SpringerWienNewYork

Progress in the Chemistry
of Organic Natural Products

Founded by L. Zechmeister

94

Progress in the Chemistry of Organic Natural Products

Authors:
S.S. Ebada, N. Lajkiewicz, J.A. Porco Jr,
M. Li-Weber, and P. Proksch
M.A.R.C. Bulusu, K. Baumann, and A. Stuetz
R.I. Misico, V.E. Nicotra, J.C. Oberti, G. Barboza,
R.R. Gil, and G. Burton

SpringerWienNewYork

Prof. A. Douglas Kinghorn, College of Pharmacy,
Ohio State University, Columbus, OH, USA

em. Univ.-Prof. Dr. H. Falk, Institut für Organische Chemie,
Johannes-Kepler-Universität, Linz, Austria

Prof. Dr. J. Kobayashi, Graduate School of Pharmaceutical Sciences,
Hokkaido University, Sapporo, Japan

SpringerWienNewYork is a part of
Springer Science + Business Media
springer.at

Typesetting: SPI, Chennai

Printed on acid-free and chlorine-free bleached paper
SPIN: 80022757

With 45 (partly coloured) Figures and 4 coloured Plates

ISSN 0071-7886
ISBN 978-3-7091-1117-8 ISBN 978-3-7091-0748-5 (eBook)
DOI 10.1007/978-3-7091-0748-5
SpringerWienNewYork

Contents

Listed in PubMed

Contributors

Gloria Barboza Departamento de Farmacia and IMBIV (CONICET), Facultad de Ciencias Químicas, Universidad Nacional de Córdoba, Ciudad Universitaria, Córdoba 5000, Argentina, gbarboza@imbiv.unc.edu.ar

Karl Baumann Novartis Institutes for BioMedical Research Vienna, Muthgasse 11/2, A-1190, Vienna, Austria

Murty A.R.C. Bulusu Novartis Institutes for BioMedical Research Vienna, Muthgasse 11/2, A-1190, Vienna, Austria

Gerardo Burton Departamento de Química Orgánica and UMYMFOR (CONICET-UBA), Facultad de Ciencias Exactas y Naturales, Universidad de Buenos Aires, Ciudad Universitaria, Pabellón 2, Buenos Aires C1428EGA, Argentina, burton@qo.fcen.uba.ar

Sherif S. Ebada Institute of Pharmaceutical Biology and Biotechnology, Heinrich-Heine University of Duesseldorf, Universitaetsstrasse 1, D-40225, Duesseldorf, Germany; Department of Pharmacognosy and Phytochemistry, Faculty of Pharmacy, Ain-Shams University, Organization of African Unity 1, 11566 Cairo, Egypt, sherif.elsayed@uni-duesseldorf.de

Roberto R. Gil Department of Chemistry, Carnegie Mellon University, 4400 Fifth Ave Pittsburgh, PA 15213, USA, rgil@andrew.cmu.edu

Neil Lajkiewicz Department of Chemistry and Center for Chemical Methodology and Library Development (CMLD-BU), Boston University, Commonwealth Avenue 590, Boston, MA 02215, USA, neiljl@bu.edu

Min Li-Weber Tumor Immunology Program (D030), German Cancer Research Center (DKFZ), Im Neuenheimer Feld 280, D-69120, Heidelberg, Germany, m.li-weber@dkfz-heidelberg.de

Rosana I. Misico Departamento de Química Orgánica and UMYMFOR (CONICET-UBA), Facultad de Ciencias Exactas y Naturales, Universidad de Buenos Aires, Ciudad Universitaria, Pabellón 2, Buenos Aires C1428EGA, Argentina, misicori@qo.fcen.uba.ar

Viviana E. Nicotra Departamento de Química Orgánica and IMBIV (CONICET), Facultad de Ciencias Químicas, Universidad Nacional de Córdoba, Ciudad Universitaria, Ciencias Químicas II, Córdoba 5000, Argentina, vnicotra@mail.fcq.unc.edu.ar

Juan C. Oberti Departamento de Química Orgánica and IMBIV (CONICET), Facultad de Ciencias Químicas, Universidad Nacional de Córdoba, Ciudad Universitaria, Ciencias Químicas II, Córdoba, Argentina, jco@mail.fcq.unc.edu.ar

John A. Porco Jr. Department of Chemistry and Center for Chemical Methodology and Library Development (CMLD-BU), Boston University, Commonwealth Avenue 590, Boston, MA 02215, USA, porco@bu.edu

Peter Proksch Institute of Pharmaceutical Biology and Biotechnology, Heinrich-Heine University of Duesseldorf, Universitaetsstrasse 1, D-40225, Duesseldorf, Germany, proksch@uni-duesseldorf.de

Anton Stuetz Novartis Institutes for BioMedical Research Vienna, Muthgasse 11/2, A-1190, Vienna, Austria, anton.stuetz@novartis.com

About the Authors

Sherif S. Ebada was born on September 1, 1978 in Cairo (Egypt). He received his B.Sc. and M.Sc. in Pharmaceutical Sciences (Pharmacognosy) from Ain-Shams University, Cairo (Egypt) under the guidance of Professors Ayoub, Singan, and Al-Azizi. In 2007, he joined the research group of Prof. Dr. Peter Proksch at the Institute of Pharmaceutical Biology and Biotechnology in the University of Duesseldorf as a doctoral candidate where he studied the isolation, structural elucidation, and structure-activity relationships of bioactive secondary metabolites from marine organisms. In 2010, he received his Ph.D. degree from the University of Duesseldorf, followed by a postdoctoral fellowship with Professor Proksch until the present.

Neil Lajkiewicz was born on November 23, 1983 in New York City, USA. He received his B.Sc. in chemistry at Boston University in 2005 and joined Lundbeck Research, USA after graduation. In 2008, he joined Sirtis Pharmaceuticals and in 2009 matriculated at Boston University for Ph.D. studies in organic synthesis. He is currently a second year graduate student in Professor John A. Porco Jr.'s laboratory studying photocycloadditions to achieve the synthesis of flavaglines and related products.

John A. Porco Jr. was born in Danbury, CT (USA) in 1963. He received his Ph.D. in 1992 from Harvard University under the direction of Professor Stuart L. Schreiber. John joined the Department of Chemistry at Boston University in 1999 as Assistant Professor after a period in industry and was promoted to Professor of Chemistry in September 2004. Professor Porco's current research is focused in two major areas: the development of new synthesis methodologies for efficient chemical synthesis of complex natural products and synthesis of complex chemical libraries.

Min Li-Weber was born on May 8, 1948 in Phnom Penh, Cambodia. She received her Master's in Biochemistry in 1975 from Peking (Beijing) University (China). From 1976 to 1979, she was a researcher at the Institute of Microbiology, Chinese Academy of Science in Beijing. From 1979 to 1980, she was a visiting scientist at the University of Utah (USA). From 1980 to 1982, she was a research assistant at the Max-Planck-Institute for cell biology (Germany). She received her Ph.D. in Biology on January 1985 from University of Heidelberg (Germany). From 1985 to 1986, she was a post-doctoral at the Max-Planck-Institute for cell biology. Since November 1986, she has been a project leader at the German Cancer Research Center (DKFZ) (Germany), where she works in the field of immunology and molecular and cellular aspects of apoptosis. She was guest professor at University of Salzburg (Austria) in 2003. Her current research is focused on the molecular mechanisms of apoptosis sensitivity and resistance in cancers and discovering and developing new anticancer drugs from natural products. She has published over 65 original research articles and several scientific review papers in the field of cancer research.

Peter Proksch was born on December 6, 1953 in Leipzig (Germany). He received his Ph.D. in Biology in 1980 from the University of Cologne. From 1980 to 1982 he was a postdoctoral at the University of California, Irvine (USA). From 1982 to 1985 he was at the University of Cologne and from 1986 to 1990 at the University of Braunschweig where he received his venia legendi for Pharmaceutical Biology. In 1990 he became Professor for Pharmaceutical Biology at the University of Wuerzburg and in 1999 he moved to his present position as Professor of Pharmaceutical Biology and Biotechnology and Head of the Institute at the University of Duesseldorf. His fields of research are bioactive natural products from marine invertebrates, higher plants and endophytic fungi. He has authored or coauthored over 300 publications and holds visiting professorships at the Universities of Beijing and Qingdao (P.R. China).

Dr. Murty Bulusu studied chemistry at Andhra University Waltair and obtained a Ph.D. degree from the Indian Institute of Technology Kanpur, India in 1983. Subsequently, he worked as Alexander von Humboldt Fellow with Prof. H. Prinzbach at the University of Freiburg i. Br., Germany, and then with Prof. A. Vasella at the University of Zürich, Switzerland. In 1989, he joined Sandoz Research Institute Vienna as a laboratory head, which later became Novartis Institutes for Biomedical Research Vienna, and then continued with its spinoff companies Sandoz AntiBiotic Research Institute (ABRI) and the New AntiBiotic Research Institute Vienna Austria (NABRIVA), and finally with the Albany Molecular Research Institute (AMRI) Hungary in 2010.

Dr. Bulusu's research interests have been on cage molecules, such as dodecahedrane, polysaccharides, such as lipid A, ascomycin and related macrolides, pleuromutilin and β-lactam antibiotics, and other low-molecular-weight classes of compounds, in various medicinal chemistry programs. He has contributed 25 research publications to peer-reviewed journals and holds five patents.

Dr. Karl Baumann studied chemistry at the Technical University in Vienna, Austria, and obtained a Ph.D. degree in organic chemistry. After a postdoctoral fellowship from 1984 to 1986 with Prof. A. Eschenmoser at the Swiss Federal Institute of Technology (ETH) in Zürich, Switzerland, he joined the Chemie Linz AG in Linz, Austria. In 1988 he joined the Sandoz Research Institute Vienna, which later became Novartis Institutes for Biomedical Research Vienna, where he worked as head of a medicinal chemistry laboratory until 2009.

Dr. Baumann invented the ascomycin derivative, SDZ 281–240, which was the first topical calcineurin inhibitor to show efficacy in patients with inflammatory skin disease. These data provided the first proof of concept and thus a milestone in the identification of this new class of topical non-steroids. He is author/coauthor of 32 publications and 20 abstracts, and the holder of 18 patents in the fields of β-lactam and quinolone-type antibiotics, natural products, labeling of organic compounds, and the development of synthetic methods.

Dr. Anton Stuetz studied chemistry and physics at the University of Vienna and obtained a Ph.D. degree in organic chemistry in 1972. After postdoctoral studies in molecular biology at the Max Planck Institute for Biophysical Chemistry, Göttingen, Germany, in 1974 he joined the Sandoz Research Institute Vienna, Austria, as head of laboratory. In 1986, he took over the responsibility of establishing dermatology research within Sandoz and became head of this new department. In 1995–1996, he served as acting head of the institute, which was renamed Novartis Research Institute Vienna after the merger of Sandoz and Ciba-Geigy. At present, he is Executive Director of Dermatology within the Disease Area Autoimmunity, Transplantation, and Inflammation as part of the Novartis Institutes for BioMedical Research, located in Vienna, Austria.

Dr. Stuetz invented terbinafine (Lamisil) in 1980, which after a worldwide launch during 1991–1997 has become the global standard for the treatment of fungal infections of the skin and nails (onychomycosis). Under his leadership a new class of anti-inflammatory agents later termed "topical calcineurin inhibitors" were pioneered, including the use of topical tacrolimus for the treatment of skin diseases, and pimecrolimus invented and its pharmacological profile established. Tacrolimus ointment (Protopic) and pimecrolimus cream (Elidel) are the first therapeutically effective and registered topical non-steroid agents for treatment of atopic dermatitis.

Dr. Stuetz is the author/coauthor of 89 publications and 170 abstracts, and holds 35 patents in the fields of synthetic and medicinal chemistry, antifungal chemotherapy, immunology, inflammation, dermatology, and translational research. He is a frequently invited speaker at international congresses and universities.

In 1994, Dr. Stuetz was appointed as professor for pharmaceutical chemistry at the University of Vienna. In 2004, he was awarded the Erwin Schrödinger Prize by the Austrian Academy of Sciences. He has served as a member of the Board of Directors of the Society for Investigative Dermatology for the period 2005–2010. In February 2011, he received the Eugene J. Van Scott Award for Innovative Therapy of the Skin and the Philipp Frost Leadership Lecture Award from the American Academy of Dermatology.

Rosana I. Misico was born in Córdoba, Argentina. She obtained her Ph.D. in chemistry (natural products) at the National University of Córdoba under the supervision of Prof. Juan C. Oberti. She then spent a postdoctoral year at the University of Illinois at Chicago working with Prof. A. Douglas Kinghorn. In 2001, she joined the group of Professor Gerardo Burton at the University of Buenos Aires. She is currently a senior researcher of the National Research Council of Argentina (CONICET). Her current research interests are on the synthesis of bioactive naphthoquinones and natural products.

Viviana E. Nicotra was born and raised in Cordoba, Argentina. She received a Pharmacy degree from the Cordoba National University, Cordoba, Argentina in 1987, a Masters in Biological Chemistry from the National University of Comahue, Neuquen, Argentina in 1988, and the Ph.D. in Chemistry (Natural Products) from Cordoba National University, under the supervision of Prof. Juan C. Oberti. She had a brief postdoctoral stay at the Instituto Universitario de Bioorganica (IUBO) at the Universidad de La Laguna, La Laguna, Canary Islands, Spain, under the supervision of Prof. Angel Gutierrez Ravelo, in 2008. Since 1999, she has been working at the Department of Chemistry of the Cordoba National University with a teaching instructor position and a senior researcher position within the research track of the National Research Council of Argentina (CONICET). Her current research interest is on the search of bioactive steroidal lactones (withanolides) from South American Solanaceae, as well as studies of montmorillonite-tetracycline interactions by circular dichroism.

Juan Carlos M. Oberti was born in the city of Paran, Entre Rios province, Argentina. He received degrees in Biochemistry (1965) and Pharmacy (1968), and a Ph.D. in 1974 under the supervision of Professor Ramón Juliani, on the topic "Alkaloids from Prosopis ruscifolia", all from Córdoba National University, Córdoba, Argentina. He spent a short postdoctoral stay at the Department of Organic Chemistry of the University of Buenos Aires with Prof. Eduardo Gros, where he also participated in the team that officially performed anti-doping tests for the soccer matches during the 1978 FIFA World Cup in Argentina. Since 1979, he has led the Natural Products research group at the Department of Chemistry of the College of Chemistry, Córdoba National University, focusing mainly on the search for sesquiterpene lactones from South American Compositae and steroidal lactones (withanolides) from South American Solanaceae. Prof. Oberti retired from Cordoba National University in 2005, where he remains as Consulting Professor, and is still active in research with a research position from the National Research Council of Argentina (CONICET). He is currently working on withanolides, as well as on sesquiterpene agarofuran alkaloids and quinones from the Celastraceae.

Gloria E. Barboza was born and raised in Salta, Argentina. She received a degree in Biology from the National University of Tucuman (Argentina) in 1985 and a Ph.D. in Biology from the National University of Cordoba (Argentina) in 1989, where she worked under the supervision of Prof. Armando T. Hunziker, a recognized specialist on the Solanaceae. Since 1990, she has held a permanent position as a researcher in the National Research Council of Argentina (CONICET) working at the Instituto Multidisciplinario de Biología Vegetal (IMBIV) in Córdoba. In 1994, she was appointed to a Professor position in Botany at the Pharmacy Department (Chemical Sciences College). Her current research interests are on the systematics of the Solanaceae, especially the South American genera, and on Argentine medicinal plants.

Roberto R. Gil was born in Catamarca, Argentina in 1961. He received the degrees of B.S./ M.S. in Organic Chemistry (1983) and Ph.D. in Natural Products Chemistry (1989) from the University of Córdoba, Córdoba, Argentina. In 1992 he received an external postdoctoral fellowship from the National Research Council of Argentina (CONICET) to work with Professors Geoffrey A. Cordell and A. Douglas Kinghorn at the University of Illinois at Chicago in the field of bioactive natural products from plants. In 1995, he returned to the University of Córdoba where he started his own research group as Assistant Professor. In 2000 he spent a year as Visiting Professor at Carnegie Mellon University working in Protein NMR with Professor Miguel Llins. In 2002, he moved to Pittsburgh, Pennsylvania, where he currently holds the position of Associate Research Professor and Director of the NMR Laboratory of the Department of Chemistry at Carnegie Mellon University. His research interest is aimed at the development and application of NMR methodologies to the analysis of the structural and physical properties of bioactive natural products, nucleic acids, peptides and synthetic polymers.

Gerardo Burton was born in Buenos Aires, Argentina. He obtained a doctoral degree in organic chemistry from the University of Buenos Aires in 1977, where he worked on the biosynthesis of steroidal lactones of animal origin with Prof. E. G. Gros. After a postdoctoral stay at the Department of Chemistry, Texas A&M University (USA) with Prof. A. Ian Scott working on porphyrin biosynthesis and biological NMR, he returned to Argentina in 1980. There he joined the faculty of the Organic Chemistry Department (Facultad de Ciencias Exactas y Naturales), University of Buenos Aires as an Assistant Professor, and started research on the design and synthesis of steroid hormone analogs. He is currently a Plenary Professor in that Department and an Investigator of the National Research Council of Argentina (CONICET). He was Chairman of the Organic Chemistry Department (University of Buenos Aires) on two occasions, and has been Director of UMYMFOR, a research institute and spectroscopic and analytical facility of CONICET, since 2001. His current research interests are in the area of organic synthesis and medicinal chemistry, specifically the design and synthesis of new bioactive steroids and their interaction with nuclear receptors.

Chemistry and Biology of Rocaglamides (= Flavaglines) and Related Derivatives from *Aglaia* Species (Meliaceae)

Sherif S. Ebada, Neil Lajkiewicz, John A. Porco Jr., Min Li-Weber, and *Peter Proksch*

Contents

John A. Porco, Min Li-Weber, and Peter Proksch contributed equally to the writing of this chapter.
Dedicated to Dr. Bambang Wahyu Nugroho, a pioneer of rocaglamide research (*9*, *14*, *16*, *17*, *27*, *54–56*, *58*, *59*, *75*, *84*, *85*) who passed away far too early.

S.S. Ebada
Institute of Pharmaceutical Biology and Biotechnology, Heinrich-Heine University of Duesseldorf, Universitaetsstrasse 1, D-40225, Duesseldorf, Germany
Department of Pharmacognosy and Phytochemistry, Faculty of Pharmacy, Ain-Shams University, Organization of African Unity 1, 11566 Cairo, Egypt
e-mail: sherif.elsayed@uni-duesseldorf.de

N. Lajkiewicz • J.A. Porco Jr.
Department of Chemistry and Center for Chemical Methodology and Library Development (CMLD-BU), Boston University, Commonwealth Avenue 590, Boston, MA 02215, USA
e-mail: neiljl@bu.edu; porco@bu.edu

M. Li-Weber
Tumor Immunology Program (D030), German Cancer Research Center (DKFZ), Im Neuenheimer Feld 280, D-69120, Heidelberg, Germany
e-mail: m.li-weber@dkfz-heidelberg.de

P. Proksch (✉)
Institute of Pharmaceutical Biology and Biotechnology, Heinrich-Heine University of Duesseldorf, Universitaetsstrasse 1, D-40225, Duesseldorf, Germany
e-mail: proksch@uni-duesseldorf.de

A.D. Kinghorn, H. Falk, J. Kobayashi (eds.), *Progress in the Chemistry of Organic Natural Products, Vol. 94*, DOI 10.1007/978-3-7091-0748-5_1,

1. Introduction

Throughout the ages, humans have relied on Nature for fulfilling their basic needs for foodstuffs, shelter, clothing, means of transportation, fertilizers, flavors and fragrances, and, last but not least, medicines. Natural products have played, for thousands of years, an important role throughout the world in treating and preventing human diseases. Natural product medicines have come from various source materials including terrestrial plants, terrestrial microorganisms, marine organisms, and terrestrial vertebrates and invertebrates (*1*). The importance of natural products in modern medicine can be assessed using three criteria: (a) the rate of introducing new chemical entities of wide structural diversity, which may serve as templates for semisynthetic and total synthetic modification, (b) the number of diseases treated or prevented by these substances, and (c) their frequency of use in the treatment of disease (*2, 3*). An analysis of the origin of drugs developed between 1981 and 2007 indicated that almost half of the drugs approved since 1994 were based on natural products (*2, 3*). Over 20

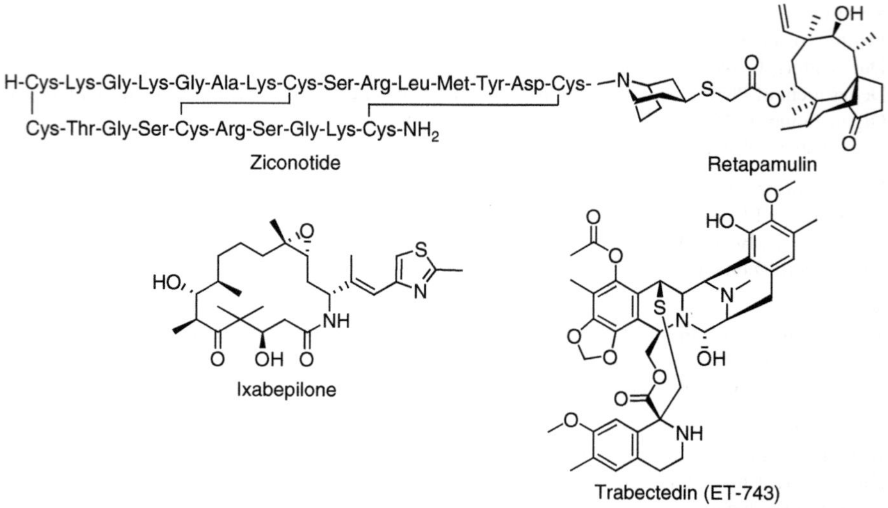

Fig. 1. Chemical structures of ziconotide, ixabepilone, retapamulin, and trabectedin (ET-743)

new drugs launched into the pharmaceutical market between 2000 and 2005 represent natural products (*2, 3*), whereas more than 13 natural-product-related drugs were approved from 2004 to 2007; four of them represent the first members of new classes of drugs: the peptide ziconotide, and the small molecules ixabepilone, retapamulin, and trabectedin (ET-743) (Fig. 1) (*3, 4*). Interestingly, over a hundred natural-product-derived compounds are currently undergoing clinical trials and at least a hundred similar substances are under preclinical development, with most of these derived from leads from plant and microbial sources (*3*). In spite of challenges facing drug discovery from plants, including the legal and logistical difficulties involved in the procurement of plant materials, and the lengthy and costly process of bioassay-guided fractionation and compound isolation, plants still provide new drug leads that prove to be of potential preclinical and/or clinical use against serious ailments such as cancer, malaria, *Alzheimer*'s disease, and AIDS (*5*).

The family Meliaceae (= Mahogany family, order Sapindales) is an angiosperm plant family of mostly trees and shrubs together with a few herbaceous plants. This family includes about 50 genera and 550 species, with a panotropical geographical distribution. Two genera, namely, *Swietenia* (Mahogany) and *Khaya* (African mahogany), are important sources of high-quality woods for building shelters and furniture due to their physical properties and also due to their resistance to insect invasion (*6*).

The genus *Aglaia* Lour. (Fig. 2) is the largest genus of the family Meliaceae, comprising about 120 woody species ranging from small to large trees up to 40 m

Fig. 2. *Aglaia* Lour. (family Meliaceae). (a): Entire tree of *Aglaia odorata*, (b): leaves of *A. tomentosa*, (c): flowers of *A. odorata*, (d): fruits of *A. forbesii* (photos by Dr. *B. W. Nugroho* and from http://dps.plants.ox.ac.uk/bol/aglaia and http://www.rareflora.com/aglaiaodo.html)

high, mainly distributed in the tropical rainforests of southeast Asia from Sri Lanka and India, through Burma, south China and Taiwan, Vietnam, Malaysia, Indonesia, the Philippines, New Guinea, the Solomon Islands, Vanuatu (New Hebrides), New Caledonia, Australia (Queensland, Northern Territory and Western Australia), Fiji, as far east as the island of Samoa in Polynesia and north to the Marianne Islands (Saipan, Roti and Guam), and the Caroline Islands (Palau and Ponape) in Micronesia (*7*). A molecular phylogeny has demonstrated that the genus is divided into three sections, section *Amoora*, section *Neoaglaia*, and section *Aglaia* (*8*). They are distinguishable morphologically, mainly on fruit characteristics and the numbers of flower parts (*8*). Like the two genera *Swietenia* (Mahogany) and *Khaya*, the timber of many *Aglaia* species is used locally for house-building, fence-posts, canoes, paddles, axe-handles, spear-shafts, and firewood. The fragrant flowers are used for scenting tea and are kept in cupboards to perfume and to protect clothing from moths. They produce sweet, fleshy fruits that are cultivated in villages in Thailand and peninsular Malaysia and are eaten in the forest by indigenous forest peoples.

The fruits of *Aglaia* (Fig. 2) are also a source of food for birds and mammals in the forests of the Indo-Malayan and Australasian regions where they occur. In West Malaysia, the fruits of species in the section *Aglaia* are indehiscent and primates break open the orange, yellow or brown, fibrous, inedible pericarp and extract the one or two seeds from within. The translucent, sweet aril adheres firmly to the seed, and the seed is often swallowed whole. Analysis of the nutrient content of the aril reveals that it contains sugars and other sweet-tasting constituents and it is thought that these are attractive to the gibbons that disperse the seeds (*7*). The fruits of sections *Amoora* and *Neoaglaia* are dehiscent and contain up to three seeds. The outer pericarp is pink or reddish-brown and contrasts with the white inner pericarp and the red aril surrounding the seed. The aril is easily detached from the testa and is removed by the action of a bird's gizzard, without destroying the rest of the seed. The aril, surrounding a relatively large seed, is rich in lipids and provides the birds that disperse the large seed with a high-calorie reward (*7*).

Several species of the genus *Aglaia*, such as *A. odorata*, are used traditionally in folk medicine for heart stimulant and febrifuge purposes, and for the treatment of coughs, diarrhea, inflammation, and injuries (*9*). Extracts have also been used as bactericides, insecticides, and in perfumery (*10*).

During the last few decades, species in the genus *Aglaia* Lour. have received an increasing scientific focus due to their bioactivity potential. Phytochemical interest in the natural constituents of *Aglaia* Lour. can be traced back to the discovery in 1982 of the first cyclopenta[*b*]benzofuran derivative, rocaglamide (**1**), from *A. elliptifolia* (*11*). To date, more than a hundred naturally occurring rocaglamide-type (= flavagline) compounds have been isolated from over 30 *Aglaia* species (*9, 12*). Rocaglamides exhibit potent insecticidal (*13–18*) and antiproliferative (*12, 19–21*) activities. In addition, antiviral (*22*), antifungal (*23*), and anti-inflammatory (*24, 25*) activities were also reported for these compounds, which are so far only known from *Aglaia* species. Other classes of natural products occurring in *Aglaia* include lignans (*13, 26–29*), flavonoids, and bisamides (*18, 22, 26, 30–36*). Some of these metabolites exhibit cytotoxic and antiviral properties as well (*22, 30*). Furthermore, many terpenoids have been reported from the genus *Aglaia* Lour. (*10, 36–51*).

The present contribution surveys the group of the rocaglamide derivatives (also known as "flavaglines" or "rocaglate derivatives") and related compounds obtained from the genus *Aglaia*, with an emphasis on their structural diversity, and highlights their potential pharmacological significance, which is the main reason for attracting a greater attention by natural product chemists and cell biologists to this class of natural products and provides a comprehensive overview on their total synthesis.

2. Structural Classification of Rocaglamides and Related Compounds

2.1. Rocaglamide Derivatives

Rocaglamide (**1**), a 1*H*-2,3,3a,8b-tetrahydrocyclopenta[*b*]benzofuran, was first structurally elucidated in 1982 by *King et al.* through single-crystal X-ray analysis (Fig. 3) (*11*). Its absolute stereochemistry was determined unambiguously to be (1*R*,2*R*,3*S*,3a*R*,8b*S*) using enantioselective synthesis in 1990 by *Trost et al.* (*52*). Comparative MS and 1D and 2D NMR spectroscopic data of rocaglamide (**1**) and

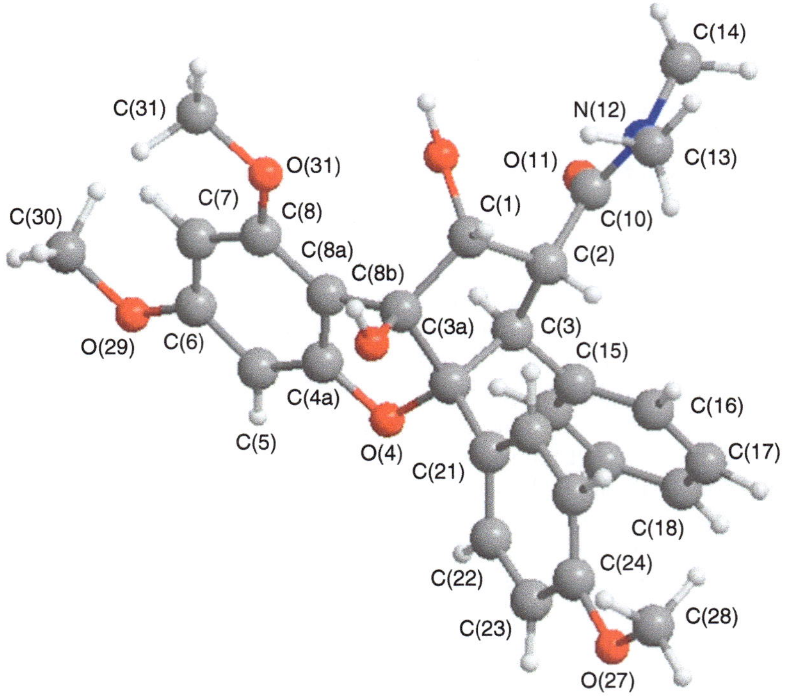

Fig. 3. X-ray crystal structure of rocaglamide (**1**) (*11*)

its analogues, desmethylrocaglamide (**7**), methyl rocaglate (**18**), and rocaglaol (**28**) were first presented in 1993 by *Ishibashi et al.* (*53*). Rocaglamide congeners differ basically with regard to their substituents at C-1, C-2, C-8b, and C-3′ at ring B. Major variations in the substitution pattern occur at C-2 while the hydroxy substituents at C-1 or C-8b can either be acetylated, methylated, or ethylated (*e.g.* congeners **4**, **5**, and **6**). The position C-3′ is either hydroxylated or methoxylated (*e.g.* congeners **2** and **3**). However, oxidation (*16*) and esterification (*17*) of the hydroxy group at C-1 have been also reported. The structures of rocaglamides known so far are summarized in Fig. 4.

The mass spectra of rocaglamide and its derivatives often show characteristic pairs of fragments at *m*/*z* 300 and 313 dependent on the substitution pattern. Plausible structures for the ions *m*/*z* 300 and 313 arising from fragmentation of rocaglamide-type compounds under EI conditions have been described (*54*), as summarized in Fig. 5. Changes in the fragmentation pattern in the range *m*/*z* 300–343 indicate the type of substitution at ring B and C-8b of the furan ring. For example, the presence of a hydroxy substituent at C-3′ shifts the characteristic pair of fragments at *m*/*z* 300 and 313 (as in rocaglamide) to *m*/*z* 316 and 329 while a methoxy substituent at the same position gives rise to fragments at *m*/*z* 330 and 343 in the EI mass spectrum of the respective derivative (*55*). Modification of the hydroxy substituent at C-8b (*e.g.* methylation) can also be determined initially by comparison of its diagnostic fragments to those of the more common structural analogues featuring a hydroxy group at that position (*56*). Rocaglamide analogues exhibit ^{1}H and ^{13}C NMR signals for aromatic protons and aromatic methoxy groups typical for those of substituted phenols. Investigation of the ^{1}H NMR spectra of several rocaglamide derivatives showed empirically that hydroxylation at C-3′ causes a deshielding effect on the aromatic protons at ring B in the following order: H-2′ > H-6′ > H-5′. Consequently, methylation of the hydroxy group at C-3′ causes a deshielding of the aromatic protons accordingly: H-6′ > H-5′ > H-2′. Moreover, substitution at C-3′ changes the symmetrical ^{1}H NMR resonance pattern for the AA′BB′ system for the *para*-substituted ring B to an ABC pattern of methines comparable to a threefold substituted phenyl ring system. Assignment of the relative configuration at C-2 has also been deduced by inspection of their ^{1}H NMR spectra. The vicinal coupling constant values of the methine protons at the C-1, C-2, and C-3 positions ($J_{1,2}$ *ca.* 5–7 Hz and $J_{2,3}$ *ca.* 13–14 Hz) indicated the 1α,2α,3β configuration as well as the *cis*-BC ring junction (*53*). NOESY experiments have been used to confirm the stereochemical relationship of the substituents from different carbon ring junctions. The NOESY spectrum showed a NOE correlation peak between H-2′ and both H-1α and H-2α but not between H-2′ and H-3β (*53*).

The CD spectra of the rocaglamides show prominent negative *Cotton* effects between 217 and 220 nm as the most characteristic feature (*54*). Their CD spectra are dominated by the nature of the cyclopenta[*b*]tetrahydrobenzofuran moiety forming the backbone of the rocaglamide derivatives with stereocenters at C-1, C-2, C-3, C-3a, and C-8b and thus by the 3D array of the main molecular chromophores, the three aromatic rings. However, the asymmetric carbon C-2 apparently can influence the CD spectra of rocaglamide congeners, as exemplified by the α-sugar-substituted

Fig. 4. Rocaglamide derivatives isolated from *Aglaia* species

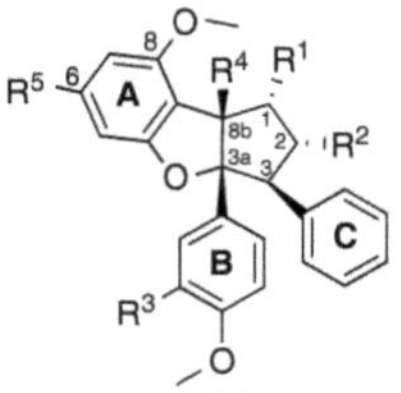

1-38

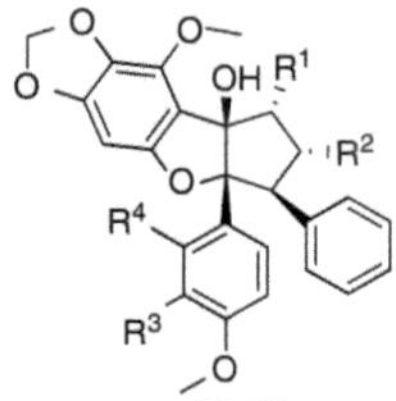

39-47

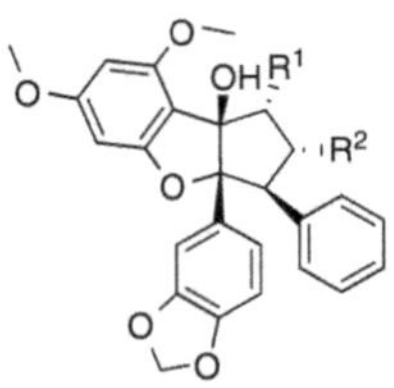

48-52

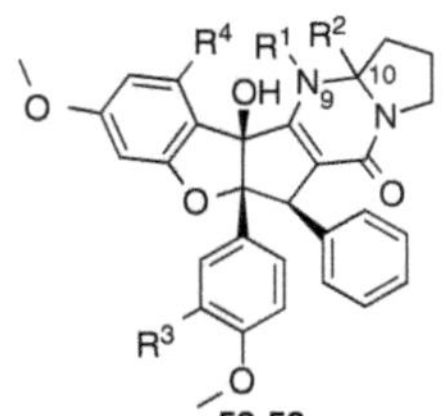

53-58

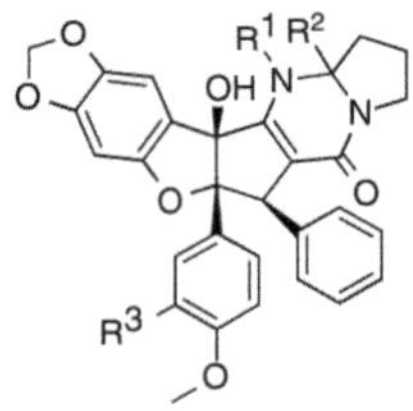

59-62

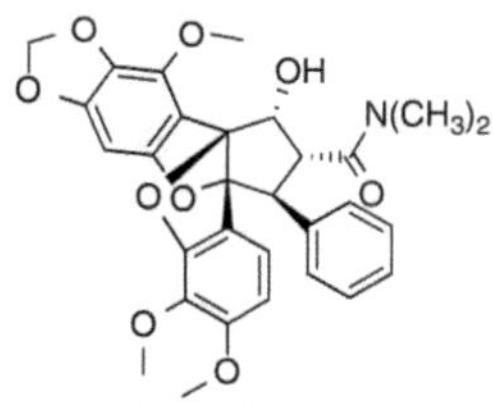

Cyclorocaglamide (**63**)

Cpd.	Trivial Name	R[1]	R[2]	R[3]	R[4]	R[5]	Ref.
1	Rocaglamide	OH	$CON(CH_3)_2$	H	OH	OCH_3	*(11, 60)*
2		OH	$CON(CH_3)_2$	OH	OH	OCH_3	*(55)*
3	Aglaroxin E	OH	$CON(CH_3)_2$	OCH_3	OH	OCH_3	*(55, 57)*
4	1-*O*-Acetylrocaglamide	$OCOCH_3$	$CON(CH_3)_2$	H	OH	OCH_3	*(58)*
5		$OCOCH_3$	$CON(CH_3)_2$	OH	OH	OCH_3	*(55)*
6		OH	$CON(CH_3)_2$	OH	OC_2H_5	OCH_3	*(56)*
7	Desmethyl-rocaglamide	OH	$CONHCH_3$	H	OH	OCH_3	*(53, 60)*
8		OH	$CONHCH_3$	OH	OH	OCH_3	*(14)*
9		$OCOCH_3$	$CONHCH_3$	H	OH	OCH_3	*(58)*
10		$OCOCH_3$	$CONHCH_3$	OH	OH	OCH_3	*(59)*
11		OH	$CONHCH_3$	OH	OC_2H_5	OCH_3	*(56)*
12	*N,N*-Didesmethyl-*N*-4-hydroxybutyl-rocaglamide	OH	$CONH(CH_2)_4OH$	H	OH	OCH_3	*(54)*
13		$OCOCH_3$	$CONH(CH_2)_4OH$	H	OH	OCH_3	*(54)*
14		OH	Ring[1]	H	OH	OCH_3	*(54)*
15	Didesmethyl-rocaglamide	OH	$CONH_2$	H	OH	OCH_3	*(60)*
16		OH	$CONH_2$	OH	OH	OCH_3	*(59)*
17		$OCOCH_3$	$CONH_2$	H	OH	OCH_3	*(56)*
18	Methyl rocaglate (Aglafoline)	OH	$COOCH_3$	H	OH	OCH_3	*(53, 60–62)*

Fig. 4. continued

Cpd.	Trivial Name	R^1	R^2	R^3	R^4	R^5	Ref.
19	3'-Hydroxyaglafoline	OH	$COOCH_3$	OH	OH	OCH_3	(*14*)
20		OH	$COOCH_3$	OCH_3	OH	OCH_3	(*17*)
21		$OCOCH_3$	$COOCH_3$	H	OH	OCH_3	(*56*)
22		$OCOCH_3$	$COOCH_3$	OH	OH	OCH_3	(*59*)
23	1-*O*-Formylaglafoline	OCHO	$COOCH_3$	H	OH	OCH_3	(*17*)
24		OCHO	$COOCH_3$	OH	OH	OCH_3	(*17*)
25		=NOH	$COOCH_3$	OCH_3	OH	OCH_3	(*14*)
26		OH	$COOCH_3$	H	OCH_3	OCH_3	(*58*)
27	Rocagloic acid	OH	COOH	H	OH	OCH_3	(*63*)
28	Rocaglaol (Aglaiastatin A)	OH	H	H	OH	OCH_3	(*53, 60, 65*)
29		OH	H	OCH_3	OH	OCH_3	(*59*)
30		OH	H	Sugar[2]	OH	OCH_3	(*54*)
31		OH	H	Sugar[3]	OH	OCH_3	(*67*)
32		OH	H	H	OC_2H_5	OCH_3	(*60*)
33		OH	H	H	OCH_3	OCH_3	(*58*)
34	1-*O*-Acetylrocaglaol	$OCOCH_3$	H	H	OH	OCH_3	(*42*)
35	Silvestrol (5'''*R*)	OH	$COOCH_3$	H	OH	Ring[4]	(*68*)
36	Episilvestrol (5'''*S*)	OH	$COOCH_3$	H	OH	Ring[4]	(*68*)
37	1-*O*-Formylrocagloic acid	OCHO	COOH	H	OH	OCH_3	(*70*)
38	3'-Hydroxyrocagloic acid	H	COOH	OH	OH	OCH_3	(*70*)
39	Aglaroxin A	OH	$CON(CH_3)_2$	H	H	-	(*16, 57*)
40	Aglaroxin B	OH	$CON(CH_3)_2$	OCH_3	H	-	(*57, 71*)
41	Aglaroxin F	OH	$CON(CH_3)_2$	OCH_3	OH	-	(*57*)
42	Aglaroxin A 1-*O*-acetate	$OCOCH_3$	$CON(CH_3)_2$	H	H	-	(*72*)
43	1-*O*-Acetyl-3'-methoxyaglaroxin A	$OCOCH_3$	$CON(CH_3)_2$	OCH_3	H	-	(*72*)
44	Pannellin	OH	$COOCH_3$	H	H	-	(*13, 71*)
45	1-*O*-Acetylpannellin	$OCOCH_3$	$COOCH_3$	H	H	-	(*13*)
46	3'-Methoxypannellin	OH	$COOCH_3$	OCH_3	H	-	(*13*)
47	Isothapsakon A	=O	Chain[5]	H	H	-	(*71*)
48	4'-Demethoxy-3',4'-methylenedioxy-methyl rocaglate	OH	$COOCH_3$	-	-	-	(*20*)
49		OH	H	-	-	-	(*20*)
50	1-Oxo-11,12-methylendioxyrocaglaol	=O	H	-	-	-	(*20*)
51		OCHO	$COOCH_3$	-	-	-	(*20*)
52		$OCOCH_3$	$COOCH_3$	-	-	-	(*17*)
53		$\Delta^{9,10}$		H	-	-	(*73*)
54		$\Delta^{9,10}$		OH	-	-	(*56*)
55	Algaroxin D (Aglaiastatin)	H	H	H	-	-	(*54, 57, 74*)
56	Marikarin	$\Delta^{9,10}$		H	OH	-	(*18*)
57	3'-Hydroxymarikarin	$\Delta^{9,10}$		OH	OH	-	(*18*)
58	Aglaiformosanin	$\Delta^{9,10}$		OCH_3	OCH_3	-	(*75*)
59	Algaroxin C	$\Delta^{9,10}$		H	-	-	(*57*)
60	Algaroxin G	$\Delta^{9,10}$		OCH_3	-	-	(*57*)
61	Algaroxin H	H	H	OCH_3	-	-	(*57*)
62	Algaroxin I	H	H	H	-	-	(*57*)

Fig. 4. continued

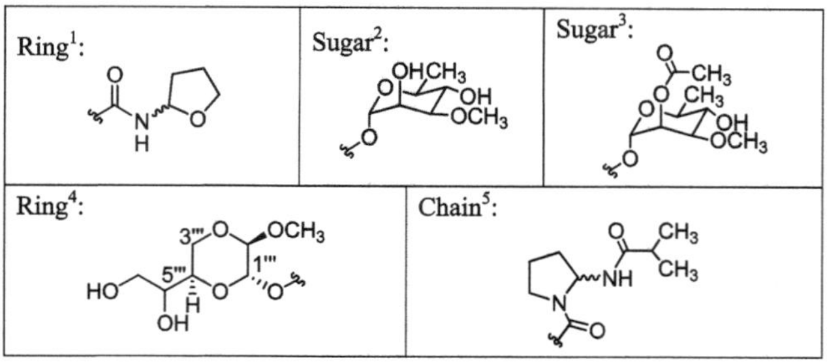

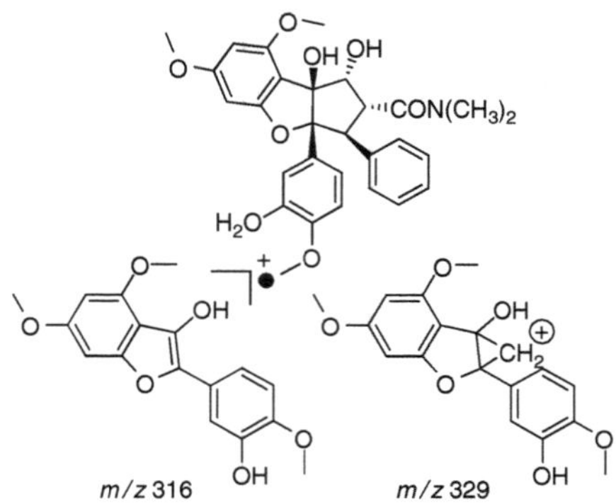

Fig. 5. Plausible structures of fragment ions m/z 316 and 329 of compound **2** under EI–MS

derivative **30** (*54*), which shows virtually the same CD spectrum as rocaglamide (**1**), but it lacks the stereocenter at C-2.

Considering rocaglamide (**1**) as the parent compound, major modifications in the substitution patterns occur at C-2, which in **1** is attached to a dimethylamino substituent characterized by two NCH_3 resonance signals at *ca.* 2.90–3.40 ppm in the ^{1}H NMR spectrum. Derivatives **2**, **5**, and **6**, with a hydroxy function at C-3′ of ring B, were isolated from the twigs (*55*) and flowers (*56*) of the Vietnamese species *Aglaia duperreana* while its methoxylated form known as aglaroxin E (**3**) was purified from the bark of the Sri Lankan species *A. roxburghiana* (*57*). Compounds with an acetoxy function at C-1 (**4** and **5**) (*55, 58*) and ethoxylated substituent at C-8b (**6**) (*56*) were obtained from the same *A. duperreana* specimen.

N-Desmethylrocaglamide congeners **7–11** feature a $-CONHCH_3$ group at C-2. *N*-Desmethylrocaglamide (**7**) was isolated from twigs and leaves of *A. odorata* (*14, 53*), whereas congeners with an acetylated hydroxy function at C-1 have been isolated from the flowers of *A. odorata* (*59*) and the roots of *A. duperreana* (*58*) collected in Vietnam. An ethylated form of substitution at C-8b occurs in compound **11**, which was obtained from the flowers of the same collection (*56*). Derivatives with an amino acyl substituent at C-2, as in congeners **12** and **13**, were isolated from *Aglaia harmsiana* (*54*). From the same species, the cyclized form of the amino acyl chain yielding the tetrahydrofuran ring, which is present in congener **14**, was isolated in its two stereoisomeric configurations. The *N*-didesmethylrocaglamide derivatives **15–17** are widely distributed among various *Aglaia* species from different geographical origins, *e.g. A. odorata* from Indonesia (*59*), *A. argentea* from Malaysia (*60*), and *A. duperreana* from Vietnam (*56*).

The methyl rocaglate congeners **18–26** were identified by their methyl ester functionality at C-2, which is indicated by a ^{13}C NMR resonance at *ca.* 170 ppm as well as by a three-proton singlet at *ca.* 3.70 ppm in its ^{1}H NMR spectrum. Methyl rocaglate (**18**) was isolated initially from *Aglaia odorata* (*53*), then later from *A. forbesii* (*60*) and *A. elaeagnoidea* (*61*). Methyl rocaglate was also named aglafoline (*62*). Compounds with acetylated substituents at C-1 (**21** and **22**) were isolated from *A. duperreana* (*56*), and also from *A. odorata* (*59*) while the formylated congeners **23** and **24** were obtained from the bark of *A. spectabilis* collected from Vietnam (*17*). An unusual C-1 oxime derivative **25** of a rocaglate was isolated from the leaves of *A. odorata* (*14*), which was exemplified by a large downfield shift of 153.0 ppm as compared to the C-1 resonance for methyl rocaglate at 80.6 ppm in addition to the loss of the H-1 resonance at 4.90 ppm. The H-2 resonance in congener **25** was observed as a doublet that coupled only with H-3, instead of a double doublet as observed in methyl rocaglate (**18**).

Rocagloic acid (**27**) is the demethylated form of methyl rocaglate or the acid congener of this series of cyclopenta[*b*]tetrahydrobenzofuran compounds. It was obtained from the leaves of the Taiwanese species *Aglaia elliptifolia* (*63*) and also from the leaves of *A. dasyclada* (*64*) collected in Yunnan Province (China). The ^{1}H NMR and ^{13}C NMR spectra of **27** are comparable to those of methyl rocaglate (**18**), with the exception of the loss of methyl ester resonance signals.

The rocaglaol derivatives **28–32** are unsubstituted at C-2. Rocaglaol (**28**) itself was first isolated from the leaves of *A. odorata* (*53*) and later proved to be identical to ferrugin, which was reported from *A. ferruginaea* (*65*) but had been initially assigned a different structure (*66*). The ^{13}C NMR spectra of compounds **28–32** exhibit no signal indicative of a carbonyl group (usually in the range of 171–175 ppm), whereas they do feature an aliphatic methylene signal at *ca.* 38 ppm for C-2, as detected from the DEPT-135 spectrum (*54*). In their ^{1}H NMR spectra, the resonances for the methylene protons at *ca.* 2.15 and 2.80 ppm appear as a pair of geminally coupled multiplets splitting as a *ddd* due to coupling with the vicinal methine protons, H-1 and H-3. Modification of the substitution pattern for compounds **28–32** occurs either at C-3′ or C-8b. Methoxylation (**29**) and glycosidation (**30**) at C-3′ have been reported for compounds isolated from the flowers of *A. odorata* (*59*) and leaves of *A. harmsiana* (*54*). Inspection of the ^{1}H NMR spectrum of the glycoside congener **30** revealed an α-linked modified rhamnose unit with a methoxy group at the C-3″ position as confirmed by N*O*E experiments (*54*). This sugar-substituted rocaglaol derivative **30** was the first rocaglamide glycoside isolated from Nature. From the leaf extract of the Malaysian species *A. laxiflora*, a similar rocaglaol rhamnoside, **31**, was isolated, which was reported to contain an additional acetyl group at the C-2″ position of the modified rhamnose unit as confirmed by HMBC (*67*). Methylation (**32**) and ethylation (**33**) of the hydroxy group at C-8b occur in compounds isolated from the roots of *A. duperreana* (*58*) and the bark of *A. forbesii* (*60*).

The three cyclopenta[*b*]tetrahydrobenzofuran derivatives **34–36** were isolated by *Kinghorn et al.* from two specimens of *Aglaia* species collected in Indonesia (*42, 68, 69*). 1-*O*-Acetylrocaglaol (**34**) was isolated from the twigs of *A. rubiginosa* (*42*). The absolute stereochemistry of **34** was deduced by a comparison of the CD spectrum with that of rocaglamide (**1**). Two methyl rocaglate congeners with an unusual dioxyanyloxy unit at C-6, silvestrol (**35**) and episilvestrol (**36**), were obtained from the fruits and twigs of *A. foveolata* (*68*). The CD spectrum of silvestrol (**35**) was very similar to that of methyl rocaglate (**18**) implying that the tricyclic cores of both molecules have the same stereochemistry. However, the relative configuration of the dioxyanyloxy unit was difficult to confirm from the available NMR data. Accordingly, the absolute configuration of **35** was established by a single-crystal X-ray analysis of its 5‴,6‴-di-*p*-bromobenzoate derivative, and was found to be (1*R*,2*R*,3*S*,3a*R*,8b*S*,1‴*S*,2‴*R*,4‴*R*,5‴*R*) (*68*). From a comparison of its 2D NMR data with those of silvestrol (**35**), compound **36** was assigned as the C-5‴ epimer of **35** (*68*). Initially, the plant material was wrongly identified as *Aglaia silvestris* (M. Roemer) Merrill, hence the name silvestrol was given to **35**. However, the species was later re-identified as *A. foveolata* Pannell (*69*).

From the fruits of *Aglaia spectabilis* (syn. *Amoora cucullata*) (Meliaceae) collected from Thailand in 2004, two rocaglamide derivatives, namely, 1-*O*-formylrocagloic acid (**37**) and 3′-hydroxyrocagloic acid (**38**) were isolated (*70*). The absolute stereochemistry of **37** was defined as having the (1*R*,2*R*,3*S*,3a*R*,8b*S*)-configuration by comparing its CD spectrum, which revealed a prominent negative *Cotton* effect at 274 nm, with that of rocaglamide (**1**) (*70*).

The group of 6,7-methyenedioxy rocaglamide analogues (**39–41**) was isolated from the stem bark of the Sri Lankan species *Aglaia roxburghiana* (*57*) and were accorded the trivial names aglaroxins A, B, and F. Compared to the fundamental structure of the rocaglamides, the ^{1}H NMR resonances for OCH_3-6 and H-7 were absent and instead replaced by a methylenedioxy singlet at *ca.* 5.90 ppm. The doublet for H-5 at *ca.* 6.30 ppm in the ^{1}H NMR spectrum of rocaglamide (**1**) was replaced by a singlet (*13*). The resonance for OCH_3-8 was also shifted downfield from δ 3.85 to δ 4.10 ppm due to the deshielding effect of the adjacent methylenedioxy function. The presence of a methylenedioxy function was also evident from a triplet resonance at *ca.* 103 ppm as revealed in its DEPT spectra (*16*). The absolute configuration of aglaroxin A (**39**) was first determined by calculation of its CD spectrum using molecular dynamics (MD) simulations (*16*). Variations for the analogues occur at ring B in which aglaroxin B (**40**) was methoxylated at C-3′ while aglaroxin F (**41**) was both methoxylated and hydroxylated at C-3′ and C-4′ (*57*). Two further aglaroxin A analogues, the 1-*O*-acetate (**42**) and the 3′-methoxy-1-*O*-acetate (**43**), were isolated from an Indonesian collection of the bark of *A. edulis* (*71*).

The pannellins **44–46** were isolated from *Aglaia elaeagnoidea* collected from Thailand (*13*). For this group of analogues, the amide function at C-2 in aglaroxins A, B, and F was replaced by a methyl ester. Pannellin-1-*O*-acetate (**45**) is the acetylated product of **44** while 3′-methoxypannellin (**46**) is characterized by an additional –OCH_3 function in ring B.

Proksch et al. described the isolation of a similar group of congeners from the twigs of a Vietnamese collection of *A. oligophylla*, including isothapsakon A (**47**), a C-1-oxo derivative of aglaroxin A (**39**), bearing a bisamide side chain at C-2 that is derived from piriferine (*16*). The ketone substituent at C-1 was identified by the carbon resonance at δ 206 ppm consequently resulting in a downfield shift of H-2, which appeared as a doublet coupling only with H-3.

Derivatives **48–52**, featuring a 3′,4′-methylenedioxy substitution in the B ring, have been first reported from *Aglaia elliptica* (*20*) collected in Thailand and the Vietnamese species *A. spectabilis* (*17*) while the congeners **39–47** possess the same 3′,4′-methylenedioxy functionality but in ring A.

The last group of rocaglamide congeners (**53–62**) is characterized by a pyrimidinone subunit fused at C-1 and C-2. The resulting pentacyclic skeleton can be considered conceptually as a rocaglamide with a 2-aminopyrrolidine amide substituent at C-2 linked to C-1 via the primary amino group. This pyrimidinone-type rocaglamide **53** was first isolated from the roots of *A. odorata* collected in Thailand and was elucidated structurally by X-ray crystallography (*72*). Later, **53** was also isolated from the leaves and twigs of the Vietnamese species *A. duperreana* (*55*) while its flowers yielded the 3′-hydroxy derivative **54** (*56*). Aglaroxin D (aglaiastatin) (**55**), the dihydro derivative of **53**, has been isolated from the leaves of *A. duperreana* (*55*) and *A. odorata* (*73*) and from the stem bark of the Sri Lankan species *A. roxburghiana* (*57*). The latter collection yielded four further pyrimidinone analogues with an additional 6,7-methylenedioxy substituent in ring A, known as aglaroxins C (**59**) and G–I (**60–62**) (*57*).

Three further pyrimidinone-type congeners, marikarin (**56**) and 3′-hydroxymarikarin (**57**), were isolated from the root bark of *Aglaia gracilis* collected in Fiji (*18*), while aglaiformosanin (**58**) was obtained from the stem bark of *A. formosana* collected in Taiwan (*74*). In 2003, aglaroxin F (**41**) was isolated from *A. oligophylla* twigs collected in Vietnam together with its 8b,10-anhydro analogue, cyclorocaglamide (**63**) (*75*). Cyclorocaglamide (**63**) was identified as the first bridged cyclopenta[*b*]benzofuran between C-8b and C-2′ of ring B, whereas to the best of our knowledge aglaroxin F (**41**) represents the only rocaglamide derivative with three oxygen functions in the B ring, bearing an additional hydroxy group at C-2′.

2.2. Aglain Derivatives

Aglains (see Fig. 6) are characterized by a cyclopenta[*bc*]benzopyran(2,5-methano-1-benzoxepin) skeleton, thought to be derived biogenetically from the addition of a flavonoid precursor and a bisamide such as odorine (**103**), odorinol (**104**), or piriferine (**105**) (Fig. 6) (*14, 60*). Formally, rocaglamides can also be considered as derivatives from aglains by cleaving the C-C bond between C-10 and C-5a, and linking C-10 and C-5a instead. The nature of the ring system of aglains was proven unambiguously by X-ray crystallography of the first congener of this group, aglain

Fig. 6. Aglain derivatives isolated from *Aglaia* species

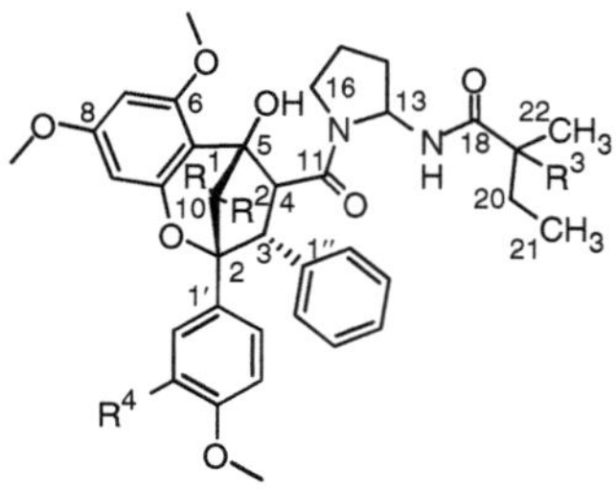

64-80

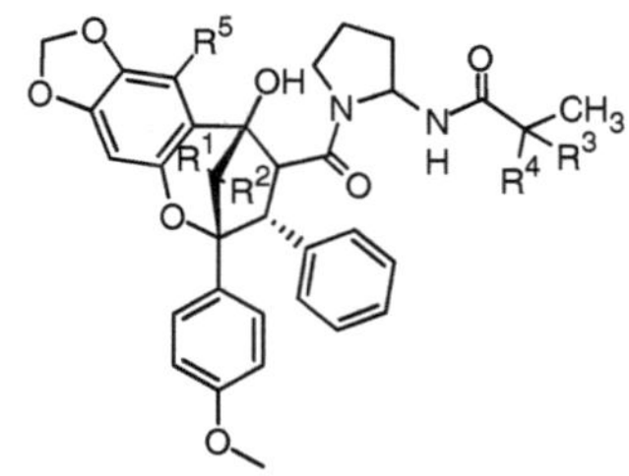

81-89

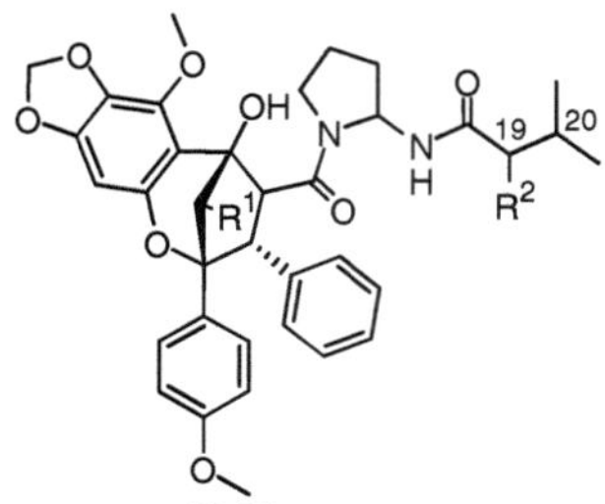

90-92

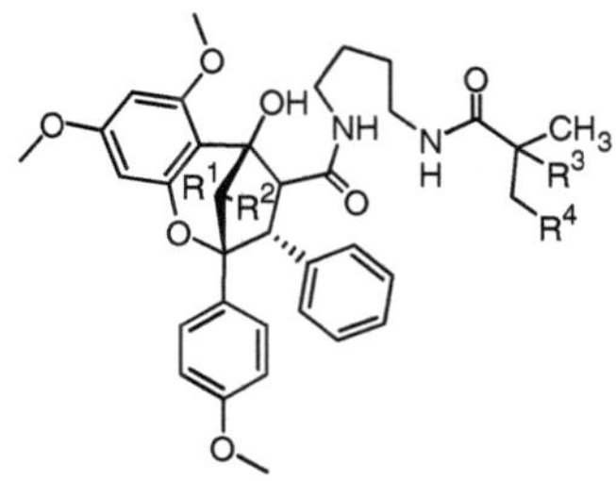

93, 94

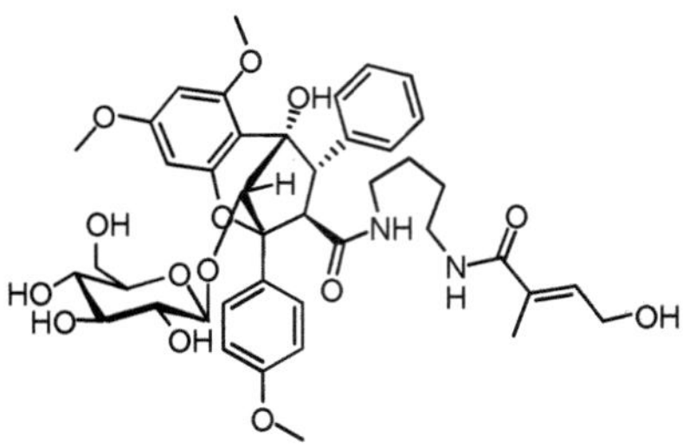

95

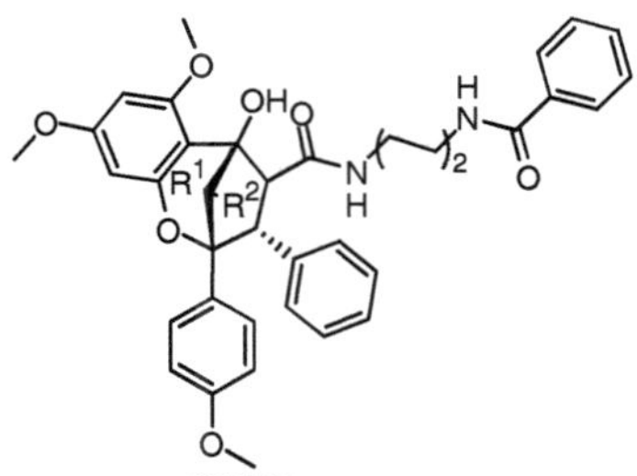

96-100

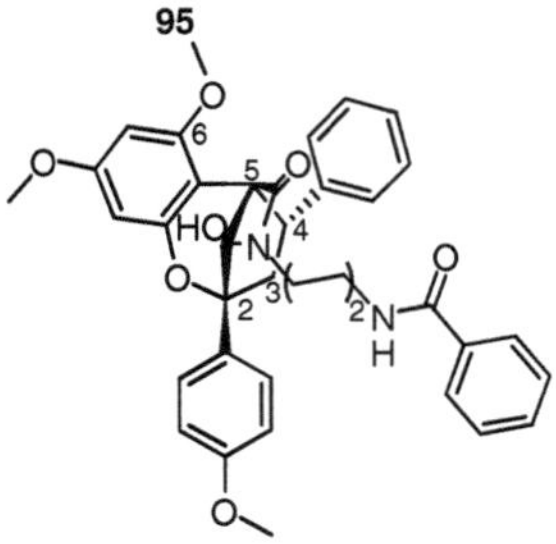

Cyclofoveoglin (**101**)

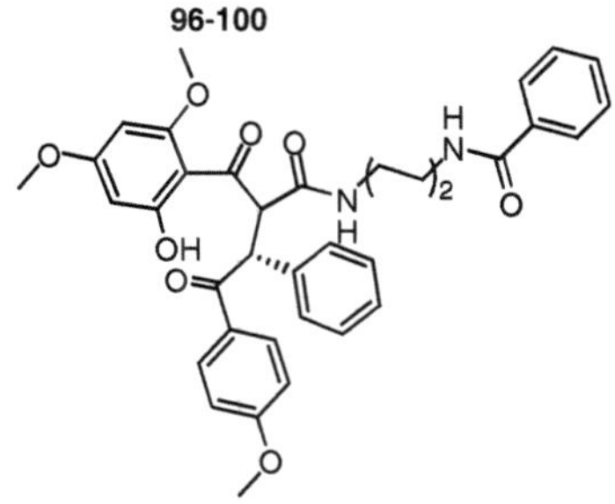

Secofoveoglin (**102**)

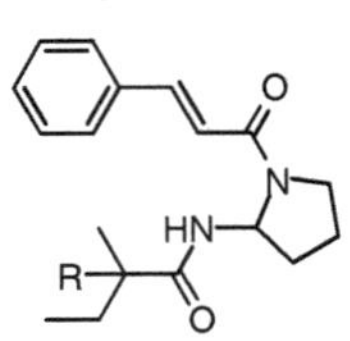

R = H: Odorine (**103**)
R = OH: Odorinol (**104**)

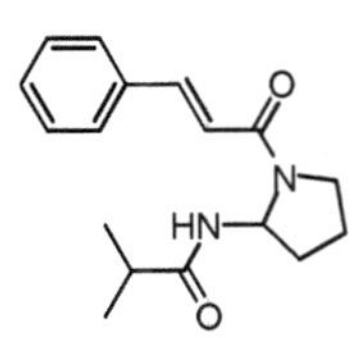

Piriferine (**105**)

Fig. 6. continued

Cpd.	Trivial Name	R^1	R^2	R^3	R^4	R^5	H-3, H-4	Ref.
64	Aglain A	$OCOCH_3$	H	H	H	-	*β,α*	(*60*)
65	4-Epiaglain A	$OCOCH_3$	H	H	H	-	*β,β*	(*80*)
66	Desacetylaglain A	OH	H	H	H	-	*β,α*	(*18*)
67	Aglain B	H	OH	H	H	-	*β,α*	(*60*)
68	10-*O*-Acetylaglain B	H	$OCOCH_3$	H	H	-	*β,α*	(*80*)
69	Aglain C	H	OH	H	H	-	*α,β*	(*60*)
70		H	OH	H	OH	-	*α,β*	(*14*)
71		H	OH	OH	OH	-	*α,β*	(*14*)
72		H	OH	OH	OCH_3	-	*α,β*	(*14*)
73		H	OH	H	OH	-	*β,α*	(*14*)
74	Aglaxiflorin A	$OCOCH_3$	H	OH	H	-	*β,α*	(*67*)
75	Aglaxiflorin B	$OCOCH_3$	H	OH	H	-	*α,β*	(*67*)
76	Aglaxiflorin D	H	OH	OH	H	-	*α,β*	(*67*)
77	Elliptifoline	H	OH	$\Delta^{19,20}$	H	-	*α,β*	(*63*)
78	Aglaroxin J	H	$OCOCH_3$	H	H	-	*α,β*	(*57*)
79	Aglaroxin L	H	OH	OH	H	-	*α,β*	(*57*)
80		H	OH	OH	H	-	*β,α*	(*57*)
81		H	OH	OH	C_2H_5	H	*β,α*	(*57*)
82		H	$OCOCH_3$	H	C_2H_5	H	*β,α*	(*57*)
83	(13*S*)-Thapsakin B	H	OH	H	CH_3	OCH_3	*β,α*	(*15*)
84	(13*R*)-Thapsakin B	H	OH	H	CH_3	OCH_3	*β,α*	(*15*)
85	Isothapsakin B	OH	H	H	CH_3	OCH_3	*β,α*	(*15*)
86	Homothapsakin A	H	OH	H	C_2H_5	OCH_3	*α,β*	(*15*)
87	10-*O*-Acetylthapsakin A	H	$OCOCH_3$	H	CH_3	OCH_3	*α,β*	(*15*)
88	Thapsakon A	=O		H	CH_3	OCH_3	*α,β*	(*15*)
89	Thapsakon B	=O		H	CH_3	OCH_3	*β,α*	(*15*)
90	Edulirin A	OH	H	-	-	-	*β,α*	(*72*)
91	10-*O*-Acetyledulirin A	$OCOCH_3$	H	-	-	-	*β,α*	(*72*)
92	19,20-Dehydroedulirin A	OH	$\Delta^{19,20}$	-	-	-	*β,α*	(*72*)
93	Grandiamide A	$OCOCH_3$	H	$\Delta^{19,20}$	CH_3	-	*α,β*	(*89*)
94		*O*-glc	H	$\Delta^{19,20}$	CH_2OH	-	*α,β*	(*75*)
96	Pyrimidaglain A	H	$OCOCH_3$	-	-	-	*β,α*	(*36*)
97	Pyrimidaglain B	$OCOCH_3$	H	-	-	-	*α α*	(*36*)
98	Desacetylpyrimidaglain A	H	OH	-	-	-	*α,β*	(*46*)
99	Desacetylpyrimidaglain C	OH	H	-	-	-	*α,β*	(*46*)
100	Desacetylpyrimidaglain D	H	OH	-	-	-	*β,α*	(*46*)

A (**64**), thus also revealing the relative configuration (*60*). The aglain skeleton was confirmed through key HMBC correlations including H-10 to C-5, C-5a, H-4 to C-11, C-5, C-5a, and H-3 to C-2″/6″, C-2, C-5 (*60*).

The aglains, aglaforbesins as well as the forbaglins contain bisamide side chains that are derived from a cinnamic acid bisamide. These low molecular weight precursors, namely, odorine (**103**) (*76, 77*), odorinol (**104**) (*19, 76*), and piriferine (**105**) (*78*), are composed of cinnamic acid, the bifunctional amine 2-aminopyrrolidine, and 2-methylbutanoic acid (in odorine), 2-hydroxy-2-methylbutanoic acid (in odorinol) or 2-methylpropanoic acid (in piriferine). In 33 of the total of 37 aglain derivatives isolated so far, the bisamide side chain is directly analogous to a naturally occurring cinnamic acid bisamide, odorine (**103**), odorinol (**104**), or piriferine (**105**). The four remaining compounds, **77** and **92–94**, can be formally obtained by dehydration of the hydroxy group at C-19 resulting in a double bond between C-19 and C-20. Aglains differ in regard to their configuration at C-19, which can be either (*R*) or (*S*), but more often remains uncertain. This finding parallels the situation of the cinnamic acid bisamides, which also occur as diastereomers at the analogous position. Similarly, the configuration at C-13 can either be (*R*) or (*S*), which again is consistent with the occurrence of both (+)- or (−)-forms of odorine (**103**), odorinol (**104**), and piriferine (**105**) in Nature. It is noteworthy that this aminal position is prone to epimerization in low molecular weight precursors (*77*), and, frequently, aglains are isolated as diastereomeric mixtures (*15*).

In their cyclic core, aglains display structural variability at the following positions: the bridging carbon atom, C-10, nearly always carries one proton as well as one oxygen-containing substituent, the latter being either a hydroxy, an acetoxy, or a sugar moiety. The substituents can be either *endo* or *exo* with regard to ring A. Only two derivatives, **88** and **89**, are known to feature a carbonyl group at C-10. In the oxepine ring, H-3 and H-4 are mostly *trans*-oriented, but both possible diastereomeric forms, *i.e.* H-3α, H-4β as well as H-3β, H-4α, occur more or less evenly distributed in Nature. For compounds with the opposite configuration at C-10, N*O*E correlation peaks are observed between the β-protons, H-3 or H-4, and OH-10 (in aprotic solvent) or $OCOCH_3$-10. Furthermore, these N*O*Es have also been used to assign the relative configurations of the H-3 and H-4 stereocenters. Additionally, the vicinal coupling constant between H-3 and H-4 can be utilized to confirm their configurations. For the H-3β, H-4α configuration, the ^{1}H NMR vicinal coupling constant varies between 5 and 6 Hz, while for the H-3α, H-4β configuration, the coupling constant amounts to 9–11 Hz (*14, 60, 67*). One exception is 4-epiaglain A (**65**), which features the β-configuration for both H-3 and H-4, and displays a coupling constant of 7.4 Hz (*79*).

As in the case of rocaglamides, ring A of aglains is usually substituted by two *m*-positioned methoxy groups at C-6 and C-8, but is also known to carry a 7,8-methylenedioxy substituent, mostly in addition to the methoxy group at C-6 except for congeners **81** and **82**, which feature no methoxy group at C-6. Ring B always carries a 4′-methoxy substituent, in some cases accompanied by a hydroxy or a methoxy group at C-3′, while ring C is always unsubstituted. These substitution patterns are again parallel to those of rocaglamide, whereas a methylenedioxy substituent in ring B has not been encountered in aglains so far.

In spite of the numerous structural analogies between rocaglamides and aglains, and the postulated similar biogenetic pathways leading to both classes of compounds, it is interesting to note that bisamide-derived side chains occur mainly in aglains (and in aglaforbesins as well as in forbaglins, see below), but are rarely encountered in rocaglamides such as in isothapsakon A (**47**). It may be speculated that bulky substituents, such as those present in odorine (**103**), odorinol (**104**), or piriferine (**105**), cannot easily be incorporated into rocaglamides, and thus are usually replaced by simpler amide or nitrogen-free side chains.

The assignment of the relative configuration of aglain A (**64**), the parent compound of this series of cyclopenta[*bc*]benzopyran derivatives, was determined from N*O*ESY NMR correlations (*60*). In 2000, the first X-ray structure of this type of compounds was obtained for aglaxiflorin A (**74**), thus confirming the relative stereochemistry (*67*). Previously, the relative configuration of aglains had been assigned from 2D N*O*E data, while the absolute configuration was deduced on the grounds of biogenetic comparison with rocaglamide (**1**). According to *Greger* and colleagues, formal conversion of cyclopenta[*bc*]benzopyran into cyclopenta[*b*] benzofuran would leave the absolute configuration at C-2 (C-3a in rocaglamides) unchanged, as was deduced by inspecting *Dreiding* models (*15*). Thus, the structures of aglain derivatives are commonly drawn with the methylene bridge (C-10) oriented upwards, while the aromatic ring B and OH-5 are oriented downwards (*9*).

Aglains A (**64**), B (**67**), and C (**69**) were isolated from the leaves of *Aglaia argentea* collected in Malaysia (*60*), while 4-epiaglain A (**65**) and 10-*O*-acetylaglain B (**68**) were obtained from an Indonesian collection of *A. elliptica* leaves (*79*). The relative configurations of **65** and **68** were solved using N*O*ESY NMR data. Deacetylaglain A (**66**), isolated from the leaves of *A. gracilis* collected in Fiji (*18*), is very similar to aglain A (**64**), except for the hydroxy group at C-10. Recently, ponapensin, the only congener featuring a methoxy group at C-13 instead of the amide side chain in aglain B (**67**), was isolated from the Micronesian species *Aglaia ponapensis* (*80*).

In thapsakones A (**88**) and B (**89**), obtained from the root bark of *Aglaia edulis* (southwest Thailand), which lack a proton at C-10, the stereochemistry of H-3 and H-4 was deduced in an elegant manner by observing a shift stronger than a lanthanide-induced shift (LIS) to the respective β-proton (4 in **88**, 3 in **89**) (*15*). The configuration of the aminal proton H-13 was assigned as being (13*S*) by observing N*O*Es between H-4 and H-13 as well as between the terminal methyl group(s) H-21 (and H-20 in the case of piriferine-derived side chains) and H-2″/6″, while no such N*O*E correlations were detected for (13*R*)-derivatives as confirmed by close inspection of *Dreiding* models (*15, 60*).

Edulirin A (**90**), 10-*O*-acetyledulirin A (**91**), and 19,20-dehydroedulirin A (**92**), together with aglaroxin A analogues **42** and **43**, were reported from an Indonesian collection of the bark of *Aglaia edulis* (*71*).

The two glycosidic derivatives **94** and **95** have been isolated from the leaves of *Aglaia dasyclada* collected in Yunnan Province, People's Republic of China (*64*). These two compounds have a hydroxytiglic amidic putrescine moiety instead of the cinnamic acid bisamides previously found as the amine substituents in other aglain derivatives.

The last group of aglain congeners with compounds **96–100** exhibits a benzoyl-1,4-butanebisamide moiety at C-4 along with the open oxepine ring congener, secofoveoglin (**101**). Pyrimidaglain A (**96**) and B (**97**) were the first congeners of this group isolated from the leaves of *Aglaia andamanica* collected in Thailand (*36*). Recently, three further congeners, desacetylpyrimidaglains A, C, and D (**98–100**), have been reported from the leaves of *A. forbesii* collected also in Thailand (*46*). The latter has been given the trivial name, isofoveoglin, and was isolated together with the open oxepine ring congener, secofoveoglin (**102**), from the leaves and stem bark of *A. foveolata* (Indonesia) (*81*). The only difference between the pyrimidaglains **96** and **97** and the desacetylpyrimidaglains **98–100** is the lack of acetylation of the OH-10 function in the latter compounds. The relative configurations of the deacetyated pyrimidaglains has been proven through the observation of the characteristic N*O*ESY cross peaks H-3 to H-4, N*H*-12, H-2′/6′, and H-2″/6″, H-4 to H-3, OH-10, N*H*-12, and H-2″/6″, and H-10 to H-2′/6′, and the most important cross peak between OH-10 and H-4, which directly proved the relative configuration at C-3, C-4, and C-10 (*46*). Cyclofloveoglin (**101**), isolated from the leaves and stem bark of *A. foveolata* (Indonesia) (*81*), represents a hitherto unprecedented five membered-cyclic amide moiety among the rocaglamide-type compounds isolated from the genus *Aglaia* so far (*9, 12*). The structure of **101** was proposed through the DEPT NMR spectrum, which revealed a quaternary carbon resonance at δ 90.6 ppm that replaced the signal of a hydroxymethine carbon at position C-10 in **100**. Furthermore, a HMBC spectrum confirmed the structure of cyclofloveoglin through correlations between the quaternary carbon, C-10, with H-4 and H-13, indicating that N-12 is bonded to C-10 (*81*).

2.3. Aglaforbesin Derivatives

The aglaforbesins are closely related to the aglains, but with a cinnamic acid bisamide-derived side chain at C-3 and the unsubstituted phenyl ring C at C-4 mutually interchanging (as in congener **95**). This structural feature was evidenced by HMBC correlations from H-3 to C-11 as well as H-4 to C-2″/6″ (*60*). To date, only ten aglaforbesin derivatives (see Fig. 7) have been described from Nature, which differ with regard to the substitution pattern of ring A as well as in the stereochemistry at C-3, C-4, and C-13. Unlike the aglains, no structural variants from the 4′-methoxy substituted ring B are known, however, in ring A, a methylendioxy functionality between C-7 and C-8 has been reported in the three congeners **109–111** (*16, 71*). Side chains are derived from odorine (**103**) (in **106** and **107**) (*60*), odorinol (**104**) (in **108**) (*67*), and piriferine (**105**) (in **109**) (*16*). However, foveoglins A (**112**) and B (**113**) feature a benzoyl-1,4-butanebisamide moiety at C-3 (*71, 81*) unlike the pyrimidaglains **96–100**, which exhibit the same moiety at C-4 (*36, 46*).

Assignment of the stereochemistry of aglaforbesins is based on the same principles as for aglains. Consequently, the configuration of the aminal proton H-13 was deduced as being (*R*) in aglaforbesins A (**106**) and B (**107**) due to

Fig. 7. Aglaforbesin derivatives isolated from *Aglaia* species

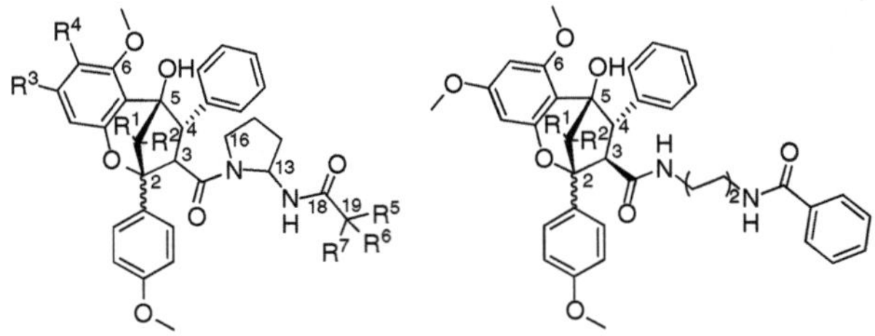

106-111　　112, 113

Cpd.	Trivial Name	R^1	R^2	R^3	R^4	R^5	R^6	R^7	H-3, H-4	Ref.
106	Aglaforbesin A	OH	H	OCH_3	H	C_2H_5	H	CH_3	*α,β*	*(60)*
107	Aglaforbesin B	H	OH	OCH_3	H	C_2H_5	H	CH_3	*α,β*	*(60)*
108	Aglaxiflorin C	H	$OCOCH_3$	OCH_3	H	C_2H_5	OH	CH_3	*α,β*	*(67)*
109		H	OH	$-OCH_2O-$		CH_3	H	CH_3	*β,α*	*(71)*
110	Isoedulirin A	H	OH	$-OCH_2O-$		H	H	$CH(CH_3)_2$	*α,β*	*(72)*
111	Edulirin B	H	OH	$-OCH_2O-$		H	H	$CH(CH_3)_2$	*β,α*	*(72)*
112	Foveoglin A	OH	H	-	-	-	-	-	*α,β*	*(82)*
113	Foveoglin B	H	OH	-	-	-	-	-	*α,β*	*(82)*

NOEs observed between H-3 and H-13 as well as between H-21 and H-2″/6″ *(60)*. Interestingly, the H-3α/H-4*β* configuration leads to a pronounced upfield shift of OCH_3-6 (*δ* approx. 3.1 ppm), since in this case the methoxy group is placed inside the shielding zone of the unsubstituted benzene ring at C-4α *(60, 67)*, while a normal chemical shift (*δ* approx. 4.1 ppm) is observed in the case of reversed stereochemistry at C-3 and C-4 *(16)*. By analogy to the aglains, configurations at their respective positions are also reflected by the magnitude of the vicinal coupling constant: $^3J_{(H\text{-}3,\ H\text{-}4)}$ amounts to 10–11 Hz when H-3 is α and H-4 is *β* *(60, 67)*, while the coupling constant is 6–7 Hz when in the opposite configuration *(16)*.

2.4. *Forbagline Derivatives*

Forbaglines are benzo[*b*]oxepines naturally occurring in the genus *Aglaia*, in which the pyran ring of the aglains is replaced by an oxepine ring. The benzo[*b*]oxepine skeleton of the forbagline derivatives can be formally obtained from the aglains by oxidative cleavage at the methylene bridge between C-5 and C-10 *(60)*. As for the aforementioned groups of rocaglamide-type compounds, the aromatic rings A, B, and C share common characteristics with their benzofuran and benzopyran counterparts. The aromatic ring A can carry either an 8-methoxy or a 7,8-methylenedioxy substituent in addition to a 6-methoxy group, while ring B may show a *p*-methoxy (as in **114–124**) or a *p*-hydroxy (as in **125**) substituent, and ring C is unsubstituted. The benzo[*b*]oxepine core is conserved in all but derivative **125**, which has a carboxylic acid functional group instead of the methyl ester group at C-10 (*64*). The only major variation in the skeleton occurs in the type of the bisamide side chain substituent at C-4.

The structure of the first derivative, forbaglin A (**114**), was established by X-ray crystallographic analysis, thus revealing the relative stereochemistry (*60*). The configurations at H-3 and H-4 of the forbagline derivatives reflect those of the aglains and aglaforbesins with only *trans* isomers having been isolated so far. By analogy to the benzopyran series, the magnitude of the vicinal coupling constant $^{3}J_{(H\text{-}3,\ H\text{-}4)}$ can be used to determine the relative stereochemistry at C-3 and C-4.

To date, 12 forbagline derivatives (see Fig. 8) have been isolated, including the 7 derivatives **114–120** with an odorine or a piriferine side chain. The other 5 analogues **121–125** revealed bisamide side chains derived from substituents other than odorine or piriferine (*64, 82*). Both derivatives **124** and **125** have a hydroxytiglic amidic putrescine moiety similar to that of **94** and **95**, and all of them were isolated from the same *Aglaia* species (*64*). Compound **124** is the only forbagline glucoside derivative isolated so far, with the glucose attached to C-21 (the bisamide side chain), whereas compound **125** has a very similar structure to **124** except for

Fig. 8. Forbagline derivatives isolated from *Aglaia* species

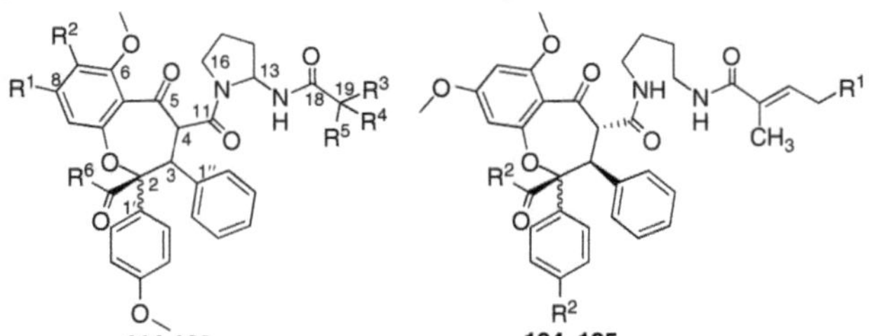

114-123 **124, 125**

Cpd.	Trivial Name	R1	R2	R3	R4	R5	R6	H-3, H-4	Ref.
114	Forbaglin A (13*R*)	OCH_3	H	CH_3	H	C_2H_5	CH_3	*α,β*	(*60*)
115	Forbaglin B (13*S*)	OCH_3	H	CH_3	H	C_2H_5	CH_3	*α,β*	(*60*)
116	(13*S*)-Thapoxepine A	-OCH_2O-		CH_3	H	CH_3	CH_3	*α,β*	(*15*)
117	(13*R*)-Thapoxepine A	-OCH_2O-		CH_3	H	CH_3	CH_3	*α,β*	(*15*)
118	Homothapoxepine A	-OCH_2O-		CH_3	H	C_2H_5	CH_3	*α,β*	(*15*)
119	(13*S*)-Thapoxepine B	-OCH_2O-		CH_3	H	CH_3	CH_3	*β,α*	(*15*)
120	(13*R*)-Thapoxepine B	-OCH_2O-		CH_3	H	CH_3	CH_3	*β,α*	(*15*)
121	Edulisone A (13*R*)	-OCH_2O-		H	H	$CH(CH_3)_2$	OCH_3	*α,β*	(*83*)
122	Edulisone B (13*S*)	-OCH_2O-		H	H	$CH(CH_3)_2$	OCH_3	*α,β*	(*83*)
123	19,20-Dehydroedulisone	-OCH_2O-		$\Delta^{19,20}$		$C(CH_3)_2$	OCH_3	*α,β*	(*72*)
124		*O*-glc	OCH_3	-	-	-	-	*α,β*	(*64*)
125		OH	OH	-	-	-	-	*α,β*	(*64*)

the absence of the sugar moiety at C-21, and the presence of a carboxylic acid and a hydroxy functional group at positions C-10 and C-4′ (*64*).

Edulisones A (**121**), B (**122**), and 19,20-dehydroedulisone A (**123**) were isolated from the bark of *Aglaia edulis* collected in Indonesia (*71, 82*). The relative stereochemistry of edulisone A (**121**) was determined by single-crystal X-ray diffraction analysis, revealing the (*R*) configuration at C-13 (*82*). Furthermore, the two epimers **121** and **122** showed different ^{1}H NMR chemical shifts for protons close to the C-13 epimeric site, which may be used to assign the relative stereochemistry at C-13. For the (13*R*)-epimer **121**, H-14a and H-14b displayed two signals in the ^{1}H NMR spectrum, while for the (13*S*)-epimer, these two protons were overlapped in a relatively upfield region (*82*). For H-16a and H-16b of the (13*R*)-epimer, the two protons overlapped in the ^{1}H NMR spectrum, while for the (13*S*)-epimer, these two protons were clearly separated, one at a higher field and one at a lower field relative to those of its (13*R*)-counterparts (*82*). The same phenomenon was also observed in forbaglins A ((13*R*), **114**) and B ((13*S*), **115**) (*60*).

3. Biosynthesis of Rocaglamides and Related Metabolites

The cyclopenta[*b*]benzofurans (rocaglamides), and the two structurally related groups, the cyclopenta[*bc*]benzopyrans (including the aglains and aglaforbesins), and the benzo[*b*]oxepines (known also as the "forbaglines"), are considered characteristic secondary metabolites of the genus *Aglaia*, because they have been only isolated from this taxon (*9*). Therefore, the collective name "flavagline" has been proposed for these compounds because their mutual biogenetic origin has been postulated to arise from common structurally related precursors that include cinnamic acid amides and the flavonoid nucleus (*9, 13–15*). A postulated biosynthetic origin was firstly proposed by *Nugroho et al.* in 1999 as depicted in Fig. 9 (*14*). According to this hypothesis, the initial *C-C*-connecting step (step A) between C-2 of the flavonoid **I** and C-3 of the cinnamic acid amide **II** is a *Michael*-type 1,4-addition of the enolate subunit of **I** to the α,β-unsaturated amide **II**. The C-2 atom of the resulting amide enolate of **III** can now attack C-4 of the previous flavonoid, which has now become a strongly activated carbonyl group, to yield a five-membered ring, giving rise to **IV** (step B). According to this concept, **IV** constitutes the biosynthetic key intermediate and precursor both to aglain and rocaglamide derivatives. Moreover, **IV** can already be considered as a dehydroaglain derivative, and a simple reduction step (*e.g.* with $[H]^-$ possibly through NADPH or a related H-nucleophile), will yield the corresponding aglain derivative **V'** (step C').

This reduction to give **V** stabilizes the strained molecule **IV**, which, as the key intermediate, may otherwise undergo a rearrangement by an intramolecular migration of the electron-rich substituted (phloroglucinol-type) aromatic ring from the previous C-4 to C-3 of the flavonoid. Mechanistically, this can be considered as an electrophilic aromatic *ipso*-substitution via the cyclopropyl derivative **V** as the

Fig. 9. Postulated joint biosynthesis scheme of rocaglamide (= flavagline)-type compounds isolated from *Aglaia* species (*14*)

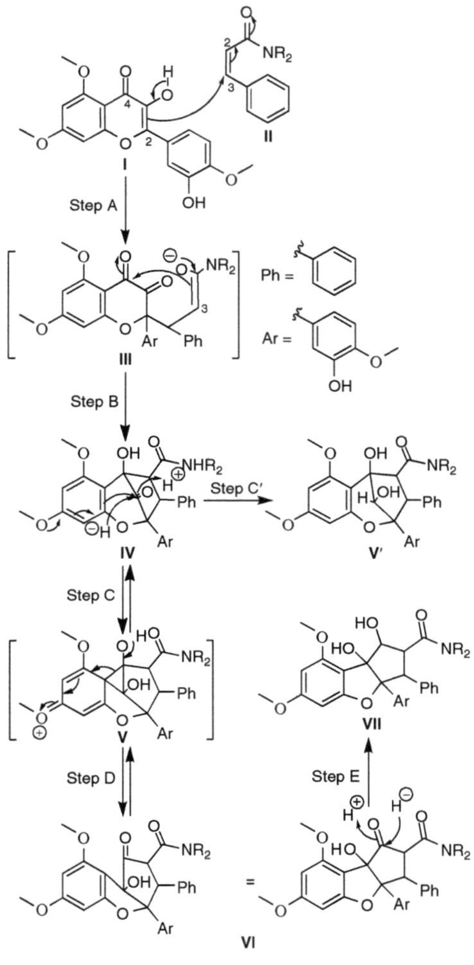

σ-complex (steps C and D), thus ultimately transforming the hydroxyketone **IV** into the isomeric hydroxyketone **VI**, which is already a dehydrorocaglamide derivative. Again, this is possibly a reversible process, which becomes definite by a stabilizing final reduction step (step E), to give rise to rocaglamide derivatives **VII**.

Although aglaforbesin derivatives are not depicted in Fig. 9, they also fit into the biogenetic scheme proposed, but differ in comparison to the aglains by the opposite orientation of the cinnamic acid amide **II** with respect to flavonoid **I**. In addition, forbaglines can be proposed as being biosynthesized through oxidative cleavage between C-5 and C-10 of hydroxyketone **IV** (numbering as in aglains and aglaforbesins). Apparently, the addition of **II** to **I** is neither regio- nor stereoselective, since all four possible stereoisomers do exist in Nature, *i.e.* both (H-3α,H-4β) and (H-3β, H-4α) derivatives have been reported.

From these aforementioned considerations, a search for the probably unstable, possibly interconverting, intermediates **IV** and **VI** of the postulated biosynthesis pathway of dehydroaglain and rocaglamide derivatives, would provide further support for the hypothetical biogenetic scheme and be rewarding, if successful.

For the bisamide-containing flavaglines, a proposed biosynthesis pathway is depicted in Fig. 10, in which a similar initial *C*-*C*-addition occurs between C-2 of

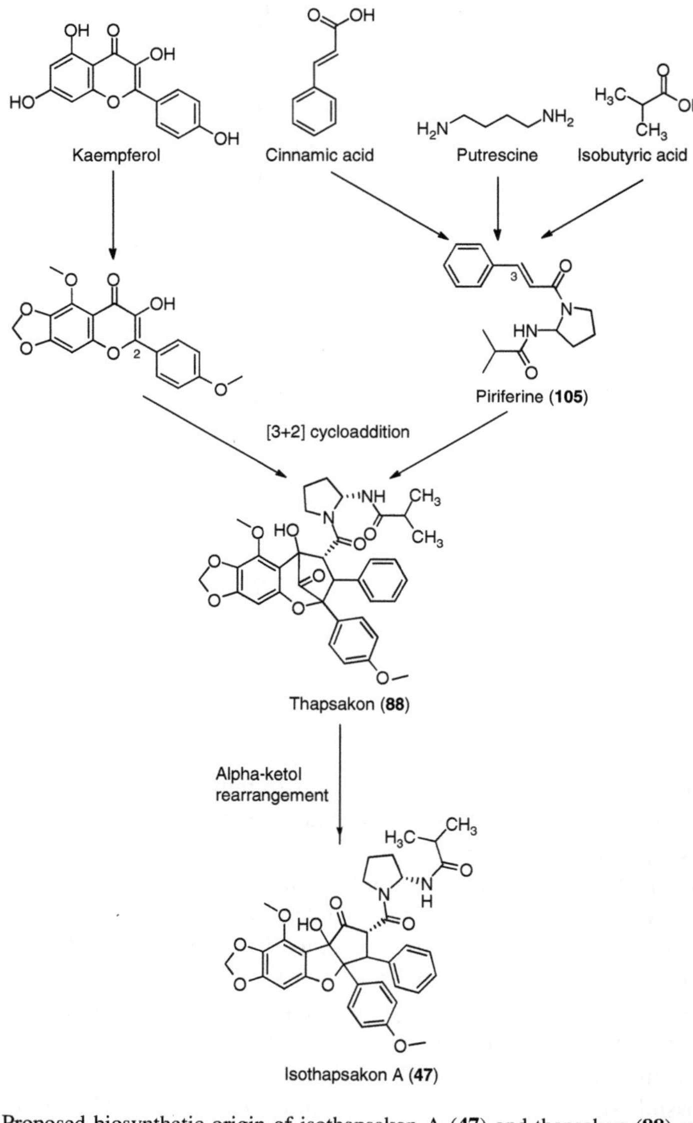

Fig. 10. Proposed biosynthetic origin of isothapsakon A (**47**) and thapsakon (**88**) as bisamide-containing rocaglamide-type derivatives

the flavonoid and C-3 of piriferine (**105**), odorine (**103**), or odorinol (**104**). This step leads to the formation of the aglain derivative, thapsakon A (**88**), which can be transformed into isothapsakon A (**47**) through an α-ketol rearrangement similar to steps (C–E) in Fig. 9.

4. Pharmacological Significance of Rocaglamides and Related Compounds

4.1. Insecticidal Activity

The first report on the remarkable insecticidal activity of extracts from *Aglaia* species in the literature originates from 1985, when Chiu *et al.* (*83*) described the antifeedant properties of a crude extract derived from *A. odorata* towards larvae of the cabbage worm, *Pieris rapae*. The active principles of *A. odorata* responsible for the strong antifeedant properties of the respective crude extracts against *P. rapae* and against other insects, however, were only identified in 1993 when *Ishibashi et al.* (*53*) reported on the bioassay-guided isolation of rocaglamide (**1**) and three of its congeners using larvae of the polyphagous noctuid, *Peridroma saucia*, as experimental insects. Interestingly, rocaglamide (**1**) had already previously been isolated in 1982 from *A. elliptifolia* and described as having inhibitory properties against a human epidemoid carcinoma of the nasopharynx (KB) cell line *in vitro* (IC_{50} = 8.7 μ*M*) (*11*), thereby providing an early hint to the parallelism between insecticidal and antiproliferative properties of rocaglamide and its derivatives that became even more obvious in subsequent studies (*84, 85*).

Sparked by these findings, a more directed search for new rocaglamide derivatives from Nature aiming also at a better understanding of their pronounced insecticidal properties commenced, which up to now has led to the isolation of more than 60 naturally occurring rocaglamide congeners as well as numerous other biogenetically related compounds isolated from over 30 different *Aglaia* species collected mainly in Indonesia, the People's Republic of China, Thailand, and Vietnam.

The majority of the 63 naturally occurring rocaglamide derivatives known so far were analyzed for their insecticidal activity employing various Lepidopteran larvae including *Peridroma saucia* (*53*), *Ostrinia nubilalis* (*86*), *Helicoverpa armigera* (*87*), or *Spodoptera littoralis* as experimental insects. Most of these studies, however, have been performed with larvae of the polyphagous pest insect *Spodoptera littoralis* (*14, 16, 17, 54–56, 58, 59*) thereby permitting the formulation of preliminary structure-activity relationships based on a larger set of data obtained with the same insect species. In the studies with *S. littoralis*, different rocaglamide derivatives were usually added at a range of concentrations to artificial diet, which was subsequently offered to newly hatched larvae in a no choice chronic feeding bioassay. From the larval survival rates, LC_{50} values can be calculated for

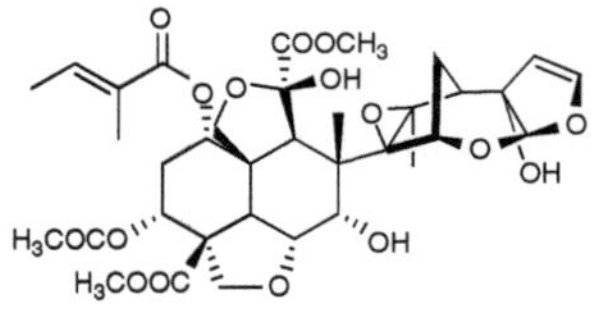

Fig. 11. Chemical structure of azadirachtin from *Azadirachta indica* (family Meliaceae)

the various rocaglamide congeners and compared to other insecticides of natural origin, such as azadirachtin from *Azadirachta indica* (family Meliaceae) (Fig. 11). With only a few exceptions, all naturally occurring rocaglamide congeners evaluated exhibited potent insecticidal activity toward the larvae of *S. littoralis*. The most active compounds, including the parent compound rocaglamide (**1**) itself or its didesmethyl analogue **15**, exhibited LC_{50} values ranging between 1 and 2 ppm, so their insecticidal activity was thus comparable to that of azadirachtin (*14, 54, 55, 59*).

Acylation of the OH group at C-1 (*e.g.* with formic or acetic acid) led to a reduction of insecticidal activity, as exemplified by comparison of the LC_{50} values of compound **2** (1.5 ± 0.7 ppm) and its acetyl derivative **5** (8.0 ± 1.4 ppm) or of congener **12** (1.1 ± 0.6 ppm) and its acetyl derivative **13** (14.7 ± 2.8 ppm) (*14, 54, 55*). On the other hand, the nature of the amide substituent present at C-2 showed little or no influence on the insecticidal activity of the resulting rocaglamide congeners, even when the dimethylamino group present in the parent compound rocaglamide (**1**) (LC_{50} of 0.9 ± 0.4 ppm) was exchanged for a rather bulky group, as, for example, in compound **14** (LC_{50} of 1.6 ± 0.6 ppm). The same findings were true when the amide group was exchanged to an ester substituent, which likewise had no significant effect on the resultant insecticidal activity of the respective compounds. However, a diminution in insecticidal activity by a factor of 5 or 6 was usually observed for rocaglamide derivatives featuring an unsubstituted C-2 when compared to analogues with an amide or carboxylic acid ester substituent at this particular position (*14, 54, 55, 59*).

In rings A or B, the occurrence of an additional oxygen substituent, when compared to the substitution pattern of the parent compound, rocaglamide (**1**), revealed only marginal influences on the insecticidal activity of the respective products (*54*). However, a dramatic effect with regard to structure-activity relationships of rocaglamide derivatives was observed for analogues with a substituted OH group at C-8b. For example, compounds featuring a methoxy group at C-8b that were isolated from the roots of *A. duperreana* proved to be completely inactive as insecticides even when tested at concentrations of more than 100 ppm. Thus, these results point out the presence of a free OH group at C-8b as being the most important structural prerequisite for insecticidal activity of rocaglamide analogues determined so far (*56, 58*).

It is difficult to decide whether the mortality of the *S. littoralis* larvae observed in chronic feeding bioassays described above is mainly caused by starvation due to feeding deterrence or by a direct toxicity of the rocaglamide derivatives evaluated or from a combination of both effects. When neonate larvae of *S. littoralis* were given the choice between an artificial diet treated with rocaglamide (**1**) and a control diet, they avoided the former and showed a clear preference for the latter (the IC_{50} values in these experiments varied between 0.2 and 0.25 ppm), indicating that **1** and its congeners have strong antifeedant properties (*54, 55*). The toxicity of rocaglamide (**1**) was proven by injecting known amounts of this cyclopenta[*b*] benzofuran derivative into the hemolymph of last instar larvae of *S. littoralis*. In these experiments, the LC_{50} of rocaglamide varied between 5.6 and 7.5 ppm (*54, 55*). Further proof for the effects of **1** on a cellular level was obtained using *in vitro* cultures of *Spodoptera frugiperda* cells. Addition of rocaglamide (**1**) to the *in vitro* cultures resulted in an arrest of cellular division as indicated by the severely reduced incorporation of [^{3}H]-thymidine. The IC_{50} of **1** amounted to 1.9 μg/cm^3 (= 3.8 μ*M*) (*84, 85*).

A more recent study was conducted to assess the insecticidal activity of rocaglamide (**1**) isolated from *Aglaia elaeagnoidea* against the gram pod borer, *Helicoverpa armigera* (Hübner) (*87*). In this study, rocaglamide was added to an artificial diet and this led to growth retardation of neonate larvae in a dose-dependent manner with an IC_{50} value of 0.76 ppm, which could be compared to that of azadirachtin (IC_{50} = 0.23 ppm). However, azadirachtin (Fig. 11) was determined as being more potent than rocaglamide in inducing growth inhibition via oral administration to the first stadium larvae utilized. By topical application, rocaglamide (**1**) was found to have LD_{50} and LD_{95} values of 0.40 and 1.02 μg per larvae against third instar larvae 96 h post-treatment, whereas the analogous values for azadirachtin were 8.16 and 25.8 μg per larva for the same period (*87*).

In spite of the severe morphological larval deformities observed in azadirachtin-treated larvae during the process of ecdysis, the cytotoxic nature of rocaglamide (**1**) was established by evaluating dietary utilization, for which the results did not implicate any antifeedant effect, but rather a toxicity-mediated effect due to the reduced efficiency of the conversion of ingested food. Conclusively, feeding deterrence was deduced not to be the primary mode of insecticidal activity of rocaglamide, with this compound having instead a centrally mediated effect, which could be due to induced cytotoxicity at non-specific cellular levels (*87*).

However, the molecular target of rocaglamide (**1**) and its congeners in insects is still unknown, and the insecticidal activity of these compounds can be linked to distinct structural features, such as the OH group at C-8b, which is an indispensable prerequisite for bioactivity. Interestingly, rocaglamide derivatives often co-occur in *Aglaia* species with biogenetically closely related compounds of the aglain, aglaforbesin, or forbagline type (*14, 16, 60*). In these latter compounds, the oxygen heterocycle of the dihydrobenzofuran nucleus in rocaglamides is replaced by a bridged pyran or by an oxepine ring. These structural differences, however, lead to a complete loss of insecticidal activity for aglain, aglaforbesin, or forbagline derivatives (*14, 16*). The putative biogenetic precursors of rocaglamides as well as aglains, aglaforbesins, and forbaglines – methylated flavonoids and 2-aminopyrrolidines, such as odorine (**103**) – are likewise devoid of any significant insecticidal activity (*54, 55*), implying that the integrity of the cyclopenta[*b*]-tetrahydrobenzofuran moiety of the rocaglamide skeleton is essential for the insecticidal activity of this unique group of natural products.

4.2. Anti-inflammatory Activity

In several countries of Southeast Asia (*e.g.* Vietnam), the leaves and flowers of *A. duperreana* and *A. odorata* are used in traditional medicine for the treatment of asthma and inflammatory skin diseases. Inflammatory diseases arise from inappropriate activation of the immune system, leading to abnormal expression of genes encoding pro-inflammatory cytokines and tissue-destructive enzymes (*88*). Overproduction of cytokines such as TNF-α and IFN-γ has been shown to be tightly associated with autoimmune and inflammatory diseases, whereas, uncontrolled expression of the cytokine IL-4 causes allergic disorders including asthma (*88, 89*). It has been shown that rocaglamide (**1**), 1-*O*-acetylrocaglamide (**4**), and 1-oxo-11,12-methylendioxyrocaglaol (**50**) can inhibit TNF-α, IFN-γ, and IL-4 production in human peripheral blood T cells at very low doses (25–50 n*M*) (*25*). This effect may partially explain the anti-inflammatory and anti-asthmatic activities of *Aglaia* species. Importantly, at the concentrations required for inhibition of cytokine expression, these compounds do not show obvious toxicities on primary blood T cells (*21, 25*).

Many inflammatory cytokine genes including IL-4 are regulated at the transcriptional level by pro-inflammatory transcription factors, such as NF-κB and AP-1 and by nuclear factor of activated T cells (NF-AT) (*89–91*). Several rocaglamide derivatives were reported to have an inhibitory effect on the activity of NF-κB. It was shown that at n*M* concentrations, rocaglamide (**1**), desmethyl-rocaglamide (**7**), *N,N*-didesmethyl-*N*-4-hydroxybutyl-rocaglamide (**12**), and didesmethyl-rocaglamide (**15**) inhibited NF-κB-mediated transcription induced by TNF-α or phorbol 12-myristate 13-acetate (PMA) in Jurkat leukemic T cells by more than 90%, as determined by a stably transfected NF-κB-regulated luciferase reporter gene (*24*). This effect, however, was not seen in non-lymphoid cells transiently transfected with the NF-κB-regulated luciferase reporter gene (*24*). The authors concluded that rocaglamide derivatives may be potent inhibitors of NF-κB in T lymphocytes but not in other types of cells. The NF-κB pathway can be activated by various stimuli. In resting T cells, NF-κB is sequestered in an inactive state by the cytoplasmic inhibitor of NF-κB (IκB). Stimulation of T cells, *e.g.* through the T cell receptor, the TNF receptor, or using PMA, leads to rapid activation of the IκB kinases (IKKs) and results in phosphorylation, ubiquitylation, and subsequent degradation of IκB proteins, which allows the nuclear translocation of NF-κB (*92*). Rocaglamide (**1**) probably interferes with the NF-κB activation pathway upstream of the IKK complex but downstream of the TNF receptor-associated proteins (*24*).

Recently, several rocaglamide derivatives were reported to inhibit NF-AT activity in a more sensitive and selective way than NF-κB (*25*). It was shown that rocaglamide (**1**), 1-*O*-acetylrocaglamide (**4**), and 1-oxo-11,12-methylendioxyrocaglaol (**50**) inhibit NF-AT-dependent transcription at doses that did not impair NF-κB- and AP1-mediated transcription in Jurkat T cells stimulated with PMA and ionomycin (*25*). Rocaglamide (**1**), which was previously reported to inhibit PMA-induced NF-κB activation in Jurkat T cells at concentrations of 25–100 n*M* (*24*), did not show inhibition of PMA-induced IκB degradation and also did not block PMA-induced nuclear translocation of p65 (a subunit of NF-κB) (*21, 25*). Instead, at concentrations <100 n*M*, rocaglamide (**1**), as well as 1-*O*-acetylrocaglamide (**4**) and 1-oxo-11,12-methylendioxy-rocaglaol (**50**), even substantially increased NF-κB-mediated transcription (*25*). In addition, using an enzyme-based NF-κB activity readout it was shown that rocaglamide (**1**) inhibited NF-κB activity only at a high dose (IC_{50} = 2 μ*M*) (*80*). Since rocaglamide also inhibits protein synthesis (see below), it is, therefore, unclear whether the observed inhibition of NF-κB activity at high concentrations of this compound is due to inhibition of NF-κB activation signalling pathway or is rather the consequence of translation inhibition. Apparently, the latter studies do not support rocaglamide as a potent NF-κB inhibitor. Nevertheless, a newly identified compound of the group of aglain derivatives named ponapensin, which is the only congener featuring a methoxy group at C-13 instead of an amide side chain in aglain B (**67**), was shown to inhibit NF-κB activity with an IC_{50} of 60 n*M* determined by the NF-κB ELISA method (*80*). So far, there are no further studies on whether and how ponapensin inhibits NF-κB activity in cells.

Members of the NF-AT family of proteins are calcium- and calcineurin-regulated transcription factors. In resting T cells, NF-AT proteins are phosphorylated and reside in the cytoplasm. T cell activation leads to activation of the calcium-dependent phosphatase calcineurin, resulting in rapid dephosphorylation of NF-AT which leads to its nuclear translocation and the induction of NF-AT-mediated gene transcription. NF-AT activities are negatively controlled by several kinases including glycogen-synthase kinase 3 (GSK3) and casein kinase 1 (CK1), which maintain NF-AT in a phosphorylated state in the cytosol (maintenance kinases), and inducible MAPKs p38 (the mitogen-activated protein kinases) and JNK (the stress-activated c-Jun N-terminal protein kinase), which induce the re-phosphorylation of nuclear NF-AT to expose a nuclear-export signal and translocate NF-AT back to the cytosol (export kinases) (*91*). Investigation of the molecular mechanism by which rocaglamide (**1**), 1-*O*-acetylrocaglamide (**4**), and 1-oxo-11,12-methylendioxyrocaglaol (**50**) inhibit NF-AT activation in Jurkat T cells revealed that these compounds can enhance T-cell-activation-induced p38 and JNK activity (*25*). Increase of p38 and JNK activity resulted in acceleration of nuclear export of NF-AT. Prevention of NF-AT nuclear translocation in activated Jurkat T cells was visualized by confocal laser scan microscopy after rocaglamide (**1**) treatment (*25*). These data suggest that rocaglamide derivatives may function as immunosuppressive agents by targeting NF-AT activity in T cells. However, later studies revealed that, in contrast to malignant T cells, rocaglamides do not activate p38 and JNK in normal T cells and have apparently little effect on the NF-AT-dependent transcription in normal T cells (*21*). How the expression of TNF-α, IFN-γ, and IL-4 in normal T cells are suppressed by rocaglamide remains unknown.

4.3. Anticancer Activity

4.3.1. Antitumor Activities *In Vivo* in Mouse Tumor Models

In 1982, rocaglamide (**1**) was shown for the first time to increase lifespan of tumor-bearing mice in a leukemic model using P388 murine lymphocytic leukemia cells (*11*). In this study, administration of rocaglamide (**1**) at a dose of 1 mg/kg/day was shown to prolong survival with a T/C value (median survival time of treated *vs.* control group) of 156% (*11*). This observation has attracted the attention of scientists and more and more newly isolated rocaglamide derivatives have been evaluated for their antitumor potential in different cancer cell lines. A battery of rocaglamide derivatives has been found to have potent inhibitory effects on proliferation in different tumor cell lines *in vitro* (summarized in Ref. (*12*)). The IC_{50} values of the antiproliferative activities of most of the rocaglamide derivatives range from 1 to 200 n*M* depending on the cell line investigated (*12*). Several rocaglamide derivatives were further tested *in vivo* in mice tumor models (*21, 68, 93, 94*). For instance, intraperitoneal treatment with 4′-demethoxy-3′,4′-methylenedioxy-methyl rocaglate (**48**) at 10 mg/kg body weight three times per week was shown to lead to delayed growth of the xenografted human breast cancer cell line BC1 in athymic mice (*93*). Treatment with silvestrol (**35**) at 2.5 mg/kg/injection, when given intraperitoneally for five consecutive days in the ip P388 murine leukemia model, afforded a lifespan increase corresponding to a T/C of 150% (*68*). Silvestrol (**35**) was also shown to inhibit growth of the xenografted human breast cancer cells MDA-MB-231 at doses of 0.5 mg/kg/day for 8 days (*94*). Significant delay in growth and tumor size (*P* value = 0.021) of the mouse lymphoma RMA was observed after 16 days intraperitoneal treatment with desmethyl-rocaglamide (**7**) at 5 mg/kg three times per week (*21*). These animal studies have confirmed the anticancer activities of rocaglamide derivatives. All animal studies showed that these compounds had no toxicity to liver evaluated by glutamate pyruvate transaminase (GPT) activity and showed also no body weight loss.

4.3.2. Cytostatic Activity and Inhibition of Translation

Since rocaglamide (**1**) and its derivatives have been found to have anticancer activities in experimental animal models, efforts have been taken to explore the molecular mechanisms of their actions. In several early publications, rocaglamide derivatives were reported to function mainly in a cytostatic rather than a directly cytotoxic manner (*11, 84, 85, 93*). It was shown that treatment of the human lung carcinoma cells Lu1 with 4′-demethoxy-3′,4′-methylenedioxy-methyl rocaglate (**48**) resulted in inhibition of tumor cell proliferation with cell accumulation in the G1/G0 phase of the cell cycle with only marginal cell death (*93*). Thereafter, didesmethyl-rocaglamide (**15**), aglaroxin D (aglaiastatin) (**55**), and silvestrol (**35**) were shown to inhibit cell proliferation of different human malignant cell lines with the cell cycle blocked at the G2/M phase with negligible death (*85, 95, 96*). Furthermore, 4′-demethoxy-3′,4′-methylenedioxy-methyl rocaglate (**48**) was found to strongly inhibit protein biosynthesis in tumor cells as determined by ^{3}H-leucine incorporation (*93*). Based on these observations, it was thought that inhibition of translation is the key mode of action by which cyclopenta[*b*]benzofurans exert their antitumor activities.

Although rocaglamide (**1**) was found to inhibit translation more than 10 years ago, its mode of function was only recently explored. Translational control involves a highly regulated process (schematically depicted in Fig. 12).

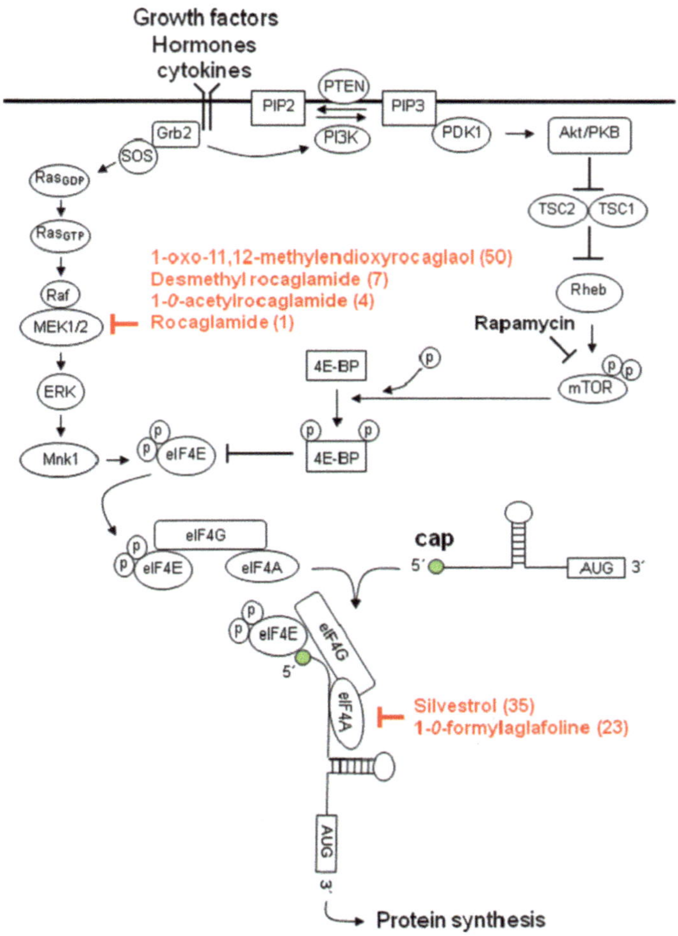

Fig. 12. Inhibition of protein synthesis by rocaglamides. The Ras-Raf-MEK-ERK and the PI3K-AKT-mTOR pathways play a central role in regulation of cap-dependent translation by activation of the eukaryotic translation initiation factor 4E (eIF4E). eIF4E binds to the 5′ cap structure and assembles eIF4G and eIF4A to initiate translation. Rocaglamide congeners **23** and **35** directly bind to and inhibit eIF4A function. Rocaglamides **1, 4, 7** and **50** block the MEK-REK pathway and inhibit eIF4E cap binding activity

Translation is initiated by binding of the initiation factor eIF4E to the mRNA 5′ cap structure. After binding to the 5′ cap structure, eIF4E interacts with eIF4G, which serves as a scaffold protein for the assembly of eIF4E and eIF4A to form the eIF4F complex. The eIF4F complex is then directed to the 5′ terminus of the mRNA and unwinds the mRNA 5′ secondary structure to facilitate ribosome binding and promotes ribosome recruitment and translation (*97*). Using a small molecule screening approach, 1-*O*-formylaglafoline (**23**) and silvestrol (**35**) were found to inhibit translation by interfering with eIF4A activity (*98*). It is thought that eIF4A exists as a free form or as part of the eIF4F complex and recycles through the eIF4F complex during translation initiation. 1-*O*-Formylaglafoline (**23**) and silvestrol (**35**) were shown to stimulate the RNA-binding activity of eIF4A and this action prevents incorporation of free eIF4A into the eIF4F complex (*98*).

In eukaryotes, most mRNAs are translated in a cap-dependent manner. The rate-limiting step of translation is largely controlled by binding of the initiation factor eIF4E to the mRNA 5′ cap structure. The activity of eIF4E is regulated by two major signalling pathways: the Ras-ERK-Mnk1 pathway and the PI3K-mTOR (mammalian target of rapamycin) pathway (*97*) (Fig. 12). In the Ras-ERK-Mnk1 pathway, the extracellular signal-regulated kinase ERK, activated by growth factors through Ras, phosphorylates the kinase Mnk1, which, in turn, phosphorylates eIF4E. Phosphorylation of eIF4E increases its affinity for the 5′ cap structure. The activity of eIF4E is also regulated by the PI3K-mTOR pathway. The assembly of the translation initiation complex eIF4F is inhibited by the translational repressor of eIF4E-binding proteins (4E-BP1). 4E-BP1 interacts with eIF4E in its hypophosphorylated state and prevents the recognition and binding of eIF4E to the 5′ cap mRNA structure. PI3K activates mTOR, which in turn phosphorylates 4E-BP1 leading to disruption of the interaction between 4E-BP1 and eIF4E and allowing eIF4E binding to the 5′ cap structure (Fig. 12). Most recently, 1-oxo-11,12-methylendioxyrocaglaol (**50**) was found to strongly inhibit protein synthesis ($IC_{50} = 30$ n*M*) in living cells, but had no inhibitory effect on protein synthesis *in vitro* in the rabbit reticulocyte cell free system (*99*). This finding indicates **50** does not directly inhibit the translational machinery, but rather acts through inhibition of the signaling pathway that is required for translation. Further investigation revealed **50** suppresses ERK phosphorylation and thereby inhibits ERK-Mnk1-mediated phosphorylation of eIF4E (*99*) (Fig. 12). Indeed, several rocaglamide derivatives, *e.g.* rocaglamide (**1**), 1-*O*-acetylrocaglamide (**4**), and 3′-hydroxyaglafoline (**19**), have been shown to inhibit ERK phosphorylation (*100*). Interestingly, these compounds do not inhibit ERK activity in normal lymphocytes (*100*). It is well known that the Ras-ERK-Mnk1 and the PI3K-mTOR pathways are frequently over-activated in many types of cancers due to gain of function mutations. Also the eIF4E, eIF4G, and eIF4A expression or activities are up-regulated in many cancers (*97*). Thus, targeting translational pathways is one of the mechanisms by which rocaglamide derivatives exert their antitumor functions.

4.3.3. Apoptosis Induction

Induction of apoptosis is now known to be also an important mechanism of rocaglamide (**1**)-mediated anticancer activity. Apoptotic cell death can be triggered through two main pathways: the extrinsic (also termed receptor-mediated) and the intrinsic (also termed mitochondrial-mediated) pathway (Fig. 13). The extrinsic apoptotic pathway is initiated by binding of ligands (*e.g.* CD95L and TRAIL) to their specific death receptors on the cell surface, which leads to the formation of the death inducing signaling complex (DISC) containing the FAS-associated death domain adaptor protein FADD and pro-caspase-8. Activation of caspase-8 at the DISC leads to activation of the effector caspase, caspase-3, which cleaves a number of target death proteins such as poly (ADP-ribose) polymerase (PARP) leading to apoptosis (*101*). The intrinsic pathway is initiated by various stimuli that directly or indirectly activate the mitochondrial pathway by inducing the release of

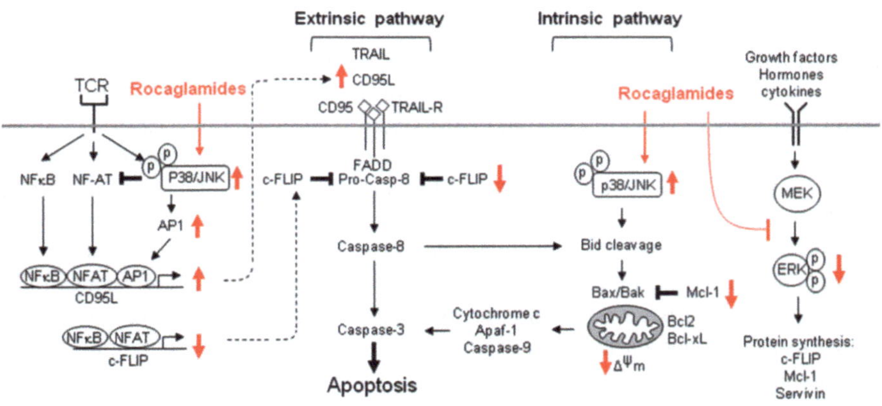

Fig. 13. Apoptosis pathways affected by rocaglamides. Apoptotic cell death is regulated by two main pathways: extrinsic (receptor-mediated) and intrinsic (mitochondria-mediated) pathways. The extrinsic pathway involves ligation of death receptors (*e.g.* CD95 and TRAIL-R) with their ligands (*e.g.* CD95L and TRAIL) resulting in a sequential activation of caspase-8 and -3, which cleave target proteins leading to apoptosis. This pathway is negatively regulated by the anti-apoptotic protein c-FLIP. Intrinsic stimuli (*e.g.* anticancer drugs) directly or indirectly activate the mitochondrial pathway by inducing release of cytochrome c and activation of caspase-9. Caspase-9, in turn, activates caspase-3. This death pathway is largely controlled by the pro-apoptotic (*e.g.* Bax and Bak) and anti-apoptotic (*e.g.* Mcl-1, Bcl-2 and Bcl-xL) proteins. Activated caspase-8 may induce cleavage of Bid, which induces translocation of Bax and/or Bak to the mitochondrial membrane and amplifies the mitochondrial apoptosis pathway. Bid cleavage can be also induced by activated p38 and JNK. Several rocaglamides can activate p38 and JNK leading to Bid cleavage. Rocaglamides can also directly inhibit protein synthesis by interfering with eIF4A (see Fig. 9) or indirectly through inhibition of the MEK-ERK-eIF4E pathway. Protein synthesis inhibition will lead to down-regulation of short-lived anti-apoptotic proteins such as c-FLIP and Mcl-1. Furthermore, rocaglamides may further increase T-cell-receptor (TCR)-mediated activation of p38 and JNK leading to down-regulation of NF-AT activity and up-regulation of AP-1 activity. This event results in up-regulation of CD95L promoter activity and suppression of c-FLIP promoter activity leading to enhancing activation-induced-cell-death

cytochrome c and the formation of the apoptosome complex with Apaf-1 and pro-caspase-9. Caspase-9 is then activated at the apoptosome and, subsequently, activates pro-caspase-3 (*102*). Activated caspase-8 may induce cleavage of Bid, which links the extrinsic death pathway to the intrinsic death pathway by induction of the translocation of the proapoptotic Bcl-2 family proteins Bax and/or Bak to the mitochondrial membrane. One major negative regulator of receptor-mediated apoptosis is the cellular caspase-8 (FLICE)-inhibitory protein (c-FLIP), which blocks processing and activation of caspase-8 at the DISC level (*101*). In the intrinsic apoptosis pathway, death and life of cells are largely controlled by pro-apoptotic, *e.g.* Bax and Bak, and anti-apoptotic proteins, *e.g.* Bcl-2, Bcl-X_L, XIAP, and myeloid cell leukemia-1 (Mcl-1) (*102*).

The first observation that rocaglamide derivatives could induce apoptotic cell death in cancer cells was reported in 2002 (*24*). It was shown that a fraction of the human leukemic CEM T cells die by apoptosis after treatment with didesmethyl-rocaglamide (**15**). It was also shown that didesmethyl-rocaglamide (**15**) could enhance apoptotic cell death induced by other apoptotic stimuli such as TNF-α and cisplatin (*24*). Thereafter, silvestrol (**35**) was also reported to have a pronounced cytotoxic rather than cytostatic effect on the human prostate carcinoma LNCaP cells (*103*). It was shown that treatment of LNCaP cells with silvestrol (**35**) resulted in cytochrome c release and consequently activation of caspase-9. However, no caspase-3 and caspase-7 activation were detected although the apoptotic cell death induced by silvestrol (**35**) could be inhibited by a pan-caspase inhibitor, Boc-D-Fmk (*103*).

Solid evidence supporting rocaglamide derivatives as being potent apoptosis inducers was provided by the study of rocaglamide (**1**) and 1-*O*-acetylrocaglamide (**4**) in leukemic cells (*100*). In this investigation, rocaglamide (**1**) treatment was shown to induce depolarization of the mitochondrial membrane potential and to trigger caspase-mediated apoptosis involving caspase-9, -8, -3, and -2 in different leukemic cell lines, *e.g.* the human acute T cell leukemia Jurkat, the human T cell lymphoma Hut78, the EBV-transformed human B lymphoblast SKW, and also primary tumor cells freshly isolated from patients having acute myeloid (AML), acute lymphoblastic (ALL), and chronic myeloid (CML) leukemias. Interestingly, rocaglamide (**1**) showed no or very low toxicities to normal peripheral blood T and B lymphocytes and had also very little toxicity to human bone marrow stem cells (*100*). Investigation of the molecular mechanisms by which rocaglamide (**1**) and 1-*O*-acetylrocaglamide (**4**) killed tumors but not normal cells revealed that these compounds preferentially induce apoptosis in malignant cells by differential modulation of the activities of ERK, p38, and JNK. It has been shown that p38 is involved in mitochondria-mediated apoptosis by promoting mitochondrial translocation of the pro-apoptotic Bcl2 family protein Bax *via* inducing Bid cleavage (*104, 105*). The protein kinase p38 has been also linked to activation of the mitochondrial pathway by directly phosphorylating Bax (*106*), Bim (*107*), and Bcl-2 (*108*) and by down-regulation of Bad phosphorylation (*109*). Consistent with those studies, activation of p38 and JNK by rocaglamide (**1**) and 1-*O*-acetylrocaglamide (**4**) was shown to correlate with Bid cleavage (*100*).

Extensive studies in the past have established that ERK is an important survival factor and, particularly, ERK has been found to be constitutively activated in the majority of cancers. Thus, targeting the ERK pathway has been considered to be one of the important strategies in anticancer therapy (*110*). Rocaglamide (**1**) and 1-*O*-acetylrocaglamide (**4**) were shown to induce activation of p38 and JNK accompanied by a long-term suppression of ERK activity (*100*). The critical role of p38/JNK and ERK in apoptosis induction in tumor cells was further demonstrated by the rocaglamide derivative 3′-hydroxylafoline (**19**), which differs from rocaglamide (**1**) and 1-*O*-acetylrocaglamide (**4**) only in having -OH at the C-3′ position. 3′-Hydroxylafoline (**19**) had a very weak effect on the activities of p38 and ERK in tumor cells and showed only a slight apoptosis induction. In addition, malignant cells that lost their ability to respond to rocaglamide-induced activation of p38/JNK or suppression of ERK were resistant to apoptosis induction by rocaglamide (*110*). Furthermore, rocaglamide (**1**) and 1-*O*-acetylrocaglamide (**4**) do not affect the activities of p38, JNK, and ERK in normal lymphocytes and therefore have no or only small effects on apoptosis induction in normal lymphocytes (*100*).

Inhibition of protein synthesis may affect expression levels of pro- and antiapoptotic proteins, in particular, those having a short half-life. Therefore, rocaglamide-mediated translational inhibition may cause imbalances of pro- and anti-apoptosis protein levels in a tumor cell. For instance, reduced expression of the anti-apoptotic proteins Mcl-1 and survivin were observed in human breast cancer MDA-MB-231 cells after exposure to silvestrol (**35**) (*94*). Also, down-regulation of c-FLIP, the major inhibitor of caspase-8, was detected in tumor cells treated with rocaglamide (**1**), desmethyl-rocaglamide (**7**), 1-*O*-acetylrocaglamide (**4**), and 1-oxo-11,12-methylendioxyrocaglamide (**50**) (*21, 99*). Expression of the c-FLIP proteins is controlled at multiple levels. The ERK and the PI3K signaling pathways have been shown to regulate c-FLIP expression (*111–113*). In addition, c-FLIP protein turnover is actively regulated by ubiquitin-mediated proteasomal degradation (*114*). Therefore, c-FLIP proteins are short-lived proteins and are required to be constitutively synthesized in tumor cells. As described before, rocaglamide derivatives can inhibit translation (Fig. 12). Inhibition of *de novo* protein synthesis will lead to down-regulation of the expression levels of short-lived anti-apoptotic proteins such as c-FLIP. This effect is well demonstrated by a recent study showing that 1-oxo-11,12-methylendioxyrocaglamide (**50**) inhibits expression of c-FLIP at the translational level by blocking the ERK signalling pathway and thereby breaks CD95 and TRAIL resistance in human T-cell leukemia virus type-1 (HTLV-1)-associated adult T-cell leukemia/lymphoma (ATL) (*99*) (Fig. 13). One can predict that 1-*O*-acetylrocaglamide (**4**), desmethyl-rocaglamide (**7**), 1-*O*-formylaglafoline (**23**), silvestrol (**35**), 4′-demethoxy-3′,4′-methylenedioxy-methyl rocaglate (**48**), and 1-oxo-11,12-methylendioxyrocaglamide (**50**) may be also capable of sensitizing receptor-mediated apoptosis by down-regulation of c-FLIP expression.

CD95L and its receptor CD95 are the most intensively studied apoptotic system of the extrinsic pathway (Fig. 13). A well-known phenomenon is that a fraction of T cells die after activation by a so called activation-induced-cell-death (AICD). AICD in T cells is mediated predominantly by CD95L and CD95 (*101*). CD95 is abundantly expressed in many types of cells. In contrast, expression of CD95L is more restricted to certain types of cells and predominantly expressed in activated T cells. In resting T cells, CD95L is expressed at an undetectable level and its expression is rapidly up-regulated upon T cell activation. Thus, the rate of CD95-mediated apoptosis is largely determined by the levels of CD95L expression. Expression of CD95L in activated T cells is regulated by multiple inducible transcription factors such as NF-AT, NF-κB, and AP-1 (Fos/Jun) (*115*). However, expression of c-FLIP in T cells is also regulated by NF-κB and NF-AT (*116, 117*). Therefore, T cell activation leads not only to induction of CD95L but also c-FLIP expression. Up-regulation of c-FLIP expression during T cell activation correlates with resistance to CD95/CD95L-mediated apoptosis and rescue of T cells from AICD. Rocaglamide (**1**) and desmethyl-rocaglamide (**7**) were shown to enhance CD95L-mediated AICD in malignant T cells by up-regulation of CD95L but down-regulation of c-FLIP expression (*21*). As mentioned in Sect. 4.2., rocaglamide (**1**) and several derivatives, such as 1-*O*-acetylrocaglamide (**4**) and 1-oxo-11,12-methylendioxyrocaglaol (**50**), can enhance T-cell-activation-induced p38 and JNK activity and thereby suppress NF-AT activity by promoting NF-AT nuclear export (*25*). The CD95L promoter is strongly regulated by JNK/AP-1 (*115*). Increase of p38/JNK activity leads to increase of AP-1 activity and consequently enhanced CD95L promoter activity (*21*). In contrast, c-FLIP expression in T cells is strongly regulated by NF-AT (*117*). Rocaglamide-mediated suppression of NF-AT resulted in down-regulation of c-FLIP expression (*21*) (Fig. 13). These studies suggest that rocaglamide derivatives may serve as adjuvants for death receptor-based anticancer therapies in the future.

5. Chemical Synthesis of Cyclopenta[*b*]benzofurans

Rocaglamide (**1**) and congeners pose an intriguing synthetic chemistry challenge: multiple functionalities, a hepta-substituted cyclopenta[*b*]benzofuran scaffold, and five contiguous, stereogenic centers. The following syntheses, presented largely in chronological order, are significant because of the daring synthetic maneuvers employed, efficiency of the transformations, and development of asymmetric processes.

5.1. First Approaches to the Synthesis of Rocaglamides

Before (−)-rocaglamide was synthesized, there were two significant approaches published. In 1987, *Richard J. K. Taylor* and coworkers first accessed the tricyclic rocaglate core employing the benzofuranone intermediate **126** (Fig. 14) (*118, 119*). The rocaglate skeleton was subsequently assembled *via* an intramolecular, 1,3-dithianyl anion-carbonyl condensation. However, in this case the 1,3-dithiane

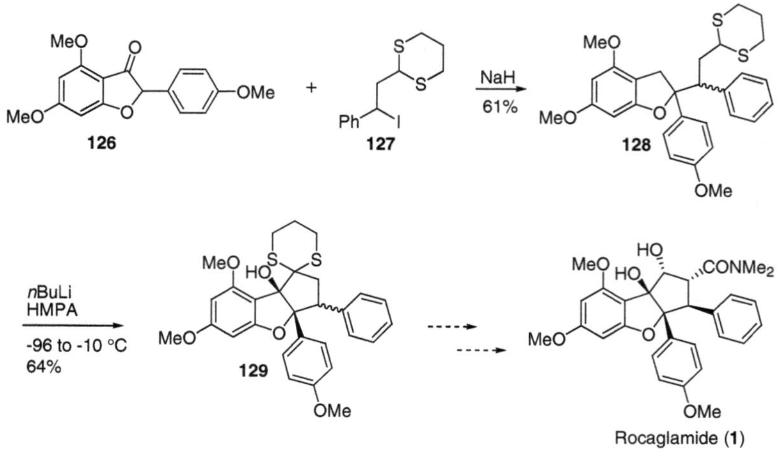

Fig. 14. *Taylor*'s approach to the rocaglate skeleton (*119*)

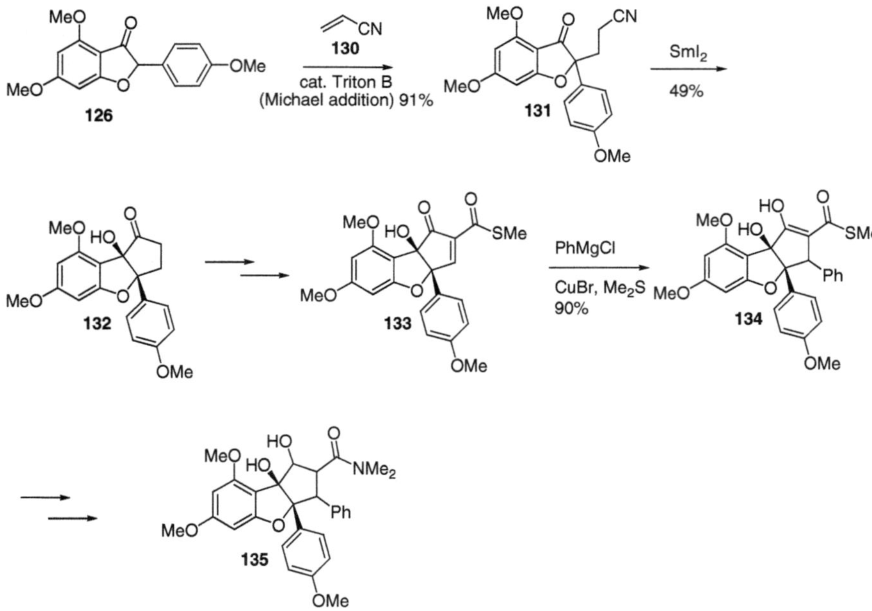

Fig. 15. *Kraus*'s synthesis of a rocaglamide di-*epi*-analogue (*120*)

group proved difficult to remove. Later syntheses, including *Taylor*'s total synthesis, have adapted the use of key intermediate **126** as starting material.

Taylor's work towards the rocaglate skeleton was then followed by *Kraus* and coworkers' synthesis of the di-*epi*-analogue of rocaglamide (Fig. 15) (*120*). Like *Taylor*'s approach, the investigators began their synthesis with the known intermediate **126**. The first significant step in this synthesis is a *Michael* addition of acrylonitrile (**130**) with **126**. Next, a samarium-mediated reductive cyclization was performed to afford rocaglate skeleton **132**. Later in the synthesis, a cuprate addition was conducted on unsaturated thioester **133** to provide the enol intermediate **134**, which through X-ray crystal structure analysis was shown to be the *trans*-phenyl-aryl isomer. Accordingly, the cuprate had added unexpectedly to the concave face of the bicyclic ring system. Subsequent amidation and reduction of **134** yielded rocaglamide isomer **135** that did not possess the same NMR spectra as natural rocaglamide. Although the natural product was not synthesized, *Kraus* established two key steps (*Michael* addition of acrylonitrile and the reductive cyclization using SmI_2) that paved the way for future investigations in the field.

5.2. The First Total Synthesis of Rocaglamide

The first total synthesis of rocaglamide (**1**) was accomplished by *Barry Trost* and coworkers in 1990 (Fig. 16) (*52*). Asymmetric [3 + 2] cycloaddition of trimethylsilylmethyl precursor **136** and chiral oxazepinedione **137** afforded cyclopentene **138**, a transformation which established the absolute configuration at C_4. Following removal of the chiral auxiliary and ozonolysis (*not shown*), cyclopentanone **139**

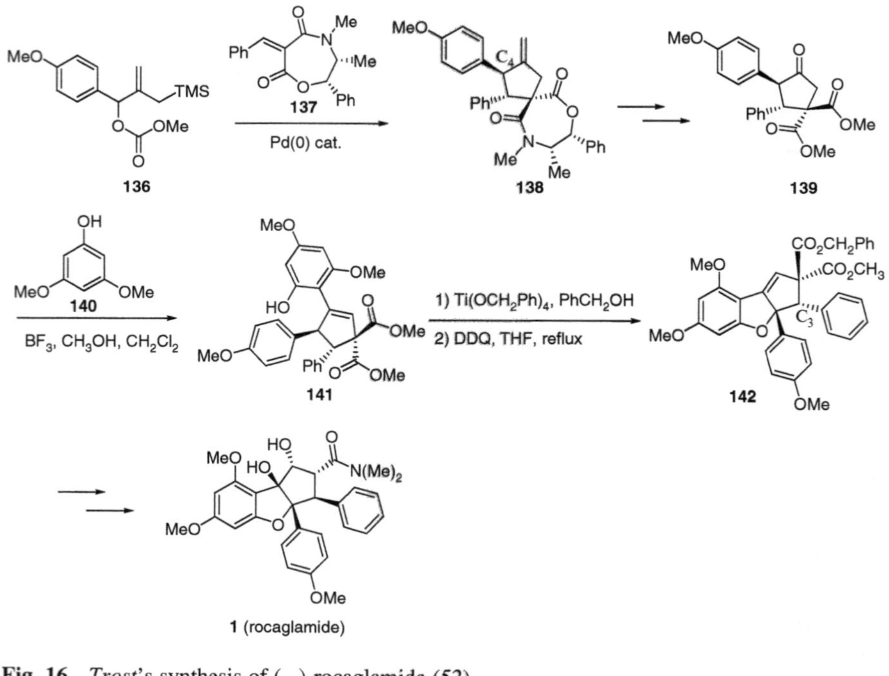

Fig. 16. *Trost*'s synthesis of (−)-rocaglamide (*52*)

was next condensed with dimethylphloroglucinol **140** to furnish intermediate **141**. Regioselective transesterification of **141** with benzyl alcohol was then performed that was followed by oxidative cyclization with DDQ to furnish intermediate **142**. With the rocaglate skeleton in hand, several reactions were performed to fully functionalized intermediate **142** and also invert stereochemistry at C_3. The synthesis was completed in 16 steps and fully established the absolute configuration of the natural product.

5.3. *Syntheses of Rocaglamide and Related Natural Products*

Shortly after the *Trost* synthesis, in 1991 *Taylor* and coworkers reported their synthesis of racemic rocaglamide (**1**) in eight steps from the benzofuran intermediate **126** (Fig. 17) (*121*). The key steps utilized were (1) *Michael* addition of *trans*-cinnamaldehyde (**144**) and benzofuran **126**; and (2) intramolecular, keto-aldehyde pinacolic coupling (reductive cyclization) mediated by SmI_2. This step afforded high diastereoselectivity (6:1) in construction of the rocaglate scaffold. Advanced intermediate **146** was then oxidized *via Swern* oxidation and was then transformed to a dimethylamide. The synthesis was completed with a hydroxy-directed, diastereoselective reduction of the α-hydroxy ketone **149** using $NMe_4BH(OAc)_3$ to afford (±)-rocaglamide (**1**). In 1992, *Taylor* and coworkers published another synthesis of rocaglamide involving hydrolysis of the dithiane group used in his original approach (*122*). These syntheses by *Taylor* are a beautiful fusion of his original

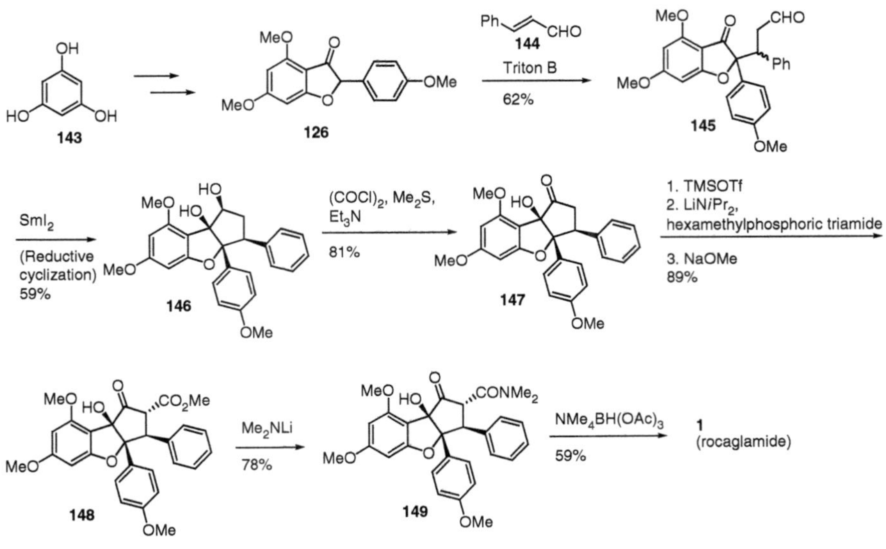

Fig. 17. *Taylor*'s racemic synthesis of rocaglamide (**1**) (*121*)

approach and the *Kraus* approach to the rocaglamide skeleton and are also frequently cited and used in later syntheses of rocaglamide and related natural products.

Inspired by the approaches of *Kraus*, *Taylor*, and *Trost*, *Takumi Watanabe* and coworkers achieved the racemic synthesis of aglaiastatin, a natural product related to rocaglamide (Fig. 18) (*123*). As the synthesis of ABC ring system had already been developed by *Taylor*, *Watanabe* began with advanced intermediate **150** (*122*). Carboxylic acid **150** was subsequently coupled with 4,4-diethoxybutanamine to form amide **151**. After the secondary alcohol was selectively oxidized, anhydrous HCl was used to form an acyl iminium species, in which case excess ammonium formate in formic acid served to form the enamine leading to cyclization with an iminium ion intermediate to afford (±)-aglaiastatin (**154**) ($dr = 2{:}1$).

In 2001, *Dobler* and coworkers reported a total synthesis of racemic rocaglamide using the *Michael* addition performed by *Taylor*. However, instead of using SmI_2, *Dobler* converted aldehyde **145** to a cyanohydrin intermediate in quantitative yield (Fig. 19) (*124*). Acyloin (**147**) was then initiated *via* addition of LDA, followed by deprotection with K_2CO_3, a marked improvement from preexisting methods.

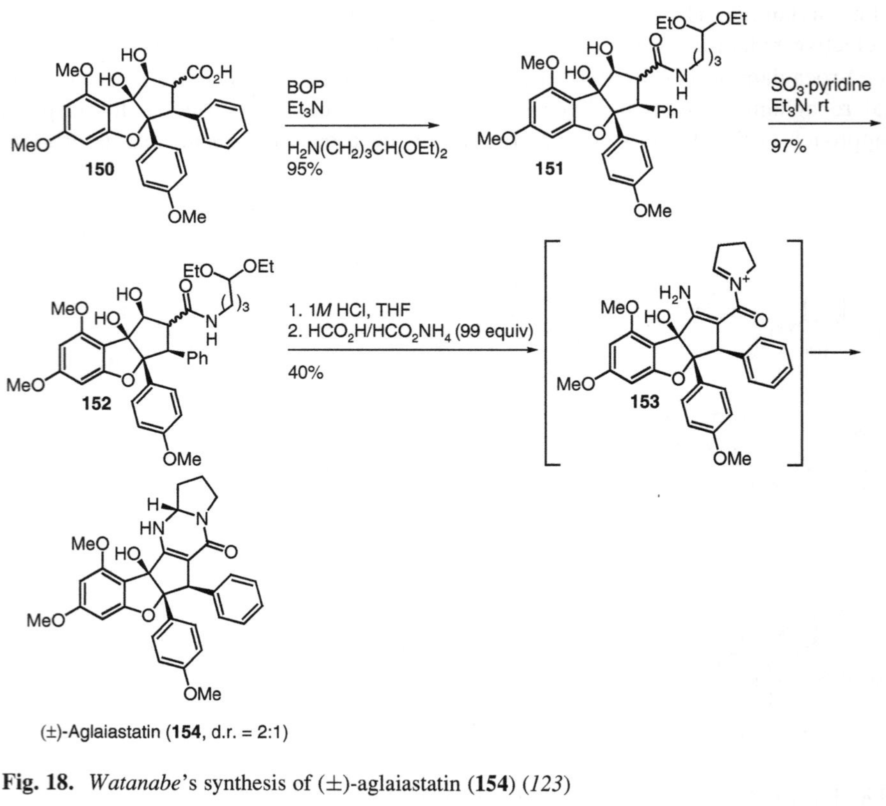

Fig. 18. *Watanabe*'s synthesis of (±)-aglaiastatin (**154**) (*123*)

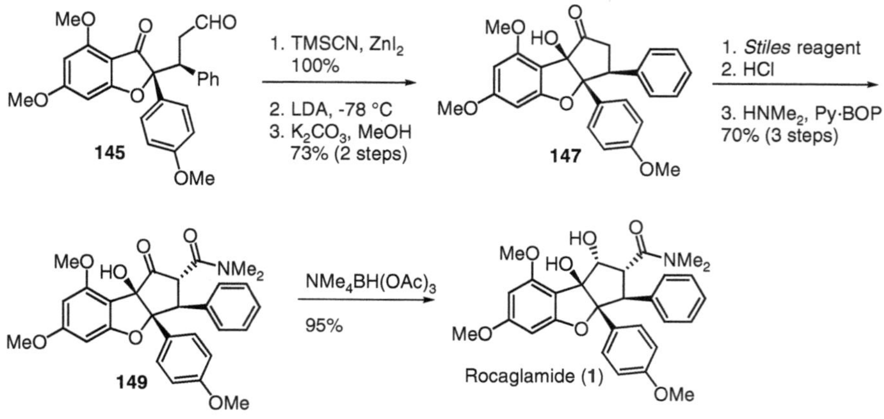

Fig. 19. *Dobler*'s synthesis of racemic rocaglamide (**1**) (*124*)

Additionally, *Stiles* carboxylation following the keto-aldehyde acyloin ring closure made the synthesis more efficient than previously reported.

5.4. New Approaches to Rocaglamide and Related Natural Products

The year 2004 saw a new synthetic approach to rocaglamide and related natural products employing 3-hydroxyflavones, *Porco* and coworkers found that upon photoirradiation of 3-hydroxyflavones (*e.g.* intermediate **155**), an oxidopyrylium species derived from excited-state intramolecular proton transfer (ESIPT) could be trapped with a dipolarophile (*e.g. trans*-methyl cinnamate **156**) (Fig. 20) (*125*) resulting in an overall [3 + 2] photocycloaddition to afford aglain intermediate **157**. The approach may mimic the proposed biosynthesis of the rocaglamides (*cf. 13–15*). Forbaglin natural products (**158**) were also accessed through oxidation of the aglain intermediate, while rocaglamide and related natural products such as methyl rocaglate were obtained through an α-ketol shift, followed by an hydroxyl-directed reduction of the α-hydroxy ketone **159**. The methodology was then used to synthesize (±)-methyl rocaglate (Fig. 21) (*125*). Photoirradiation of 3-hydroxyflavone (**161**) and methyl cinnamate (**156**) afforded the aglain (**162**) as well as a benzo[*b*]-cyclobutapyran-8-one (**163**), which were then subjected to basic conditions to provide α-ketol rearrangement product **148**. Hydroxyl-directed reduction of α-hydroxyketone **148** yielded racemic methyl rocaglate (**18**).

To further improve their methodology, *Porco* and coworkers developed an enantioselective [3 + 2] photocycloaddition that was then used to synthesize the natural enantiomer (−)-methyl rocaglate, (−)-rocaglamide, and (−)-rocaglaol (Fig. 22) (*126*). Using functionalized TADDOL (*127*) derivatives, chiral *Brønsted* acids to enhance excited state intramolecular proton transfer, asymmetry was

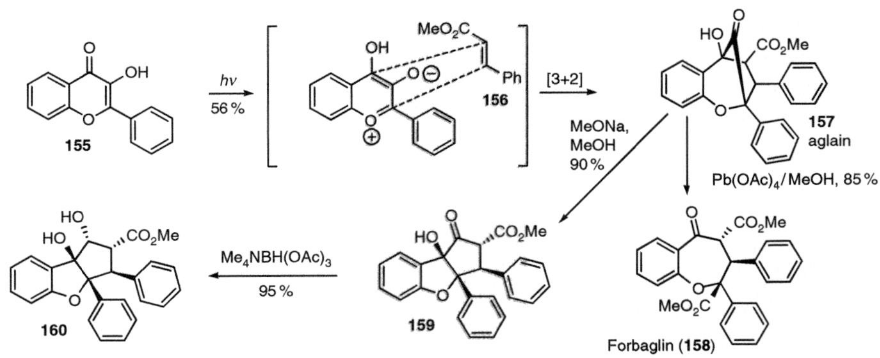

Fig. 20. *Porco*'s unified approach to the aglains, forbaglins, and rocaglamides (*125*)

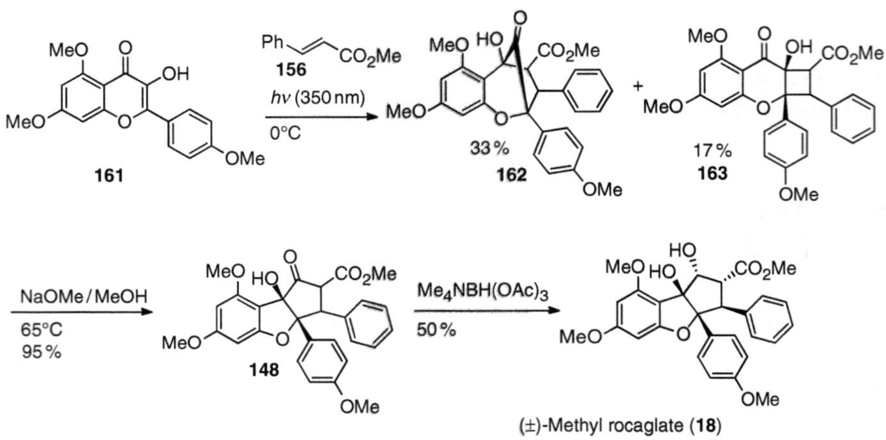

Fig. 21. *Porco*'s synthesis of (±)-methyl rocaglate (**18**) (*125*)

successfully induced with an enantiomeric excess (*ee*) of 86% using the TADDOL **164**. Recrystallization of methyl rocaglate (**18**) enhanced the *ee* to 94%, with 86% recovery. (−)-Rocaglaol (**28**) was also obtained through decarboxylation followed by diastereoselective reduction of intermediate **148**. Likewise, (−)-rocaglamide (**1**) was synthesized from **148** through a reduction/saponification/amide coupling sequence.

The work of *Thede* and *Ragot* in 2004 and 2005 focused on the synthesis of rocaglaol and analogues. While *Taylor's* approach to rocaglate natural products was quite viable, access to very electron-rich benzofuranones derived from *Hoesch* or *Friedel Crafts* reactions was limited. Using methods for unsymmetrical 3,4-diaryl-cyclopent-2-enones (*128*), *Thede* and *Ragot* published an imaginative approach to the molecule, which was highlighted by intramolecular epoxide ring opening (Fig. 23) (*129*). Although the method did not afford the natural product, it did allow for preparation of 3-*epi*-rocaglaol (**170**) and analogues through variation of aryl boronic acids on the western portion of the molecule. In 2005, *Thede* and *Ragot*

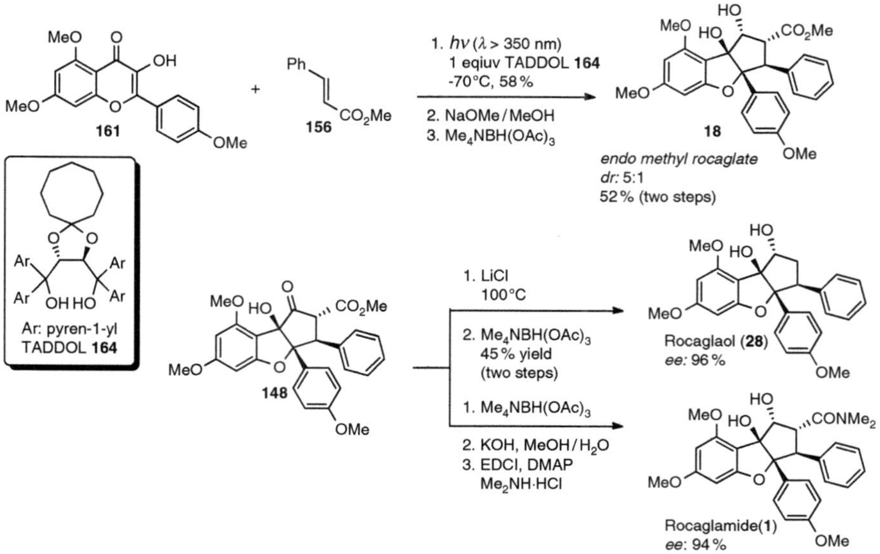

Fig. 22. *Porco*'s enantioselective methodology for the synthesis of (−)-methyl rocaglate (**18**), (−)-rocaglaol (**28**), and (−)-rocaglamide (**1**) (*126*)

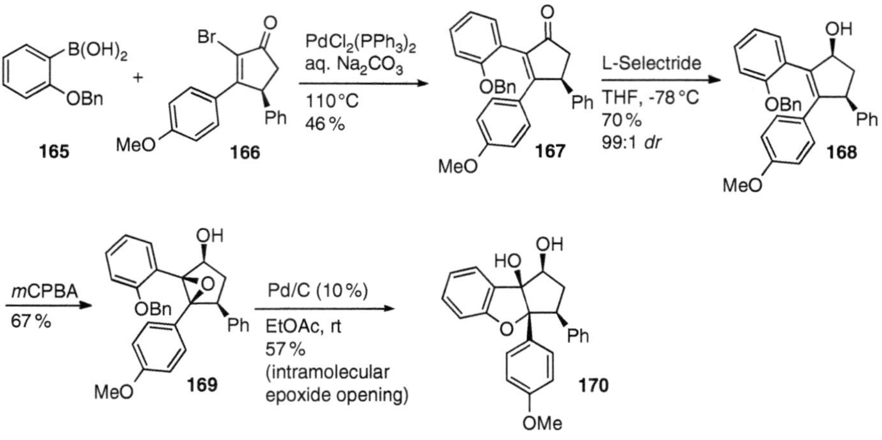

Fig. 23. *Thede* and *Ragot*'s stereoselective synthesis of (±)-rocaglaol analogues (*129*)

developed an alternative method to synthesize rocaglaol (**28**) and alteration of the benzofuran moiety to create rocaglaol analogues (Fig. 24) (*130*). This was achieved through the α-arylation of ketones using *Taylor*'s method to synthesize the skeleton. The α-arylation entailed *Suzuki*-type reaction of brominated silyl enol ethers with aryl boronic acids, allowing broad variation of the benzofuran moiety.

Control of stereochemistry of the adjacent phenyl and *para*-methoxyphenyl (PMP) groups is a difficulty frequently encountered in the synthesis of rocaglates and analogues. *Philip Magnus* and coworkers synthesized (±)-1,2-anhydro methyl

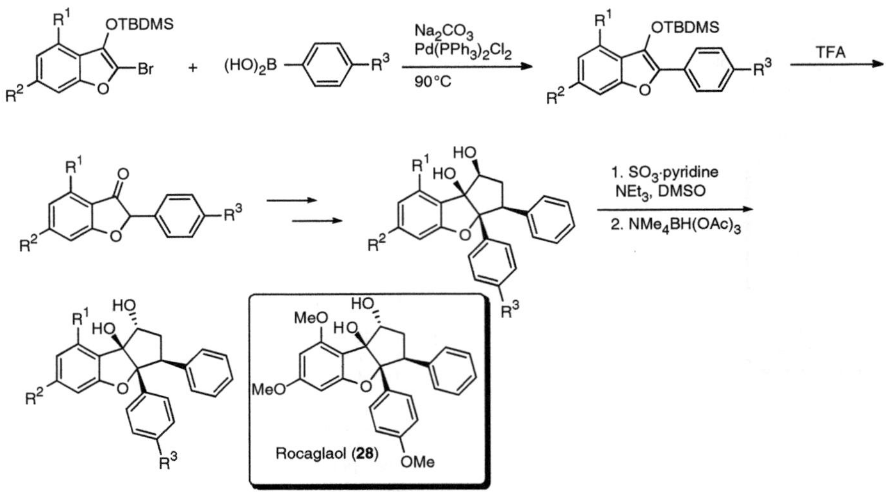

Fig. 24. *Thede* and *Ragot*'s stereoselective synthesis of (±)-rocaglaol (**28**) and analogues (*130*)

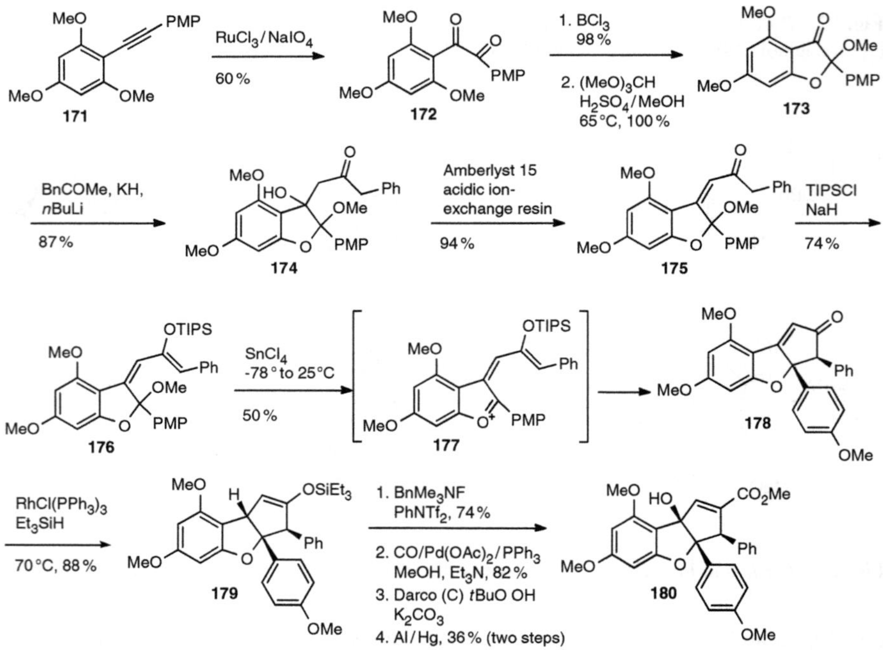

Fig. 25. *Magnus*'s stereospecific synthesis of (±)-1,2-anhydro methyl rocaglate (**180**) (*131*)

rocaglate stereospecifically in which the stereocontrol was generated through an *Isler-Mukaiyama* aldol reaction (conrotatory *Nazarov* reaction) (Fig. 25) (*131*). The synthesis commenced with coupling of 2,4,6-trimethoxyiodobenzene with

4-methoxyphenylacetylene under *Sonogashira* coupling conditions to afford intermediate **171**. Next, an oxidation was carried out with catalytic ruthenium chloride/sodium periodate to provide dione **172**, which was then treated with boron trichloride to selectively demethylate and simultaneously form a hemiketal. The hemiketal was subsequently methylated using trimethoxymethane and sulfuric acid to afford benzofuranone (**173**). Benzofuranone (**173**) was then alkylated to give substrate **174**, which was then exposed to Amberlyst 15 ion-exchange resin to afford **175**. Treatment with base and TIPSCl produced the substrate used in the conrotatory *Nazarov* cyclization. Since the reaction is stereospecific, only *cis*-aryl-phenyl diastereomer **178** was formed. Using *Wilkinson*'s catalyst, **178** was hydrosilated to give **179**, which was then desilylated and trapped with *N*-phenyltriflamide. This intermediate was then subjected to *Ortar* reaction conditions to afford a methyl ester intermediate that was further oxidized to a *tert*-butyl peroxide and subsequently reduced to yield (±)-1,2-anhydro methyl rocaglate (**180**).

In 2008, *Qin* and coworkers reported the shortest synthesis of (±)-rocaglamide and its 2,3-di-*epi*-analogue to date (Fig. 26) (*132*). A methoxycarbonyl group was introduced into *Michael* acceptor **181**, thus making the synthesis immensely more efficient. Samarium (II)-mediated reductive cyclization afforded keto-methyl rocaglate (**148**) which was then converted to the amide **149** and reduced with tetramethylammonium triacetoxyborohydride to afford the natural product **1**.

The most recent synthesis of rocaglamide was accomplished by *Alison Frontier* and coworkers (Fig. 27) (*133*). Benzofuran (**126**) was first alkylated with vinyl magnesium bromide and then oxidized to form aldehyde **183**. Subsequent alkylation with

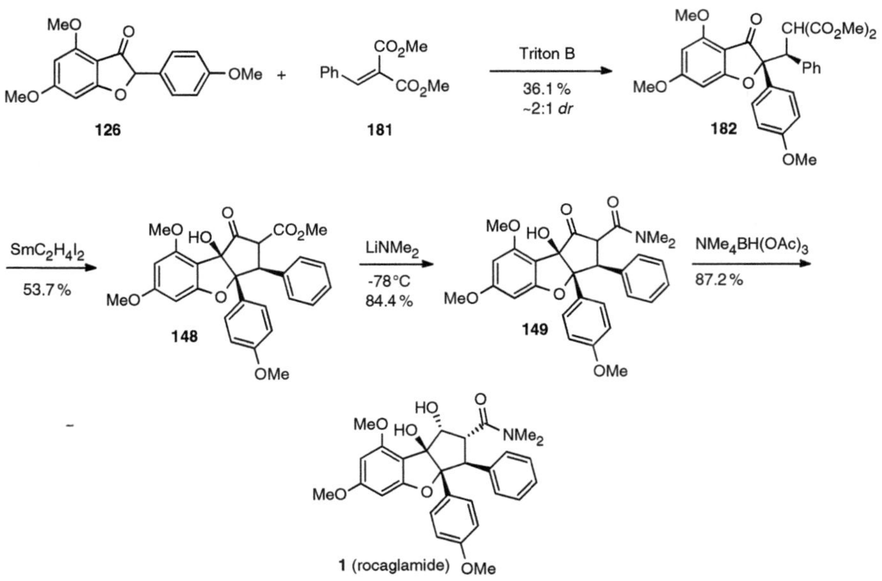

Fig. 26. *Qin*'s racemic synthesis of rocaglamide (**1**) (*132*)

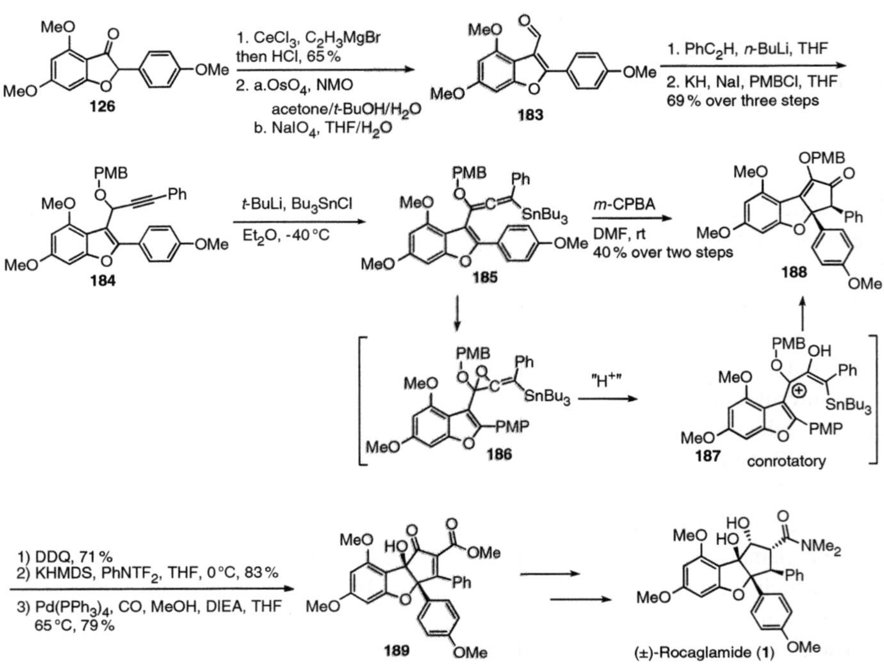

Fig. 27. *Frontier*'s synthesis of (±)-rocaglamide (1) *via Nazarov* cyclization (*133*)

phenylacetylene and protection of the propargyl alcohol afforded propargyl ether **184**. Next, **184** was deprotonated with *tert*-butyllithium and the resulting allenyl anion trapped with tri-*n*-butyltin chloride to afford stannyl alkoxyallene **185**. In a key step, the allenol ether was oxidized with *meta*-chloroperbenzoic acid (*m*-CPBA) forming an epoxide intermediate **186**, which opened to form the pentadienyl cation **187** necessary for the *Nazarov* cyclization under acidic conditions. Treatment of intermediate **188** with excess DDQ led to a production of a diosphenol intermediate, which was subsequently converted to a triflate, and the latter was then carbonylated to afford intermediate **189** and further advanced to racemic rocaglamide.

5.5. *Syntheses of Silvestrol*

The enantioselective synthesis of the complex rocaglate silvestrol was reported by *Porco* and coworkers in 2007 (Fig. 28) (*134*). Through a convergent strategy, the rocaglate intermediate **191** and dioxanyloxy fragment **193** were connected and subsequently deprotected to form the natural product. Methoxy methyl ether protected 3-hydroxyflavone **190** was photo-excited to its oxidopyrylium tautomer

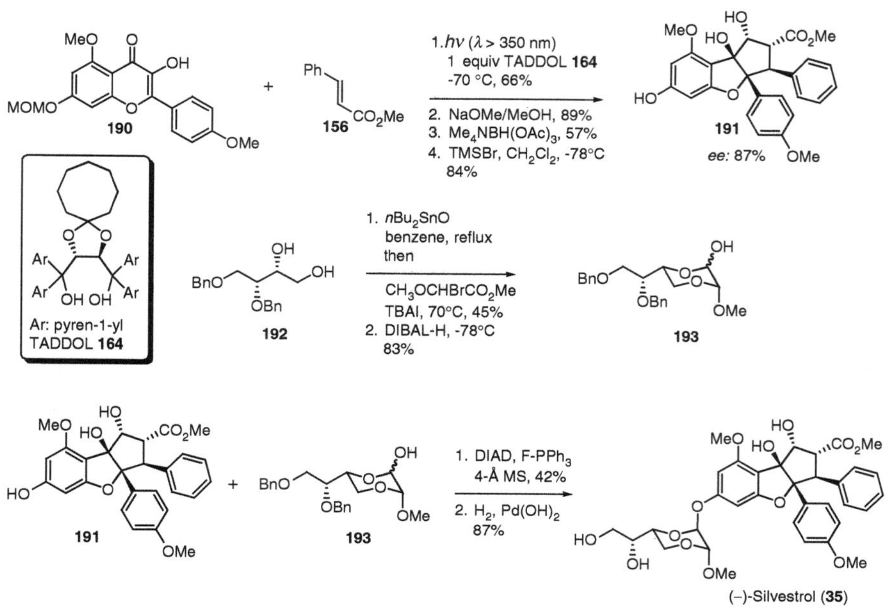

Fig. 28. *Porco*'s enantioselective synthesis of (−)-silvestrol (**35**) (*134*)

and then trapped with methyl cinnamate (**156**) mediated by TADDOL (**164**). After a ketol rearrangement/hydroxyl directed reduction sequence, the resulting cyclopenta[*b*,*c*]benzofuran intermediate was deprotected using TMSBr to remove the methoxymethyl ether (MOM), which afforded rocaglate intermediate **191** in 87% *ee* after recrystallization. Intermediate **192** was used to form an *O*-stannylene acetal, which was then combined with 2-bromo-2-methoxy acetate to afford a dioxanylone intermediate. Reduction of the dixoanylone intermediate with diisobutylaluminum hydride (DIBAL-H) afforded the dioxanyloxy fragment **193**. Using *Mitsunobu* conditions, fragments **191** and **193** were successfully coupled. Finally, hydrogenation of the coupled product gave (−)-silvestrol (**35**).

Rizzacasa and coworkers synthesized *epi*-silvestrol and silvestrol in a convergent fashion employing a method similar to their published pilot studies (*135*). Their plan was to connect the 1,4-dioxanyloxy fragment **199** with the cyclopentabenzofuran **191** core *via Mitsunobu*-type coupling to access *epi*-silvestrol (**36**), which was subsequently converted to silvestrol (**35**) using a *Mitsunobu* reaction (Fig. 29) (*136*). The synthesis of the 1,4-dioxanyloxy fragment **199** commenced with glycosylation of bromide **194** using *para*-methoxybenzyl alcohol, followed by removal of the acetates *via* methanolysis. Subsequent benzylidene formation afforded acetal **195**, which was treated with $BH_3{\cdot}THF$ and $Cu(OTf)_2$ to selectively cleave the benzylidene and afforded the protected sugar **196**. Periodate cleavage, followed by selective reduction of the newly formed aldehyde-lactol, yielded the

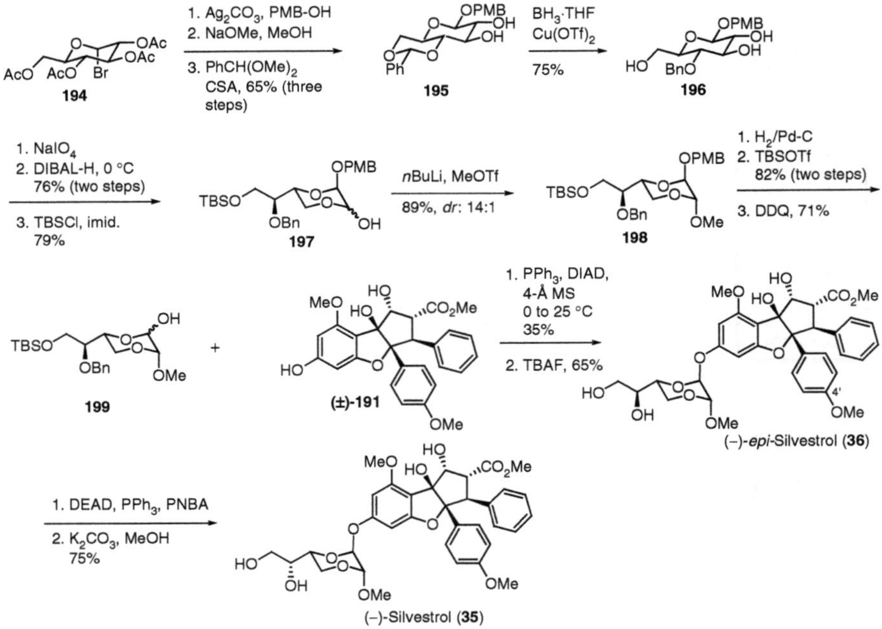

Fig. 29. *Rizzacasa*'s syntheses of (−)-silvestrol (**35**) and (−)-*epi*-silvestrol (**36**) (*136*)

lactol as a 1:3 mixture of anomers. The resultant lactol was then selectively protected at the primary alcohol to afford TBS ether **197**, which was then methylated with high selectivity to provide axial product **198**. DDQ was then used to remove the *para*-methoxybenzyl protecting group to afford the 1,4-dioxanyloxy fragment **199**. Racemic rocaglate core **191** was synthesized using the methods developed by *Porco* and coworkers, which was then coupled with optically pure fragment **199** to afford two major axial diastereomers in equal amounts, and two equatorial diastereomers in equal amounts (2:1 ratio axial:equatorial products) in 35% overall yield. The diastereomers were then deprotected to afford (−)-*epi*-silvestrol (**36**). (−)-Silvestrol (**35**) was then achieved through a selective "double *Mitsunobu*" reaction of (−)-*epi*-silvestrol (**36**). Future studies by *Rizzacasa* and coworkers disclosed a method to resolve racemic **191** using (−)-menthol, as well as an improved *Mitsunobu* coupling of **191** and **199** using (di-2-methoxyethyl azodicarboxylate) (*137*).

Aside from their complexity and synthetic challenge, silvestrol and *epi*-silvestrol displayed potent anticancer activity in A549 lung cancer proliferation assays with IC_{50} values of 33 n*M* and 30 n*M* (*137*). The synthetic analogue 4′-desmethoxy-*epi*-silvestrol was also found to be highly potent against LIM1215 colon cancer cells proliferation with an IC_{50} of 10 n*M*, foreshadowing the development of cytotoxic flavagline analogues that are more potent than their parent counterparts.

5.6. *Development of Rocaglates and Analogues as Therapeutic Agents*

Recently, *Laurent Désaubry* and coworkers discovered the first synthetic rocaglamide-type compound that inhibits cell proliferation and viability at lower doses than the parent compound, rocaglaol (**28**) (Fig. 30) (*138*). Using *Dobler*'s

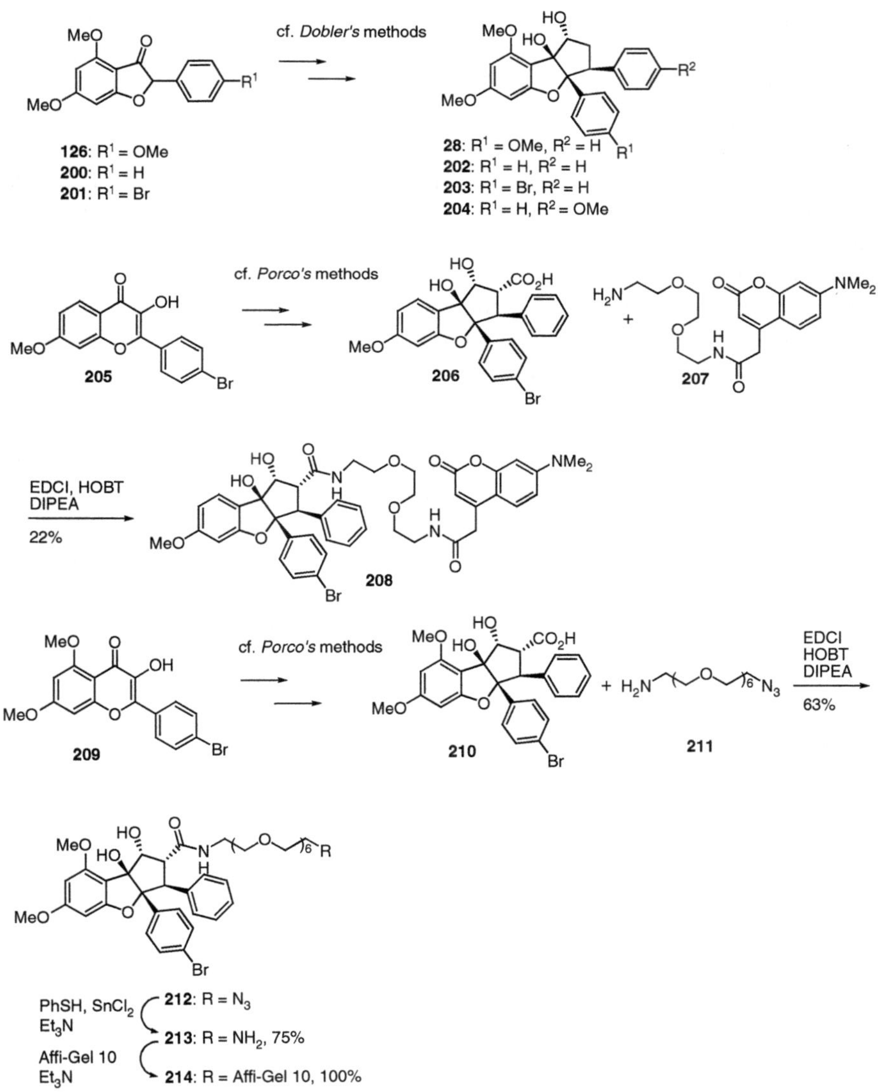

Fig. 30. *Désaubry*'s synthetic rocaglaol derivatives **28** and **202–204**, fluorescent probe **208**, and affinity ligand **214** (*138*)

procedure for the preparation of rocaglamides, the synthesis of racemic rocaglaol (**28**) and rocaglaol derivatives **202**, **203**, and **204** were achieved. These compounds were tested on a variety of human cancer cell lines. In all cases, brominated rocaglaol **203** was the most potent, having IC_{50} values lower than 1 n*M* for KB (nasopharynx), MCF7R (breast), HCT116 (colon), and HL60 (neutrophil) human cancer cell lines. No aryl substituents manifested a decrease in potency, while rocaglaol derivative **204** was highly inactive (methoxy group on the other aryl ring). Fluorescent probe **208** was synthesized from an amide coupling between fluorescent coumarin intermediate **207** and rocagloic acid intermediate **206**. The synthesis of affinity ligand **214** was initiated by amide coupling between fragments **210** and **211** to afford azide **212**, which was converted to amine **213** using $SnCl_2$ and benzene thiol. Then, **213** was conjugated to Affi-Gel 10 to afford the affinity ligand **214** used in pull down experiments that did not reveal any interaction between **214** and eukaryotic initiation factor 4A (eiF4A). Fluorescent probe **208** was shown to accumulate in the endoplasmic reticulum (ER), implying that rocaglamides bind to their target in the ER, inducing the death of cancer cells *via* activation of the AIF and caspase-12 pathways.

Porco and coworkers, using optimized reaction conditions and sequences, displayed the broad scope of their photocycloaddition methodology in the syntheses of rocaglate analogues (Fig. 31) (*139*). Using a variety of dipolarophiles, novel cycloadducts were obtained including thioester **216b**, *Weinreb* amide **216e**, and amide **216f**. The aglain intermediates were then converted to cyclopenta[*b*,*c*]-benzofurans using two methods. Cycloadducts bearing a methyl ester moiety were readily converted into rocaglate isomers, but in many cases a *Lewis* acid-catalyzed method was found to be necessary. When 25 compounds were tested for *in vitro* potency as inhibitors of eukaryotic translation, six compounds had IC_{50} values lower than 10 μ*M*. Compounds **217e** and **217f** were found to be the most potent with IC_{50} values of 300–400 n*M*. The positive control, silvestrol, had an IC_{50}

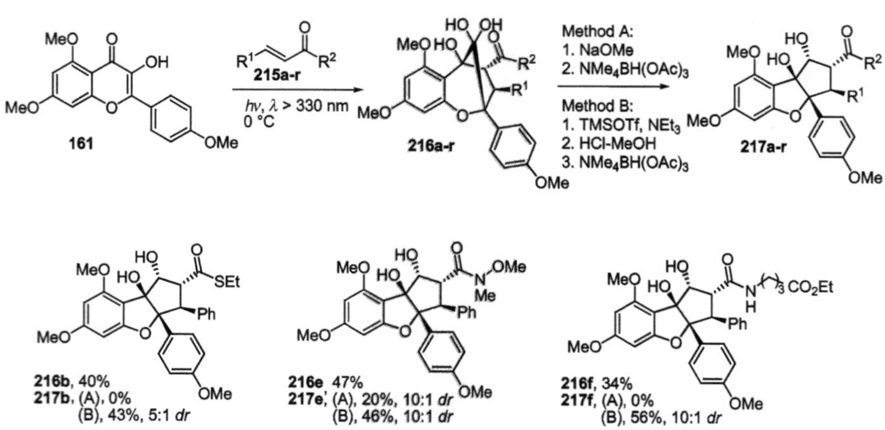

Fig. 31. Scope of [3 + 2] photocycloaddition to produce analogues of rocaglamide (*139*)

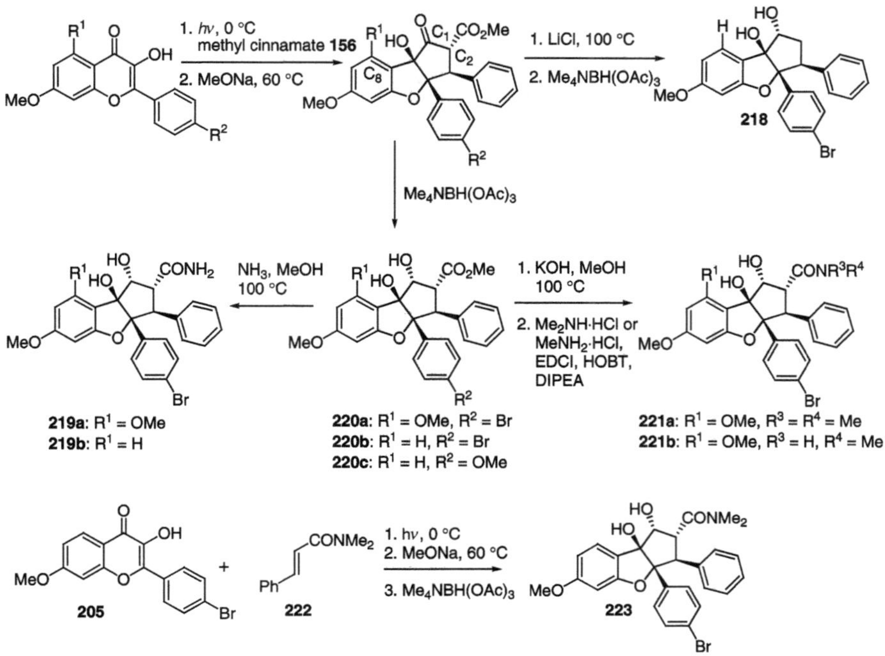

Fig. 32. *Désaubry*'s rocaglamide and rocaglaol analogues (*140*)

of 100 n*M* in the same assay, demonstrating excellent potency from the synthetic analogues. When tested *in vivo*, hydroxamate **217e** was the most potent, inhibiting 85% of protein synthesis over the course of an hour, similar to silvestrol.

With knowledge that bromine substituted at the C-4′ position improves cancer cell cytotoxicity, *Désaubry* and coworkers studied structure activity relationships (SAR) when C-1 and C-2 substituents are varied, and how the C-8 methoxy affected activity. The rocaglamide core was constructed through [3 + 2] photocycloaddition, followed by acyloin rearrangement (Fig. 32) (*140*). Compound **218** was prepared by decarboxylation and subsequent reduction. Methyl rocaglate analogues **220a-220c** were tested, and from **220a** and **220b**, amide analogues **219a** and **219b** were achieved through nucleophilic displacement using ammonia. Compounds **221a** and **221b** were prepared from **220a** and **220b** by KOH hydrolysis followed by amide coupling. Using modified 3-hydroxyflavone **205** and cinnamic amide **222**, synthetic rocaglamide **223** was efficiently prepared using a [3 + 2] photocycloaddition, ketol rearrangement, and subsequent reduction. With the C-2 analogues in hand, C-1 analogues were constructed from α-hydroxyketone **224** (Fig. 33) (*140*). Reduction with $Me_4NBH(OAc)_3$ afforded rocaglaol analogue **225**, which was esterified to afford analogues **226**, **227**, and **228**. When **224** was treated with sodium borohydride, *Désaubry* observed opposite selectivity from $Me_4NBH(OAc)_3$, forming *cis*-diol **229**, which was esterified to afford **230**. Oxime methyl ether **231** was reduced

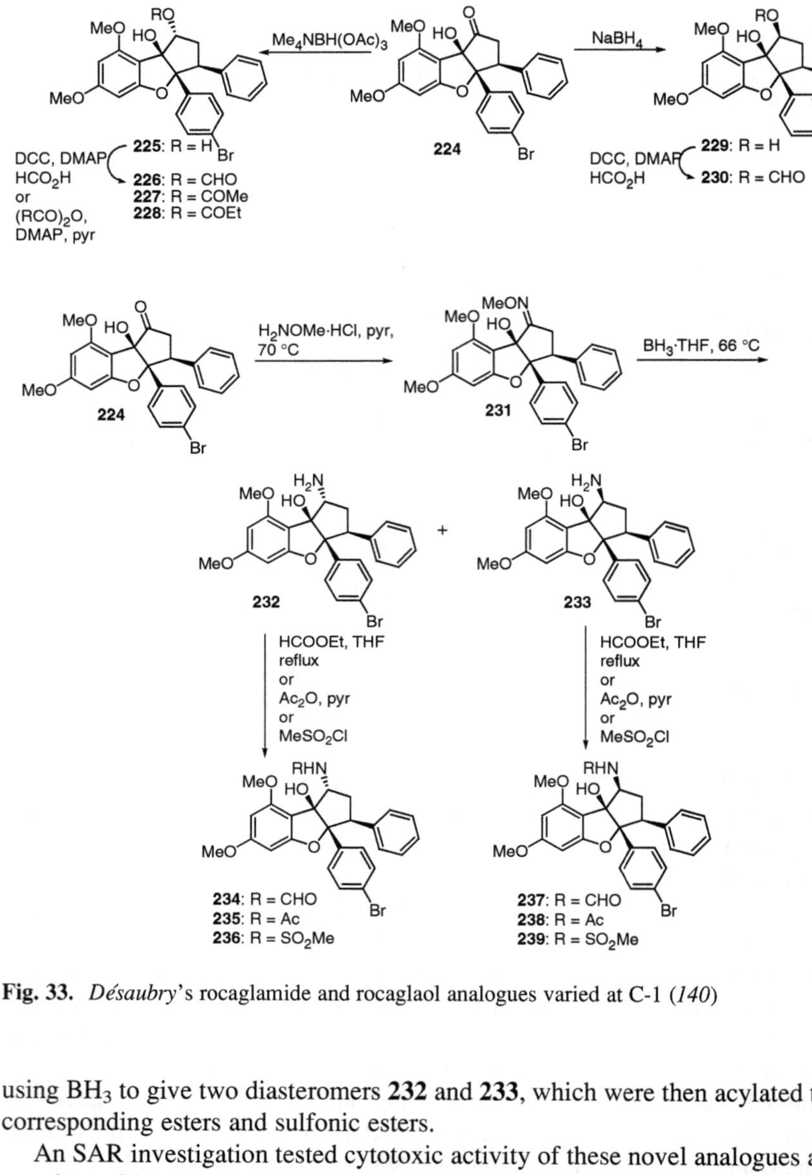

Fig. 33. *Désaubry*'s rocaglamide and rocaglaol analogues varied at C-1 (*140*)

using BH_3 to give two diasteromers **232** and **233**, which were then acylated to give corresponding esters and sulfonic esters.

An SAR investigation tested cytotoxic activity of these novel analogues against a variety of human cancer cell lines, and found that 8-demethoxy compounds **218**, **219b**, **220b**, **220c**, and **223** were less active than their 8-methoxy counterparts, suggesting a preference of a methoxy group for cytotoxicity. It was also determined that having an amide or ester substituent at C-2 was detrimental for activity against HL60R cells, suggesting a detriment to multidrug resistance as well. Although configuration of the hydroxy at C-1 was not crucial for activity, formamide analogue **237** had better activity in comparison to diastereomer **234**. Compound **237** also proved to be the best therapeutic candidate from *in vitro* studies. *In vivo* studies were then performed on mice with xenografted 3LL tumors using **237**, and reduced the growth of the tumor by 65% without any visible toxicity in the mice.

6. Concluding Remarks

The present review surveys and summarizes the rocaglamides (= flavaglines) and related derivatives isolated from the genus *Aglaia* (family Meliaceae). Both the unique unprecedented chemical skeleton and the interesting pharmacological properties of rocaglamides have attracted the attention of natural product chemists, cell biologists, and pharmacologists alike. The overwhelming research interests placed on the genus *Aglaia* have led to the isolation of more than 120 congeners since the first report of the prototype compound, rocaglamide (**1**), in 1982 by King *et al.* (*11*).

In addition to their insecticidal activity, rocaglamide derivatives revealed several other bioactivities of interest, such as anti-inflammatory, *in vitro* and *in vivo* antiproliferative, and apoptosis induction activities. While rocaglamides are unlikely to be developed into a commercial insecticide, their anti-tumor and apoptosis induction activities make them potential oncology drug candidates.

Acknowledgements Preparation of this chapter was supported by a grant of BMBF (to *P.P.*). A scholarship granted and financed by the Egyptian government (predoctoral fellowship for *S.S.E.*) is gratefully acknowledged.

References

1. Newman DJ, Cragg GM, Snader KM (2000) The Influence of Natural Products upon Drug Discovery. Nat Prod Rep **17**: 215
2. Chin Y-W, Balunas MJ, Chai HB, Kinghorn AD (2006) Drug Discovery from Natural Sources. AAPS J **8**: e239
3. Butler MS (2008) Natural Products to Drugs: Natural Product-Derived Compounds in Clinical Trials. Nat Prod Rep **25**: 475
4. Altmann K-H, Höfle G, Müller R, Mulzer J, Prantz K (2009) The Epothilones: an Outstanding Family of Anti-Tumor Agents – From Soil to the Clinic. Prog Chem Org Nat Prod 90, 1
5. Balunas MJ, Jones WP, Chin Y-W, Mi Q, Farnsworth NR, Soejarto DD, Cordell GA, Swanson SM, Pezzuto JM, Chai H-B, Kinghorn AD (2006) Relationships between Inhibitory Activity against a Cancer Cell Line Panel, Profiles of Plants Collected, and Compound Classes Isolated in an Anticancer Drug Discovery Project. Chem Biodiversity **3**: 897
6. Isman MB, Gunning PJ, Spollen KM (1997) Tropical Timber Species as Sources of Botanical Insecticides. In: Hedin PA, Hollingworth RM, Masler EP, Miyamoto J, Thompson DG (eds.) Phytochemicals for Pest Control, Symposium Series 658, p. 27. ACS Books, Washington, DC
7. Pannell CM (2007) *Aglaia*. In: Soepadmo E, Saw LG, Chung RCK, Kiew R (eds.) Tree Flora of Sabah and Sarawak, p. 24. Ampang Press Sdn Bhd, Kuala Lumpur

8. Muellner AN, Samuel R, Chase MW, Pannell CM, Greger H (2005) *Aglaia* (Meliaceae): An Evaluation of Taxonomic Concepts Based on DNA Data and Secondary Metabolites. Am J Bot **92**: 534
9. Proksch P, Edrada RA, Ebel R, Bohnenstengel FI, Nugroho BW (2001) Chemistry and Biological Activity of Rocaglamide Derivatives and Related Compounds in *Aglaia* species (Meliaceae). Curr Org Chem **5**: 923
10. Janaki S, Vijayasekaran V, Viswanathan S, Balakrishna K (1999) Anti-Inflammatory Activity of *Aglaia roxburghiana* var. *beddomei* Extract and Triterpenes Roxburghiadiol A and B. J Ethnopharmacol **67**: 45, and references cited therein
11. King ML, Chiang C-C, Ling H-C, Fujita E, Ochiai M, McPhail AT (1982) X-Ray Crystal Structure of Rocaglamide, a Novel Antileukemic 1*H*-Cyclopenta[*b*]benzofuran from *Aglaia elliptifolia*. J Chem Soc Chem Commun: 1150
12. Kim S, Salim AA, Swanson SM, Kinghorn AD (2006) Potential of Cyclopenta[*b*]benzofurans from *Aglaia* Species in Cancer Chemotherapy. Anticancer Agents Med Chem **6**: 319
13. Brader G, Vajrodaya S, Greger H, Bacher M, Kalchhauser H, Hofer O (1998) Bisamides, Lignans, Triterpenes, and Insecticidal Cyclopenta[*b*]benzofurans from *Aglaia* Species. J Nat Prod **61**: 1482
14. Nugroho BW, Edrada, RA, Wray V, Witte L, Bringmann G, Gehling M, Proksch P (1999) An Insecticidal Rocaglamide Derivative and Related Compounds from *Aglaia odorata* (Meliaceae). Phytochemistry **51**: 367
15. Bacher M, Hofer O, Brader G, Vajrodaya S, Greger H (1999) Thapsakins: Possible Biogenetic Intermediates towards Insecticidal Cyclopenta[*b*]benzofurans from *Aglaia edulis*. Phytochemistry **52**: 253
16. Dreyer M, Nugroho BW, Bohnenstengel FI, Ebel R, Wray V, Witte L, Bringmann G, Mühlbacher J, Herold M, Hung PD, Kiet LC, Proksch P (2001) New Insecticidal Rocaglamide Derivatives and Related Compounds from *Aglaia oligophylla*. J Nat Prod **64**: 415
17. Schneider C, Bohnenstengel FI, Nugroho BW, Wray V, Witte L, Hung PD, Kiet LC, Proksch P (2000) Insecticidal Rocaglamide Derivatives from *Aglaia spectabilis* (Meliaceae). Phytochemistry **54**: 731
18. Greger H, Pacher T, Brem B, Bacher M, Hofer O (2001) Insecticidal Flavaglines and Other Compounds from Fijian *Aglaia* Species. Phytochemistry **57**: 57
19. Hayashi N, Lee K-H, Hall IH, McPhail AT, Huang H-C (1982) Structure and Stereochemistry of (−)-Odorinol, an Antileukemic Diamide from *Aglaia odorata*. Phytochemistry **21**: 2371
20. Cui B, Chai H, Santisuk T, Reutrakul V, Farnsworth NR, Cordell GA, Pezzuto JM, Kinghorn AD (1997) Novel Cytotoxic 1*H*-Cyclopenta[*b*]benzofuran Lignans from *Aglaia elliptica*. Tetrahedron **53**: 17625
21. Zhu JY, Giaisi M, Köhler R, Müller WW, Mühleisen A, Proksch P, Krammer PH, Li-Weber M (2009) Rocaglamide Sensitizes Leukemic T Cells to Activation-Induced Cell Death by Differential Regulation of CD95L and c-FLIP Expression. Cell Death Differen **16**: 1289
22. Saifah E, Suttisri R, Shamsub S, Pengsuparp T, Lipipum V (1999) Bisamides from *Aglaia edulis*. Phytochemistry **52**: 1085
23. Engelmeier D, Hadacek F, Pacher T, Vajrodaya S, Greger H (2000) Cyclopenta[*b*] benzofurans from *Aglaia* Species with Pronounced Antifungal Activity against Rice Blast Fungus (*Pyricularia grisea*). J Agric Food Chem **48**: 1400
24. Baumann B, Bohnenstengel F, Siegmund D, Wajant H, Weber C, Herr I, Debatin K-M, Proksch P, Wirth T (2002) Rocaglamide Derivatives are Potent Inhibitors of NF-κB Activation in T-Cells. J Biol Chem **277**: 44791
25. Proksch P, Giaisi M, Treiber MK, Palfi K, Merling A, Spring H, Krammer PH, Li-Weber M (2005) Rocaglamide Derivatives are Immunosuppressive Phytochemicals that Target NF-AT Activity in T cells. J Immunol **174**: 7075
26. Greger H, Pacher T, Vajrodaya S, Bacher M, Hofer O (2000) Infraspecific Variation of Sulfur-Containing Bisamides from *Aglaia leptantha*. J Nat Prod **63**: 616

27. Wang B-G, Ebel R, Nugroho BW, Prijono D, Frank W, Steube KG, Hao X-J, Proksch P (2001) Aglacins A–D, First Representatives of a New Class of Aryltetralin Cyclic Ether Lignans from *Aglaia cordata*. J Nat Prod **64**: 1521
28. Wang B-G, Ebel R, Wang C-Y, Wray V, Proksch P (2002) New Methoxylated Aryltetrahydronaphthalene Lignans and a Norlignan from *Aglaia cordata*. Tetrahedron Lett **43**: 5783
29. Wang B-G, Ebel R, Wang C-Y, Edrada RA, Wray V, Proksch P (2004) Aglacins I–K, Three Highly Methoxylated Lignans from *Aglaia cordata*. J Nat Prod **67**: 682
30. Saifah E, Puripattanavong J, Likhitwitayawuid K, Cordell GA, Chai H, Pezzuto JM (1993) Bisamides from *Aglaia* Species: Structure Analysis and Potential to Reverse Drug Resistance with Cultured Cells. J Nat Prod **56**: 473
31. Saifah E, Suparakchinda N (1998) Bisamide from *Aglaia rubiginosa*. Planta Med **64**: 682
32. Inada A, Shono K, Murata H, Inatomi Y, Darnaedi D, Nakanishi T (2000) Three Putrescine Bisamides from the Leaves of *Aglaia grandis*. Phytochemistry **53**: 1091
33. Seger C, Pacher T, Greger H, Saifah E, Hofer O (2002) Aglairubine: Structure Revision of a Chemotaxonomically Interesting Bisamide in *Aglaia* (Meliaceae). Monatsh Chem **133**: 97
34. Duong TN, Edrada RA, Ebel R, Wray V, Frank W, Duong AT, Lin WH, Proksch P (2007) Putrescine Bisamides from *Aglaia gigantea*. J Nat Prod **70**: 1640
35. Greger H, Hofer M, Teichmann K, Schinnerl J, Pannell CM, Vajrodaya S, Hofer O (2008) Amide-Esters from *Aglaia tenuicaulis* – First Representatives of a Class of Compounds Structurally Related to Bisamides and Flavaglines. Phytochemistry **69**: 928
36. Puripattanavong J, Weber S, Brecht V, Frahm AW (2000) Phytochemical Investigation of *Aglaia andramanica*. Planta Med **66**: 740
37. Mohamad K, Martin M-T, Leroy E, Tempête C, Sévenet T, Awang K, Païs M (1997) Argenteanones C–E and Argenteanols B–E, Cytotoxic Cycloartanes from *Aglaia argentea*. J Nat Prod **60**: 81
38. Roux D, Martin M-T, Adeline M-T, Sevenet T, Hadi AHA, Païs (1998) Foveolins A and B, Dammarane Triterpenes from *Aglaia foveolata*. Phytochemistry **49**: 1745
39. Mohamad K, Sévenet T, Dumontet V, Païs M, Tri MV, Hadi H, Awang K, Martin M-T (1999) Dammarane Triterpenes and Pregnane Steroids from *Aglaia lawii* and *A. tomentosa*. Phytochemistry **51**: 1031
40. Mohamad K, Martin M-T, Najdar H, Gaspard C, Sévenet T, Awang K, Hadi H, Païs M (1999) Cytotoxic 3,4-Secoapotirucallanes from *Aglaia argentea* Bark. J Nat Prod **62**: 868
41. Weber S, Puripattanavong J, Brecht V, Frahm AW (2000) Phytochemical Investigation of *Aglaia rubiginosa*. J Nat Prod **63**: 636
42. Rivero-Cruz JF, Chai H-B, Kardono LBS, Setyowati FM, Afriastini JJ, Riswan S, Farnsworth NR, Cordell GA, Pezzuto JM, Swanson SM, Kinghorn AD (2004) Cytotoxic Constituents of the Twigs and Leaves of *Aglaia rubiginosa*. J Nat Prod **67**: 343
43. Cai X-H, Luo X-D, Zhou J, Hao X-J (2005) Dolabellane Diterpenoids from the Higher Plant *Aglaia odorata*. Helv Chim Acta **88**: 2938
44. Su B-N, Chai H, Mi Q, Riswan S, Kardono LBS, Afriastini JJ, Santarsiero BD, Mesecar AD, Farnsworth NR, Cordell GA, Swanson SM, Kinghorn AD (2006) Activity-Guided Isolation of Cytotoxic Constituents from the Bark of *Aglaia crassinervia* Collected in Indonesia. Bioorg Med Chem **14**: 960
45. Xie B-J, Yang S-P, Chen H-D, Yue J-M (2007) Agladupols A–E, Triterpenoids from *Aglaia duperreana*. J Nat Prod **70**: 1532
46. Joycharat N, Greger H, Hofer O, Saifah E (2008) Flavaglines and Triterpenoids from the Leaves of *Aglaia forbesii*. Phytochemistry **69**: 206
47. Joycharat N, Greger H, Hofer O, Saifah E (2008) Flavaglines and Triterpenes as Chemical Markers of *Aglaia oligophylla*. Biochem Syst Ecol **36**: 584
48. Pointinger S, Promdang S, Vajrodaya S, Pannell CM, Hofer O, Mereiter K, Greger H (2008) Silvaglins and Related 2,3-Secodammarane Derivatives – Unusual Types of Triterpenes from *Aglaia silvestris*. Phytochemistry **69**: 2696

49. Seger C, Pointinger S, Greger H, Hofer O (2008) Isoeichlerianic Acid from *Aglaia silvestris* and Revision of the Stereochemistry of Foveolin B. Tetrahedron Lett **49**: 4313
50. Hofer O, Pointinger S, Brecker L, Peter K, Greger H (2009) Silvaglenamin – A Novel Dimeric Triterpene Alkaloid from *Aglaia silvestris*. Tetrahedron Lett **50**: 467
51. Cai X-H, Wang Y-Y, Zhao P-J, Li Y, Luo X-D (2010) Dolabellane Diterpenoids from *Aglaia odorata*. Phytochemistry **71**: 1020
52. Trost BM, Greenspan PD, Yang BV, Saulnier MG (1990) An Unusual Oxidative Cyclization. A Synthesis and Absolute Stereochemical Assignment of (−)-Rocaglamide. J Am Chem Soc **112**: 9022
53. Ishibashi F, Satasook C, Isman MB, Towers GHN (1993) Insecticidal 1*H*-Cyclopentatetrahydro[*b*]benzofurans from *Aglaia odorata*. Phytochemistry **32**: 307
54. Nugroho BW, Güssregen B , Wray V, Witte L, Bringmann G, Proksch P (1997) Insecticidal Rocaglamide Derivatives from *Aglaia elliptica* and *A. harmsiana*. Phytochemistry **45**: 1579
55. Nugroho BW, Edrada RA, Güssregen B, Wray V, Witte L, Proksch P (1997) Insecticidal Rocaglamide Derivatives from *Aglaia duperreana*. Phytochemistry **44**: 1455
56. Chaidir, Hiort J, Nugroho BW, Bohnenstengel FI, Wray V, Witte L, Hung PD, Kiet LC, Sumaryono W, Proksch P (1999) New Insecticidal Rocaglamide Derivatives from Flowers of *Aglaia duperreana* (Meliaceae). Phytochemistry **52**: 837
57. Molleyres L-P, Rindlisbacher A, Winkler T, Kumar V (1999) Insecticidal Natural Products: New Rocaglamide Derivatives from *Aglaia roxburghiana*. Pestic Sci **55**: 494
58. Hiort J, Chaidir, Bohnenstengel FI, Nugroho BW, Schneider C, Wray V, Witte L, Hung PD, Kiet LC, Proksch P (1999) New Insecticidal Rocaglamide Derivatives from the Roots of *Aglaia duperreana*. J Nat Prod **62**: 1632
59. Güssregen B, Fuhr M, Nugroho BW, Wray V, Witte L, Proksch P (1997) New Insecticidal Rocaglamide Derivatives from Flowers of *Aglaia odorata*. J Biosci (Z Naturforsch) **52c**: 339
60. Dumontet V, Thoison O, Omobuwajo OR, Martin M-T, Perromat G, Chiaroni A, Riche C, Païs M, Sévenet T, Hadi AHA (1996) New Nitrogenous and Aromatic Derivatives from *Aglaia argentea* and *A. forbesii*. Tetrahedron **52**: 6931
61. Fuzzati N, Dyatmiko W, Abdul Rahman, Achmad F, Hostettman K (1996) Triterpenoids, Lignans and a Benzofuran Derivative from the Bark of *Aglaia elaeagnoidea*. Phytochemistry **42**: 1395
62. Ko F-N, Wu T-S, Liou M-J, Huang T-F, Teng C-M (1992) PAF Antagonism In Vitro and In Vivo by Aglafoline from *Aglaia elliptifolia* Merr. Eur J Pharmacol **218**: 129
63. Wang S-K, Cheng Y-J, Duh C-Y (2001) Cytotoxic Constituents from Leaves of *Aglaia elliptifolia*. J Nat Prod **64**: 92
64. Chaidir, Lin WH, Ebel R, Edrada RA, Wray V, Nimtz M, Sumaryono W, Proksch P (2001) Rocaglamides, Glycosides, and Putrescine Bisamides from *Aglaia dasyclada*. J Nat Prod **64**: 1216
65. Mulholland DA, Naidoo N (1998) A Revision of the Structure of Ferrugin from *Aglaia ferruginaea*. Phytochemistry **47**: 1163
66. Dean FM, Monkhe TV, Mulholland DA, Taylor DAH (1993) An Isoflavanoid from *Aglaia ferruginaea*, an Australian Member of the Meliaceae. Phytochemistry **34**: 1537
67. Xu Y-J, Wu X-H, Tan BKH, Lai Y-H, Vittal JJ, Imiyabir Z, Madani L, Khozirah KS, Goh SH (2000) Flavonol–Cinnamate Cycloadducts and Diamide Derivatives from *Aglaia laxiflora*. J Nat Prod **63**: 473
68. Hwang BY, Su B-N, Chai H, Mi Q, Kardono LB, Afriastini JJ, Riswan S, Santarsiero BD, Mesecar AD, Wild R, Fairchild CR, Vite GD, Rose WC, Farnsworth NR, Cordell GA, Pezzuto JM, Swanson SM, Kinghorn AD (2004) Silvestrol and Episilvestrol, Potential Anticancer Rocaglate Derivatives from *Aglaia silvestris*. J Org Chem **69**: 3350
69. Hwang BY, Su B-N, Chai H, Mi Q, Kardono LB, Afriastini JJ, Riswan S, Santarsiero BD, Mesecar AD, Wild R, Fairchild CR, Vite GD, Rose WC, Farnsworth NR, Cordell GA, Pezzuto JM, Swanson SM, Kinghorn AD (2004) Silvestrol and Episilvestrol, Potential Anticancer Rocaglate Derivatives from *Aglaia silvestris*. J Org Chem **69**: 6156

70. Chumkaew P, Kato S, Chantrapromma K (2006) Potent Cytotoxic Rocaglamide Derivatives from the Fruits of *Amoora cucullata*. Chem Pharm Bull **54**: 1344
71. Kim S, Chin Y-W, Su B-N, Riswan S, Kardono LBS, Afriastini JJ, Chai H, Farnsworth NR, Cordell GA, Swanson SM, Kinghorn AD (2006) Cytotoxic Flavaglines and Bisamides from *Aglaia edulis*. J Nat Prod **69**: 1769
72. Kokpol U, Venaskulchai B, Simpson J, Weavers RT (1994) Isolation and X-ray Structure Determination of a Novel Pyrimidinone from *Aglaia odorata*. J Chem Soc Chem Commun (6):773
73. Ohse T, Ohba S, Yamamoto T, Koyano T, Umezawa K (1996) Cyclopentabenzofuran Lignan Protein Synthesis Inhibitors from *Aglaia odorata*. J Nat Prod **59**: 650
74. Wang S-K, Duh C-Y (2001) Cytotoxic Cyclopenta[*b*]benzofuran Derivatives from the Stem Bark of *Aglaia formosana*. Planta Med **67**: 555
75. Bringmann G, Mühlbacher J, Messer K, Dreyer M, Ebel R, Nugroho BW, Wray V, Proksch P (2003) Cyclorocaglamide, the First Bridged Cyclopentatetrahydro-benzofuran, and a Related "Open Chain" Rocaglamide Derivative from *Aglaia oligophylla*. J Nat Prod **66**: 80
76. Shiengthong D, Ungphakorn A, Lewis DE, Massy-Westropp RA (1979) New Nitrogenous Compounds – Odorine and Odorinol. Tetrahedron Lett **24**: 2247
77. Purushothaman KK, Sarada A, Connolly JD, Akinniyi JA (1979) The Structure of Roxburghilin, a Bis-amide of 2-Aminopyrrolidine from the Leaves of *Aglaia roxburghiana* (Meliaceae). J Chem Soc Perkin Trans I :3171
78. Saifah E, Jongbunprasert V, Kelley CJ (1988) Piriferine, a New Pyrrolidine Alkaloid from *Aglaia pirifera* leaves. J Nat Prod **51**: 80
79. Inada A, Sorano T, Murata H, Inatomi Y, Darnaedi D, Nakanishi T (2001) Diamide Derivatives and Cycloartanes from the Leaves of *Aglaia elliptica*. Chem Pharm Bull **49**: 1226
80. Salim AA, Pawlus AD, Chai H-B, Farnsworth NR, Kinghorn AD, Carcache-Blanco EJ (2007) Ponapensin, a Cyclopenta[*bc*]benzopyran with Potent NF-κB Inhibitory Activity from *Aglaia ponapensis*. Bioorg Med Chem Lett **17**: 109
81. Salim AA, Chai H-B, Rachman I, Riswan S, Kardono LBS, Farnsworth NR, Carcache-Blanco EJ, Kinghorn AD (2007) Constituents of the Leaves and Stem Bark of *Aglaia foveolata*. Tetrahedron **63**: 7926
82. Kim S, Su B-N, Riswan S, Kardono LBS, Afriastini JJ, Gallucci JC, Chai H, Farnsworth NR, Cordell GA, Swanson SM, Kinghorn AD (2005) Edulisones A and B, Two Epimeric Benzo-[*b*]oxepine Derivatives from the Bark of *Aglaia edulis*. Tetrahedron Lett **46**: 9021
83. Chiu SF (1985) Recent Research Findings on Meliaceae and Other Promising Botanical Insecticides in China. J Plant Diseases and Protection **92**: 310
84. Bohnenstengel FI, Steube KG, Meyer C, Nugroho BW, Hung PD, Kiet LC, Proksch P (1999) Structure Activity Relationships of Antiproliferative Rocaglamide Derivatives from *Aglaia* species (Meliaceae). J Biosci (Z Naturforsch) **54c**: 55
85. Bohnenstengel FI, Steube KG, Meyer C, Quentmeier H, Nugroho BW, Proksch P (1999) 1 *H*-Cyclopenta[*b*]benzofuran Lignans from *Aglaia* species Inhibit Cell Proliferation and Alter Cell Cycle Distribution in Human Monocytic Leukemia Cell Lines. J Biosci (Z Naturforsch) **54c**: 1075
86. Ewete F, Nicol RW, Hengsawad V, Sukumalanand P, Satasook C, Wiriyachitra P, Isman MB, Kahn Y, Duval F, Philogène BJR, Arnason JT (1996) Insecticidal Activity of *Aglaia odorata* Extract and the Active Principle, Rocaglamide, to the European Corn Borer *Ostrinia nubilalis* Hübn. (Lep., Pyralidae). J Appl Ent **120**: 483
87. Koul O, Kaur H, Goomber S, Wahab S (2004) Bioefficacy and Mode of Action of Rocaglamide from *Aglaia elaeagnoidea* (syn. *A. roxburghiana*) against Gram Pod Borer, *Helicoverpa armigera* (Hübner). J Appl Ent **128**: 177
88. O'Shea JJ, Ma A, Lipsky P (2002) Cytokines and Autoimmunity. Nat Rev Immunol **2**: 37
89. Li-Weber M, Krammer PH (2003) Regulation of IL4 Gene Expression by T Cells and Therapeutic Perspectives. Nat Rev Immunol **3**: 534

90. Yamamoto Y, Gaynor RB (2001) Therapeutic Potential of Inhibition of the NF-kappaB Pathway in the Treatment of Inflammation and Cancer. J Clin Invest **107**:135
91. Macian F (2005) NFAT Proteins: Key Regulators of T-Cell Development and Function. Nat Rev Immunol **5**: 472
92. Li Q, Verma IM (2002) NF-kappaB Regulation in the Immune System. Nat Rev Immunol **2**: 725
93. Lee SK, Cui B, Mehta RR, Kinghorn AD, Pezzuto JM (1998) Cytostatic Mechanism and Antitumor Potential of Novel 1*H*-Cyclopenta[*b*]benzofuran Lignans Isolated From *Aglaia elliptica*. Chem Biol Interact **115**: 215
94. Cencic R, Carrier M, Galicia-Vázquez G, Bordeleau ME, Sukarieh R, Bourdeau A, Brem B, Teodoro JG, Greger H, Tremblay ML, Porco JA Jr, Pelletier J (2009) Antitumor Activity and Mechanism of Action of the Cyclopenta[*b*]benzofuran, Silvestrol. PLoS One **4**: e5223
95. Hausott B, Greger H, Marian B (2004) Flavaglines: a Group of Efficient Growth Inhibitors Block Cell Cycle Progression and Induce Apoptosis in Colorectal Cancer Cells. Int J Cancer **109**: 933
96. Mi Q, Kim S, Hwang BY, Su BN, Chai H, Arbieva ZH, Kinghorn AD, Swanson SM (2006) Silvestrol Regulates G2/M Checkpoint Genes Independent of p53 Activity. Anticancer Res **26**: 3349
97. Silvera D, Formenti SC, Schneider RJ (2010) Translational Control in Cancer. Nat Rev Cancer **10**: 254
98. Bordeleau ME, Robert F, Gerard B, Lindqvist L, Chen SM, Wendel HG, Brem B, Greger H, Lowe SW, Porco JA Jr, Pelletier J (2008) Therapeutic Suppression of Translation Initiation Modulates Chemosensitivity in a Mouse Lymphoma Model. J Clin Invest **8**: 2651
99. Bleumink M, Köhler R, Giaisi M, Proksch P, Krammer PH, Li-Weber M (2011) Rocaglamide Breaks TRAIL Resistance in HTLV-1-associated Adult T-cell Leukemia/ Lymphoma by Translational Suppression of c-FLIP Expression. Cell Death Differ **18**: 362
100. Zhu JY, Lavrik IN, Mahlknecht U, Giaisi M, Proksch P, Krammer PH, Li-Weber M (2007) The Traditional Chinese Herbal Compound Rocaglamide Preferentially Induces Apoptosis in Leukemia Cells by Modulation of Mitogen-Activated Protein Kinase Activities. Int J Cancer **121**: 1839
101. Krammer PH, Arnold R, Lavrik IN (2007) Life and Death in Peripheral T Cells. Nat Rev Immunol **7**: 532
102. Galluzzi L, Larochette N, Zamzami N and Kroemer G (2006) Mitochondria as Therapeutic Targets for Cancer Chemotherapy Oncogene **25**: 4812
103. Kim S, Hwang BY, Su BN, Chai H, Mi Q, Kinghorn AD, Wild R, Swanson SM (2007) Silvestrol, a Potential Anticancer Rocaglate Derivative from *Aglaia foveolata*, Induces Apoptosis in LNCaP Cells Through the Mitochondrial/Apoptosome Pathway Without Activation of Executioner Caspase-3 or −7. Anticancer Res **27**: 2175
104. Zhuang S, Demirs JT, Kochevar IE (2000) p38 Mitogen-Activated Protein Kinase Mediates bid Cleavage, Mitochondrial Dysfunction, and Caspase-3 Activation During Apoptosis Induced by Singlet Oxygen but not by Hydrogen Peroxide. J Biol Chem **275**: 25939
105. Yoshino T, Kishi H, Nagata T, Tsukada K, Saito S, Muraguchi A (2001) Differential Involvement of p38 MAP Kinase Pathway and Bax Translocation in the Mitochondria-Mediated Cell Death in TCR- and Dexamethasone-Stimulated Thymocytes. Eur J Immunol **31**: 2702
106. Kim BJ, Ryu SW, Song BJ (2006) JNK- and p38 Kinase-Mediated Phosphorylation of Bax Leads to Its Activation and Mitochondrial Translocation and to Apoptosis of Human Hepatoma HepG2 Cells. J Biol Chem **281**: 21256
107. Cai B, Chang SH, Becker EB, Bonni A, Xia Z (2006) p38 MAP Kinase Mediates Apoptosis Through Phosphorylation of BimEL at Ser-65. J Biol Chem **281**: 25215
108. De Chiara G, Marcocci ME, Torcia M, Lucibello M, Rosini P, Bonini P, Higashimoto Y, Damonte G, Armirotti A, Amodei S, Palamara AT, Russo T, Garaci E, Cozzolino F

(2006) Bcl-2 Phosphorylation by p38 MAPK: Identification of Target Sites and Biologic Consequences. J Biol Chem **281**: 21353

109. Grethe S, Coltella N, Di Renzo MF, Porn-Ares MI (2006) p38 MAPK Downregulates Phosphorylation of Bad in Doxorubicin-Induced Endothelial Apoptosis. Biochem Biophys Res Commun **347**: 781

110. Balmanno K, Cook SJ (2009) Tumour Cell Survival Signalling by the ERK1/2 Pathway. Cell Death Differ **16**: 368

111. Yeh JH, Hsu SC, Han SH, Lai MZ (1998) Mitogen-Activated Protein Kinase Kinase Antagonized fas-Associated Death Domain Protein-Mediated Apoptosis by Induced FLICE-Inhibitory Protein Expression. J Exp Med **188**: 1795

112. Wang W, Prince CZ, Mou Y, Pollman MJ (2002) Notch3 Signaling in Vascular Smooth Muscle Cells Induces c-FLIP Expression via ERK/MAPK Activation. Resistance to Fas ligand-Induced Apoptosis. J Biol Chem **277**: 21723

113. Panka DJ, Mano T, Suhara T, Walsh K, Mier JW (2001) Phosphatidylinositol 3-Kinase/Akt Activity Regulates c-FLIP Expression in Tumor Cells. J Biol Chem **276**: 6893

114. Poukkula M, Kaunisto A, Hietakangas V, Denessiouk K, Katajamäki T, Johnson MS, Sistonen L, Eriksson JE (2005) Rapid Turnover of c-FLIPshort is Determined by its Unique C-Terminal Tail. J Biol Chem **280**: 27345

115. Li-Weber M, Krammer PH (2003) Function and Regulation of the CD95 (APO-1/Fas) Ligand in the Immune System. Semin Immunol **15**: 145

116. Kreuz S, Siegmund D, Scheurich P and Wajant H (2001) NF-kappaB Inducers Upregulate cFLIP, a Cycloheximide-Sensitive Inhibitor of Death Receptor Signaling. Mol Cell Biol **21**: 3964

117. Ueffing N, Schuster M, Keil E, Schulze-Osthoff K and Schmitz I (2008) Up-regulation of c-FLIP short by NFAT Contributes to Apoptosis Resistance of Short-Term Activated T Cells. Blood **112**: 690

118. Katamna C (1970) Synthesis of 2-Arylbenzofuran-3-ones. Bull Soc Chim Fr **6**: 2309

119. Davey AE, Taylor RJK (1987) A Novel 1,3-Dithiane-based Cyclopenta-annulation Procedure: Synthesis of the Rocaglamide Skeleton. J Chem Soc, Chem Commun **1**: 25

120. Kraus GA, Sy JO (1989) A Synthetic Approach to Rocaglamide via Reductive Cyclization of δ-Keto Nitriles. J Org Chem **54**: 77

121. Davey AE, Schaeffer MJ, Taylor RJK (1991) Synthesis of the Novel Antileukemic Tetrahydrocyclopenta[*b*]benzofuran, Rocaglamide. J Chem Soc, Chem Commun **16**: 1137

122. Davey AE, Schaeffer MJ, Taylor RJK (1992) Synthesis of the Novel Anti-leukemic Tetrahydrocyclopenta[*b*]benzofuran, Rocaglamide and Related Synthetic Studies. J Chem Soc, Perkin Trans 1 **20**: 2657

123. Watanabe T, Takeuchi T, Kohzuma S, Umezawa K, Otsuka M (1998) Total Synthesis of (±)-Aglaiastatin, a Novel Bioactive Alkaloid. Chem Commun **10**: 1097

124. Dobler MR, Bruce I, Cederbaum F, Cooke NG, Diorazio LJ, Hall RG, Irving E (2001) Total Synthesis of (±)-Rocaglamide and Some Aryl Analogues. Tetrahedron Lett **42**: 8281

125. Gerard B, Jones G, Porco JA (2004) A Biomimetic Approach to the Rocaglamides Employing Photogeneration of Oxidopyryliums Derived from 3-Hdroxyflavones. J Am Chem Soc **126**: 13620

126. Gerard B, Sangji S, O'Leary DJ, Porco JA (2006) Enantioselective Photocycloaddition Mediated by Chiral Brnsted Acids: Asymmetric Synthesis of the Rocaglamides. J Am Chem Soc **128**: 7754

127. McDougal NT, Schaus SE (2003) Asymmetric Morita-Baylis-Hillman Reactions Catalyzed by Chiral Brnsted Acids. J Am Chem Soc **125**: 12094

128. Schoop A, Helmut G, Axel G (2000) A New Analogue of Rocaglamide by an Oxidative Dihydrofuran Synthesis. Tetrahedron Lett **41**: 1913

129. Thede K, Diedrichs N, Ragot JP (2004) Stereoselective Synthesis of (±)-Rocaglaol Analogues. Org Lett **6**: 4595

130. Diedrichs N, Ragot JP, Thede K (2005) A Highly Efficient Synthesis of Rocaglaols by a Novel α-Arylation of Ketones. Eur J Org Chem **9**: 1731
131. Magnus P, Stent MAH (2005) Stereospecific Synthesis of (±)-1,2-Anhydro Methyl Rocaglate. Org Lett **7**: 3853
132. Li H, Fu B, Wang MA, Li N, Liu WJ, Xie ZQ, Ma YQ, Qin Z (2008) Total Synthesis and Biological Activity of (±)-Rocaglamide and its 2,3-Di-*epi* Analogue. Eur J Org Chem **10**: 1753
133. Malona JA, Cariou K, Frontier JA (2009) Nazarov Cyclization Initiated by Peracid Oxidation: the Total Synthesis of (±)-Rocaglamide. J Am Chem Soc **131**: 7560
134. Gerard B, Cencic R, Pelletier J, Porco JA (2007) Enantioselective Synthesis of the Complex Rocaglate (−)-Silvestrol. Angew Chem Int Ed **46**: 7831
135. El Sous M, Rizzacasa MA (2005) Biomimetic Synthesis of the Novel 1,4-Dioxanyloxy Fragment of Silvestrol and Episilvestrol. Tetrahedron Lett **46**: 293
136. El Sous M, Khoo M, Holloway G, Owen DJ, Scammells PJ, Rizzacasa MA (2007) Total Synthesis of (−)-Episilvestrol and (−)-Silvestrol. Angew Chem Int Ed **46**: 7835
137. Adams TE, El Sous M, Hawkins BC, Hirner S, Holloway G, Khoo ML, Owen DJ, Savage GP, Scammells PJ, Rizzacasa MA (2009) Total Synthesis of the Potent Anticancer *Aglaia* Metabolites (−)-Silvestrol and (−)-Episilvestrol and the Active Analogue (−)-4′-Desmethoxyepisilvestrol. J Am Chem Soc **131**: 1607
138. Thuaud F, Bernard Y, Turkeri G, Dirr R, Aubert G, Cresteil T, Baguet A, Tomasetto C, Svitkin Y, Sonenberg N, Nebigil CG, Désaubry L (2009) Synthetic Analogue of Rocaglaol Displays a Potent and Selective Cytotoxicity in Cancer Cells: Involvement of Apoptosis Inducing Factor and Caspase-12. J Med Chem **52**: 5176
139. Roche SP, Cencic R, Pelletier J, Porco JA (2010) Biomimetic Photocycloaddition of 3-Hydroxyflavones: Synthesis and Evaluation of Rocaglate Derivatives as Inhibitors of Eukaryotic Translation. Angew Chem Int Ed **49**: 6533
140. Thuaud F, Ribeiro N, Gaiddon C, Cresteil T, Désaubry L (2011) Novel Flavaglines Displaying Improved Cytotoxicity. J Med Chem **54**: 411

Chemistry of the Immunomodulatory Macrolide Ascomycin and Related Analogues

Murty A.R.C. Bulusu, Karl Baumann, and *Anton Stuetz*

Contents

1. Introduction

This chapter, after giving a short overview of the natural product ascomycin and related analogues, provides a summary of the biological properties and the clinical use of the ascomycin derivative pimecrolimus, a calcineurin inhibitor, developed and registered for topical treatment of the inflammatory skin disease atopic dermatitis. This is followed by an in-depth description of the structural features of ascomycin and synthesis aspects including specific modifications in the binding region, cleavage reactions, semisynthetic approaches, and modifications in the effector and cyclohexyl regions. The literature covers journal articles up to July 2010 with emphasis on chemistry. Structure-activity relationships are mentioned only qualitatively. No attempt has been made to review the patent literature in depth.

M.A.R.C. Bulusu and K. Baumann are former employees of Novartis Institutes for BioMedical Research Vienna.

M.A.R.C. Bulusu • K. Baumann • A. Stuetz (✉)
Novartis Institutes for BioMedical Research Vienna, Muthgasse 11/2, A-1190, Vienna, Austria
e-mail: anton.stuetz@novartis.com

A.D. Kinghorn, H. Falk, J. Kobayashi (eds.), *Progress in the Chemistry of Organic Natural Products, Vol. 94*, DOI 10.1007/978-3-7091-0748-5_2,

1.1. Ascomycin and Related Natural Products

The successful application of cyclosporin A (CyA, Sandimmune®) in the prevention of transplantation rejection stimulated a worldwide search for new selective T-cell inhibitors resulting in the discovery of the macrolactam tacrolimus (**1**, FK506, Prograf®, Scheme 1) (*1–3*). Ascomycin (**2**), a compound initially referred to as FR-900520 and isolated earlier from *Streptomyces hygroscopicus* var. *ascomyceticus*, (Fig. 1) due to its antifungal activities, and for which the structure was not elucidated originally, was later shown to be a close structural analogue of tacrolimus (ethyl instead of allyl on C-21) (*4–7*). Analogues bearing other substituents, such as methyl (*5*), or propyl (*8*), on C-21, or, homologues in which several methyl groups on the macrocycle are replaced by a higher alkyl (*i.e.* ethyl) have been isolated as well (*9, 10*).

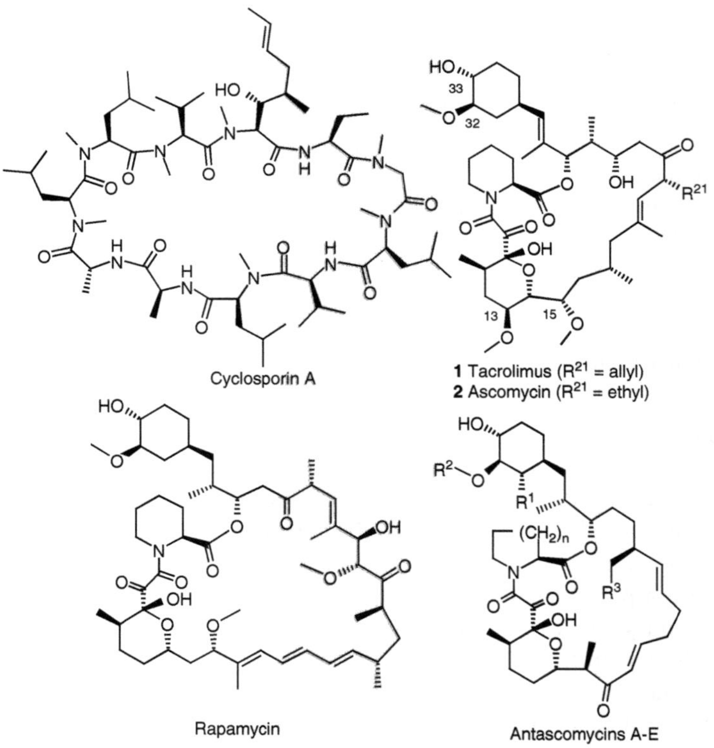

	R^1	R^2	R^3	n
A	H	H	H	2
B	OH	H	H	2
C	OH	Me	H	2
D	H	H	H	1
E	H	H	OH	2

Scheme 1. Structures of cyclosporin A, tacrolimus (**1**), ascomycin (**2**), and related macrolactams

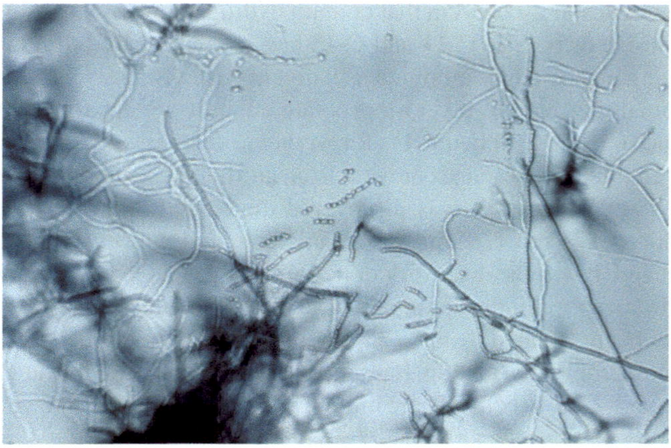

Fig. 1. Photograph of *Streptomyces* sp. grown in tap water agar. Branching filaments, abundant aerial mycelia, and long chains of small spores are visible, all of which are characteristic of the organism. Taken from en.wikipedia.org

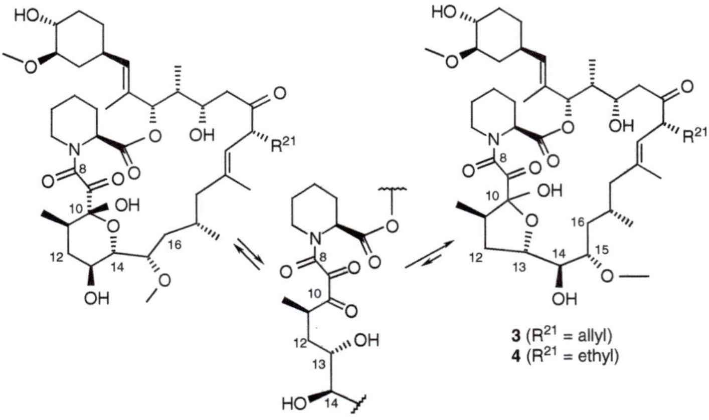

Scheme 2. Equilibria of 13-*O*-desmethyl derivatives **3** and **4** of tacrolimus (**1**) and ascomycin (**2**)

Macrolactams featuring a proline unit instead of the pipecolic acid moiety have been found as by-products in the fermentation broth of tacrolimus producer strains (*11*). Numerous 32-, 13-, and/or 15-*O*-desmethyl derivatives of ascomycin (**2**) and tacrolimus (**1**) are readily available through biotransformation (*11–15*). Interestingly, the 13-*O*-desmethyl-derivatives **3** and **4** exist in solution as highly complex mixtures of isomers (Scheme 2). Thus, seven different isomers of 13-*O*-desmethyl-FK506 (**3**) could be differentiated in COSY, HMBC, and HMQC experiments. The formation of these isomers could be explained by an epimerization at C-10, formation of hemiketal rings between C-10 and C-13, or, C-9 and C-13, and the occurrence of (*Z*)/(*E*)-isomers at the amide bond (*16*).

Other analogues of ascomycin that contain hydrogen, methyl, or ethyl instead of a methoxy at one or both of C-13 and C-15 have been prepared through fermentation employing a genetically modified ascomycin gene cluster (*17*). Related natural products containing an α,β-diketo-pipecolate subunit are rapamycin (Sirolimus) and the antascomycins (Scheme 1). Rapamycin, another macrolactam, had also been discovered like CyA and ascomycin in an antifungal screening of fermentation broths (*18, 19*). Although it was discovered quite early that rapamycin inhibits immune responses in rats (*20*), its therapeutic potential, however, was initially not fully recognized. It was the discovery of tacrolimus (**1**) that led to a renewed interest in rapamycin as an immunosuppressant (*21, 22*). The antascomycins A, B, C, D, and E have been isolated from a strain of *Micromonospora* (*23*). They bind strongly to macrophilin (FK506 binding protein, FKBP12) and antagonize the effects of **1** and **2** on T-cells. Their biological properties have not yet been fully explored.

1.2. Ascomycin Derivatives, a Novel Class of Anti-inflammatory Compounds

The discovery of this novel class of anti-inflammatory compounds started with reports of the high efficacy of oral cyclosporin A in the treatment of psoriasis, atopic dermatitis, and other inflammatory skin diseases. However, efforts failed to render this calcineurin inhibitor effective topically, in order to minimize its systemic immunosuppressive side effects. Using a newly developed model of allergic contact dermatitis (ACD) in pigs, calcineurin inhibitors of the tacrolimus- and ascomycin-type were shown to be highly effective after topical application. These findings provided the first pharmacological evidence of the potential of this novel class of topical agents for the treatment of inflammatory skin diseases (*24*). Topical application of the first representative of this class, SDZ 281–240 (Scheme 3), in

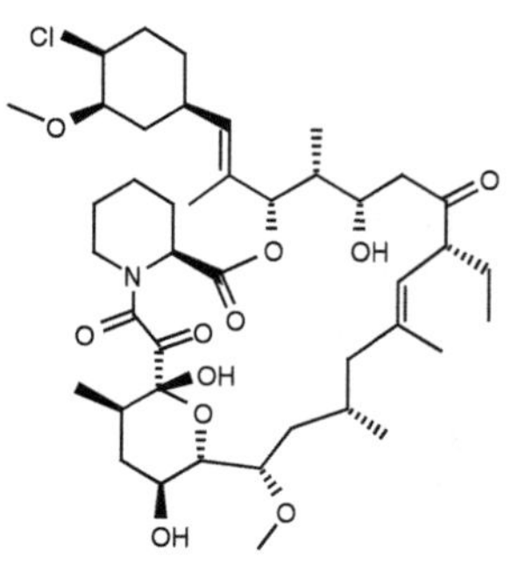

2a : Pimecrolimus (SDZ ASM 981)

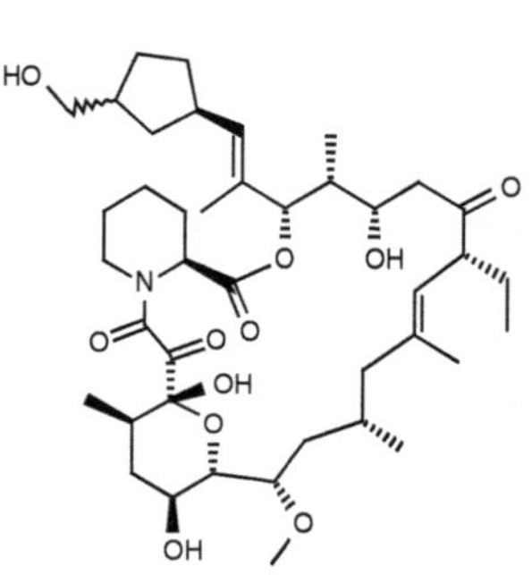

SDZ 281-240

Scheme 3. Structures of the 33-*epi*-chloro-derivative of ascomycin (**2**) pimecrolimus (**2a**), and SDZ 281–240

chronic plaque psoriasis under *Finn*-chamber occlusion confirmed the validity of this concept in man (*25*). Intensive studies on structure-activity relationships and comparative pharmacological evaluations among a large number of newly synthesized derivatives to identify a compound combining high anti-inflammatory activity with minimal side effects finally resulted in the discovery and development of pimecrolimus (**2a**, SDZ ASM 981, Scheme 3) (*26, 27*). A detailed review of the discovery and development of the then new class of topical calcineurin inhibitors has been published (*28*).

1.2.1. Pimecrolimus

1.2.1.1. Pharmacology *In Vitro* and *In Vivo*

Pimecrolimus (**2a**) binds with high affinity to the cytosolic receptor macrophilin-12 and inhibits the phosphatase calcineurin, an enzyme required for the dephosphorylation of the cytosolic form of the nuclear factor of activated T-cells (NF-AT). As a consequence, it prevents in T-cells the transcription and release of both T-helper type 1 cell (TH1) and T-helper type 2 cell (TH2) inflammatory cytokines such as interleukin-2 (IL-2), interferon-γ (IFN-γ), interleukin-4 (IL-4), interleukin-5 (IL-5), interleukin-10 (IL-10), tumor necrosis factor alpha (TNF-α), and granulocyte macrophage colony-stimulating factor (GM-CSF) as well as T-cell proliferation (*29*). A graphical representation of the biological mechanism of action is shown in Fig. 2. The inhibitory effect has been shown using the Jurkat human T-cell

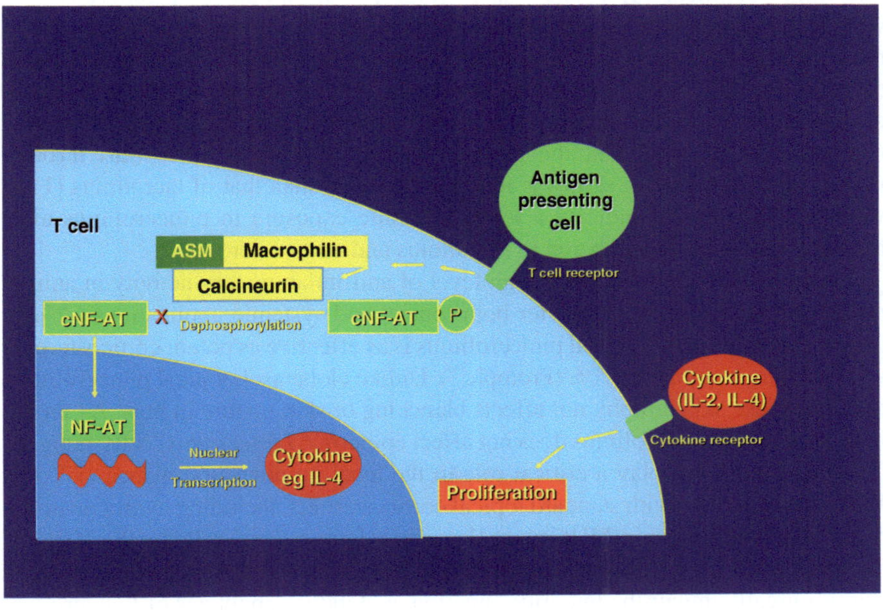

Fig. 2. Mechanism of action of pimecrolimus (ASM) on T-cells

line, peripheral blood mononuclear cells from healthy subjects, as well as human T-helper cell clones isolated from the skin of an atopic dermatitis patient. In these T-cell clones, pimecrolimus (**2a**) inhibits cytokine production at (sub)nanomolar concentrations and as potently as tacrolimus. Furthermore, pimecrolimus shows selectivity for antigen-primed memory T-cells, an effect not seen with tacrolimus (*30*). Pimecrolimus (**2a**) also prevents the production of TNF-α and the release of pro-inflammatory mediators like histamine, hexosaminidase, and tryptase in activated primary human skin mast cells and rodent mast cell lines (*31*). Pimecrolimus does not affect the proliferation of keratinocyte, endothelial, and fibroblast cell lines and has, in contrast to corticosteroids, no effect on the differentiation, maturation, functions, and viability of human dendritic cells (*32*). A recent study revealed that pimecrolimus (**2a**) increases at low nanomolar concentrations innate immune functions of human keratinocytes, such as expression of Toll-like receptors 2 and 6, as well as the production of antimicrobial peptides (cathelicidin, human beta defensin-2 and 3). Pimecrolimus (**2a**) also enhances the functional capacity of keratinocytes to inhibit the growth of *Staphylococcus aureus*. These data suggest that **2a** can amplify the cutaneous innate host defense (*33*).

Topical pimecrolimus penetrates similarly into, but permeates less through the skin *in vitro*, when compared to corticosteroids or tacrolimus. In comparison with bethamethasone, clobetasol, and difluorcortolone used as 1% solutions, pimecrolimus (**2a**) permeates less through the human skin by factors of 60–110. When comparing pimecrolimus (**2a**) and tacrolimus (**1**) in the same vehicle at the same concentration, the same skin concentrations were found with both compounds, but permeation rates of pimecrolimus through human and pig skin were lower by a factor of 9–10 (*34*). When comparing pimecrolimus cream 1% (Elidel®) and tacrolimus ointment 0.1% and 0.03% (Protopic®), similar skin concentrations were determined with pimecrolimus and tacrolimus. However, the permeation rates through the skin were found to be lower with pimecrolimus (**2a**) than those of tacrolimus (**1**), with both ointment preparations by factors of about 6 and 4, despite higher drug concentrations in Elidel® cream. In agreement with the results obtained with human skin, the permeation rate of pimecrolimus (**2a**) through normal and inflamed pig skin was found to be lower than that of tacrolimus (**1**) as well (*35*). These data indicate a lower systemic exposure to pimecrolimus after topical application as compared to tacrolimus and corticosteroids.

Pimecrolimus (**2a**) exhibits a high level of anti-inflammatory activity in animal models of skin inflammation after both topical and systemic applications (*36*). In the pig model of ACD, topical pimecrolimus is as effective as potent corticosteroids and tacrolimus ointment 0.1% (Protopic®). Unlike clobetasol, topical pimecrolimus does not cause skin atrophy nor affects blanching or skin texture in pigs. As shown in mice, topical pimecrolimus does not affect epidermal *Langerhans'* cells, antigen-presenting cells that play a critical role in the local immunosurveillance (*37, 38*). While the treatment with standard topical corticosteroids, including hydrocortisone, resulted in a reduction in MHC class II-positive *Langerhans'* cells by 96–100% in the treated skin, no effect on *Langerhans'* cells was noted. In contrast, corticosteroids greatly impair the integrity, function, and induce apoptosis of *Langerhans'* cells (LC) in mice. A recent analysis of skin biopsies of atopic dermatitis patients has confirmed that treatment for 3 weeks with the corticosteroid ß-methasone 0.1%, but not Elidel® cream 1%, resulted in depletion of *Langerhans*' cells, while both drugs significantly reduced T-cells (*39*). These results indicate that topically applied pimecrolimus is unlikely to interfere with the function of *Langerhans*'/ dendritic cells to differentiate naïve T-cells into effector T-cells, which is key for the developing immune system and maintenance of specific immunocompetence.

Pimecrolimus (**2a**) proved to be highly anti-inflammatory effective also after systemic administration to rodents. Oral and subcutaneous treatment of mice reveals pimecrolimus to be as potent as tacrolimus and more potent than cyclosporin A in inhibiting the elicitation phase, which is the clinically apparent inflammatory phase of ACD (*40*). In contrast to cyclosporin A and tacrolimus (**1**), oral treatment of mice with **2a** neither impairs the induction phase of ACD (sensitization) nor decreases weight and cellularity of draining lymph nodes, indicating that the primary immune response in ACD is not impaired by pimecrolimus (**2a**). In rat ACD, oral pimecrolimus is more potent than cyclosporin A by a factor of 4 and more potent than tacrolimus by a factor of 2 in inhibiting the elicitation phase of ACD. In contrast to tacrolimus, pimecrolimus has no effect on ongoing immune responses in the lymph nodes draining the application site of the hapten (*41*).

In comparison to cyclosporin A and tacrolimus, pimecrolimus (**2a**) has a lower potential to affect systemic immune responses. In rats, subcutaneous injections of cyclosporin A and tacrolimus suppress the localized graft-versus-host reaction 8-fold and 66-fold more potently than pimecrolimus. In the same species, the potency of tacrolimus to inhibit antibody formation against sheep red blood cells is 48-fold higher than that of pimecrolimus. Oral cyclosporin A and tacrolimus are immunosuppressive at lower doses than pimecrolimus in the rat kidney transplantation model by factors of 3 and 15, which correlates with exposure to lymph nodes (*42*).

Pimecrolimus (**2a**) may have therapeutic potential in inflammatory conditions beyond dermatological disorders as well. Results from ophthalmic studies in dogs with chronic keratokonjunctivitis sicca treated locally with experimental pimecrolimus eye drops indicate that pimecrolimus has therapeutic potential in inflammatory eye diseases in man (*43*). Studies in standard rat models of arthritis show that oral pimecrolimus (**2a**) exerts dose-dependent anti-inflammatory and disease-modifying efficacy indicating therapeutic potential for the treatment of human rheumatoid arthritis (*44*). Studies in a SCID model of inflammatory bowel disease indicate that oral pimecrolimus has therapeutic potential, superior to those of cyclosporin A and tacrolimus (*45*).

Taken together, the data suggest that pimecrolimus (**2a**) has favorable pharmacological profiles *in vitro* and *in vivo*:

- When applied topically, it has a high and selective anti-inflammatory activity in the skin, minimal percutaneous resorption, and a low potential to affect local and systemic immunosurveillance. It differs from corticosteroids by its selective action on T cells and mast cells, by a lack of effects on *Langerhans*' cells/ dendritic cells, by the lack of induction of skin atrophy, and by much less permeation through the skin. It differs from tacrolimus by less permeation

through skin and by a lower potential to affect systemic immune responses, thus specifically targeting skin inflammation.
- When applied systemically, it exerts a high anti-inflammatory activity, but has a lower potential for immunosuppression and/or is better tolerated than tacrolimus or cyclosporin A (*28*).

1.2.1.2. Clinical Profile

Therapeutic efficacy and safety of topical pimecrolimus (**2a**) has been established in short-term and long-term management of atopic dermatitis in extensive double-blind, randomized, vehicle-controlled studies with patients including adults, children, and infants. In short-term studies with children, considerable efficacy was already evident at the first evaluation on day 8; significant relief from pruritus was observed also within the first week of treatment. In infants, results were similar to those obtained in the studies with children. Long-term studies were performed in children and infants (1-year treatment) and in adults (6 months). In summary, clinical trials have shown Elidel® to be highly effective in relieving the signs and symptoms of atopic dermatitis in adults, children, and infants. Clinically significant improvement was seen within 3 days of the first application. In long-term studies, Elidel® has demonstrated a unique ability to prevent disease progression if applied at the first signs or symptoms of disease.

In addition to its therapeutic efficacy, pimecrolimus (**2a**) has proven to be safe and well tolerated, as derived from animal and human studies. Topical application led to consistently low systemic exposure, irrespective of age, disease severity, or body surface treatment. Pimecrolimus cream 1% (Elidel®) was approved in the USA at the end of 2001 and in European and other countries in the autumn of 2002. Experience with more than 10 million patients treated so far in clinical practice has confirmed the high efficacy and safety elaborated in the controlled studies. Detailed reviews on clinical studies with pimecrolimus 1% cream (Elidel®) in patients with atopic dermatitis and other inflammatory skin diseases have been published (*46–51*).

In addition to topical application, pimecrolimus (**2a**) was shown to be highly effective and safe after oral treatment. In psoriasis patients, pimecrolimus down-regulated the expression of genes associated with leukocyte activation/proliferation, lymphocyte chemotaxis, and trafficking as well as inflammation. No changes in gene expression were observed that might be linked with drug-related side effects (*52*). Multicenter studies with 3-month treatments of psoriasis and atopic dermatitis patients have proven the efficacy and safety of this compound, thus confirming preclinical studies.

1.3. Structural Features of Ascomycin

As confirmed by X-ray crystal structure analysis and NMR-studies the left hand parts of ascomycin (**2**) and tacrolimus (**1**) mediate binding to their common immunophilin macrophilin (FK506-binding protein, FKBP-12) and have therefore been termed "binding domains" (Scheme 4). The right hand parts of the macrolactams, together with elements of the immunophilin, interact with the protein-phosphatase calcineurin, which plays a key role in the Ca^{2+} dependent activation of lymphocytes, and are called "effector domains" (*53–58*). The X-ray crystal structure of pimecrolimus (**2a**) is shown in Fig. 3. A model of the complex of pimecrolimus and macrophilin derived from the binding complex of L-685,818 and macrophlin is presented in Fig. 4 and also as cover picture of this volume.

Ascomycin (**2**), tacrolimus (**1**), and related analogues represent highly functionalized 23-membered macrocycles, containing a pipecolate residue in an amide (C-8) and an allylic ester linkage (C-26–C-29) with a polyketide backbone. Both macrocycles feature fourteen chiral centers, an endocyclic trisubstituted double bond (C-19–C-20) with (*E*)-geometry and located in an allylic position to a ketone functionality (C-22), three methoxy groups (at C-13, C-15, and C-32), and three secondary hydroxy groups at C-33, C-14, and C-24, with one of these part of a β-hydroxy ketone unit (C-22–C-24). Most notably, within the binding domain the macrolactams feature a unique pattern of three adjacent carbonyl groups (C-8–C-10, tricarbonyl portion, α,β-diketo-amide moiety), of which one carbonyl group (C-10) is masked as a hemiketal with the secondary C-14–OH, resulting in a tetrahydropyran unit (C-10–C-14). Biosynthetically, the C-9 carbonyl is introduced at a late stage *via* C-9–hydroxylation of the corresponding 9-deoxo precursor followed by its oxidation (*59*). In $CDCl_3$-solution, ascomycin (**2**) and tacrolimus (**1**) exist as mixtures of amide bond rotamers ((*E*):(*Z*)=approx. 2:1) (*60*). In the crystalline state, tacrolimus and **2** adopt an (*E*)-amide configuration, whereas the (*Z*)-diastereomer is observed in the tacrolimus/macrophilin complex (*54, 61*). Interestingly, due to electronic repulsion, the planes of the conjugated carbonyls (C-8/C-9) are almost orthogonally oriented. The cyclohexyl side chain and the pyran unit adopt chair conformations, whereas all substituents except the C-10–OH are oriented equatorially.

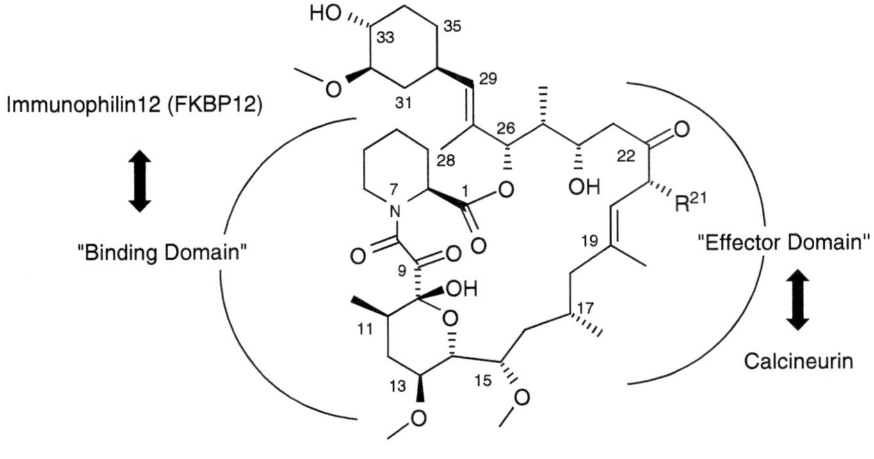

Scheme 4. Dual domain model of tacrolimus (**1**) and ascomycin (**2**)

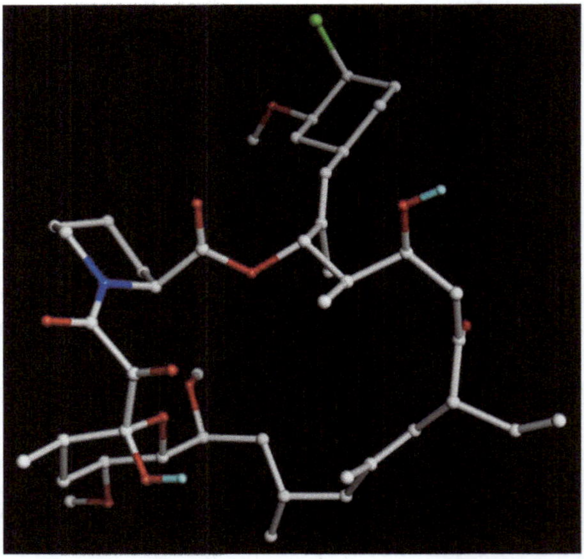

Fig. 3. View of the X-ray crystal structure of pimecrolimus (Weber HP, Sandoz, unpublished results): The ring of pimecrolimus adopts a *cis*-amide conformation and the structure is identical to the X-ray structure of ascomycin (*113*)

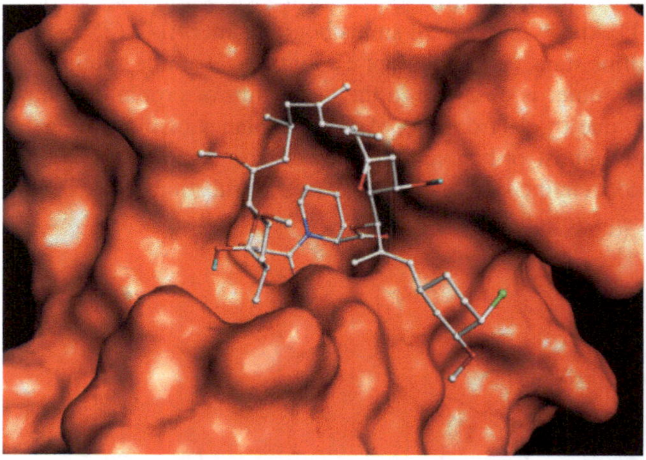

Fig. 4. Model of the complex of pimecrolimus (**2a**) and macrophilin – derived based on the X-ray structure of the native macrophilin (Burkhard P, Taylor P, Walkinshaw MD (2000) J Mol Biol **295**: 953; PDB: 1d60), and the conformation of L-685,818 (18-hydroxyascomycin) as observed in the binding complex with macrophilin (Becker JW, Rotonda J, McKeever BM, Chan HK, Marcy AI, Wiederrecht G, Hemes JD, Springer JP (1993) FK-506-binding protein: three-dimensional structure of the complex with the antagonist L-685,818 J Biol Chem **268**: 11335; PDB: 1fkd) in the trans-amide conformation. The surface of macrophilin is marked in red and pimecrolimus is represented by sticks. Contacts with macrophilin are formed by the pipecolinyl and pyranose rings and the dicarbonyl groups. The pipecolinyl ring is embedded in a deep cavity. The chlorine atom does not form strong contacts to the protein

1.3.1. Structural Flexibility of Ascomycin

Although the structure of ascomycin (**2**) shown in Scheme 4 is the main isomeric form adopted in organic solution, the close proximity of the tricarbonyl unit to C-14–OH potentially allows the formation of numerous alternative isomers (Scheme 5). Thus, liberation and enolization of the tricarbonyl portion followed by re-hemiketalizations could give rise to the four six- or seven-membered hemiketal forms **A**-**D** and their C-11–isomers 11-*epi*-**A**-**D** (not shown in the Scheme). Furthermore, anticipating a 1,4-addition of C-14–OH to the enolized tricarbonyl form **E** allows the generation of a set of isomeric "furano-ascomycins", **F1**-**F4**. In addition, in an aqueous environment, the formation of a hydrate form, **H**, could also be anticipated. Finally, each of the above potential equilibrium products could exist as a mixture of the amide bond rotamers.

Despite the numerous equilibrium products that could be formed, only the isomeric forms **B** and **C** (Scheme 5) of ascomycin (**2**) and tacrolimus (**1**) have been identified and characterized so far (*62–64*). The existence of the tricarbonyl form **T** and its hydrate form **H** has not yet been established, but is suggested by the following findings. Addition of minor amounts of water to a colorless solution of **2** in acetonitrile causes a yellow coloration indicating the formation of the free tricarbonyl form. Further addition of water to the yellow solution leads back to a colorless solution, indicating the conversion of **T** into its hydrate form **H**. An equilibrium among the tricarbonyl form **T**, the hemiketal form **A,** and an alternative hemiketal form (most probably **C**), has also been suggested by the results from reversed-phase LC/MS experiments. The major equilibrium product of ascomycin

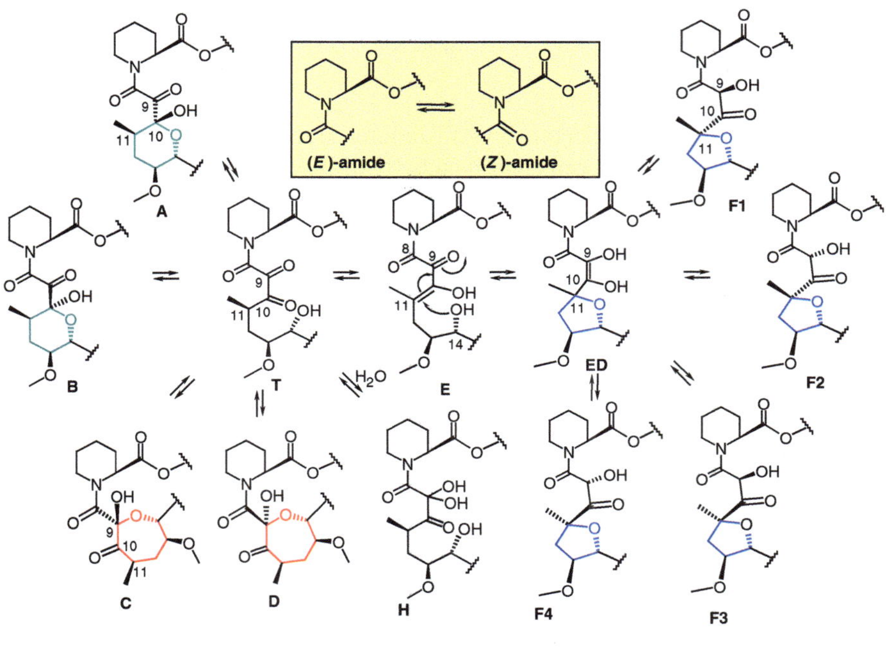

Scheme 5. (Hypothetical) structural flexibility of ascomycin (**2**) and tacrolimus (**1**)

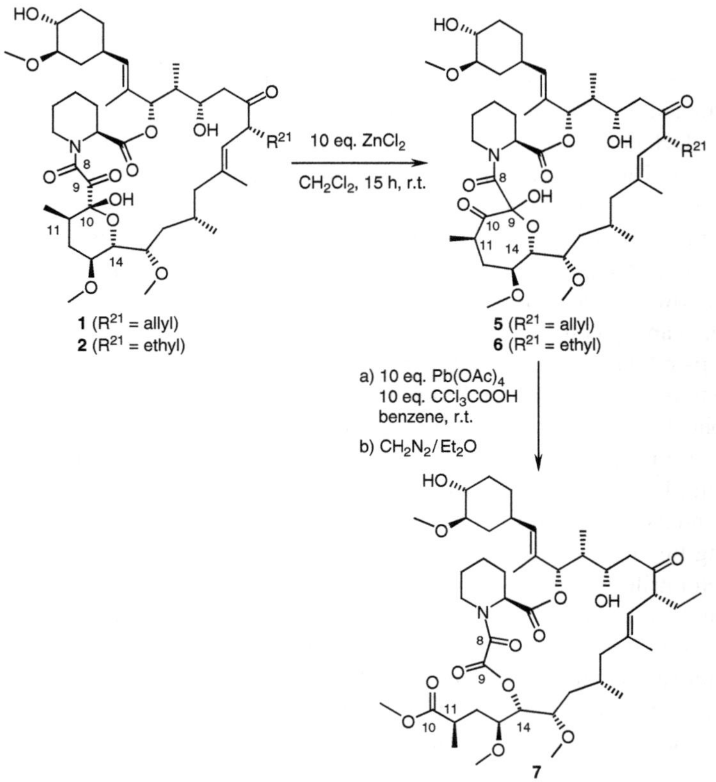

Scheme 6. Conversion of tacrolimus (**1**) and ascomycin (**2**) to their hemiketal forms **5** and **6**, and the chemical degradation of **6** to **7**

(**C**) has been isolated *via* selective crystallization and reversed-phase chromatography (*63*). Its structure has been confirmed by NMR spectroscopy and its synthesis followed by oxidative degradation (Scheme 6) (*64*). Thus, the action of *Lewis* acids (*i.e.* zinc halides) in non-protic organic solvents converts ascomycin (**2**) and tacrolimus (**1**) into the corresponding seven-membered C-9–hemiketal forms, **6** and **5**, almost quantitatively. Lead tetraacetate-mediated chemoselective α-ketol cleavage of **6**, followed by esterification of the crude product and chromatography provided the ester **7** in a 80% yield, thus confirming the C-14–O–C-9 linkage.

2. Synthesis Aspects

2.1. Synthesis of the Four Diastereomeric "Furano-Ascomycins"

An unexpected and interesting reaction occurring in the binding domain of ascomycin (**2**) could be used to synthesize the "furano-ascomycins" **F1-F4** (Scheme 5) from **2** (*65*). Thus, bis-silylation of **2** gave **8**, which upon action of diiodo-triphenyl-phosphorane in the presence of imidazole in refluxing acetonitrile, gave a mixture of the silyl protected 9- and 10-deoxo-furano-ascomycins, **9a**, **9b**, **10a**, and **10b** (Scheme 7). The reaction probably proceeds through equilibration to the ene-diol form **ED**, replacement of either of the OH groups by iodide, followed by iodide ion-mediated deiodination. Starting from **10a** and **10b**, simple functional group manipulations afforded the (11*S*)-furano-ascomycins **13a** and **13b** and their (11*R*)-isomers **13c** and **13d** (Scheme 8). Thus, oxidation of the activated methylene groups in **10a** and **10b** with *Dess-Martin* periodinane in the presence of pyridine yielded the yellow tricarbonyl derivatives **11a** and **11b**, which, after desilylation, provided the derivatives **12a** and **12b**, in high yields. Chemoselective reduction of the highly activated C-9–carbonyl group of **12a** and **12b** with zinc/glacial acetic acid, followed by chromatography afforded the individual isomers **13a-13d** in high yields.

The furano-ascomycins **13a** and **13b** differ only at the configuration of C-9, as could be shown by equilibration under basic conditions. Analogous results

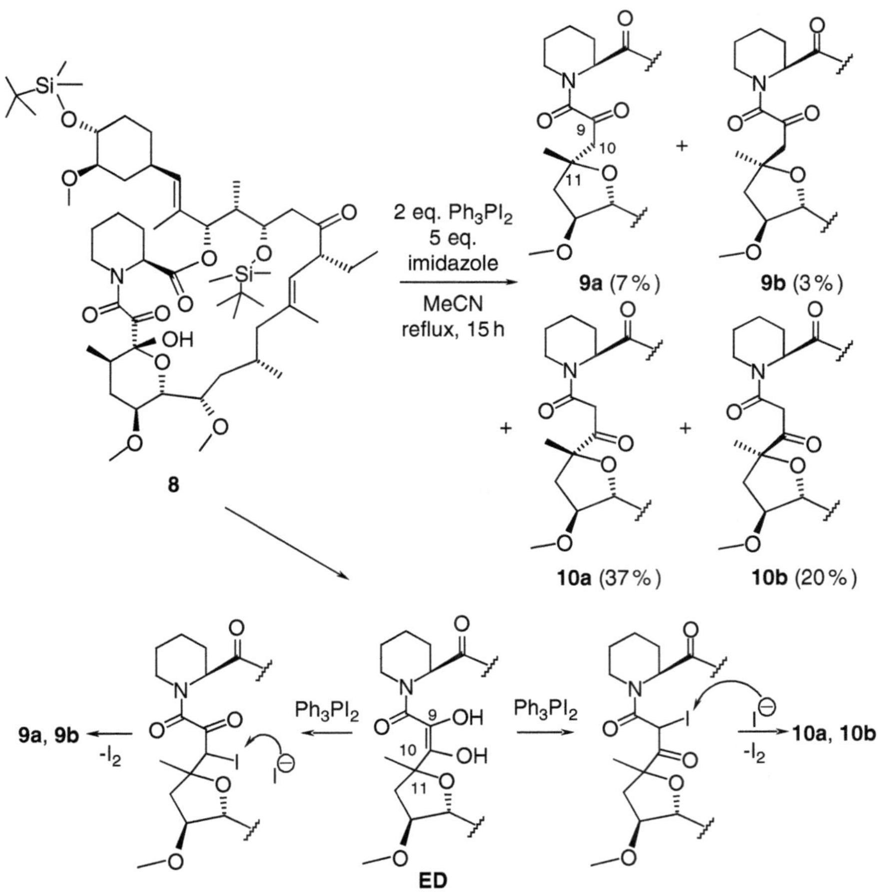

Scheme 7. Transformation of **8** to the 9-, and 10-deoxo-furano-ascomycins **9a**, **9b**, **10a**, and **10b** and a possible mechanism

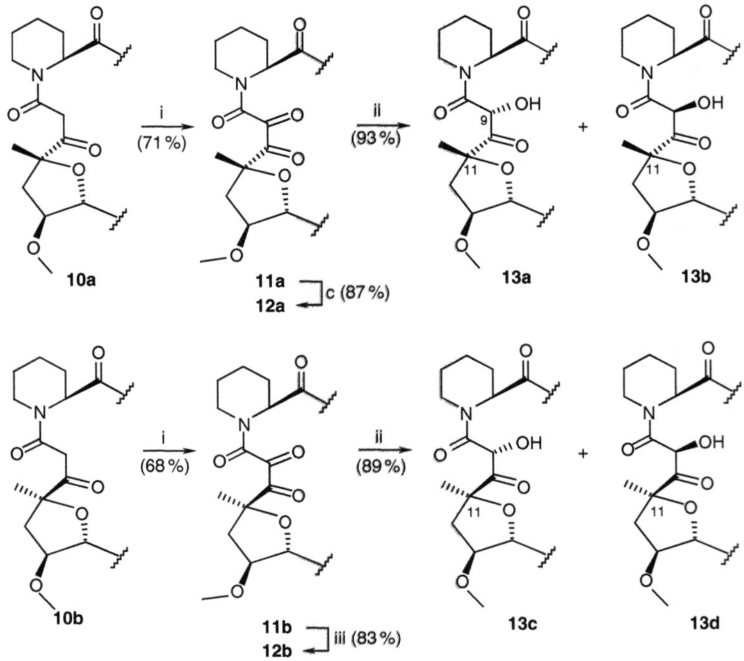

i. Periodinane, DBU, *t*-BuOH; ii. Zn, AcOH, MeCN; iii. HF, MeCN

Scheme 8. Transformation of the deoxo-furano-ascomycins **10a** and **10b** to the furano-ascomycins **13a**, **13b**, **13c**, and **13d**

were obtained starting from the 11-*epi*-isomers **13c** and **13d**. The isolated pure furano-ascomycins are remarkably stable at room temperature for several months. Furthermore, no reconversion into ascomycin (**2**) in protic or aprotic solutions under neutral, basic, or acidic conditions could be seen. Also, starting from ascomycin (**2**) no formation of **13a-13d** could be demonstrated in solution. Thus, there is no evidence for equilibrium between the furano-ascomycins and the parent compound ascomycin. No biological activities of these compounds have been reported.

2.2. *Synthesis of* ^{13}*C Labeled Ascomycin*

It is essential to have appropriate tools to establish the purity of a drug substance (DS) unambiguously. In the case of an ascomycin-derived DS it is important to be able to distinguish between the "real by-products" and DS-related inherent equilibrium compounds. Researchers at Novartis succeeded in labeling **2** at the diagnostically most relevant C-9 or C-10–carbons in the binding domain (*66*). ^{13}C-Labeled ascomycin (^{13}C-n–**2**), or drug substances derived thereof, serve as versatile tools for studying equilibrium phenomena in more complex mixtures, such as galenical formulations. For the synthesis of ^{13}C-9–**2**, a ring contraction/ring expansion strategy has been applied (Scheme 9). Thus, ascomycin (**2**) was silylated to furnish the yellow 14,24,33-tris-*O*-TBDMS- derivative **14**, bearing the unmasked tricarbonyl unit. Treatment of **14** with excess calcium hydroxide in THF-water afforded, *via* an irreversible benzilic acid-type rearrangement reaction, the ring-contracted α-hydroxy acid **15** as a >95:5 mixture of diastereoisomers in favor of the (10*S*)-enantiomer. Oxidative decarboxylation of the latter with lead tetraacetate furnished the ring contracted ketoamide **16** quantitatively, setting the stage for ring expansion. The ^{13}C-label was introduced through reaction of **16** with ^{13}C-methylene iodide and butyllithium. Further functional group manipulations

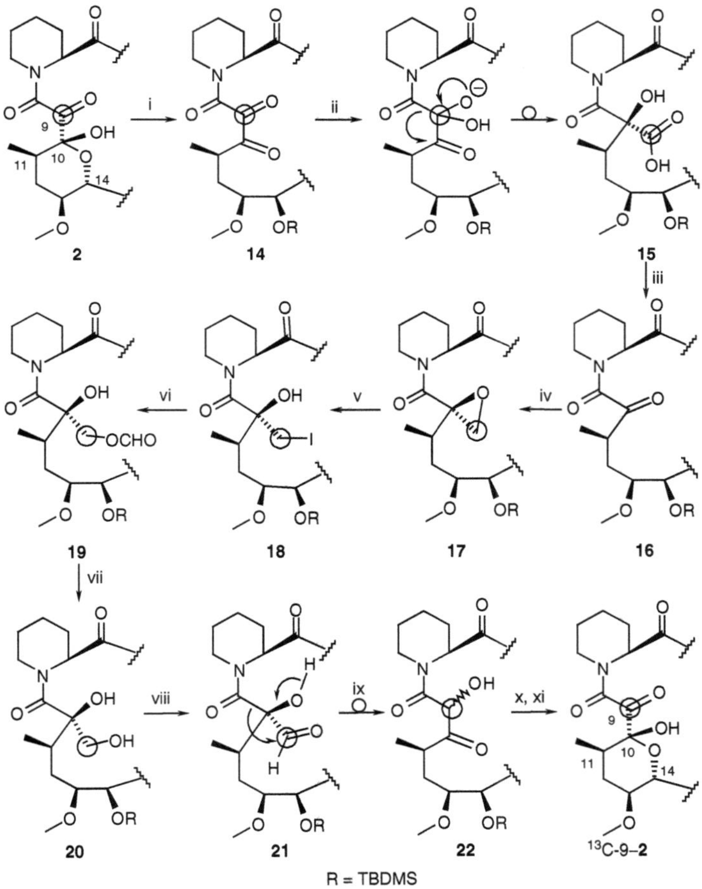

i) TBDMSOTf, 2, 6-lutidine, CH_2Cl_2, r.t.; ii) $Ca(OH)_2$, THF, water, r.t.; iii) $Pb(OAc)_4$, C_6H_6, r.t.; iv) (^{13}C)-CH_2I_2, BuLi, THF, –78°; v) MgI_2, CH_2Cl_2, r.t.; vi) $AgBF_4$, DMF, r.t.; vii) $ZnCl_2$, MeOH, r.t.; viii) Oxalyl chloride, DMSO, Et_3N; ix) $ZnCl_2$,CH_2Cl_2 , r.t., or Florisil, THF, reflux; x) Oxalyl chloride, DMSO, Et $_3$N; xi) aq. HF, MeCN, r.t.

Scheme 9. Synthesis of ^{13}C-labeled ascomycin (^{13}C-9–**2**)

provided the α-hydroxyaldehyde **21** as the key intermediate for the subsequent ring expansion protocol. Thus, exposure of **21** to zinc chloride in methanol at room temperature, or to magnesium silicate on silica gel (Forisil®) in refluxing THF solution, provided the ^{13}C-9–labelled 9(*R*/*S*)-dihydroascomycin **22** regioselectively. Oxidation of the secondary hydroxy group in **22**, followed by deprotection, completed the synthesis of ^{13}C-9–**2**.

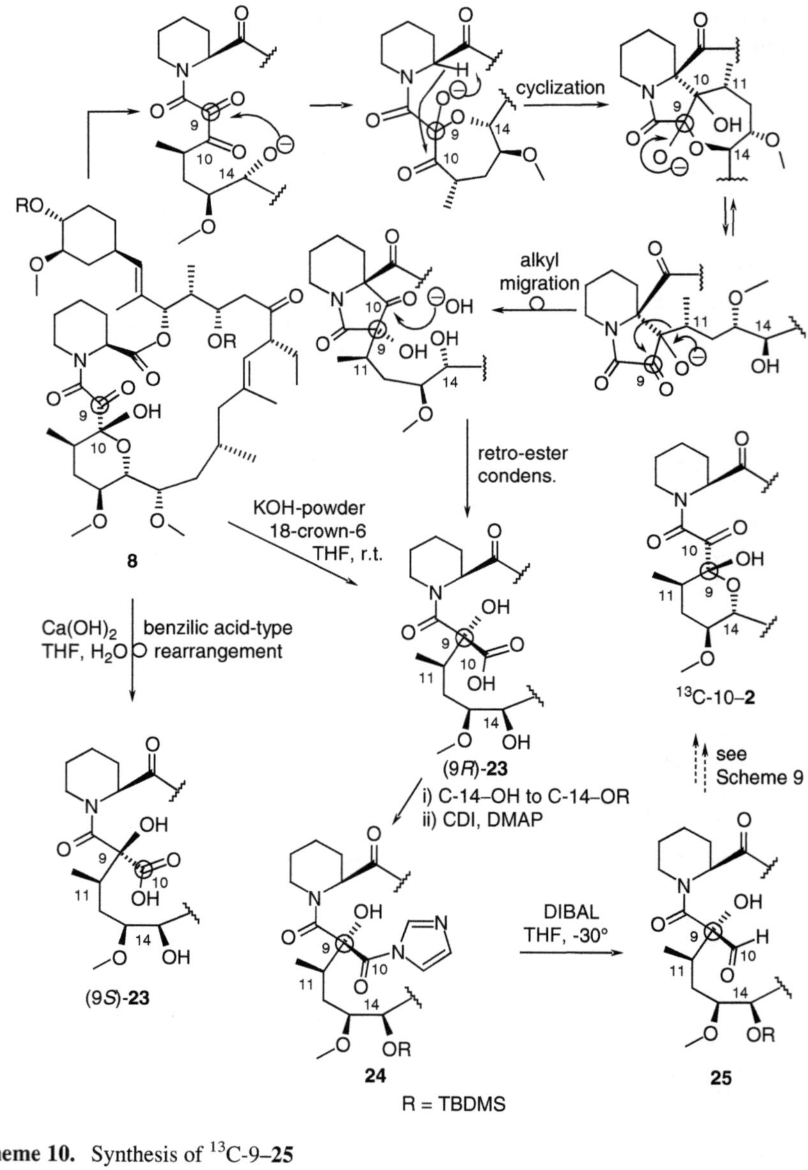

Scheme 10. Synthesis of ^{13}C-9–**25**

Potentially, ^{13}C-9–ascomycin (^{13}C-9–**2**) could be converted into ^{13}C-10–ascomycin (^{13}C-10–**2**) applying a cascade of diastereoselective rearrangement reactions (Scheme 10). As described above, treatment of 24,33-bis-*O*-TBDMS-ascomycin **8** with calcium hydroxide in THF-water solution results in a facile benzilic acid-type rearrangement to give diastereoselectively the (9*S*)-α-hydroxy acid, (9*S*)-**23**, in a quantitative yield. Remarkably, changing to anhydrous reaction conditions (powdered KOH, 18-crown-6, THF, room temperature) afforded almost exclusively (>97:3) the (9*R*)-isomer, (9*R*)-**23**, in a high yield. Quenching and trapping experiments revealed that its formation proceeds not *via* a benzilic acid-type rearrangement, but *via* a cyclization followed by an α-ketol-rearrangement involving the migration of C-11, which is then followed by a diastereoselective (retention at C-2) *retro*-ester condensation. Thus, the newly created quaternary carbon atom of the α-hydroxy-acid (9*R*)-**23** is the former C-9 carbon of ascomycin (**2**). Silylation of the C-14–OH, followed by reduction of the acid *via* its imidazolide **24** provided the (9*R*)-α-hydroxy-aldehyde **25**, which potentially could be transformed to **2** as described in Scheme 9 for the (9*S*)-isomer. Thus, provided that C-8 migrates in the ring enlargement reaction (as observed for the (9*S*)-isomer), this sequence would allow the conversion of ^{13}C-9–ascomycin (^{13}C-9–**2**) into ^{13}C-10–ascomycin (^{13}C-10–**2**).

Tritium labeled tacrolimus was prepared by a metal-catalyzed hydrogen isotope exchange procedure (*67*). The radiochemical purity was 98% and ^{3}H-NMR spectroscopy was used to identify the tritium incorporation in about 10–15 different positions. When administered to rats, HTO-formation turned out to be less than 6% of the dose, thus indicating a reasonable biological stability.

2.3. *Reactivity of the Binding Domain*

The structural flexibility of ascomycin (**2**) and related macrolactams in the binding domain imparts a high reactivity to this region towards a broad variety of reaction conditions. Thus, selective transformations on other parts of the molecule are often difficult to achieve without provoking concomitant changes in this unit.

2.3.1. Reactivity of Ascomycin and Tacrolimus Towards Diazomethane

Reaction of ascomycin (**2**) with excess ethereal diazomethane led to the diastereoisomeric 9-epoxides **26** and **27**, as expected, together with eight minor by-products **28–35** (Scheme 11; for clarity only the relevant portions of the macrocycle are drawn) (*68–70*). Compounds **26–32** result through insertion of a methylene unit, while compounds **33–35** arise from **31** and **32** through a second methylene insertion reaction. The formation of **26–31** could be rationalized through the intermediacy of the betaines **A** (keto-form) and **B** (hemiketal form) (Scheme 11). Thus, ring closure with C-9–OH leads to the formation of the oxiranes **26** and **27**. Alternatively, ring

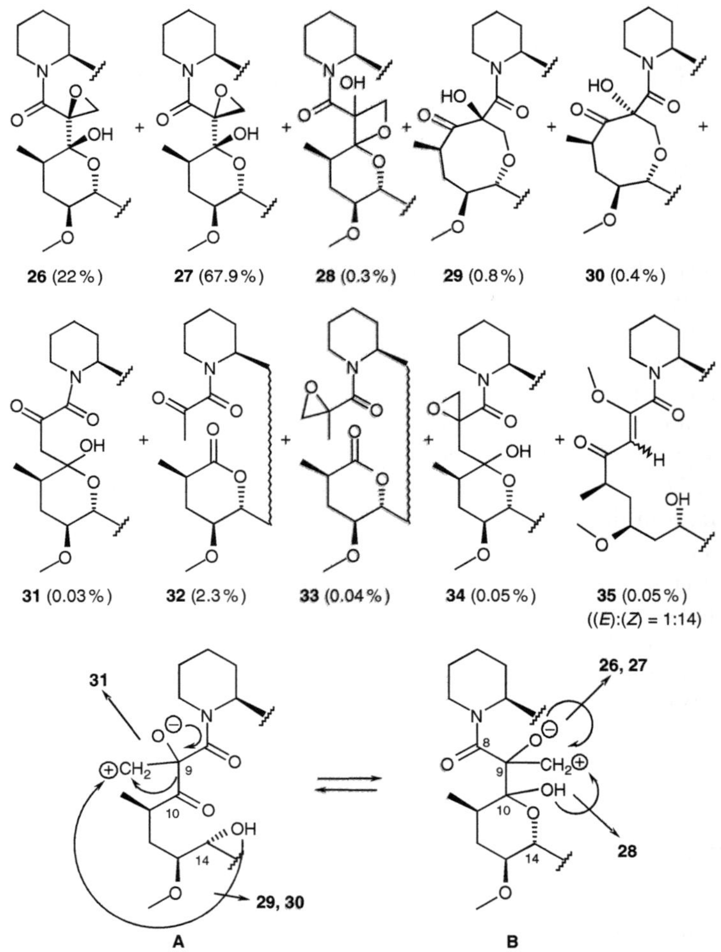

Scheme 11. Products from the reaction of ascomycin (**2**) with diazomethane and a possible mechanistic picture for the formation of the products **26–31**. The *long wavy lines* indicate partial structures

closure with C-10–OH or C-14–OH, leads, respectively, to the unusual *spiro*-oxetane **28**, or the oxocanone isomers **29** and **30**. The ring-enlarged derivative **31** could be formed through migration of C-10 to the positively charged carbon of the betaines **A** or **B**. Two distinct pathways could be proposed for the formation of the unusual *seco*-compound **32** (Scheme 12). Thus, starting from the betaine **B**, a *Grob*-type fragmentation would provide the enol form **C** of **32**. Alternatively, starting from the betaine **A**, ring closure to the hydroxy-epoxide **D** followed by a similar fragmentation could lead to the enol ester **E**, which after an intramolecular transesterification, leads to the *seco*-compound **32**. In fact, the enol ester **36** was isolated as a by-product from the reaction of 14,24,33-tris-*O*-TBDMS-ascomycin with diazomethane. This supports the intermediacy of compounds of type **E** in these

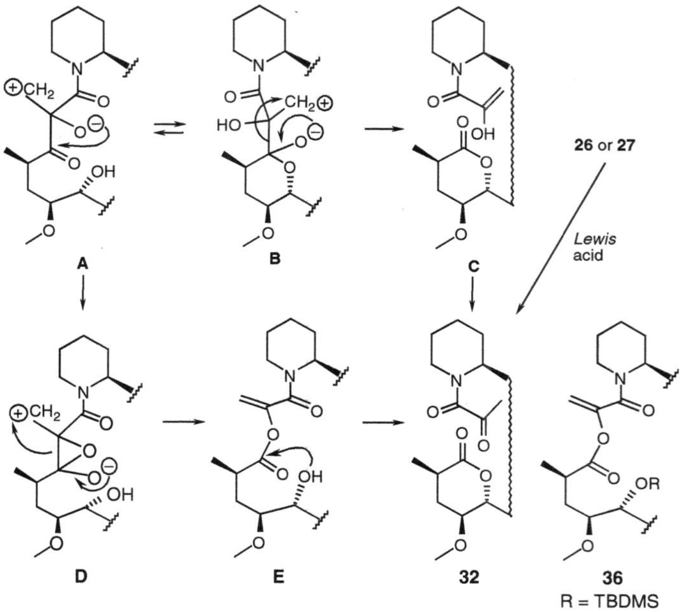

Scheme 12. Possible pathways for the formation of the product **32** in the reaction of ascomycin (**2**) with diazomethane, and the structure of the isolated intermediate **36**. The *long wavy lines* indicate partial structures

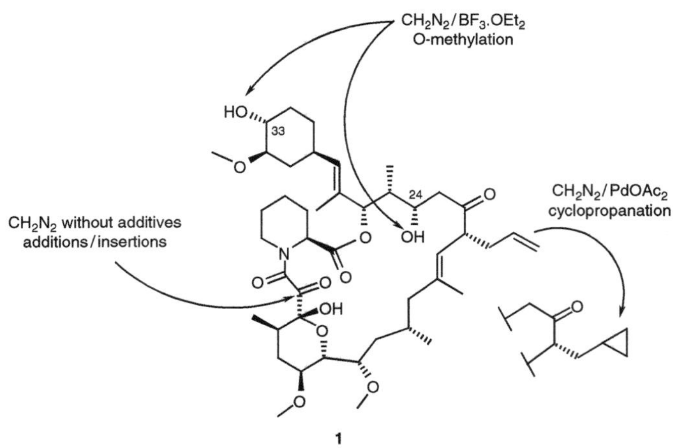

Scheme 13. Selective reactions of diazomethane with tacrolimus (**1**)

reactions. Interestingly, treatment of the oxiranes **26** and **27** with *Lewis* acids leads to the *seco-* compound **32** in an almost quantitative yield (*69, 71*).

Although the tricarbonyl portions of ascomycin (**2**) and tacrolimus (**1**) are highly sensitive towards diazomethane, selective O-methylations on C-24–OH and C-33–OH could be achieved in the presence of catalytic amounts of $BF_3.OEt_2$ (Scheme 13). On the other hand, reaction with excess of diazomethane in the presence of $Pd(OAc)_2$ led selectively to cyclopropanation of the double bond on the allylic side chain of tacrolimus (**1**), leading to 37,38-cyclopropano-FK506, which showed excellent macrophilin binding but somewhat weaker T-cell inhibitory activities (*26, 72–74*).

2.3.2. Rearrangement Reactions in the Binding Domain

2.3.2.1. Benzilic Acid-Type Rearrangement Reactions

Early attempts to cleave the endocyclic pipecolic ester bond by a base-catalyzed saponification procedure did not lead to the expected 1,26-*seco*- derivative, but instead to a 22-membered lactone with a rearranged C-9–C-10 region (Scheme 14). Based on labeling studies, a benzilic acid-type rearrangement occurring on the free tricarbonyl form was proposed as a mechanism for this ring-contraction reaction. More detailed studies performed at Novartis showed that the reaction conditions originally applied (*i.e.* 1.1 eq. LiOH, THF-H_2O) (*73, 75, 76*) provided not only the hydroxy acid (9*R*)-**37** but also its 9-*epi*-derivative (9*S*)-**37** (*77, 78*). Oxidative degradation of both isomers led to **38,** which after deprotection, led to the common ring-contracted *nor*-C-9 ascomycin derivative **39**, thus showing that all other chiral positions remained unchanged. Further studies showed that the derivative (9*S*)-**37** was formed *via* a distinct reaction pathway and could be prepared selectively by applying anhydrous reaction conditions. An almost exclusive formation of the 24,33-bis-*O*-TBDMS protected (9*S*)-hydroxy esters **40a** (R′=methyl) and **41a**, **42a** and **43a** (R′=ethyl, propyl, *iso*-propyl), or their unprotected congeners **40b**, **41b**, **42b** and **43b,** could be accomplished by reacting 24,33-bis-*O*-TBDMS-ascomycin (**8**) or unprotected ascomycin (**2**) in the appropriate alcohol with *Lewis* acids (*i.e.* ZnX_2, MgX_2, X=Cl, Br, I; $Ti(O\text{-}^{i}Pr)_4$) (*71*).

2.3.2.2. Rearrangement Reactions in Aprotic Media

Treatment of unprotected **2** or 24,33-bis-*O*-TBDMS protected ascomycin (**8**) with various non-nucleophilic bases in aprotic solvents led to a set of unusual rearranged derivatives, which have been termed "cyclo-ascomycins" (Scheme 15). Thus, refluxing ascomycin with excess triethylamine and catalytic amounts of powdered KOH in acetonitrile yielded (9*S*,11*S*),2,10-cyclo-ascomycin **44** (SDZ ASD 732), together with traces of the stereoisomers **45–47** (*79, 80*). Separation of the diastereomers **44–47** was accomplished using stimulated moving bed technology or centrifugal counter-current chromatography (*81, 82*).

The isomer **44** could be transformed to 11-*epi*-ascomycin in a series of steps (Scheme 15) (*83*). Thus, silylation of **44** afforded 14,24,33-tris-*O*-TBDMS-protected **44** that on treatment with powdered KOH and 18-crown-6 under aprotic

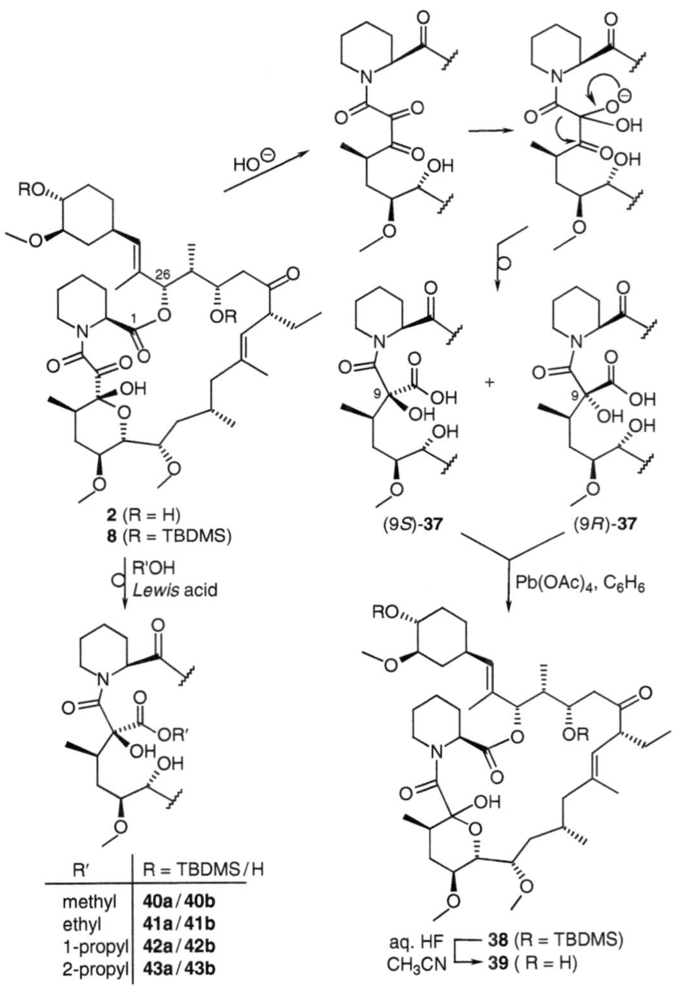

Scheme 14. Benzilic acid-type rearrangement reactions of ascomycin (**2**)

conditions led to 11-*epi*-(9*R*)-**23a**. This compound, upon further transformations analogous to those described in Scheme 10, led to 11-*epi*-ascomycin (11-*epi*-**2**). In contrast to ascomycin (**2**), which exists in $CDCl_3$ solution as a mixture of six- and seven-membered hemiketal forms (ratio 15:1), 11-*epi*-**2** adopts exclusively two diastereomeric (10*S*)- and (10*R*)-six-membered hemiketal forms in the ratio 4:1. Notably, 11-*epi*-**2** is stable to acidic conditions, but is converted completely to **2** under weakly basic conditions, whereas no conversion of **2** to 11-*epi*-**2** was observed under the same conditions or on storage under protic, aprotic, organic, or aqueous solutions. Interestingly, treatment with DBU converts 11-*epi*-**2** to **44** as the sole product. Therefore, it is likely that the formation of **44** from **2** involves

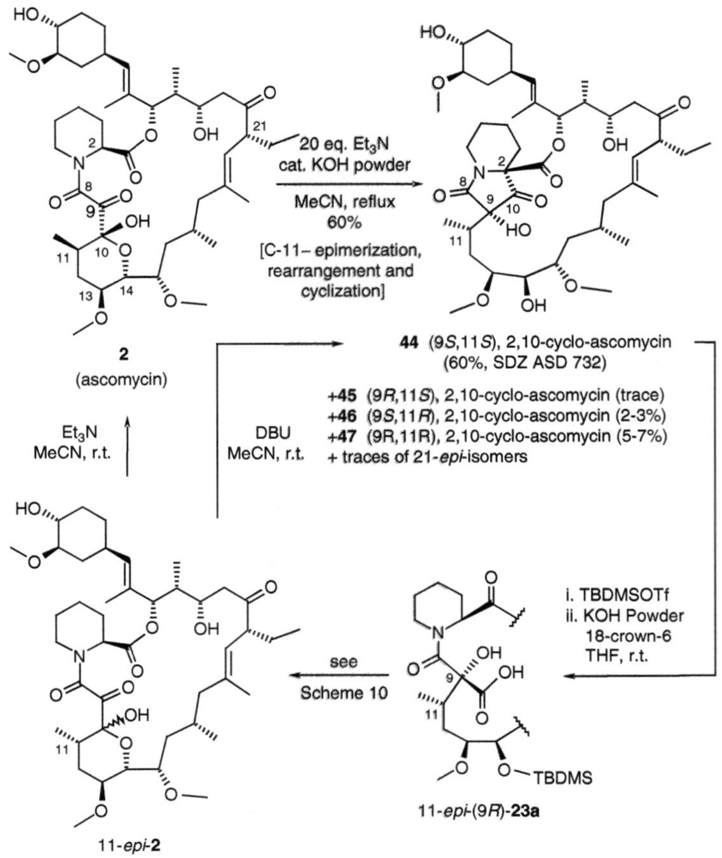

Scheme 15. Conversion of ascomycin (**2**) to the cyclo-ascomycin **44** (SDZ ASD732) and other isomers, and, transformation of **44** to 11-*epi*-ascomycin (11-*epi*-**2**) and *vice versa*

epimerization at C-11 as the first step, which is then followed by rearrangement and ring closure.

Starting from 24,33-bis-*O*-TBDMS-protected ascomycin (**8**), the isomeric cyclo-ascomycins **45–47** could be prepared as the sole products (*79, 84, 85*) (Scheme 16). Thus, treatment of **8** with the sodium salt of tosylamide provided with high stereoselectivity the cyclized compound **48** ((10*R*)-configuration), which, upon treatment with cesium carbonate, rearranged to **49**, and, after desilylation, afforded the (9*R*,11*R*),2,10-cyclo-ascomycin **47**. Alternatively, one could first carry out the rearrangement and then esterify to give **40b**, and then cyclize it to **49** (compare Scheme 14).

The isomer **46** can be prepared with a high stereoselectivity (Scheme 17). Thus, 24,33-bis-*O*-TBDMS-ascomycin (**8**) is transformed to (10*S*)-**48** with an (*S*)-configuration at C-10 using CsF in a 68% yield. Alternatively, **8** is converted

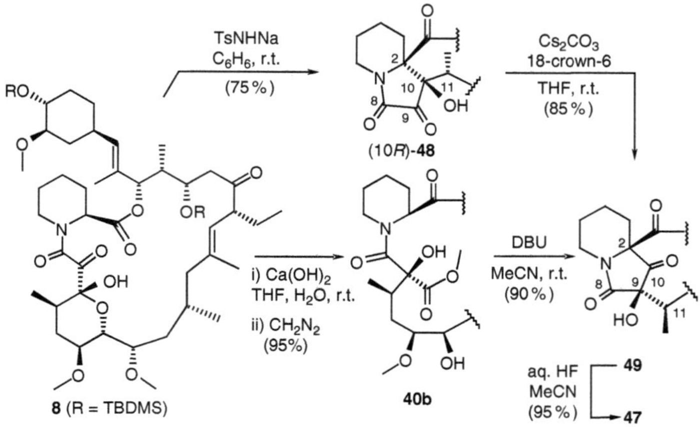

Scheme 16. Transformation of **8** to the cyclo-ascomycin **47**

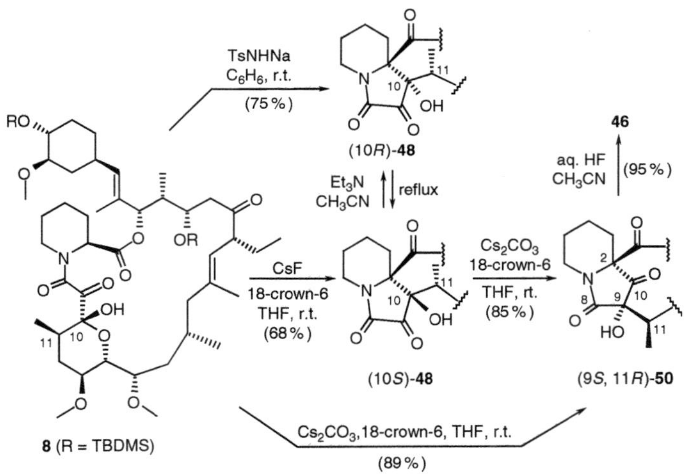

Scheme 17. Transformation of **8** to the cyclo-ascomycin **46**

first to (10*R*)-**48** and then epimerized to (10*S*)-**48**. Treatment of isolated (10*S*)-**48** with Cs_2CO_3 in the presence of 18-crown-6 in tetrahydrofuran initiates a diastereoselective alkyl migration to give the 24,33-bis-*O*-TBDMS-derivative **50**, which after desilylation affords the (9*S*,11*R*),2,10-cyclo-ascomycin **46**. More conveniently, compound **50** is also obtained in a one-pot reaction starting from **8**.

No methods are available so far for the preparation of the fourth stereoisomer (9*R*,11*S*),2,10-cyclo-ascomycin (**45**), in a good yield. The only available method involves silylation of **44** to give **51**, which is then transformed to its 9-chloro- derivative **52** by routine methods (Scheme 18). An unusual halogen/hydride exchange furnished the deoxo- derivative **53** in a low yield. Due to

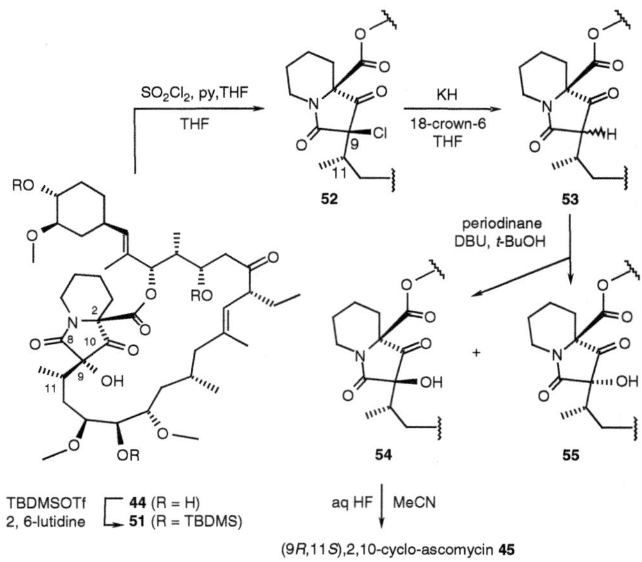

Scheme 18. Transformation of the cyclo-ascomycin **44** to the isomer **45**

keto/enol tautomerism, **53** exists in $CDCl_3$-solution as a mixture of diastereomers. Oxidation of **53** resulted in the reintroduction of the hydroxy group to give the tris-*O*-TBDMS-protected cyclo-ascomycin-derivatives **54** and **55**. Finally, desilylation of **54** afforded the desired (9*R*,11*S*),2,10-cyclo-ascomycin **45**.

In summary, four diastereoisomeric cyclo-ascomycins (**44**–**47**) can be prepared stereoselectively by simple treatment of protected or unprotected ascomycin (**2**) in aprotic media with an appropriate base. Although C-2 is clearly involved during their formation, no C-2–epimeric cyclo-ascomycins have been found so far under any of the reaction conditions, thus indicating a powerful remote stereocontrol. Interestingly, it could be shown that upon treatment with powdered KOH in the presence of 18-crown-6, the 14,24,33-tris-*O*-TBDMS-cyclo-ascomycins **54**–**57** undergo ring cleavage to the corresponding α-hydroxy acids **58**–**61**, thereby leading to, in all the four isomers, C-2 in its natural (*S*)-configuration (Scheme 19). Furthermore, **58**–**61** can be transformed to esters **62**–**65** that may then be cyclized back diastereoselectively to give **54**–**57**. The α-hydroxy acids **58**–**61** have served as versatile tools for the determination of relative configurations. Thus, oxidative decarboxylation resulted in the formation of the epimeric 14,24,33-tris-*O*-TBDMS-α-ketoamides **66** and **67** differing in their configuration at C-11, as could be shown by equilibration experiments and by their conversion to the silyl enol ether **68**. Desilylation of **67** led to the already known derivative **39**, thus giving evidence for the outlined C-11–configuration and the confirmed (*S*)-configuration at C-2 for all the α-hydroxy acids. The stereochemistry at C-2 of the cyclo-ascomycins **54**–**57** (and thus also of their deprotected congeners **44**–**47**) has been established by another chemoselective oxidative degradation protocol (Scheme 20). Thus, reaction

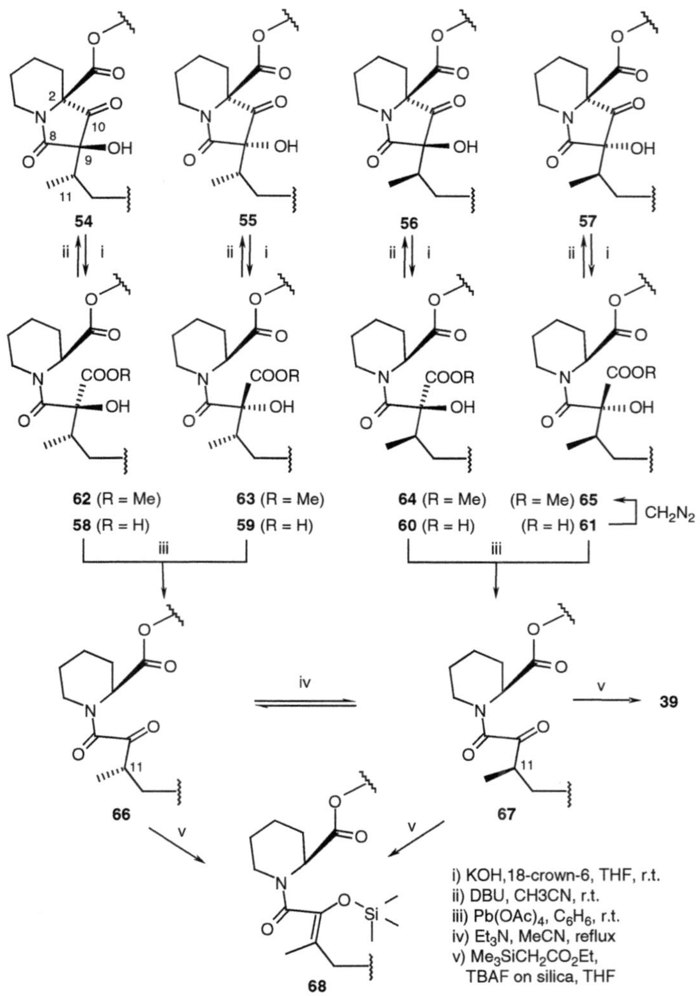

Scheme 19. Ring-cleavage reactions of the 24,33-bis-TBDMS-cyclo-ascomycins **54–57** and their further transformations for establishing the stereochemistry

of **54–57** with excess tetrapropylammonium perruthenate (TPAP) provided the two C-11–epimeric α-ketoamides **69** and **70**, bearing a carboxylic acid functionality at C-2. After treatment with diazomethane they provided the corresponding esters **71** and **72**, which in an intramolecular base-mediated transesterification event provided the enol ester derivative **73**. The latter can only exhibit the intact original C-2–configuration of the starting cyclo-ascomycins **54–57**. Having ascertained the relative stereochemistry at C-2, C-9, and C-11 of all cyclo-ascomycins, the absolute configurations were established by X-ray crystal structure analysis of the

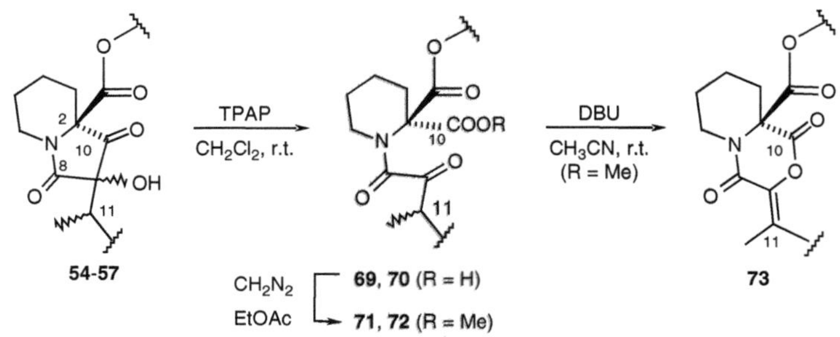

Scheme 20. Degradation reactions of the 24,33-bis-TBDMS-cyclo-ascomycins **54–57** towards establishing the C-2–configuration

unprotected (9*S*,11*S*)-derivative **44** (*71, 84, 85*). The biological activities of these compounds have not been reported.

2.3.3. Reduction, Deoxygenation, Imine- and Aminal-Formation in the Binding Domain

9-Dihydro-FK506 (**74**, tsukubamycin I), together with its 9-dihydro-9-*O*-methyl-congener **75** (tsukubamycin A), was first isolated as a by-product from the fermentation broth of a tacrolimus producer strain (*8*) (Scheme 21). Synthetically, chemo- and stereoselective reductions of the 9-carbonyl group of ascomycin (**2**), tacrolimus (**1**), and 24,33-bis-O-protected congeners thereof, have been accomplished using either DIBAH or *Evans*' reagent (tetramethylammonium-triacetoxyborohydride) (*74, 86*). The absolute stereochemistry of the 9-dihydro- derivatives could be deduced by NMR spectroscopy from their corresponding rigid thiocarbonates **79** and **80**, which were prepared readily by the action of thiocarbonyl diimidazole. The latter served as versatile starting materials for further modifications. Thus, radical deoxygenation/elimination of the thiocarbonates **79** and **80** gave the (*E*)- and (*Z*)-9,10-unsaturated compounds (*E*)-**81** and (*Z*)-**81**. Different (*E*)/(*Z*) ratios were obtained when Bu_3SnH or $(Me_3Si)_3SiH$ in combination with AIBN or triethylborane were used, allowing the preparation of either isomer selectively. Hydration of the enolether **81** provided the 9-deoxo-derivative **83** as a single diastereomer, whereas hydrogenation led to the two diastereomers, (10*S*/*R*)-9-deoxo-10-deoxy-**82**. Interestingly, the unusual reduction- and deoxygenation-product **85** was obtained when ascomycin (**2**) was reacted with diphenylsilane in the presence of cesium fluoride.

Action of ammonia in methanol converts ascomycin (**2**) to a mixture of the C-10–aminal **86** and the C-9–imine **87**, whereas reaction with methylamine led to the C-9–imine **88** exclusively (Scheme 21) (*87*).

9-Deoxoascomycin (**83**), together with minor amounts of the 10-deoxo- derivative **84**, could be prepared from ascomycin (**2**) directly through reaction with hydrogen sulfide in the presence of pyridine (Scheme 22) (*88*). Notably, the

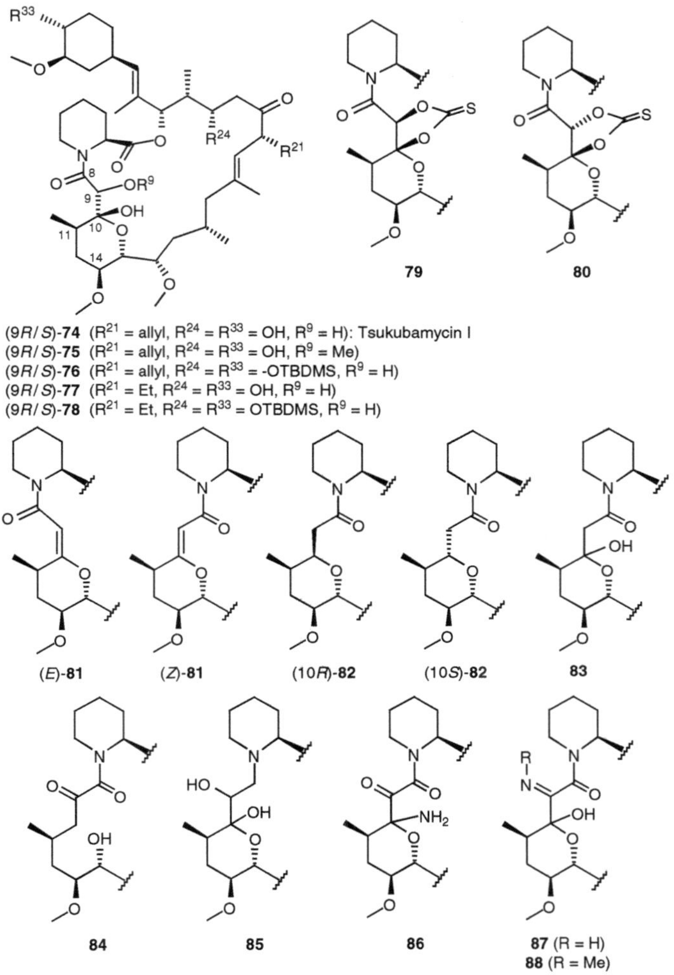

Scheme 21. Derivatives of ascomycin (**2**) and tacrolimus (**1**) with diverse modifications in the binding domain

exclusive formation of either the 9- or 10-deoxo- derivative was observed when the imine **87** or the aminal **86** was treated with hydrogen sulfide in the absence of base.

Semi-rigid derivatives of ascomycin (**2**), featuring heterocyclic structural elements in the binding region, were synthesized from the aminal **86** (*89*). The carbonyl C atoms, for example, were integrated into an oxazole ring giving **89**, and the tetrahydropyran ring of **86** was transformed into the pyridine unit in **90**.

Several of the analogues in this section have also been prepared with tacrolimus, but are omitted here for simplicity. Structure-activity relationships (SARs) have been reported (*74*) and are summarized as follows. The 9-dihydro- derivatives **74** and **77** showed good macrophilin binding but reduced inhibition of T-cell activation

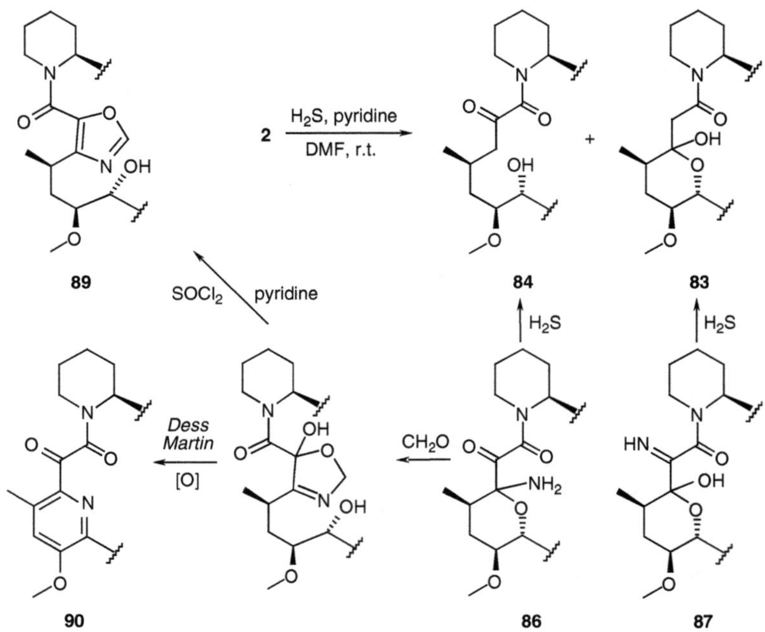

Scheme 22. Derivatives of ascomycin (**2**) featuring desoxygenations and heterocyclic rings in the binding domain

in cellular assays. The 9-desoxo derivative **83** showed lower binding affinity and reduced inhibition of T-cell activation. Interestingly, the 10-amino and 9-imino analogues **86** and **87** had higher macrophilin binding affinities than tacrolimus; the tacrolimus analogue of **86** was highly active in cellular assays. The pyridine analogue **90** was totally inactive on T-cells.

2.3.4. Modification in the Binding Domain Through Photochemistry

It has been reported that tacrolimus (**1**), on exposure to visible light (30,000 Lux) for 10 days, undergoes a rearrangement in the binding domain, leading to the oxazolidinone **91**, and degradation, affording the diastereomeric cleavage products (11*R*/*S*)-**92** (Scheme 23) (*90, 91*). The formation of **91** represents a modification on the amino acid unit, the type of which could not be achieved easily through routine chemical methods. This prompted Novartis researchers to investigate photochemical reactions in more detail.

Thus, irradiation of ascomycin (**2**) in acetonitrile using >280 nm light led to the [1,3]-sigmatropic shift product **94**, the cleavage product **96** and the oxazolidinone **97** (Scheme 24) (*92–94*). On the other hand, irradiation in methanol under the same conditions led to the methoxy derivative **93** in addition to the other products **94–96**; none of the intramolecular cyclization product **97** was observed under these

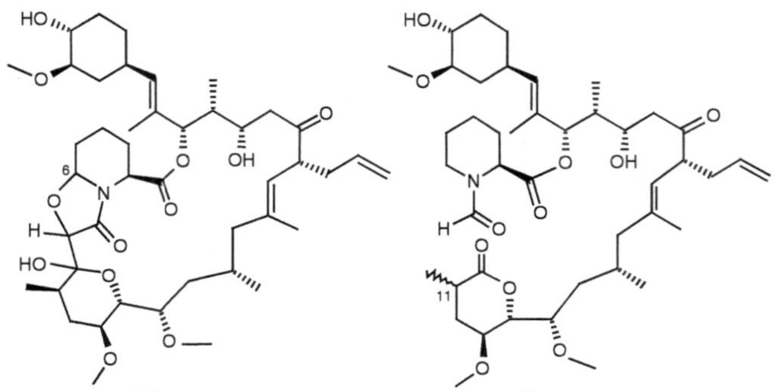

91 92

Scheme 23. Structures of the products formed after exposure of tacrolimus (**1**) to visible light

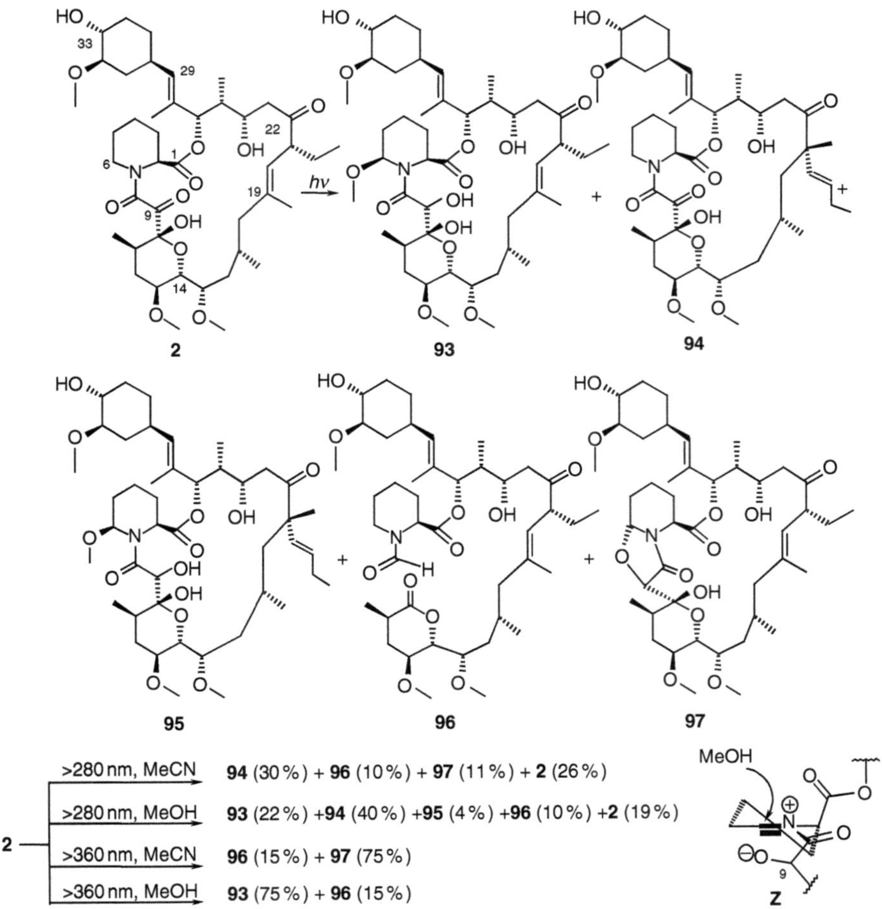

Scheme 24. Products formed upon irradiation of ascomycin (**2**) with ultraviolet light of different wavelengths in acetonitrile and methanol

reaction conditions. Interestingly, by employing light at >360 nm only the tricarbonyl chromophore is selectively excited and, hence, the formation of the [1,3]-shift product **95** could be suppressed leading to high selectivities. Thus, using these longer wavelengths the reaction in acetonitrile afforded the oxazolidinone **97**, whereas the reaction in methanol furnished the methoxy derivative **93** as the main product, hence providing preparatively useful protocols. The formation of these products could be explained through the intermediacy of the zwitterion **Z**.

The photoproduct **93** could be oxidized selectively using catalytic amounts of Cu $(OAc)_2$, pyridine, and oxygen in dichloromethane to afford 9-methoxyascomycin (**98a**) in excellent yield (Scheme 25). The ethoxy and propoxy analogues **98b** and **98c** could also be prepared analogously through irradiation of **2** in ethanol or propanol, followed by oxidation of the resulting 9-alkoxy photoproducts. On the

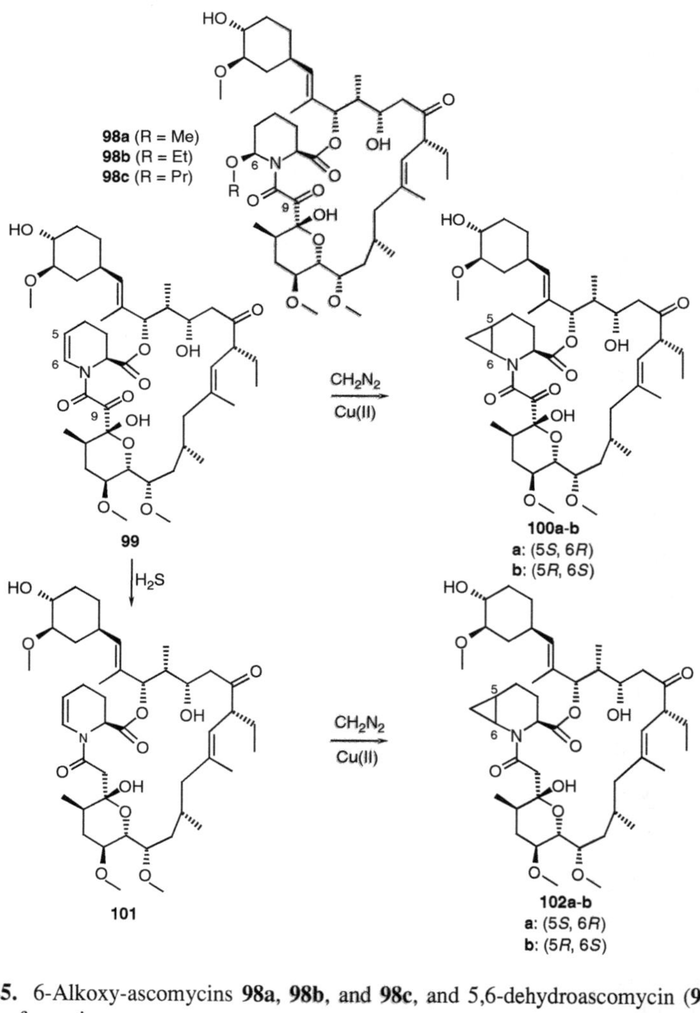

Scheme 25. 6-Alkoxy-ascomycins **98a**, **98b**, and **98c**, and 5,6-dehydroascomycin (**99**) and its further transformations

other hand, **93**, upon elimination of MeOH using ammonium chloride in DMF, gave 5,6-dehydro-9-dihydroacomycin, which after Cu(II)-catalyzed oxidation of the C-9–OH, afforded 5,6-dehydroascomycin (**99**), a close bioisostere of ascomycin in an excellent yield (*95*). Compound **99** served as a useful starting material for the preparation of 5,6-tritium labeled ascomycin in a single step through catalytic tritiation (*96*). Cu(II)-catalyzed cyclopropanation of **99** using diazomethane furnished both the α- and β- isopropano analogues **100a** and **100b** as further rigid derivatives of ascomycin (**2**). Further, **99** could be desoxygenated using hydrogen sulfide/pyridine giving the 9-desoxo analogue **101**, from which the two isomers of the cyclopropano analogues **102a** and **102b** could be obtained in low yields, in an analogous manner to the reactions described above. Starting from the proline analogue of ascomycin **103**, which is available as a side product in the fermentation of ascomycin (**2**), the 5-methoxy analogues **104** and **105**, and the 4,5-dehydro analogue **106** were prepared analogously (Scheme 26).

The hemiacetal photoproduct **93** can be hydrolyzed giving the aldehyde **107** in an excellent yield (Scheme 27). The reactivity of the aldehyde functionality can be used for preparing interesting derivatives. Thus, employing $NaClO_2$ and Cu(II)-catalyzed oxidations, *Wittig* and esterification reactions, **107** could be transformed to the linear amino acid analogues **108a**, **108b**, **109a**, and **109b** in a few steps and good yields. Furthermore, Cu(II)- catalyzed oxidation of **107** provided the corresponding C-9=O derivative, which underwent acid-catalyzed cyclization leading to the bicyclic analogues **110a** and **110b**.

For the purpose of broader derivatization in the amino acid region, the suitably protected key aldehyde **113** was synthesized starting from the easily available protected ascomycin derivative **111**, using the photoreaction described above (Scheme 28) (*97*). Thus, starting from **111** photolysis afforded the methoxy derivative **112**, which was hydrolyzed, and the C-33–*O*-silyl group that fell off was reintroduced giving **113**. Aldehyde **113** was transformed to the allyl carbonate **114** through a series of transformations. Interestingly, **114** exists exclusively as a seven-membered hemiketal in $CDCl_3$. Pd-Catalyzed cyclization of **114**, followed by desilylation and selective oxidation of the C-22–OH led to the 6-vinyl analogue **115** as an inseparable mixture of C-6–epimers. The two C-5–epimers of the 5-vinylproline analogue **116** were also synthesized starting from **113**, as individual isomers, using a similar strategy.

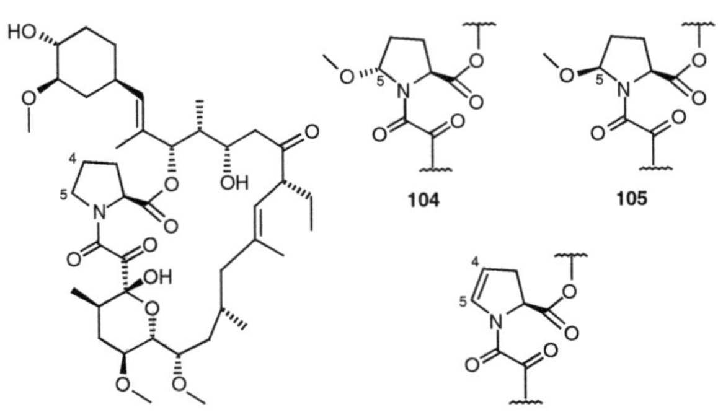

103 104 105 106

Scheme 26. 5-Alkoxy- and 4,5-dehydro- derivatives **104**, **105** and **106** of the proline analogue of ascomycin **103**

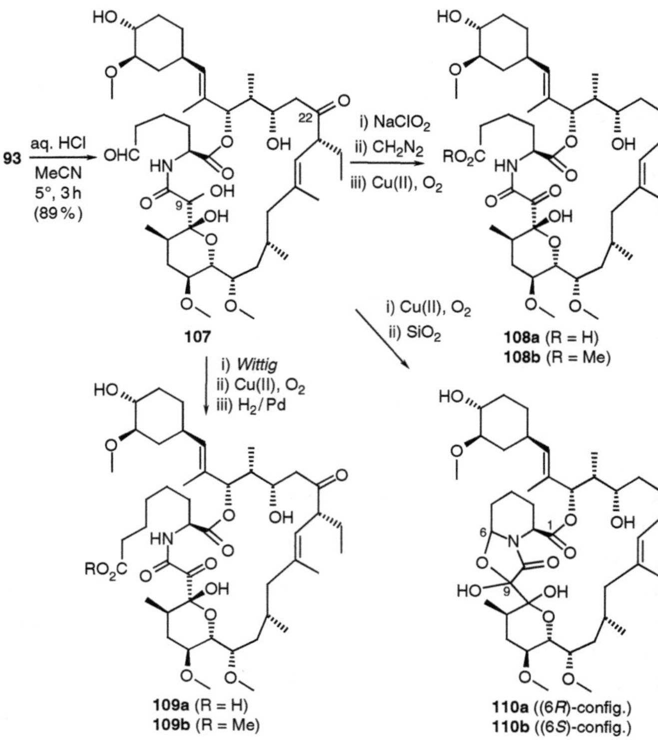

Scheme 27. Hydrolysis of the photoproduct **93** to the aldehyde **107** and its further transformations

The biological activities of these compounds resulting from photochemistry were discussed in detail in the light of the possible conformations (*92*). Briefly, the observed activities of the three 6-alkoxy derivatives **98a**, **98b**, and **98c** showed that larger alkoxy groups result in weaker binding to macrophilin. Among the cyclopropano analogues, those bearing the cyclopropano group in the α-face of the piperidine ring **100a** and **102a** showed higher activity than the β-counterparts **100b** and **102b**. The 9-oxo analogues **99**, **100a** and **100b** were more active than the 9-desoxo counterparts **101**, **102a**, and **102b**, indicating the importance of the C-9−carbonyl group to the binding with macrophilin. Other analogues, such as the vinyl analogues **115**, **116a**, **116b**, and the bridged analogues **110a** and **110b** were not active. The acyclic amino acid analogues **108a-110a** and **108b-110b** did not show any activity, reflecting on the importance of the necessity of the pipecolic acid ring structure, and hence the conformation of the macrolide. It is noteworthy that the ring-contracted butenyl analogue **94** and its saturated butyl derivative (not shown in the scheme), featuring unaltered binding domains, showed excellent binding to macrophilin, but were not active in T-cell assays. Thus, in these cases, modifications in the effector domain do not favor further interactions of the initially formed macrophilin-binding complex with calcineurin A and calcineurin B, and thus do not result in any T-cell modulatory activities. Of all these derivatives, the 5,6-dehydro-ascomycin **99** and the dehydroproline **106** showed activities close to

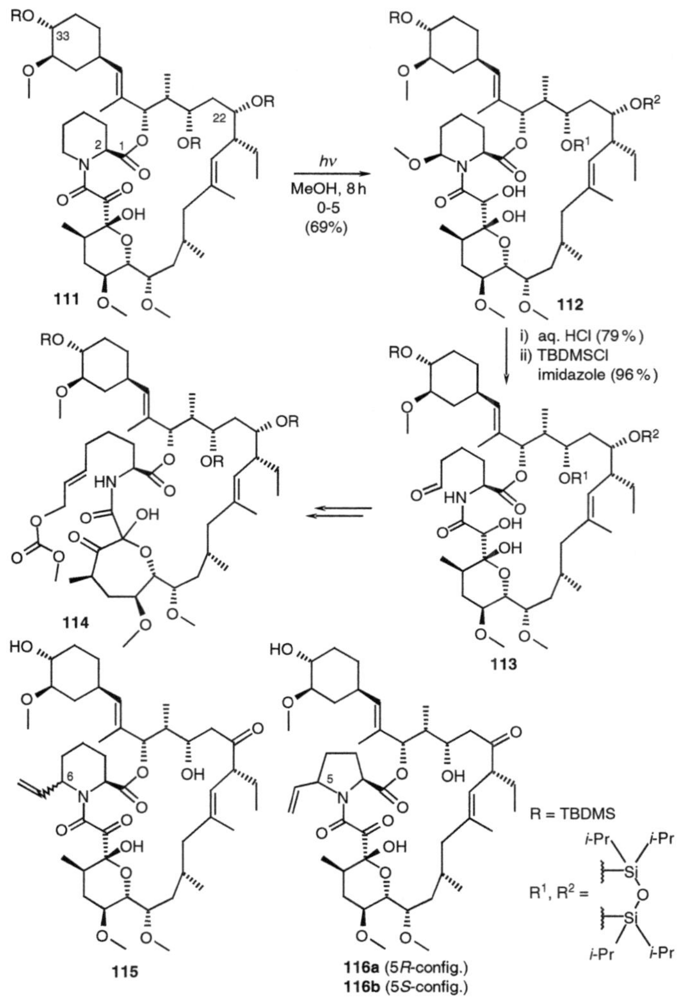

Scheme 28. Syntheses of the 6-vinyl-ascomycin **115** and the 5-vinylproline-analogues **116a** and **116b**

the parent ascomycin. Compound **99** (SDZ 283–871) had been under detailed investigations at Novartis, but its development was terminated later in preference to pimecrolimus.

2.3.5. Cleavage Reactions Within the Binding Domain: Amino Acid Exchange

In addition to the already mentioned photochemical decarbonylative cleavage and the *Lewis* acid assisted cleavage of 9-oxirane derivatives (see Scheme 12), several other cleavage protocols have been elaborated. Thus, reaction of ascomycin (**2**) or tacrolimus (**1**) derivatives with lead tetraacetate in methanol, or with *in situ* generated LiOOH or $Ca(OOH)_2$ in THF/aq. hydrogen peroxide solution, led to a facile oxidative cleavage of the C-9–C-10 bond, affording, depending on the work-up method, *seco-* structures of the general formula **A** or **B** (Scheme 29) (*64, 73, 98–100*).

Two additional methods, allowing a reductive cleavage at the binding domain, have been described as well. Thus, reaction of tacrolimus with benzyl amine and sodium cyanoborohydride in methanol afforded the diketopiperazine **117** and a mixture of the C-10–C-34 fragments **118** and **119** (Scheme 30) (*73*). The hydroxyketone fragment **119** exists in equilibrium with its hemiketal tautomer **120**. Mechanistically, this interesting fragmentation might involve reductive amination of the C-9=O followed by an intramolecular aminolysis of the ester, and, retroaldol cleavage of the C-9–C-10 bond yielding the observed products.

Alternatively, reductive cleavage of 24,33-bis-*O*-TBDMS-ascomycin (**8**) has been achieved by the action of 9-BBN (9-bora-bicyclo[3.3.1]nonane) in THF giving the hydroxy-acetamide **121** in an excellent yield (Scheme 31) (*100*). Treatment of the latter with sodium hydride furnished the C-10–C-34 fragment **122** through an intramolecular transesterification reaction. Interestingly, **122** could also be obtained by a cyanide-induced cleavage. In the event, after addition of cyanide to the C-9–carbonyl, cleavage of the C-9–C-10 bond occurs to give the cyanohydrin **123** (*99, 101–103*). Depending on the cyanide source used and the work-up conditions, either the cyanohydrin itself and/or the fragment **122** could be isolated.

The cleavage products having lost the pipecolic acid moiety could be elaborated to new analogues incorporating new amino acid units such as sarcosine, proline, D-pipecoline, *N*-methyl-alanine, 2-methyl-alanine, and β-alanine in place of the original amino acid (Schemes 32 and 33) (*100, 104*). For example, coupling of a suitable N-protected amino acid to the most reactive, allylic C-26–OH of **122**, followed by a chemoselective reduction of the lactone unit, afforded the lactol **124**. Addition of a C_2 unit to the lactol through a *Wittig* reaction, protection of the C-14–OH and deprotection of the amine and the carboxylic acid groups followed by a macrolactamization reaction provided the enamide **125** as a key

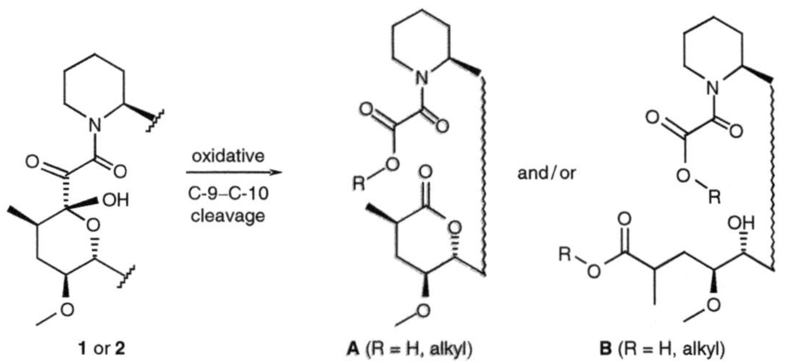

Scheme 29. Oxidative cleavage of derivatives of ascomycin (**2**) and tacrolimus (**1**). The *long wavy lines* indicate partial structures

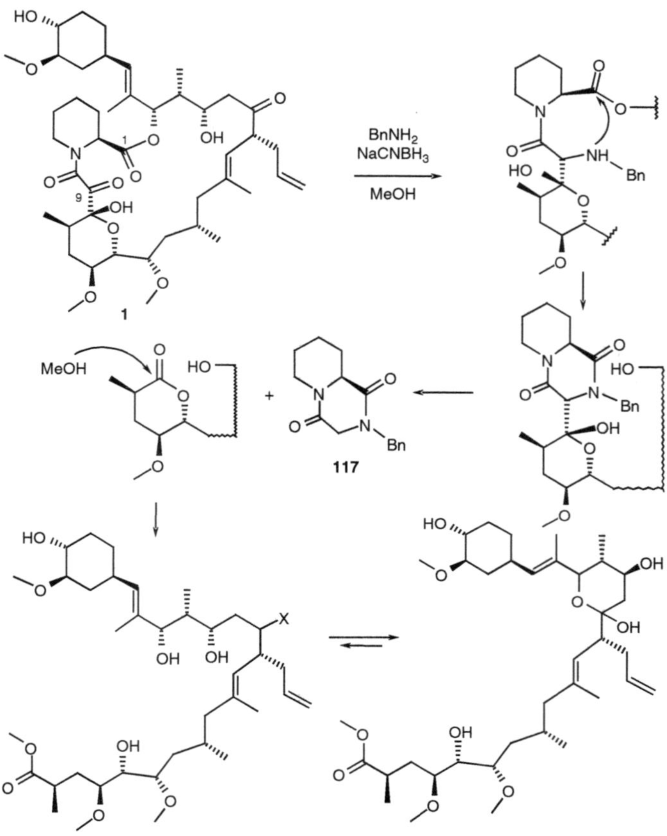

Scheme 30. Reductive cleavage of the C-1–C-9 unit of tacrolimus (**1**). The *long wavy lines* indicate partial structures

intermediate. Bishydroxylation of the C-9–C-10 double bond followed by *Dess-Martin* periodinane oxidation led to the tricarbonyl product **126**. Finally, removal of all protecting groups led to the 23-membered macrocycles **127–130** incorporating the appropriate amino acid. Interestingly, the analogues featuring 2-methyl-alanine or β-alanine exist preferentially in the seven-membered hemiketal-form.

Finally, mention must be made here of a novel photochemical amide cleavage reaction (*92, 105*). Thus, the photoproduct **112** upon oxidation of the C-9–OH gave the protected 6-methoxyascomycin analogue **131** in an excellent yield (Scheme 34). Irradiation of **131** in MeOH afforded the amide cleavage product **132** in a good yield. The reaction most probably proceeds through the zwitterionic intermediate $\mathbf{Z_3}$, which is attacked by MeOH on C-8=O instead of the usually observed attack on C-6. It is interesting to note that similar cleavage reactions have also been reported on other protected and, surprisingly, also on the unprotected ascomycin

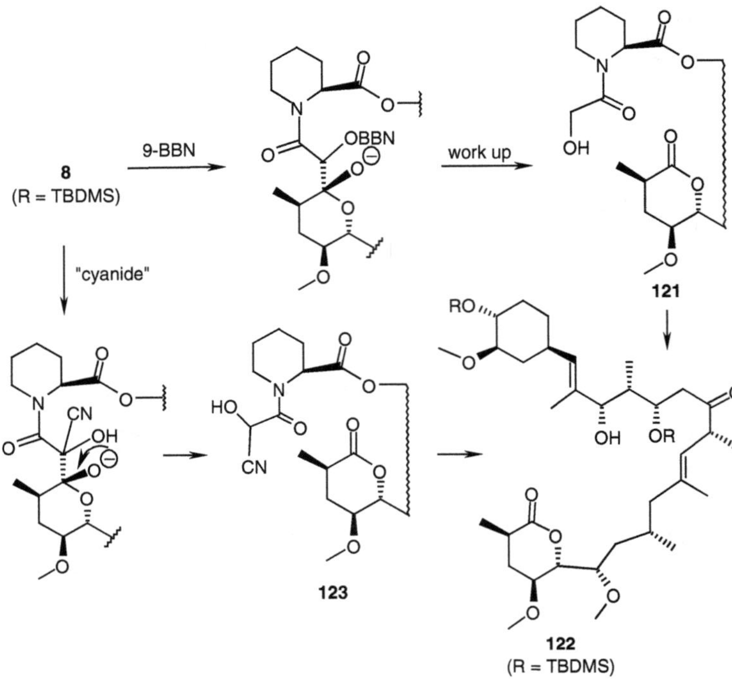

Scheme 31. Cleavage reactions in the binding domain of 24,33-bis-*O*-TBDMS-ascomycin (**8**). The *long wavy lines* indicate partial structures

analogues. The product **132** is a potentially useful intermediate for semisynthetic modifications, but this has not been explored further. Biological activities of the compounds in this section have not been reported.

2.4. Modifications in the Effector and Cyclohexyl Domains

Several modifications have been done on the effector parts of tacrolimus (**1**) and ascomycin (**2**). Whereas treatment of tacrolimus (**1**) with strong bases led to degradation of the molecule, treatment with 1,5-diazabicyclo[4,3,0]nonane (DBN) in CH_2Cl_2 at room temperature led to isomerization at C-21, affording a 1:2 equilibrium mixture of **1** and the C-21−epimer **133** (Scheme 35) (*106*). The isomer **133** was isolated and the structure established by X-ray crystal structure analysis. Compound **133** binds well to macrophilin but showed very weak immunosuppressive activities in cells (*74*). Acid-catalyzed dehydration of ascomycin (**2**) gave 23,24-dehydroascomycin (**134**), which was hydrogenated to give 24-desoxyascomycin (**135**) (*74, 107*). Whereas the dehydro derivative **134** showed only weak *in vitro* and *in vivo* activities, the desoxyascomycin **135** was as active as

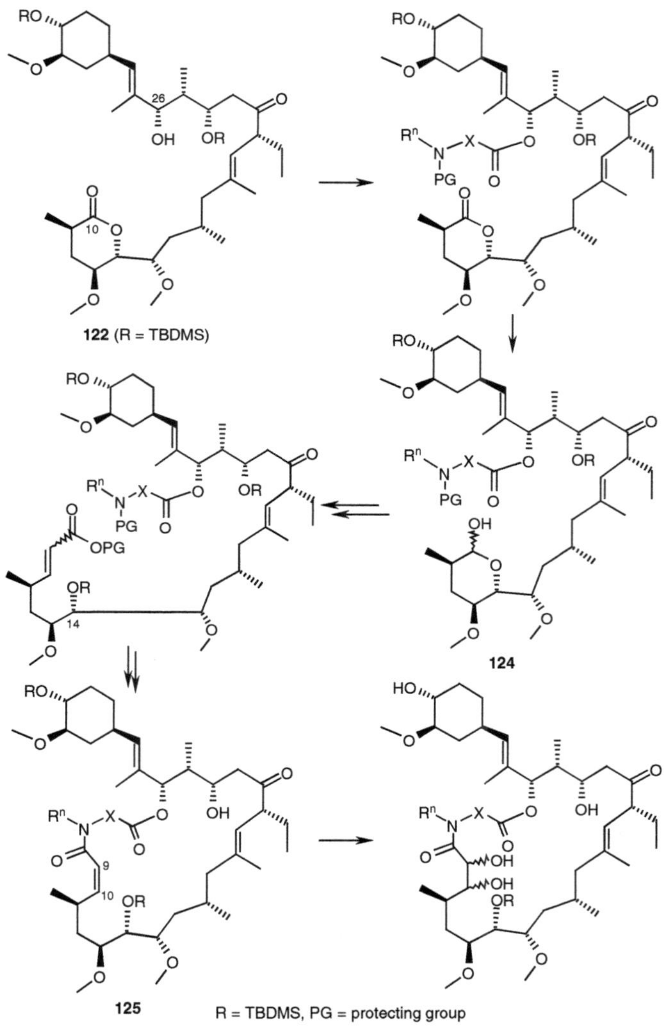

Scheme 32. Synthesis of amino acid analogues of ascomycin (**2**) starting from fragment **122**

ascomycin (**2**). NMR structures of the complexes formed between macophilin and **2** or **135** were found to be very similar, suggesting that hydrogen-bonding interactions with the C-24–OH are not important for complex formation.

Numerous derivatives of the formula **136** with modifications on the terminal carbon of the allyl group of tacrolimus (**1**) have been prepared employing the olefin cross metathesis reaction (Scheme 35). These were employed as modules for the assembly of chemical inducers of dimerization (CIDs) (*108, 109*).

Several derivatives of **2** with modifications at C-18 were prepared through allylic oxidation (Scheme 36) (*110*). Thus, oxidation of **2** by selenium oxide and *t*-butyl

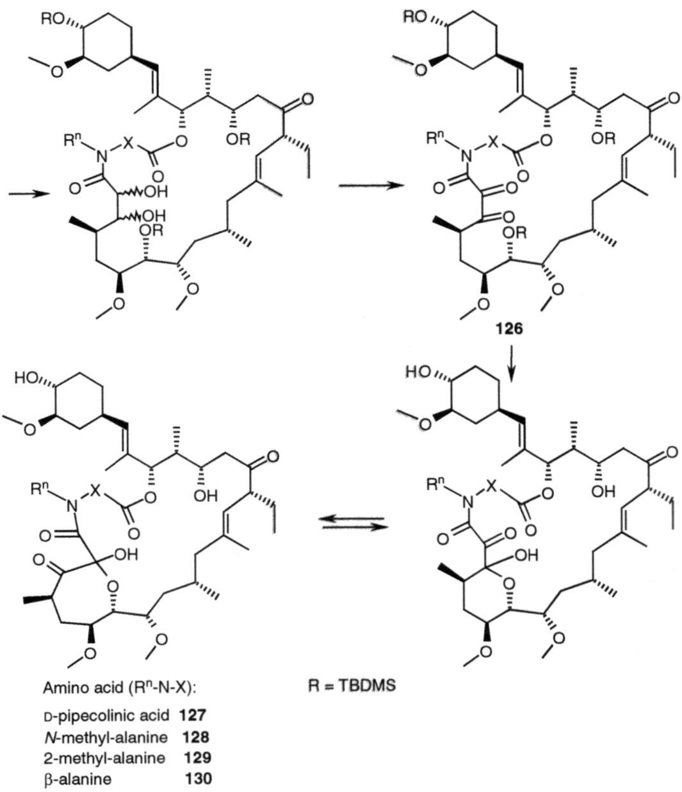

Scheme 33. Synthesis of amino acid analogues of ascomycin (**2**) starting from fragment **122**, continued from Scheme 32

hydroperoxide in dichloromethane afforded the hydroxyascomycin **137** (R=(18*S*)-OH), also referred to in literature as L-685818, in addition to side products arising through dehydration at C-23–C-24. None of the (18*R*)-isomer was formed in this reaction. On the other hand, under the same reaction conditions, oxidation of the 24,33-bis-silyl derivative **8** gave the corresponding 18-hydroxy derivatives as a mixture of isomers from which the (18*R*)-isomer **137** (R=(18*R*)-OH) could be obtained after its separation and desilylation. The 24,33-bis-silyl protected (18*R*/*S*)-OH derivatives could be oxidized and desilylated to the oxo analogue **138**. The 24,33-bis-silyl-protected (18*S*)-OH derivative could be acylated and desilylated to give (18*S*)-acetoxy, -iodoacetoxy, and -benzoyloxy derivatives of the formula **137**. Whereas 24,33-bis-silyl-ascomycin could be deprotected in excellent yield, the corresponding 24,33-bis-silyl-18-hydroxy analogues were, because of the allylic nature of the additional hydroxy group, sensitive to the usual desilylation conditions and resulted in the cyclic derivatives **139** and **140** as side products. (18*R*/*S*)-Hydroxyascomycins bind tightly to macrophilin and are potent inhibitors of rotamase (PPIase) activity, but do not show any immunomodulatory activity

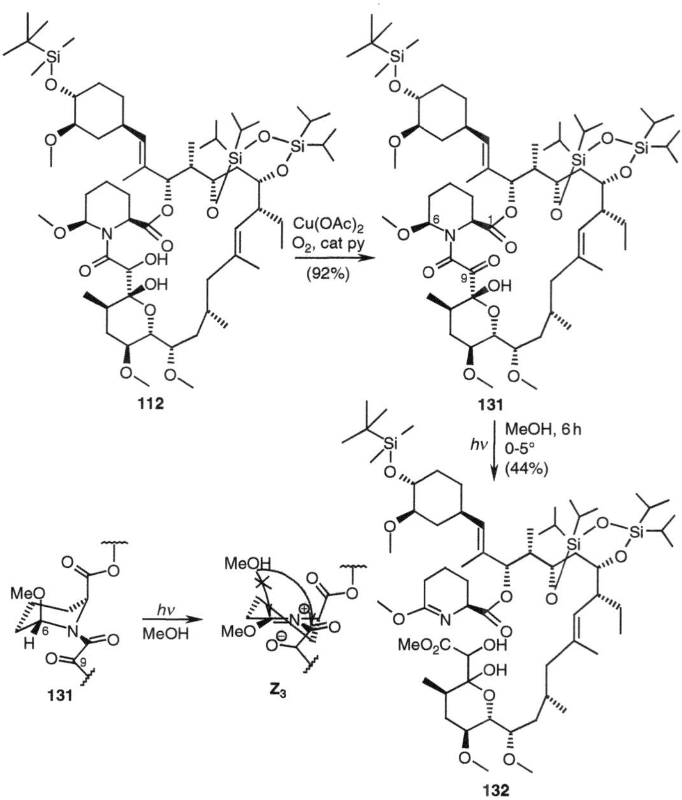

Scheme 34. A novel photochemical amide cleavage reaction of derivative 131

in vitro or *in vivo* despite their good bioavailability. Further, they reversed the inhibition of calcineurin caused by the tacrolimus/macrophilin complex. This indicated clearly that the immunomodulatory activity has nothing to do with the inhibition of the PPIase activity of tacrolimus/macrophilin. Later, it was demonstrated that binding of the initially formed tacrolimus/macrophilin complex to calcineurins A and B is necessary for immunomodulatory activity.

Reaction of 33-*O*-TBDMS-ascomycin (**141b**) with *t*-butyldimethylsilyloxy-1-methoxyethene in the presence of $BF_3.Et_2O$ resulted in the addition of a methoxycarbonylmethyl unit to the C-22=O (Scheme 37) (*111*). The resulting aldol product upon treatment with 1 *N* HCl underwent 33-*O*-desilylation and concomitant intramolecular esterification leading to the lactone **142**. The product **142** then underwent facile acid-catalyzed dehydration to give the unsaturated lactone **143**. The enone **144**, which could be prepared through treatment of **141b** with methanesulfonyl chloride and excess of 4-dimethylaminopyridine, also underwent 1,2-addition upon treatment with *t*-butyldimethylsilyloxy-1-methoxyethane in the presence of $BF_3.Et_2O$, resulting in the corresponding silyl derivative, which could be desilylated to afford **145** as a single isomer with an unknown

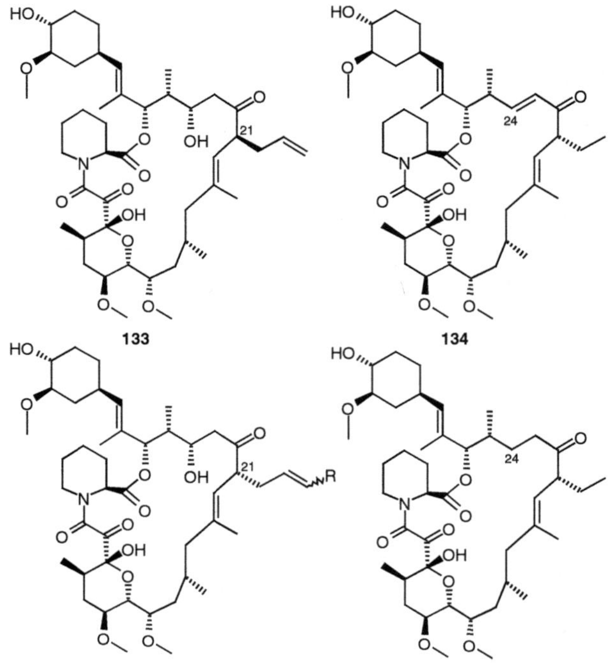

R = -CHOCH$_2$, -CO$_2$H, -CONH$_2$, -C$_6$H$_5$, -C$_6$H$_5$-(p)O-C$_6$H$_5$, -C$_6$H$_4$(p)-CO$_2$H, -CH$_2$CH$_2$CH$_2$CN, -CH$_2$NH$_2$

Scheme 35. 21-*epi*-FK506 (**133**), 24-desoxyascomycin (**135**) and olefin cross-metathesis derivatives of tacrolimus (**1**) **136**

C-22–configuration. On the other hand, the enone **144** underwent predominant 1,4-addition upon reaction with *t*-butyldimethylsilyloxy-1-methoxyethane in the presence of $LiClO_4$ in dichloromethane, providing the methyl ester **146** (as a single isomer of unknown stereochemistry) after desilylation (Scheme 38). Further, hydrosilylation of **144** gave the enol ether **147**, which could be employed in aldol reactions leading to **148a** and **148b**, as single isomers of unknown configuration at the newly formed stereocenters, featuring modifications at C-23.

Novel cyclopropano dervatives could be prepared using carbene chemistry. Thus, both tacrolimus (**1**) and ascomycin (**2**) were transformed to their hydrazones **149a** (62%) and **149b** (56%) (Scheme 39) (*112*). Treatment of the hydrazones **149a** and **149b** with excess manganese dioxide in dichloromethane resulted in several products. However, reaction in methanol under the same conditions led to the cyclopropano derivatives **150a** and **150b** as the main products; no biological activities are reported.

Numerous pyrazole analogues of ascomycin (**2**) have been prepared (Scheme 40) (*113*). Thus, the 33-*O*-TBDMS derivative **141b**, upon oxidation with DMSO and oxalyl chloride followed by desilylation with HF/aceteonitrile, gave a 91% yield of

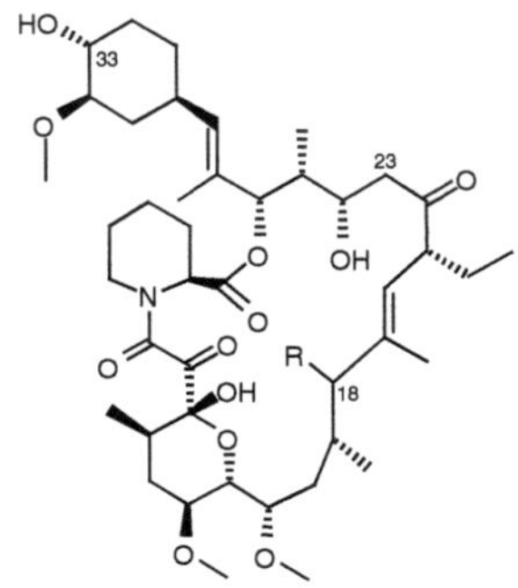

137

R = (18*R*)-OH, (18*S*)-OH, (18*S*)-OAc, (18*S*)-OBz, (18*S*)-OCOCH$_2$I

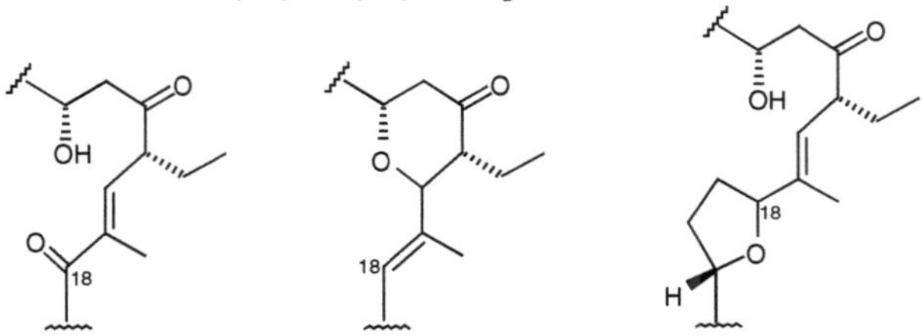

138 **139** **140**

Scheme 36. 18-Hydroxyascomycin (**137**) and its derivatives **138–140**

151, which upon treatment with hydrazine, afforded a mixture of the isomeric pyrazoles **152** (R=H) and **153** (R=H). The substituted pyrazoles **152** and **153** were prepared either through alkylation of the 33-*O*-TBDMS-protected unsubstituted pyrazoles, or through reaction of **151** with substituted hydrazines followed by chromatographic separation of the isomers. No biological activities were reported.

Several modifications in the cyclohexyl domain have been reported. Thus, activation of C-33–OH of the 24-*O*-TBDMS-ascomycin **154** as its triflate, followed by treatment with triethyl amine in dichloromethane at 45°C for 2 days and at room temperature for 2 days, resulted in an inseparable 3:1 mixture of the two isomers **155** and **156** in a 30% yield and the enol ether **157** (3% yield) (Scheme 41) (*114*). Osmium tetroxide-catalyzed *cis*-hydroxylation of the mixture **155** + **156** using *N*-methylmorpholine *N*-oxide in THF at room temperature for 2 days, followed by separation and desilylation of the individual isomers, afforded (34*R*)-hydroxyascomycin (**158**) and the 31,32-dihydroxy-33-methoxy analogue **159**. Interestingly, the introduction of an additional hydroxy group increased the solubility of compounds **158** and **159** by 300- and 150-fold compared to ascomycin (**2**) at pH 6.5. Compound **158** showed *in vitro* and *in vivo* activities comparable to those of ascomycin.

Several 33-*epi*-amino derivatives of ascomycin and its 24-desoxy analogue have been synthesized (Scheme 42) (*115*). Thus, the C-33–OH of 24-*O*-TBDMS-ascomycin (**154**) was activated as its *o*-nitrobenzenesulfonate (*o*-nitrobenzenesulfonyl

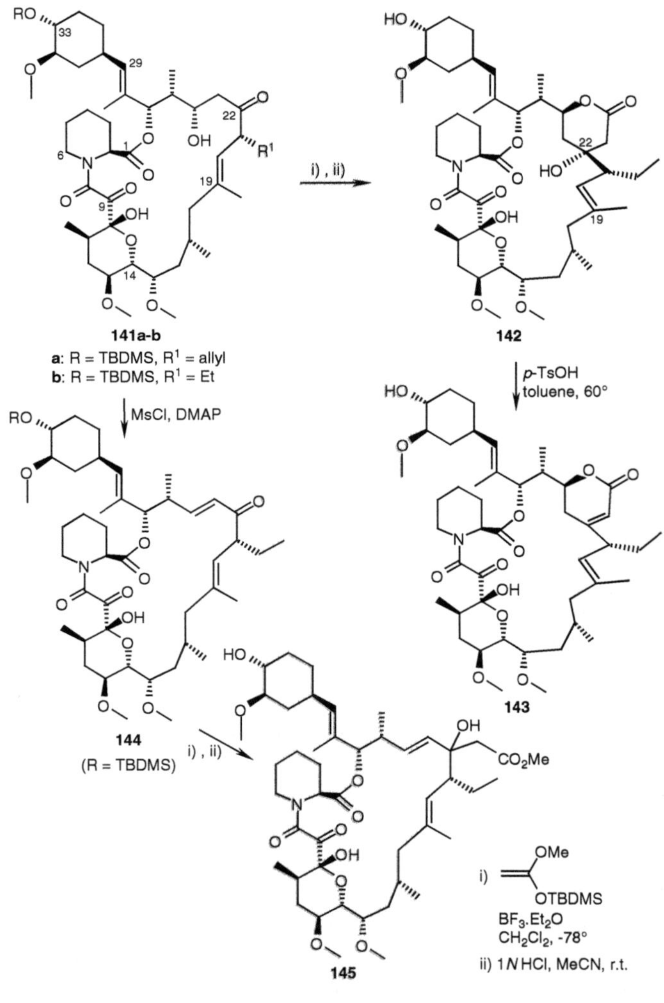

Scheme 37. Derivatives of ascomycin (**2**) with modifications in the C-22–C-24 region

chloride, triethyl amine, DMAP, CH_2Cl_2) and transformed to the 33-*epi*-azide through treatment with sodium azide in DMF at 60°C. Reduction of the azide with triphenyl phosphine in a mixture of THF and water at refluxing temperature afforded the 33-*epi*-amino-ascomycin **160** (R=H) in a 73% yield over three steps. Starting from **160** (R=H) several amides and carbamates have been synthesized. Similarly, from the 24-desoxyascomycin **135** the 33-*epi*-amino derivative **161** (R=NH_2) was synthesized, from which several amino analogues were prepared through condensation with aldehydes in THF, followed by reduction with sodium cyanoborohydride. The activities of these compounds in inhibiting T-cell proliferation were reported. Briefly, the 33-*epi*-amino analogue **160** (R=H) was tenfold less

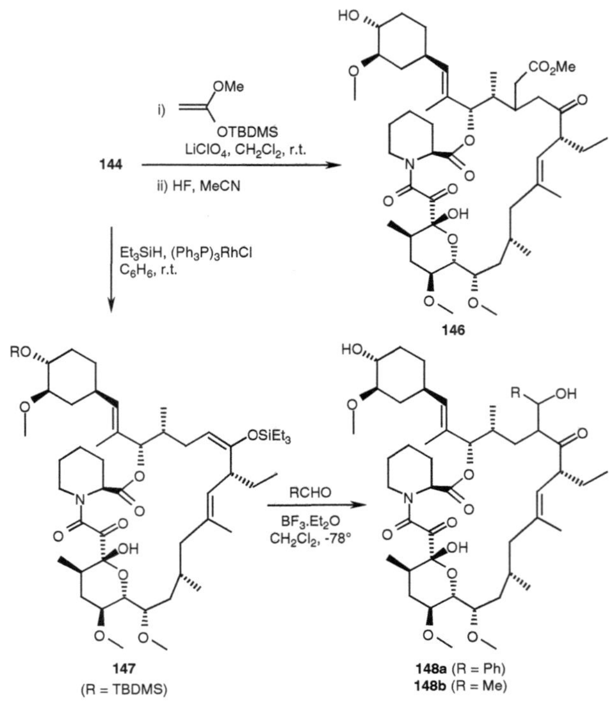

Scheme 38. Derivatives of ascomycin (**2**) with modifications in the C-22–C-24 region; continued from Scheme 37

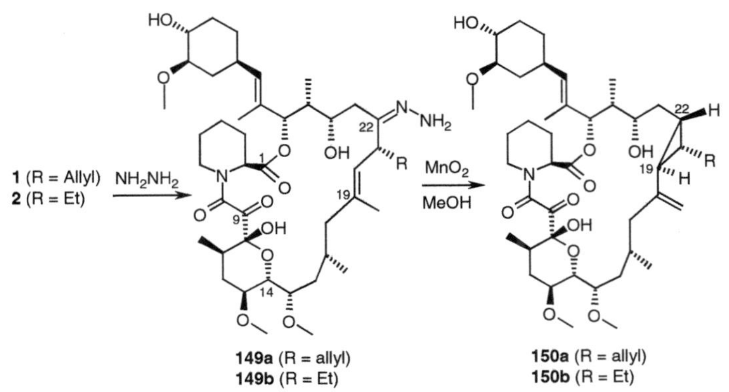

Scheme 39. Transformation of tacrolimus (**1**) and ascomycin (**2**) to the cyclopropano- analogues **150a** and **150b**

active than ascomycin; this loss was partially recovered by removing the 24-OH group. Among the carbamate derivatives, the methyl carbamate **160** (R=MeOCO-) showed activity closest to that of ascomycin.

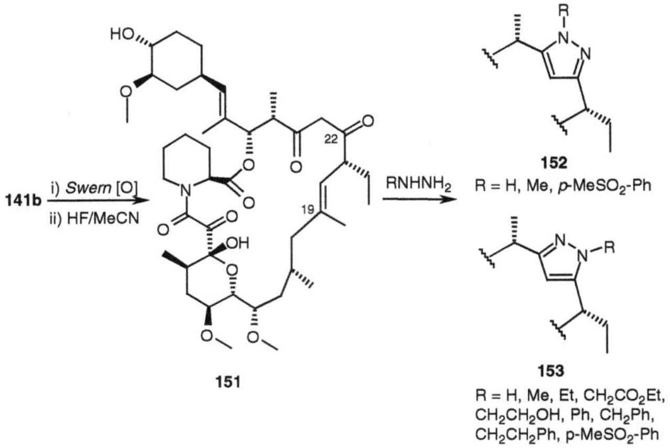

Scheme 40. Transformation of 33-*O*-TBDMS-ascomycin (**141b**) to the pyrazole analogues **152** and **153**

Pimecrolimus (**2a**) is prepared through selective chlorination of **2** using dichloro-triphenyl-phosphorane in a good yield (Scheme 43) (*116*). Other replacements of the C-33−OH of ascomycin by halogens or pseudohalogens were also reported (*116, 117, 118*). Further, the (33*R* or *S*)-*O*-cyano derivative **162** is easily accessible from **2** (*117*). Starting from **162** the derivatives **163–169** (SDZ 281–240 is a mixture of **169a** and **169b**) could be prepared easily (*117, 118*). Of these derivatives, 33-*epi*-ascomycin (**163**) and 33-desoxy-32-oxo-ascomycin (**167**) showed good activities in a MLR (mixed lymphocyte reaction), but were, however, only weakly active in the animal models of ACD. Noteworthy is the observation that the *cis*-isomer **169b** is by a factor of 2 more active than the *trans*-isomer **169a**, in *in vitro* and *in vivo* models.

Several carbamate derivatives at C-33 have been prepared starting from 24-*O*-TBDMS-ascomycin (**154**) (Scheme 44) (*119*). Thus, reaction of **154** with the appropriate acyl isocyanate or acyl isothiocyanate followed by desilylation afforded the acylated carbamate **170a** and the acylated thiocarbamates **170b-h**. The hydrazide derivative **170i** was prepared through reaction of ascomycin (**2**) with triphosgene giving the corresponding 33-*O*-chloroformate selectively, followed by its reaction with phenyl hydrazine. Binding to macrophilin and the *in vitro* immunomodulatory activities were demonstrated; derivatives **170a** and **170i** showed activities similar to ascomycin (*119*).

Numerous C-33−ether derivatives were prepared by O-alkylation of 24-*O*-TBDMS-ascomycin (Scheme 44). Thus, reaction of **154** with allyl 2,2,2-trichloroacetimidate in the presence of a catalytic amount of trifluoromethane sulfonic acid gave **171a**, which after desilylation using HF in acetonitrile, afforded the allyl ether **171b** in a 75% yield (*120*). Similarly, the unsubstituted cinnamyl ether **171c** and several analogues of **171c** with electron-withdrawing and -releasing substituents on the phenyl ring have been synthesized. Further, using the same method, several analogues of **171d** featuring alkyl, alkenyl, alkynyl, aralkyl, aralkenyl, and aralkynyl units, were prepared and their binding to macrophilin

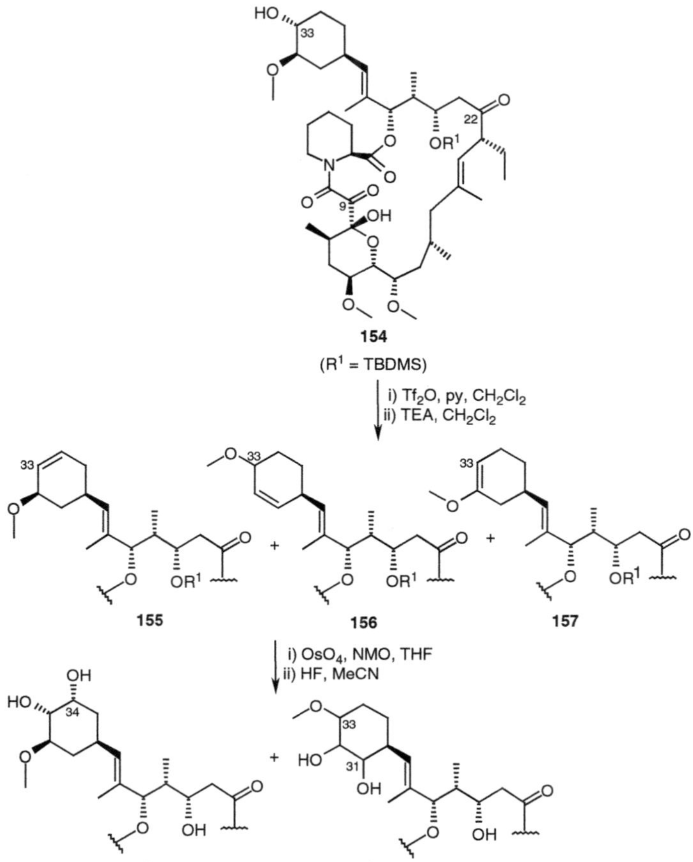

Scheme 41. Synthesis of the hydroxylated derivatives **158** and **159**

and inhibition of T-cells studied. It is interesting to note that **171c** and the corresponding *m*- and *p*-hydroxy cinnamyl ethers, in spite of their relatively weaker binding, showed inhibitory activities on T cells comparable to those of ascomycin. However, they were reported to be less efficacious *in vivo* especially upon oral administration (*121*). In further studies aimed at optimizing the linker between the C-33–O and the phenyl groups, the 24-silyl protected allyl ether **171a** was transformed to the aldehyde **171e** by osmium tetroxide-catalyzed hydroxylation followed by cleavage using sodium periodate in aqueous THF, in an overall yield of 64%. Starting from the valuable intermediate **171e**, using the sequences reductive amination followed by desilylation, or, reduction to alcohol, displacement of a trichloro acetimidate, followed by desilylation, ether derivatives of the type **172** (R=alkyl, aryl; tether=secondary amine or ether) with varying tethers were synthesized. Further, the aldehyde **171e** was transformed to several imidazol-2-yl-methyl ether derivatives of type **173** (R^1=H, Et; R^2=H, Me, Ph, substituted-Ph)

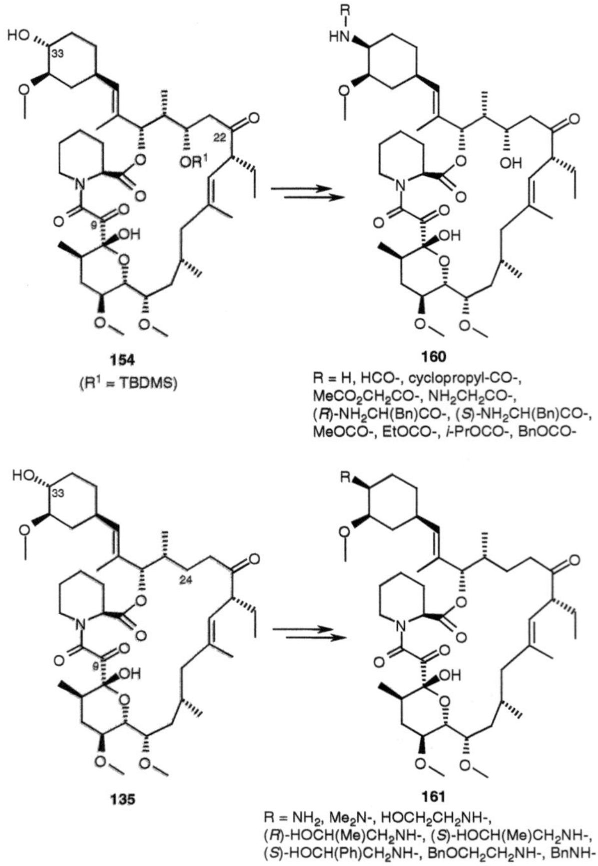

Scheme 42. Synthesis of the 33-*epi*-amino analogues **160** and **161** of ascomycin (**2**) and 24-desoxyascomycin (**135**)

with different substituents on the imidazole ring. Among the compounds investigated, derivative **173** (R^1=H, R^2=3,5-dimethoxyphenyl), referred to as L-733,725, was reported to have *in vivo* activities comparable to those of tacrolimus and a better therapeutic index. A convergent practical synthesis was developed and was used for synthesizing multi-kilogram quantities of L-733,725 of consistently high purity (*122*). Furthermore, labelled L-733,725 carrying a ^{14}C-label on the CH_2 of the side chain was also synthesized for utilization in animal and human drug metabolism studies (*123*).

In a similar study, several 33-*O*-ether derivatives featuring a phenyl group connected through a carbon tether of varying length were synthesized (Scheme 45) (*124*). The synthesis started from the aldehyde **171e** and proceeded through addition of an organomagnesium bromide to the aldehyde, followed by transformation of the resulting alcohol to the corresponding trifluoro acetate, its elimination, selective hydrogenation of the resulting double bond and, finally, deprotection of the C-24−O−silyl group leading to the compounds **174a** (R=$PhCH_2CH_2$, $PhCH_2CH_2CH_2$); the biological data were compared with **174a** (R=Ph, $PhCH_2$)

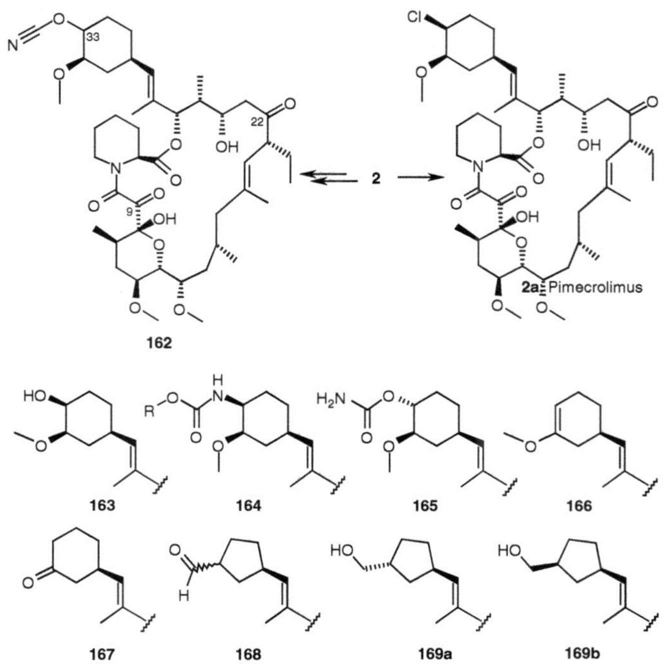

Scheme 43. Transformation of ascomycin (**2**) to several analogues *via* the 33-*O*-cyano-derivative **162**, and, synthesis of pimecrolimus (**2a**) and the individual isomers **169a** and **169b** of SDZ 281–240

(*125*). The derivatives **174b** bearing hydroxy or keto groups on the carbon tether have also been synthesized using similar strategies. A two-carbon tether provided optimum *in vitro* activity. The acetophenone derivatives **174b** (R=$PhCOCH_2$) showed efficacy in models of immunosuppression. However, it was found that they were rapidly converted to the arylhydroxy ether products when incubated in rat blood. Hence, an extensive series of substituted arylhydroxy ethers **174c** were synthesized using analogous chemistry (*126*). Several of these compounds showed potent *in vitro* and *in vivo* immunosuppressive activities. The derivative **174c** (R=2-naphthyl-$CHOHCH_2$-) was claimed to have an improved therapeutic index compared to tacrolimus (**1**) (*126*).

Using pentavalent bismuth derivatives ($Ar_3Bi(OAc)_2$, CH_2Cl_2, THF) the C-33–OH of ascomycin (**2**) could be preferentially arylated leading to compounds of type **177a** in yields ranging from 14% to 72% (Scheme 45) (*125, 127*). Similarly, starting from the C-33–O-desmethyl analogue **176**, derivatives **177b** and **177c** could be synthesized. Among the derivatives of the type **177a** (R^1=Me, R^2=Ph), **177b** and **177c**, those with electron-donating substituents such as HO and Me_2N in the *para-* position of the phenyl group, showed better *in vitro* immunosuppressive activities. The two indole ether derivatives of the type **177a** turned out to be the best in the series, being even superior to the parent macrolide **2** by a factor of 3 in *in vitro* immunosuppression. The two derivatives **175a** and **175b**, prepared from

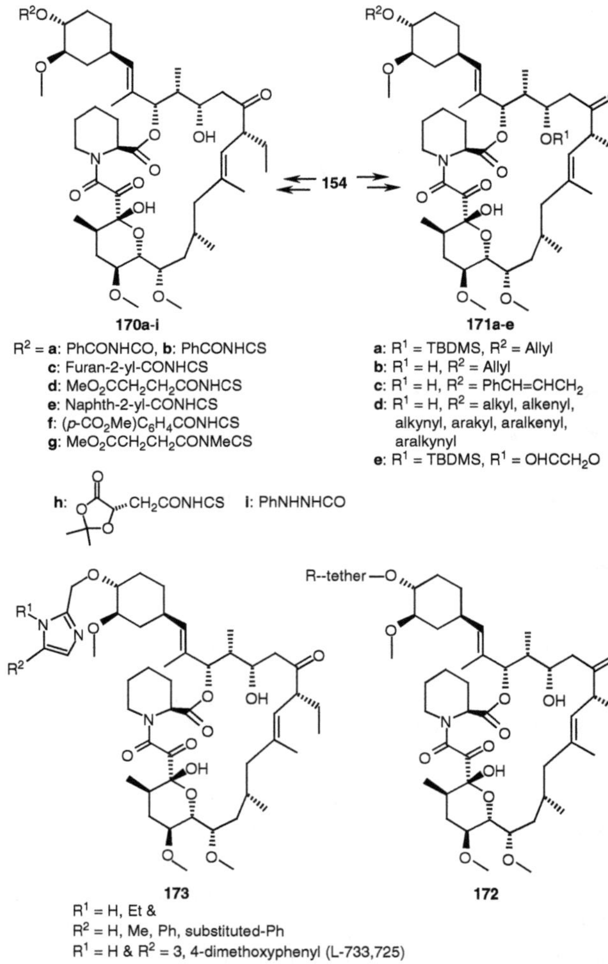

Scheme 44. Synthesis of 33-*O*-acyl, 33-*O*-thioacyl, and 33-*O*-ether derivatives of ascomycin (**2**)

(18*S*)-hydroxyascomycin (**137** (R=(18*S*)-OH) using a similar method, showed 8- to 15-fold weaker immunosuppressive activities compared to **2**.

The C-33−OH group of **2** could be etherified selectively by rhodium (II) acetate-catalyzed insertion of benzyl diazoacetate in dichloromethane affording **178** (R=Bn), from which the corresponding acid **178** (R=H) could be obtained by hydrogenolysis of the benzyl ester (Scheme 46) (*128*). Starting from the acid, ester and amide derivatives of types **178** and **179** were prepared. While the carboxylic acid **178** (R=H) did not show any *in vitro* activity, the ester and amide derivatives showed potent T-cell inhibitory activities. Surprisingly, hydrophobic amides were reported to have weakened macrophilin binding by several orders of magnitude, while maintaining potent activity to inhibit T-cell activation (*128*).

Scheme 45. Synthesis of ether derivatives at the C-31 and C-33 positions of ascomycin (**2**)

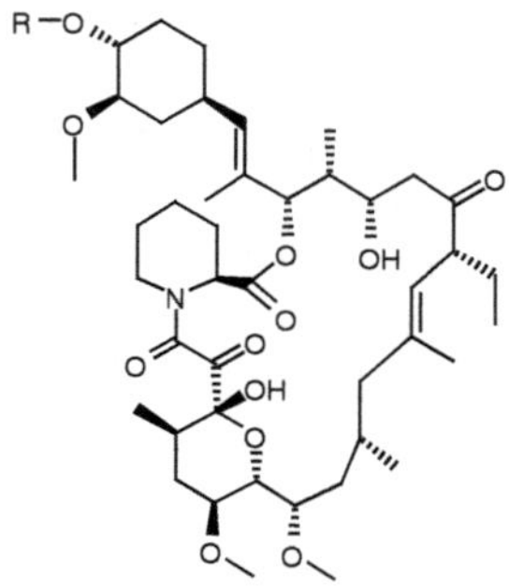

174a-c

a: R = Ph, $PhCH_2$, $PhCH_2CH_2$, $PhCH_2CH_2CH_2$
b: R = $OHCH_2CH_2$, $PhCHOHCH_2$, $PhCOCH_2$, $PhCH_2CO$, $PhC(=CH_2)CH_2$
c: R = $ArCHOHCH_2$

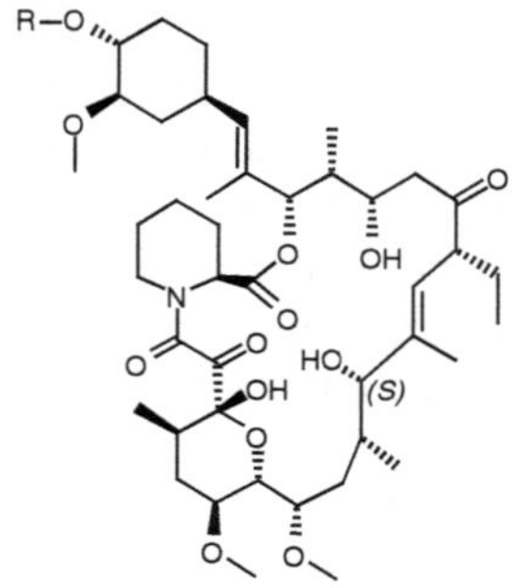

175a-b

a: R = 2-naphthyl
b: R = *N*-methyl-5-indolyl

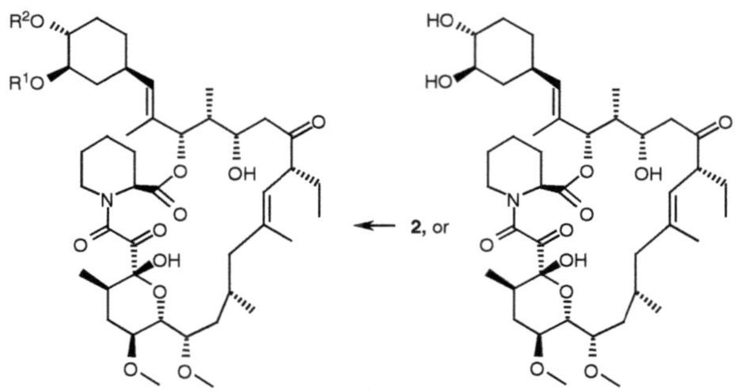

177a-c **176**

Prepared from **2:**
a: R^1 = Me, R^2 = substituted Ph, 2-naphthyl, 5-indolyl, *N*-methyl-5-indolyl
Prepared from **176:**
b: R^1 = H, R^2 = substituted Ph
c: R^1 = substituted Ph, R^2 = H

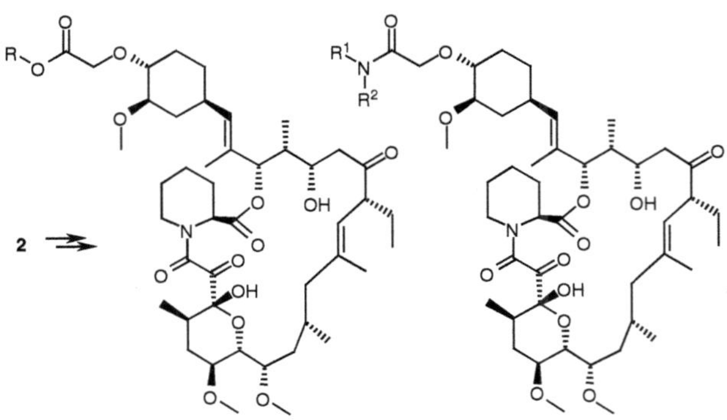

178
R = H, Et, Bn

179a-b
a: R^1 = H, and R^2 = cyclohexyl, Ph, substituted aryl, Bn, CH_2CH_2Ph
b: R^1 = Me, and R^2 = Bn, CH_2CH_2Ph

Scheme 46. Diverse esters and amides of the ascomycin-33-*O*-acetic acid skeleton

Derivatives like **180** featuring a heterocycle at C-33 have been prepared through selective activation of C-33–OH as its triflate, followed by its displacement by nitrogen of a small heterocycle (Scheme 47) (*129*). Further, selective activation of C-33–OH as a *p*-nitrophenylcarbonate, followed by reaction with nitrogen nucleophiles afforded the carbamates **181**. The tetrazole **180** (R=tetrazol-1-yl), also referred to as ABT-281, was reported to be equipotent to tacrolimus (**1**) in the swine contact hypersensitivity model after topical application, despite its several-fold lower potency for inhibiting swine T cells *in vitro*. The high activity was consistent with its superior skin penetration. Further, it was found to exhibit a three- to fivefold more rapid pharmacokinetic elimination in the rat, swine, and monkey, in addition to showing a substantial reduction in potency for immunologic responses in all three species after systemic administration. These features of ABT-281 were claimed to result in both efficacy and a high margin of safety for topical therapy of skin disease as compared to tacolimus (*129*).

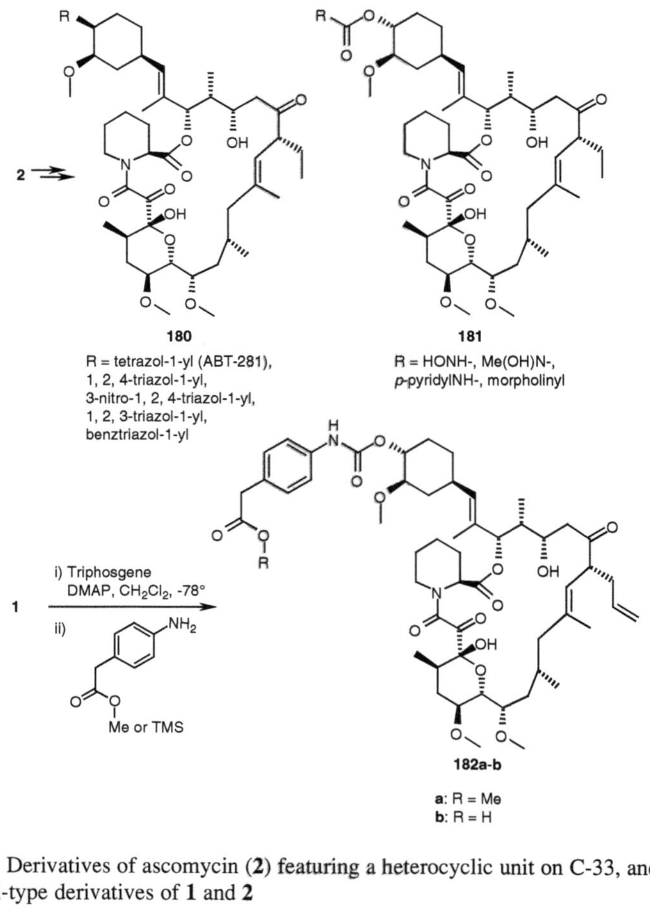

Scheme 47. Derivatives of ascomycin (**2**) featuring a heterocyclic unit on C-33, and diverse 33-*O*-carbamoyl-type derivatives of **1** and **2**

Tacrolimus was transformed to the carbamates **182a** and **182b** through activation of the C-33–OH as its chloroformate followed by reaction with *p*-amino-phenylacetic acid methyl ester, leading to **182a** in a 70% yield, or, with *p*-amino-phenylacetic acid trimethylsilyl ester followed by hydrolysis providing the acid **182b** in a 26% yield (*130*). Compound **182a** could also be prepared through activation of C-33–OH as its *p*-ntrophenyl carbonate followed by quenching with the amine. The carbamate **182a**, also referred to as MLD987, inhibited the activation, proliferation and release of cytokines from T-cells with IC_{50} values in the low nanomolar range. Furthermore, in a brown Norway rat model of allergic asthma, **182a**, when given into the airways by intertracheal administration (ED_{50} = 1 mg/kg) or by inhalation (ED_{50} = 0.4 mg/kg), potently reduced the influx of leucocytes into the bronchoalveolar fluid. Interestingly, **182a** had an appreciably

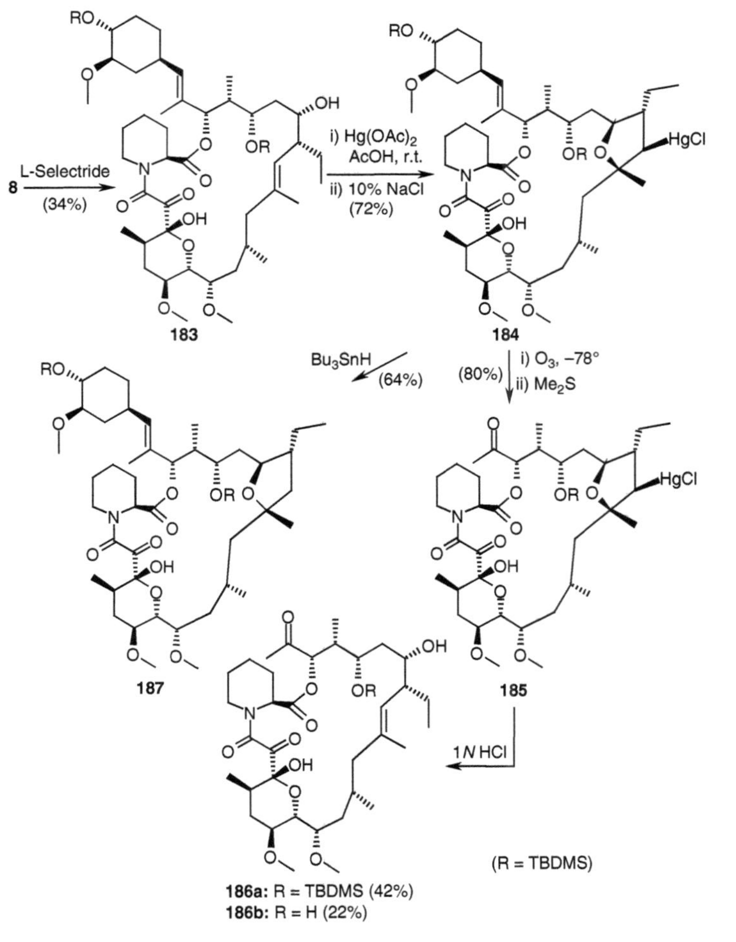

Scheme 48. Transformations using oxymercuration and ozonolysis resulting in the derivative **187** and the cleavage product **186b**

weaker activity when given orally and intravenously. Furthermore, pharmacokinetic evaluations showed that **182a** had a low oral and pulmonary bioavailability. In addition, **182a** was metabolized in the blood to the much less potent acid **182b**. These aspects indicated a potential of **182a** to serve as a soft drug after local application for therapy of asthma.

With a view to synthesizing derivatives with new substituents in place of the cyclohexyl moiety, a strategy was designed for chemoselective cleavage of the C-28 = C-29 double bond. Thus, the bis-silyl-protected ascomycin **8** was reduced stereo- and regioselectively at C-22=O giving the alcohol **183** in a 34% yield, which was then subjected to oxymercuration leading to the tetrahydrofuran **184**, thus protecting the C-19 = C-20 double bond against ozone (Scheme 48) (*131*). Ozonolysis of **184** afforded the ketone **185** in an excellent yield. Further, acidic hydrolysis of the organomercurial chloride afforded the required derivatives **186a** and **186b**. Finally, tributyl tin hydride reduction of **184** afforded the analogue **187**. The derivatives **188** and **189** could also be synthesized in an analogous manner (Scheme 49).

Starting from **8**, using K-selectride instead of L-selectride, the isomer **190** was prepared analogously to the synthesis of **186** (Scheme 50) (*132*). The C-27 = O keto group was then used for preparing derivatives of the types **191** and **192**. The required C-22 = O carbonyl could be re-established through oxidation leading to the 24-*O*-silyl derivatives **193** featuring novel replacements for the cyclohexylvinylidene unit. Several attempts to establish a carbon-carbon double bond at C-28 of **190** *via* a *Wittig* or *Wittig-Horner* olefination were unsuccessful and no reaction was observed. However, reaction of **190** with acetylmethylidenephosphorane ($CH_3COCH = PPh_3$) under forced conditions (65°C, 10 days) led to furano derivatives through reaction at the tricarbonyl region.

From the fermentation broth of *Steptomyces tsukubaensis* 9993, the novel ring-contracted metabolite *iso*-FK506 (**197a**) was isolated (Scheme 51) (*133*). A synthetic pathway for the transformation of tacrolimus to **197a** has been established (*134*). Thus, *Evans*' reduction of 33-*O*-TBMDS-FK506 (**141a**) afforded the (22*S*)-dihydro derivative **194a** in a good yield. Treatment of **194a** with DMAP in DMF

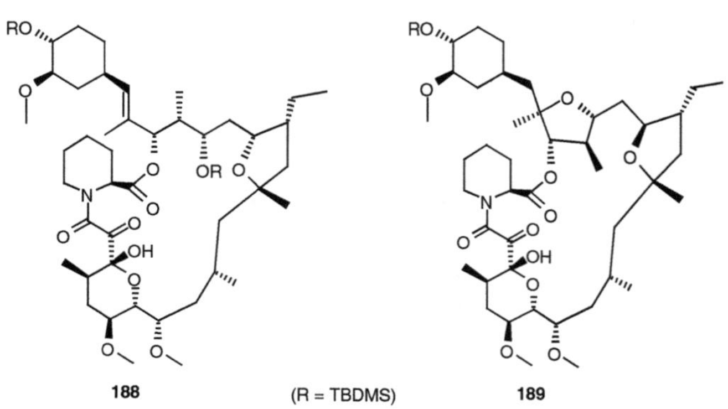

188 189

Scheme 49. Derivatives **188** and **189** prepared in an analogous manner to those in Scheme 48

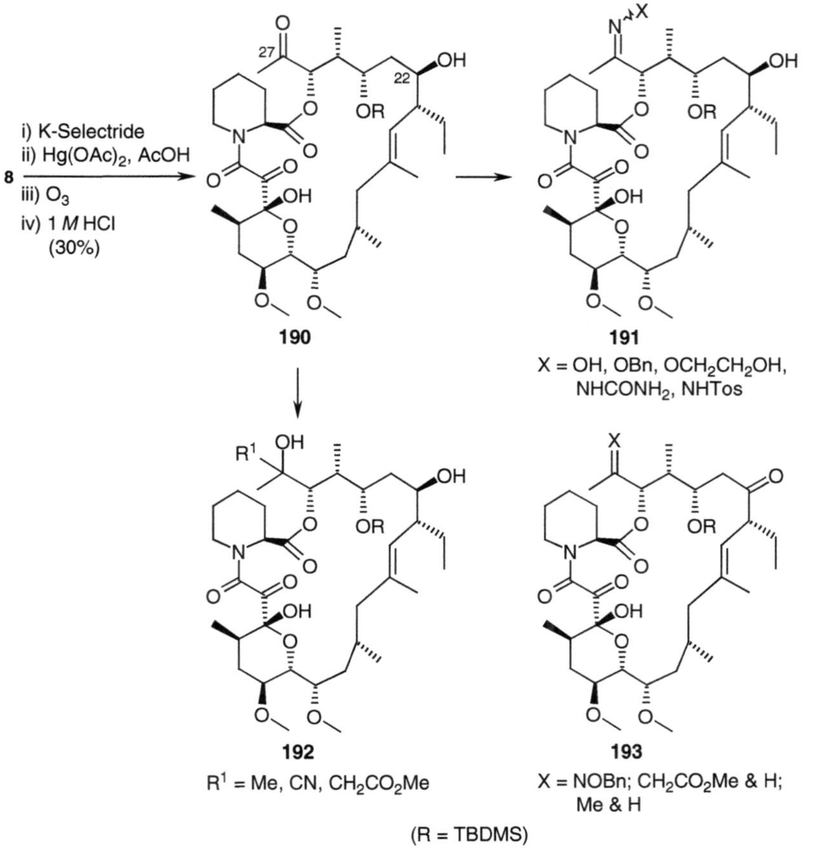

Scheme 50. Derivatives of ascomycin (**2**) featuring novel replacements for the cyclohexylvinylidene subunit

brought about acyl migration, giving **195a**. Attempts to oxidize the C-22–OH group selectively were not successful. However, after protection of the C-26–OH as its TMS derivative **196a**, oxidation using TPAP followed by desilylation afforded the *iso*-FK506 **197a**. The chemistry of the ethyl analogue **197b** followed in an analogous manner.

Under the usual desilylation conditions employing aqueous HF, the 22-dihydro *iso*-ascomycin derivative **198b** underwent intramolecular cyclization through attack of the C-22–OH on C-26 leading to the tetrahydropyran **199** in a 36% yield (Scheme 52) (*135*). On the other hand, **198a** could be desilylated using 1*N* HCl and transformed to the epoxide **200**. Interestingly, **200** upon treatment with $BF_3.Et_2O$, underwent acyl migration from 24-*O* to 22-*O*, followed by cyclization involving the released C-24–OH and the epoxide ring, affording the ring-contracted 19-membered macrolide **201** in a 38% yield.

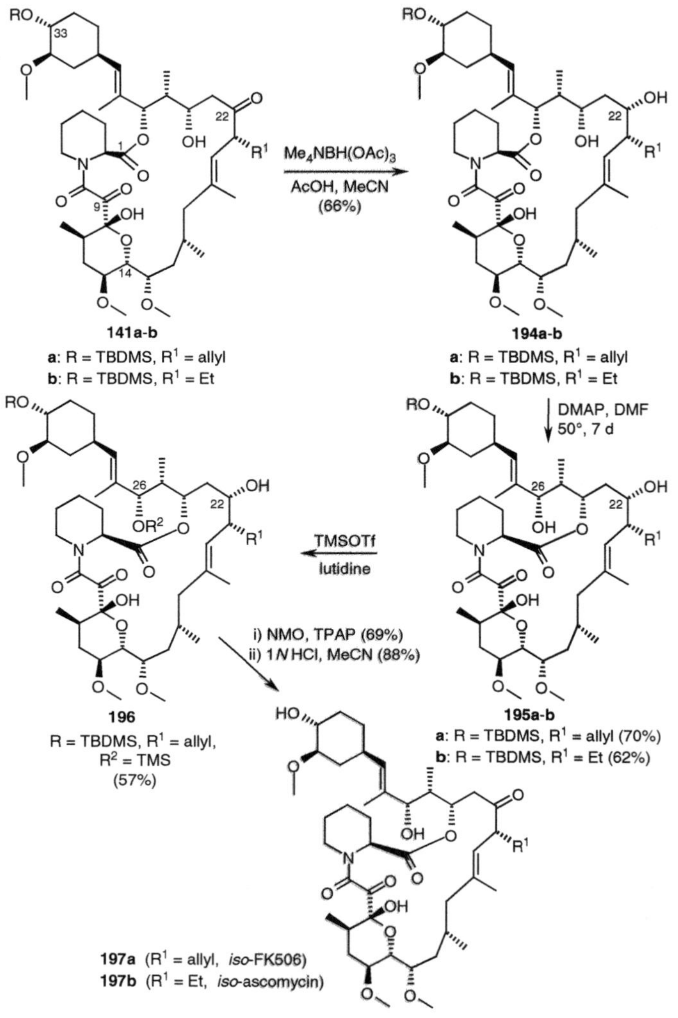

Scheme 51. Synthesis of *iso*-FK506 (**197a**) and *iso*-ascomycin (**197b**)

Several derivatives of *iso*-ascomycin featuring broad modifications in the cyclohexyl region have been synthesized. Thus, selective *cis*-hydroxylation of the C-27=C-29 double bond in a 33-*O*-TBDMS-*iso*-ascomycin (**195b**) with osmium tetraoxide led to **202** as a mixture of diastereomers, which were cleaved with periodate to give the aldehyde **203** as a valuable intermediate (Scheme 53) (*136*). Compound **203** exists predominantly as a 1:1 mixture of the anomeric hemiacetals formed through intramolecular addition of the C-22–OH to the C-26–aldehyde. The aldehyde group of **203** could be transformed to **204a** through *Wittig* reaction and to the oxime and hydrazone derivatives **204b-d** through condensation reactions.

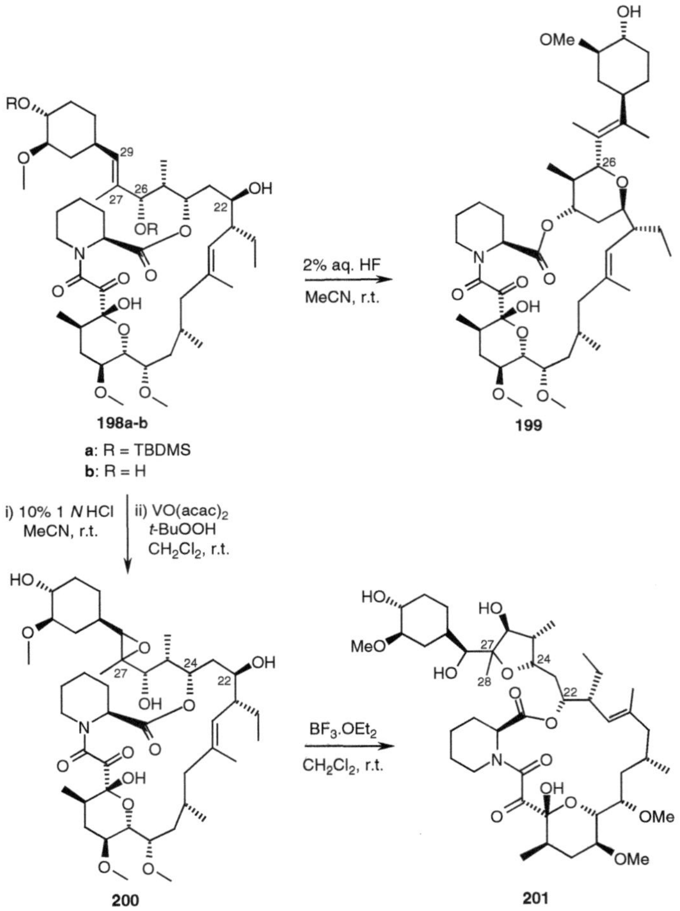

Scheme 52. Intramolecular cyclization, and ring-contraction reactions of 22-dihydro-*iso*-ascomycin (**198b**) leading to **199** and **201**

The derivatives **204a** and **204b** could be transformed to the analogues **205** (76%) and **206** (49%) through oxidation of the C-22–OH with *Dess-Martin's* periodinane.

In contrast to the chemistry observed with the (22*S*)-OH series (Scheme 52), osmium tetraoxide catalyzed dihydroxylation of the (22*R*)-dihydro-*iso*-ascomycin (**198b**) led to hydroxylation of both double bonds giving a 50% yield of the tetraol **207** as an isomeric mixture (Scheme 54) (*137*). Diol-cleavage using excess of sodium periodate led to the bis-*seco*-derivative **208**. Alternatively, **198b** could be transformed to **208** in a single step by ozonolysis followed by treatment with dimethyl sulfide.

Tacrolimus (**1**) is poorly soluble in water, and hence the conventional i.v. dosage forms contain surfactants such as cremophor EL or hydrogenated polyoxy 60 castor oil (HCO-60). Thus, towards increasing the water solubility of **1**, the methoxy-

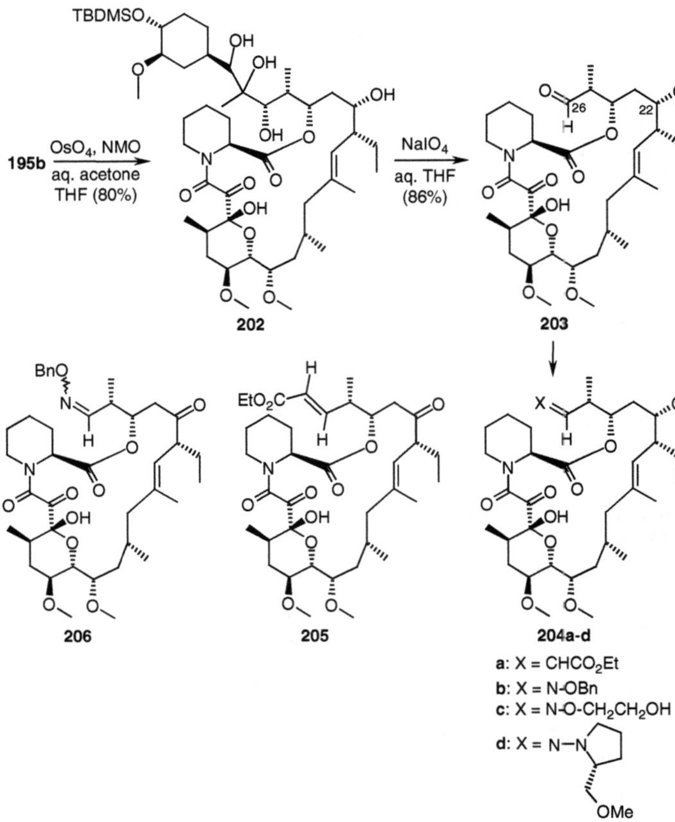

Scheme 53. Transformations of (22*S*)-diydro-33-*O*-TBDMS-*iso*-ascomycin (**195b**)

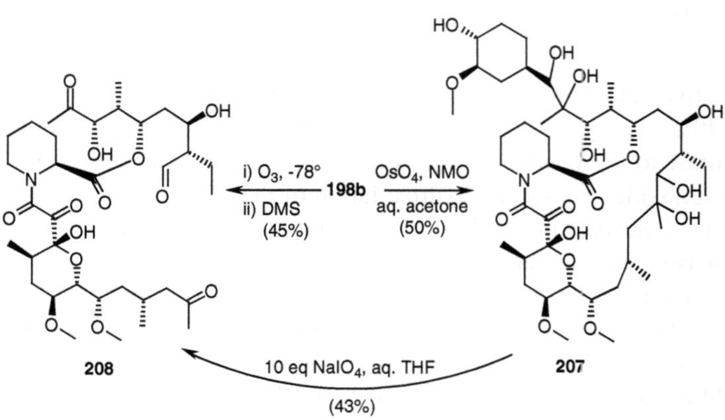

Scheme 54. Transformations of the (22*R*)-dihydro-*iso*-ascomycin **198b**

(polyethyleneglycol) (mPEG) conjugates **209d**, **209e**, and **209f** were synthesized by esterification of **1** with iodoacetic acid using dicyclohexylcarbodiimide followed by chromatographic separation giving the individual iodoacetates **209a** and **209b** and **209c**, and displacement of the iodide with mPEG-SH in a mixture of acetonitrile and 0.1 *M* aqueous sodium hydrogen carbonate (Scheme 55) (*138*). The half-life of these conjugates in phosphate buffer (pH 7.4, 37°C) was *ca.* 20 h, showing acceptable stability; in human liver homogenates they decomposed readily with a half-life of 10 min and released the drug tacrolimus (**1**). This indicates their potential as useful water-soluble prodrugs. The effects of the derivative **209d** on the proliferation of T-cells, B-cells and mast cells, and on IgG production in human B-cells, and histamine release, were investigated *in vitro* (*139*). Further, *in vivo*

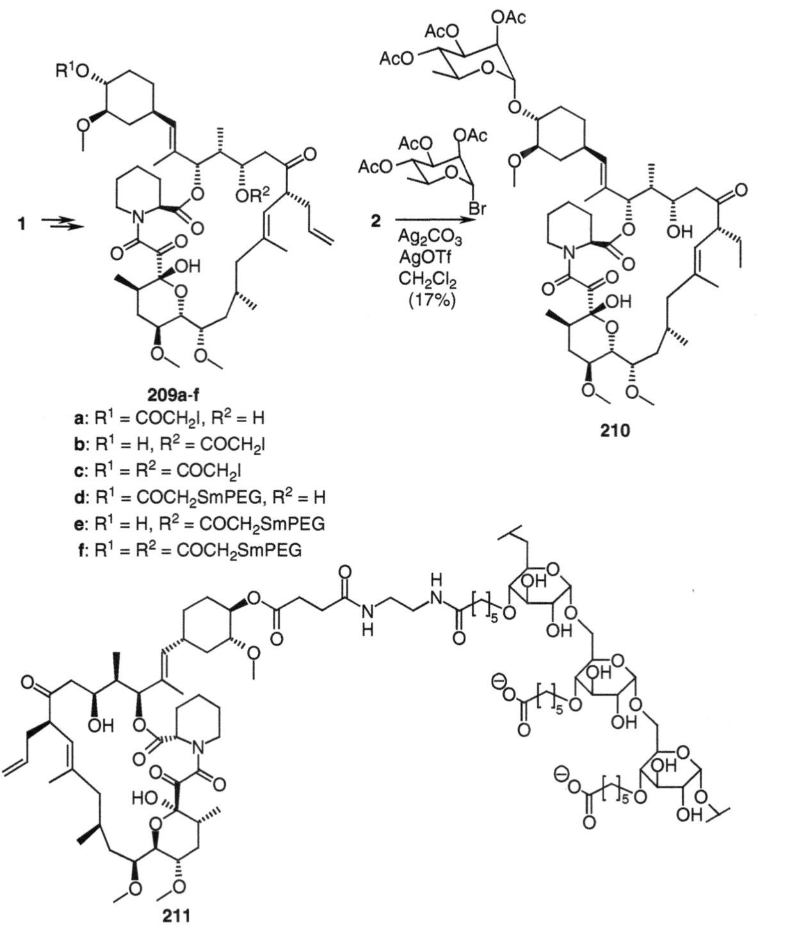

Scheme 55. Water-soluble prodrugs of tacrolimus **209d**, **209e**, **209f**, featuring methoxy-poleyethyleneglycol units attached through an ester linkage, a rhamnose derivative of ascomycin (**210**), and tacrolimus-C-6-dextran conjugate (**211**)

studies were performed on skin graft rejection in mice and in *Freund*'s adjuvant arthritis in rats. The results demonstrated that the mPEG derivative **209d** and tacrolimus have similar effects both *in vitro* an *in vivo*, thus indicating the former to be a useful prodrug with its advantageous physicochemical properties for preparing formulations for different applications.

With a view to developing a macromolecular prodrug of tacrolimus (**1**) with modified pharmacokinetic properties for systemic administration, the dextran conjugate **211** has been prepared (Scheme 55) (*140*). Thus, dextran (T-70) was O-alkylated with 6-bromohexanoic acid to give carboxy-*n*-pentyl-dextran (C6D); the free carboxylic acid groups were then modified through coupling with ethylene diamine to give C6D-ED. The free amino groups were in turn coupled through an amide bond to FK506-33-*O*-hemisuccinate, affording the conjugate **211**. The conjugate was estimated to contain 0.45% of tacrolimus (**1**) and the coupling molar ratio was approximately 1:1 (dextran:tacrolimus) (*141*). After incubation of the conjugate **211** in phosphate buffer (pH 7.4), tacrolimus was released and the half-life of the conjugate was 150 h. The *in vitro* immunosuppressive activity was reported to be almost comparable to that of free tacrolimus. *In vivo* biodistribution studies demonstrated that conjugation dramatically changes the pharmacokinetic properties of tacrolimus. After i.v. injection in rats, the AUC of the conjugate was reported to be almost 2000 times higher than that of free tacrolimus and organ uptake and clearances of the conjugate were significantly smaller than those of the free drug. However, no further activities have been reported.

Ascomycin (**2**) can be glycosylated at C-33−OH selectively using triacetyl-L-rhamnosyl bromide giving a 17% yield of the derivative **210** (Scheme 55) (*142*). Compound **210** showed good efficacy in *in vitro*, and in animal models including rat heart allograft rejection and rat adjuvant arthritis, and indicated a lower potential for neurotoxicity compared to tacrolimus.

Tacrolimus has been linked at C-33−OH through an ester or carbamate bond to several polymers for the purpose of affinity chromatography (*143, 144*). Finally, several total syntheses (*145–155*) of tacrolimus and ascomycin, and also segments thereof, have been reported, and biosynthetic pathways discussed (*156*).

3. Summary

The discovery of the potent anti-inflammatory activity of the natural compounds ascomycin and tacrolimus in a new animal model of skin inflammation provided the first pharmacological evidence that calcineurin inhibitors of this type may represent a novel class for topical treatment of inflammatory skin diseases. A subsequent extensive medicinal chemistry effort finally resulted in the selection of the ascomycin derivative pimecrolimus for development, due to its favorable pharmacology and safety profile. Since 2001/2 the new class of topical calcineurin inhibitors represented by pimecrolimus cream and tacrolimus ointment has become the first and only alternative to topical corticosteroids for treatment of inflammatory skin diseases such as atopic dermatitis.

Biological properties and clinical data of pimecrolimus are summarized. This compound differs from corticosteroids by selective action on T- and mast cells only, and by the lack of induction of skin atrophy and by less permeation through the skin. Pimecrolinus differs from other calcineurin inhibitors by specifically targeting skin inflammation, by a lower potential to affect systemic immune responses, and by its favorable skin penetration properties. Pimecrolimus cream 1% (Elidel®) has proven to be well tolerated, safe, and highly effective in extensive clinical studies in patients with atopic dermatitis and other inflammatory skin diseases.

Medicinal chemistry efforts aiming at understanding the structure-activity relationships required a detailed study of the chemical properties of the highly complex macrolactam structure of ascomycin and its derivatives, and elaboration and establishment of methodologies for selective transformations in various regions of the molecule. This chapter summarizes for the first time the chemical investigations in our laboratories and discusses the work published by others in this field as well. Furthermore, certain structure-activity relationships are described qualitatively.

Ascomycin (**2**) and tacrolimus (**1**) both exist as a mixture of several isomers in the macrophilin-binding region. Potential equilibrium products including the "furano-ascomycins" **13a**, **13b**, **13c**, and **13d** have been synthesized. The tricarbonyl moiety is highly sensitive and undergoes a benzilic acid-type of rearrangement, giving two diastereomers through two different pathways depending on the reaction conditions. Based on this, protocols were developed for the synthesis of labeled ascomycins. A new class of derivatives termed "cyclo-ascomycins" **44–47**, arising through cyclization in the binding domain, has been synthesized. In addition to other selective transformations in the binding region, photochemical investigations and further elaboration of the intermediates led to the biologically active 5,6-dehydro-ascomycin **99** and several other analogues. Using cyanide or 9-BBN, selective cleavage of C-1–C-9 was accomplished and semisynthetic strategies starting from the resulting fragment **122** led to the new analogues **127–130** featuring new amino acids. In the effector side of the molecule, several selective transformations such as allylic oxidation on C-18, epimerization of C-21, enumeration of the C-21–allyl side chain through *Grubbs*' cross metathesis reactions, and dehydration of C-24–OH leading to the enone 23,24-dehydroascomycin (**134**) were achieved. The protected enone **144** could be further transformed through selective addition reactions. Whereas the 21-*epi* derivative **133** did not show useful activities, 24-desoxyascomycin (**135**) was as active as ascomycin (**2**). On the cyclohexyl part of the molecule, demethylations, or introduction of additional hydroxy groups led to the more water-soluble derivatives **158** and **159**. Extensive derivatization of the C-33–OH has been achieved. Noteworthy is the ring contraction leading to SDZ 281–240, which was the first topical calcineurin inhibitor to demonstrate clinical proof of concept in patients with inflammatory skin disease. The cyclohexyl-methylidene group was cleaved off and new moieties could be incorporated instead. Furthermore, *iso*-tacrolimus (**197a**), a ring-contracted derivative, also isolated from the fermentation broths, has been synthesized and its further chemistry summarized.

In conclusion, this chapter summarizes the extensive chemistry and biology studies on a natural product, which have resulted in a novel therapy approved worldwide. This helps underscore the importance of natural products as a versatile source of novel structures with unique biological activities.

Acknowledgments The authors thank Dr. *Anthony Winiski* for his assistance with the literature search and Dr. *Philipp Floersheim* for his support in the molecular modeling.

References

1. Goto T, Kino T, Hatanaka H, Okuhara M, Kohsaka M, Aoki H, Imanaka H (1991) FK506: Historical Perspectives. Transplant Proc **23**: 2713

2. Goto T, Kino T, Hatanaka H, Nishiyama M, Okuhara M, Kosaka M, Aoki H, Imanaka H (1987) Discovery of FK-506, a Novel Immunosuppressant Isolated from *Streptomyces tsukubaensis*. Transplant Proc **19**: 4

3. Tanaka H, Kuroda A, Marusawa H, Hashimoto M, Hatanaka H, Kino T, Goto T, Okuhara M (1987) Physicochemical Properties of FK-506, a Novel Immunosuppressant Isolated from *Streptomyces tskubaensis*. Transplant Proc **19** (Suppl 6): 11

4. Arai T, Koyama Y, Suenaga T, Honda H (1962) Ascomycin as Antifungal Antibiotic. J Antibiotics **15** (Ser. A): 231

5. Hatanaka H, Iwami M, Kino T, Goto T, Okuhara M (1988) FR-900520 and FR-900523, Novel Immunosuppressants Isolated from a *Streptomyces*. I. Taxonomy of the Producing Strain. J Antibiotics **41**: 1586

6. Hatanaka H, Kino T, Miyata S, Inamura N, Kuroda A, Goto T, Tanaka H, Okuhara M (1988) FR-900520 and FR-900523, Novel Immunosuppressants Isolated from a *Streptomyces*. II. Fermentation, Isolation and Physicochemical and Biological Characteristics. J Antibiotics **41**: 1592

7. Morisaki M, Arai T (1992) Identity of Immunosuppressant FR-900520 with Ascomycin. J Antibiotics **45**: 126

8. Borris RP, Wicker LS, Zink DL, Lien T (1990) Eur Pat Appl EP 358508; Chem Abstr **113**: 38913

9. Fleissner G, Haecker H, Kuesters E, Penn G (2001) PCT Int Appl WO 0190110; Chem Abstr **136**: 4783

10. Junker B, Reddy J, Olewinski R, Gailliot P, Byrne K, Gbewonyo K (1998) Secondary Metabolite Scale-Up to Minimize Homolog Impurity Levels. Biotechnol Bioengineer **59**: 595

11. Hatanaka H, Kino T, Asano M, Goto T, Tanaka H, Okuhara M (1989) FK-506 Related Compounds Produced by *Streptomyces tsukubaensis* No. 9993. J Antibiotics **42**: 620

12. Chen TS, Arison BH, Wicker LS, Inamine ES, Monaghan RL (1992) Microbial Transformation of Immunosuppressive Compounds. I. Desmethylation of FK 506 and Immunomycin (FR 900520). J Antibiotics **45**: 118

13. Haag M, Baumann K, Billich A, Bulusu M, Fehr T, Grassberger M, Haidl E, Schulz G, Sanglier JJ (1998) Bioconversion of Ascomycin and its 6-Alkoxy Derivatives. J Mol Catal B: Enzymat **5**: 389

14. Arison BH, Inamine ES, Chen SST, Wicker LS (1990) Eur Pat Appl EP 378321; Chem Abstr **115**: 112812

15. Shafiee A, Kaplan L, Dumont F, Colwell LF, Arison BH (1992) Can Pat Appl CA 2071066; Chem Abstr **118**: 253402
16. Schueler W, Christians U, Schmieder P, Schiebel HM, Holze I, Sewing KF, Kessler H (1993) 159. Structural Investigations of 13-*O*-Desmethyl-FK506 and its Isomers Generated by *in vitro* Metabolism of FK506 Using Human Liver Microsomes. Helv Chim Acta **76**: 2288
17. Revill WP, Voda J, Reeves CR, Chung L, Schirmer A, Ashley G, Carney JR, Fardis M, Carreras CW, Zhou Y, Feng L, Tucker E, Robinson D, Gold BG (2002) Genetically Engineered Analogs of Ascomycin for Nerve Regeneration. J Pharmacol Exp Ther **302**: 1278
18. Vezina C, Kudelski A, Sehgal SN (1975) Rapamycin (AY-22,989), a New Antifungal Antibiotic. I. Taxonomy of the Producing Streptomycete and Isolation of the Active Principle. J Antibiotics **28**: 721
19. Sehgal SN, Baker H, Vezina C (1975) Rapamycin (AY-22,989), a New Antifungal Antibiotic. II. Fermentation, Isolation and Characterization. J Antibiotics **28**: 727
20. Martel RR, Klicius J, Galet S (1977) Inhibition of the Immune Response by Rapamycin, a New Antifungal Antibiotic. Can J Physiol Pharmacol **55**: 48
21. Meiser BM, Wang J, Morris RE (1989) Rapamycin: A new and highly active immunosuppressive macrolide with an efficacy superior to cyclosporine. In: Melchers F (ed) Progress in Immunology, vol VII: Proceedings of the 7th International Congress of Immunology, Springer-Verlag Berlin Heidelberg, p 1195
22. Morris RE, Meiser BM, Wu J, Shorthouse R, Wang J (1991) Use of Rapamycin for the Suppression of Alloimmune Reactions *in vivo*: Schedule Dependence, Tolerance Inductions, Synergy with Cyclosporine and FK 506, and Effect of Host-Versus-Graft and Graft-Versus-Host Reactions. Transplant Proc **23**: 521
23. Fehr T, Sanglier JJ, Schuler W, Gschwind L, Ponelle M, Schilling W, Wioland C (1996) Antascomycins A, B, C, D and E. Novel FKBP12 Binding Compounds from a *Micromonospora* Strain. J Antibiotics **49**: 230
24. Meingassner JG, Stuetz A (1992) Immunosuppressive Macrolides of the Type FK506: A Novel Class of Topical Agents for Treatment of Skin Diseases? J Invest Dermatol **98**: 851
25. Rappersberger K, Meingassner JG, Födinger D, Putz E, Fialla R, Tong D, Stuetz A, Wolff K (1996) Clearing of Psoriasis by a Novel Macrolide. J Invest Dermatol **106**: 701
26. Stuetz A, Grassberger MA, Baumann K, Edmunds AJF, Hiestand P, Meingassner JG, Nussbaumer P, Schuler W, Zenke G (1993) Immunophilins as Drug Targets. In: Testa B, Kyburz E, Fuhrer W, Giger R (eds) Perspectives in Medicinal Chemistry. Verl Helvetica Chimica Acta, Basel, p 427
27. Grassberger M, Meingassner J, Stuetz A, Wolff K (1998) Ascomycins. In: Roenigk HH, Maibach HI (eds) Psoriasis, 3rd edn. Marcel Dekker, Inc. New York, Basel, Hong Kong, Chapter 66, p 769
28. Stuetz A, Baumann K, Grassberger M, Wolff K, Meingassner JG (2006) Discovery of Topical Calcineurin Inhibitors and Pharmacological Profile of Pimecrolimus. Int Arch Allergy Immunol **141**: 199
29. Grassberger M, Baumruker T, Enz A, Hiestand P, Hultsch T, Kalthoff F, Schuler W, Schulz M, Werner FJ, Winiski A, Wolff B, Zenke G (1999) A Novel Anti-inflammatory Drug, SDZ ASM 981, for the Treatment of Skin Diseases: *in vitro* Pharmacology. Br J Dermatol **141**: 264
30. Kalthoff FS, Winiski A, Fichtinger P, Schwendinger B, Wang S, Weishaeupl C, Stuetz A (2007) Differential Inhibition of Primary versus Pre-activated T Cells by Pimecrolimus but not by Tacrolimus *in vitro*. Int Arch Allergy Immunol **142**: 255
31. Hultsch T, Müller KD, Meingassner JG, Grassberger M, Schopf RE, Knop J (1998) Ascomycin Macrolactam Derivative SDZ ASM 981 Inhibits the Release of Granule Associated Mediators and of Newly Synthesized Cytokines in RBL 2 H3 Mast Cells in an Immunophilin-Dependent Manner. Arch Dermatol. Res **290**: 501
32. Kalthoff F, Chung J, Musser P, Stuetz A (2003) Pimecrolimus does not Affect the Differentiation, Maturation and Function of Human Monocyte-Derived Dendritic Cells, in Contrast to Corticosteroids. Clin Exp Immunol **133**: 350

33. Büchau A, Schauber J, Hultsch T, Stuetz A, Gallo R (2008) Pimecrolimus Enhances Antimicrobial Peptide Expression in Keratinocytes. J Invest Dermatol **128**: 2646
34. Billich A, Aschauer H, Aszody A, Stuetz A (2004) Percutaneous Absorption of Drugs Used in Atopic Eczema: Pimecrolimus Permeates Less Through Skin than Corticosteroids and Tacrolimus. Int J Pharmaceutics **269**: 29
35. Meingassner JG, Aschauer H, Stuetz A, Billich A. (2005) Pimecrolimus Permeates Less than Tacrolimus Through Normal, Inflamed or Corticosteroid Pre-treated Skin. Exp Dermatol **14**: 752
36. Meingassner JG, Grassberger M, Fahrngruber H, Moore HD, Schuurman H, Stütz A (1997) A Novel Anti-inflammatory Drug, SDZ ASM 981, for the Topical and Oral Treatment of Skin Diseases: *in vivo* Pharmacology. Br J Dermatol **137**: 568
37. Meingassner JG, Kowalsky E, Schwendinger H, Elbe-Buerger A, Stuetz A (2003) Pimecrolimus does not Affect Langerhans Cells in Murine Epidermis, in Contrast to Corticosteroids. Br J Dermatol **149**: 853
38. Hoetzenecker W, Meingassner JG, Ecker R, Stingl G, Stuetz A, Elbe-Buerger A (2004) Corticosteroids but not Pimecrolimus Affect Viability, Maturation and Immune Function of Murine Epidermal Langerhans Cells. J Invest Dermatol **122**: 673
39. Hoetzenecker W, Ecker R, Kopp T, Stuetz A, Stingl G, Elbe-Bürger A (2005) Pimecrolimus Leads to an Apoptosis-induced Depletion of T Cells but not Langerhans Cells in Patients with Atopic Dermatitis: Results from a Randomized, Double-blind, Vehicle-controlled Clinical Trial. J Allergy Clin Immunol **115**: 1276
40. Meingassner JG, Fahrngruber H, Bavandi A (2003) Pimecrolimus Inhibits the Elicitation Phase but does not Suppress the Sensitization Phase in Murine Contact Hypersensitivity, in Contrast to Tacrolimus and Cyclosporin A. J Invest Dermatol **121**: 77
41. Bavandi A, Fahngruber H, Aschauer A, Hartmann B, Meingassner JG, Kalthoff FS (2006) Pimecrolimus and Tacrolimus Differ in their Inhibition of Lymphocyte Activation During the Sensitization Phase of Contact Hypersensitivity. J Dermatol Sci **43**: 117
42. Stuetz A, Grassberger M, Meingassner JG (2001) Pimecrolimus (Elidel®, SDZ ASM 981) – Preclinical Pharmacological Profile and Skin Selectivity. Sem Cutan Med Surg **20**: 233
43. Nell B, Walde I, Billich A, Vit P, Meingassner JG (2005) The Effect of Topical Pimecrolimus on Keratoconjunctivitis Sicca and Chronic Superficial Keratitis in Dogs: Results from an Exploratory Study. Vet Ophthal **8**: 39
44. Hiestand PC, Feifel R, Stuetz A (2003) Pimecrolimus (SDZASM981, Elidel®) is Highly Effective in Two Rat Models of Rheumatoid Arthritis. Inflammat Res (Suppl. **2**): S 157
45. Moore H, Tyldesley M, Martland A, Edjlalipour M (2003) Pimecrolimus Protects Mice from Developing T Cell-induced Inflammatory Bowel Disease without Systemic Immunosuppression, in Contrast to Tacrolimus or Cyclosporine A. J Invest Dermatol **121**, Abstr. 868
46. Wolff K, Stuetz A (2004) Pimecrolimus for the Treatment of Inflammatory Skin Disease. Expert Opin Pharmacother **5**: 643
47. Grassberger M, Steinhoff M, Schneider D, Luger TA (2004) Pimecrolimus – an Anti-inflammatory Drug Targeting the Skin. Exp Dermatol **13**: 721
48. Spergel JM (2009) Pimecrolimus Cream in the Management of Patients with Atopic Eczema. Clin Cos Invest Dermatol **2**: 85
49. Erchen J, Sunderkoetter C, Luger T, Steinhoff M (2008) Calcineurin Inhibitors for Treatment of Atopic Dematitis. Expert Opin Pharmacother **9**: 3009
50. Luger T, Paul C (2007) Potential New Indications of Topical Calcineurin Inhibitors. Dermatology **215** (Suppl 1): 45
51. Hebert AA (2006) Review of Pimecrolimus Cream 1% for Treatment of Mild to Moderate Atopic Dermatitis. Clin Ther **28**: 1972
52. Rappersberger K, Komar M, Ebelin ME, Scott G, Burtin P, Greig G, Kehren J, Chibout SD, Holter W, Richter L, Oberbauer R, Cordier A, Stuetz A, Wolff K (2002) Pimecrolimus Identifies a Common Genomic Anti-inflammatory Profile, is Clinically Highly Effective in Psoriasis and is Well Tolerated. J Invest Dermatol **119**: 876
53. Van Duyne GD, Standaert RF, Karplus PA, Schreiber SL, Clardy J (1991) Atomic Structure of FKBP-FK506, an Immunophilin-Immunosuppressant Complex. Science **252**: 839

54. Lepre CA, Thomson JA, Moore JM (1992) Solution Structure of FK506 bound to FKBP-12. FEBS Lett **302**: 89
55. Van Duyne GD, Standaert RF, Karplus P, Schreiber SL, Clardy J (1993) Atomic Structures of the Human Immunophilin FKBP-12 Complexes with FK506 and Rapamycin. J Mol Biol **229**: 105
56. Griffith JP, Kim J, Kim EE, Sintchak MD, Thomson JA, Fitzgibbon MJ, Fleming MA, Caron PR, Hsiao K, Navia MA (1995) X-Ray Structure of Calcineurin Inhibited by the Immunophilin-Immunospressant FKBP12-FK506 Complex. Cell **82**: 507
57. Kissinger CR, Parge HE, Knighton DR, Lewis CT, Pelletier LA, Tempczyk A, Kalish VJ, Tucker KD, Showalter RE, Moomaw EW (1995) Crystal Structures of Human Calcineurin and the Human FKBP12-FK506-Calcineurin Complex. Nature **378**: 641
58. Liu J, Farmer JD, Lane WS, Friedman J, Weissman I, Schreiber SL (1991) Calcineurin is a Common Target of Cyclophilin-Cyclosporin A and FKBP-FK506 Complexes. Cell **66**: 807
59. Motamedi H, Shafiee A, Cai SJ, Streicher SL, Arison BH, Miller RR (1996) Characterization of Methyltransferase and Hydroxylase Genes Involved in the Biosynthesis of the Immunosuppressants FK506 and FK520. J Bacteriol **178**: 5243
60. Karuso P, Kessler H, Mierke DF (1990) Solution Structure of FK506 from Nuclear Magnetic Resonance and Molecular Dynamics. J Am Chem Soc **112**: 9434
61. Meadows RP, Nettesheim DG, Xu RX, Olejniczak ET, Petros AM, Holzman TF, Severin J, Gubbins E, Smith H, Fesik SW (1993) Three-Dimensional Structure of the FK506 Binding Protein/Ascomycin Complex in Solution by Heteronuclear Three- and Four-Dimensional NMR. Biochemistry **32**: 754
62. Namiki Y, Kihara N, Koda S, Hane K, Yasuda T (1993) Tautomeric Phenomenon of a Novel Immunosuppressant (FK506) in Solution. I. Isolation and Structure Determination of Tautomeric Compounds. J Antibiotics **46**: 1149
63. Gailliot FP, Natishan TK, Ballard JM, Reamer RA, Kuczynski D, McManemin GJ, Egan RS, Buckland BC (1994) Isolation and Characterization of the Major Equilibrium Product of FK-520. J Antibiotics **47**: 806
64. Baumann K, Oberhauser B, Grassberger MA, Haidl G, Schulz G (1995) Synthesis and Oxidative Cleavage of the Major Equilibrium Products of Ascomycin and FK506. Tetrahedron Lett **36**: 2231
65. Baumann K, Oberhauser B, Strnadt G, Knapp H, Schulz G, Grassberger MA (1999) Conversion of Ascomycin into its Furano-Isomers. Synlett **S1**: 877
66. Baumann K, Damont A, Högenauer K, Steck A (2002) Poster presented at the 1st International SPECS and BioSPECS Symposium, Delft, the Netherlands, September 18–21, Abstract book: 64
67. Shevchenko VP, Nagaev IYu, Myasoedov NF, Andres H, Moenius T, Susan A (2004) Synthesis of Tritiated Cyclosporin A and FK506 by Metal-Catalyzed Hydrogen Isotope Exchange. J Lab Comp Radiopharm **47**: 407
68. Baumann K, Haidl E, Högenauer K, Kollenz G, Schulz G, Weber HP (1996) Poster presented at the 17th European Colloquium on Heterocyclic Chemistry, Regensburg, Germany, October 6–9, Abstract book: 211
69. Högenauer K, Diploma work (1997) Department of Organic Chemistry, Karl Franzens University, Graz, Austria
70. Baumann K, Högenauer K, Knapp H, Bacher M, Steck A, Wagner T (2005) On the Reactivity of Ascomycin at the Binding Domain. Part 3: Reactivity of the Binding Domain towards Diazomethane. Tetrahedron **61**: 4819
71. Baumann K, *et al.* Novartis Institutes for BioMedical Research Vienna, Austria, unpublished results
72. Edmunds AJF, Baumann K, Grassberger M, Schulz G (1991) Highly Selective Reactions of FK506 with Diazomethane. Tetrahedron Lett **32**: 7039
73. Fisher MJ, Chow K, Villalobos A, Danishefsky SJ (1991) On the Remarkable Propensity for Carbon-Carbon Bond Cleavage Reations in the C_8-C_{10} Region of FK-506. J Org Chem **56**: 2900

74. Grassberger MA, Baumann K (1993) Macrolide Immunosuppressants. Exp Opin Ther Pat **3**: 931
75. Askin D, Reamer RA, Joe D, Volante RP, Shinkai I (1989) Chemistry of FK-506: Benzilic Acid Rearrangement of the Tricarbonyl System. Tetrahedron Lett **30**: 671
76. Askin D, Jones TK, Reamer RA, Volante RP, Shinkai I (1990) Eur Pat Appl EP 364031; Chem Abstr **113**: 152155
77. Baumann K, Bacher M, Damont A, Högenauer K, Steck A (2003) On the Reactivity of Ascomycin at the Binding Domain. Part 1: Liberation of the Tricarbonyl Portion of Ascomycin. Tetrahedron **59**: 10075
78. Baumann K, Bacher M, Steck A, Wagner T (2004) On the Reactivity of Ascomycin at the Binding Domain. Part 2: Hydroxide Mediated Rearrangement Reactions. Tetrahedron **60**: 5965
79. Baumann K (1993) Eur Pat Appl EP 569337; Chem Abstr **122**: 187265
80. Koch G, Jeck R, Hartmann O, Kuesters E (2001) Selective Synthesis of a New Ascomycin Rearrangement Product (SDZ ASD732) on a Pilot Plant Scale. Org Proc Res Devel **5**: 211
81. Wang-Fan W, Kusters E, Mak CP, Wang Y (2000) Application of Centrifugal Counter-Current Chromatography to the Separation of Macrolide Antibiotic Analogues. II. Determination of Partition Coefficients in Comparison with the Shake-Flask Method. J Liq Chomatogr Rel Technol **23**: 1365
82. Kusters E, Heuer C, Wieckhusen D (2000) Purification of an Ascomycin Derivative with Simulated Moving Bed Chromatography. A Case Study. J Chromatogr A **874**: 155
83. Baumann K, Bacher M, Damont A, Steck A (2004) Selective Transformation of Ascomycin into 11-*epi*-Ascomycin. Tetrahedron Lett **45**: 549
84. Baumann K, Strnadt G, Knapp H, Schulz G, Haidl E, Weber HP (1996) Poster presented at the 17th European Colloquium on Heterocyclic Chemistry, Regensburg, Germany, October 6–9, Abstract book: 167
85. Baumann K, Meingassner J, Schulz G, Steck A, Grassberger M (2000) Lecture given at the XVIth International Symposium on Medicinal Chemistry, Bologna, Italy, September 18–22, Abstract book: 103
86. Emmer G, Weber-Roth S (1992) Synthesis of Derivatives of FK506 and FR900520: Modifications at the Binding Domain. Tetrahedron **48**: 5861
87. Nussbaumer P, Grassberger M, Schulz G (1992) C9-Imino and C10-Amino Derivatives of Ascomycin (21-Ethyl FK506). Tetrahedron Lett **33**: 3845
88. Baumann K, Knapp H, Strnadt G, Schulz G, Grassberger MA (1999) Carbonyl to Methylene Conversions at the Tricarbonyl Portion of the Ascomycin Derivatives. Tetrahedron Lett **40**: 7761
89. Nussbaumer P, Grassberger MA, Hiestand P, Schuler W, Schulz G, Zenke G (1992) Poster presented at the XIIth International Symposium on Medicinal Chemistry, Basel, Switzerland, September 13–17, Abstract book: 347
90. Namiki Y, Kihara N (1992) PCT Int Appl WO 9213862; Chem Abstr **117**: 233717
91. Namiki Y, Kihara N (1992) Brit. Pat. Appl. GB 2249787; Chem. Abstr. **117**: 150803
92. Bulusu ARCM, Waldstaetten P, Tricotet T, Rochais C, Steck A, Bacher M, Schulz G, Meingassner JG, Hiestand P, Zenke G, Schuler W, Wagner T (2009) New Derivatives of Ascomycin with Modifications in the Amino Acid Region - Synthesis and Biological Activities and X-Ray Crystal Structure of 5,6-Dehydroascomycin. Helv Chim Acta **92**: 839
93. Bulusu ARCM, Haidl EH, Schulz G, Waldstaetten P, Grassberger M (1999) Novel Derivatives of Ascomycin through Photochemistry. Ind J Chem **38B**:1159
94. Bulusu, A.R.C.M. (1994) Eur Pat Appl EP 626385; Chem Abstr **122**: 213857
95. Bulusu M, Waldstaetten P, Schulz G, Grassberger M (2004) Novel Analogs of Ascomycin with Modifications in the Amino Acid through Photochemistry: the Synthesis of 5,6-Dehydroascomycin. Tetrahedron Lett **45**: 2523
96. Moenius T, Baumann K, Bulusu M, Schweitzer A, Voges R (2001) Labelling of Pharmacologically Active Macrocycles. In: Pleiss U, Voges R (eds) Synthesis and Applications of Isotopically Labelled Compounds, vol 7. John Wiley Sons, Chichester, UK, p 424

97. Bulusu M, Waldstaetten P, Tricotet T, Rochais C, Steck A, Bacher M (2004) Synthesis of 6-Vinyl and 5-Vinylproline Analogues of Ascomycin. Tetrahedron Lett **45**: 5833
98. Jones TK, Askin D, Mills SG, Reamer RA, Desmond R, Volante RP, Tschaen DM, Shinkai I (1990) Eur Pat Appl EP 378318; Chem Abstr **114**: 81436
99. Coleman RS, Danishefsky SJ (1989) Degradation and Manipulations of the Immunosuppressant FK506: Preparation of Potential Synthetic Intermediates. Heterocycles **28**: 157
100. Horvath A, Grassberger MA, Haidl E, Schulz G (2001) Lecture given at the Fall Meeting of the Swiss Chemical Society, October 12, Zürich, Switzerland
101. Goulet MT, Mills SG, Parsons WH, Rupprecht KM, Wyvratt MJ (1993) Chemistry of FK506. In Lukacs G (ed) Recent Progress in the Chemical Synthesis of Antibiotics and Related Microbial Products. Springer, Berlin, p 141
102. Luengo JI, Rozamus LW, Holt DA (1993) Efficient Removal of Pipecolinate from Rapamycin and FK506 by Reaction with n-$Bu_4N^+CN^-$. Tetrahedron Lett **34**: 4599
103. Cooper ME, Donald DK, Tanaka H (1991) Eur Pat Appl EP 444829; Chem Abstr **115**: 279490
104. Askin D, Reamer RA, Joe D, Volante RP, Shinkai I (1990) Synthesis of the Novel Sarcosine and Proline (FK525) Analogs of FK-506: Rearrangement of the FK506 Allylic System. J Org Chem **55**: 5448
105. Bulusu M, Waldstaetten P, Tricotett T, Schulz G (2004) Selective Photochemical Cleavage of an α-Ketoamide in a Highly Functionalised Macrolide Ascomycin. Tetrahedron Lett **45**: 2527
106. Skytte DM, Frydenvang K, Hansen L, Nielsen PG, Jaroszewski JW (2010) Synthesis and Characterization of an Epimer of Tacrolimus, an Immunosuppressive Drug. J Nat Prod **73**: 776
107. Wiedeman PE, Fesik SW, Petros AM, Nettesheim DG, Mollison KW, Lane BC, Or YS, Luly JR (1999) Retention of Immunosuppressant Activity in an Ascomycin Analogue Lacking a Hydrogen-Bonding Interaction with FKBP 12. J Med Chem **42**: 4456
108. Clemons PA, Gladstone BG, Chao ED, Seth A, Chao ED, Foley MA, Schreiber SL (2002) Synthesis of Calcineurin-Resistant Derivatives of FK506 and Selection of Compensatory Receptors. Chem Biol **9**: 49
109. Marinec PS, Evans CG, Gibbons GS, Tarnowski MA, Overbeek DL, Gestwicki JE (2009) Synthesis of Orthogonally Reactive FK506 Derivatives via Olefin Cross Metathesis. Bioorg Med Chem **17**: 5763
110. Kawai M, Lane BC, Hsieh GC, Mollison KW, Carter GW, Luly JR (1993) Structure-Activity Profiles of Macrolactam Immunosuppressant FK506 Analogues. Fed Eur Biochem Soc **316**: 107
111. Horvath A, Grassberger MA, Schulz G, Haidl E, Sperner H, Steck A (2000) Regioselective Carbon-Carbon Bond Formation in the Effector Domain of Ascomycin. Tetrahedron **56**: 7469
112. Nussbaumer P, Grassberger MA, Schulz G (1999) Synthesis of Ascomycin and FK 506 Derivatives with Modified Effector Domain. Tetrahedron Lett **40**: 3869
113. Or YS, Clark RF, Xie Q, McAlpine J, Whittern DN, Henry R, Luly JR (1993) The Chemistry of Ascomycin: Structure Determination and Synthesis of Pyrazole Analogues. Tetrahedron **49**: 8771
114. Kawai M, Gunawardana IWK, Mollison KW, Hsieh GC, Lane BC, Luly JR (1998) Studies on an Immunosuppressive Macrolactam, Ascomycin: Synthesis of a C-33 Hydroxyl Derivative. Bioorg Med Chem Lett **8**: 935
115. Ok HO, Szumiloski JL, Beattie TR, Goulet MT, Strauch MJ, Dumont FJ, Wyvratt MJ (1997) C32-Amino Derivatives of the Immunosuppressant Ascomycin. Bioorg Med Chem Lett **7**: 2199
116. Baumann K, Emmer G (1991) Can Pat. Appl CA 2029694; Chem. Abstr 116; 128509
117. Baumann K, Emmer G (1994) US Pat Appl US 5352671; Chem Abstr 122: 187265
118. Baumann K, Emmer G (1991) Eur Pat Appl EP 427680; Chem Abstr 115: 231991

119. Hersperger R, Schuler W, Zenke G (1999) Preparation and Immunosuppressive Activity of 32-(*O*)-Acylated and 32-(*O*)-Thioacylated Analogues of Ascomycin. Bioorg Med Chem Lett **9**: 227
120. Goulet MT, Hodkey DW, Staruch MJ, Dumont FJ, Cryan JG, Parsons WH, Wyvratt MJ (1994) Alkyl Ether Analogs of the FK506 Related Immunosuppressive Macrolide L-683,590 (Ascomycin). Bioorg Med Chem Lett **4**: 921
121. Goulet MT, McAlpine SR, Staruch MJ, Koprak S, Dumont FJ, Cryan JG, Wiederrecht GJ, Rosa R, Wilusz MB, Peterson LB, Wyvratt MJ, Parsons WH (1998) C(32)-*O*-Imidazol-2-yl-methyl Ether Derivatives of the Immunosuppressant Ascomycin with Improved Therapeutic Potential. Bioorg Med Chem Lett **8**: 2253
122. Song Z, DeMarco A, Zhao M, Corley EG, Thompson AS, McNamara J, Li Y, Rieger D, Sohar P, Mathre DJ, Tschaen DM, Reamer RA, Huntington MF, Ho G-J, Tsay F-R, Emerson K, Shuman R, Grabowski EJJ, Reider PJ (1999) Highly Chemoselective Trichloroacetimidate-Mediated Alkylation of Ascomycin: A Convergent, Practical Synthesis of the Immunosuppressant L-733,725. J Org Chem **64**: 1859
123. Egan MAMcE, Dean DC, Marks TM, Song Z, Melillo DG (2000) Carbon-14 Labeling of a Potential New Immunoregulant Agent. J Lab Comp Radiopharm **43**: 1095
124. Goulet MT, Sinclair PJ, Wong F, Staruch MJ, Dumont FJ, Cryan JG, Wiederrecht GJ, Wyvratt MJ, Parsons WH (1999) C32-*O*-Phenalkyl Ether Derivatives of the Immunosuppressant Ascomycin: A Tether Length Study. Bioorg Med Chem Lett **9**: 2085
125. Sinclair PJ, Wong F, Wyvratt M, Staruch MJ, Dumont F (1995) Preparation and *in vitro* Activity of Aryl Ether Derivatives of the FK506 Related Immunosuppressive Macrolides Ascomycin and L-683,742. Bioorg Med Chem Lett **5**: 1035
126. Armstrong HM, Wong F, Holmes MA, Sinclair PJ, Goulet MT, Dumont FJ, Staruch MJ, Koprak S, Peterson LB, Rosa R, Wilusz MB, Wiederrecht GJ, Cryan JG, Wyvratt MJ, Parsons WH (1999) Potent Immunosuppressive C32-*O*-Arylether Derivatives of Ascomycin with Reduced Toxicity. Bioorg Med Chem Lett **9**: 2089
127. Sinclair PJ, Wong F, Staruch MJ, Wiederrecht G, Parsons WH, Dumont F, Wyvratt M (1996) Preparation and *in vitro* Activities of Naphthyl and Indolyl Ether Derivatives of the FK506 Related Immunosuppressive Macrolide Ascomycin. Bioorg Med Chem Lett **6**: 2193
128. Wagner R, Rhoades TA, Or YS, Lane BC, Hsieh G, Mollison KW, Luly JR (1998) 32-Ascomycinyloxyacetic Acid Derived Immunosuppressants. Independence of Immunophilin Binding and Immunosuppressive Potency. J Med Chem **41**: 1764
129. Mollison KW, Fey TA, Gauvin DM, Sheets MP, Smith ML, Pong M, Krause R, Miller L, Or YS, Kawai M, Wagner R, Wiedeman PE, Clark RF, Gunawardana IWK, Rhoades TA, Henry CL, Tu NP, BaMaung NY, Kopecka H, Liu L, Xie Q, Lane BC, Trevillyan JM, Marsh K, Carter GW, Chen Y-W, Hsieh GC, Luly JR (1998) Discovery of Ascomycin Analogs with Potent Topical but Weak Systematic Activity for Treatment of Inflammatory Skin Diseases. Curr Pharm Design **4**: 367
130. Hersperger R, Buchheit KH, Cammisuli S, Enz A, Lohse O, Ponelle M, Schuler W, Schweitzer A, Walker C, Zehender H, Zenke G, Zimmerlin AG, Zollinger M, Mazzoni L, Fozard JR (2004) A Locally Active Antiinflammatory Macrolide (MLD987) for Inhalation Therapy of Asthma. J Med Chem **47**: 4950
131. Zimmer R, Grassberger MA, Baumann K, Schulz G, Haidl E (1994) Synthetic Modifications of Ascomycin – I. A Chemoselective Removal of the Cyclohexyl Residue of Ascomycin. Tetrahedron **50**: 13655
132. Zimmer R, Baumann K, Sperner H, Schulz G, Haidl E, Grassberger MA (2005) Synthetic Modifications of Ascomycin. V. Access to Novel Ascomycin Derivatives by Replacement of the Cyclohexylvinylidene Subunit. Croatica Chem Acta **78**: 17
133. Grassberger MA, Schulz G, Fehr T (1990) Ger Offen DE 3938754; Chem Abstr 115: 157124
134. Grassberger MA, Fehr T, Horvath A, Schulz G (1992) Isolation of an Isomer of FK-506 from Fermentation of *Streptomyces tsukubaensis* and its Chemical Synthesis from FK-506. Tetrahedron **48**: 413

135. Zimmer R, Baumann K, Sperner H, Schulz G, Haidl E, Grassberger MA (1999) Ring Contraction of an Ascomycin Derivative to a 19-Membered Macrolactam. Helv Chim Acta **82**: 1038

136. Zimmer R, Grassberger MA, Baumann K, Horvath A, Schulz G, Haidl E (1995) Synthetic Modifications of Ascomycin – II. A Simple and Efficient Way to Modified *iso*-Ascomycin Derivatives. Tetrahedron Lett **36**: 7635

137. Zimmer R, Grassberger MA, Baumann K, Zenke G, Schuler W (1999) Synthetic Modifications of Ascomycin: Part III – A Concise Transformation of *iso*-Ascomycin to 19,20-*seco*- Derivatives. Ind J Chem **38B**: 831

138. Chung Y, Cho H (2004) Preparation of Highly Water Soluble Tacrolimus Derivatives: Poly (Ethylene Glycol) Esters as Potential Prodrugs. Arch Pharm Res **27**: 878

139. Choi DB, Cho H (2009) Effect of Tacrolimus Derivatives on Immunosuppression. Arch Pharm Res **32**: 549

140. Yura H, Yoshimura N, Oka T, Takakura Y, Hashida M (1998) Development of a Macromolecular Prodrug of FK506: I. Synthesis of FK506-Dextran Conjugate. Transplant Proc **30**: 3598

141. Yura H, Yoshimura N, Hamashima T, Akamatsu K, Nishikawa M, Takakura Y, Hashida M (1999) Synthesis and Pharmacokinetics of a Novel Macromolecular Prodrug of Tacrolimus (FK506), FK506-dextran Conjugate. J Cont Release **57**: 87

142. Koch K, Newborg MF, Hanson DC, Cooper K, Shepard RM, Biehl ML, Biggers MS, Ramchandani M, Schulte G, Snyder JR, Ferraina RA, Donovan C, Guadliana MA, Kostek GJ, Cole SH, Connolly MJ, Sawyer PS, I T-P, Blocker LW, Meiser BM, Melvin LS (1995) The C-32 Triacetyl-L-rhamnose Derivative of Ascomycin: A Potent, Orally Active Macrolactone Immunosuppressant. J Med Chem **38**: 1255

143. Takahashi T, Shiyama T, Mori T, Hosoya K, Tanaka A (2006) Isolating the Whole Complex of Target Proteins of FK506 Using Affinity Resins from Novel Solid Phases. Anal Bioanal Chem **385**: 122

144. Takahashi T, Shiyama T, Hosoya K, Tanaka A (2006) Development of Chemically Stable Solid Phases for the Target Isolation with Reduced Nonspecific Binding Proteins. Bioorg Med Chem Lett **16**: 447

145. Fujisawa; Klinge Pharma (1992) FK506, Fujimycin, Tacrolimus, Prograf. Drugs Fut **17**: 732

146. Ireland RE, Liu L, Roper TD (1997) Total Synthesis of FK506. Part 1: Construction of the C (16)-C(34) Fragment. Tetrahedron **53**: 13221

147. Ireland RE, Liu L, Roper TD, Gleason JL (1997) Total Synthesis of FK506. Part 2: Completion of the Synthesis. Tetrahedron **53**: 13257

148. Ireland RE, Gleason JL, Gegnas LD, Highsmith TK (1996) A Total Synthesis of FK506. J Org Chem **61**: 6856

149. Smith III AB, Condon SM, McCauley JA (1998) Total Synthesis of Immunosuppressants: Unified Strategies Exploiting Dithiane Couplings and σ-Bond Olefin Constructions. Acc Chem Res **31**: 35

150. Smith III AB, Chen K, Robinson DJ, Laakso LF, Hale KJ (1994) Formal Total Synthesis of FK506. Concise Construction of the C(10)-C(34) Segment via an Effective Coupling Tactic. Tetrahedron Lett **25**: 4271

151. Batchelor MJ, Gillespie RJ, Golec JMC, Hedgecock CJR, Jones SD, Murdoch R (1994) Total Synthesis of Close Analogues of the Immunosuppressant FK506. Tetrahedron **50**: 809

152. Nakatsuka M, Ragan JA, Sammakia T, Smith DB, Uehling DE, Schreiber SL (1990) Total Synthesis of FK506 and an FKBP Probe Reagent, (C_8,C_9-$^{13}C_2$)-FK506. J Am Chem Soc **112**: 5583

153. Jones TK, Mills SG, Reamer RA, Askin D, Desmond R, Volante RP, Shinkai I (1989) Total Synthesis of Immunosuppressant (−)-FK506. J Am Chem Soc **111**: 1157

154. Jones TK, Reamer RA, Desmond R, Mills SG (1990) Chemistry of Tricarbonyl Hemiketals and Application of Evans Technology to the Total Synthesis of the Immunosuppressant (−)-FK506. J Am Chem Soc **112**: 2998

155. Jones AB, Villalobos A, Linde RG, Danishefsky SJ (1990) A Formal Synthesis of FK506: Exploration of Some Alternatives to Macrolactamisation. J Org Chem **55**: 2786
156. Reynolds KA, Demain AL (1997) Rapamycin, FK506 and Ascomycin-related Compounds. In: Strohl WR (ed) Drugs and the Pharmaceutical Sciences, vol 82 (Biotechnology of Antibiotics, 2nd ed), Marcel Dekker, Inc., New York, p 497

Withanolides and Related Steroids

Rosana I. Misico, Viviana E. Nicotra, Juan C. Oberti, Gloria Barboza, Roberto R. Gil, and *Gerardo Burton*

Contents

R.I. Misico • G. Burton (✉)
Departamento de Química Orgánica and UMYMFOR (CONICET-UBA), Facultad de Ciencias Exactas y Naturales, Universidad de Buenos Aires, Ciudad Universitaria, Pabellón 2, Buenos Aires C1428EGA, Argentina
e-mail: misicori@qo.fcen.uba.ar; burton@qo.fcen.uba.ar

V.E. Nicotra • J.C. Oberti
Departamento de Química Orgánica and IMBIV (CONICET), Facultad de Ciencias Químicas, Universidad Nacional de Córdoba, Ciudad Universitaria, Ciencias Químicas II, Córdoba 5000, Argentina
e-mail: vnicotra@mail.fcq.unc.edu.ar; jco@mail.fcq.unc.edu.ar

G. Barboza
Departamento de Farmacia and IMBIV (CONICET), Facultad de Ciencias Químicas, Universidad Nacional de Córdoba, Ciudad Universitaria, Córdoba 5000, Argentina
e-mail: gbarboza@imbiv.unc.edu.ar

R.R. Gil
Department of Chemistry, Carnegie Mellon University, 4400 Fifth Ave, Pittsburgh, PA 15213, USA
e-mail: rgil@andrew.cmu.edu

A.D. Kinghorn, H. Falk, J. Kobayashi (eds.), *Progress in the Chemistry of Organic Natural Products, Vol. 94*, DOI 10.1007/978-3-7091-0748-5_3,

1. Introduction

Since the isolation of the first withanolides in the mid 1960s (*1, 2*), *ca.* 650 members of this family of compounds have been described, with most of these from genera of the Solanaceae. The basic structure of withaferin A (**1**), a C_{28} ergostane with a modified side chain forming a δ-lactone between carbons 22 and 26, was considered for many years the basic template for withanolides (Fig. 1). Nowadays this has given place to a considerable number of related structures that are considered part of the withanolide family. Withanolides have shown to possess many biological activities that include anti-inflammatory, antitumor, cytotoxic, immunomodulating, and cancer chemopreventive as well as antibacterial and antifungal properties. Extracts from *Withania somnifera*, known by its Sanskrit name "Ashwagandha" or "Indian ginseng", have been used for over 3,000 years in traditional medicine in India (Ayurvedic system). Many of the curative properties of this extract may be

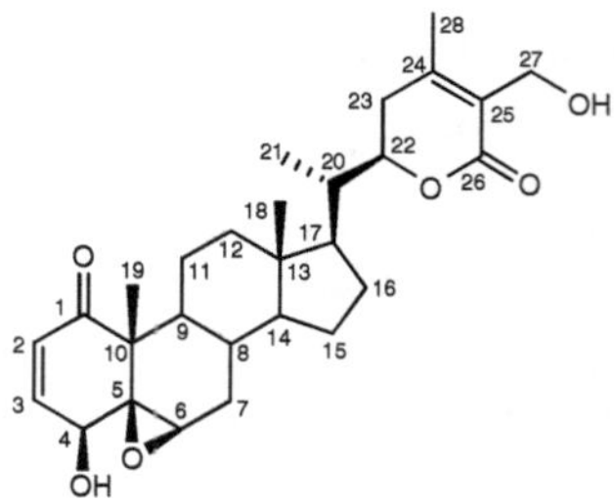

1 (withaferin A)

Fig. 1. Structure of withaferin A (**1**) and numbering scheme

associated with the presence of withanolides. Several withanolides also exhibit insecticidal activities, mostly as feedant deterrents, and selective phytotoxicity. A number of review articles have dealt with structural aspects of this group of compounds (*3–5*) as well as with their biological activities (*6*, *7*). The present chapter covers developments and findings in the chemistry and bioactivity of withanolides and related compounds since 1994.

2. Withanolides in the Plant Kingdom

2.1. Solanaceous Genera Containing Withanolides

The Solanaceae, also known as the nightshade or potato family, is one of the largest flowering plant families in containing nearly 100 genera and *ca.* 2,500 species, with a worldwide distribution in temperate and tropical continents (*8–10*). However, they are much more diversified in the Andean/Amazonian regions of South America in habitats that vary dramatically. The Solanaceae is also the third most important plant taxon economically and the most valuable in terms of vegetable crops, including the tuber-bearing potato, a number of fruit-bearing vegetables (tomato, eggplant, peppers), ornamental plants (petunias, *Nicotiana*, *Nierembergia*), plants with edible leaves (*Solanum aethiopicum*, *S. macrocarpon*), and medicinal plants (*e.g. Atropa*, *Capsicum*, *Datura*).

A modern traditional classification of the Solanaceae, predominantly based on morphological evidence with a minor emphasis on chemistry, recognizes six subfamilies (*9*). However, a recently proposed phylogenetic classification of the Solanaceae provides a different framework to the morphological system, with the inclusion of genera traditionally excluded from the family and with a greater resolution among lineages within the subfamily Solanoideae (*10*).

Among the nearly 100 genera included in Solanaceae (*10*), the occurrence of withanolides is restricted to the subfamily Solanoideae. Table 1 summarizes all genera and species known to contain withanolides to date. The four major contributors of withanolide structures are the genera *Jaborosa* Juss., *Datura* L., and *Physalis* L. from the North and South American continents, and *Withania* Pauq., native to the Old World. Up to the present, *ca.* 50% of the species in these genera have been investigated. *Jaborosa* is an interesting South American genus with a much varied corolla odor, color, and morphology as an adaptation to different pollinators growing from southern Peru to Argentina in very diverse habitats. *Datura* comprises annual or short-lived perennial herbs, mostly with erect large flowers, common in semiarid habitats in Mexico and the southwest of the United States, but introduced in many countries. *Physalis* includes mostly American herbs (Central America, Mexico and United States, except for the Euroasian *P. alkekengi*), with solitary pendant yellow or white flowers, and fruiting calyxes that become enlarged and inflated. Finally, *Withania* is a well-known genus

Table 1. Genera and species of the Solanaceae containing withanolides

Genera[a]	Species	Previous names[b]
Acnistus (1/1)	*arborescens* (L.) Schltdl.[c]	*Acnistus ramiflorus* Miers
Brachistus (2/4)	*stramoniifolius* (Kunth) Miers[c]	
	hunzikeri (D'Arcy) Sousa-Peña[c]	*Witheringia hunzikeri*
Datura (6/*ca.* 11)	*ferox* L.[c]	
	inoxia Mill.[c]	
	metel L.[c]	
	metel var. *fastuosa* (L.) Saff.[c]	*Datura fastuosa* L.
	quercifolia Kunth[c]	
	stramonium L.	
	stramonium var. *tatula* (L.) Torr.[c]	*Datura tatula* L.
Deprea (1/7)	*orinocensis* (Kunth) Raf.[c]	
Discopodium (1/1)	*penninervium* Hochst.[c]	
Dunalia (2/5)	*brachyacantha* Miers[c]	
	solanacea Kunth[c]	
Eriolarynx (1/3)	*lorentzii* (Dammer) Hunz.[c]	*Vassobia lorentzii* Dammer; *Acnistus lorentzii* (Dammer) Hunz.
Exodeconus (1/6)	*maritimus* (Benth.) D'Arcy[c]	
Hyoscyamus (1/17)	*niger* L.[c]	
Iochroma (4/*ca.* 25)	*australe* Griseb.[c]	*Acnistus australis* (Griseb.) Griseb.; *Dunalia australis* (Griseb.) Sleumer
	coccineum Scheid.	
	fuchsioides Miers	
	gesneroides Miers[c]	
Jaborosa (13/23)	*araucana* Phil.[c]	
	bergii Hieron.[c]	
	caulescens var. *bipinnatifida* (Dunal) Reiche[c]	
	caulescens Gillies & Hook. var. *caulescens*[c]	
	integrifolia Lam.	
	kurtzii Hunz. & Barboza[c]	
	laciniata (Miers) Hunz. & Barboza[c]	*Trechonaetes laciniata* Miers
	lanigera (Phil.) Hunz. & Barboza[c]	
	leucotricha (Speg.) Hunz. [c]	
	magellanica (Griseb.) Dusén	
	odonelliana Hunz.[c]	
	rotacea (Lillo) Hunz. & Barboza[c]	
	runcinata Lam.[c]	
	sativa (Miers) Hunz. & Barboza[c]	
Larnax (2/*ca.* 30)	*glabra* (Standl.) Sawyer[c]	
	subtriflora (Ruiz & Pav.) Miers[c]	*Deprea subtriflora* Ruiz & Pav.
Lycium (2/*ca.* 80)	*chinense* Mill.	
	barbarum L.	*Lycium halimifolium* Mill.

(continued)

Table 1. (continued)

Genera[a]	Species	Previous names[b]
Nicandra (1/2)	*physalodes* (L.) Gaertn.	
Margaranthus (1/1)	*solanaceous* Schltdl.[c]	*Physalis solanaceous* (Schltdl.) Axelius
Physalis (15/*ca.* 90)	*alkenkegi* L.[c]	
	alkenkegi var. *franchetii* (Mast.) Makino[c]	
	angulata L.[c]	
	chenopodifolia Lam.[c]	
	cinerascens (Dunal) Hitchc.[c]	
	coztomatl Dunal[c]	
	divaricata D. Don[c]	
	lanceifolia Nees	
	minima L.[c]	
	peruviana L.[c]	
	philadelphica Lam.[c]	*Physalis ixocarpa* Brot. ex Hornem.
	pruinosa L.	
	pubescens L.	
	virginiana Mill.[c]	
	viscosa L.	*Physalis curassavica* L.
Salpichroa (1/16)	*origanifolia* (Lam.) Thell.[c]	
Schraderanthus (1/1)	*viscosus* (Schrad.) Averett[c]	*Saracha viscosa* Schrad. *Leucophysalis viscosa* (Schrad.) Hunz.
Solanum (2/*ca.* 1,500)	*ciliatum* Lam.[c]	*Solanum cilistum* Lam.
	sisymbriifolium Lam.[c]	
Tubocapsicum (1/2)	*anomalum* (Franch. & Sav.) Makino[c]	
Vassobia (1/2)	*breviflora* (Sendtn.) Hunz.[c]	*Acnistus breviflorus* Sendtn.
Withania (6/20)	*adpressa* Cors.[c]	
	aristata (Aiton) Pauq.[c]	
	coagulans (Stocks) Dunal[c]	*Withania coagulance*
	frutescens (L.) Pauq.	
	obtusifolia V. Tackh.	
	somnifera (L.) Dunal[c]	
Witheringia (2/12)	*coccoloboides* (Dammer) Hunz.[c]	
	solanacea L'Hér.[c]	

[a]In parenthesis are the number of species with withanolides/number of total species. [b]Previous names cited in literature. [c]Species included in this chapter

of perennial herbs or shrubs with flowers that are perfect or functionally imperfect, occurring in Europe and Asia.

Table 1 also includes four monotypic genera, *Acnistus* Schott (Southern Mexico to Eastern Brazil and Paraguay), *Margaranthus* Schltdl. (Southern United States to the Antilles), *Schraderanthus* Averett (Mexico to Guatemala), and *Discopodium* Hochst. (tropical Africa). In the remaining genera containing withanolides, only isolated species have been characterized phytochemically.

2.2. *Non-Solanaceous Genera Containing Withanolides*

Withanolides have been detected in six species belonging to different families (Table 2), including the rhizomes of species of *Tacca* J.R. Forst. & G. Forst. (family Dioscoreaceae, formerly the family Taccaceae), in the aerial parts of *Senna siamea* (family Leguminosae), in the bark of *Eucalyptus globulus* (family Myrtaceae), and in species of *Ajuga* L. (family Lamiaceae).

3. Classification of Withanolides

The withanolides are polyoxygenated steroids with a C_{28} ergostane skeleton. A common feature is the presence of oxygen atoms at C-1, C-22, and C-26 although a few exceptions with a non-functionalized C22 are included. They may be classified into two major groups depending on the arrangement of the side chain, those with a δ-lactone or δ-lactol comprising C-22 and C-26 and those with a γ-lactone usually involving C-23 and C-26.

3.1. *Withanolides with a δ-Lactone or δ-Lactol Side Chain*

Most of the known withanolides belong to this group, which may be further divided into 13 subgroups: withanolides with the parent skeleton of withaferin A (**1**), withaphysalins, physalins, acnistins, withajardins, withametelins, sativolides, subtriflora-δ-lactones, spiranoid-δ-lactones, norbornane-type withanolides, ring-D aromatic withanolides, ring-A aromatic withanolides, and taccalonolide-δ-lactones (Fig. 2). The δ-lactone formed between a carboxyl group at C-26 and a hydroxy at

Table 2. Genera containing withanolides from outside the Solanaceae

Family and genera	Species	Name cited[a]
Monocotyledoneae	*chantrieri* André[b]	
Fam. Dioscoreaceae		
Tacca	*paxiana* H. Limpr.[b]	
	plantaginea (Hance) Drenth[b]	
	subflabellata P.P. Ling & C.T. Ting[b]	*Tacca subflaellata*
Dicotyledoneae	*siamea* (Lam.) Irwin & Barneby	*Cassia siamea* Lam.
Fam. Leguminosae		
Senna		
Fam. Myrtaceae	*globulus* Labill.[b]	
Eucalyptus		
Fam. Lamiaceae	*bracteosa* Wall. ex Benth.[b]	
Ajuga	*parviflora* Benth.[b]	

[a]Names cited in literature. [b]Species included in this chapter

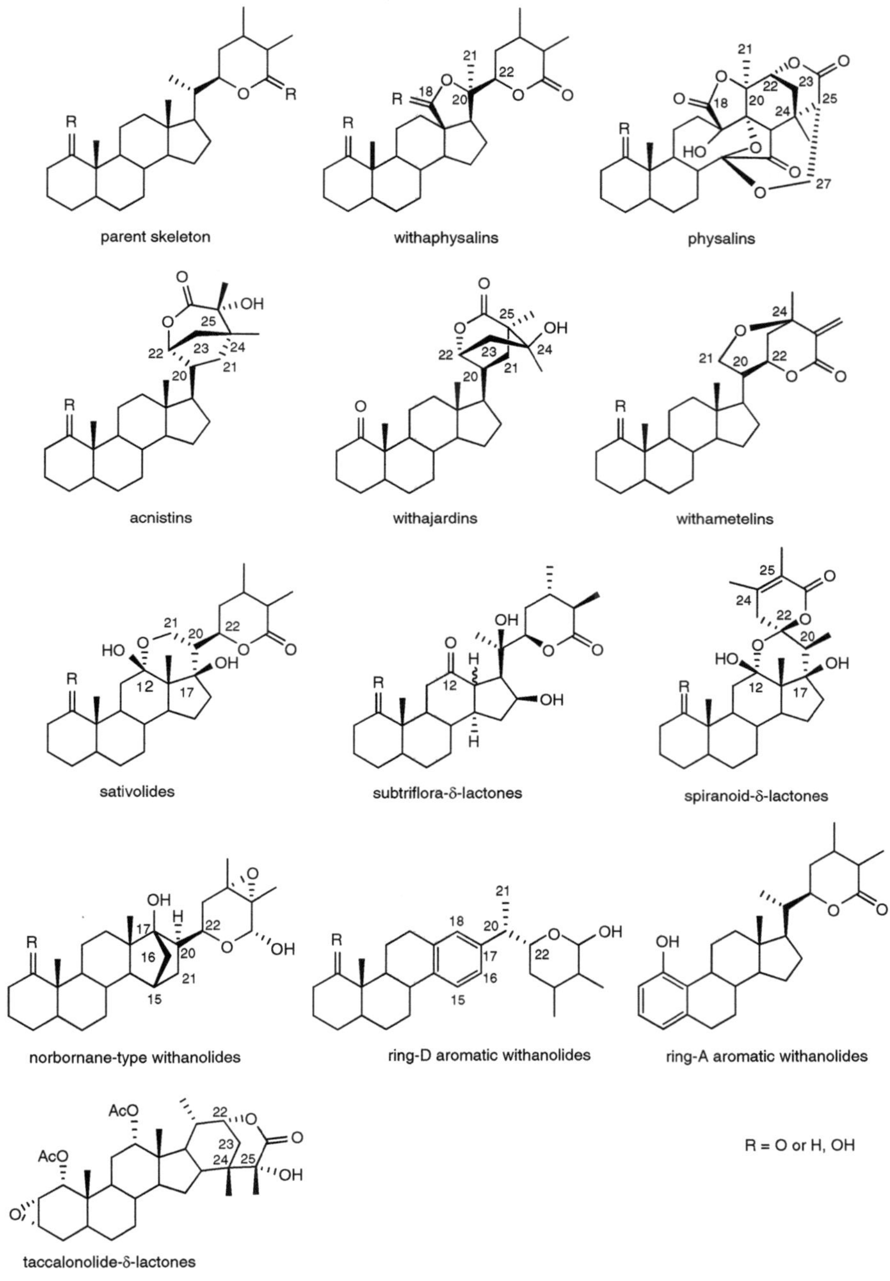

Fig. 2. General structures of withanolides with a δ-lactone or δ-lactol side chain. Numbering of relevant positions has been added for clarity in some structures

C-22 is the most common arrangement; other types include lactone formation with a hydrated carbonyl at C-22 and δ-lactols between an aldehyde at C-26 and a hydroxy at C-22. All known withanolides have the same stereochemistry at C-22, which corresponds to (22*R*) except when substituents at C-23 or C-22 change the relative priorities of groups around the asymmetric center.

Withanolides with an unmodified skeleton are the most abundant and are regarded as possible precursors of most of the other compounds in this group. A further subdivision of the withanolides is usually made according to the orientation of the side chain; those with the "normal" 17*β*-oriented side chain as well as those with the less usual 17α-oriented side chain are known. In the latter case a 17*β*-hydroxy (either free or involved in a cyclic ether) is generally present although several exceptions are known.

3.2. Withanolides with a γ-Lactone Side Chain

The presence of an oxygenated function at C-23 allows the formation of a γ-lactone with a carboxyl group at C-26; these withanolides may be divided in five subgroups, spiranoid withanolides, trechonolides, subtriflora-γ-lactones, ixocarpalactones, and taccalonolide-γ-lactones. A sixth subgroup with a γ-lactone side chain involving C-26 and C-28 corresponds to the perulactones (Fig. 3).

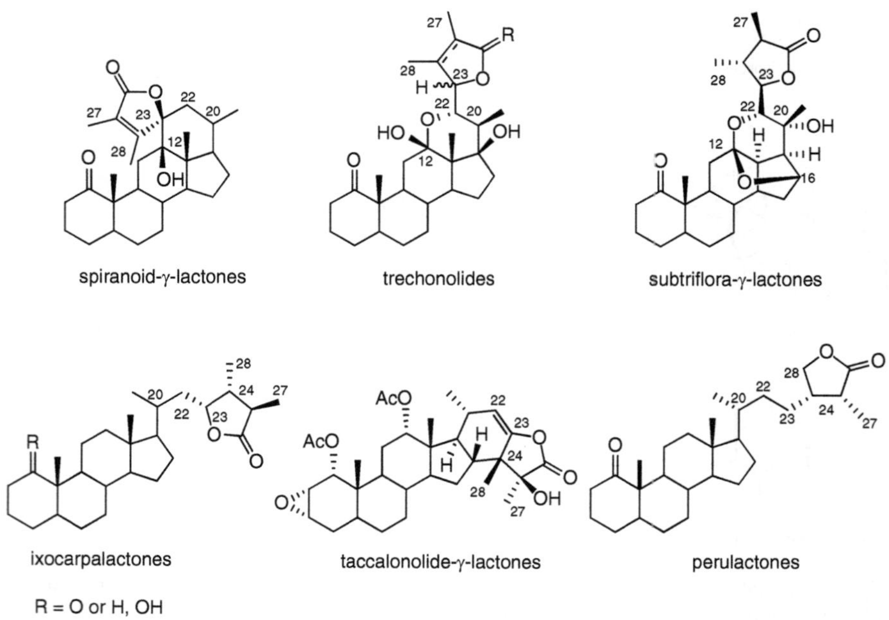

Fig. 3. General structures of withanolides with a γ-lactone side chain. Numbering of relevant positions has been added for clarity

4. Withanolides with an Unmodified Skeleton

Despite the large number of withanolides with the parent skeleton of withaferin A (**1**) that are already known, many new entities have been described with minor variations. These correspond mostly to different combinations of hydroxylated substituents and the occurrence of glycosidated derivatives. Among the many structures that fall within this group, the *Withania* withanolides comprise a major subgroup and are presented separately. Then, other representative substitutions on the parent skeleton will be described.

4.1. The Withania *Withanolides*

The *Withania* genus, although studied extensively in the past, has continued to provide new withanolide structures. Leaves, roots, and fruits of *W. somnifera* and *W. coagulans* have been investigated, with a total of 45 and 29 new structures reported, respectively. New withanolides have also been isolated from *W. adpressa* and *W. aristata*. Unfortunately, with several groups working simultaneously on the same plant, some structures have been reported as new more than once. As mentioned above, all of them conform to the classical withanolide skeleton with few unusual features.

4.1.1. 5*β*,6*β*-Epoxywithanolides and Related Compounds

The basic structure of withaferin A (**1**) may be found in several new structures isolated from *W. somnifera*, and simple variations are the 17α-hydroxy derivative **2** (*11*) and the Δ^{16} analogues **3** (*12*) and witharistatin (**4**) (*13*), with the latter isolated from *W. aristata*. The same basic structure may be found in **5** where the 2,3-double bond has been hydrated (*14*) and in **6** in which the double bond is reduced (*15*). The latter compound also has an unusual fragment etherifying the 4*β*-hydroxy group. One of the most common variations in the withaferin A substitution pattern is the hydrolytic cleavage of the 5,6-epoxide according to the *Fürst-Plattner* rule (*16*), to give the *trans*-diaxial 5α,6*β*-diol as in coagulin H (**7**) and coagulin S (**8**) isolated from *W. coagulans* (*17, 18*). A less common cleavage is that occurring in a *trans*-diequatorial manner, as found in **9** and **10** (*14*), later reported as new by *Kuroyanagi et al.* (*19*), and in 4-deoxywithaperuvin (**11**) (*20*). Compound **12** represents an unusual variant of the above, with a 3α,6α-epoxy bridge (*21*), which could derive from the cyclization of **10** also present in the same extract. While *Michael*-type addition of alcohols to the Δ^2-1-keto system of withanolides is well known, intramolecular addition is rare. Nevertheless, the same arrangement has been found previously in withaperuvins D (*22*) and F (*23*), although the authors did not directly compare the spectroscopic data of these substances.

Compounds **13** and **14** are two glycosides closely related to withaferin A (**1**). In the case of **13** the configuration of the hydroxy group at C-16 was not determined (*15*).

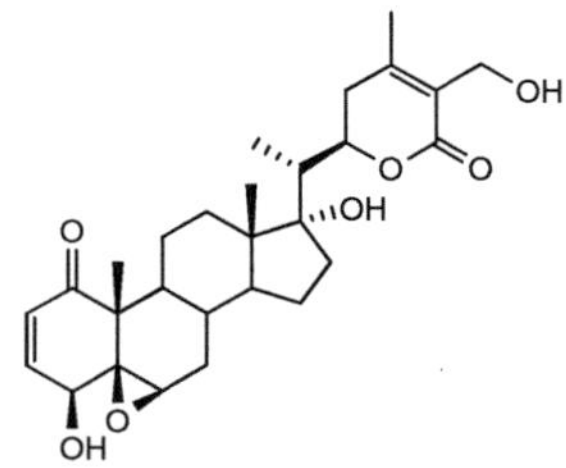

2 (17α-hydroxywithaferin A)

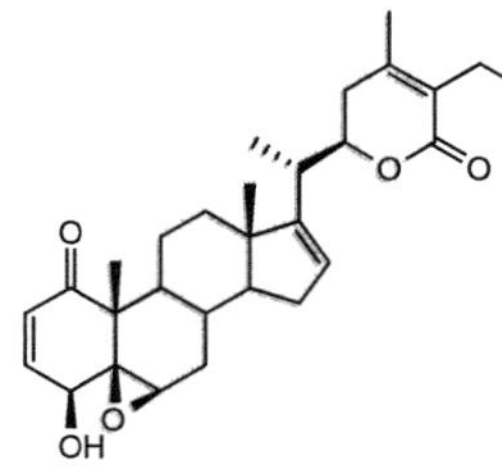

3 R = H
4 R = OH (witharistatin)

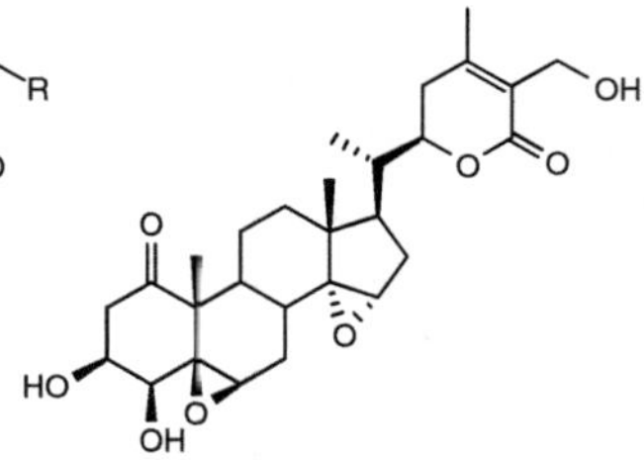

5 (withaoxylactone)

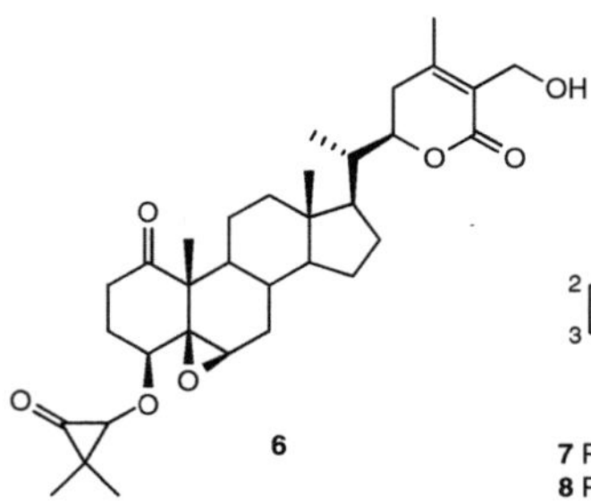

6

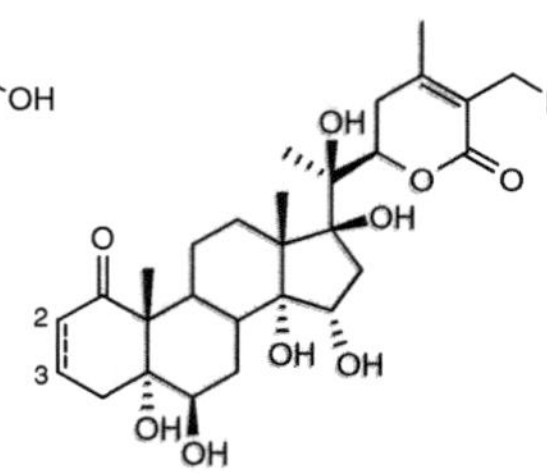

7 R = H, Δ^2 (coagulin H)
8 R = OH, 2,3-dihydro (coagulin S)

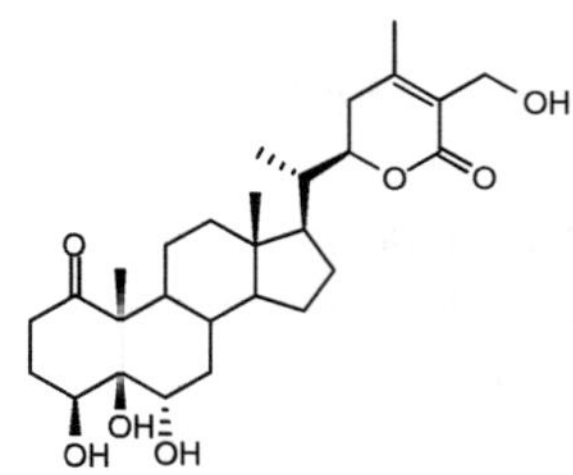

9 (somnifericin)

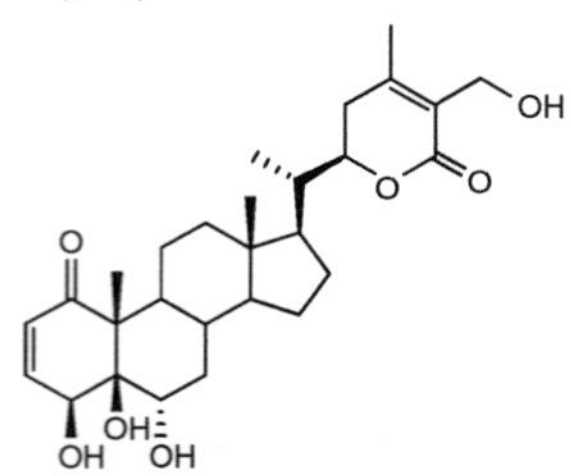

10 (2,3-dehydrosomnifericin)

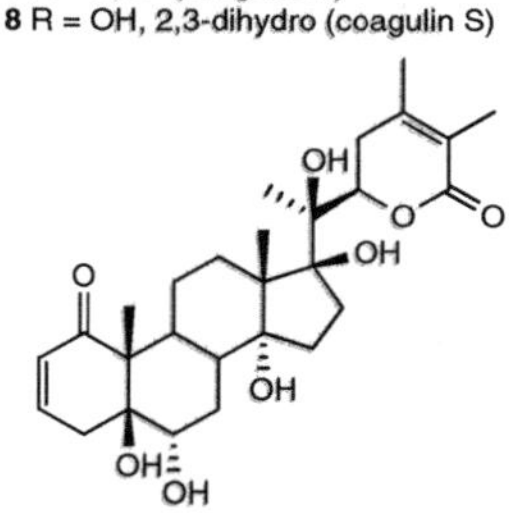

11 (4-deoxywithaperuvin)

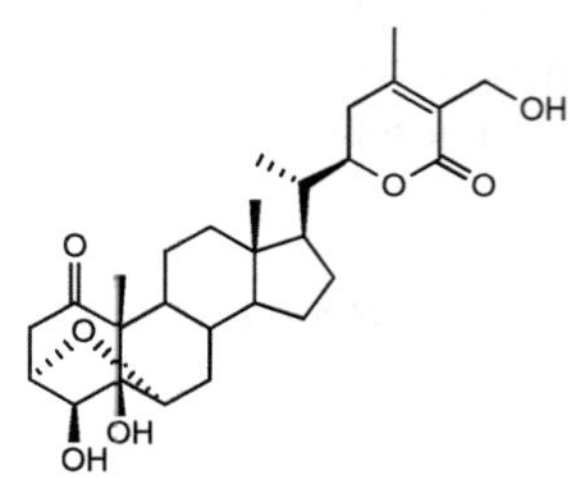

12

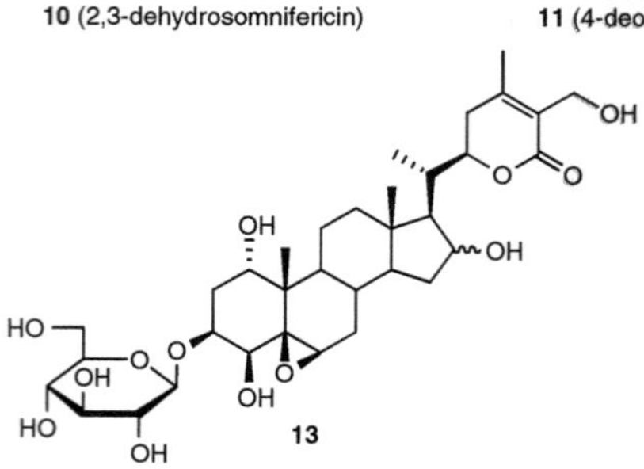

13

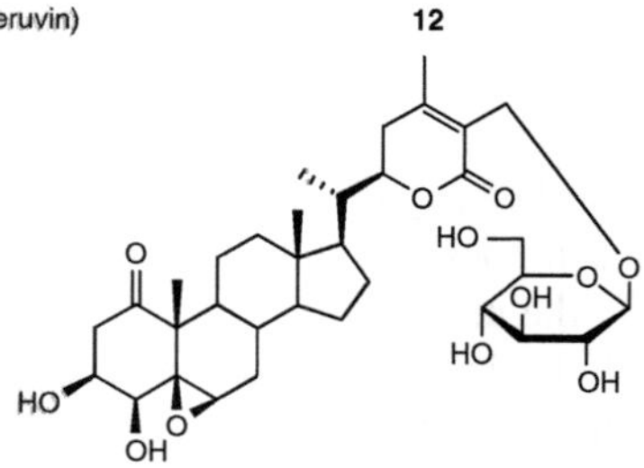

14

4.1.2. 5α-Hydroxy-6α,7α-Epoxywithanolides and Related Compounds

The 5α-hydroxy-6α,7α-epoxy substitution pattern is also a common arrangement found in *W. somnifera* withanolides. Simple variations arising from the combination of hydroxy substituents at both typical positions (C-14, C-17, C-20) and some less common ones such as C-16 or C-23 are observed in withasomniferols A (**15**) and B (**16**) (*24*), 14α,17α-dihydroxywithanolide R (**17**) (*20*), iso-withanone (**18**) (*25*), and the 16β-acetate **19** (*26*). Withasomniferol B (**16**) is a stereoisomer of the known ixocarpanolide, although the configurations at positions 24 and 25 were not specified. Hydration of the 2,3-double bond was found in compounds **20** (*11*), **21**, and **22** (*12*), and the latter was isolated as the 3-sulfate, an unusual feature in withanolides (see Sect. 4.2.5.). Other variations found were the reduced 1α-alcohol **23** (*25*) and the related glycosides withanoside II (**24**), withanoside I (**25**), and withanoside III (**26**) (*27*), with the 1α,3β-dihydroxy arrangement being fairly common among withanolides (see next section). Also, the chlorinated withanolide Z (**27**) (*28*), arising from *trans*-diaxial cleavage of the 6,7-epoxide and the Δ^7-withanolide **28** (*24*) are known. Compound **29** contains an unusual 5α,7α-epoxy bridge that would result from rearrangement of the 5α-hydroxy-6α,7α-epoxide (*26*).

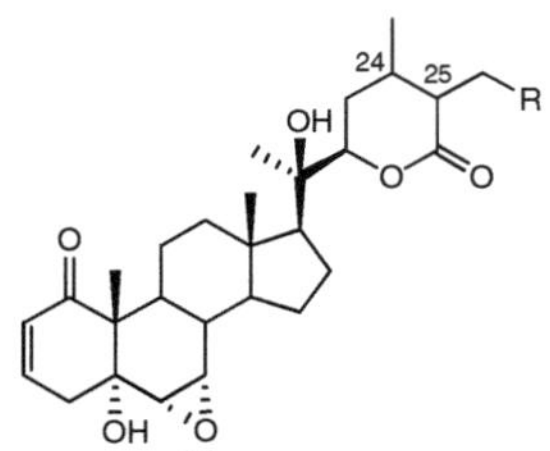

15 R = OH, Δ^{24} (withasomniferol A)
16 R = H (withasomniferol B)

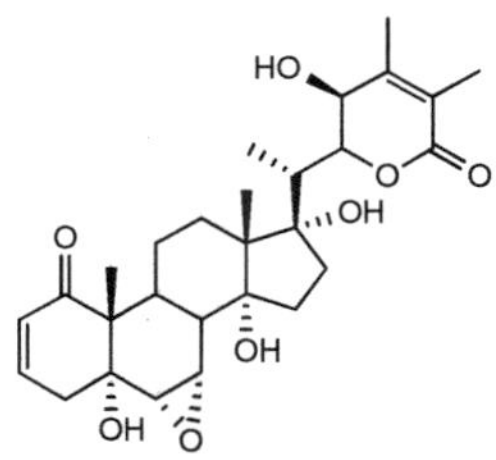

17 (14α,17α-dihydroxywithanolide R)

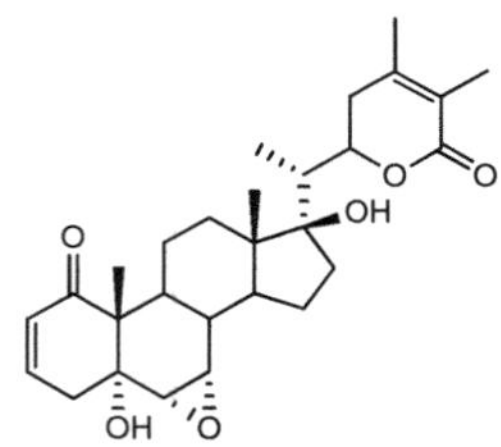

18 (iso-withanone)

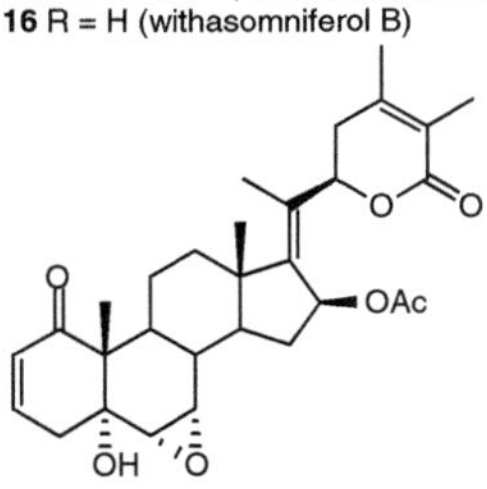

19

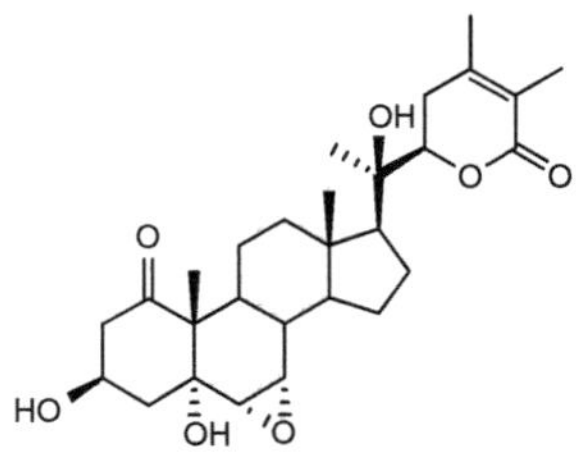

20

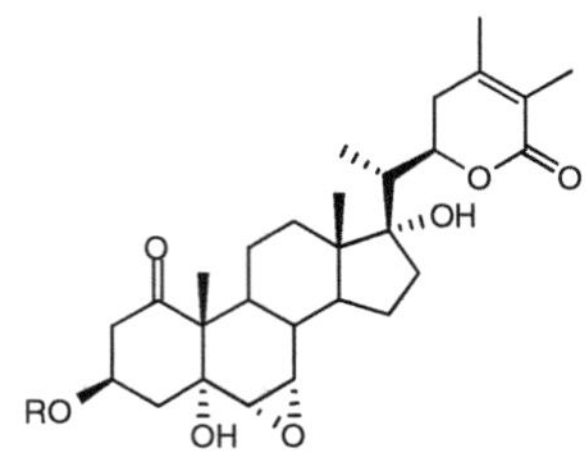

21 R = H
22 R = SO_3H

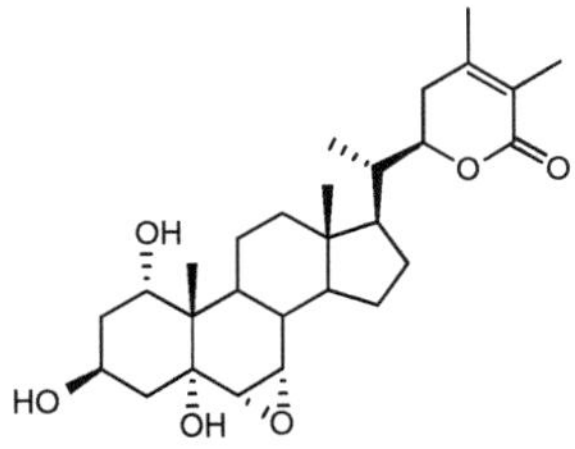

23

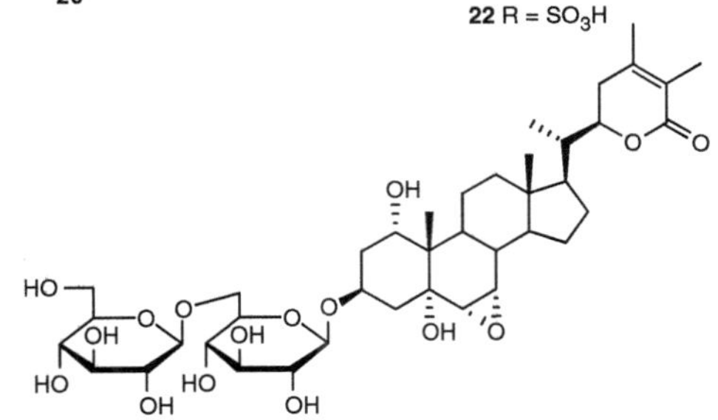

24 (withanoside II)

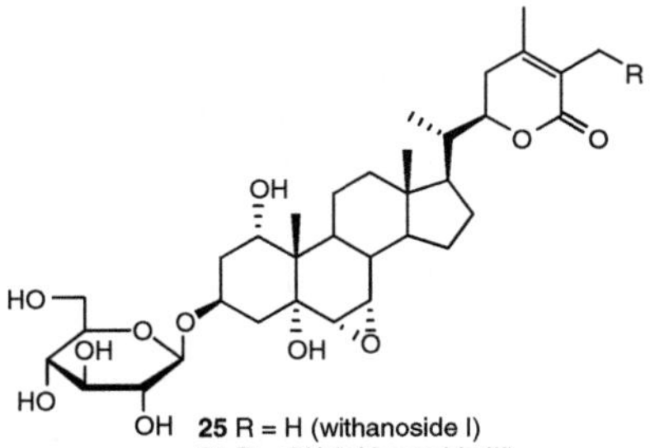

25 R = H (withanoside I)
26 R = OH (withanoside III)

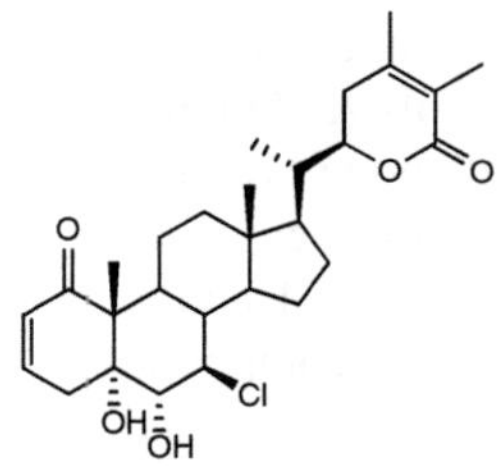

27 (withanolide Z)

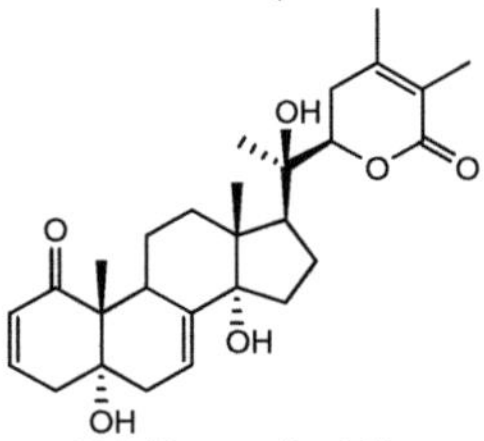

28 (withasomniferol C)

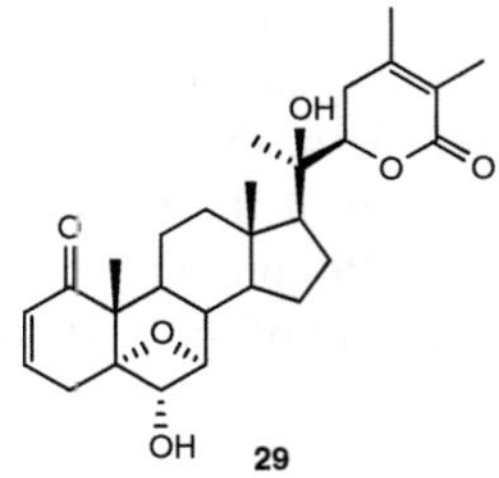

29

4.1.3. 1*α*,3*β*-Dihydroxy-Δ^5-Withanolides and Related Compounds

As mentioned above, the 1α,3*β*-dihydroxy arrangement is a well-known structural variation among the withanolides and several examples have been shown in combination with 5*β*,6*β*- or 6α,7α-epoxides. Eleven 3*β*-*O*-glycosides isolated from *W. somnifera* contained the 1α,3*β*-dihydroxy arrangement combined with a 5,6-double bond in ring B. These include compound **30**, later reported as withanoside V (*19, 27*), withanosides IV (**31**), VI (**32**) and VII (**33**) (*27*), and 24,25-dihydrowithanoside VI (**34**) containing a disacharide at position 3 (*29*). The configurations at positions 24 and 25 of the latter compound were not elucidated. Also within this group are three withanolides with an additional sugar moiety at C-27 (*21*), withanosides VIII (**35**), IX (**36**), and X (**37**) (this structure was reported again as new, one year later (*15*)), an analogue of withanoside VI with a monosacharide at C-3 named withanoside XI (**38**) (*21*), and an analogue of withanoside IV with a trisacharide moiety at C-3 (**39**) (*15*). Coagulin Q (**40**) isolated from *W. coagulans* (*30*), has the same aglycone as withanoside VI (**32**), but with a monosacharide unit at C-3. In all cases the carbohydrate units are *β*-D-glucose.

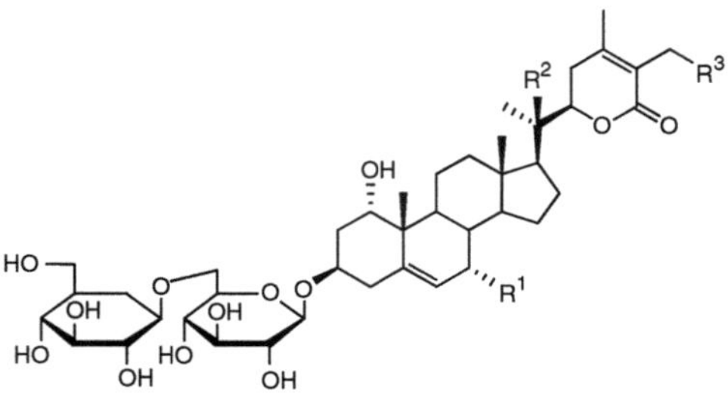

30 $R^1 = R^2 = R^3 = H$ (withanoside V)
31 $R^1 = R^2 = H$, $R^3 = OH$ (withanoside IV)
32 $R^1 = R^3 = H$, $R^2 = OH$ (withanoside VI)
33 $R^1 = OH$, $R^2 = R^3 = H$ (withanoside VII)
35 $R^1 = R^2 = H$, $R^3 = O$-β-gluc (withanoside VIII)
36 $R^1 = R^2 = H$, $R^3 = O$-β-gluc(1-6)-β-gluc (withanoside IX)

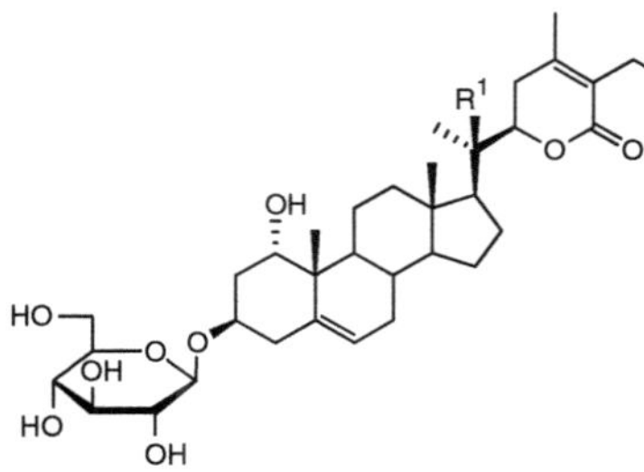

37 $R^1 = H$, $R^2 = O$-β-gluc (withanoside X)
38 $R^1 = R^2 = OH$ (withanoside XI)
40 $R^1 = OH$, $R^2 = H$ (coagulin Q)

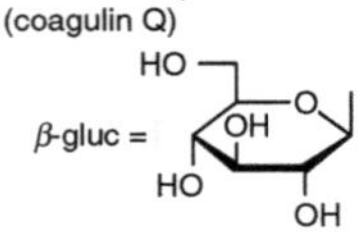

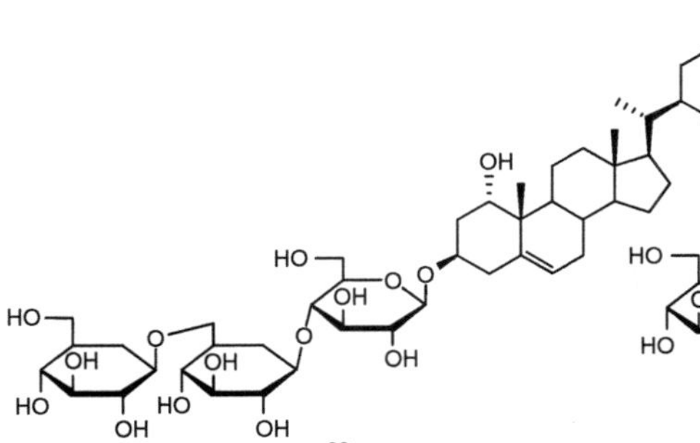

39

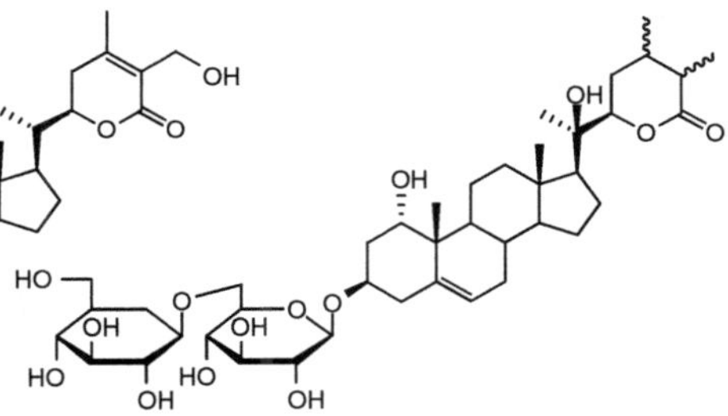

34 (24,25-dihydrowithanoside VI)

4.1.4. Other Δ^5-Withanolides

Withanolides containing a 5,6-double bond in combination with either a Δ^2- or Δ^3-1-ketone or a 3β-hydroxy-1-ketone are quite common and are the biosynthetic precursors of the 5β,6β-epoxywithanolides. Several new withanolides with these arrangements have been isolated from *Withania* species, mostly from *W. coagulans*. Compound **41** was isolated originally from *W. adpressa* (*31*) and later reported as new from *W. coagulans* (*32*). Compound **42** had been synthesized by *Lavie* and coworkers (*33*) but was isolated for the first time as a natural product by *Atta-ur-Rahman et al.* from *W. coagulans* (*34*). Ten withanolides with closely related structures were also isolated from this plant, withacoagulin (**43**) (*34*), withacoagulins A–F (**44**–**49**) (*35*), the 14β,15β-epoxide **50**, the 14α-alcohol **51** (*36*), and the 17-epimer of withacoagulin D, coagulansin A (**52**) (*37*). The 3β-hydroxy-Δ^5 arrangement is present in coagulansin B (**53**) (*37*) and in the three 3β-*O*-glycosides, coagulin L (**54**) (*17*), coagulin O (**55**) (*38*), and coagulin P (**56**) (*30*).

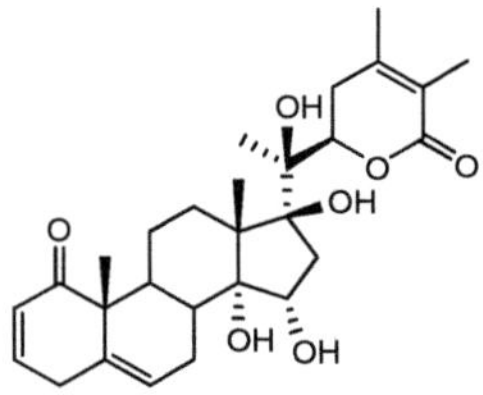

41

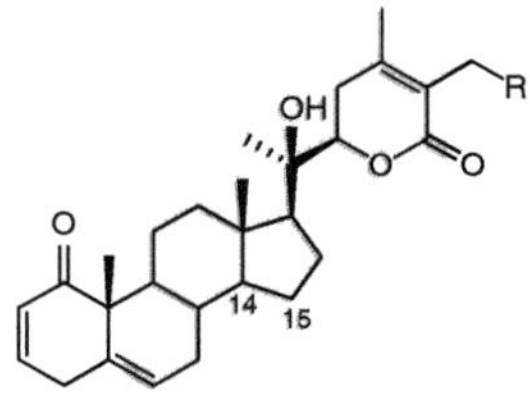

42 R = H
43 R = OH, Δ^{14} (withacoagulin)

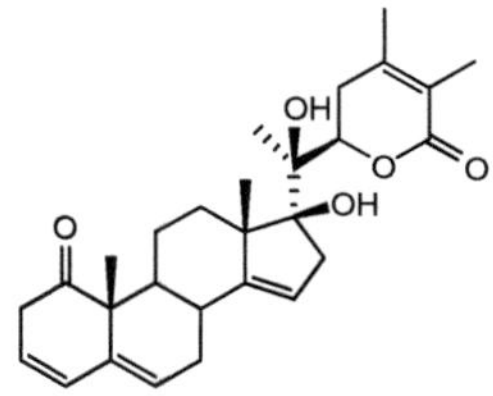

44 (withacoagulin A)

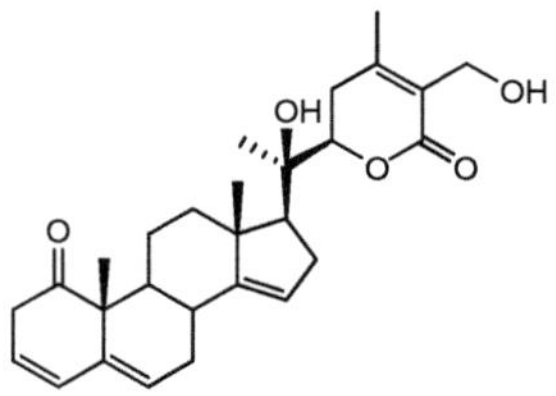

45 (withacoagulin B)

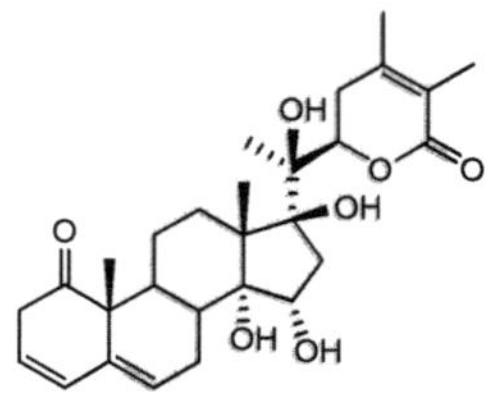

46 (withacoagulin C)

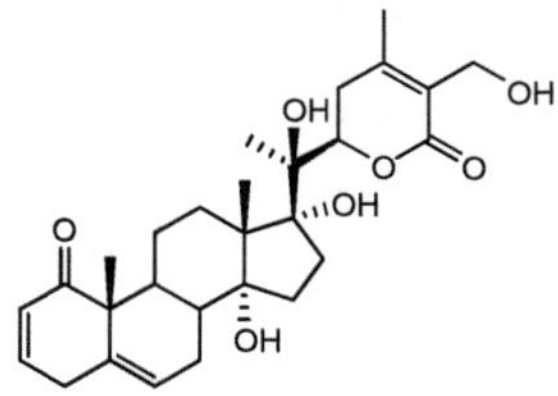

47 (withacoagulin D)

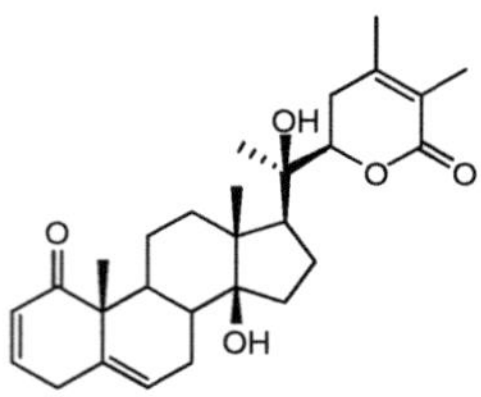

48 (withacoagulin E)

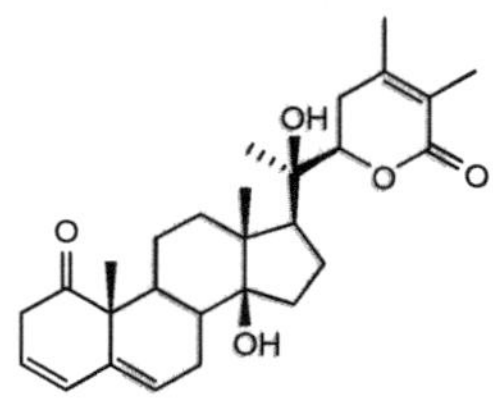

49 (withacoagulin F)

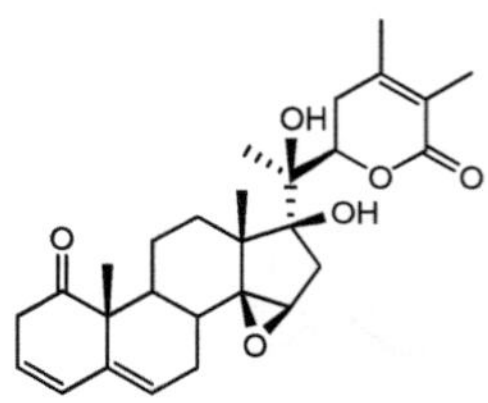

50 (14,15β-epoxywithanolide I)

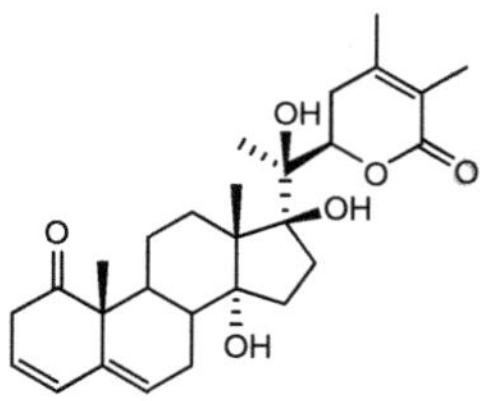

51 (17β-hydroxywithanolide K)

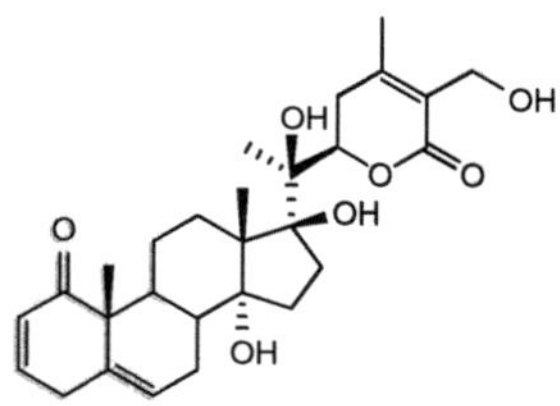

52 (coagulansin A)

Glucosomniferanolide (**57**), isolated from *W. somnifera*, contains a glucose unit at the tertiary hydroxy group at C-20 (*39*). This compound was described as having a (22*S*) configuration, opposite to that found in all withanolides with a δ-lactone side chain. However, spectroscopic observations supporting this assumption only indicate that H-22 is equatorial instead of axial (broad signal with $W_{1/2} = 5$ Hz), an orientation that can result from a simple conformational inversion of the lactone ring half chair, probably due to the presence of the bulky substituent at C-20. Thus, this structure should be revised.

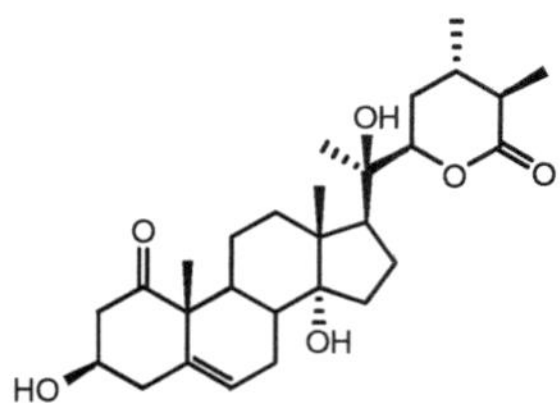

53 (coagulansin B)

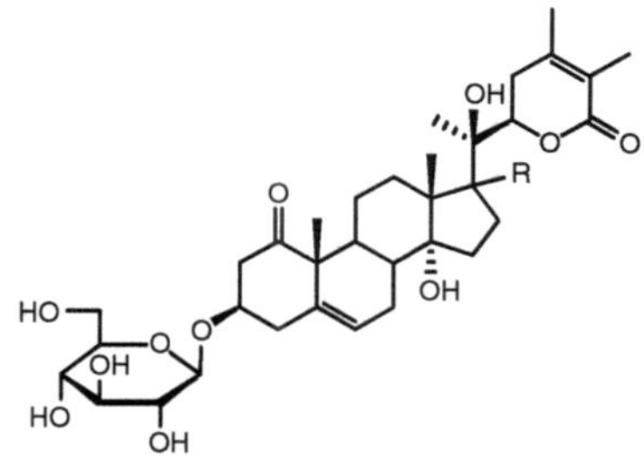

54 R = β-OH (coagulin L)
55 R = α-H (coagulin O)

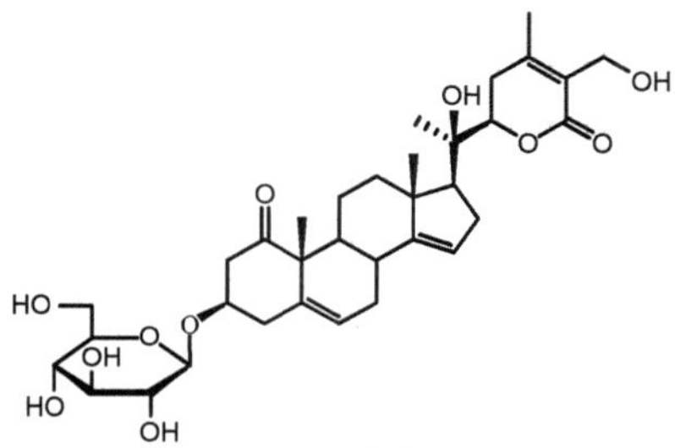

56 (coagulin P)

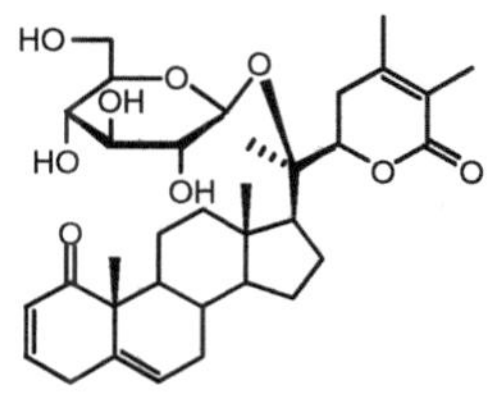

57 (glucosomniferanolide)

Several 14α,20-epoxywithanolides closely related to the known coagulin (**58**) (*40*), were isolated from *W. coagulans*. These included coagulins B–E (**59–62**) (*41*), coagulins F and G (**63**, **64**) (*42*), coagulin R (**65**) (*30*), the diol coagulin M (**66**) and the glycoside coagulin N (**67**) (*38*), coagulin J (**68**), the diol coagulin I (**69**) and the glycoside coagulin K (**70**) (*17*), and compound **71** (*34*). *Malik* and coworkers had previously reported the isolation of **71** and the 3-*O*-glycoside of coagulin R from *Physalis peruviana* (*43, 44*) and of ajugin, identical to coagulin R from *Ajuga parviflora* (see Sect. 4.2.) (*45*).

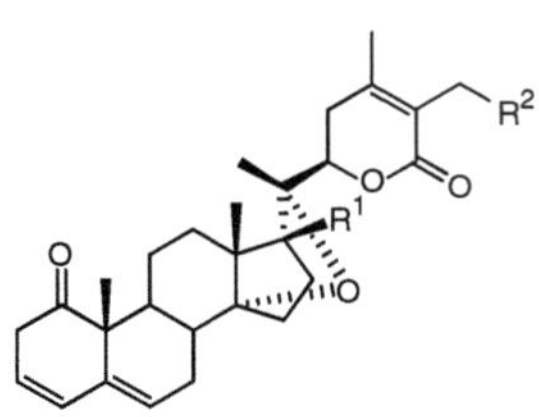

58 R^1 = OH, R^2= OH (coagulin)
62 R^1 = R^2 = H (coagulin E)
63 R^1 = H, R^2 = OH (coagulin F)
71 R^1 = OH, R^2 = H

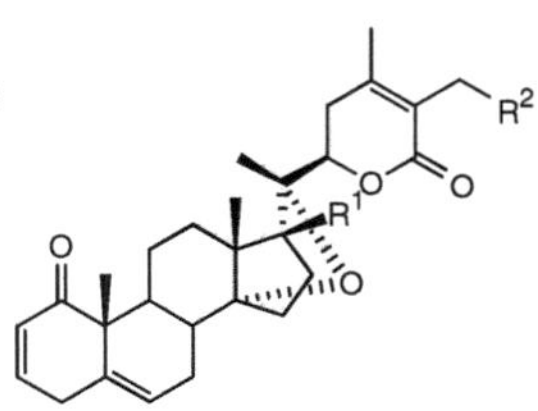

59 R^1 = H, R^2 = OH (coagulin B)
60 R^1 = OH R^2 = H (coagulin C)
61 R^1 = R^2 = H (coagulin D)
64 R^1 = R^2 = OH (coagulin G)

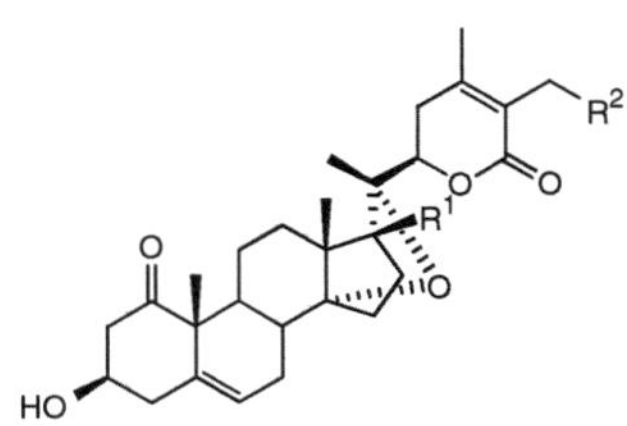

65 R^1 = OH R^2 = H (coagulin R)
68 R^1 = H R^2 = OH (coagulin J)

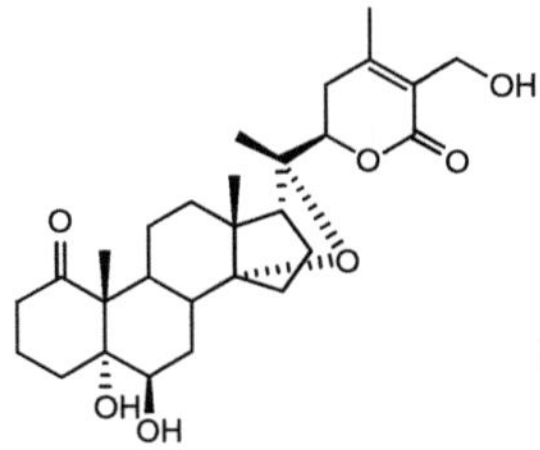

66 (coagulin M)

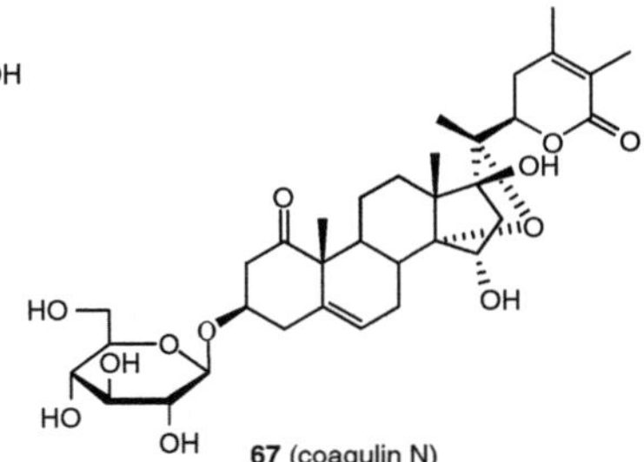

67 (coagulin N)

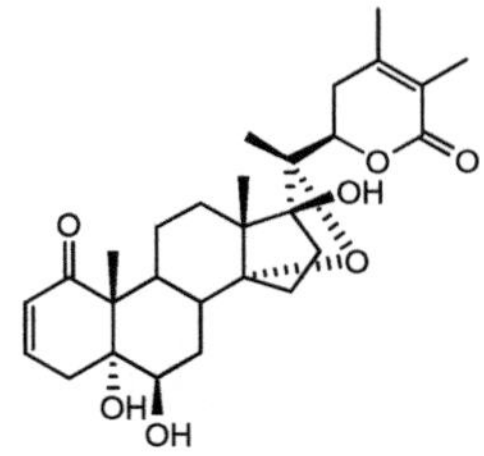

69 (coagulin I)

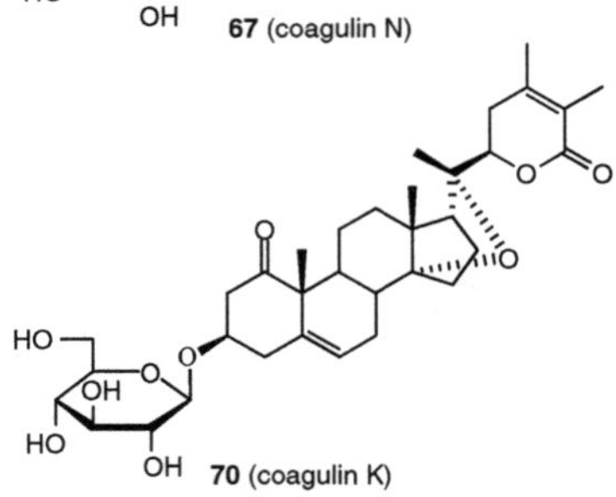

70 (coagulin K)

4.1.5. Other Substitution Patterns

Some less common substitution patterns isolated from *W. somnifera* include a $\Delta^{1,4}$-3-keto withanolide (**72**) (*12*), a series of four 8*β*-hydroxywithanolides (**73–76**), a 7*β*-hydroxywithanolide with a $\Delta^{2,4}$-1-keto arrangement in ring A (**77**) (*46*), and two dimeric withanolides bound by a thioether linkage (**78**) (*47*) or a sulfoxide (**79**) (*48*). Other less common features found in these compounds are the presence of an 11*β*-hydroxy (**74**, **76**) or an 18-hydroxy group (**75**, **77**) (*46*). The 3*β*-*O*-sulfate of 2,3-dihydrowithaferin A (**80**) was isolated from aeroponically grown *W. somnifera* plants (*49*).

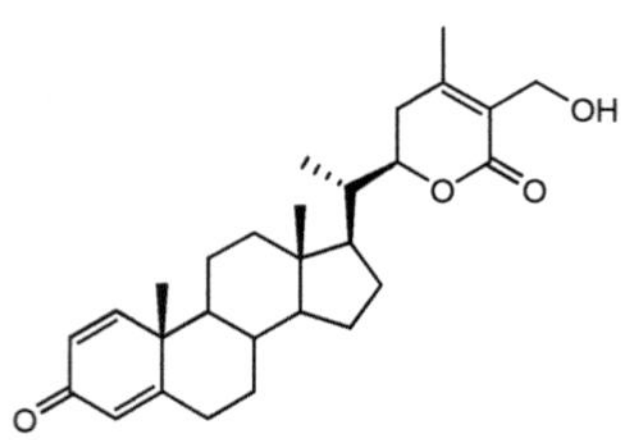

72

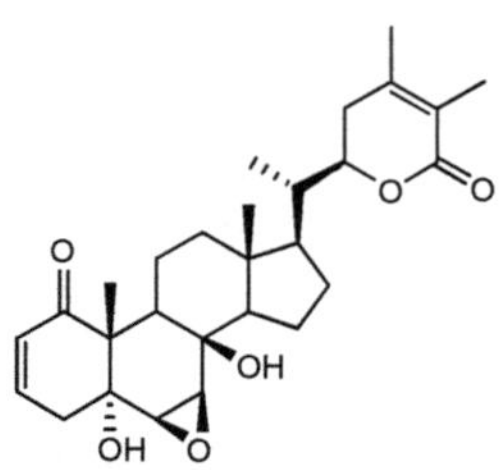

73 (withasomnilide)

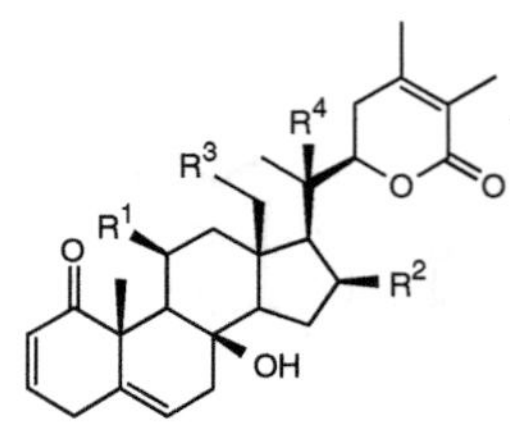

74 $R^1 = R^2 = OH$, $R^3 = R^4 = H$ (withasomniferanolide)
75 $R^1 = R^2 = H$, $R^3 = R^4 = OH$ (somniferawithanolide)

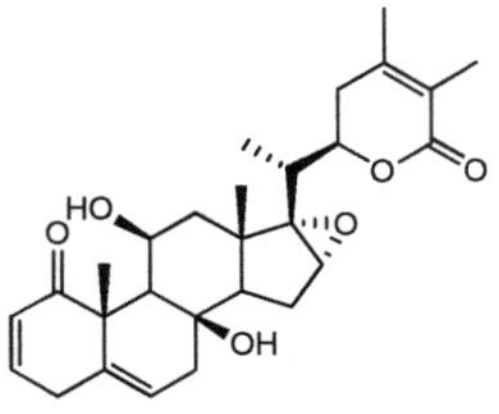

76 (somniferanolide)

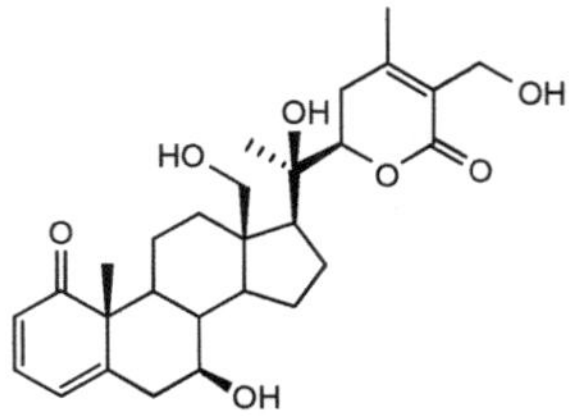

77 (somniwithanolide)

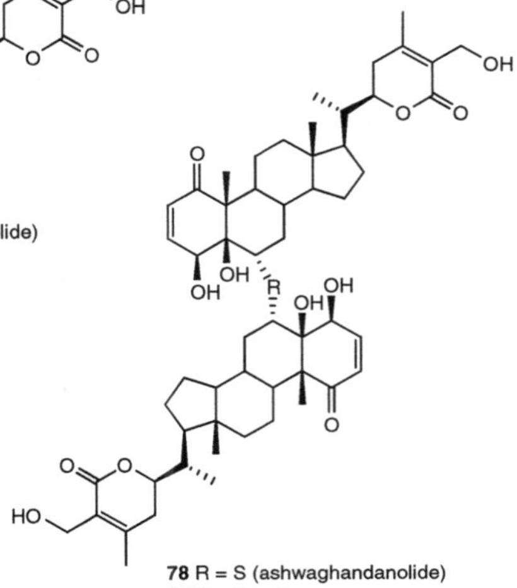

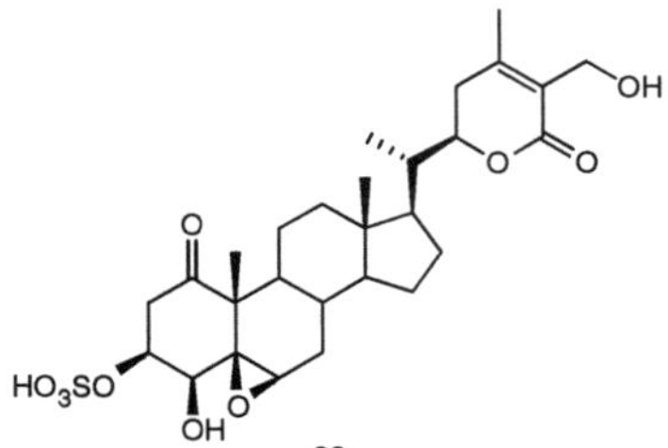

80

78 R = S (ashwaghandanolide)
79 R = S = O

4.2. Other Withanolides with an Unmodified Skeleton

As mentioned above, a considerable number of withanolides with the parent skeleton of withaferin A (**1**) have been isolated. With few exceptions, in new withanolides the substitution patterns of rings A and B correspond to those described in the previous section for the *Withania* withanolides. Structural variations consist mainly of combinations of oxygenated functions (hydroxy or carbonyl groups) at different positions of the steroid nucleus (most commonly at positions 12, 14, 16, 17, and 18) and the side chain (mostly at C-20, C-21, and C-27). These functionalities may also be involved in cyclic entities as lactones, lactols, or cyclic ethers. These withanolides are presented in Sect. 5., with the exception of 14α,20-epoxywithanolides that are included in Sect. 4.2.1.

4.2.1. C-14, C-17, and C-20 Hydroxylated Withanolides and Related Compounds

Hydroxylation at C-14, C-17 and C-20 is common in many withanolides. Usually the 14-hydroxy group has the α-orientation (see Sect. 4.1.), although there is a growing number of 14β-hydroxywithanolides. As already mentioned, hydroxy substitution at C-17 occurs with either the $\alpha-$ or $\beta-$orientation, of which the former is more common. New structures with different combinations of hydroxy groups at the above-mentioned positions have been reported, occasionally combined with hydroxy groups at positions 15, 16, and 18. The genus *Physalis* is particularly rich in 14-hydroxywithanolides, with these probably being the biosynthetic precursors of polyoxyfunctional structures such as the physalins, present in many *Physalis* species (see Sect. 5.2.). Both 14α- and 14β-hydroxywithanolides are present in *Physalis* plants. A series of 14α,17β,20-trihydroxywithanolides was isolated from *P. peruviana* (**81–85**) (*50*). Phyperunolides B (**81**) and C (**82**) have the unusual feature of a free hydroxy group at C-28; another 28-hydroxywithanolide, **86**, was reported by *Dinan et al.* from the same plant (*51*). Hydroxylation at C-28 is a prerequisite in the formation of perulactones (Fig. 3), common components of *P. peruviana*. The 3-ethoxy withanolide **85** is probably an artifact formed during isolation. Also from *P. peruviana*, *Ahmad* and coworkers isolated the closely related **87** (*52*), the glycosides **88–91** (*53*), and a glycoside of coagulin R (**92**) (*44*). As already mentioned (Sect. 4.1.4.), another withanolide with a 14α,20-ether bridge (**71**), was also isolated from this plant (*43*). From *P. cinerascens* collected in Mexico, *Maldonado et al.* isolated 24,25-dihydrowithanolide S (**93**) with a saturated lactone side chain, together with the known withanolide S (*54*).

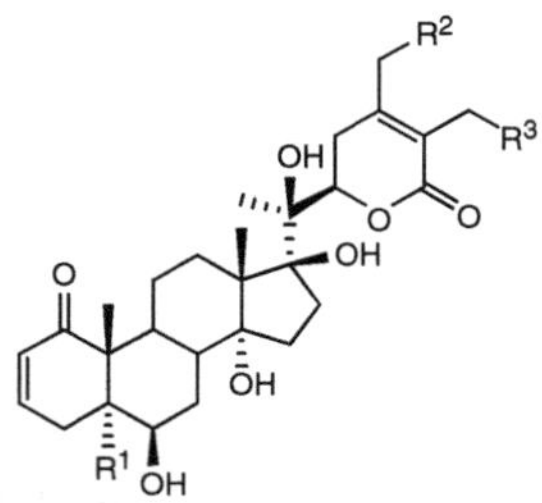

81 $R^1 = R^2 =$ OH, $R^3 =$ H (phyperunolide B)
82 $R^1 =$ Cl, $R^2 =$ OH, $R^3 =$ H (phyperunolide C)
83 $R^1 = R^3 =$ OH, $R^2 =$ H (phyperunolide D)

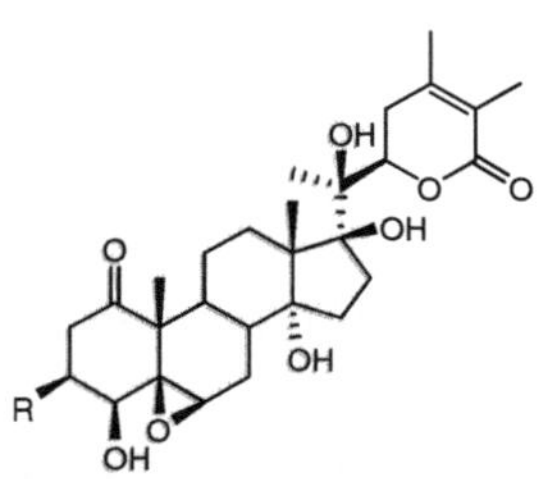

84 R = OH (phyperunolide E)
85 R = OCH_2CH_3 (phyperunolide F)

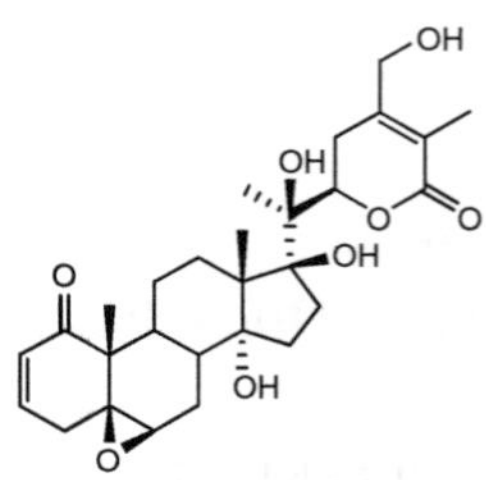

86 (28-hydroxywithanolide E)

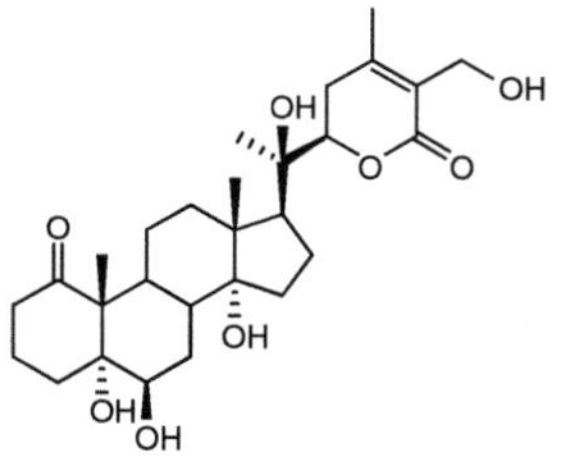

87

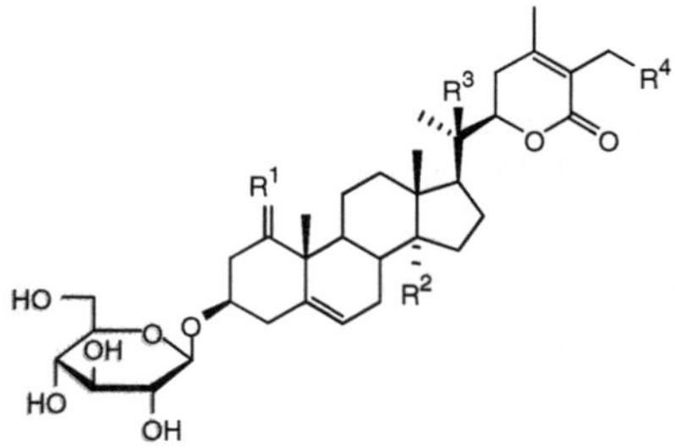

88 $R^1 =$ H, α-OAc, $R^2 = R^3 =$ OH, $R^4 =$ H
89 $R^1 =$ H, α-OAc, $R^2 = R^3 =$ H, $R^4 =$ OH
90 $R^1 =$ O, $R^2 =$ H, $R^3 = R^4 =$ OH
91 $R^1 =$ O, $R^2 = R^3 = R^4 =$ OH

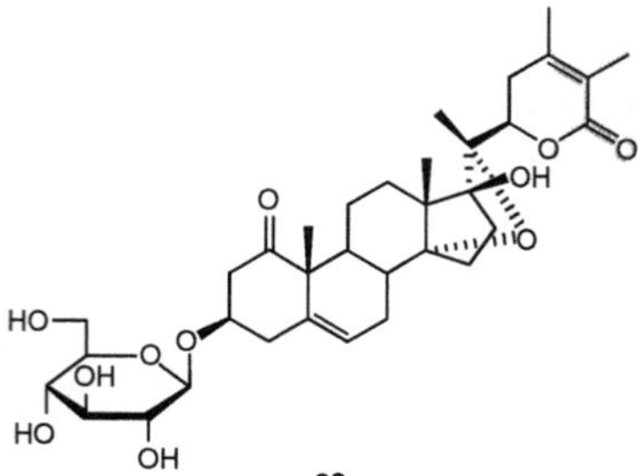

92

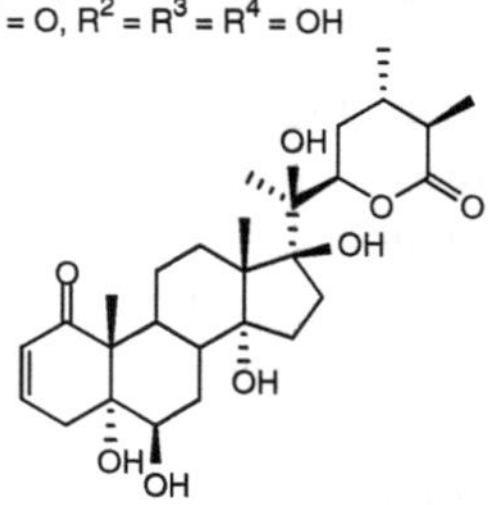

93 (24,25-dihydrowithanolide S)

An investigation of *P. angulata* growing in Taiwan gave the 15-oxygenated withangulatins B–D (**94–96**) and the 16-hydroxylated withangulatins G (**97**) and H (**98**) together with withangulatin E (**99**) and a Δ^{16} 14α-hydroxywithanolide, withangulatin F (**100**) (*55*). Several other Δ^{16} 14α-hydroxywithanolides have been reported from *Physalis* species. These include phyperunolide A (**101**) isolated from *P. peruviana* (*50*) and four 15-acetyloxy withanolides isolated from *P. angulata*, withangulatin I (**102**) (*56*), physagulin M (**103**) with the unusual feature of a free hydroxy group at C-23 (*57*), physagulin O (**104**), and compound **105** (*58*). The latter withanolide was incorrectly named physagulin L, as this name had already been assigned (see below). Physagulin N (**106**), the methanol addition product of physagulin A, was probably formed during its isolation (*57*). Δ^{16}-Withanolides with oxygen substituents at C-14 and C-15 have the appropriate functionalities for cleavage of the 13,14-bond, and the occurrence of such withanolides in physalin-rich plants strongly suggests that they are either precursors or shunt products in the biosynthesis of physalins (*59*). A Δ^{14}-withanolide, **107**, was also reported from *P. minima* collected in Pakistan (*60*).

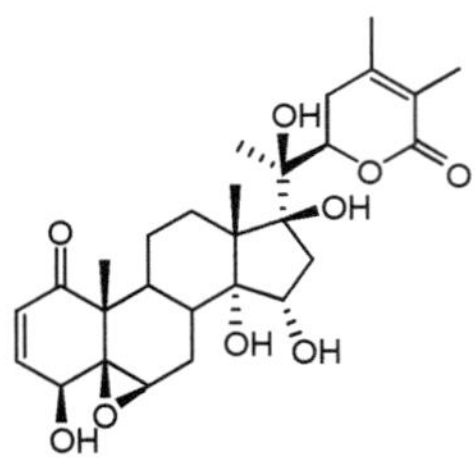
94 (withangulatin B)

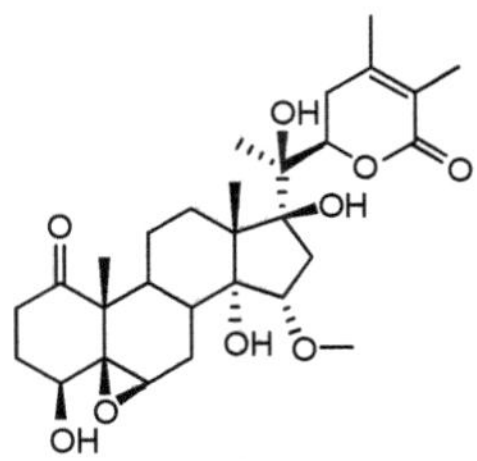
95 (withangulatin C)

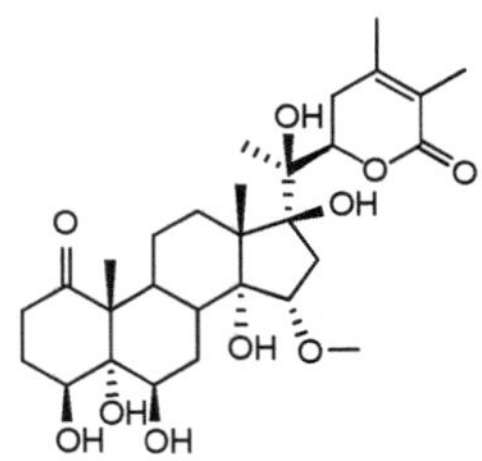
96 (withangulatin D)

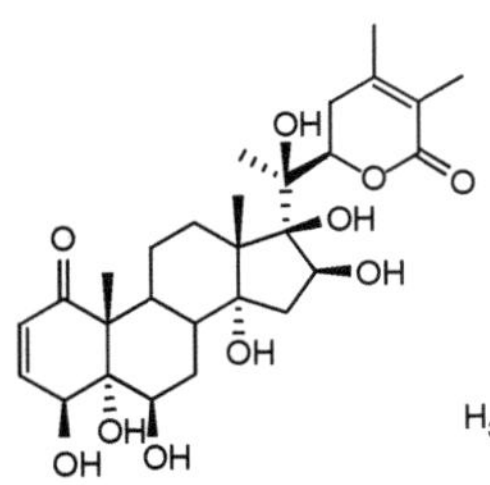
97 (withangulatin G)

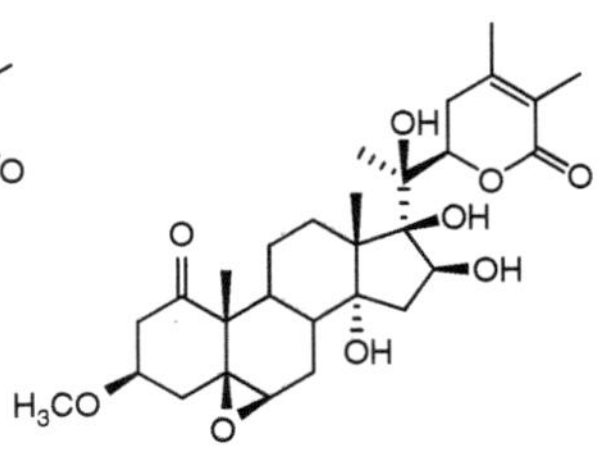
98 (withangulatin H)

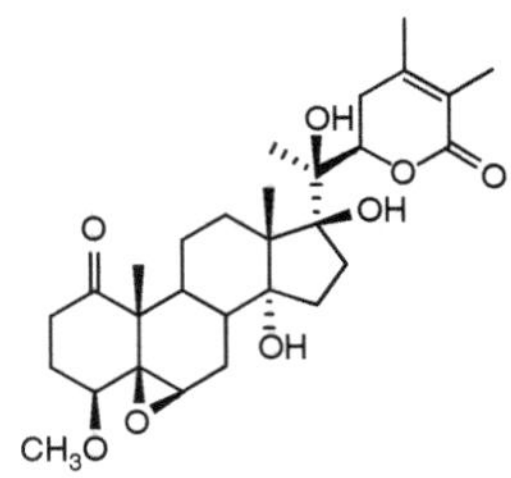
99 (withangulatin E)

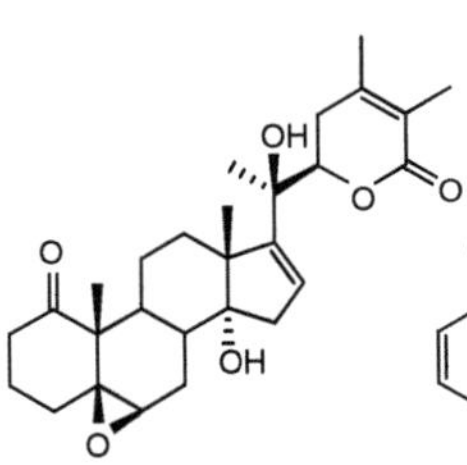
100 (withangulatin F)

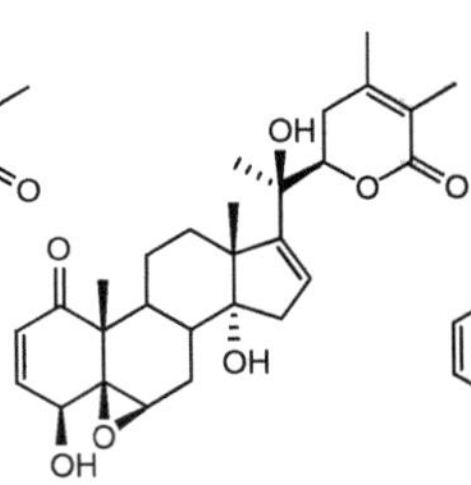
101 (phyperunolide A)

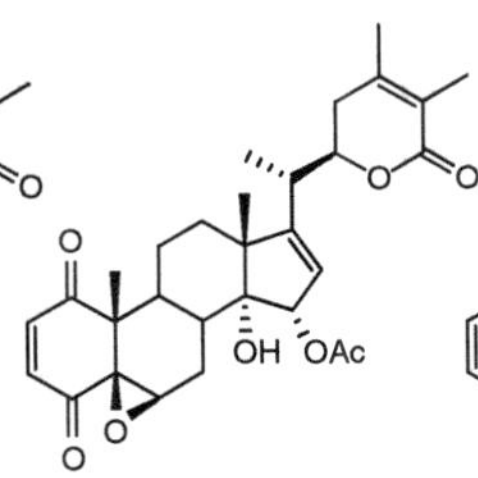
102 (withangulatin I)

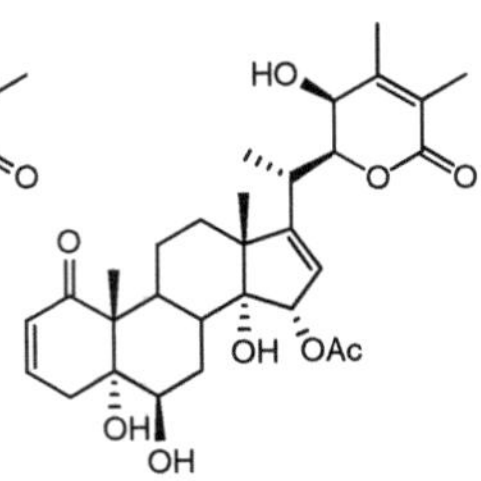
103 (physagulin M)

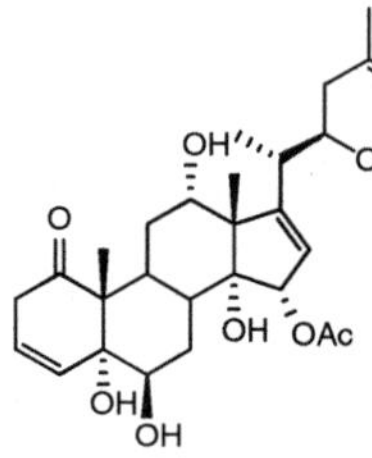

104 (physagulin O)

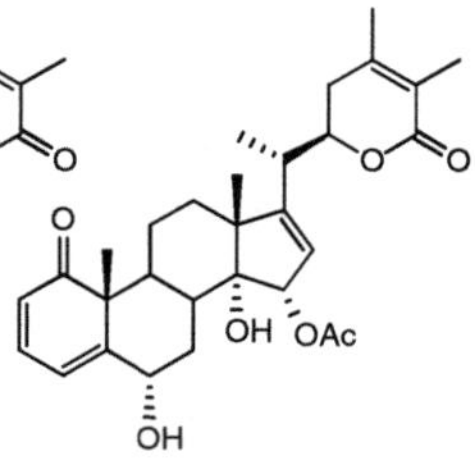

105

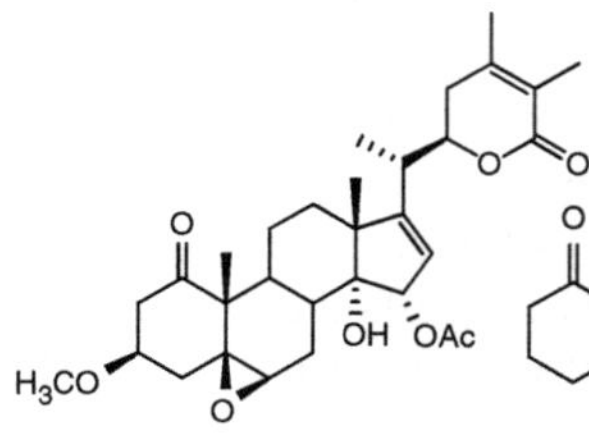

106 (physagulin N)

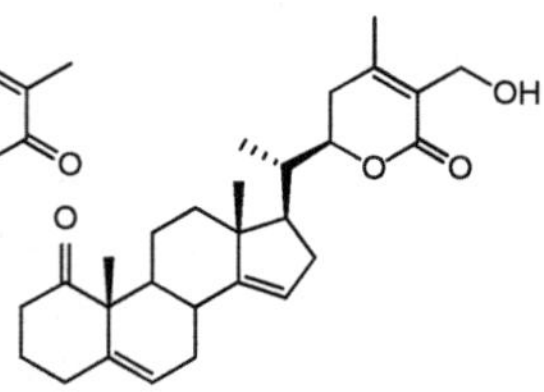

107

Ajuga (Lamiaceae) is one of the few genera outside the Solanaceae that contains withanolides, most of which are closely related to the coagulins (Sect. 4.1.). From *A. parviflora*, *Malik* and coworkers have reported seven 14α-hydroxywithanolides, compound **108** (isolated together with coagulin J) (*61*), ajugins A (**109**), B (**110**) (*62*), C (**111**), D (**112**) (*63*), E (**113**), and F (**114**) (*64*), a Δ^{14} withanolide (**115**), and a 14α,20-epoxywithanolide (**116**) (*65*). As already mentioned, ajugin, identical to coagulin R (**65**), was first isolated from this plant (*45*).

Several 14β-hydroxywithanolides have been isolated from *Physalis* species, and all of them also have an α−oxygenated function (hydroxy or acetate) at position 15. These include physagulins H–K (**117–120**) (*66*), physagulin L (**121**) (*57*), and compounds **122** and **123** (*58*) from *P. angulata*. The latter two compounds were

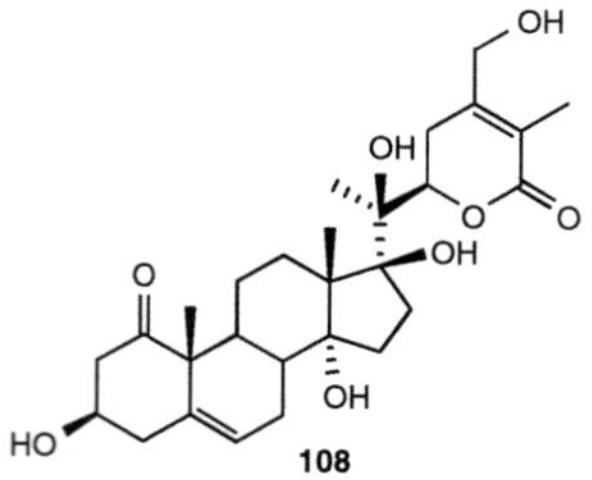

108

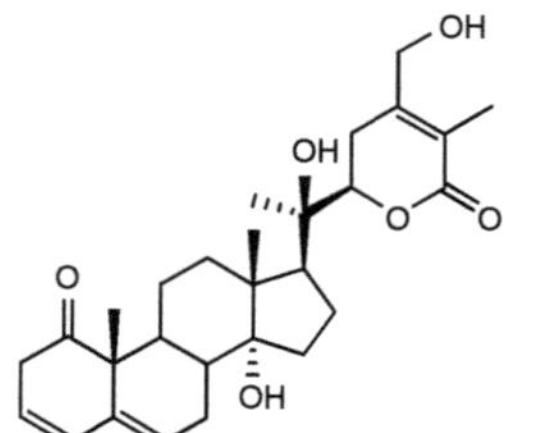

109 (ajugin A)

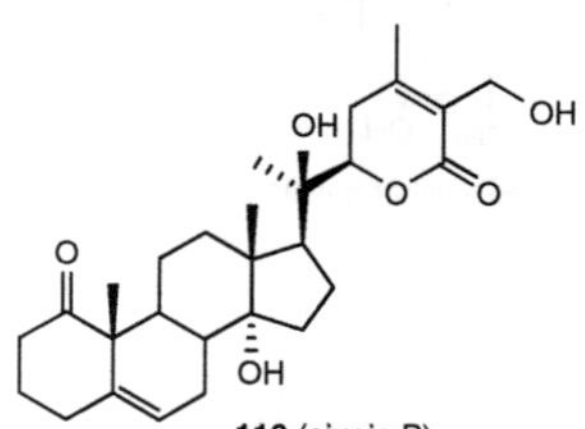

110 (ajugin B)

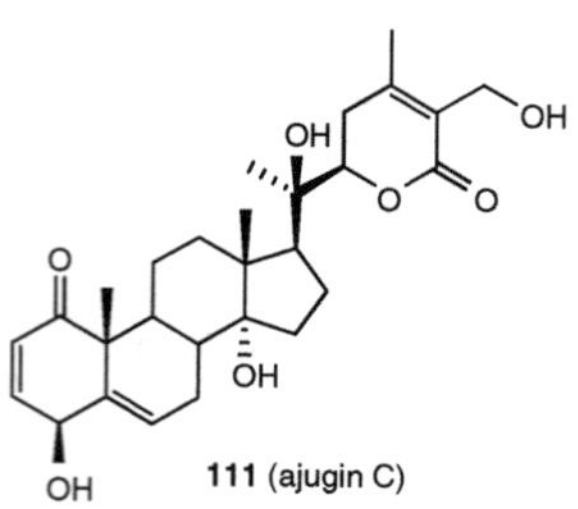

111 (ajugin C)

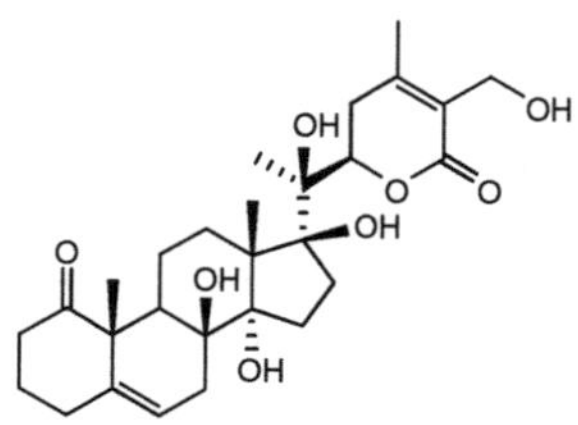

112 (ajugin D)

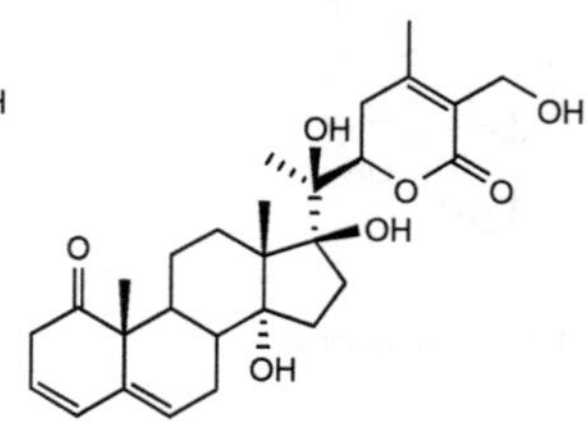

113 (ajugin E)

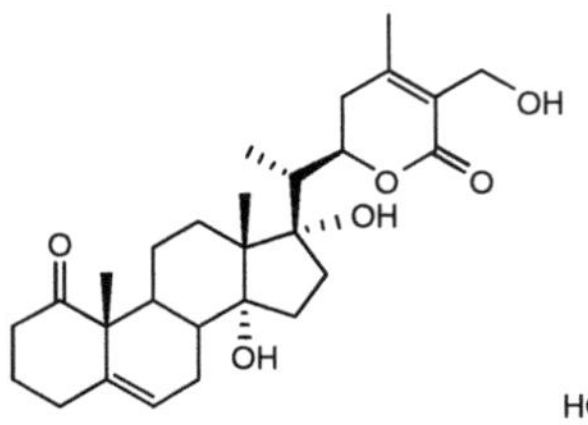

114 (ajugin F)

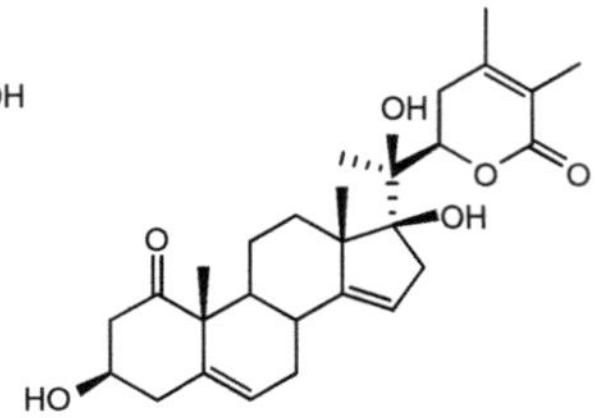

115

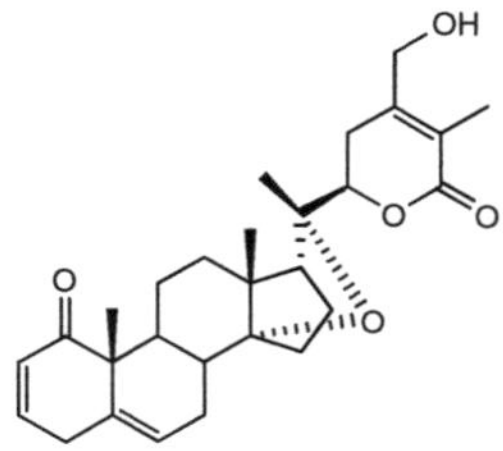

116

incorrectly named physagulins M and N, as these names had already been assigned (see above). The chlorohydrin **124** was reported from *P. alkekengi* var. *franchetii* (*67*) and the deacetylated analogue of physapubenolide (**125**) from *P. peruviana* (*52*). Outside the *Physalis* genus, new 14β-hydroxywithanolides have been isolated from *Jaborosa leucotricha* (jaborosalactone 8 (**126**)) (*68*) and *J. bergii* (jaborosalactol 23 (**127**)) (*69*).

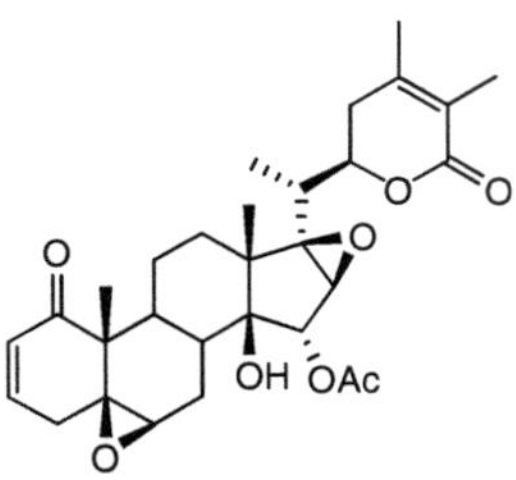

117 (physagulin H)

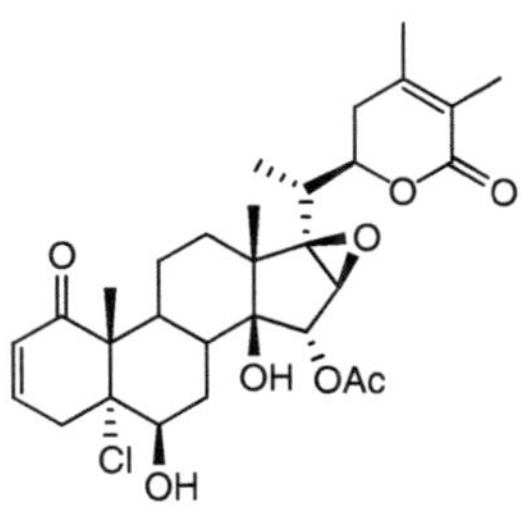

118 (physagulin I)

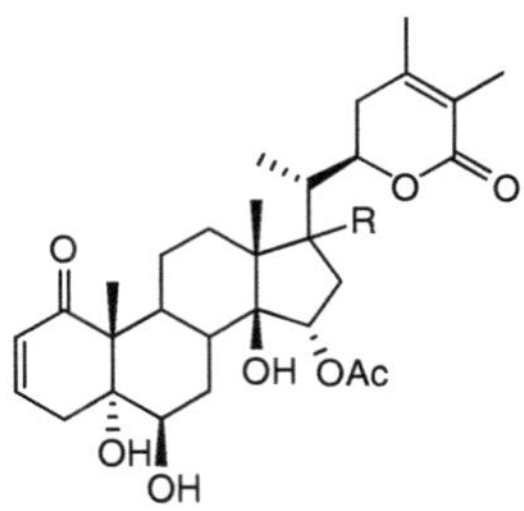

119 R = α-H (physagulin J)
120 R = β-OH (physagulin K)

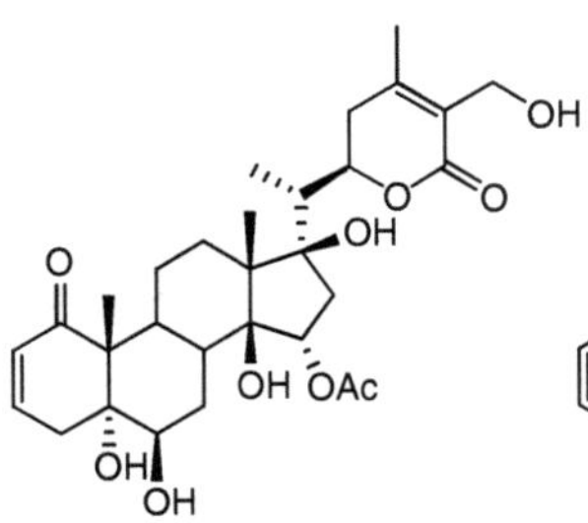

121 (physagulin L)

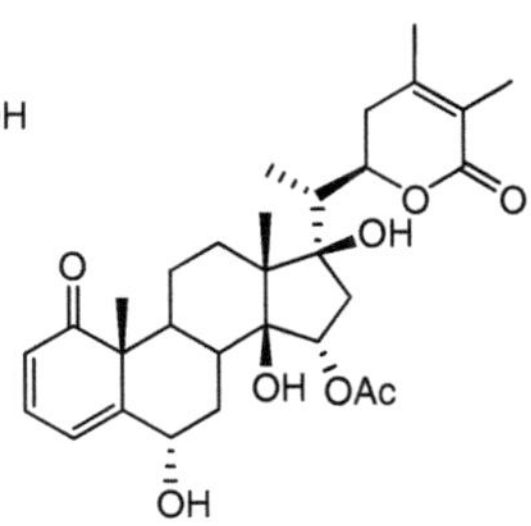

122

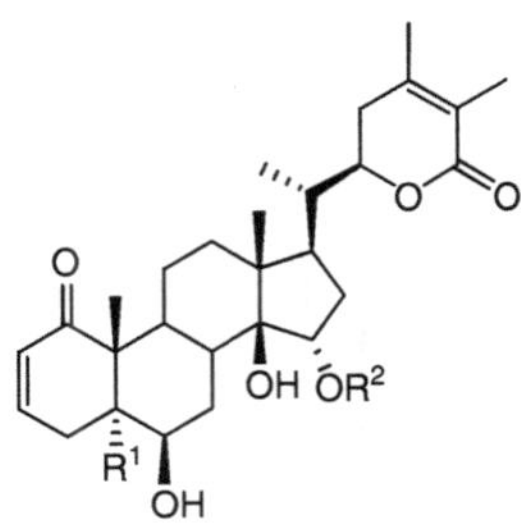

123 R^1 = OH, R^2 = H
124 R^1 = Cl, R^2 = Ac

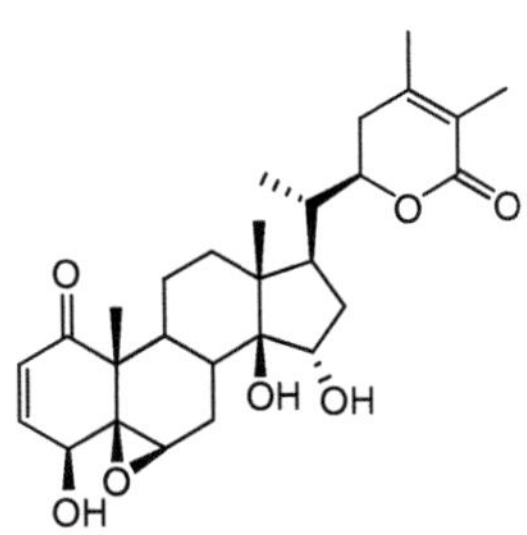

125 (deacetylphysapubenolide)

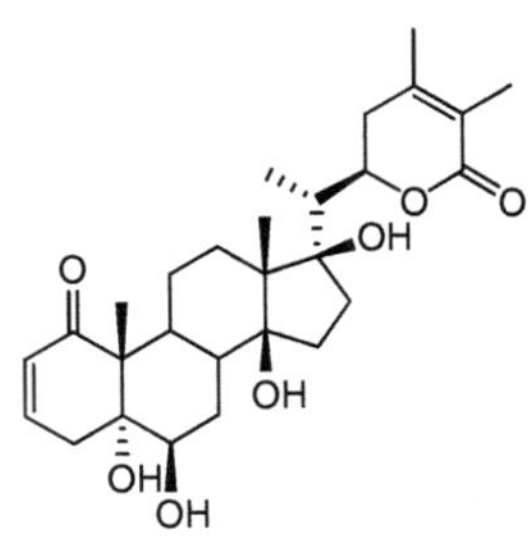

126 (jaborosalactone 8)

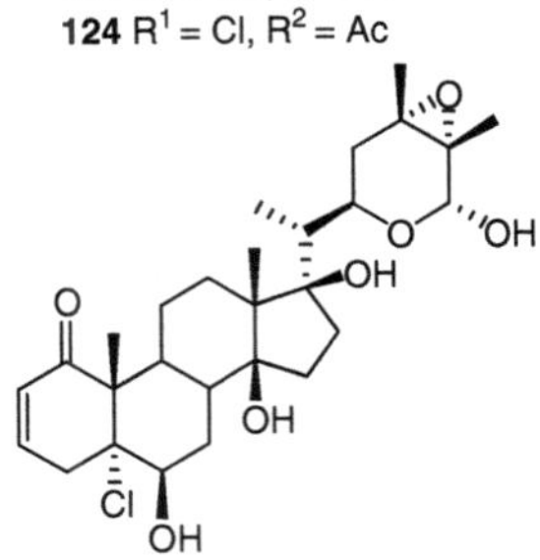

127 (jaborosalactone 23)

4.2.2. C-18 Hydroxylated Withanolides

Withanolides with a functionalized C-18 at various oxidation levels (alcohol, aldehyde, and lactone carbonyl) have been isolated from plants of the genera *Withania* (see Sect. 4.1.5.), *Acnistus*, *Dunalia*, *Eriolarynx*, *Iochroma*, and *Physalis* (*4*, *5*). Hydroxylation at C-18 is usually combined with hydroxy groups occurring at positions 14, 17, or 20. Thus, 18-hydroxywithanolide D (**128**) was isolated from *Eriolarynx lorentzii* (synonym *Vassobia lorentzii*) (*70*), and the related 18-hydroxywithanolide **129** and the corresponding 18-aldehyde **130** were isolated from *Dunalia brachyacantha* (*71*) (both plants collected in Argentina), while the 18-acetoxywithanolide **131** was reported from *Iochroma gesneroides* together with several 3-methoxylated derivatives formed during the extraction procedure (*72*). As part of a systematic study of Mexican *Physalis* species, five 18-acetoxywithanolides, the physachenolides A–E (**132–136**) were isolated from the leaves, flowers, and stems of *Physalis chenopodifolia* (*73*). Physachenolide B (**133**) has a 28-hydroxy group while physachenolide E (**136**) is a Δ^{14}-withanolide. The aerial parts of *Physalis coztomatl*, also collected in Mexico,

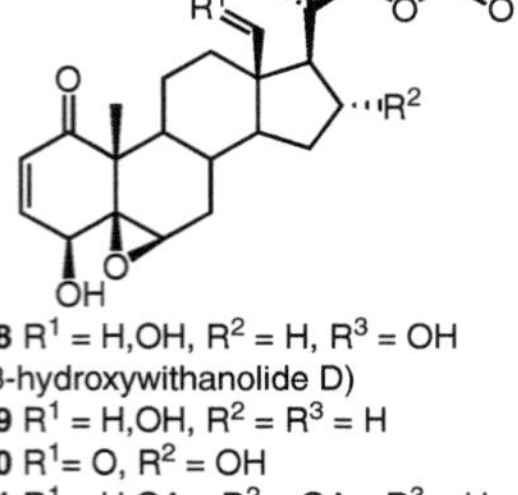

128 R^1 = H,OH, R^2 = H, R^3 = OH (18-hydroxywithanolide D)
129 R^1 = H,OH, R^2 = R^3 = H
130 R^1= O, R^2 = OH
131 R^1 = H,OAc, R^2 = OAc, R^3 = H

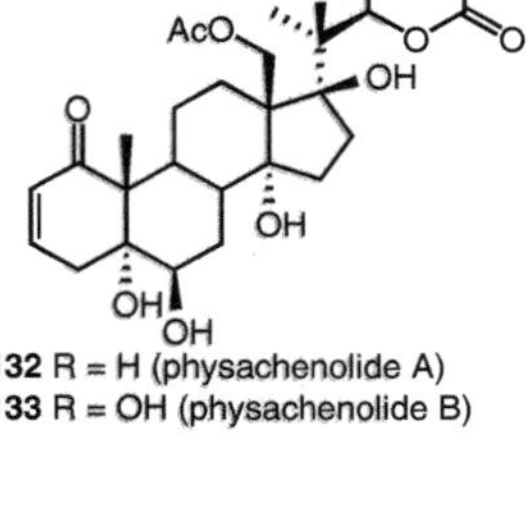

132 R = H (physachenolide A)
133 R = OH (physachenolide B)

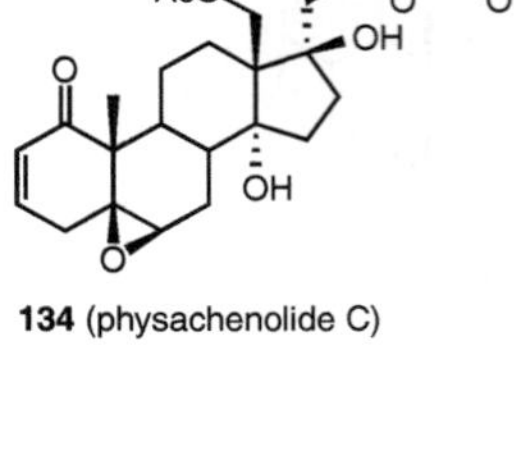

134 (physachenolide C)

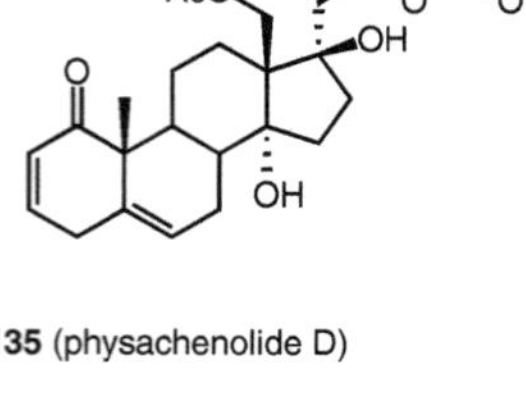

135 (physachenolide D)

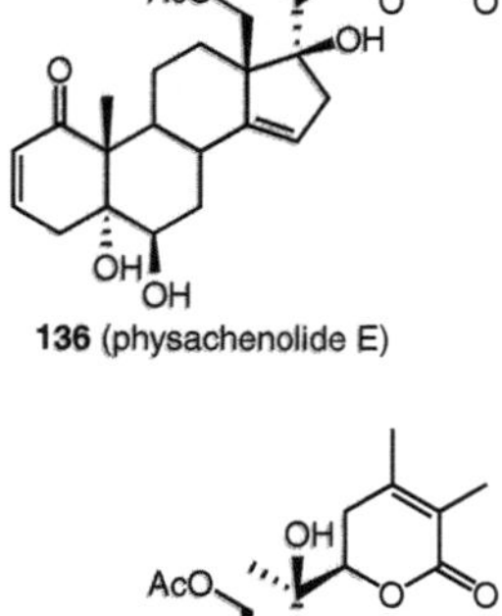

136 (physachenolide E)

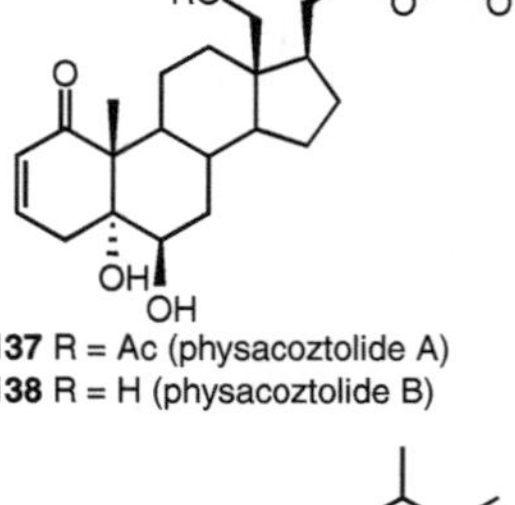

137 R = Ac (physacoztolide A)
138 R = H (physacoztolide B)

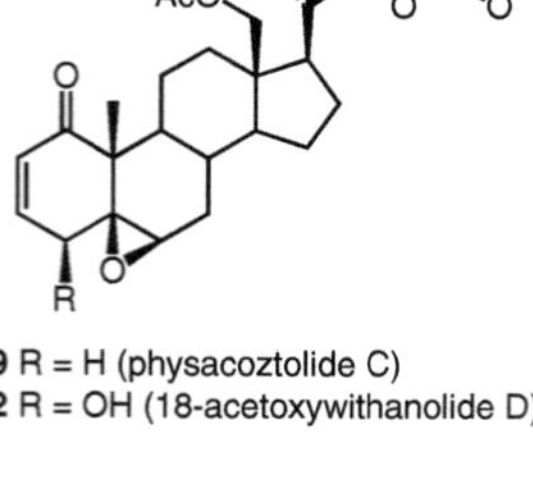

139 R = H (physacoztolide C)
142 R = OH (18-acetoxywithanolide D)

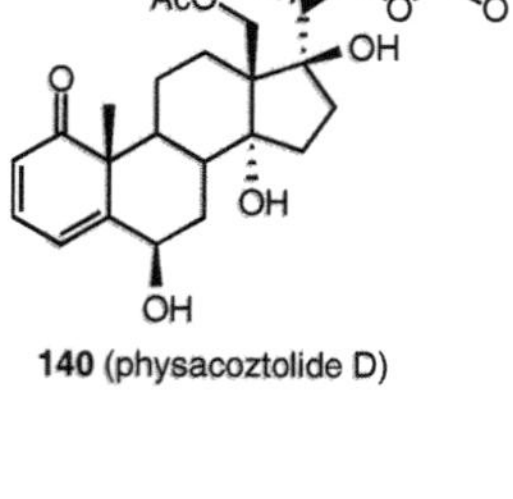

140 (physacoztolide D)

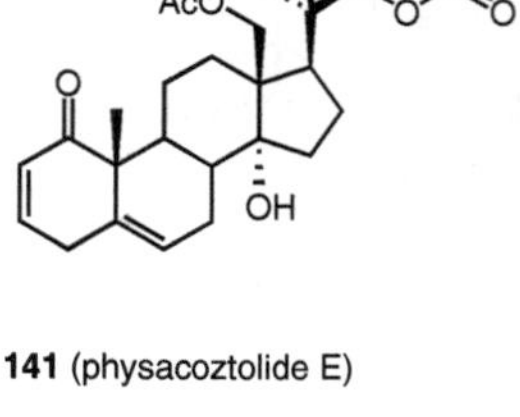

141 (physacoztolide E)

rendered five new withanolides functionalized at C-18, physacoztolides A–E (**137–141**), together with physachenolide C (**134**), 18-acetoxywithanolide D (**142**), and 18-hydroxywithanolide D (**128**) (*74*).

4.2.3. C-12 and C-21 Oxygenated Withanolides

Withanolides with a free hydroxy or keto group at C-12 are mostly restricted to the *Datura* genus. In the genus *Jaborosa*, 12-ketowithanolides are most probably involved as precursors in the formation of additional rings with the side chain, giving rise to several modified skeletons (see Sect. 5.9.). However, only two withanolides with unmodified skeletons containing a free ketone at C-12 have been reported from these plants, (−)-jaboromagellonine (**143**) from *J. magellanica* (*75*) and jaborosalactone 44 (**144**) from *J. kurtzii* (*76*). Four new 12-oxygenated withanolides were isolated from plants of *Datura ferox* collected in Argentina, together with other known daturalactones. 15*β*-Hydroxynicandrin B (**145**) was found to have the common 5*α*-hydroxy-6*α*,7*α*-epoxy substitution pattern in ring B (see Sect. 4.1.2.) (*77*), while daturalactones 5–7 (**146–148**) exhibit related arrangements resulting from hydrolytic cleavage of the epoxide or rearrangement of the epoxyalcohol (*78*). The 12*β*-epimer of **145**, baimantuololine A (**149**) (*79*) and the closely related glycosides baimantuoluosides A–C (**150–152**) (*80*), were isolated from the dry flowers of *Datura metel* (used in Chinese medicine). The 1*β*-alcohol **153** from *Datura quercifolia* collected in India is closely related to the compounds mentioned above (*81*). 12-Oxygenated withanolides have also been reported from the stem bark of the Ethiopian shrub *Discopodium penninervium* (**154**) (*82*), from *Iochroma gesneroides* (**155**) (*72*), and from *Acnistus arborescens* (**156**) (*83*), with the latter two compounds bearing a 12*β*-acetoxy group.

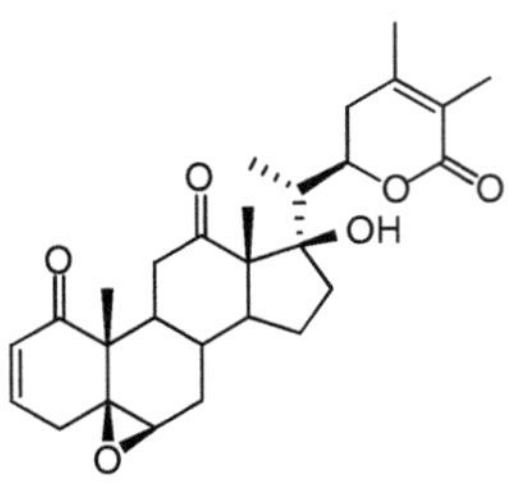

143 ((-)jaboromagellonine)

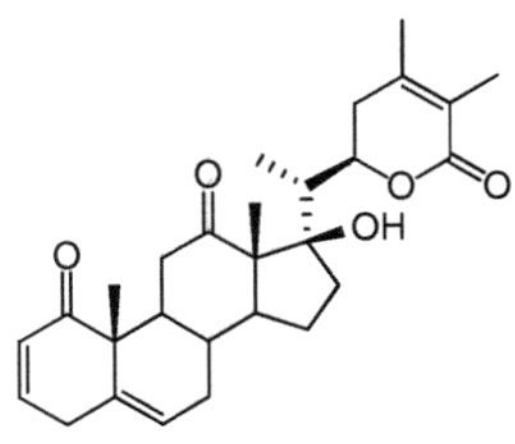

144 (jaborosalactone 44)

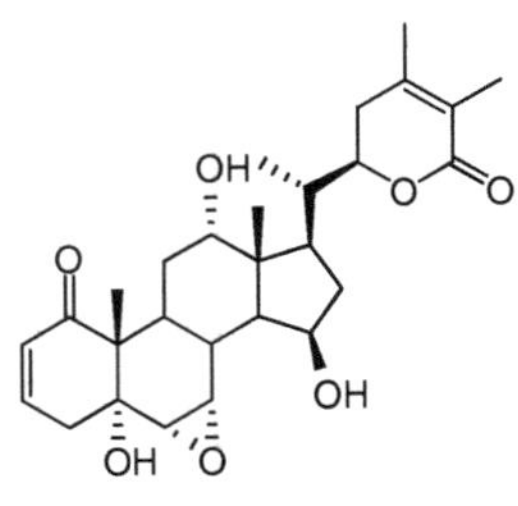

145 (15*β*-hydroxynicandrin B)

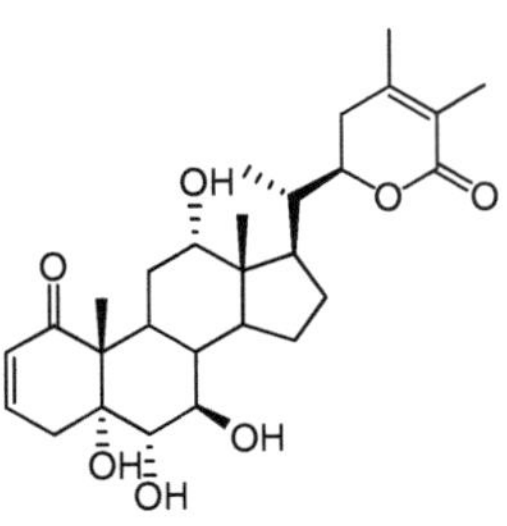

146 (daturolactone 5)

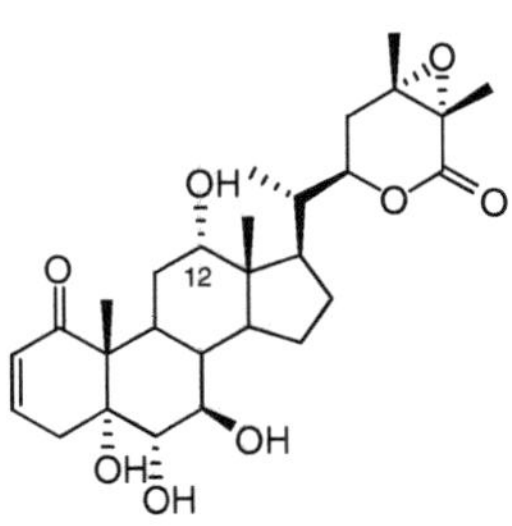

147 (daturolactone 6)

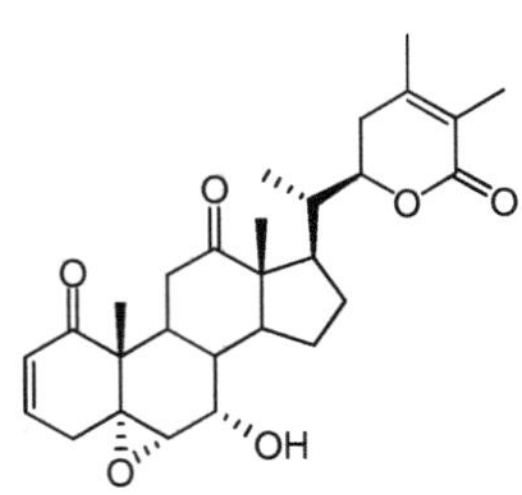

148 (daturolactone 7)

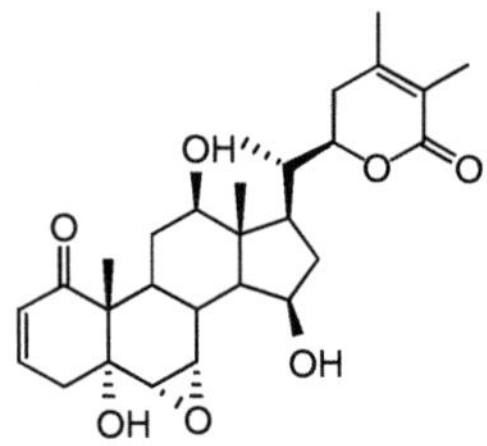

149 (baimantuoluoline A)

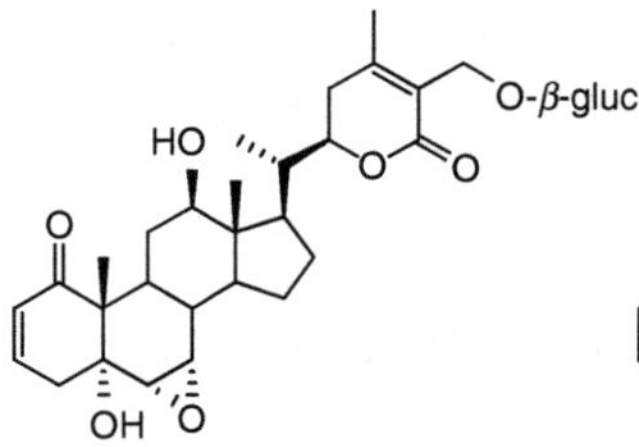

150 (baimantuoluoside A)

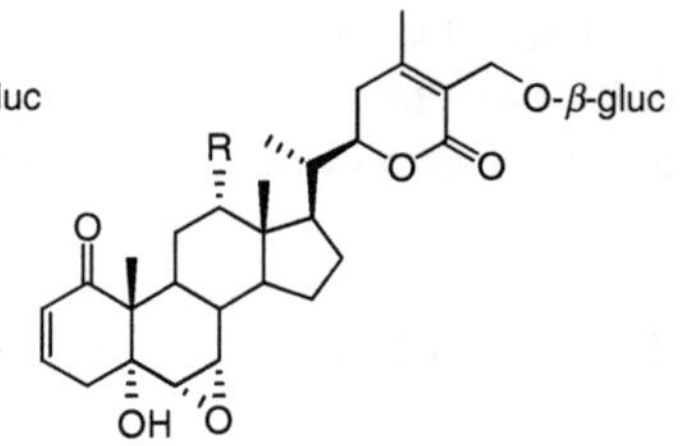

151 R = OH (baimantuoluoside B)
152 R = H (baimantuoluoside C)

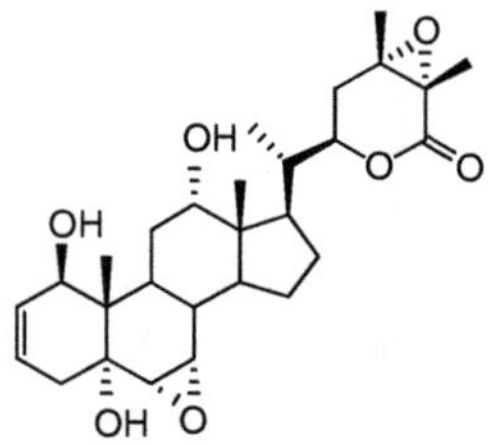

153

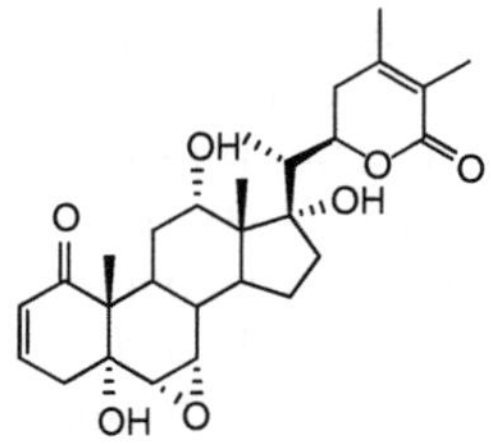

154

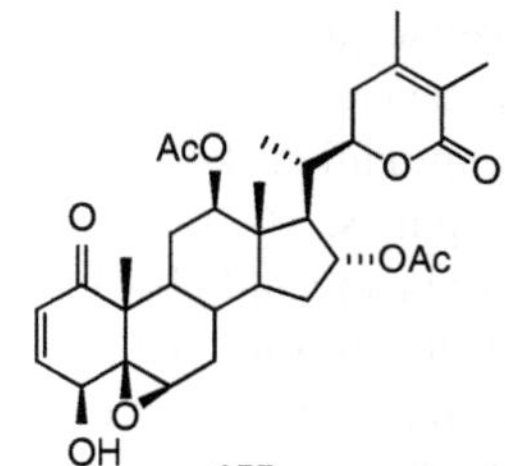

155

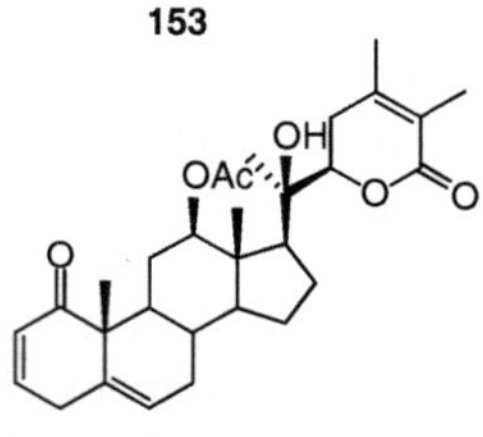

156

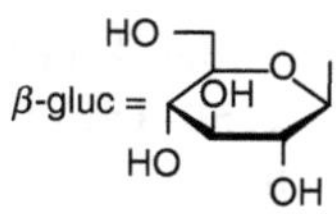

From *Dunalia brachyacantha* collected in Bolivia, *Bravo et al.* isolated the glycosides dunawithanine G (**157**) and dunawithanine H (**158**) (*84*), closely related to the known dunawithanine F (**159**) (*85*).

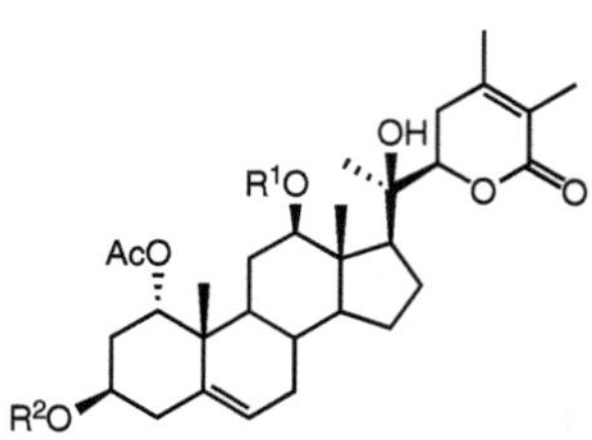

157 R^1 = Ac, R^2 =
(dunawithanine G)

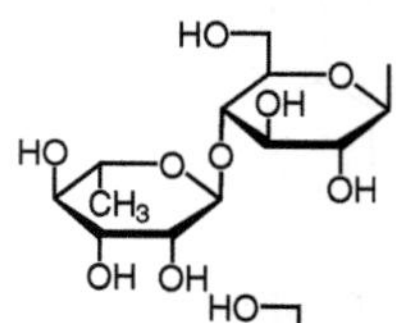

158 R^1 = H, R^2 =
(dunawithanine H)

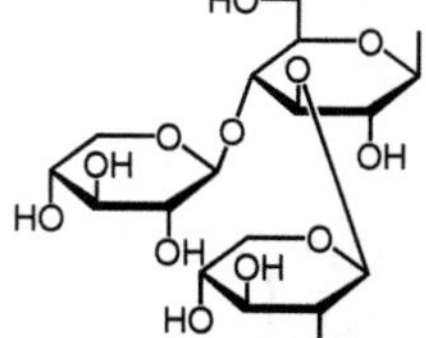

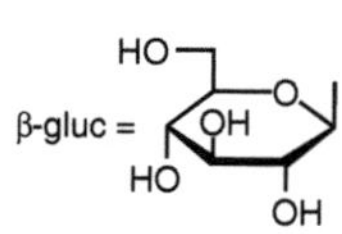

159 R^1 = H, R^2 = β-gluc(1-3)-[β-gluc(1-4)]-β-gluc-
(dunawithanine F)

Outside the Solanaceae, most withanolides from *Tacca* species bear a 12α-acetoxy group. *Tacca* species originate in the tropical and subtropical regions, mostly in Asia and Africa. They contain highly oxygenated withanolides, usually with modified skeletons (see Sect. 5.4.4.) characterized by a 1α-acetoxy group and a 2α,3α-epoxide in ring A, although several have been reported with the unmodified parent skeleton. From the rhizomes of *Tacca chantrieri*, *Yokosuka et al.* isolated two glycosides, chantriolides A (**160**) and B (**161**), with a glucose unit at C-27 (*86*). From *Tacca plantaginea*, *Liu et al.* isolated plantagiolides A–E (**162–166**) (*87*). *Tacca* species are also rich in glycosidated sterols not included in this review; the withanolide **167** with a steroid nucleus that resembles a sterol was also isolated from the rhizomes of *T. chantrieri* (*88*).

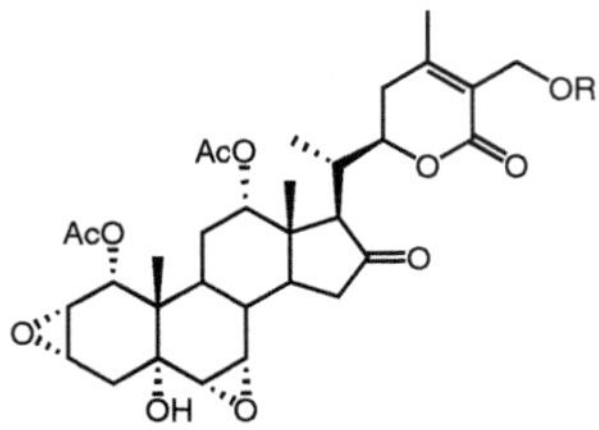

160 R = β-gluc (chantriolide A)
162 R = H (plantagiolide A)

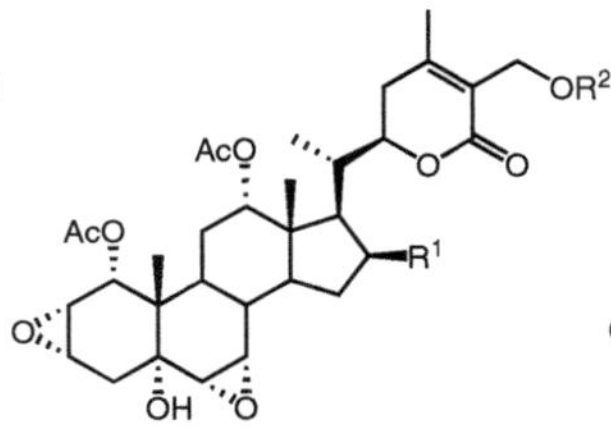

161 R^1 = OH, R^2 = β-gluc (chantriolide B)
163 R^1 = OH, R^2 = H (plantagiolide B)
164 R^1 = R^2 = H (plantagiolide C)

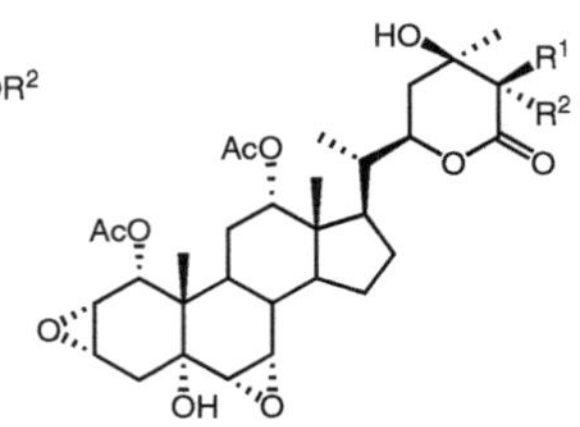

165 R^1 = CH_3, R^2 = H (plantagiolide D)
166 R^1 = OH, R^2 = CH_3 (plantagiolide E)

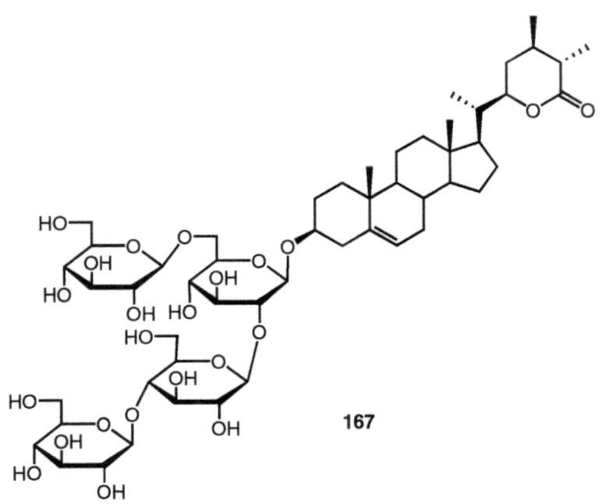

167

Although 21-oxygenated withanolides are present in *Datura*, *Jaborosa*, and *Tacca* species, 21-hydroxywithanolides with an unmodified skeleton have only been reported in the genus *Datura*, in many cases combined with a 12β-hydroxy group. Withafastuosin D (**168**) is the major withanolide of *D. metel* var. *fastuosa* (synonym *D. fastuosa*), and was isolated from the leaves together with withafastuosin E (**169**) (*89*); withafastuosin F (**170**) was isolated from the flowers of the plant (*90*). Withametelin H (**171**) isolated from the leaves of *D. metel* has the unusual feature of a methoxy group at C-27 (*91*). From flowers of *D. metel*, *Pan et al.* isolated the 27-glycoside withametelin P (**172**) (*92*) and *Yang et al.*

isolated baimantuoluolines C, B, and F (**173–175**), with the former also having a 27-methoxy group (*79, 93*). It should be noted that the name “withametelin” is usually reserved for withanolides possessing an oxygen bridge between C-21 and C-24 (see Sect. 5.1.1.), and 21-hydroxywithanolides are the most probable biosynthetic precursors of these compounds. Another five 21-hydroxywithanolides, withatatulin (**176**) (*94*) and withatulins B–E (**177–180**), were isolated by *Ray* and coworkers from *D. stramonium* var. *tatula* (synonym *D. tatula*), which grows in the sub-Himalayan tracts of India and is often cultivated as an ornamental plant (*95, 96*).

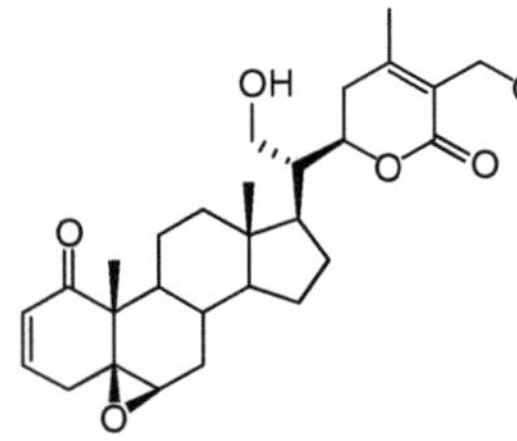

168 (withafastuosin D)

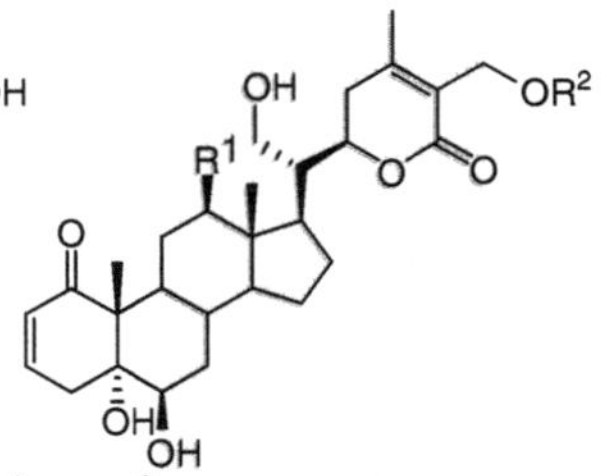

169 R^1 = H, R^2 = H (withafastuosin E)
170 R^1 = OH, R^2 = H (withafastuosin F)
171 R^1 = H, R^2 = CH_3 (withametelin H)
172 R^1 = H, R^2 = β-gluc (withametelin P)
173 R^1 = OH, R^2 = CH_3 (baimantuoluoline C)

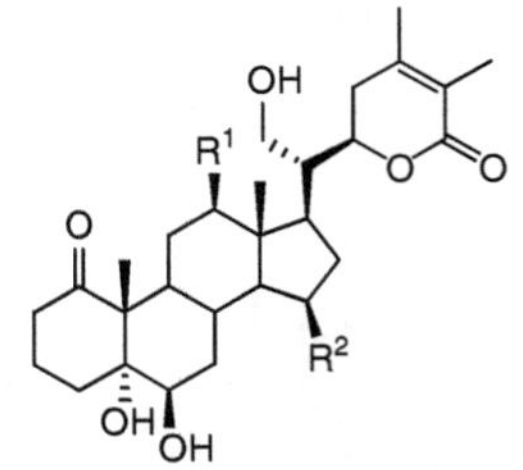

174 R^1 = H, R^2 = OH (baimantuoluoline B)
175 R^1 = OH, R^2 = H (baimantuoluoline F)

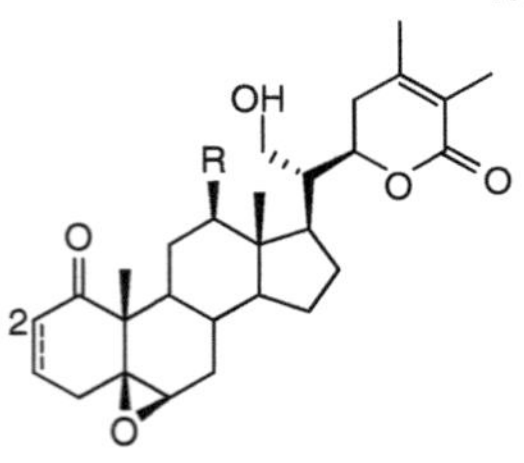

176 R = H (withatatulin)
177 R = OH, Δ^2 (withatatulin B)
180 R = H, Δ^2 (withatatulin E)

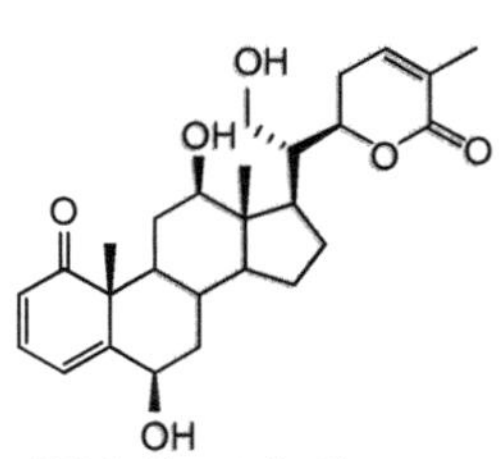

178 (withatatulin C)

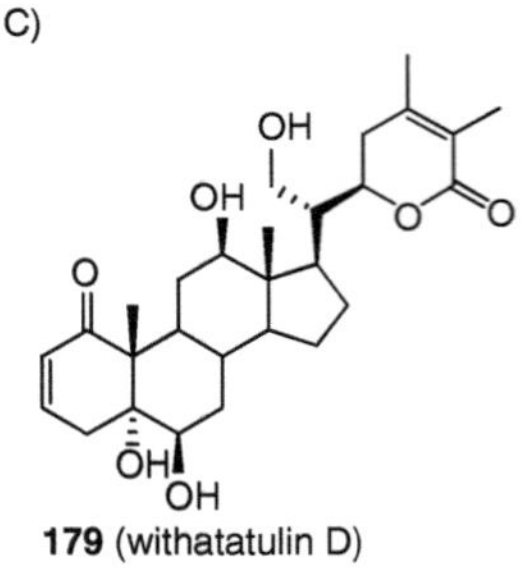

179 (withatatulin D)

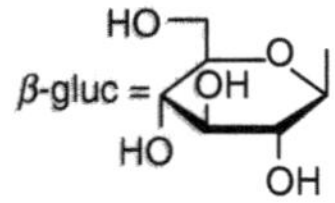

4.2.4. Other Hydroxylated Withanolides

Besides those already mentioned above, several other 16-oxygenated withanolides have been reported. Exodeconolides A–C (**181–183**) were isolated from *Exodeconus maritimus* collected in Peru (*97*), the 16-acetates **184** and **185** were reported from *Acnistus arborescens* collected in Brazil (*98*), 16α-acetoxyhyoscyamilactol (**186**) from the seeds of *Hyoscyamus niger* used in

Chinese medicine (*99*), **187** from *Dunalia brachyacantha* collected in Argentina (*71*), **188–190** from the leaves of *Discopodium penninervium* (*100*), and virginols A (**191**) and C (**192**) from *Physalis virginiana* (*101*). Virginol B (**193**) was also reported from *P. virginiana* (*101*) and **194** was isolated from the roots of *D. penninervium* (*102*). The known withaphysacarpin (**195**) and its 3-methoxy derivative, **196**, with a 16*β*-hydroxy group, were isolated from the fruits of *Physalis philadelphica* and the configuration of the lactone methyl groups was established as shown, for both compounds (*103*). LC-MS analysis of an ethyl acetate extract of the plant suggested that **196** may occur naturally although in minor amounts, being generated to a larger extent as a result of the extraction procedure. From leaves and stems of *P. philadelphica*, *Kinghorn* and coworkers isolated philadelphicalactones A and B (**197**, **198**) and compound **199**, all of them with a saturated δ-lactone side chain (*104*, *105*).

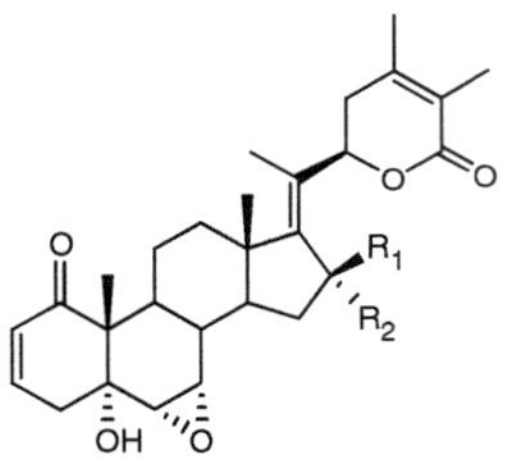

181 R^1 = OH, R^2 = H (exodeconolide A)
182 R^1 = H; R^2 = OH (exodeconolide B)

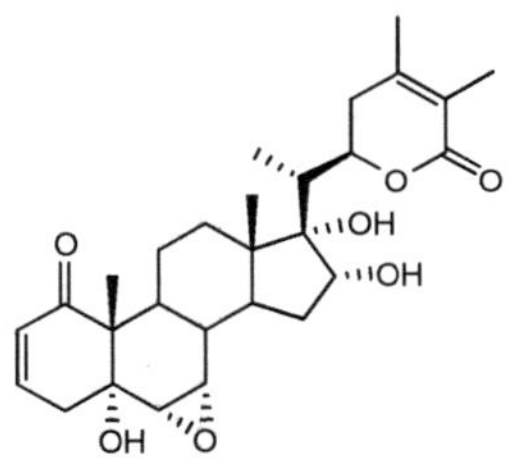

183 (exodeconolide C)

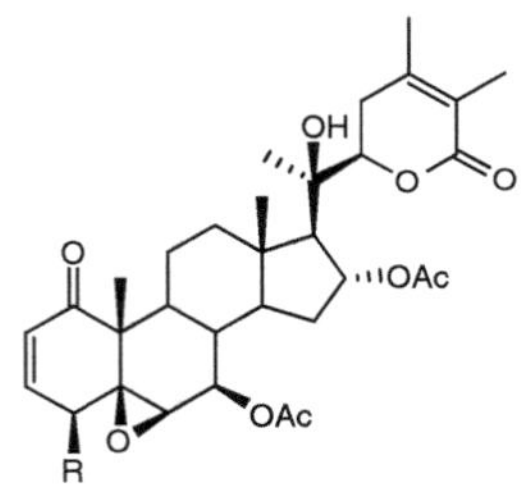

184 R = OH
185 R = H

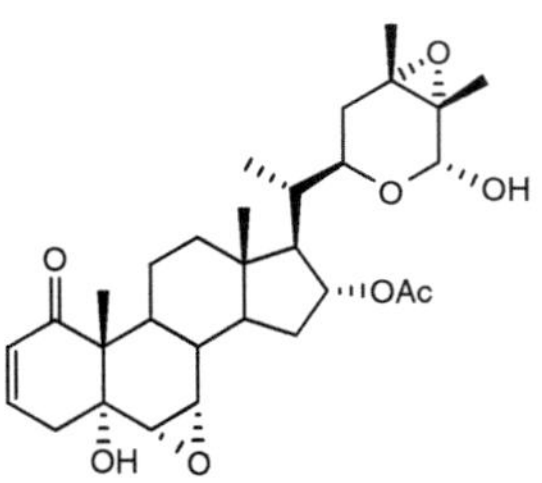

186 (16*α*-acetoxyhyoscyamilactol)

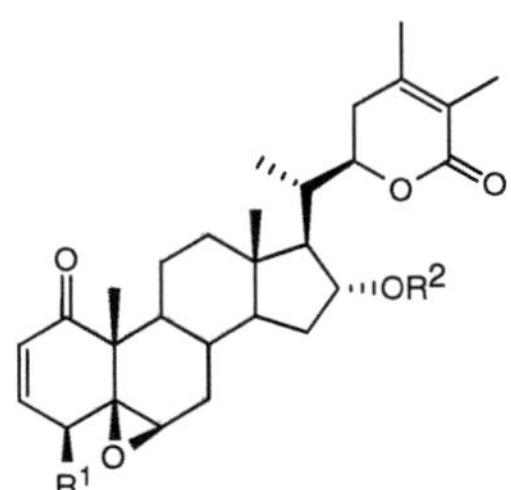

187 R^1 = OH, R^2 = H
188 R^1 = H, R^2 = Ac

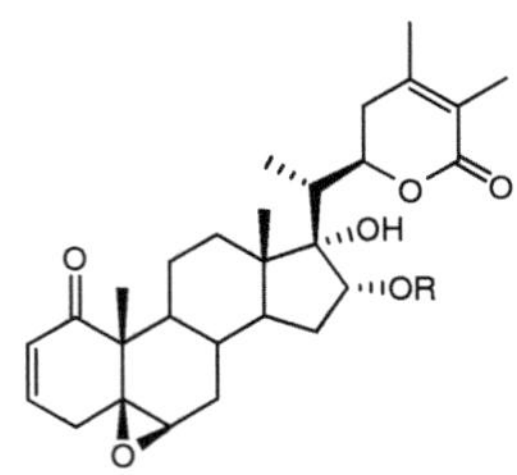

189 R = Ac
190 R = H

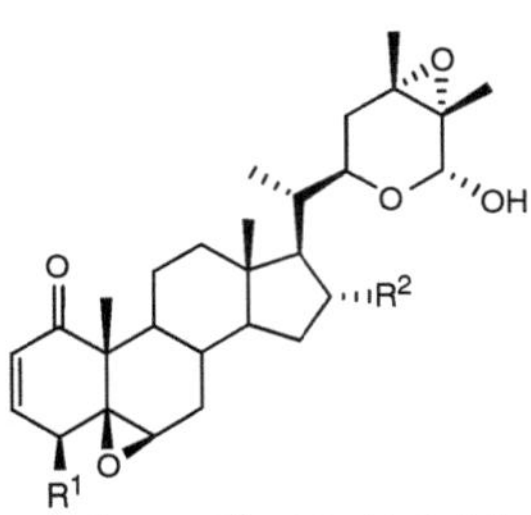

191 R^1 = OH, R^2 = OAc (virginol A)
192 R^1 = H, R^2 = OAc (virginol C)
193 R^1 = OH, R^2 = H (virginol B)

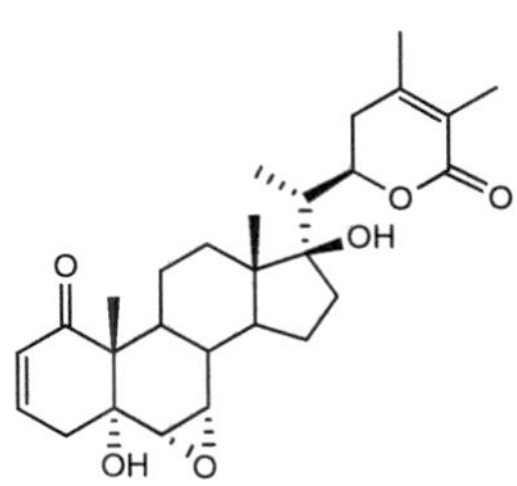

194

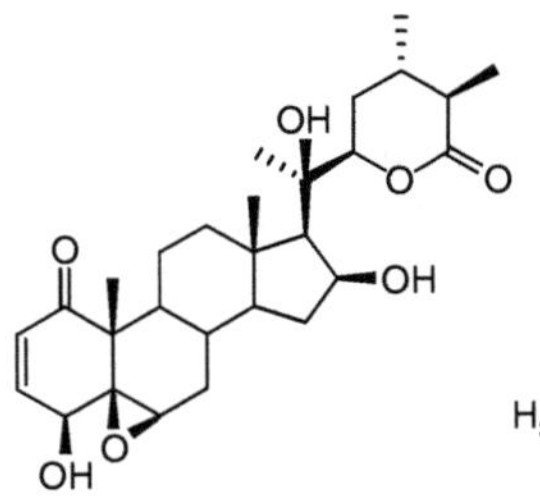

195 (withaphysacarpin)

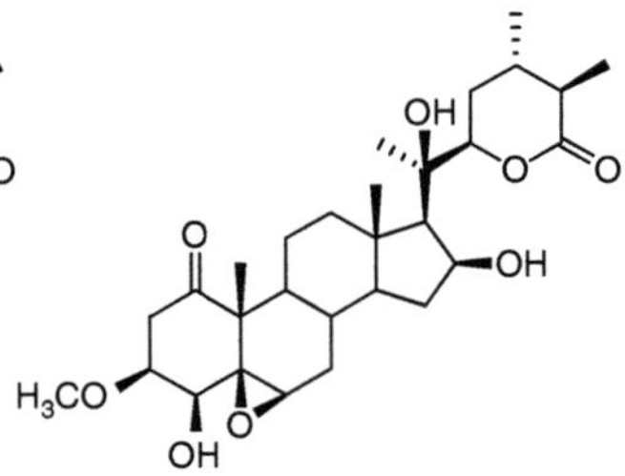

196

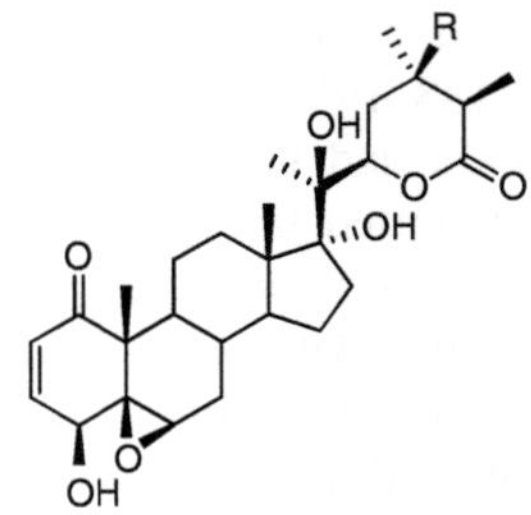

197 R = H (philadelphicalactone A)
198 R = OH (philadelphicalactone B)

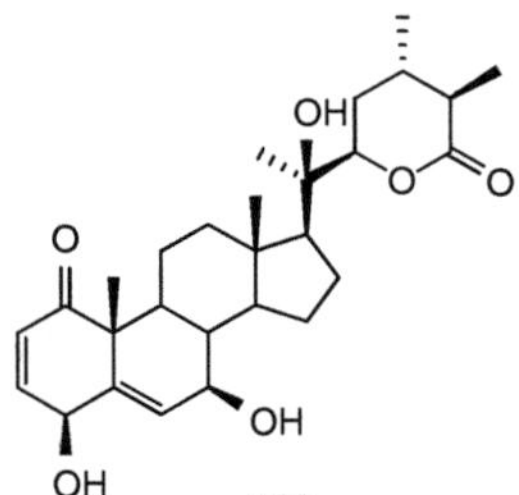

199

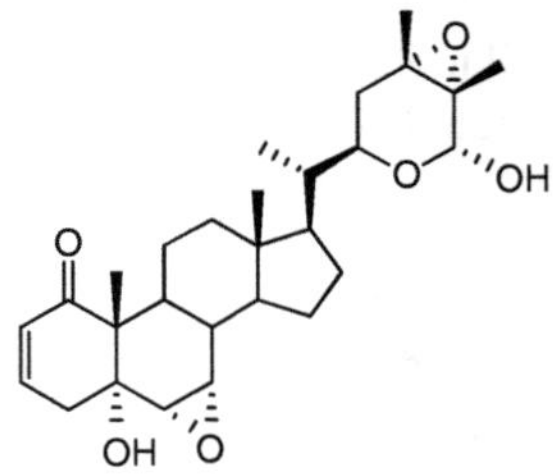

200 (Nic-3)

A 24,25-diol related to Nic-3 was reported from *Larnax glabra* (*106*), but, however, comparison of the NMR data indicates that this compound (larnaxolida A) is actually Nic-3 (**200**) (*3*). An epimer at C-5 was also reported (larnaxolida B), but the spectroscopic data do not correspond to the proposed structure.

Tubocapsanolides A (**201**), D (**202**), and F (**203**) and the 20-hydroxy and the 23-hydroxy derivatives of tubocapsanolide A **204** and **205** were isolated from *Tubocapsicum anomalum* (*107*). From *Ajuga bracteosa* collected in the north of Pakistan, *Malik* and coworkers isolated two 28-hydroxywithanolides, bracteosins A (**206**) and B (**207**), with C-19 in the latter compound oxidized to a carboxylic acid, and bracteosin C (**208**) (*108*). As in previous cases, these compounds probably derive from the corresponding Δ^2-1-ketones, upon reaction with methanol during their isolation.

From the aerial parts of *Datura metel* collected in China, *Ma et al.* isolated three new 27-glycosides, daturametelins H, I, and J (**209–211**), together with the known daturataturin A (**212**) and the aglycone **213** (*109*). The closely related 7-acetate **214** was isolated from *Iochroma gesneroides* (*72*).

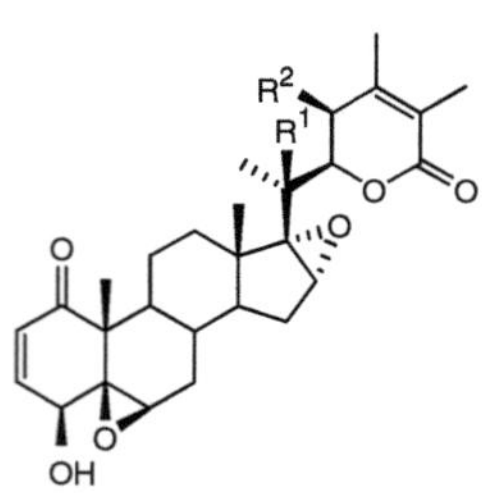

201 $R^1 = R^2 = H$ (tubocapsanolide A)
204 $R^1 = OH, R^2 = H$
205 $R^1 = H, R^2 = OH$

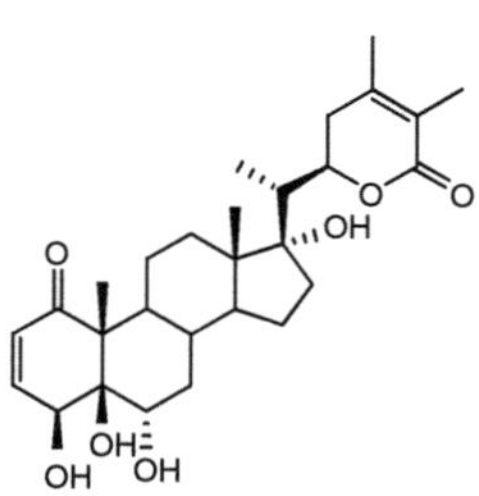

202 (tubocapsanolide D)

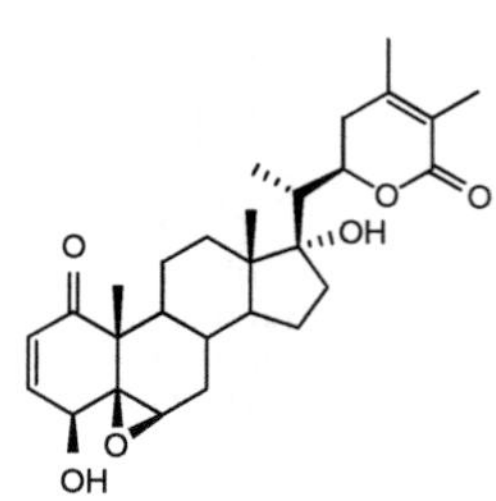

203 (tubocapsanolide F)

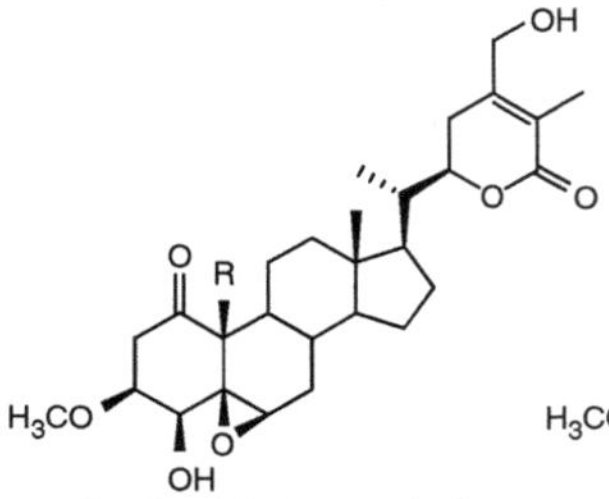

206 R = CH_3 (bracteosin A)
207 R = CO_2H (bracteosin B)

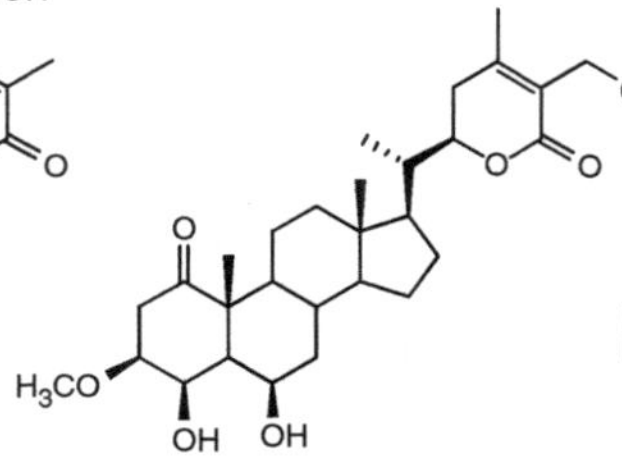

208 (bracteosin C)

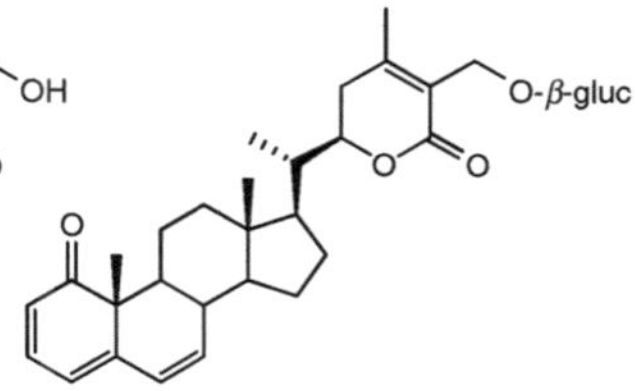

209 (daturametelin H)

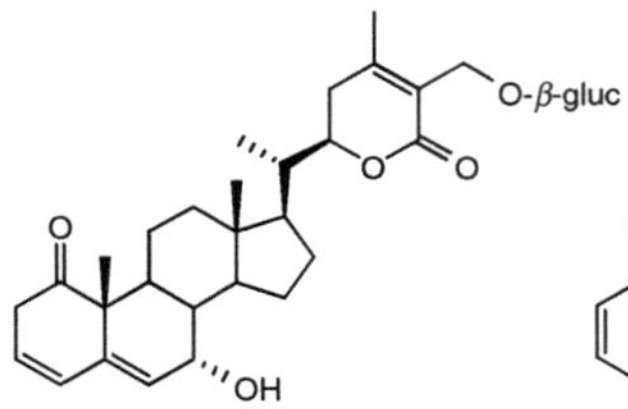

210 (daturametelin I)

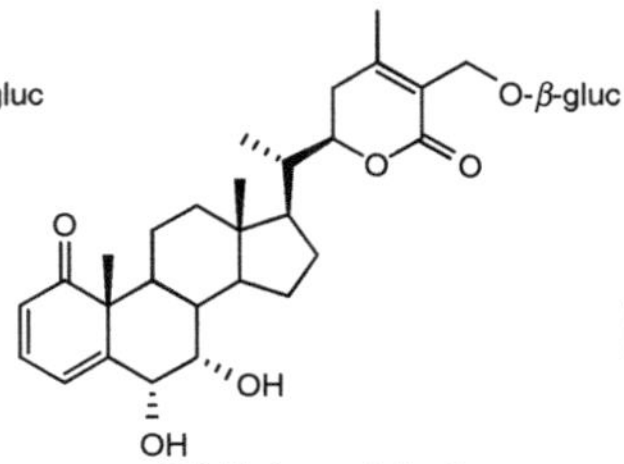

211 (daturametelin J)

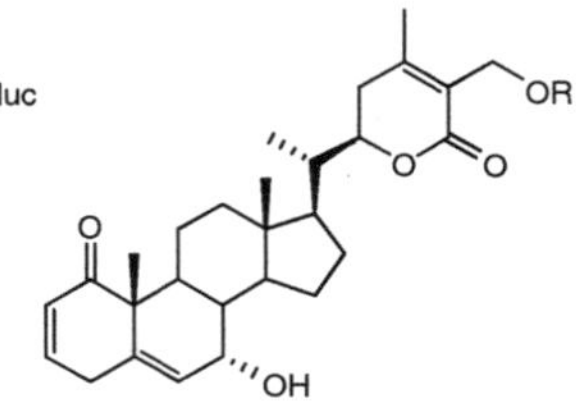

212 R = β-gluc (daturataturin A)
213 R = H

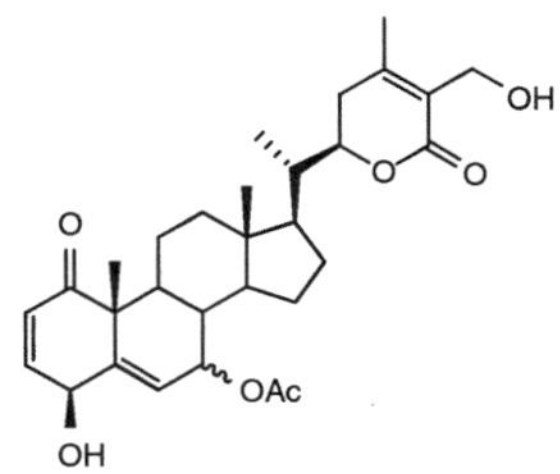

214

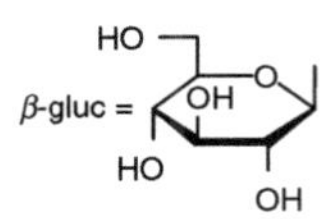

4.2.5. Cilistols and Related Withanolides

Nohara and coworkers investigated *Solanum ciliatum* (synonym *Solanum cilistum*) and isolated a series of withanolides with a δ-lactol side chain, which included several Δ^5-3β-*O*-sulfates and 6-substituted 3,5-cyclowithanolides. Variations at the side chain comprise a 24,25-epoxide or the corresponding 24,25-diol from hydrolytic cleavage, 26-*O*-glycosides, or reaction products of the epoxide or the hemiketal with methanol. Cilistols a, b, d, q, g, and f (**215–220**) contain a $\Delta^{2,5}$-1-ketone in rings A/B, and the latter two compounds (with the cleaved epoxide) exist as equilibrium mixtures of the epimeric C-26 hemiketals (*110*). The C-26 glycosides, cilistols t, i, and j (**221–223**) present a Δ^5-1α,3β-diol arrangement in rings A/B and cilistols y (**224**) and w (**225**) have the unusual feature of a 3β-*O*-sulfate group (*111*). The 3,5-cyclosteroid moiety found in cilistols u, p, pm, and p1 (**226–229**) (*112*) has no precedent among natural products but is easily formed from Δ^5-steroids with a good leaving group at C-3 (*e.g.* a sulfate); this rearrangement occurs even under very mild conditions (traces of methanol or water in the extraction or purification solvents may be enough), thus the assumption that these compounds are natural products should be taken with caution. The ergostane glycoside, cilistol v (**230**), or a closely related sterol, is probably the biosynthetic precursor of cilistols in this plant (*111*). Cilistepoxide (**231**) and cilistediol (**232**) are two closely related withanolides isolated from *Solanum sisymbriifolium* collected in Brazil (*113*).

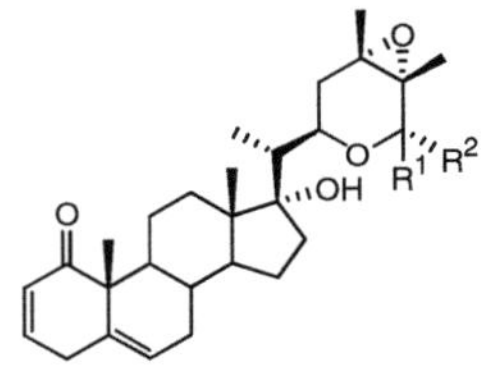

215 R^1 = H, R^2 = OH (cilistol a)
216 R^1 = OCH_3, R^2 = H (cilistol b)
217 R^1 = H, R^2 = OCH_3 (cilistol d)
218 R^1 = H, R^2 = O-β-gluc (cilistol q)

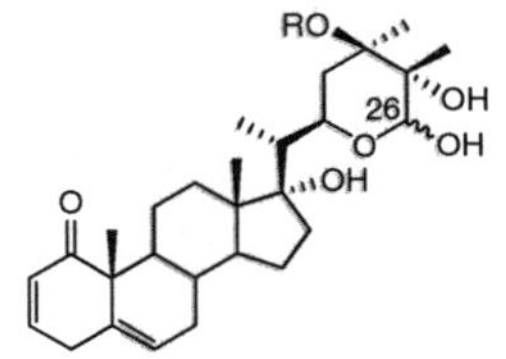

219 R = H, (26*R*/*S*) (cilistol g)
220 R = CH_3, (26*R*/*S*) (cilistol f)

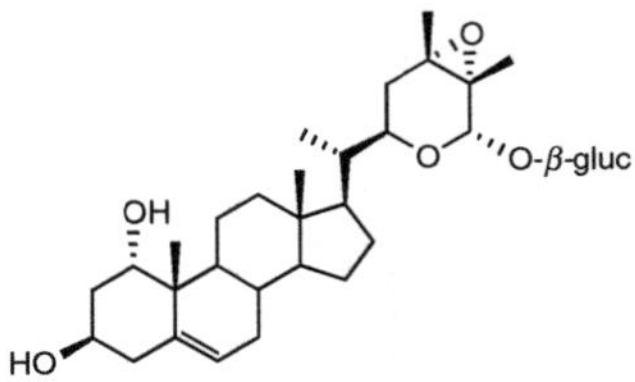

221 (cilistol t)

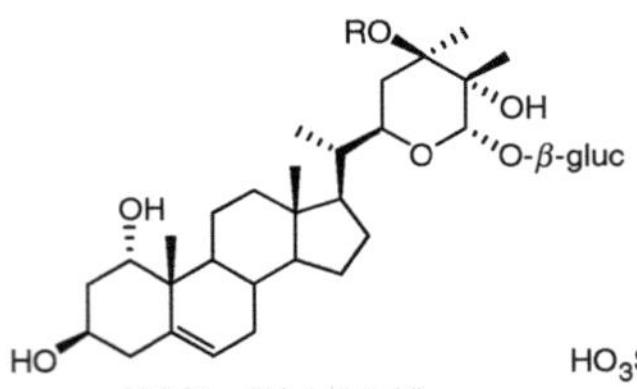

222 R = CH_3(cilistol i)
223 R = H (cilistol j)

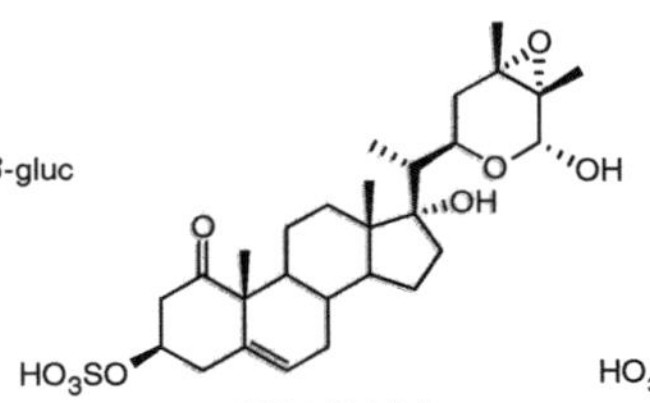

224 (cilistol y)

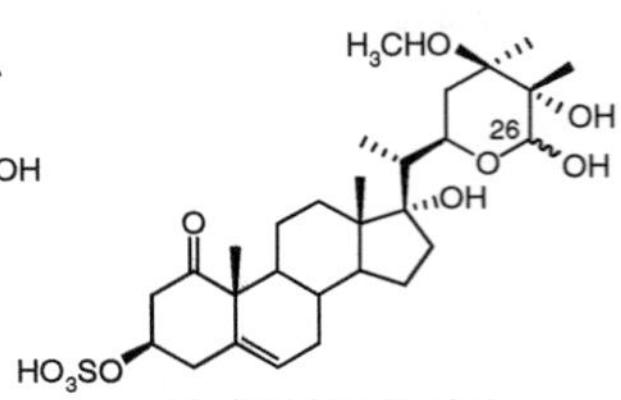

225 (26*R*/*S*) (cilistol w)

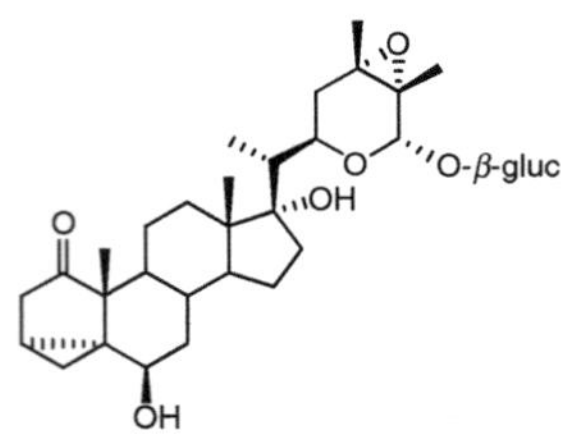

226 (cilistol u)

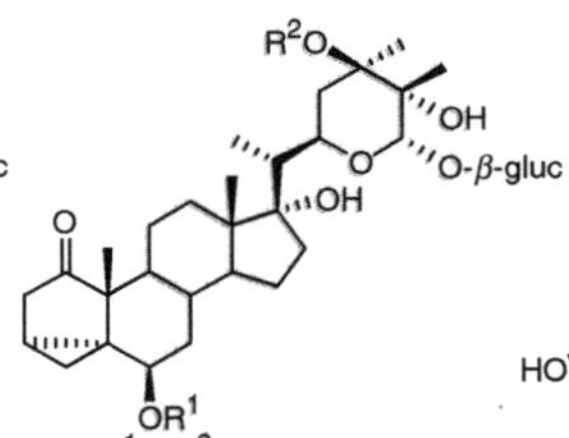

227 R^1 = R^2 = H (cilistol p)
228 R^1 = CH_3, R^2 = H (cilistol pm)
229 R^1 = R^2 = CH_3 (cilistol p1)

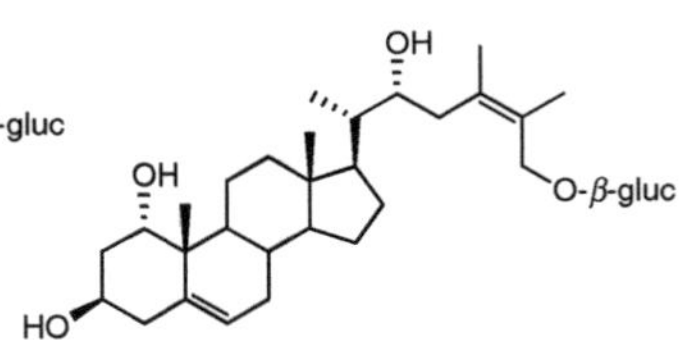

230 (cilistol v)

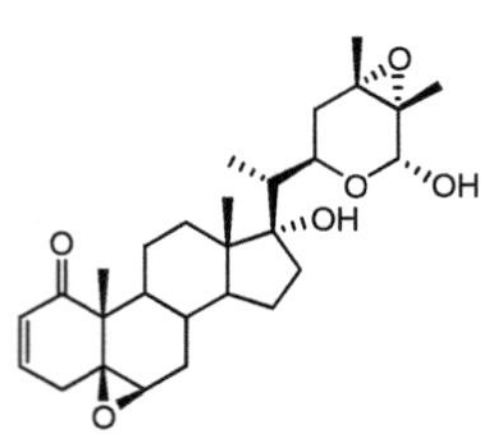

231(cilistepoxide)

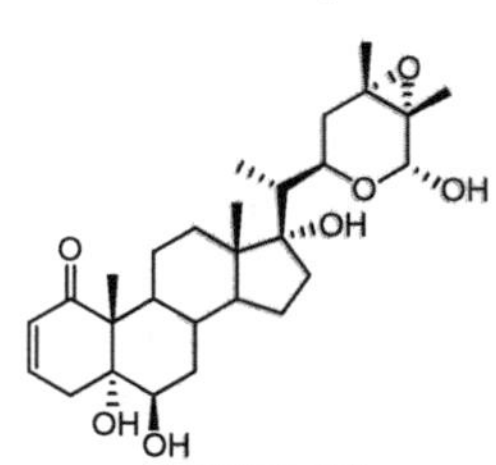

232 (cilistediol)

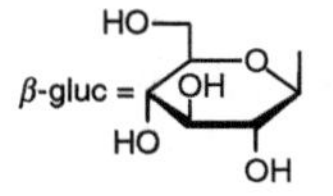

5. Withanolides with Modified Skeletons

A large number of withanolides with structures that depart from the classical withaferin A (**1**) parent structure are known nowadays. Modifications usually found are additional rings formed by direct C–C bonds, cyclic ethers and hemiketals, or lactones that may involve carbons from the steroid nucleus or from the side chain. Other modifications include formation of *seco*-steroids (as in physalins), a rearranged steroid nucleus, aromatic rings, and lack of an angular methyl as among the most important. One particularly interesting example is that present in a C_{29} withanolide from the bark of *Eucalyptus globulus* (**233**) that has an ethyl substituent at C-25 instead of the usual methyl group (*114*).

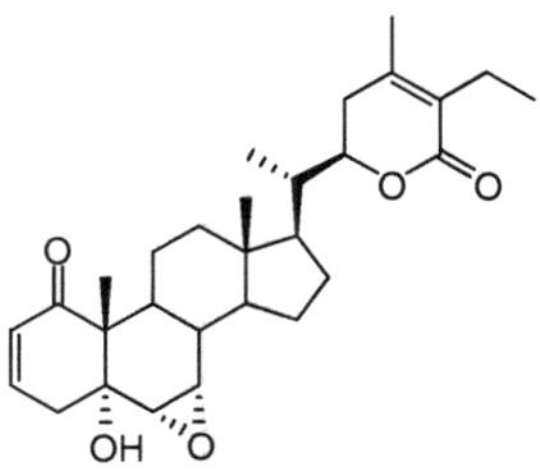

233

5.1. Withanolides with Additional Rings Involving C-21

Carbon-21 may participate in additional rings by forming either direct C–C bonds or ether bridges with other carbons in the side chain or in the steroid nucleus. Several new withanolide types with these arrangements have been reported.

5.1.1. Withametelins and Sativolides

Withametelin (**234**) exhibits an oxygen bridge between C-21 and C-24 resulting in a bicyclic side chain; it was isolated by *Ray* and coworkers from the dried leaves of *Datura metel* (*3*). Several withanolides with this functionality were subsequently isolated; some have also been named as withafastuosins, daturametelins, and baimantuoluolines on different occasions. The flowers of *D. metel*, used for centuries in traditional Chinese medicine, have been shown to contain several withametelins. *Pan et al.* isolated the 12*β*-hydroxylated derivatives withametelins I–M (**235–239**), together with withametelins N (**240**) and O (**241**), and the *seco*-withametelins **242** and **243** (*92*). Also from the dry flowers, *Yang et al.* isolated baimantuoluolines D (**244**) and E (**245**) (*93*). The closely related structures withametelinone (**246**), withametelinol (**247**), withametelinols A (**248**) and B (**249**), witharifeen (**250**), and daturalicin (**251**) were isolated by *Siddiqui et al.*

from the aerial parts of *Datura inoxia* collected in Pakistan (*115–117*). Daturacin (**252**) was reported also from *D. inoxia* as the first withanolide with a (22*S*) configuration based on a negative *Cotton* effect at 249.4 nm (*118*). The proposed configuration inversion at C-22 and C-24 compared to other withametelins (*92*), would require H-20 to occupy an *axial* instead of *equatorial* position. However, the almost perfect coincidence of the ^{1}H and ^{13}C NMR data of the side chain with those of other "normal" withametelins and the assignment of the configuration at C-5 based exclusively on CD data, suggest that this structure should be extensively revised.

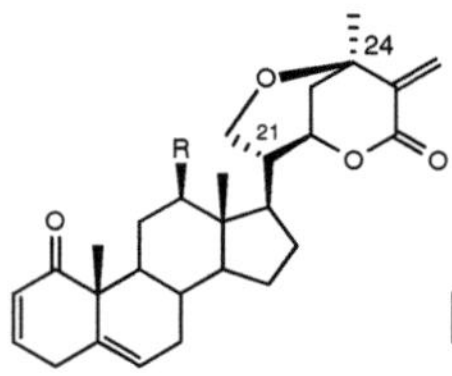

234 R = H (withametelin)
238 R = OH (withametelin L)

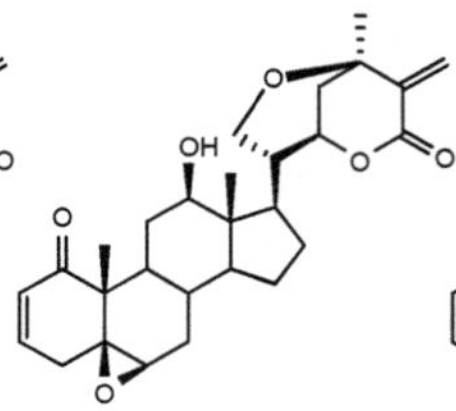

235 (withametelin I)

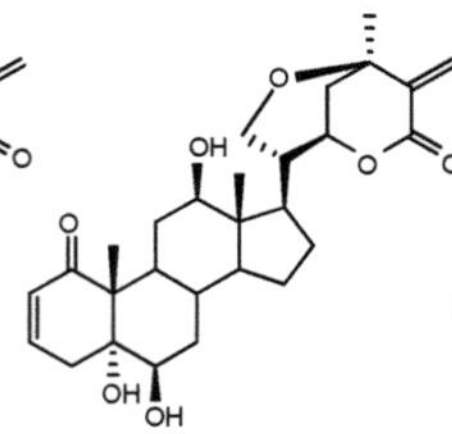

236 (withametelin J)

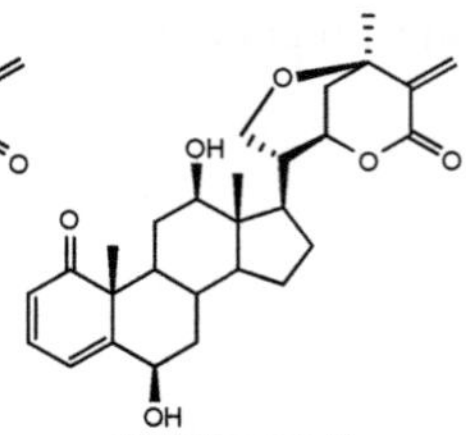

237 (withametelin K)

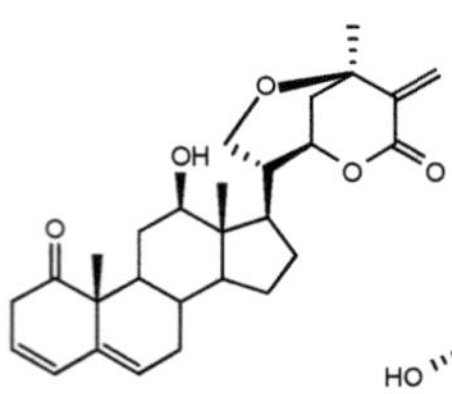

239 (withametelin M)

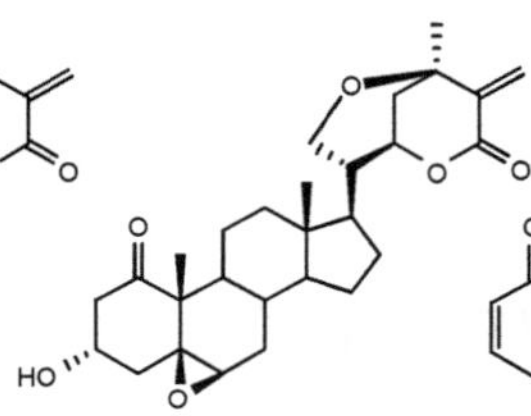

240 (withametelin N)

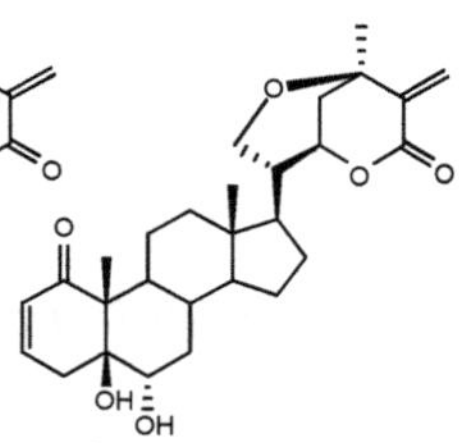

241 (withametelin O)

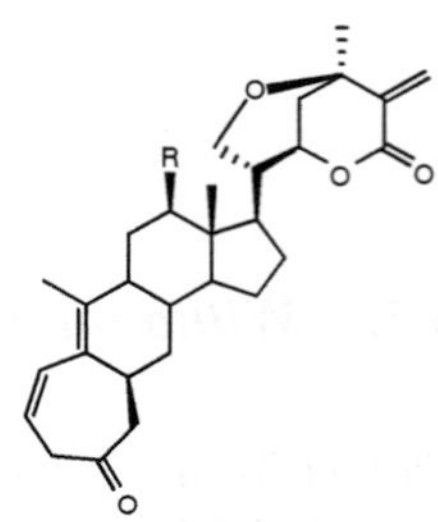

242 R = H
243 R = OH

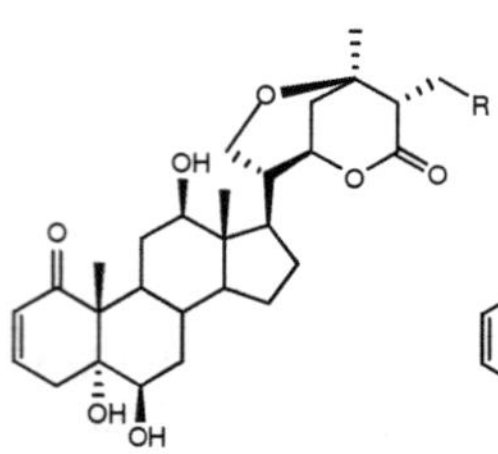

244 R = OCH_3 (baimantuoluoline D)
245 R = OH (baimantuoluoline E)

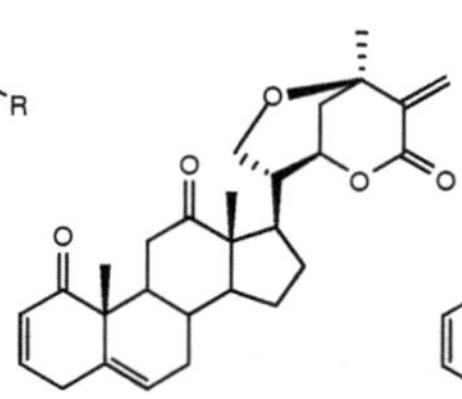

246 (withametelinone)

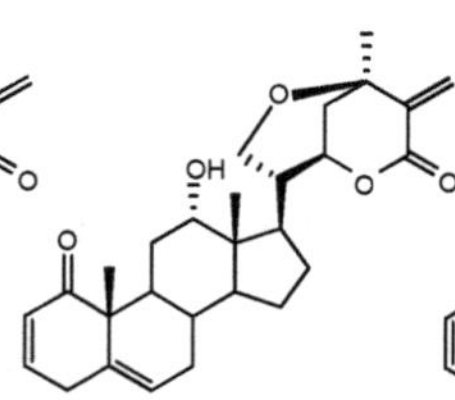

247 (withametelinol)

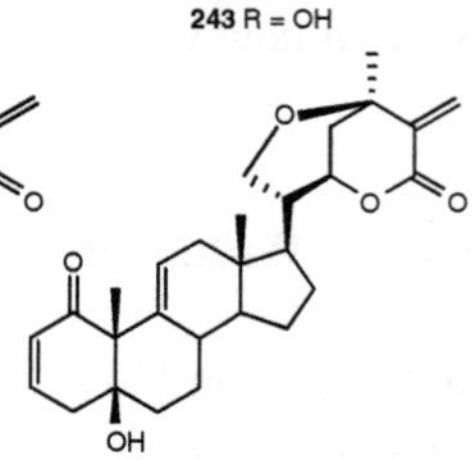

248 (withametelinol A)

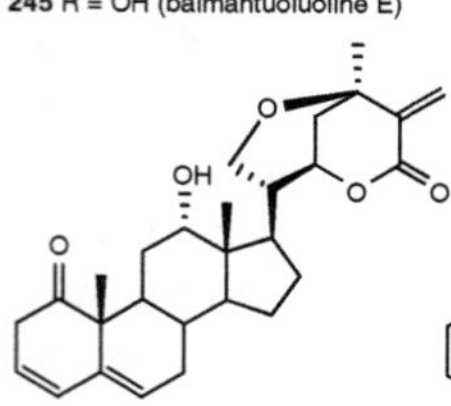

249 (withametelinol B)

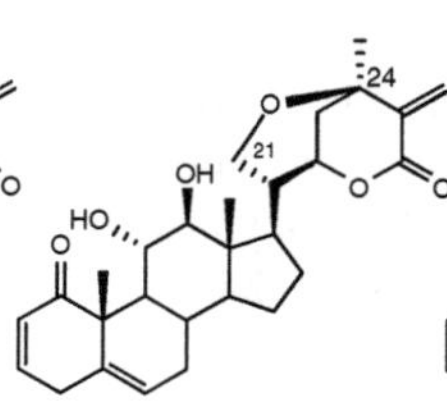

250 (witharifeen)

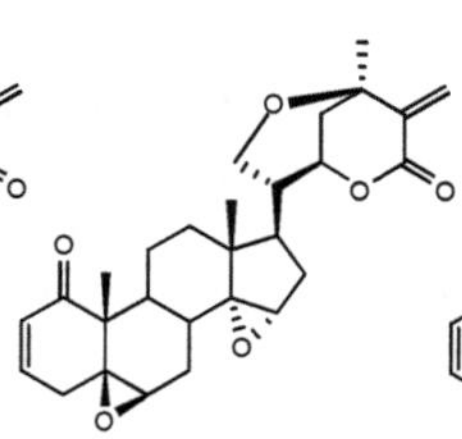

251 (daturalicin)

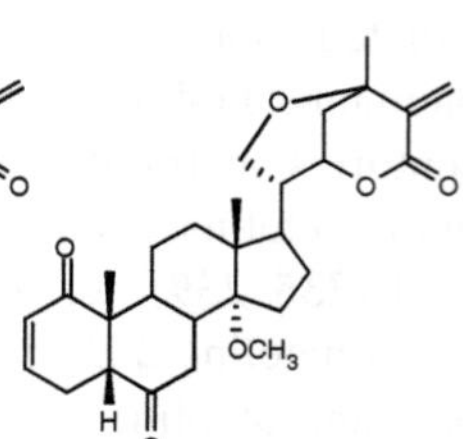

252 (daturacin)

At variance with the above, sativolides have an oxygen bridge between C-21 and C-12 in the steroid nucleus. The additional six-membered hemiketal (or ketal) ring, results from what must have been originally a C-12 ketone and a C-21 hydroxy group. Jaborosalactones R (**253**), S (**254**), and T (**255**), were isolated from *Jaborosa sativa* (synonym *Trechonaetes sativa*) collected in Argentina (*119*). C-12-Hemiketals are highly reactive towards simple alcohols (see Sect. 5.4.2.), thus the methyl ketal **255** is probably formed during isolation. *Nicotra et al.* reported the isomeric clorohydrin, jaborosalactone 37 (**256**), from *Jaborosa rotacea* (*120*). Interestingly, the Δ^2-withanolide **257** was present in *Jaborosa caulescens* var. *caulescens* (isolated together with the 12-*O*-methyl derivative) while its 2,3-dihydro derivative **258** was present in *J. caulescens* var. *bipinnatifida* (*121*).

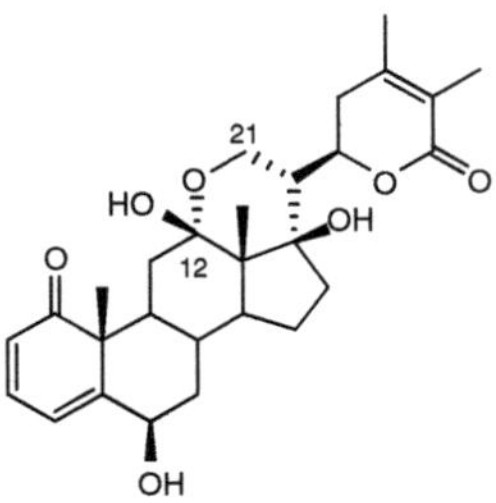

253 (jaborosalactone R)

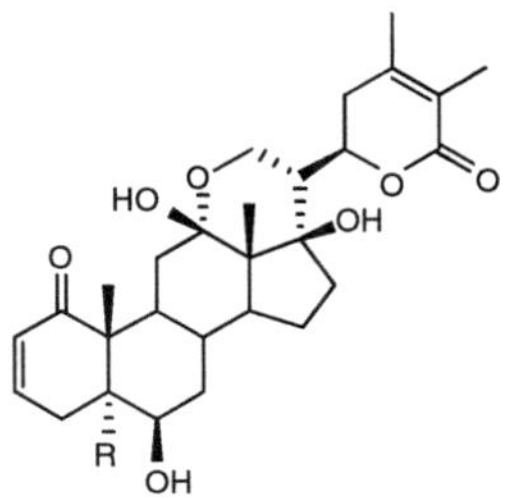

254 R = OH (jaborosalactone S)
256 R = Cl (jaborosalactone 37)

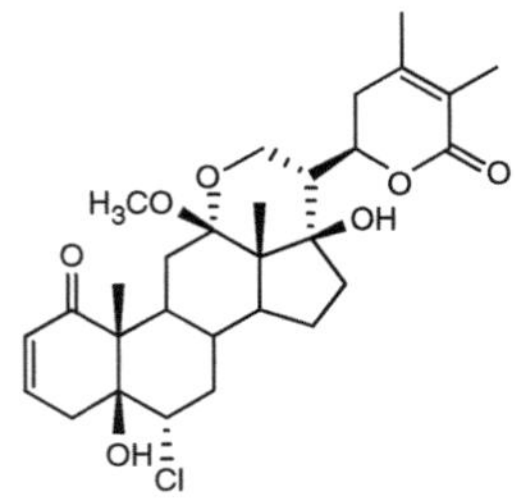

255 (jaborosalactone T)

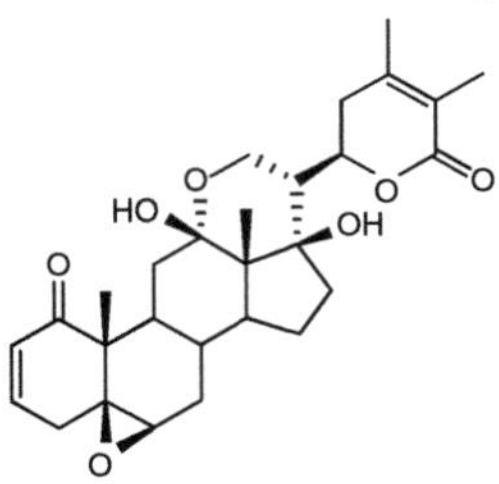

257 (jaborosalactone 38)

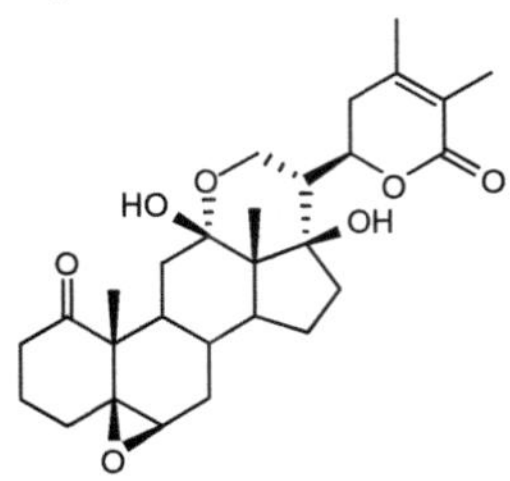

258 (jaborosalactone 39)

5.1.2. Acnistins

The acnistins also exhibit a bicyclic side chain involving C-21 and the lactone ring but, at variance with withametelins, C-21 is directly bonded to C-24 via a C–C bond instead of an ether bond (*3*, *4*). It has been proposed that the 21,24 bond is probably formed via a SN_2 type reaction in withanolides having a good leaving group at C-21. The first examples of this family, acnistins A (**259**) and E (**260**), were isolated by *Usubillaga et al.* from plants of *Acnistus arborescens* (synonym *Acnistus ramiflorus*) collected in Venezuela, with their stereochemical and spectroscopic assignments later revised by *Luis et al.* (*122*, *123*). Withanolides with this bicyclic side chain have been reported also from *Tubocapsicum anomalum* (*3*). *Luis* and coworkers isolated acnistins A and E, together with the new acnistins B (**261**),

C (**262**), D (**263**), F (**264**), G (**265**), and H (**266**), from the leaves of *Dunalia solanacea* collected in Medellin (Colombia) (*124–126*).

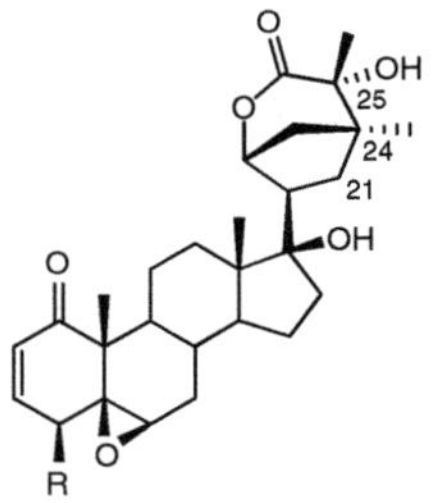

259 R = H (acnistin A)
260 R = OH (acnistin E)

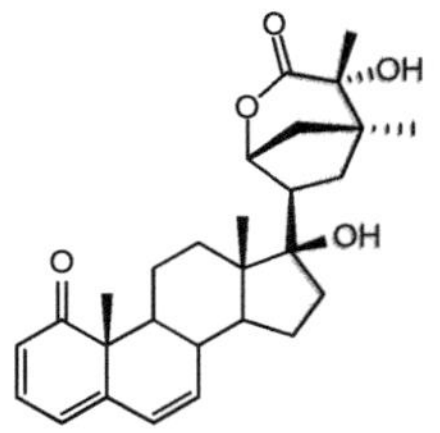

261 (acnistin B)

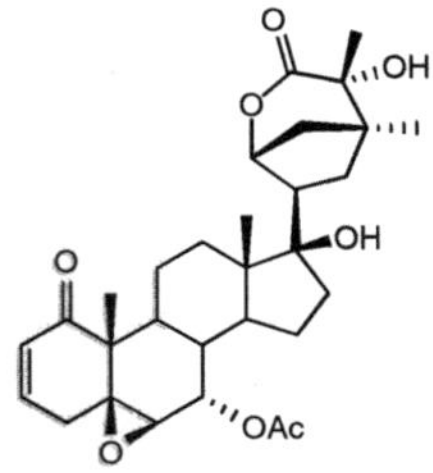

262 (acnistin C)

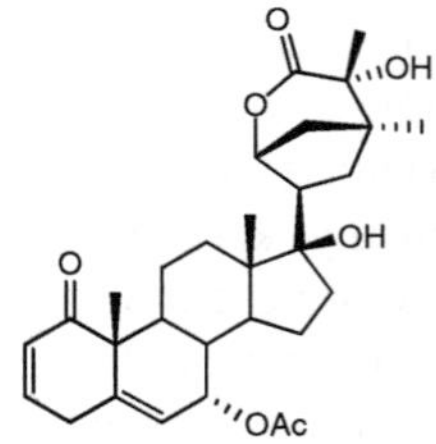

263 (acnistin D)

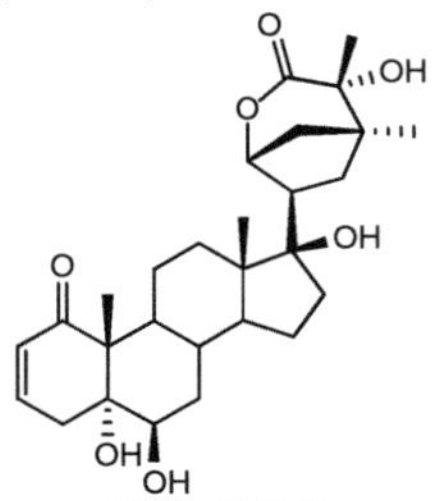

264 (acnistin F)

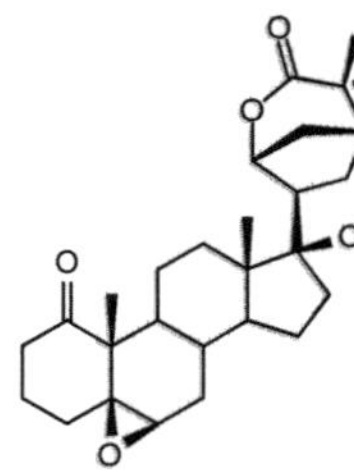

265 (acnistin G)

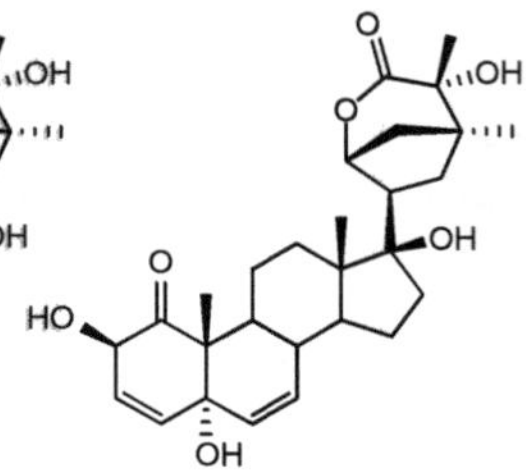

266 (acnistin H)

An epimer of acnistin A, 17-epiacnistin A (**267**) was isolated from *Discopodium penninervium* collected in Ethiopia (*127*). Recently, using bioassay-directed fractionation, six new 17-epiacnistins were isolated from *Tubocapsicum anomalum* collected in Taiwan (*107*). Anomanolides A (**268**) and B (**269**) were identified as the 17-epimer of acnistin E and the ring B diol resulting from diequatorial cleavage of the epoxide. Anomanolides C–F (**270–273**) had an additional 16α-hydroxy substituent. From fruits of *T. anomalum* collected in Japan, *Kiyota et al.* isolated three acnistin glycosides, isotubocaposides A–C (**274–276**) with a 1α,3β-dihydroxy substitution pattern in ring A (*128*). The distinctive feature of these acnistins was an inverted configuration at C-25 as determined by X-ray crystallography.

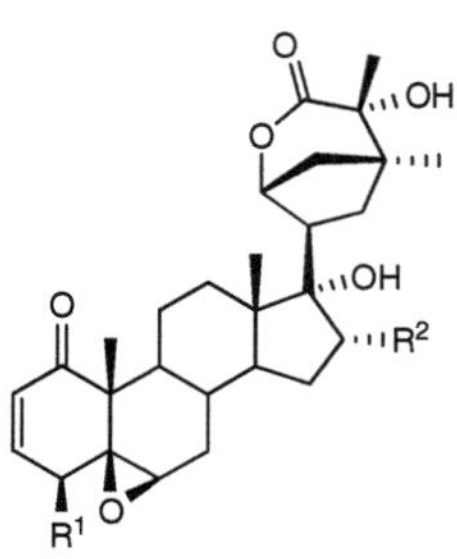

267 $R^1 = R^2 = H$ (17-epiacnistin A)
268 $R^1 = OH, R^2 = H$ (anomanolide A)
270 $R^1 = R^2 = OH$ (anomanolide C)

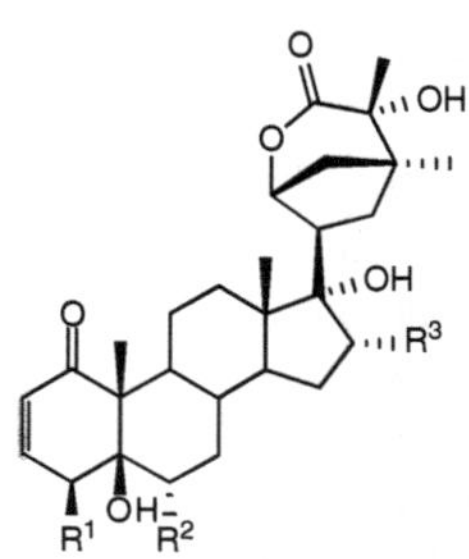

269 $R^1 = R^2 = OH, R^3 = H$ (anomanolide B)
271 $R^1 = R^3 = OH, R^2 = Cl$ (anomanolide D)
272 $R^1 = H, R^2 = R^3 = OH$ (anomanolide E)
273 $R^1 = R^3 = OH, R^2 = H$ (anomanolide F)

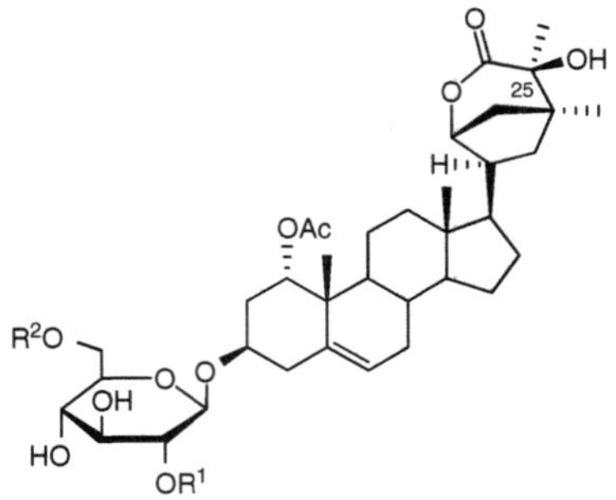

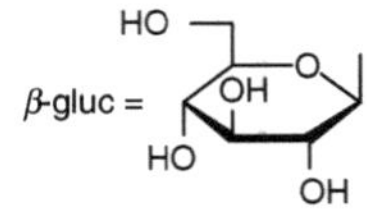

274 $R^1 = R^2 = \beta$–gluc (isotubocaposide A)
275 R^1 = H, $R^2 = \beta$–gluc (isotubocaposide B)
276 $R^1 = \beta$–gluc, R^2 = H (isotubocaposide C)

5.1.3. Withajardins

In the withajardins, C-21 is directly bonded to C-25 resulting in a bicyclic lactone side chain with a six-membered homocycle. Withajardins A–E (**277–281**) were isolated from plants of *Deprea orinocensis* collected in Colombia (*129*, *130*). A 16,17-dihydroxylated withajardin, tubonolide A (**282**), was isolated from the stems and leaves of *T. anomalum* (*107*) and the glycosides tuboanosides A and B (**283** and **284**) from the fruits of this same plant (*131*). The latter three compounds have an inverted configuration at C-24 compared to the other withajardins, and this was confirmed by X-ray diffraction in the case of tuboanosides A and B (Fig. 4). (It should be noted that in the original publication, the configuration at C-20 in the structure drawing of these compounds is incorrect; the structure shown here was taken from the X-ray data deposited at the Cambridge Crystallographic Data Centre, CCDC). A common precursor has been proposed in the biogenetic routes to acnistins, withajardins, and withametelins; the simultaneous finding of acnistins and withajardins in *T. anomalum* supports this proposal.

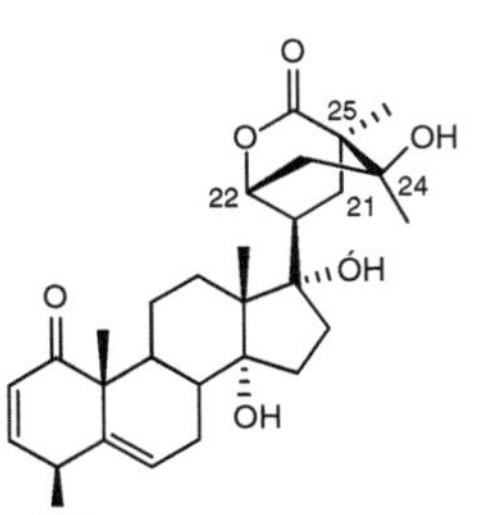

277 R = H (withajardin A)
280 R = Ac (withajardin D)

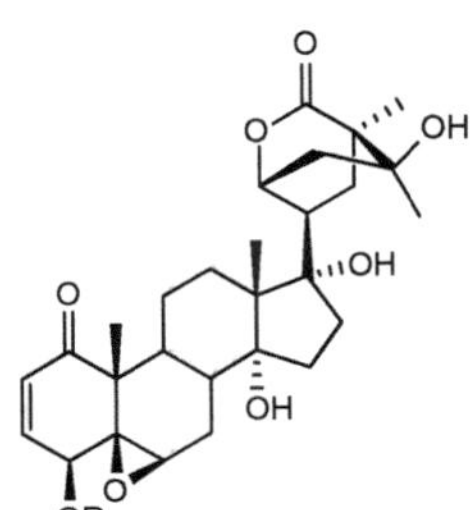

278 R = H (withajardin B)
279 R = Ac (withajardin C)

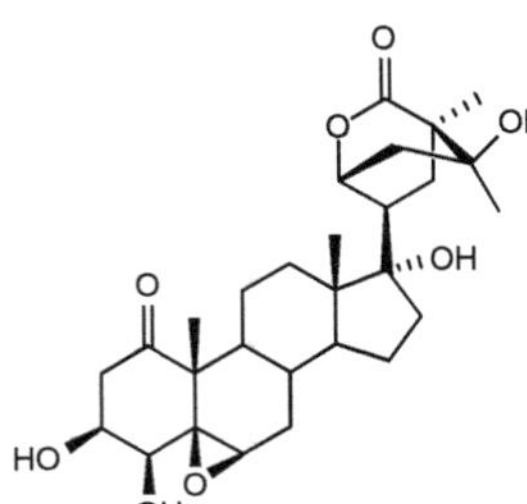

281 (withajardin E)

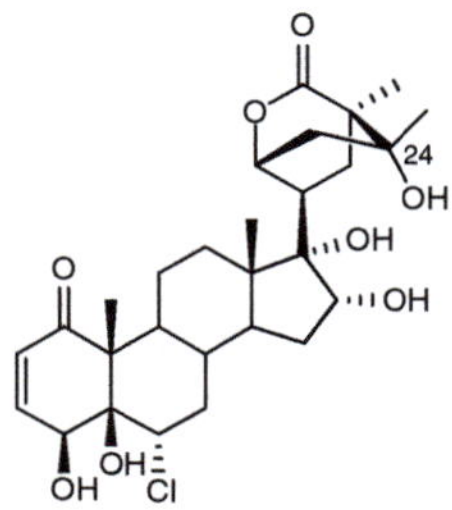

282 (tubonolide A)

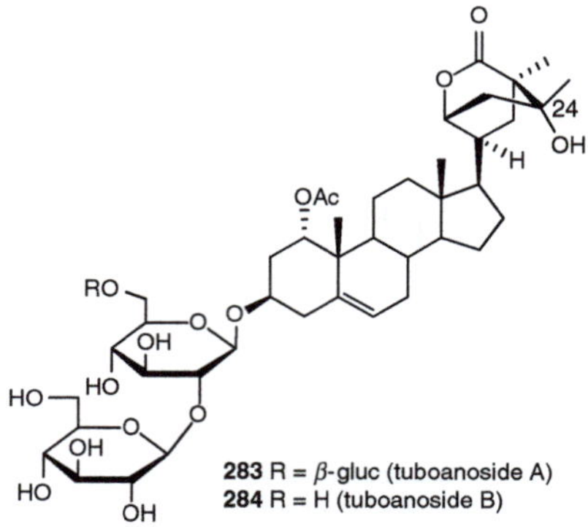

283 R = β-gluc (tuboanoside A)
284 R = H (tuboanoside B)

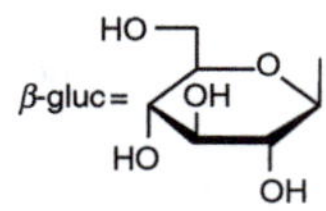

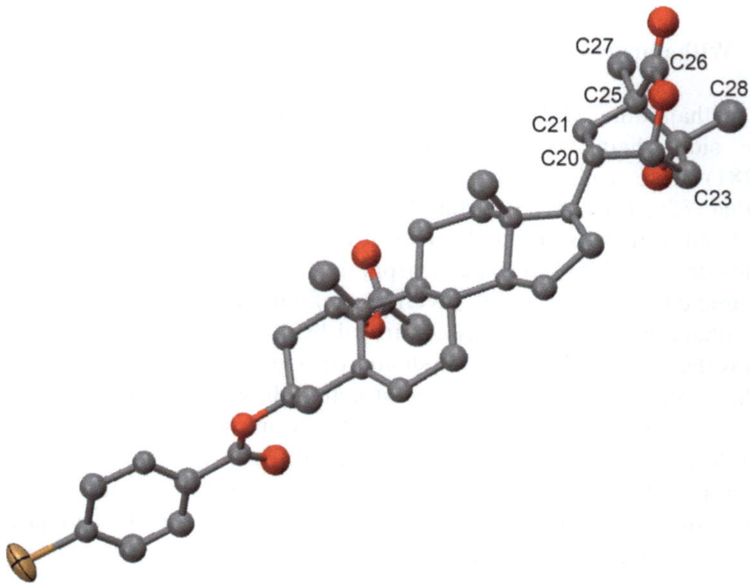

Fig. 4. X-ray crystal structure of the *p*-bromobenzoate of tuboanosigenin (CCDC 680092), aglycone of tuboanosides A (**283**) and B (**284**) (*131*). Structure drawing generated with Mercury 2.3

5.1.4. 15,21-Cyclowithanolides (Norbornane Type)

Nicotra et al. reinvestigated *Jaborosa bergii* and isolated five new withanolides with a carbon-carbon bond between C-15 and C-21, resulting in a novel norbornane-type structure in ring D (*69*). Jaborosalactols 18 (**285**) and 22 (**286**) have a 14α-hydroxy group while jaborosalactols 19–21 (**287–289**) contain a 8,14 double bond. 14α-Hydroxywithanolides are known to dehydrate easily giving a mixture of $\Delta^{8,14}$ and Δ^{14} unsaturated derivatives (*132*), and the finding that jaborosalactone 22 (**286**) spontaneously gave **289** strongly suggests that, in this case, the $\Delta^{8,14}$ unsaturated compounds are artifacts formed during isolation. The bridgehead nature of C-15 would prevent the formation of Δ^{14} derivatives (*69*).

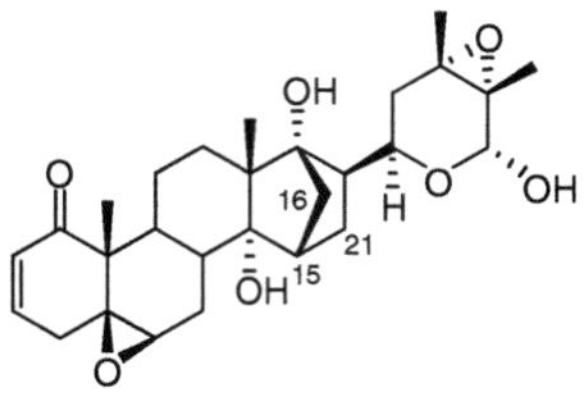

285 (jaborosalactol 18)

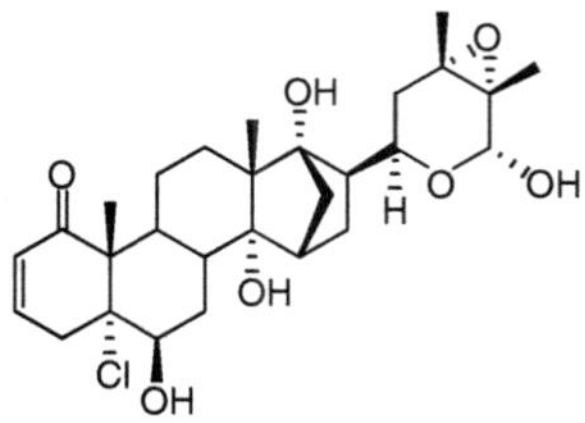

286 (jaborosalactol 22)

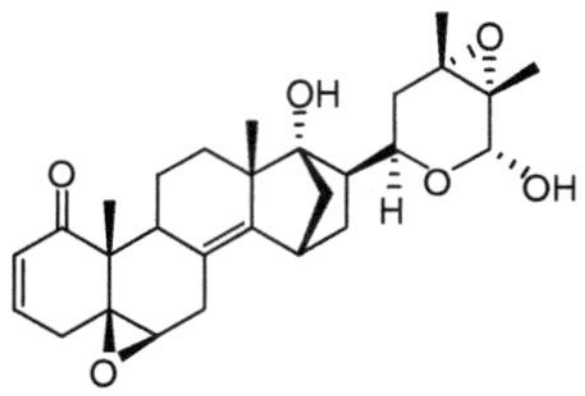

287 (jaborosalactol 19)

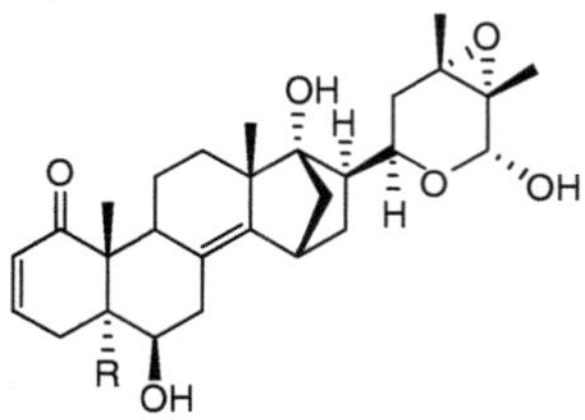

288 R = OH (jaborosalactol 20)
289 R = Cl (jaborosalactol 21)

5.2. *Physalins and Withaphysalins*

The physalins are a group of 13,14-*seco*-16,24-cycloergostane constituents of certain *Physalis* species (*3*, *133*). In recent years, some known physalins have been found in *Brachistus stramoniifolius* (physalins B, F, and H) (*134*), *B. hunzikeri* (*sub nom. Witheringia hunzikeri*) (physalin B) (*135*), *Margaranthus solanaceous* (*sub nom. Physalis solanaceous*) (physalins A, B, D, and F) (*136*), *Schraderanthus viscosus* (*sub nom. Saracha viscosa*) (physalins D, F, and H) (*137*), and *Witheringia solanacea* (physalins B, D, and F) (*138*). A total of 22 new physalins have been reported from *Physalis* species and several structures described previously have been revised. Withaphysalins, with an oxidized C-18 involved in a lactone or lactol ring with C-20 (see Fig. 2), are believed to be the biosynthetic precursors of physalins.

5.2.1. Normal Physalins

Fifteen new withanolides with the normal physalin skeleton were isolated from *Physalis* species, differing in the substitution pattern of rings A and B. From the calyces of *P. alkekengi* var. *franchetii*, *Qiu et al.* isolated three 3-hydroxy-1-ketophysalins (*139*), the new physalins Y (**290**) and Z (**291**) with a 3α-oriented hydroxy group, and compound **292** that had NMR data coincident with isophysalin G previously isolated from the same plant by *Sunayama et al.*, but was not fully characterized (*140*). NOE correlations of **292** established the β-orientation for the 3-hydroxy group of isophysalin G. The 3-methoxy analogues **293** and **294** had been isolated previously by the same authors (*141*). Physalin S (**295**) isolated from the same plant, had a 6β-hydroxy-3,5-cyclo arrangement, a common acid rearrangement product of 3-hydroxy-Δ^5 steroids (see Sect. 4.2.5.) (*142*).

Two other physalins from *P. alkekengi* var. *franchetii* corresponded to the 5α,6β-diol, physalin T (**296**) (as already mentioned derived from hydrolytic opening of a 5,6-epoxide) (*143*), and the less common 2,5-endoperoxy-Δ^3 arrangement of physalin Q (**297**) (*144*). *Choudhary et al.* isolated the four physalins **298**, **299** (*60*), **300**, and **301** (*145*) from *P. minima*, the latter three containing a 11β-hydroxy group. A reduced derivative at C-1, physalin V (**302**), was also isolated from *P. angulata* (*146*).

Several 3-alkoxy derivatives besides those mentioned above have been reported, but most probably all of these are artifacts formed by reaction of a Δ^2-1-ketone with ethanol or methanol during isolation. Thus, the 3-ethoxy derivative **303** isolated from *P. alkekengi*, was found only in trace amounts in the original extract when analyzed by HPLC (*67*). The 3-methoxy derivative **304** (physalin U) was initially isolated from *P. minima* (*60*) and later from *P. angulata* (*146*), while the 3-methoxy derivative **305** was isolated from *P. angulata* (*55*). The latter compound was named physalin W although this name had already been assigned to another physalin (see below).

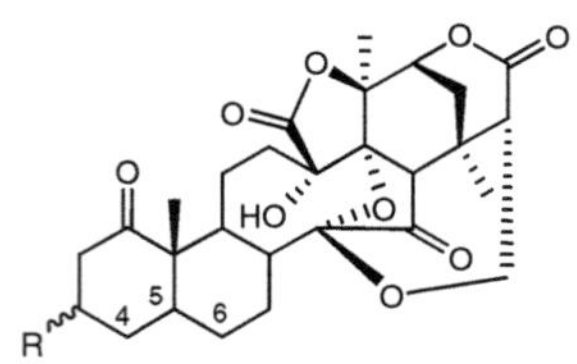

290 R = α-OH, Δ^5 (physalin Y)
291 R = α-OH, $\Delta^{4,6}$ (physalin Z)
292 R = β-OH, $\Delta^{4,6}$ (isophysalin G)
293 R = α-OCH_3, $\Delta^{4,6}$
294 R = β-OCH_3, $\Delta^{4,6}$

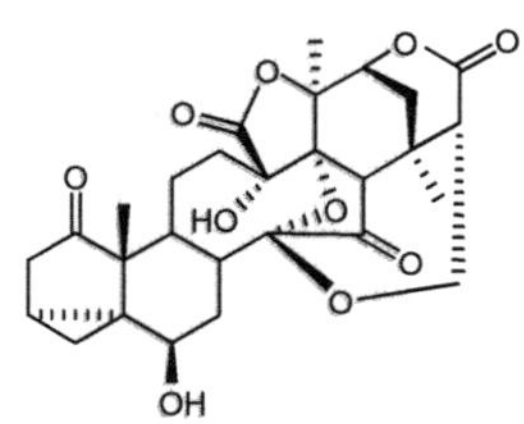

295 (physalin S)

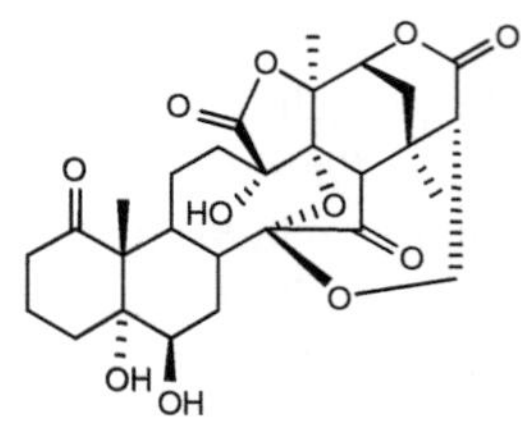

296 (physalin T)

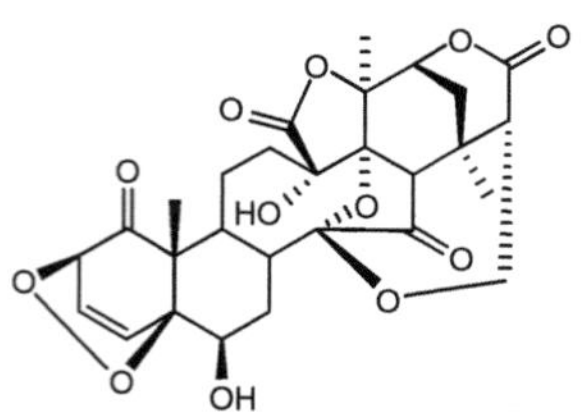

297 physalin Q

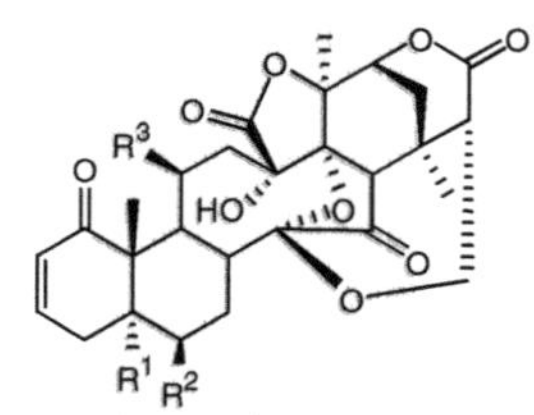

298 R^1 = OH, R^2 = R^3 = H
299 R^1 = R^2 = R^3 = OH
300 R^1 = OCH_3, R^2 = H, R^3 = OH

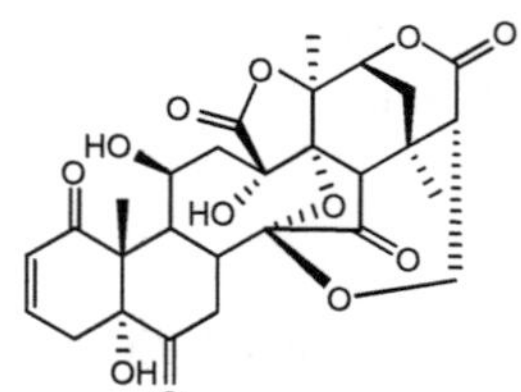

301

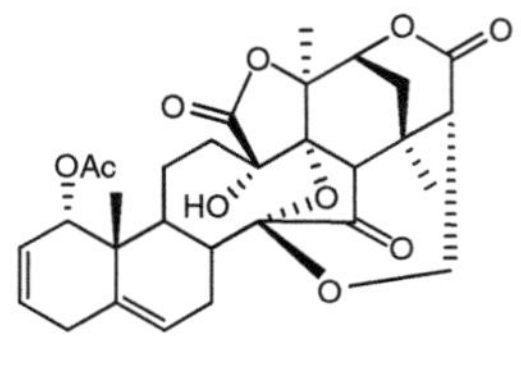

302 (physalin V)

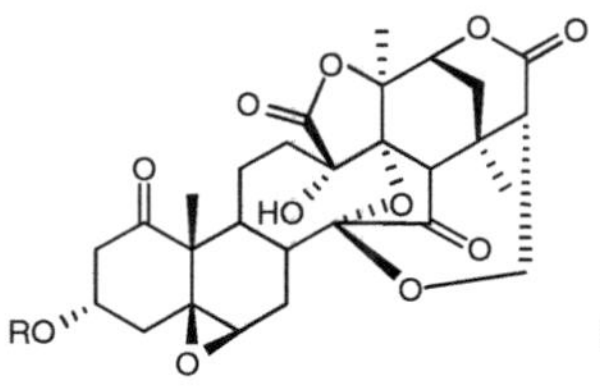

303 R = CH_2CH_3
304 R = CH_3 (physalin U)

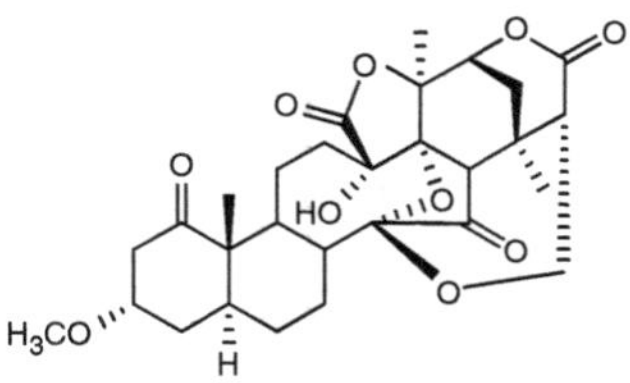

305

5.2.2. Neophysalins and Cyclophysalins

Neophysalins have a rearranged skeleton in which C-14 is directly bound to C-16 and the C-15 carbonyl forms a lactone with the oxygen atom at C-17. Four new neophysalins and a 11,16-cyclophysalin were isolated from *P. alkekengi* var. *franchetii*. Physalins W (**306**) and X (**307**) were initially isolated by *Chen et al.* (*147*) and their structures revised by *Qiu* and coworkers (*139*). The latter authors also reported the isolation of the 3-methoxy analogues physalin I (**308**) and physalin II (**309**). Physalin R (**310**) is a normal physalin with an additional bond between C-11 and C-16 (*142*). The authors showed that this cyclophysalin skeleton could be obtained from normal physalins upon irradiation with an halogen-tungsten lamp under argon.

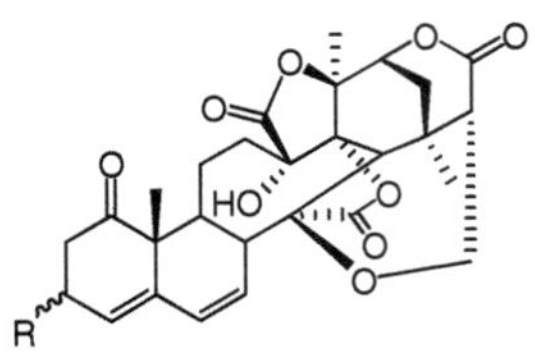

306 R = α-OH (physalin W)
307 R = β-OH (physalin X)

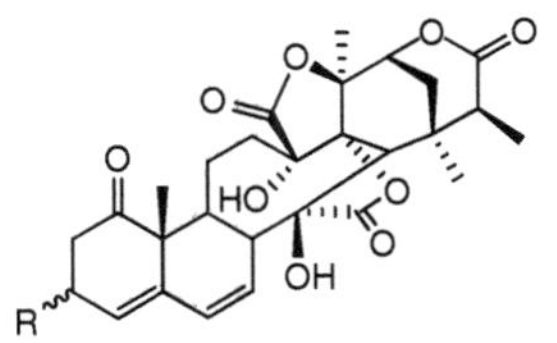

308 R = α-OCH_3 (physalin I)
309 R = β-OCH_3 (physalin II)

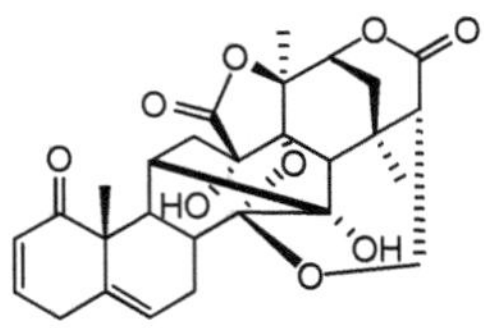

310 (physalin R)

5.2.3. Revised Physalin Structures

The structures of several physalins have been revised. Spectroscopic studies and chemical correlations revealed that the reported structure of physalin K containing a 4α,5α-epoxy-6α-hydroxy-2-en-1-one arrangement in rings A/B was incorrect, and the revised structure corresponds to a 2α,5α-epidioxy-6β-hydroxy-3-en-1-one (**311**), an isomer of physalin Q (**297**) (*144*).

Makino et al. have revised the structure of physalin H, originally reported as having a Δ^5-7β-hydroxy arrangement in ring B. The correct structure corresponded to the clorohydrin **312** (*148*). The authors also demonstrated that physalin E originally reported as a 5α,7α-dihydroxy-2-en-1-one is identical to the 5,6-diol physalin D (**313**). The acetate of physalin E also corresponds to the 6-acetate of physalin D. *Chen et al.* revised the structure of physalin G (**314**), and the NMR resonances were assigned using 2D NMR and the configuration at C-6 established as (*R*) based on NOE data and the coupling constants of H-6 with H-7α and H-7β (*149*).

Sen and *Pathak* reported a constituent of *P. minima* as "physalin L" but the proposed structure **315** (*150*) was different from that reported previously for this compound and was also inconsistent with the spectroscopic data given as shown by *Kawai et al.*, who synthesized compound **315** (*151*). The true structure of the compound isolated by *Sen* and *Pathak* has not been established although its ^{1}H NMR spectrum was comparable to the 2,3-dihydro derivative of **315**.

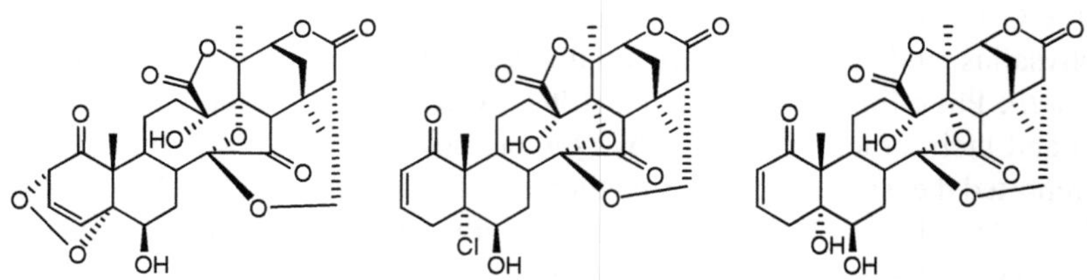

311 (physalin K) **312** (physalin H) **313** (physalin D and E)

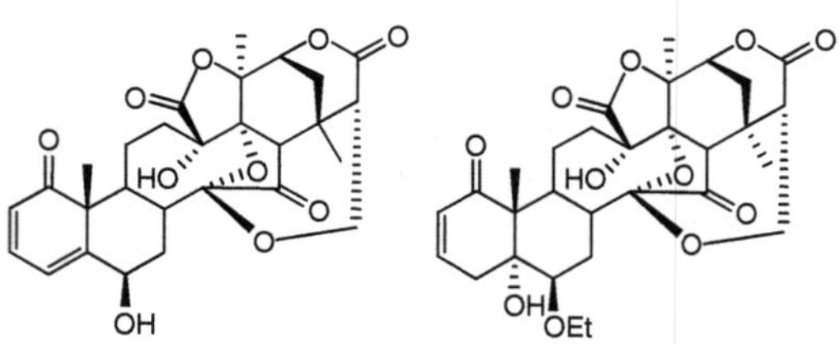

314 (physalin G) **315**

5.2.4. Withaphysalins and Related Withanolides

Withaphysalins comprise a group presenting an oxygen bridge between C-18 and C-20; depending on the oxidation state of C-18, a lactol or lactone ring may result (see Fig. 2). Hemiketals at C-18 are highly reactive towards simple alcohols and usually the methyl ketals are formed when methanol is used for extraction or purification. When the free hemiketals are isolated, they exist as an equilibrium mixture of both epimers at C-18; epimeric methyl ketals on the other hand may be separated.

Withaphysalins F–L (**316–322**) were isolated from *Eriolarynx lorentzii* (*sub nom. Vassobia lorentzii*) collected in Argentina (*70*). Compounds **318** and **319** are the corresponding methyl ketals of withaphysalin G (**317**) and most probably artifacts. The hemiketal corresponding to withaphysalins K and L was not isolated. *Veras et al.* isolated withaphysalins M (**323**), N (**324**), O (**325**), 2,3-dihydrowithaphysalin F (**326**), and withaphysalin F (**316**) from *Acnistus arborescens* collected in northeastern Brazil (*152, 153*). The ethyl ketal **325** is most probably an artifact from reaction of **316** with ethanol used for extraction. The 4-acetate of withaphysalin F (**327**) and the saturated lactone derivative **328** were isolated from *Dunalia brachyacantha* (*71*).

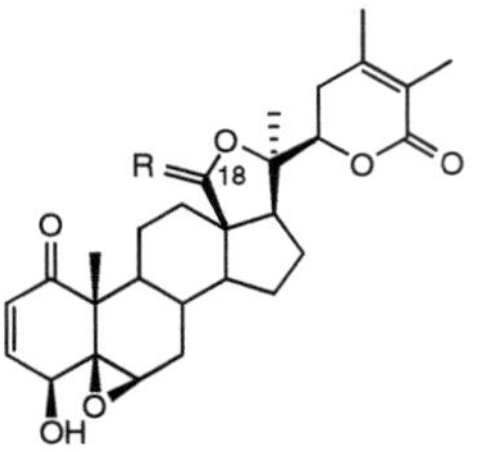

316 R = H,OH, (18 *R*/ *S*) (withaphysalin F)
323 R = O (withaphysalin M)
325 R = H,OEt (withaphysalin O)

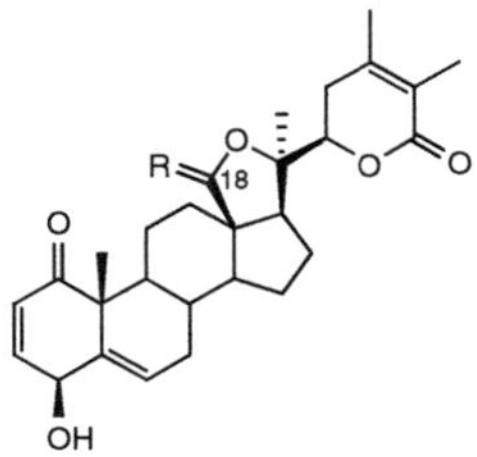

317 R = H,OH, (18 *R*/ *S*) (withaphysalin G)
318 R = H,OCH_3,(18 *R*)(withaphysalin H)
319 R= H,OCH_3, (18 *S*) (withaphysalin I)
320 R = O (withaphysalin J)

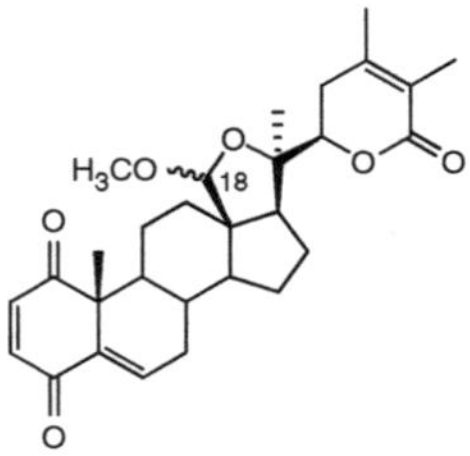

321 (18*R*) (withaphysalin K)
322 (18*S*) (withaphysalin L)

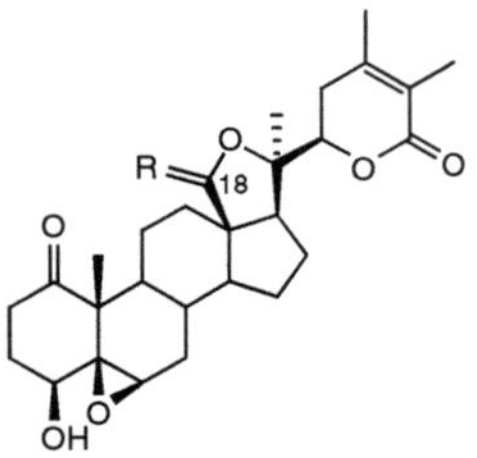

324 R = O (withaphysalin N)
326 R = H,OH, (18*R*/*S*)

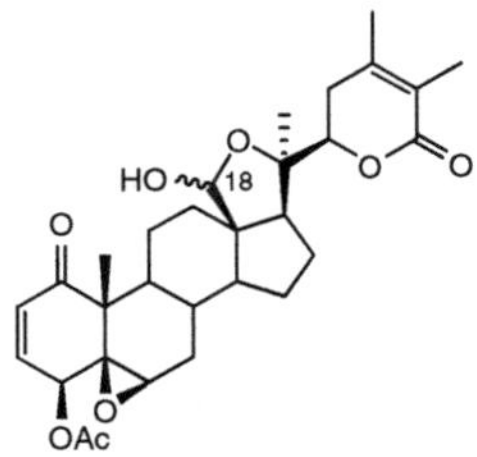

327 (18*R*/*S*)

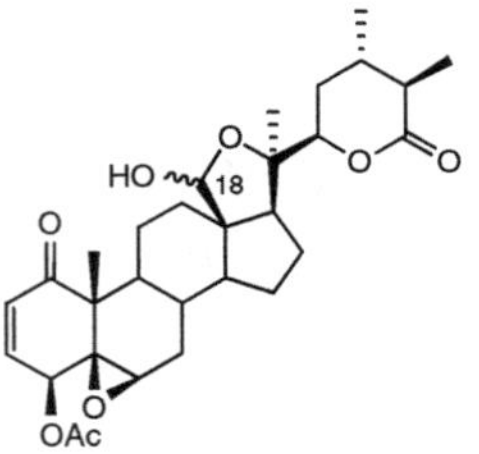

328 (18*R*/*S*)

From *Physalis minima*, *Ma et al.* reported seven new withaphysalins (*154*). Withaphysalins Q–S (**329–331**) and the 5-*O*-methyl derivative **332** were isolated as the methyl ketals at C-18; as hemiketals are highly reactive and methanol was extensively used during isolation and purification, it is possible that the actual natural products are the free hemiketals. The 3-methoxy group in **329** probably derives from the Δ^2-1-ketone and the 5-methoxy group in **332** could result from addition of methanol to a 5β,6β-epoxide, thus both compounds might be artifacts derived from the known withaphysalin B (**333**). Withaphysalin P (**334**) and the acetylated derivatives of the known withaphysalin C, **335** and **336**, appear to be biosynthetic intermediates in the conversion of withaphysalins to physalins. The authors also mention the isolation of the known 5α,6α-epoxywithaphysalin A (**337**), but this compound has been described only as a synthetic product (*155*). 5α,6α-Withanolides are rare and this would be the first report of **337** as a natural product.

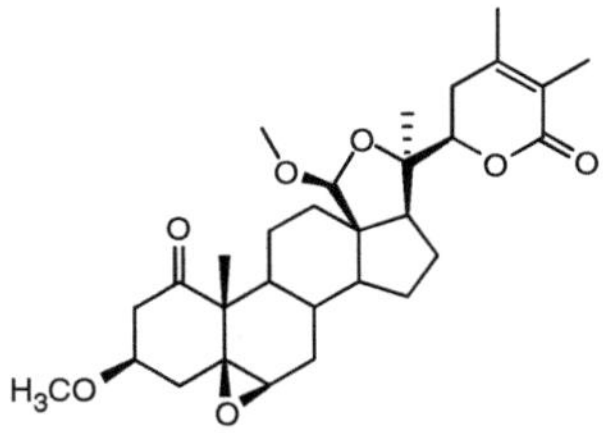

329 (withaphysalin Q)

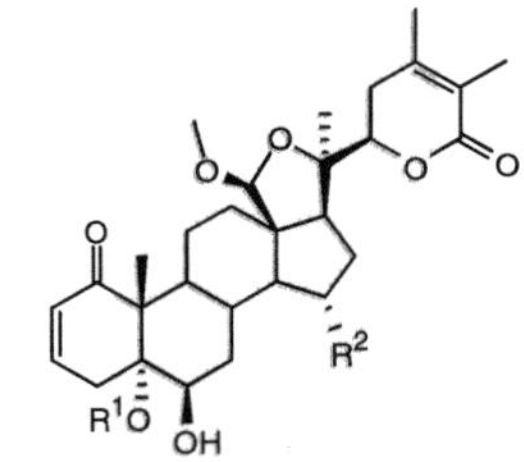

330 R^1 = H, R^2 = H (withaphysalin R)
331 R^1 = H, R^2 = OH (withaphysalin S)
332 R^1 = CH_3, R^2 = H

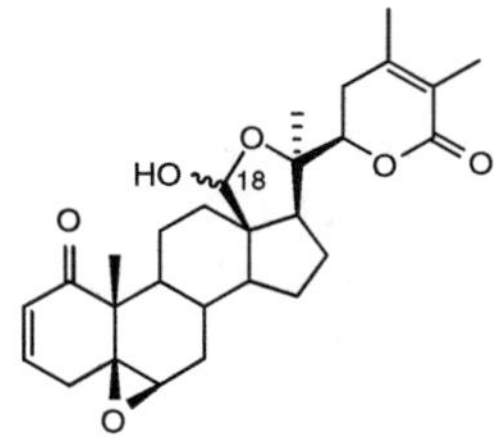

333 (18*R/S*) (withaphysalin B)

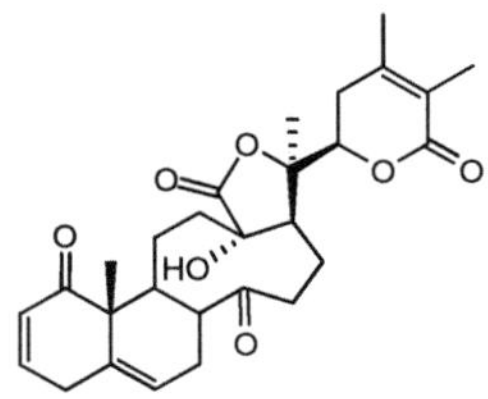

334 (withaphysalin P)

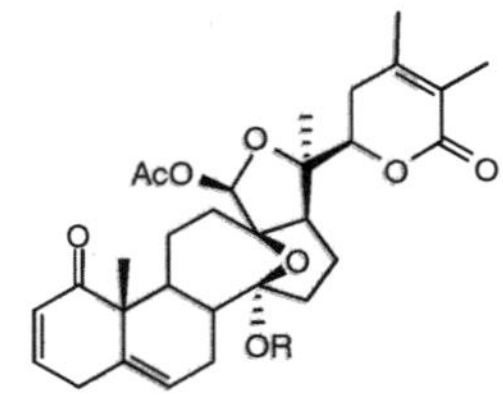

335 R = H
336 R = Ac

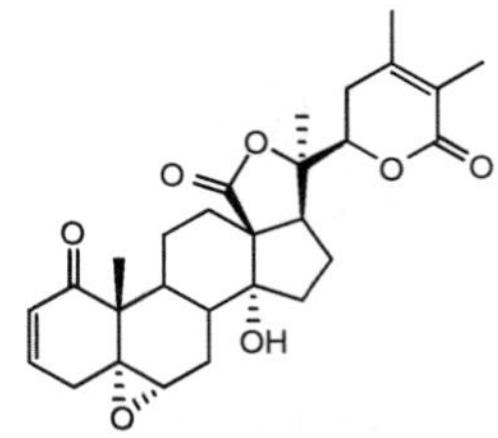

337 (5α,6α-epoxywithaphysalin A)

5.3. *Withanolides Containing an Aromatic Ring and Related Steroids*

Two distinct groups of withanolides containing aromatic rings in the steroid nucleus have been found. One of them corresponds to an *abeo*-ergostane skeleton with an expanded 6-membered ring D that incorporates C-18 (see Fig. 2) (*3*). The other group presents an aromatic A ring with loss of the angular methyl at C-10. In the discussion that follows, closely related withanolides from a biosynthesis standpoint are also included.

5.3.1. Aromatic Ring-D Withanolides and Related Steroids

A small group of withanolides and related steroids with a six-membered aromatic ring D, the nicandrenoids, were isolated in the early 1970s from the Peruvian "shoofly" plant *Nicandra physalodes* (*e.g.* Nic-1, **338**) (*59*). These compounds remained a curiosity within the withanolides for almost 20 years, until *Veleiro et al.* isolated salpichrolide A (**339**) from *Salpichroa origanifolia* (*156*). Compound **339** was also the first withanolide having a 5,6-epoxide with α-configuration, a feature that proved to be characteristic of several salpichrolides. Further studies on this plant showed that the withanolides present and the relative amounts were strongly dependent on the time of the year in which plants were collected and also on their geographical origin. The major components in *S. origanifolia* plants collected in Buenos Aires and Córdoba provinces (Argentina), were salpichrolides A (**339**) and G (**340**), with salpichrolides B (**341**) and C (**342**) being isolated as minor components (*157*, *158*).

At variance with other withanolide families, salpichrolides present limited modifications in the substitution pattern of rings A and B. On the other hand, a higher variability was observed for the side chain. Salpichrolides H (**343**) and I (**344**) were isolated from plants collected in Buenos Aires in the winter (*158*), and salpichrolides J (**345**), K (**346**), and M (**347**) from plants collected in Salta province (Argentina) in the summer (*159*). Salpichrolides H (**343**) and M (**347**) correspond to the two possible products resulting from hydrolytic (*trans*) cleavage of the side-chain epoxide. Salpichrolides J (**345**) and K (**346**) are the first examples of withanolides with a side chain in which the oxidation levels at C-22 and C-26 are reversed; salpichrolide K (**346**) slowly cyclized to salpichrolide J (**345**) in solution. Figure 5 shows a possible biosynthesis pathway for these compounds.

Plants collected in Buenos Aires in winter also contained two ergostane derivatives, salpichrolides E (**348**) and F (**349**), probably resulting from degradation of the lactone side chain of salpichrolides A and C (*160*). The configuration at C-22 was assumed to be the same as that in the salpichrolides with an intact side chain (*i.e.* (22*R*)), but the orientation of the C-24 methyl could not be determined. It is noteworthy that similar degradation products are present in *N. physalodes*, the other plant known to contain withanolides with an aromatic ring D (*59*).

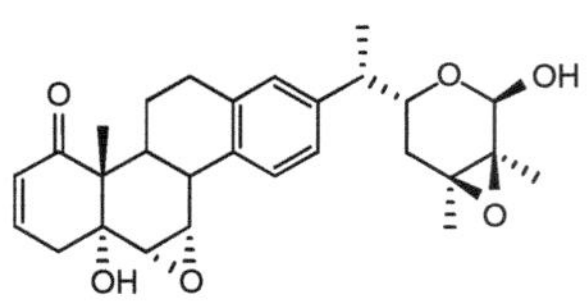

338 (Nic-1)

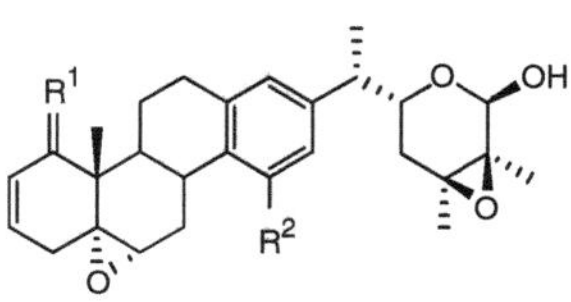

339 R^1 = O, R^2 = H (salpichrolide A)
340 R^1 = O, R^2 = OH (salpichrolide G)
341 R^1 = α-OH, β-H, R^2 = H (salpichrolide B)

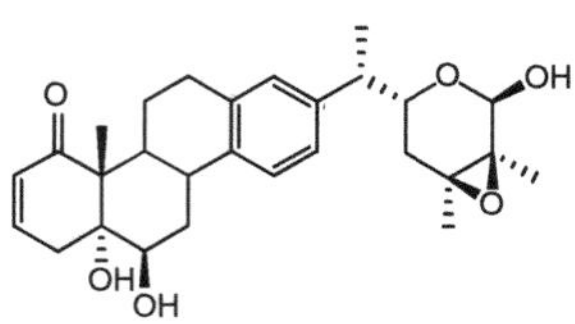

342 (salpichrolide C)

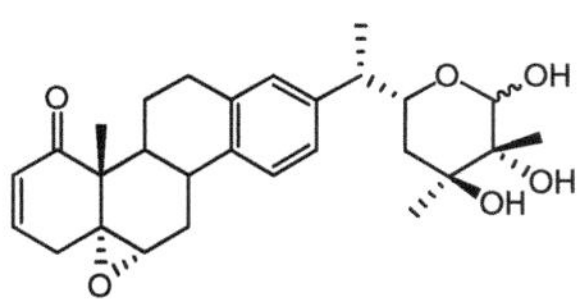

343 (salpichrolide H)

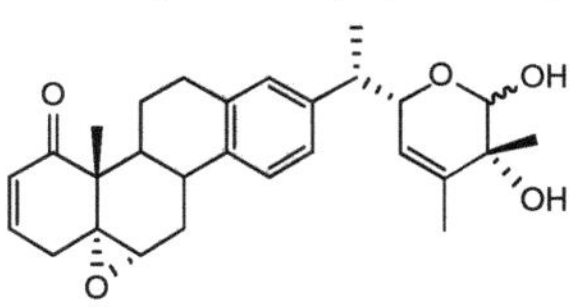

344 (salpichrolide I)

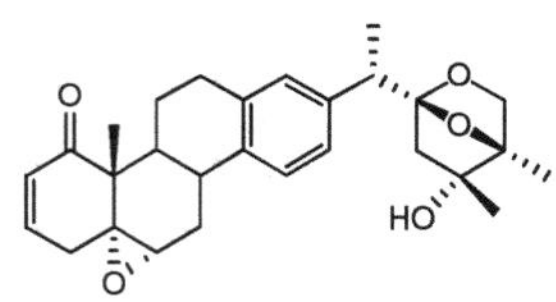

345 (salpichrolide J)

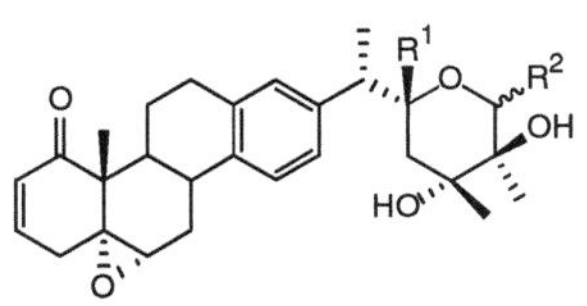

346 R^1 = OH R^2 = H (salpichrolide K)
347 R^1 = H R^2 = OH (salpichrolide M)

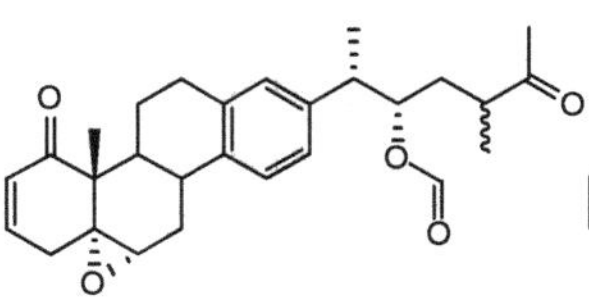

348 (salpichrolide E)

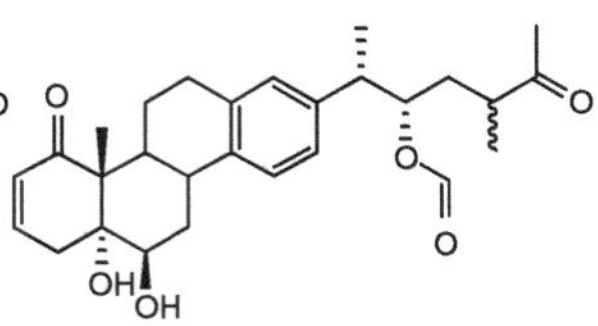

349 (salpichrolide F)

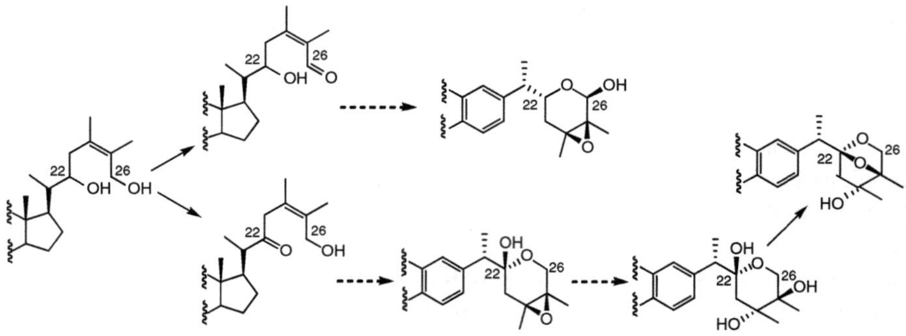

Fig. 5. Proposed biosynthetic pathway for the formation of the side chain in normal salpichrolides and in salpichrolides J (**345**) and K (**346**)

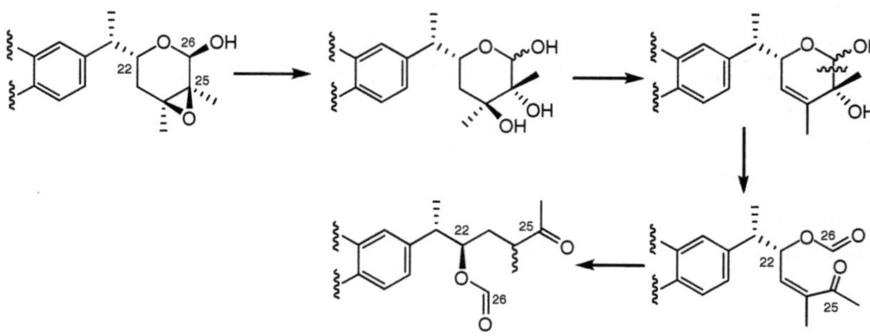

Fig. 6. Proposed degradative pathway for the formation of the side chain in salpichrolides E (**348**) and F (**349**). Starting from salpichrolide A (**339**) the first two intermediates correspond to compounds **343** and **344**

Salpichrolides H (**343**) and I (**344**) could be intermediates in the degradation pathway leading from salpichrolide A (**339**) to salpichrolide E (**348**). Oxidative cleavage of the C-25–C-26 bond would give rise to the formyloxy group (C-26) and the methyl ketone (Fig. 6).

Besides the withanolides with an aromatic D ring, salpichrolides D (**350**), (*157*) L (**351**), and N (**352**) (*159*), with a normal (5-membered) D ring were isolated from *S. origanifolia*. All of these have a characteristic 5α,6α-epoxide moiety, unique to *S. origanifolia*. A possible pathway for ring D aromatization proposed by *Whiting* involves the oxidation of C-18 followed by a 1,2-shift of C-17 to form a new six-membered ring via a cyclopropyl fused intermediate (*161*). Salpichrolide L (**351**) may be the precursor of the putative 14,16-diene intermediate; cleavage of the C-13–C-17 bond would lead to salpichrolide A and related compounds (Fig. 7, route *a*). An alternative cleavage of the cyclopropyl intermediate through the C-13–C-18 bond would result in migration of the angular methyl to give salpichrolide N (**352**) (Fig. 7, route *b*).

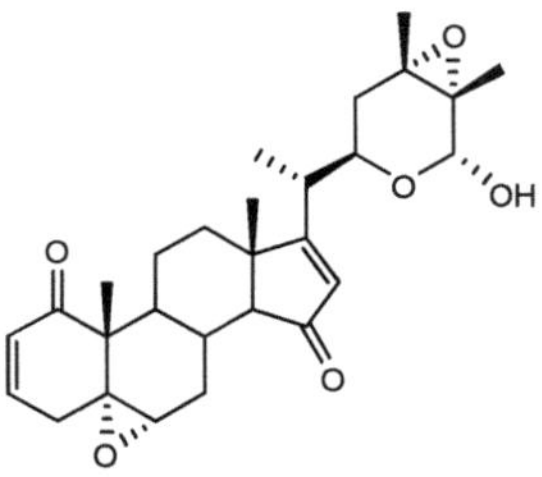

350 (salpichrolide D)

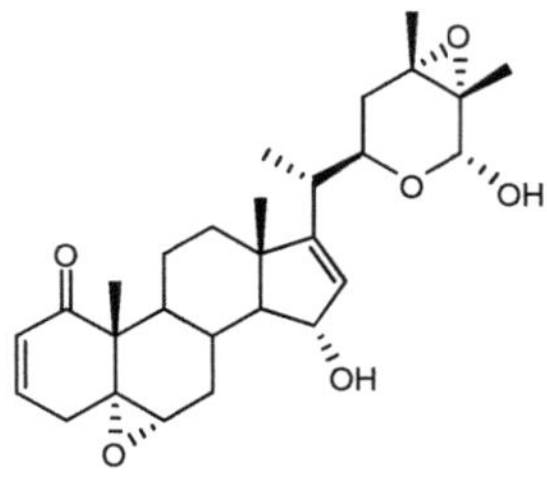

351 (salpichrolide L)

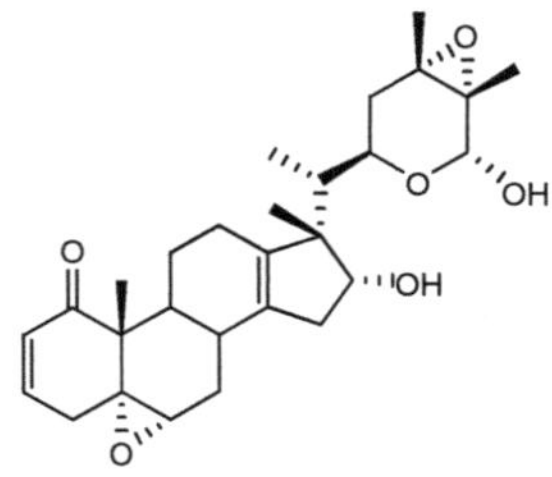

352 (salpichrolide N)

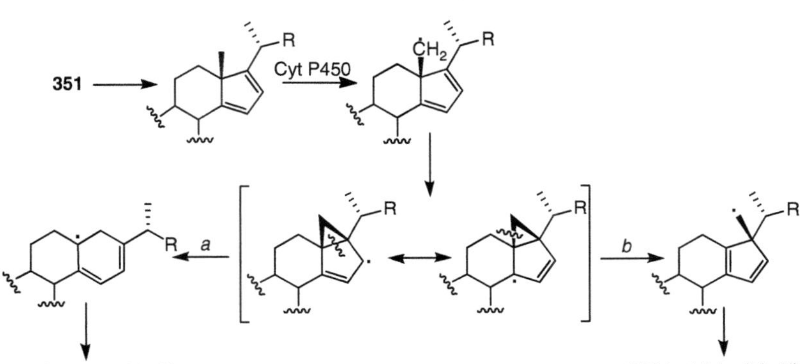

339 (salpichrolide A)

352 (salpichrolide N)

Fig. 7. Proposed biosynthesis pathways for the formation of withanolides with an aromatic D ring (*e.g.* salpichrolide A (**339**), pathway *a*) and for the rearranged skeleton in salpichrolide N (**352**) (pathway *b*)

5.3.2. Aromatic Ring-A Withanolides and 19-Hydroxywithanolides

The first 19-hydroxylated withanolide, jaborosalactone O (**353**), was isolated from *Jaborosa leucotricha* collected in late spring in Argentina (*162*). Another three 19-hydroxywithanolides, jaborosalactones V (**354**), W (**355**), and X (**356**) were isolated from plants collected in the autumn together with jaborosalactone Q (**357**) (*163*) and jaborosalactone 7 (**358**) (*68*), with the latter two containing an aromatic A ring. Compound **357** had been previously found in plants of *J. leucotricha* collected at a different location (*164*). Cinerolide (**359**), isolated from *Physalis cinerascens* collected in Mexico is the only 19-hydroxywithanolide outside the *Jaborosa* genus (*54*). The coexistence of 19-hydroxywithanolides and A-ring aromatic 19-norwithanolides in *J. leucotricha* is indicative of an oxidative degradation pathway for the loss of C-19 and aromatization. (+)-Jaborol and jaborosalactone 45 (see Sect. 5.4.2.) are the only other known withanolides with an aromatic A ring.

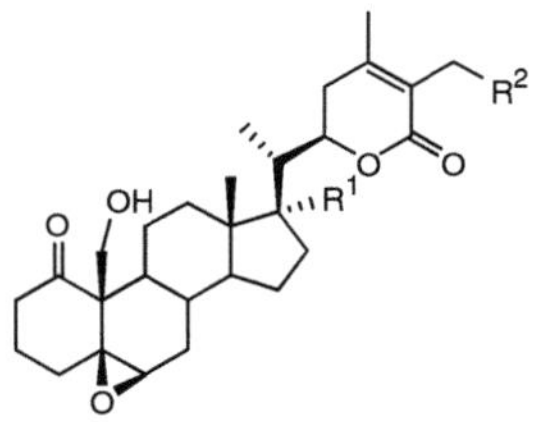

353 R^1 = OH, R^2 = H (jaborosalactone O)
355 R^1 = H, R^2 = OH (jaborosalactone W)

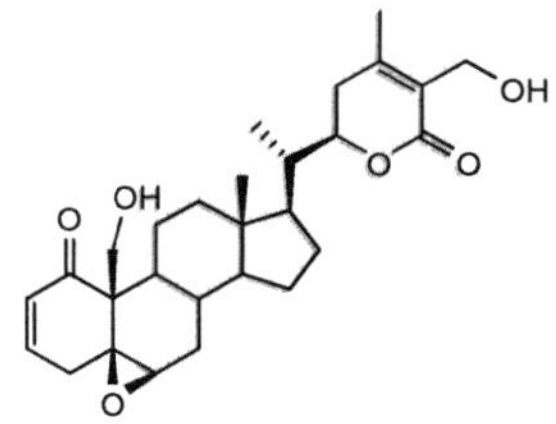

354 (jaborosalactone V)

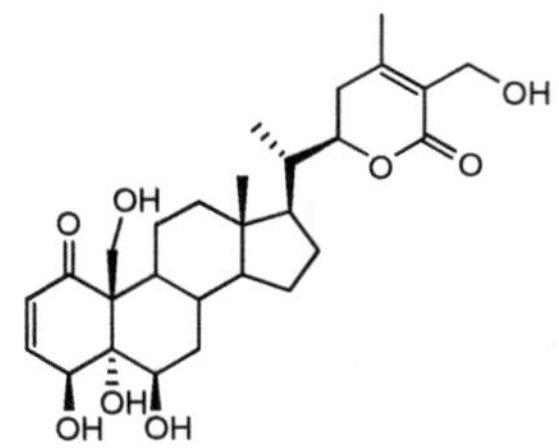

356 (jaborosalactone X)

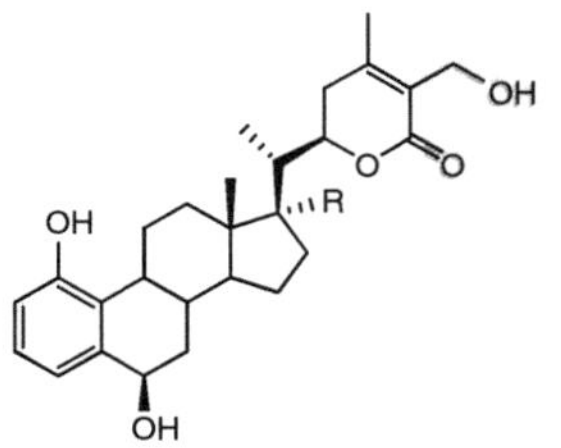

357 R = H (jaborosalactone Q)
358 R = OH (jaborosalactone 7)

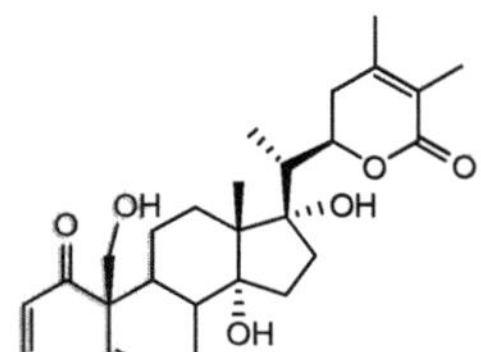

359 (cinerolide)

5.4. Withanolides with a γ-Lactone Side Chain

Ixocarpalactone A (**360**) was the first withanolide isolated with a γ-lactone side chain. It is the major withanolide of *Physalis philadelphica* (*sub nom. Physalis ixocarpa*), a plant with edible fruits (tomatillos) cultivated in Mexico and Guatemala (*165*). Nowadays several groups of withanolides containing variations of this γ-lactone moiety are known (see Fig. 3).

5.4.1. Ixocarpalactones and Perulactones

Kinghorn and coworkers reinvestigated *Physalis philadelphica* and isolated two new withanolides with a γ-lactone side chain of the ixocarpalactone type (**361**, **362**), also the configuration of ixocarpalactone A was confirmed by X-ray crystallography (*104*, *105*) (Fig. 8). The 3-methoxy derivatives of ixocarpalactones A and B, resulting from the addition of methanol were also isolated and shown to be artifacts of the isolation procedure. The 4-acetate of **361** had been reported previously from fruits of the same plant (*165*). *Huang et al.* isolated two perulactones, taccalonolides O (**363**) and P (**364**), from rhizomes and tubers of *Tacca*

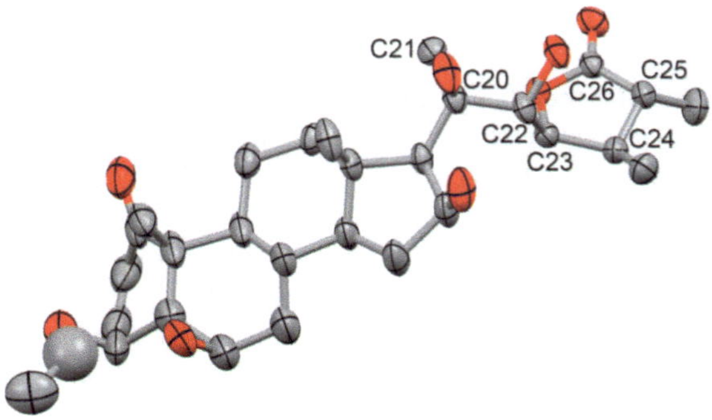

Fig. 8. X-ray crystal structure of ixocarpalactone A (**360**, CCDC 171420) (*104*). The two atoms close to the 4-hydroxy group probably correspond to a solvent molecule. Structure drawing generated with Mercury 2.3

subflabellata (*166, 167*). Recently, perulactones C (**365**) and D (**366**) were reported from *Physalis peruviana* (*168*).

Physanolide A (**367**) was isolated from *Physalis angulata* by *Kuo et al.* (*146*). This withanolide has a novel skeleton related to the perulactones (γ-lactone between C-26 and C-28) with a carbon-carbon bond between C-16 and C-25 that results in a seven-membered ring.

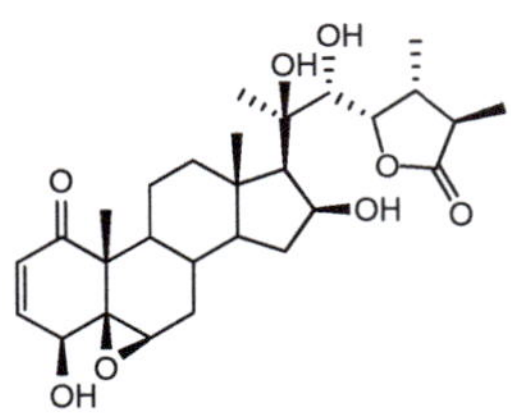

360 (ixocarpalactone A)

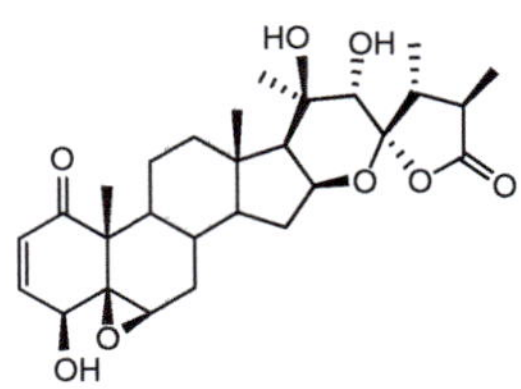

361 (ixocarpalactone B)

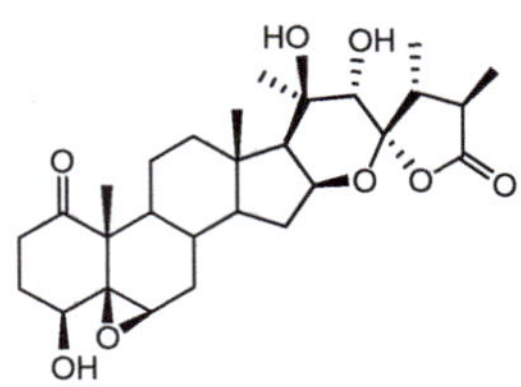

362

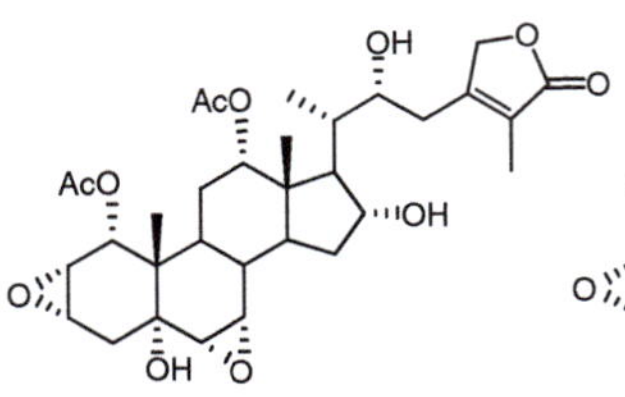

363 (taccalonolide O)

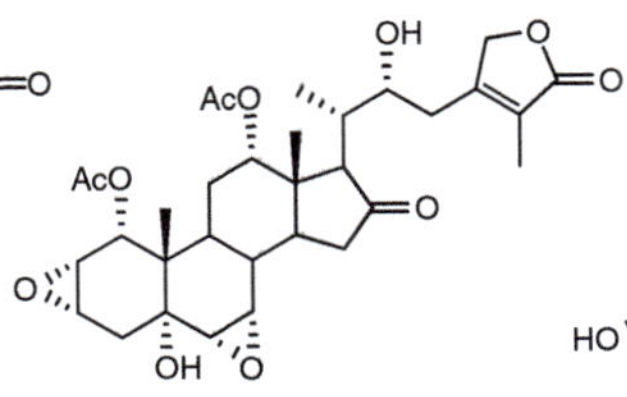

364 (taccalonolide P)

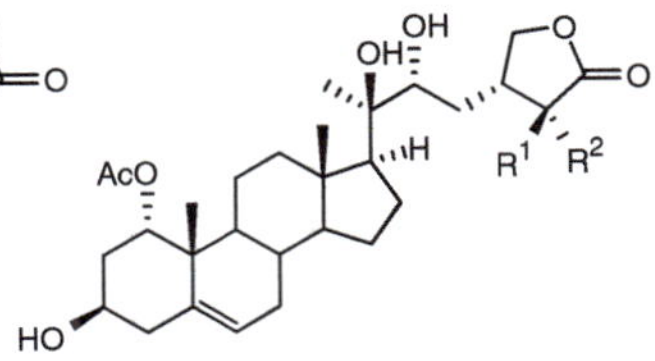

365 $R^1 = CH_3$, $R^2 = OH$ (perulactone C)
366 $R^1 = OH$, $R^2 = CH_3$ (perulactone D)

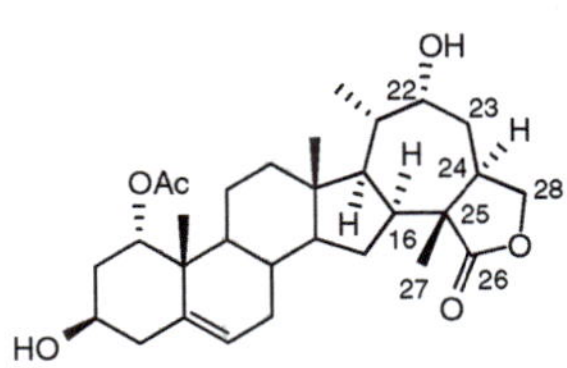

367 (physanolide A)

5.4.2. Trechonolides

The first member of this group of withanolides was trechonolide A (**368**), isolated by *Lavie et al.* from *Jaborosa laciniata* (*sub nom. Trechonaetes laciniata*) collected in Argentina (*169*). Closely related to the ixocarpalactones, an unusual characteristic feature of this compound was a hemiketal bridge formed by the 22-hydroxy group and a ketone at C-12, resulting in a six-membered ring with a β-oriented hydroxy at C-12. The same compound was later isolated by *Parvez et al.* from the Chilean *J. magellanica* (*170*). Curiously, although in both cases the structure was elucidated by X-ray crystallography, the configuration at C-23 was incorrectly assigned as (*R*). In 2006 *Nicotra et al.* reported the isolation of the C-23 epimer of trechonolide A (jaborosalactone 32, **369**) from *J. rotacea* (*120*). The configuration at C-23 was established by X-ray crystallography and shown to be (*R*) (Fig. 9). Careful inspection of the original X-ray data for trechonolide A (*169*, *170*) confirmed the (23*S*) configuration. The C-23 epimers of jaborotetrol and jaborochlorotriol (**370** and **371**) were also isolated from *J. rotacea* and shown to be (23*R*) (*120*). The chemical shift of C-23 and the sign of the *Cotton* effect at 218 nm may be used as direct indicators of the configuration at this position of trechonolides, thus the (23*S*) epimers have a negative *Cotton* effect and the C-23 resonance at δ 82.0–82.5 ppm, while the (23*R*) epimers exhibited a positive *Cotton* effect and a downfield shift for C-23 to δ 85.5–86.0 ppm. According to this, the structures of all previously known trechonolides that have been assigned the (23*R*) configuration upon comparison with trechonolide A should now be revised. For the above structures and in those that follow, the configuration at C-23 has been corrected according to *Nicotra et al.* when appropriate.

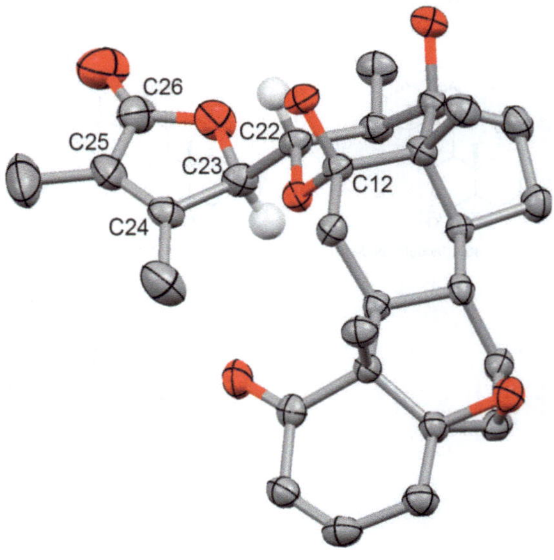

Fig. 9. X-ray crystal structure of jaborosalactone 32 (**369**, CCDC 255337), the C-23 epimer of trechonolide A showing the (23*R*) configuration (*120*). Hydrogens at positions 22 and 23 are included for clarity. Structure drawing generated with Mercury 2.3

Several withanolides structurally related to trechonolide A with the classical variations in the substitution pattern of rings A and B, have been subsequently isolated from different species of *Jaborosa*. Jaborotetrol (**372**), previously isolated from *J. magellanica*, and trechonolide A (**368**) are the most commonly found. As already observed with the sativolides, the C-12 hemiketal of the trechonolides is highly reactive towards alcohols and even small amounts of methanol or ethanol used during the isolation procedure will give the 12-*O*-methyl or ethyl derivatives. For example, when **370** was dissolved in deuterochloroform containing a few drops of deuteromethanol (to enhance solubility), the 12-*O*-trideuteromethyl derivative was formed (*120*). The clorohydrins jaborosalactone 42 (**373**) and jaborosalactone 49 (**374**) were isolated from *J. caulescens* var. *bipinnatifida* (*121*) and *J. laciniata* (*171*). Also from *J. caulescens* var. *bipinnatifida* were isolated two 21-hydroxytrechonolides epimeric at C-23, **375** and **376** (*121*). The 19-oxygenated trechonolides **377–379** were isolated from *J. laciniata* together with **380** (and its 12-*O*-methyl derivative) containing an aromatic ring A (*171*). Again, the 19-hydroxy withanolides appear as intermediates in an oxidative degradation pathway leading to the loss of the C-10 methyl and aromatization of ring A (see Sect. 5.3.2.). As already mentioned, the 12-*O*-methyl derivatives are most probably formed during the isolation procedure, with the 12-hydroxy compounds being the actual natural

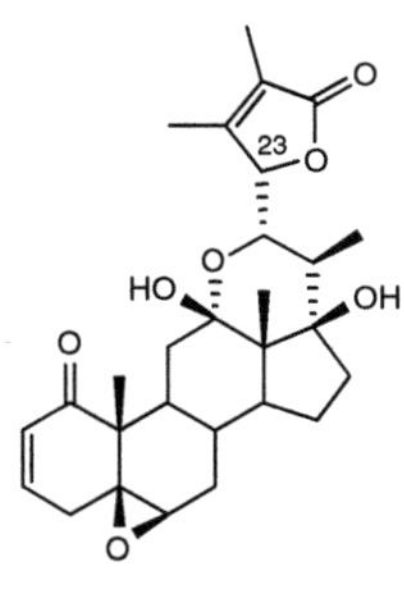

368 (23*S*) (trechonolide A)
369 (23*R*) (jaborosalactone 32)

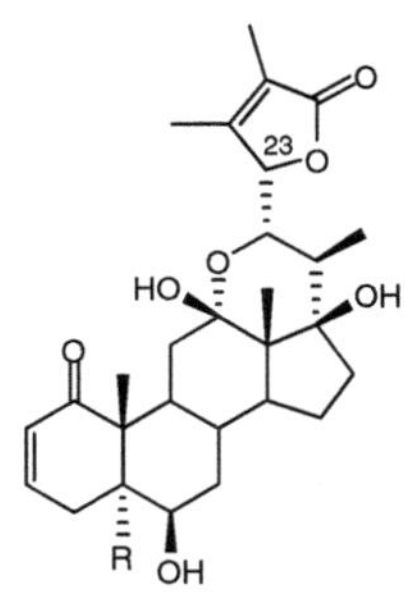

370 (23*R*), R = OH (jaborosalactone 33)
372 (23*S*), R = OH (jaborotetrol)
373 (23*R*), R = Cl (jaborosalactone 42)
374 (23*S*), R = Cl (jaborosalactone 49)

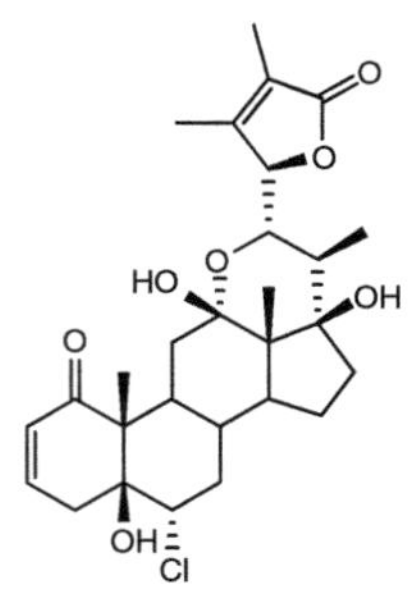

371 (jaborosalactone 34)

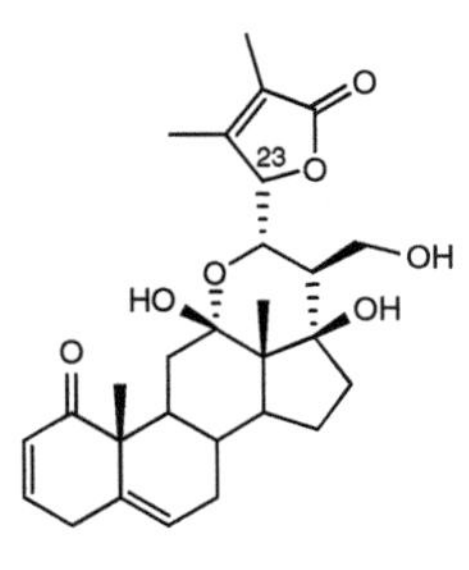

375 (23*R*) (jaborosalactone 40)
376 (23*S*) (jaborosalactone 41)

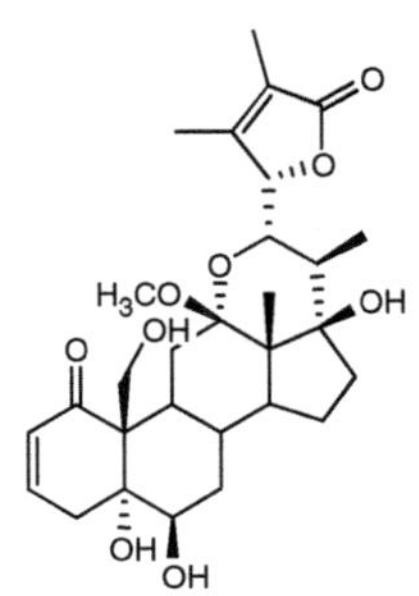

377 (jaborosalactone 46)

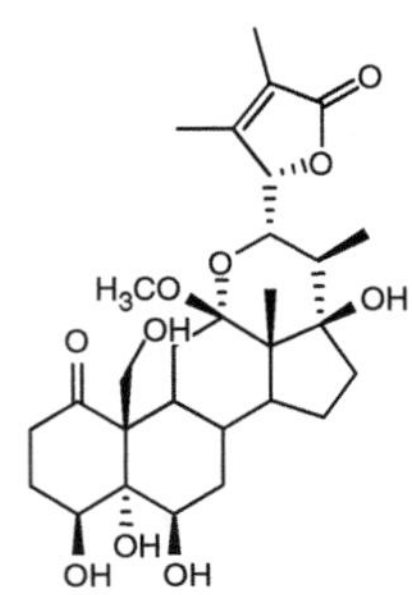

378 (jaborosalactone 47)

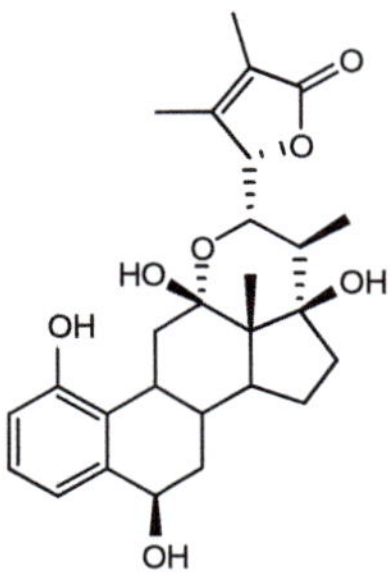

380 (jaborosalactone 45)

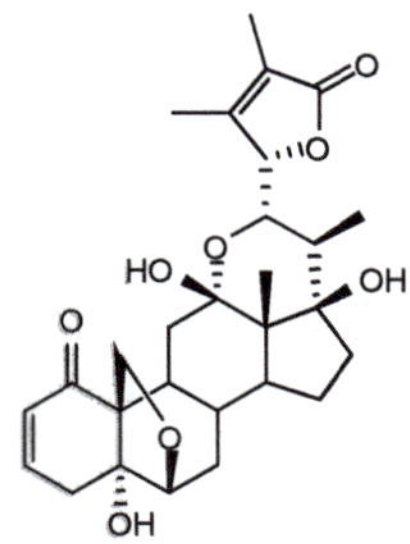

379 (jaborosalactone 48)

products. The 6,19-oxygen bridge present in **379** is an unusual functionality for a natural product. Interestingly, synthetic steroids with this moiety exhibit remarkable biological properties as selective glucocorticoid receptor modulators (*172*).

Although epoxy-δ-lactones and lactols are quite common among the withanolides, this did not appear to be the case for the γ-lactone side chains. Jaborosalactone U (**381**) isolated from *J. sativa* (Argentina), is the only known example of a 24,25-epoxy-γ-lactone (*119*). The (23*R*) configuration was originally proposed for this compound based on NOE data, however, more recently X-ray crystallography showed that the configuration is (23*S*) (Fig. 10) (*173*). Recently, several 24,25-epoxy-γ-lactols were isolated from plants of *J. parviflora* (**382–385**); some 12-*O*-ethyl derivatives were also reported (*174*). The authors used powder

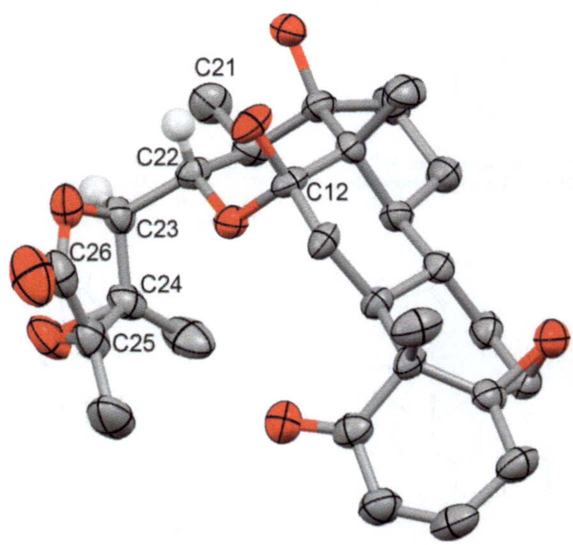

Fig. 10. X-ray crystal structure of jaborosalactone U (**381**) showing the configuration of the side chain epoxylactone (*173*). Hydrogens at positions 22 and 23 are included for clarity. Structure drawing generated with Mercury 2.3

X-ray diffraction analysis and NMR spectroscopy residual dipolar couplings to establish the absolute configuration of the epoxy-lactol side chain of **382**, confirming it was (23*S*). Both methodologies proved to be valid alternatives to single crystal X-ray diffraction. For the other compounds the same configuration was established by comparison of their ^{13}C NMR spectra.

Tettamanzi et al. had reported the isolation of a closely related epoxy-γ-lactol from *J. lanigera* for which the (23*R*) configuration was proposed (*175*); this compound had ^{13}C NMR data identical to **382** for C-12–C-18 and C-20–C-28 indicating that both compounds should have the same configuration in the side chain, hence the revised structure **386** is proposed. Thus so far, all known epoxy-γ-lactones and lactols have the same configuration at C-23, opposite to that of trechonolide A. Jaborosalactones 35 (**387**) and 36 (**388**) isolated from *J. rotacea* would result from cyclization of a 21-hydroxy-epoxy-γ-lactone (*120*). It should be noted that to date, trechonolides have been reported in nine of the thirteen *Jaborosa* species studied. A group of closely related γ-lactones isolated from *Larnax subtriflora* (*sub nom. Deprea subtriflora*) is discussed in Sect. 5.5.2.

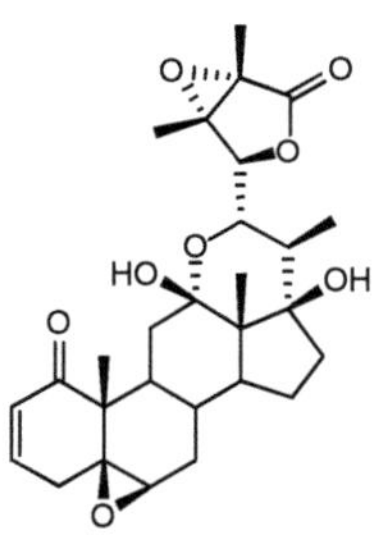

381 (jaborosalactone U)

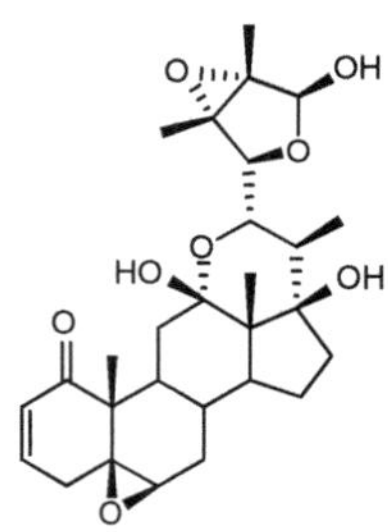

382 (jaborosalactol 24)

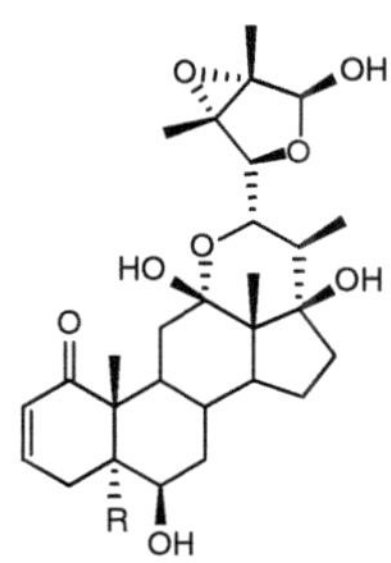

383 R = Cl (jaborosalactol 25)
385 R = OH (jaborosalactol 27)

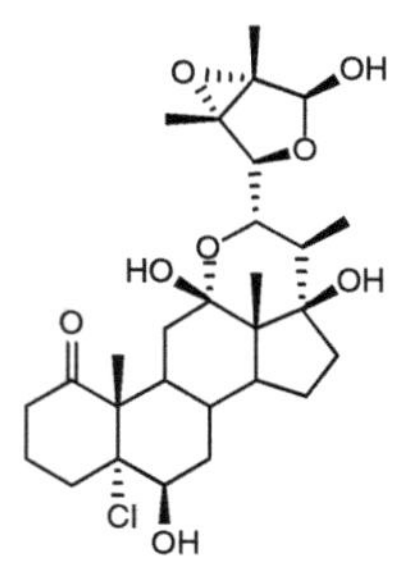

384 (jaborosalactol 26)

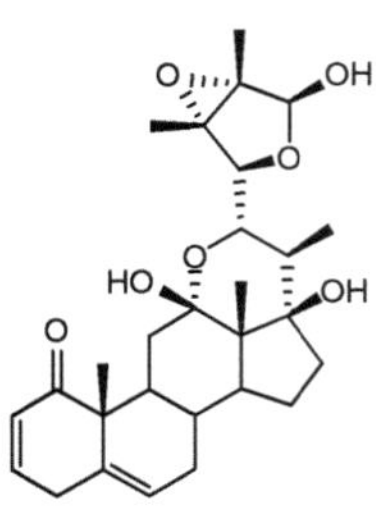

386

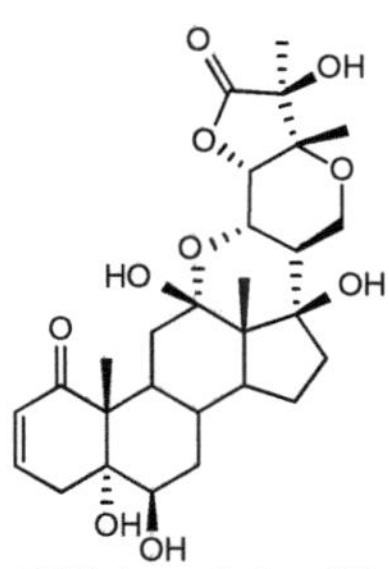

387 (jaborosalactone 35)

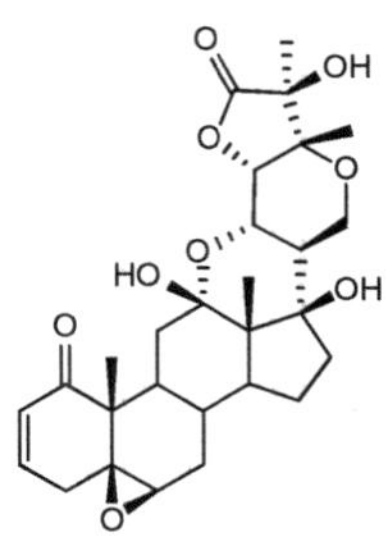

388 (jaborosalactone 36)

5.4.3. Spiranoid-γ-Lactones

The first withanolide with a spiranoid γ-lactone side chain, jaborosalactone P (**389**), was isolated by *Monteagudo et al.* from plants of *Jaborosa odonelliana* collected in Argentina (*176*). *Cirigliano et al.* reinvestigated this plant collected at different times of the year. Jaborosalactones 10 (**390**), 14 (**391**), and 15 (**392**) were found in plants collected in the summer while jaborosalactones 11 (**393**), 12 (**394**), 13 (**395**), and 14 (**391**) were present in plants collected in the autumn (*177*). Jaborosalactone P was the major component in both cases. The C-23 epimer of jaborosalactone P, jaborosalactone 24 (**396**), was isolated as a minor component from plants collected in April and December (*178*). This is the only spiranoid withanolide with a (23*R*) configuration isolated so far.

A group of six spiranoid withanolides with a 17(*20*)-ene-22-keto system, jaborosalactones 1–6 (**397–402**) was isolated from *Jaborosa runcinata* collected in Argentina. Jaborosalactone 2 (**398**) was also isolated from *Jaborosa araucana* (*179*). More recently, jaborosalactone 25 (**403**) was isolated as a minor component of *J. runcinata* (*178*).

The structural similarity of jaborosalactone 2 (**398**) and trechonolide A (**368**), both present in *J. araucana*, suggests that these compounds may have a common biosynthetic precursor. In trechonolide A, ring closure on C-12 has occurred with a C-22 hydroxy group as shown in Fig. 11 pathway *a*. Oxidation of the intermediate or a related compound to the 22-ketone would allow cyclization between C-23 and the C-12 ketone to give the 22-keto-spiranoid withanolides (Fig. 11 pathway *b*).

Jaborosalactone 31 (**404**), isolated from *J. rotacea*, is closely related to the spiranoid withanolides isolated from *J. odonelliana*, *J. runcinata*, and

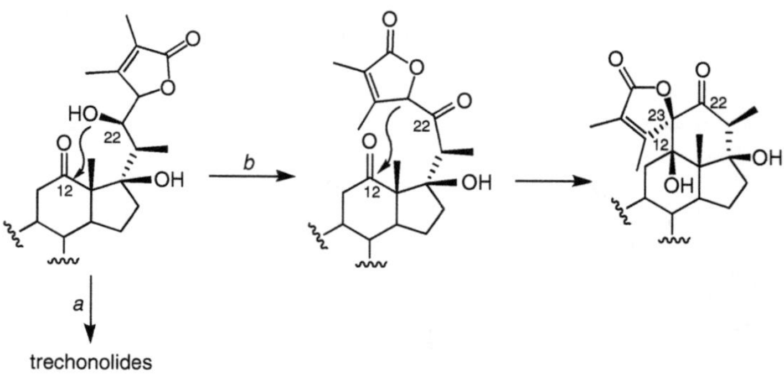

Fig. 11. Proposed biosynthesis routes to trechonolides (pathway *a*) and spiranoid withanolides (pathway *b*) via a common precursor

J. araucana. In this case, the C-12–C-23 bond is still present but instead of the spiranoid-γ-lactone arrangement (there is no oxygenated function at C-23), a δ-lactone is formed between the C-26 carboxyl and the C-12 hydroxy group (*120*).

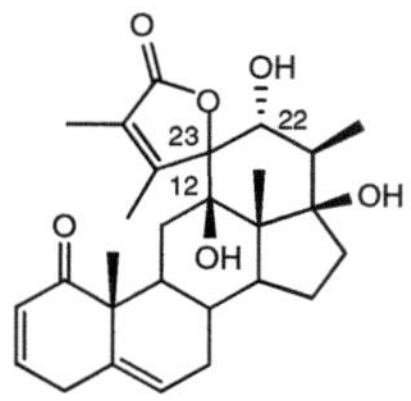

389 (23*S*) (jaborosalactone P)
396 (23*R*) (jaborosalactone 24)

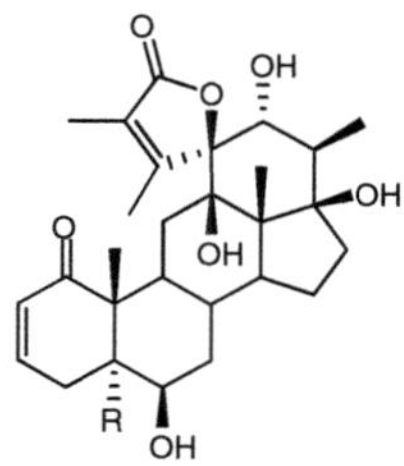

390 R = Cl (jaborosalactone 10)
391 R = OCH_3 (jaborosalactone 14)
395 R = OH (jaborosalactone 13)

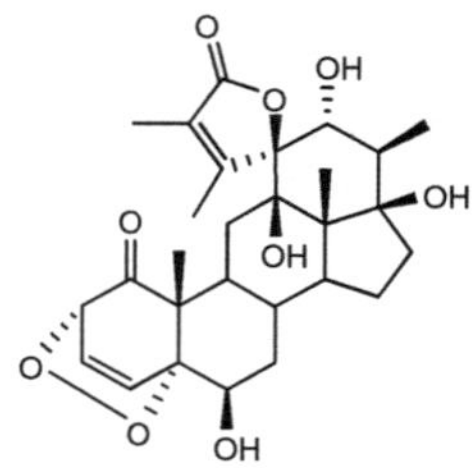

392 (jaborosalactone 15)

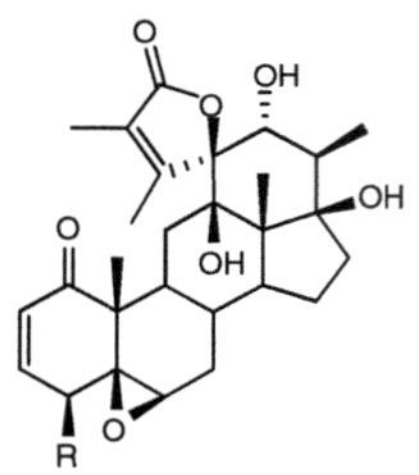

393 R = H (jaborosalactone 11)
394 R = OH (jaborosalactone 12)

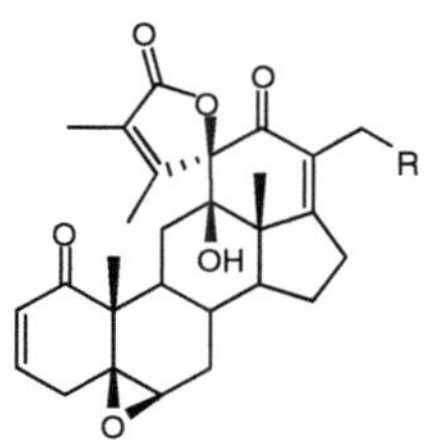

397 R =H (jaborosalactone 1)
400 R = OH (jaborosalactone 4)

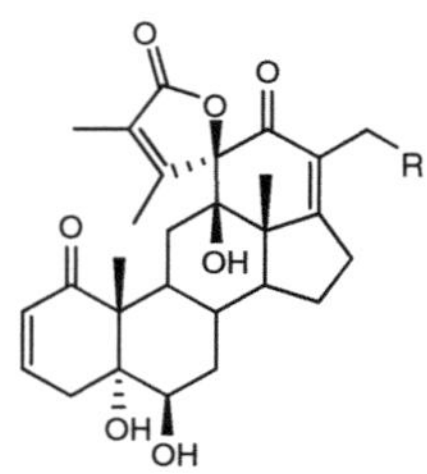

398 R = H (jaborosalactone 2)
403 R = OH (jaborosalactone 25)

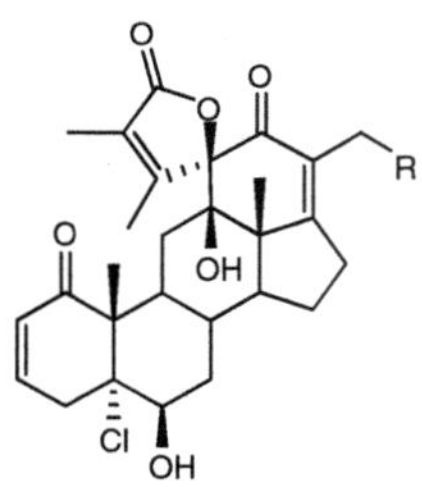

399 R = H (jaborosalactone 3)
402 R = OH (jaborosalactone 6)

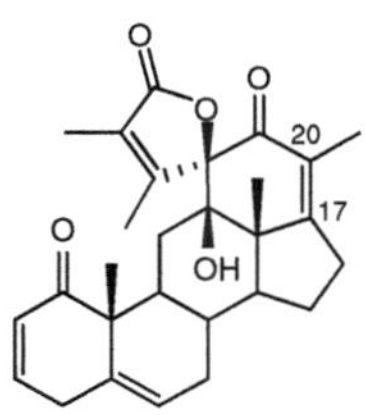

401 (jaborosalactone 5)

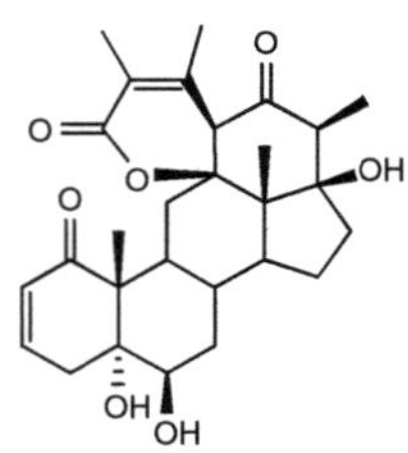

404 (jaborosalactone 31)

5.4.4. Taccalonolides

Tacca species contain highly oxygenated ixocarpalactone-type withanolides having an additional ring formed by a carbon-carbon bond between C-16 and C-24, with taccalonolide A (**405**) being the first example of these compounds (*3*). Eight new withanolides related to taccalonolide A were isolated from *Tacca plantaginea*, taccalonolides L (**406**), M (**407**) (*180*), G, H, I, J, K (**408–412**) (*181*), and W (**413**) (*182*). Taccalonolides R, S, T, U, and V (**414–418**) were isolated from the Vietnamese plant *Tacca paxiana* together with the known taccalonolides A (**405**), K (**412**), B, E, and N (**419–421**) (*183*). A distinctive feature of most taccalonolides is the absence of a hydroxy group at C-22. Taccalonolide Q (**422**) (*167*) and Y (**423**) (*182*) containing a δ-lactone side chain with a C–C bond between C-16 and C-24 were isolated from *T. subflabellata* and *T. plantaginea*.

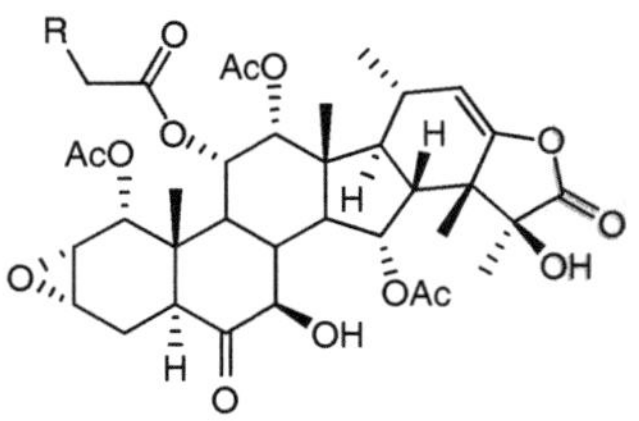

405 R = H (taccalonolide A)
406 R = OH (taccalonolide L)

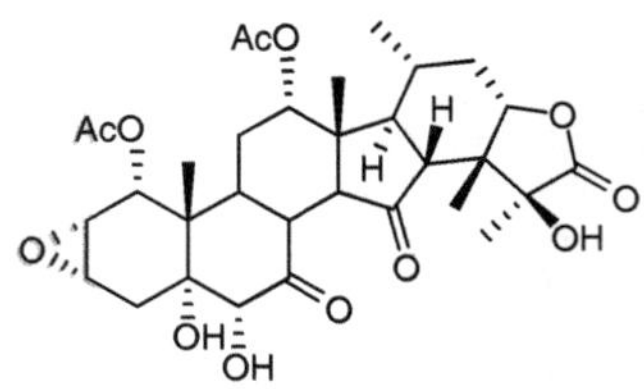

407 (taccalonolide M)

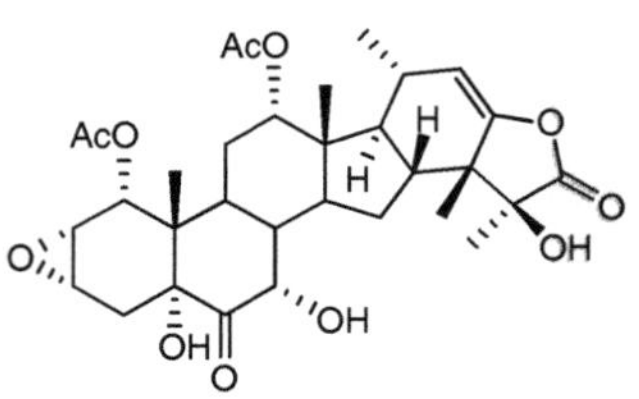

408 (taccalonolide G)

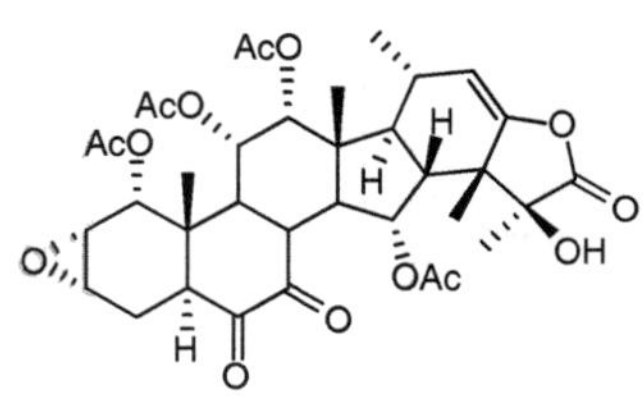

409 (taccalonolide H)

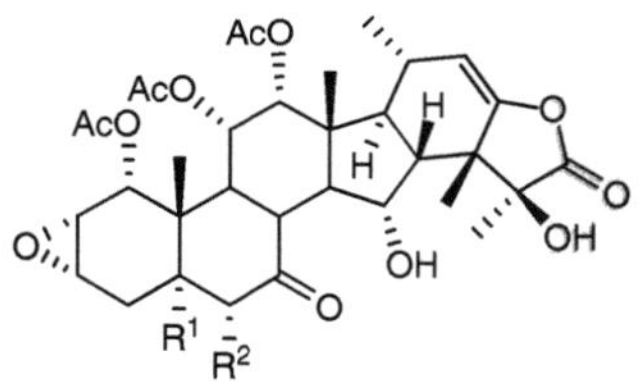

410 R^1 = H, R^2 = OH (taccalonolide I)
411 R^1 = H, R^2 = OAc (taccalonolide J)
412 R^1 = R^2 = OH (taccalonolide K)

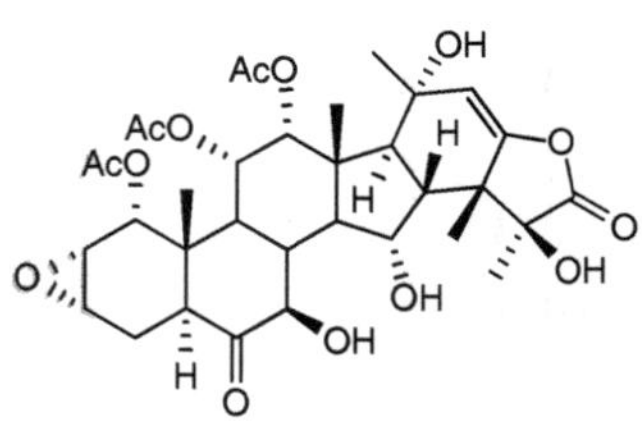

413 (taccalonolide W)

A sizeable number of steroids partially resembling withanolides have been isolated from *Tacca* species and some also dubbed taccalonolides, *e.g.* taccalonolide X (**424**) (*182*), but they are not included in this chapter. Other taccalonolides structurally related to the perulactones have been presented previously (see Sect. 5.4.1.).

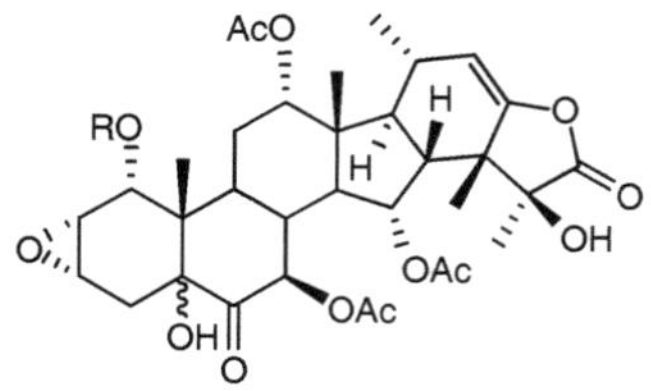

414 R = Ac (taccalonolide R)
416 R = isovaleryloxy (taccalonolide T)
417 R = H (taccalonolide U)

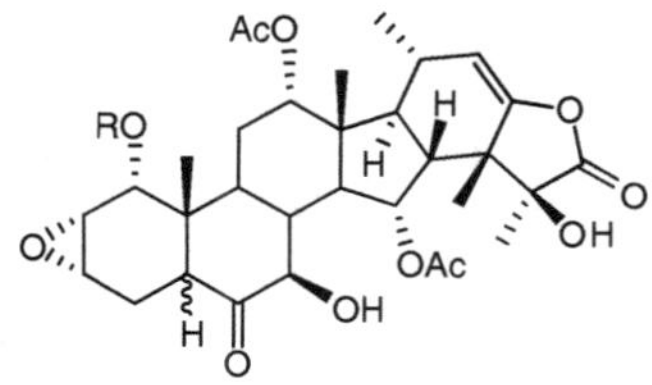

415 R = isobutyryloxy (taccalonolide S)

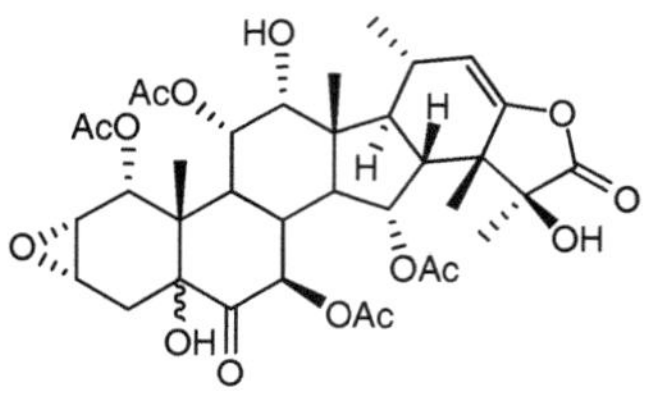

418 (taccalonolide V)

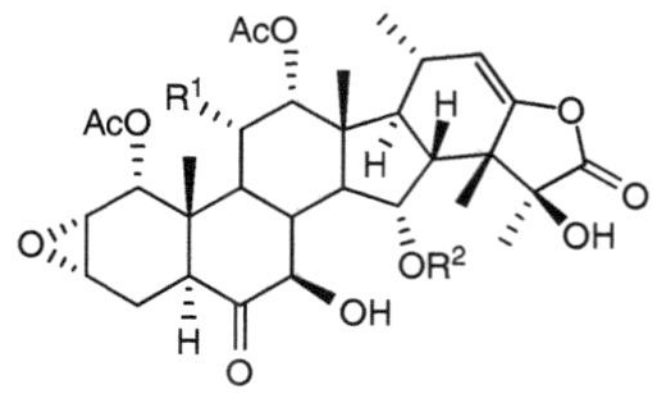

419 R^1 = OAc, R^2 = H (taccalonolide B)
420 R^1 = H, R^2 = Ac (taccalonolide E)
421 R^1 = R^2 = H (taccalonolide N)

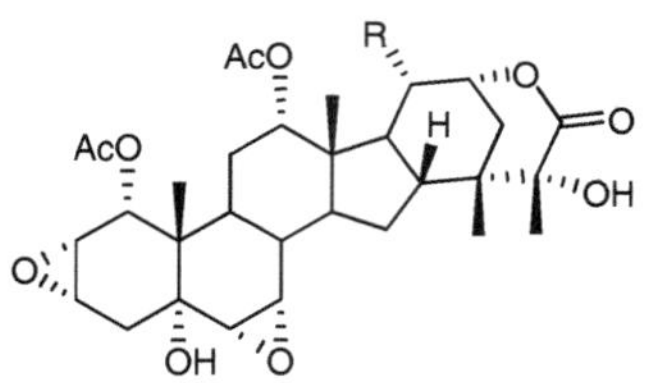

422 R = CO_2H (taccalonolide Q)
423 R = CH_2OH (taccalonolide Y)

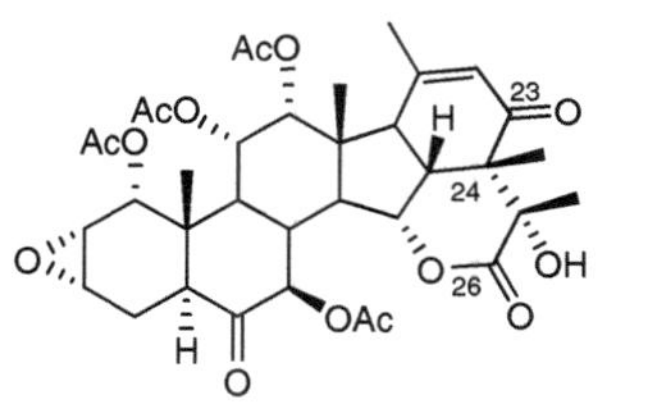

424 (taccalonolide X)

5.5. *18-Norwithanolides*

Kinghorn and coworkers used a quinone reductase induction assay for the activity-monitored fractionation of an extract of *Larnax subtriflora* (*sub nom. Deprea subtriflora*) collected in Peru. This led to the isolation of a novel group of highly oxygenated C_{27} 18-norwithanolides, dubbed subtrifloralactones. All subtrifloralactones found so far, have oxygenated functions at positions 12, 16, and 20 and a saturated lactone (γ or δ) side chain (see Figs. 2 and 3) (*184*, *185*). Another C_{27} 18-norwithanolide related to the physalins, has been reported recently from a *Physalis* species. A small group of C_{28} 17-methyl-18-norwithanolides is also included in this section.

5.5.1. Subtriflora-δ-Lactones and Related Withanolides

Subtrifloralactones D (**425**) and E (**426**) resemble the classic withanolide structure except for the lack of C-18. On the other hand, subtrifloralactones H, I, and J (**427–429**) have rings C and D *cis* fused, due to epimerization at position 13 (probably favored by the presence of the neighboring 12-ketone), allowing formation of a ketal bridge between a 20-hydroxy group and a 12-ketone (*184*). Due to the high reactivity of C-12 hemiketals towards even traces of small alcohols (see Sect. 5.4.2.), the formation of the methyl and ethyl ketals in the latter compounds probably took place during isolation. Another distinctive feature of subtrifloralactones H, I, and J is the presence of a formate group esterifying the 16-hydroxy group. The isolation of 13*β*-hydroxymethylsubtrifloralactone E (**430**) from the same plant (*185*) indicates an oxidative pathway for the loss of C-18 that could end as the 16-formate group via rearrangement of a 16,18-hemiketal (Fig. 12).

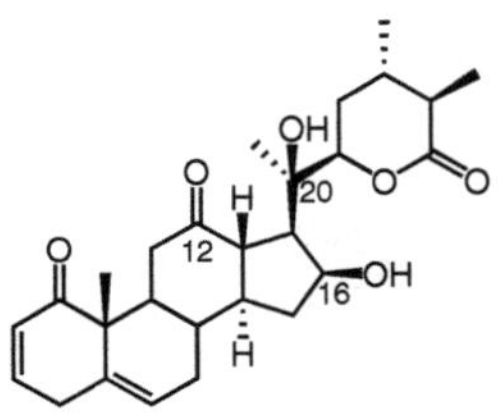

425 (subtrifloralactone D)

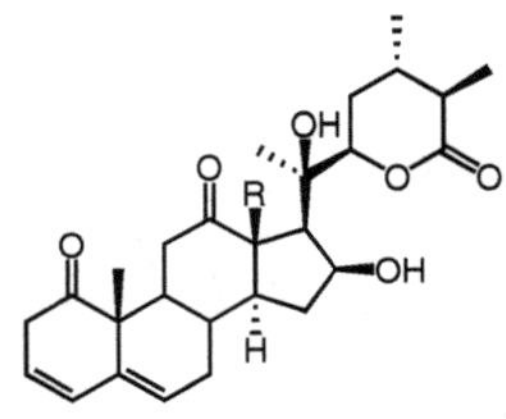

426 R = H(subtrifloralactone E)
430 R = CH_2OH

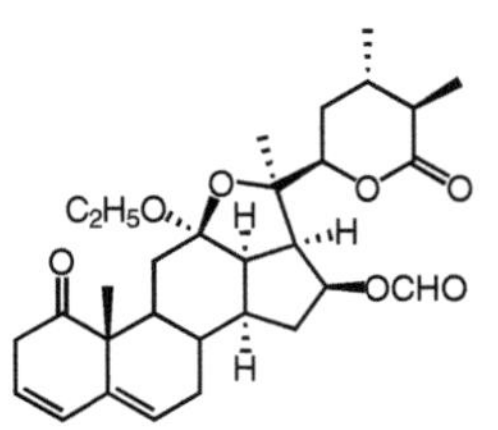

427 (subtrifloralactone H)

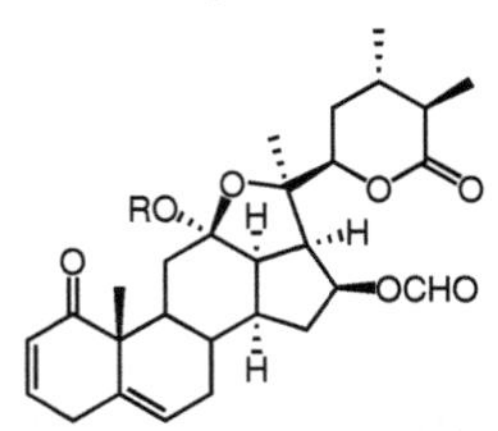

428 R = C_2H_5 (subtrifloralactone I)
429 R = CH_3 (subtrifloralactone J)

5.5.2. Subtriflora-γ-Lactones

Subtrifloralactones A (**431**), B (**432**), C (**433**), K (**434**), and L (**435**) present a side chain arrangement closely related to that observed in the trechonolides, with a γ-lactone between C-26 and C-23 and a ketal bridge between a 22-hydroxy and a 12-ketone (*184*, *185*). However, the *cis* fusion of rings C and D and the presence of a 16-hydroxy group allow an additional ketal bridge involving this hydroxy and

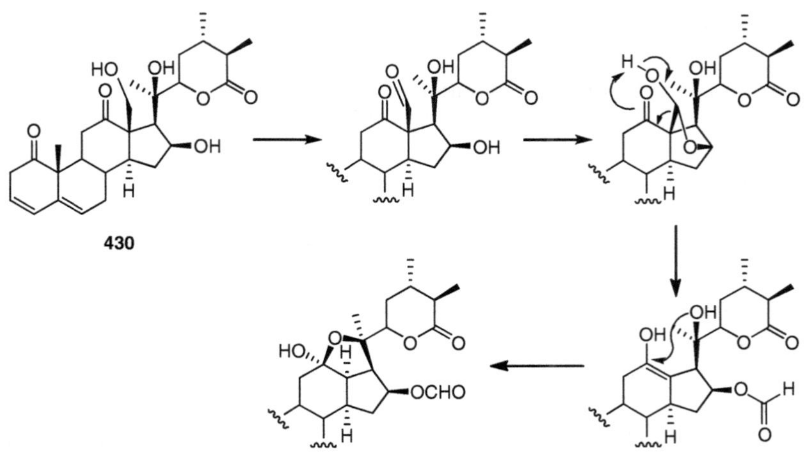

430

Fig. 12. Proposed biosynthetic pathway for the conversion of 13β-hydroxymethylsubtriflor-alactone E (**430**) to subtriflora-δ-lactones

C-12. In subtrifloralactones F (**436**) and G (**437**) the *trans* fusion of rings C and D results in an ixocarpalactone-type structure (*184*).

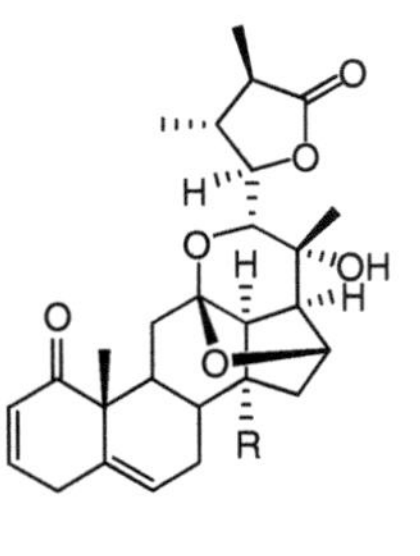

431 R = H (subtrifloralactone A)
433 R = OH (subtrifloractone C)

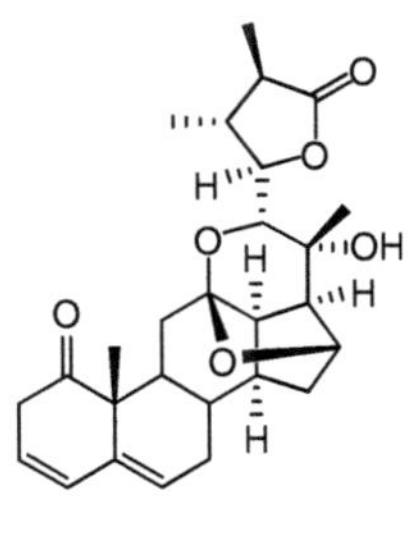

432 (subtrifloralactone B)

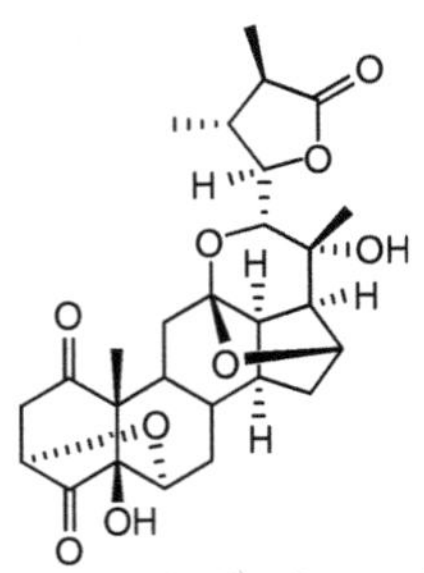

434 (subtrifloralactone K)

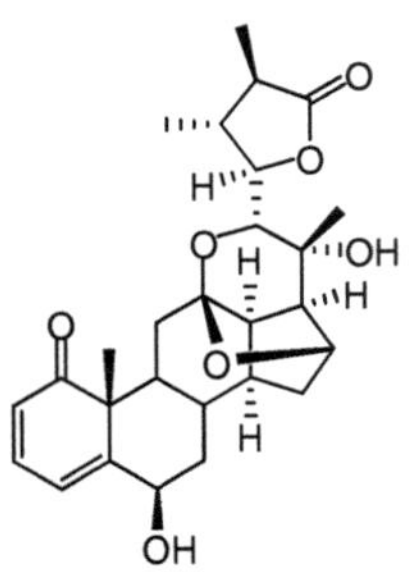

435 (subtrifloralactone L)

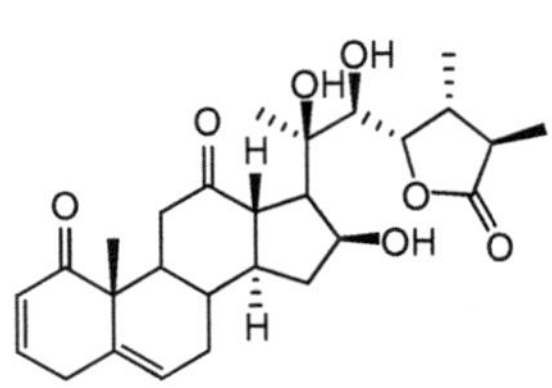

436 (subtrifloralactone F)

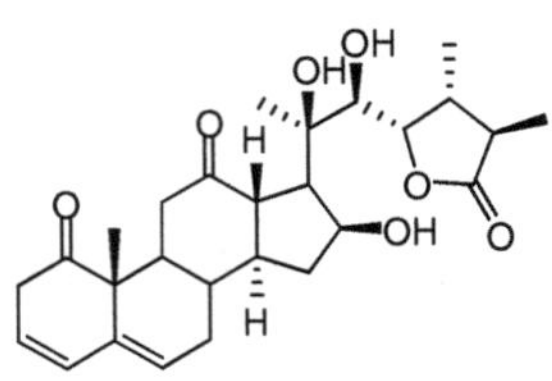

437 (subtrifloralactone G)

5.5.3. Other 18-Norwithanolides

Recently, *Ma et al.* reported the isolation of another 18-norwithanolide structurally related to the physalins, withaphysanolide A (**438**), together with several known physalins and withaphysalins from *Physalis divaricata* collected in Pakistan (*186*). The structure and configuration of withaphysanolide A was confirmed by X-ray crystallography.

TH-6 (**439**) and TH-12 (**440**) are two 17-methyl-18-nor-ergostanes isolated in 1990 by *Shingu et al.* from the acid hydrolysate of a methanolic extract of *Tubocapsicum anomalum* (*187*). The authors related these compounds to a putative precursor with a withanolide side chain that would rearrange in acid media. Recently *Hsieh et al.* isolated from the same plant three withanolides with the rearranged skeleton of TH-6 and TH-12 named tubocapsenolides A, F, and G (**441–443**) (*107*). Salpichrolide N (**352**) isolated from *Salpichroa origanifolia* (see Sect. 5.3.1.), was the first withanolide reported with this rearranged skeleton (*159*).

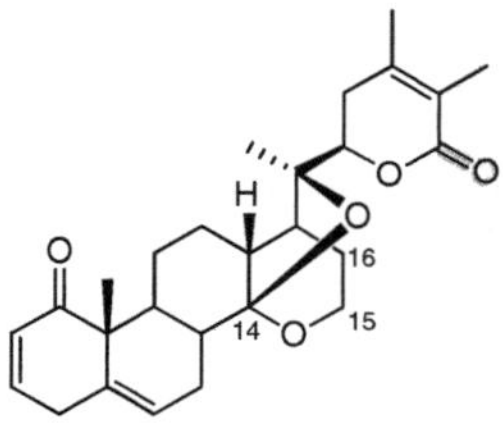

438 (withaphysanolide A)

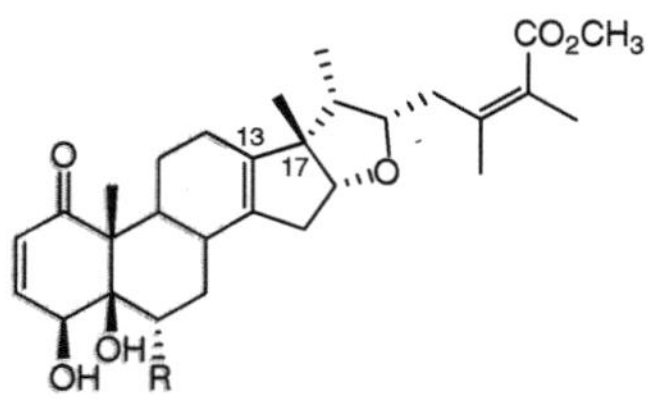

439 R = Cl (TH-6)
440 R = OH (TH-12)

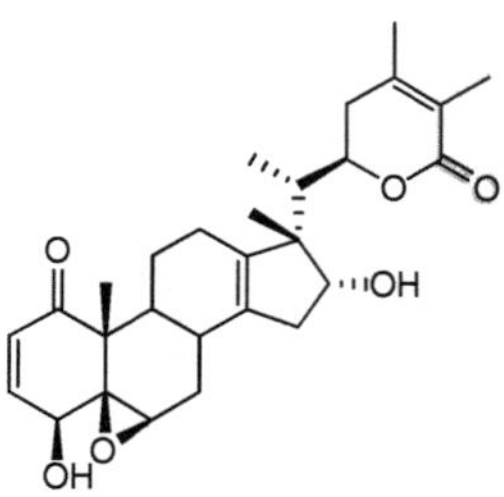

441 (tubocapsenolide A)

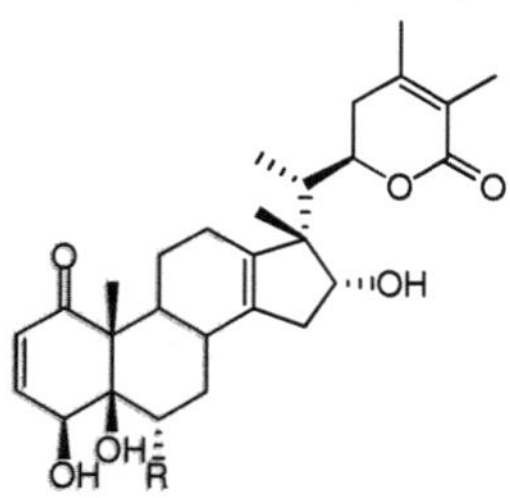

442 R = OH (tubocapsenolide F)
443 R = Cl (tubocapsenolide G)

5.6. *Spiranoid Withanolides at C-22*

This group of withanolides has a hemiketal bridge between what must have originally been ketone functions at C-12 and C-22. This gives rise to a new six-membered ring with a β-oriented hydroxy group at C-12 and a spiroketal at C-22 upon formation of the δ-lactone. Jaborosalactones 26–30 (**444–448**) were isolated from *Jaborosa rotacea* (*120*) and jaborosalactone 43 (**449**) was isolated from *J. kurtzii* (*76*) both collected in Mendoza, Argentina. The 12-ketowithanolide jaborosalactone 44 (**144**) also present in *J. kurtzii* (see Sect. 4.2.3.) has been proposed as the biogenetic precursor of **449**.

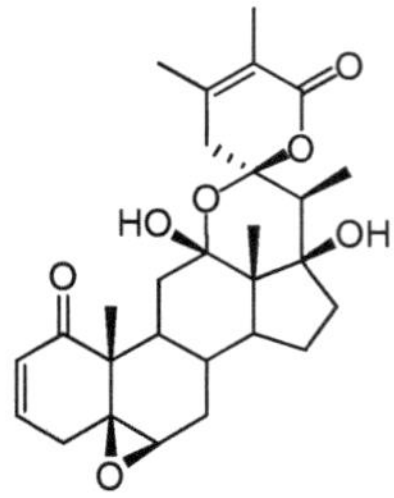

444 (jaborosalactone 26)

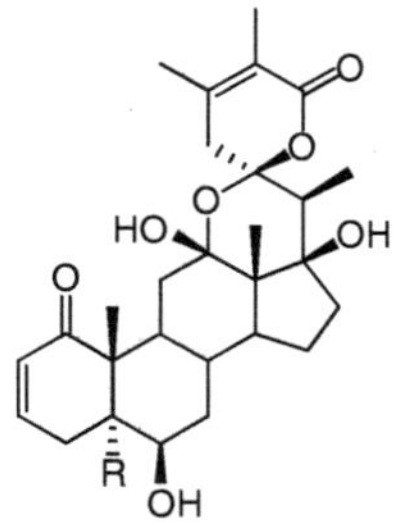

445 R = OH (jaborosalactone 27)
446 R = Cl (jaborosalactone 28)

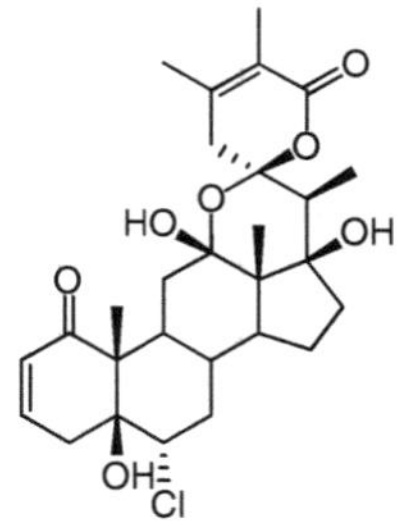

447 (jaborosalactone 29)

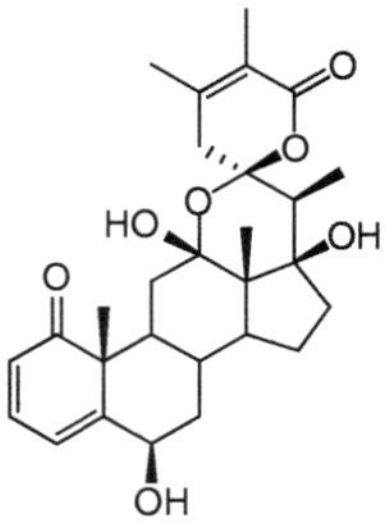

448 (jaborosalactone 30)

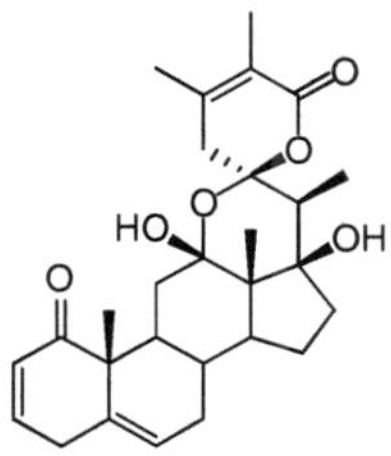

449 (jaborosalactone 43)

6. Chemical and Bio-transformations of Withanolides

The early synthesis work carried out on withanolides has been reviewed by *Glotter* (*59*), *Ray* and *Gupta* (*3*), and more recently by *Kovganko* and *Kashkan* (*188*). No new attempts on total or partial syntheses of withanolides have been reported in the last two decades. On the other hand, transformations usually involving interconversions of functional groups, of one withanolide into another or synthesis of simple derivatives, are fairly common and mostly used for structure confirmation or to study biological activities.

6.1. Chemical Transformations

Reactivity of the Δ^2-1-keto system of withanolides towards *Michael* addition (mostly of simple alcohols) has been discussed in the previous sections and several examples shown. As already mentioned, 5*β*,6*β*-epoxides are fairly common among the withanolides and also constitute a highly reactive center. Many chemical transformations involve the cleavage of the epoxide moiety to give the corresponding diols, chlorohydrins or alcohol addition products, with their configuration according to the *Fürst-Plattner* rule (*16*).

For the structure assignment of physalin H (**312**), *Makino et al.* treated physalin F (**450**) with aqueous hydrochloric acid to give a mixture of the diol **313** (physalin D) and the 5α,6*β*-clorohydrin **312** (physalin H) (Chart 1). When concentrated hydrobromic acid in THF was used the brominated analog of physalin H (**451**) was obtained. Reaction of physalin D (**313**) with phosphorus oxychloride in dry pyridine gave the isomeric chlorohydrin **452** (*148*). Cleavage of the epoxide of physalin F (**450**) with concentrated hydrochloric acid in ethanol gave the 5α-ethoxy-6*β*-hydroxy withanolide **453** (*151*).

The 5α,6α-epoxide present in the salpichrolides and a few other withanolides, reacts in a similar way. Thus, treatment of salpichrolide A (**339**) with THF containing 0.75% of 1.5 *N* sulphuric acid gave the corresponding diol **342** (salpichrolide C)

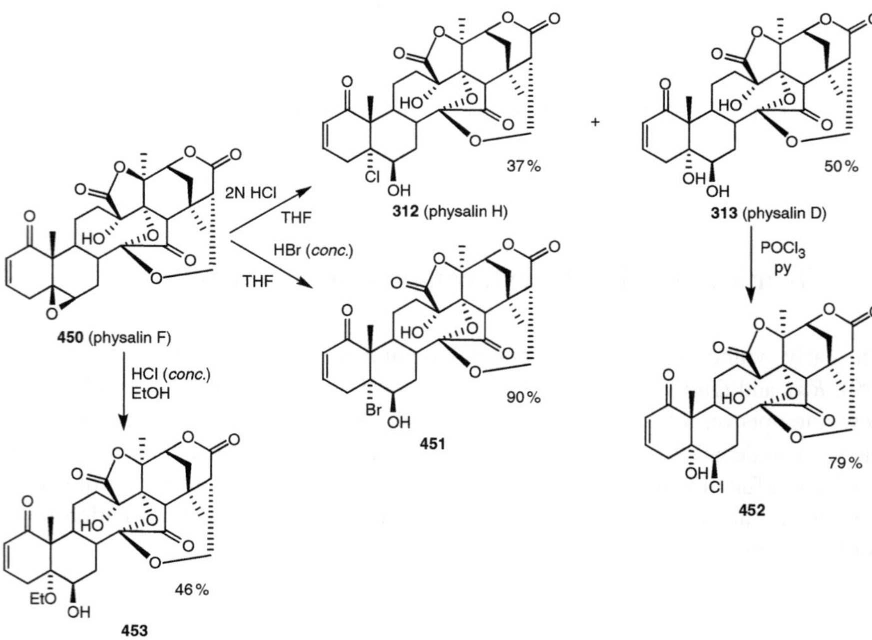

Chart 1 Cleavage reactions of the 5*β*,6*β*-epoxide of physalin F (**450**)

(*157*); increasing the amount of acid 10-fold, resulted in the concomitant cleavage of the 24,25-epoxide (compound **454**) (*159*). Reaction of salpichrolide A (**339**) with *Jones* reagent in acetone for 3 h gave the hydroxyketone **455** where oxidation of the side chain lactol to the lactone also occurred (*189*). On the other hand, cleavage of the epoxide with potassium bicarbonate in methanol gave the $\Delta^{2,4}$-6α-hydroxy withanolide **456** (Chart 2) (*190*). Reaction of physalin J (**457**) with sulfuric acid in ethanol gave the 5α-hydroxy-6β-ethoxy physalin **315** (*151*).

Misra et al. studied the reaction of 2-mercaptoethanol with several withanolides from *W. somnifera*. Withaferin A (**1**) and other 5β,6β-epoxywithanolides reacted readily under mild acid catalysis to give the 5β,6α-oxyethylenethio derivatives (Chart 3) in 47–60% yield. The additional ring would result from nucleophilic attack of the thiol at position 6, followed by acid-catalyzed condensation of the 6α-hydroxyethylthio intermediate. Withanolides with a 6α,7α-epoxide did not react

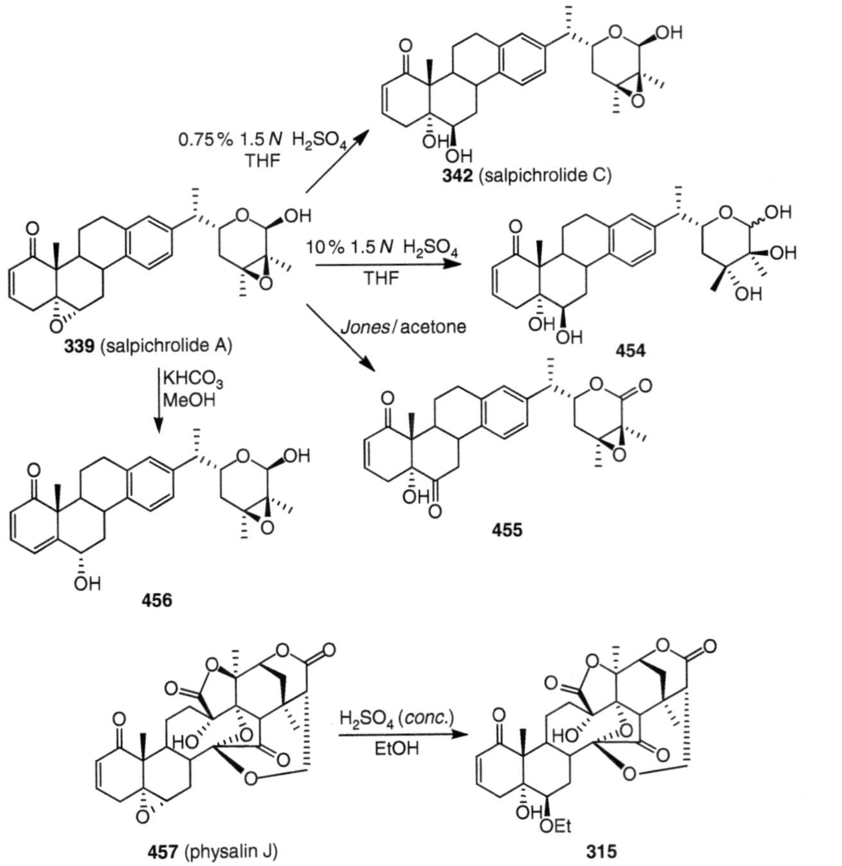

Chart 2 Cleavage reactions of 5α,6α-epoxy withanolides

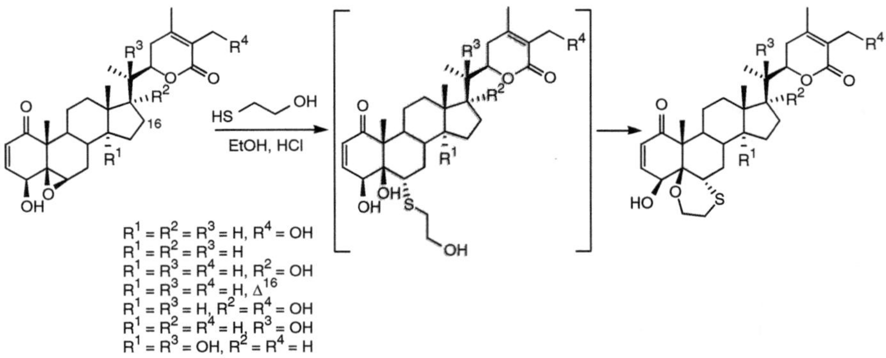

Chart 3 Reaction of withaferin A ($R^1 = R^2 = R^3 = H$, $R^4 = OH$) and related 5*β*,6*β*-epoxy withanolides with mercaptoethanol

under the same reaction conditions; attempts to force the reaction using a higher temperature and a lower pH resulted in complex mixtures (*191*).

Salpichrolide B (**341**), a minor constituent of *S. origanifolia* (see Sect. 5.3.1.), was obtained from the abundant salpichrolide A (**339**) by acetylation of the side chain hemiketal followed by stereoselective reduction of the C-1 ketone with sodium borohydride; deacetylation of the side chain occurred during workup. Following a similar protection/deprotection sequence, salpichrolide C (**342**) was oxidized to the 6-ketone **458** (Chart 4) (*190*).

Modifications at the side chain are less common, and one interesting example by *Mohan* and coworkers is the synthesis of a biotinylated analog of withaferin A (**459**) (Chart 5) for use as a probe to study angiogenesis (see Sect. 7.6.3.) (*192*).

6.2. Photochemical Transformations

Physalins containing an endoperoxy moiety in ring A were prepared from the corresponding $\Delta^{2,4}$-6-hydroxy physalin by photosensitized oxygenation. Thus, a *ca.* 1:1 mixture of physalins K (**311**) and Q (**297**) was obtained in 65% yield, by bubbling oxygen through a solution of physalin G (**314**) containing Rose Bengal and irradiated at 480 nm (Chart 6) (*144*). The same reaction carried out on 6-epiphysalin G (**460**) gave the 2*β*,5*β*-endoperoxide **461** as the major product (76% yield) and a trace amount (2%) of the 2*α*,5*α*-isomer **462**. (Note that names of physalin G and its 6-epimer have been assigned according to the revised structure of the former; see Sect. 5.2.3.). Photosensitized oxygenation of physalin B (**463**) gave a mixture of physalin K (**311**), the isomeric endoperoxide **461**, the 5*α*-hydroperoxide **464** and the corresponding alcohol **465** (Chart 6) (*144*).

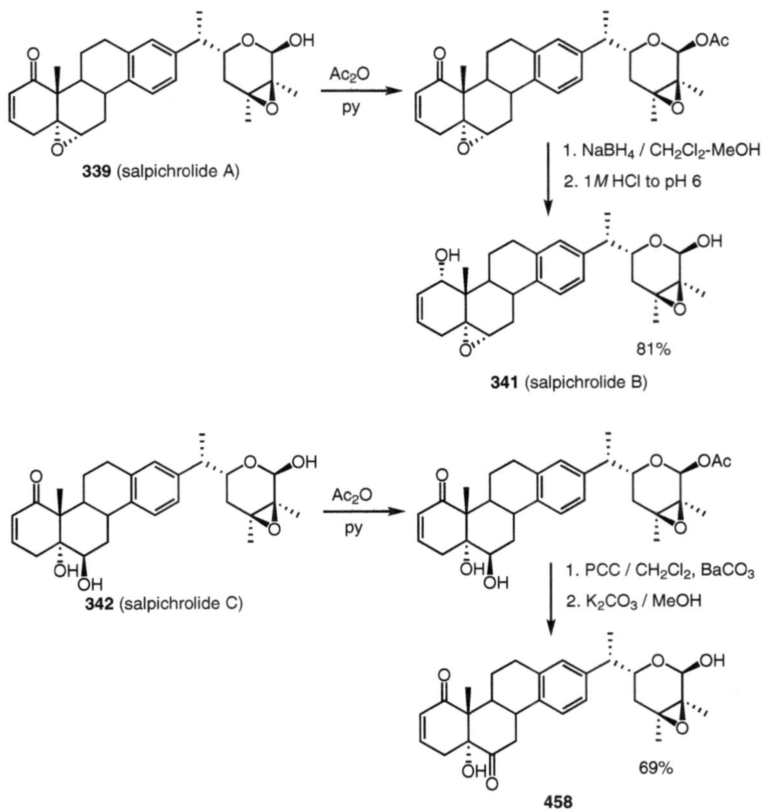

Chart 4 Chemical modification of salpichrolides A (**339**) and C (**342**)

Irradiation of an acetone solution of physalin B (**463**) with a tungsten-halogen lamp under argon, gave physalin R (**310**) in 49% yield. Irradiation of physalin F (**450**) under similar conditions gave the corresponding cyclophysalin (**466**). The 7α-hydroxy physalins, physalin N (**467**), A (**468**), and O (**469**), the latter two lacking the C-14–C-27 oxygen bridge, also gave the cyclophysalins (**470–472**) but required irradiation with a high-pressure mercury lamp (Chart 7) (*142*). The Δ^2-1-ketone system of the physalins would be involved in a self-sensitizing mechanism, as 2,3-dihydrophysalins failed to give the cyclized product.

6.3. *Biotransformations*

The modification of natural products by microorganisms may lead to new structures with potential biological activities. Application of this strategy to withanolides has been reviewed by *Anjaneyulu et al.* (*4*). Biotransformation of physalin H (**312**) by the fungus *Rhizopus stolonifer* gave the elimination product **473** in 2.1%

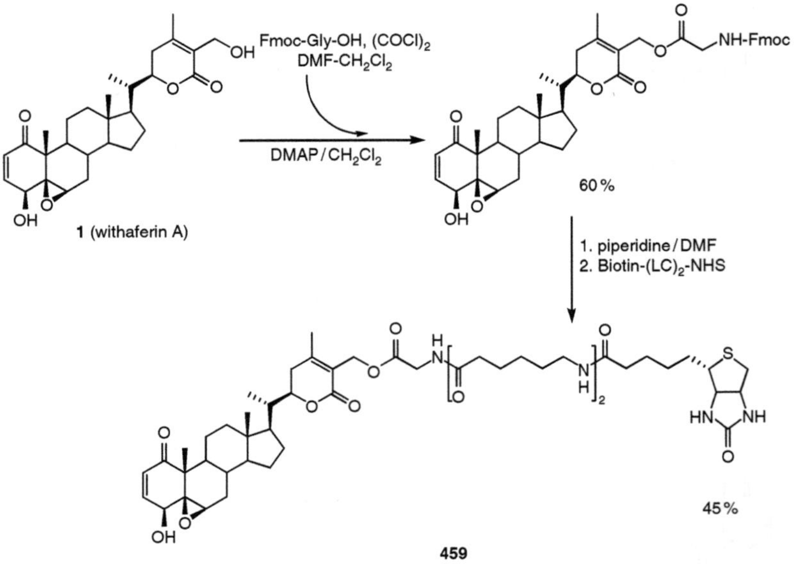

Chart 5 Synthesis of a biotinylated analog of withaferin A (**1**)

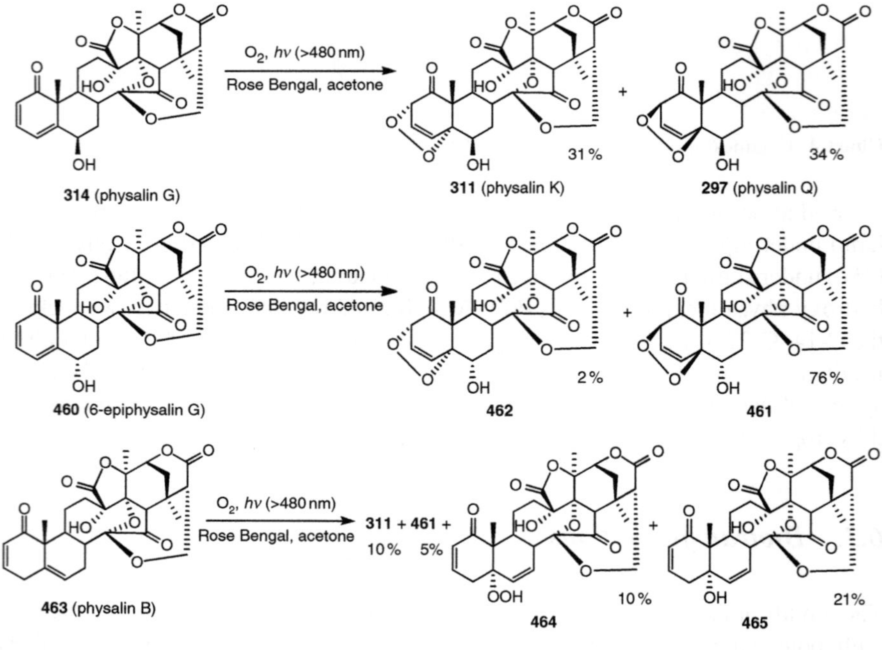

Chart 6 Photooxygenation of physalin G (**314**), 6-epiphysalin G (**460**) and physalin B (**463**)

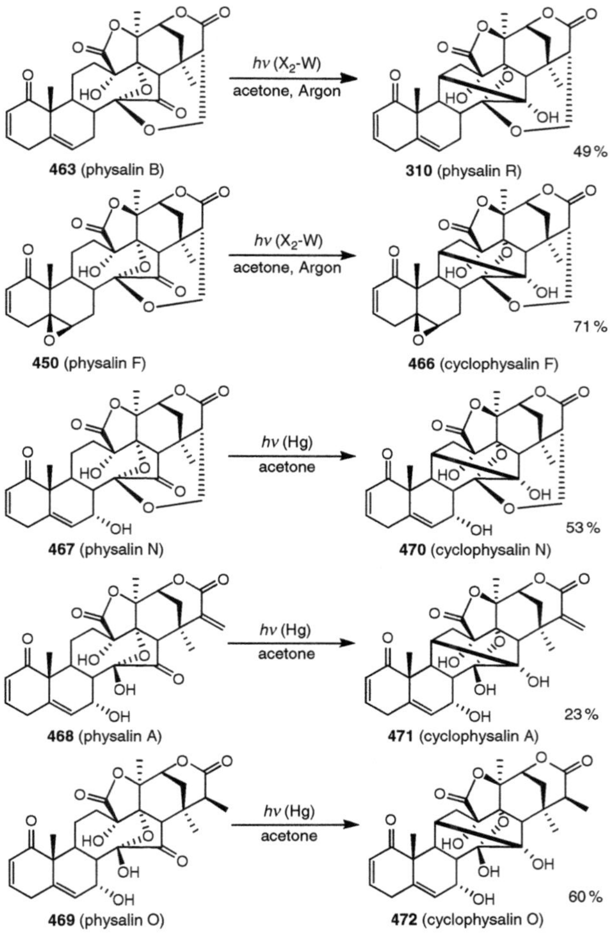

Chart 7 Photocyclization of physalins to cyclophysalins. *X_2-W* halogen-tungsten lamp; *Hg* high pressure mercury lamp

yield, while incubation with *Cunninghamella elegans* gave isophysalin B (**474**) and 6-deoxyphysalin H (**475**) in 9.4% overall yield (Chart 8) (*193*). Both **473** and **475** are new compounds. Incubation of withaferin A (**1**) with *Cunninghamella echinulata* gave 12β-hydroxywithaferin A (**476**) and 15β-hydroxywithaferin A (**477**) (*194*); the same hydroxylated products have been obtained upon incubation with *Cunninghamella elegans* (*4*). Several reports also describe the production of withanolides (mostly withaferin A) by shoot cultures of *Withania somnifera* (*195–198*).

Tuli and coworkers isolated two specific glucosyltransferases from *W. somnifera*. The cytosolic 3β-hydroxy sterol glucosyltransferase was most active on 24-methylene-cholesterol, and showed moderate activity for the 3β-*O*-monoglucosylation of the

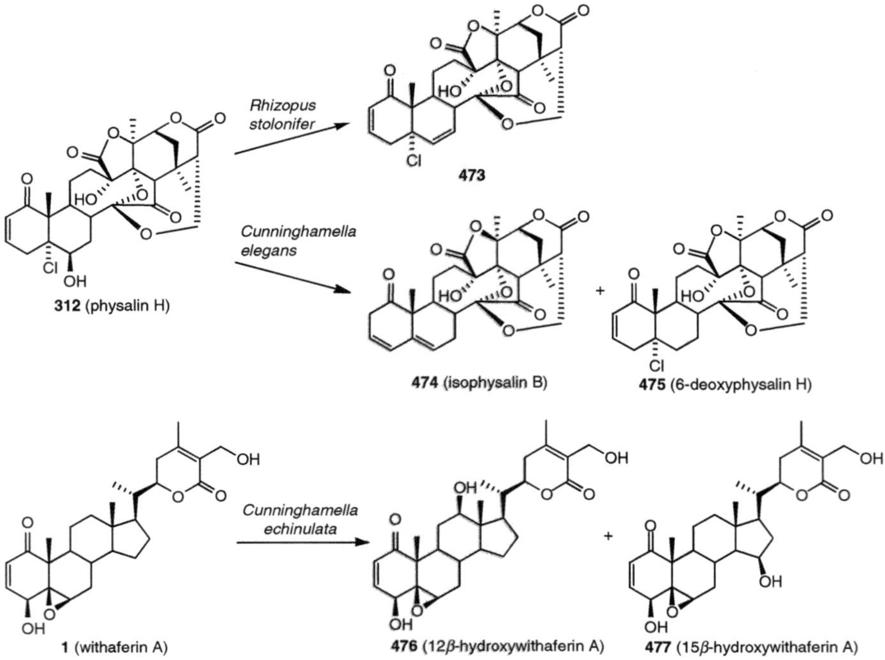

Chart 8 Biotransformations of physalin H (**312**) and withaferin A (**1**) with microorganisms

aglycone of withanoside V (**478**) (*199*). On the other hand the 27-hydroxy glucosyltransferase monoglucosylated several 27-hydroxywithanolides, provided a hydroxy group was also present at C-17 (Chart 9) (*200*).

7. Biological Activities of the Withanolides

As already mentioned, several withanolide-containing plants are used in traditional folk medicine throughout the world and many of the verified activities can be traced to their withanolide constituents. *Withania somnifera* used in Ayurvedic medicine in India since ancient times is the prototypical example, but similar uses have been accorded to *Datura metel* in traditional Chinese medicine and to several *Physalis* species in Asia and the Americas. Withanolides have proven active not only in a wide variety of assays related to human ailments, but also in potential applications as natural agrochemicals for pest and weed control. Despite the structural variety of withanolides and the many different activities they present, it is noteworthy, as will become evident in the following sections, that some structural characteristics are usually associated with biological activity (or the lack of it). One such feature is, with few exceptions, the lack of activity of 5α,6β-diols as compared to the corresponding epoxides (or the Δ^5 analogues) that are usually active.

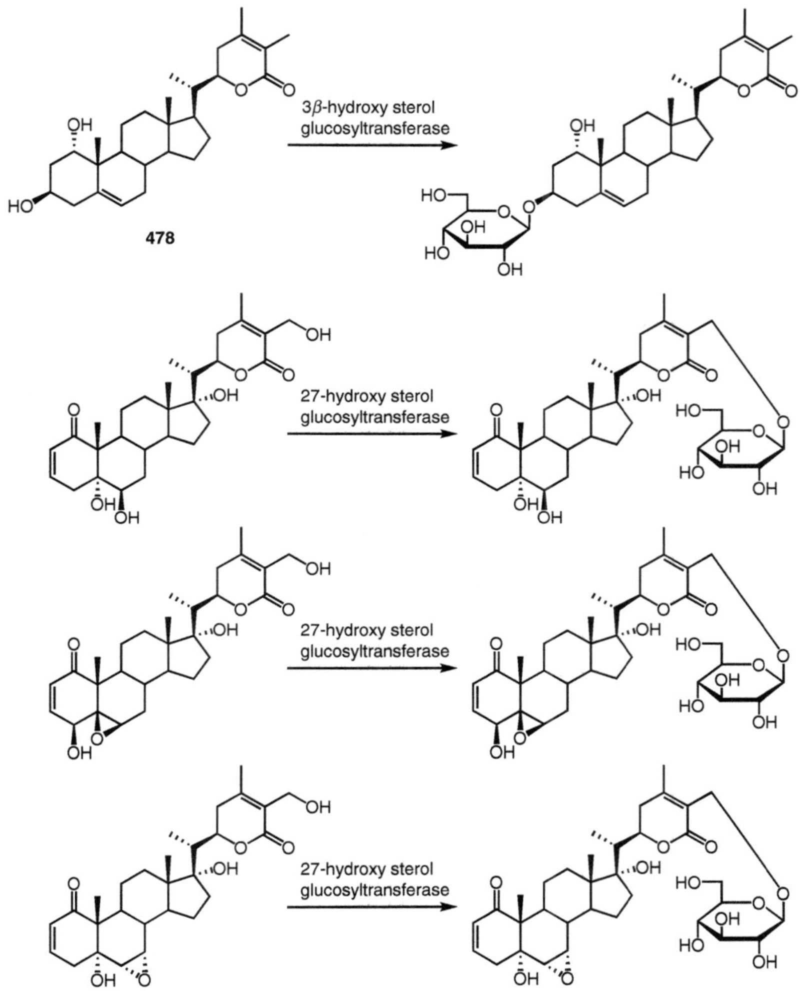

Chart 9 Enzymatic glucosylation of 3β- and 27-hydroxy withanolides

7.1. *Insecticidal Activities*

Insecticidal properties of withanolides were first noted for components isolated from the Peruvian plant *Nicandra physalodes* in the early 1960s. The major component of this plant, nicandrenone or Nic-1 (**338**), was later shown to be responsible for the insecticidal properties (*59*). Since then, several withanolides have been shown to exhibit insecticidal activity.

7.1.1. Antifeedant and Growth Inhibition

Antifeedant effects and species-specific activity were initially shown for the cotton leafworm *Spodoptera littoralis* (Boisd.) (Lepidoptera), the Mexican bean beetle, *Epilachna varivestis* Muls. (Coleoptera), and the red flour beetle, *Tribolium castaneum* (Herbst) (*59*). *Elliger* and coworkers found that some chromatographic fractions of an extract of *Physalis peruviana* leaves had a strong inhibitory effect on the development of larvae of the corn earworm *Helicoverpa zea*, an economic pest of numerous crops including tobacco and tomato (*201*). Bioassay-directed fractionation led to the isolation of a series of saccharide esters structurally related to the withanolides, with the δ-lactone side chain open and the carboxyl group esterified by mono-, di-, or trisaccharides (*202, 203*). The most active compound in inhibiting larval growth was the 11β-hydroxy diglucoside ester **479** (ED_{50} 5.4 ppm) and the least active was the closely related monoglucoside ester **480** (ED_{50} 110 ppm). No clear structure-activity relationships could be established, but the lack of toxicity of the compounds led the authors to conclude that growth inhibition was a consequence of feeding deterrence resulting in semi-starvation of the animals. Previously, the feeding deterrent withanolides 4β-hydroxywithanolide E (**481**) and withanolide E (**482**) had been found in very high concentrations in the leaves and berries of *P. peruviana* (*204*).

The antifeedant activities of the major withanolides of *Salpichroa origanifolia*, salpichrolides A (**339**), C (**342**), and G (**340**), and some synthetic derivatives, were investigated on larvae of the common fly *Musca domestica* (*189*), the stored grain pest *Tribolium castaneum* (*205*), and the Mediterranean fly *Ceratitis capitata* (*190*). Salpichrolide A (**339**) was the most active in all cases producing a significant development delay in surviving larvae, when added to the diet at 500 ppm concentration. Similar delays were observed when medium and low nutrition diets (without withanolides) were offered as food, supporting the idea that these compounds act as feeding deterrents. The effects of salpichrolides C (**342**) and G (**340**) differed from one insect species to the other, thus both compounds produced development delays only at high concentrations (2,000 ppm) in *M. domestica* and *C. capitata*, but salpichrolide G was almost as effective as salpichrolide A in *T. castaneum*. Regarding toxicity, the three salpichrolides produced significant mortality for *M. domestica* larvae (EC_{50} 200–300 ppm), but only salpichrolides A and G had this effect on *C. capitata*. The authors proposed that differences could be related to different detoxifying mechanisms.

Oxidation of the hemiketal in the side chain to give the δ-lactone **483** eliminated the activity both for *M. domestica* and *T. castaneum*. However, acetylation to give **484** drastically reduced the activity only for the latter insect. Reduction of the Δ^2 double bond had only a minor negative effect on activity (*189, 205*). The effect of functional group modifications in rings A and B on the resultant activity was evaluated on *C. capitata* larvae (*190*). Salpichrolide B (**341**), a minor component of *S. origanifolia*, was prepared from salpichrolide A in sufficient amounts for testing (see Sect. 6.1.). This compound was the most active of all salpichrolides tested, producing significant mortality when incorporated to the diet even at low doses (EC_{50} 83 ppm). It also produced clearly observed development delays in surviving larvae at 25 ppm. Reduction of the Δ^2 double bond (**485** and **486**) produced a significant decrease in activity. Oxidation of salpichrolide C to the 6-ketone (**458**) or rearrangement of the 5,6-epoxide in salpichrolide A to the $\Delta^{2,4}$-6α-alcohol **456** resulted in inactive compounds.

The content of the salpichrolides in *S. origanifolia* was monitored during plant development, reaching a maximum during summer when insect populations are higher. These results, in conjunction with the observed toxic and feeding deterrent activities, suggest that these compounds may act as a chemical defense providing protection to the plant against phytophagous insects (*189*).

Feeding-deterrent activity of the major components of *Jaborosa odonelliana*, the spiranoid withanolides jaborosalactone P (**389**) and jaborosalactone 10 (**390**), was studied against *T. castaneum* (*177*). In this case, only jaborosalactone P (**389**) produced a significant delay in the development of neonate larvae. In the case of *J. integrifolia*, only jaborosalactone A (**487**) exhibited antifeedant activity on larvae of *Spodoptera littoralis* (*206*). Rearrangement of the epoxide to the allylic alcohol as in jaborosalactone B (**488**) or cleavage to the diol as in jaborosalactone D (**489**) resulted in complete loss of the activity. Jaborosalactone S (**254**) from *J. sativa* was a feeding deterrent for *Tenebrio molitor* (*207*) and the trechonolide-type withanolide **386** isolated from *J. lanigera* produced significant development delays in *C. capitata* larvae (*175*).

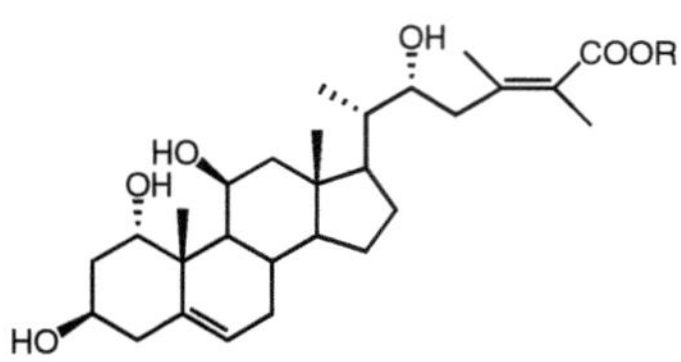

479 R = gluc-(1-2)-gluc-6'-Ac
480 R = gluc

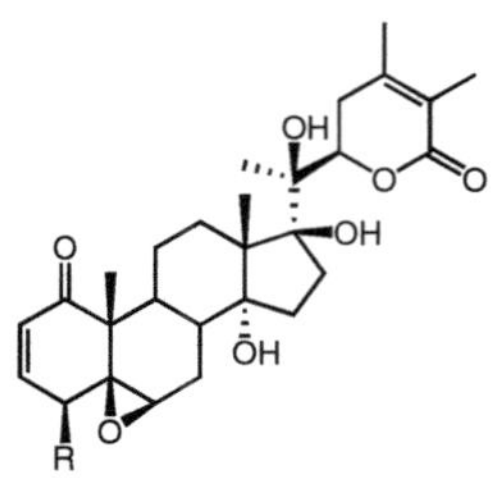

481 R = OH (4β-hydroxywithanolide E)
482 R = H (withanolide E)

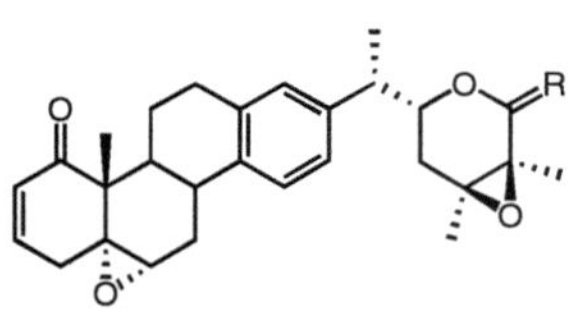

483 R = O
484 R = α-H, β-OAc

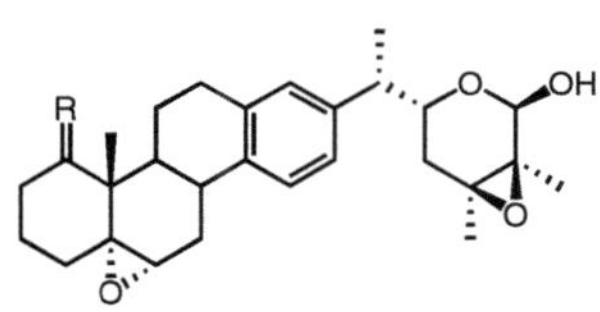

485 R = O
486 R = α-OH, β-H

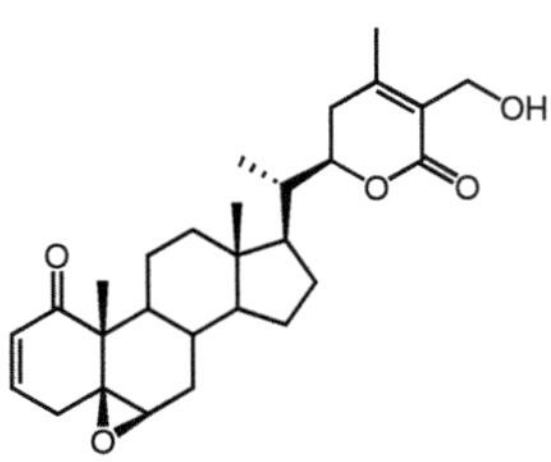

487 (jaborosalactone A)

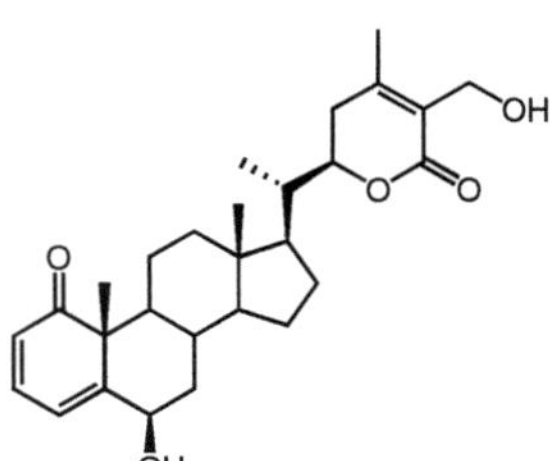

488 (jaborosalactone B)

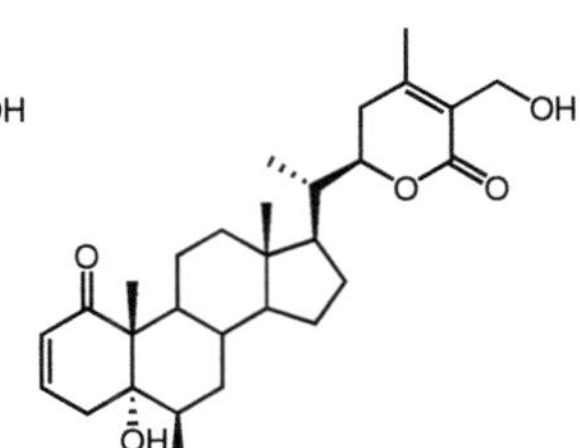

489 (jaborosalactone D)

Azambuja and coworkers studied the effect of several physalins from *Physalis angulata* on the blood-sucking insect *Rhodnius prolixus*, vector of *Trypanosoma rangeli*. Physalins B (**463**), D (**313**), F (**450**), and G (**314**) produced immune depression in *R. prolixus*, although apparently by different mechanisms. Without their defense system, insects infected with *T. rangeli* die (*208–210*).

7.1.2. Ecdysteroid Agonists and Antagonists

The ecdysteroid endocrine system is vital for insect development and a variety of secondary metabolites from plants have been shown to interfere with ecdysteroids, probably as a chemical defense mechanism. *Dinan* and coworkers developed a microplate-based bioassay with the ecdysteroid-responsive *Drosophila melanogaster* B_{11} cell line and used it to evaluate 16 withanolides isolated from *Iochroma gesneriodes*, for agonistic/antagonistic activity (*211*). Only withanolides containing an oxygenated functionality at C-3 (hydroxy or methoxy) and an α,β-unsaturated lactone in the side chain showed antagonistic activity, with 2,3-dihydro-3*β*-hydroxywithacnistine (**490**) being the most active (ED_{50} 2.5 μ*M* versus 0.05 μ*M* for 20-hydroxyecdysone). In a systematic study of 128 species of solanaceous plants including those known to contain high levels of withanolides, only a few of the methanolic extracts showed weak ecdysteroid antagonist activity (*212*). The high reactivity of the Δ^2-1-keto system present in many withanolides, raises the possibility that, even though inactive *in vitro*, they could be activated by conversion to the 2,3-dihydro-3*β*-hydroxywithanolides upon ingestion by insects (*213*). Further studies on 21 withanolides from different sources showed that withaperuvin D (**491**), with a C-3, C-6 oxygen bridge, had moderate agonistic activity (EC_{50} 25 μ*M*) (*214*).

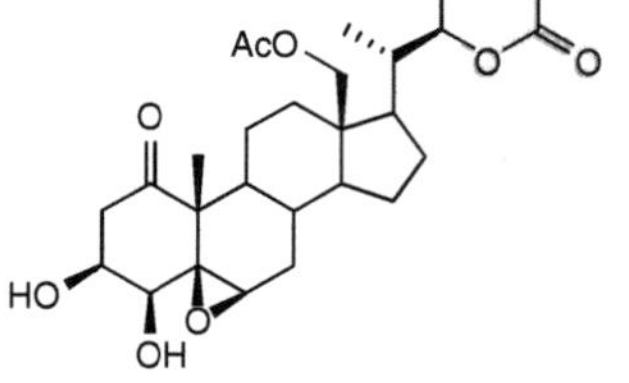

490 (2,3-dihydro-3*β*-hydroxywithacnistine)

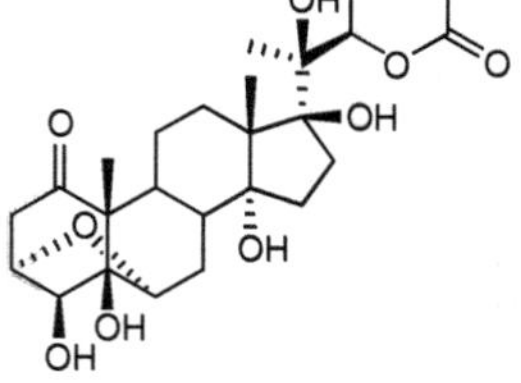

491 (withaperuvin D)

7.2. *Phytotoxic Activities*

Selective phytotoxicity has been reported for several withanolides. Three 7-oxygenated withanolides, **492–494**, isolated from *Iochroma australe*, reduced radicle growth of the weeds *Sorghum halepense* (L.) Pers. (Monocot.) and *Chenopodium album* L. (Dicot.). Compound **493** inhibited radicle growth of *Lactuca sativa* L. (lettuce) but only at high concentration (1,000 ppm) (*215*). Jaborosalactol 18 (**285**) a major constituent of *J. bergii* (see Sect. 5.1.4.) showed significant inhibition of radicle growth at 2 m*M* on the dicotyledoneous species *C. album*, *Ipomoea purpurea* (L.) Roth and *L. sativa* (phytogrowth inhibitory activity > 49%). On the other hand, **285** had a strong stimulatory effect for the monocotyledoneous species tested (*Zea mays* L. and *Sorghum halepense*) (*69*). Several withanolides isolated from *J. rotacea* also exhibited different activities towards mono- and dicotyledoneous species. Thus, jaborosalactones 29 (**447**), 30 (**448**), 31 (**404**), and 33 (**370**) had opposite effects on the dicotyledon *L. sativa* and the monocotyledon *Phalaris canariensis* L. Jaborosalactone 29 (**447**) was the most active, selectively inhibiting radicle growth, germination and the emergence rate index of *L. sativa* but with no effect on *P. canariensis*. Jaborosalactones 30 and 31 (**448**, **404**) had a strong stimulatory effect on radicle growth of *P. canariensis*. The chlorohydrin jaborosalactone 28 (**446**) was the only withanolide that inhibited the radicle growth of *P. canariensis* (*120*). A similar selectivity was exhibited by the major component of *J. kurtzii*, jaborosalactone 43 (**449**) (spiranoid type), and by withanolides of *J. caulescens*, the 12-*O*-ethyl derivative of jaborosalactone 42 (12-*O*-ethyl-**373**) (trechonolide type) and the sativolides **258**, and 12-*O*-methyl-**258**. All these compounds strongly inhibited radicle growth of *L. sativa* but exhibited a marginal effect on the monocotyledoneous *Avena sativa* L. (*76*, *121*).

7.3. Antiparasitic Activities

The first withanolides with antileishmanial and antitrypanosomial activities were isolated from *Dunalia brachyacantha* when screening extracts of Bolivian plants against *Trypanosoma cruzi* (Chagas disease), and several *Leishmania* species. Bioassay-guided fractionation of the leaf extract gave the known 18-acetoxywithanolide D (**142**) and its Δ^5 analogue **495**, with the latter compound being the most active against epimastigote forms of *T. cruzi* and promastigote cultures of *L. amazonensis*, *L. braziliensis*, and *L. donovani* (*84*).

7.3.1. Trypanocidal Activity

In a systematic study of the trypanocidal activity of secondary metabolites isolated from plants from northeastern Brazil, *Vieira et al.* found withaphysalins M and O (**323**, **325**) from *Acnistus arborescens* (see Sect. 5.2.4.) to be the most active compounds against epimastigotes of *T. cruzi*, with IC_{50} values of 100-fold less than the reference compound benznidazole (*216*). Reduction of the 2,3 double bond as in withaphysalin N (**324**) reduced the activity by an order of magnitude. Most interestingly, these withaphysalins had no cytotoxic activity against dividing normal cells. Physalin F (**450**) from *Physalis angulata* was shown to be moderately active. *Abe* and coworkers investigated the trypanocidal activity of ten withanolides from *P. angulata* growing in Japan, against *T. cruzi* epimastigotes and trypomastigotes (the infectious form of the parasite) (*57*, *66*). Physagulins A–C (**496–498**), H (**117**) and I (**118**), and withangulatin A (**499**) had activity against both forms of *T. cruzi* similar to their cytotoxicity; activity against trypomastigotes was higher for these withanolides. Physagulins F (**500**), J (**119**), K (**120**), L (**121**), and M (**103**) and withaminimin (**501**) containing the 5α,6β-diol moiety were weakly active against epimastigotes, but the first two of these compounds were moderately active against the infectious trypomastigote form. Physagulin N (**106**), the methoxy derivative of physagulin A, was also marginally active confirming that the A ring enone is required for activity.

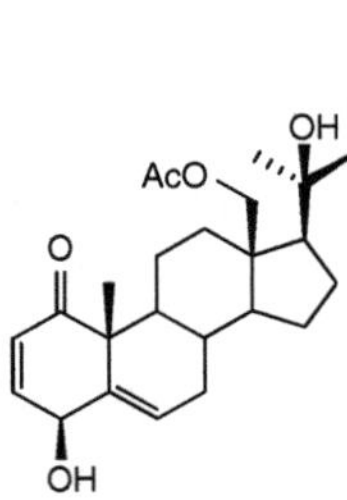

495

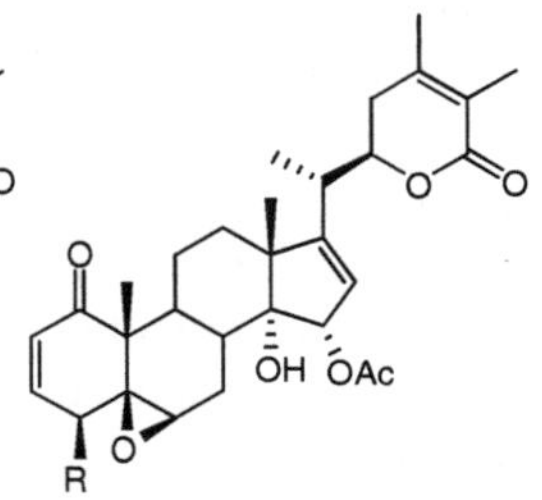

496 R = H (physagulin A)
499 R = OH (withangulatin A)

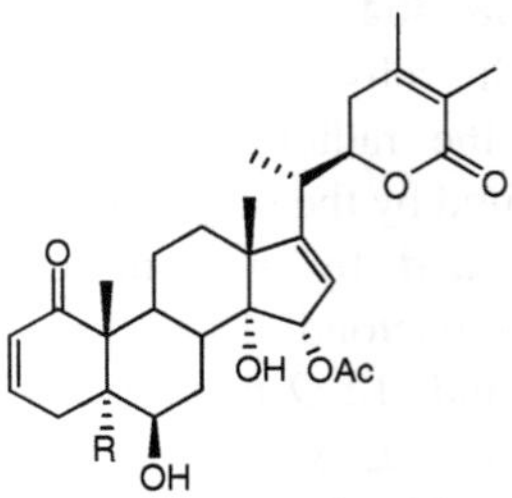

497 R = Cl (physagulin B)
501 R = OH (withaminimin)

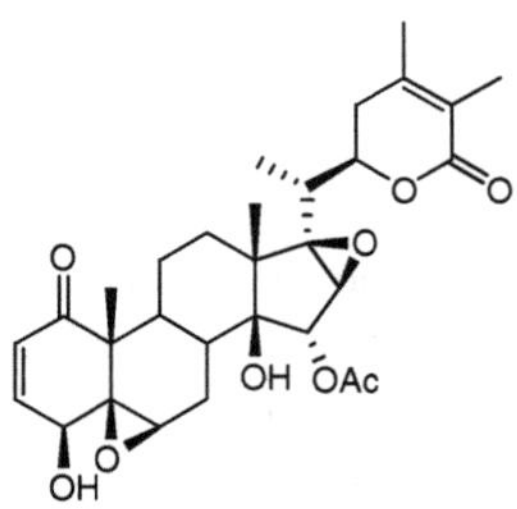

498 (physagulin C)

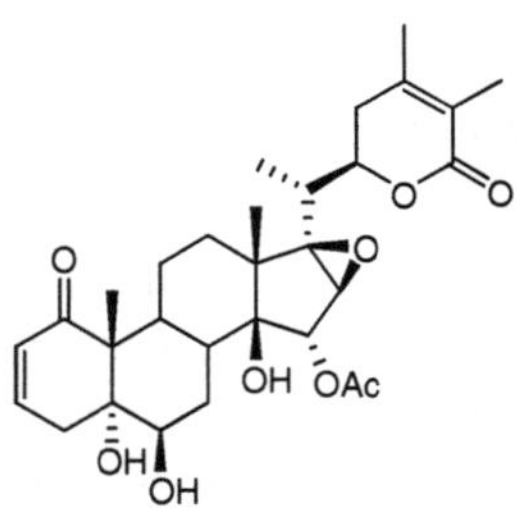

500 (physagulin F)

7.3.2. Leishmanicidal Activity

Atta-ur-Rahman and coworkers have reviewed the antileishmanial activity of withanolides from *Physalis minima* as well as some of their biotransformation products (*217*). Physalin **299** was the most active substance against *Leishmania major* promastigotes, (see Sect. 5.2.1.) while physalins **298**, **300**, **301**, physalin H (**312**), isophysalin B (**474**), and 5β,6β-epoxyphysalin B (**502**) also had significant activity.

The 3-methoxy derivative physalin U (**304**) and withanolide **107** with an unmodified skeleton were only marginally active (*60, 145*). Removal of the C-6 hydroxy group of physalin H (**312**) to give either the dehydration product **473** or the deoxygenated physalin **475** (see Sect. 6.3.) did not affect the antileishmanial activity (*193*).

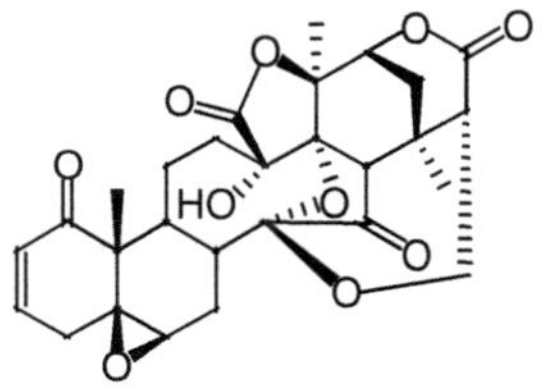

502 (5β,6β-epoxyphysalin B)

Echeverri and coworkers investigated the leishmanicidal activity of several acnistins (A, B, C, E, F, and G) and withajardins (A, B, and C) isolated from *Dunalia solanacea* and *Deprea orinocensis*, respectively (see Sects. 5.1.2. and 5.1.3.). Efficacy was evaluated using intracellular amastigotes of *Leishmania (V) panamensis* (*218*). Withajardin B (**278**) and acnistins A (**259**), C (**262**), and E (**260**) were the most active; withajardin A (**277**), and acnistins B (**261**), and F (**264**) were the least active. The 3-methoxy derivative of withajardin A was inactive. All compounds had poor selectivity, with antileishmanial activity closely paralleling cytotoxicity in all cases. The authors did a 3D-QSAR study and concluded that differences in bioactivity could be explained by the influence of steric and electrostatic fields in the vicinity of ring A. According to the model, bending of the steroid nucleus at the A/B ring junction, an increase of positive charge near positions 2, 3, and 4, or of a negative change near positions 5 and 6, increased the bioactivity.

Withaferin A (**1**) inhibited growth of *L. donovani* promastigotes. *In vitro* studies showed that withaferin A inhibits protein kinase C in the parasite, leading to apoptosis (*219*). Withanolide Z (**27**), but not withaferin A, partially inhibited *L. donovani* topoisomerase I (*28*).

7.4. Antimicrobial Activities

The antibacterial and antifungal properties of withaferin A (**1**) have been known for a long time and many other withanolides are known to display these activities (*3, 4*). However, data are scattered throughout the literature involving the action of different compounds on different microorganisms thus making it difficult to rationalize the results. Withaferin A has been shown to be strongly active against *Bacillus subtilis* and moderately active for *Escherichia coli* and *Staphylococcus aureus*, but inactive against *Pseudomonas aeruginosa*. As an antifungal it exhibited strong activity against *Aspergillus niger*, but was inactive against *Rhizopus oryzae* and *Candida albicans* (*220*). 4-Deoxywithaperuvin (**11**) was moderately active against several *Gram*-positive (*B. cereus*, *B. subtilis*, *Streptomyces* spp.), and *Gram*-negative (*Pseudomonas fluorescens*, *Serratia marcescens*) bacteria, but inactive against *Micrococcus luteus*, *M. roseus* and *S. aureus*; antifungal activity was poor (*20*). On the other hand, the 18-oxygenated withanolides 18-acetoxywithanolide D (**142**) and its Δ^5 analogue **495** from *D. brachyacantha* were active against *S. aureus* and *B. subtilis* and inactive against *E. coli* and *Shigella flexneri* (*84*).

17*β*-Hydroxywithanolide K (**51**) and the closely related 14*α*,20-epoxywithanolide, **71** (see Sect. 4.1.4.) showed antifungal activity against the human pathogens (*MIC* 300 μg/cm^3) *Aspergillus niger*, *Stachybotrys atra*, *Allescheria boydii*, *Drechslera rostrata*, *Microsporum canis*, and *Curvularia lunata* (*36*, *43*).

Among the physalins, physalin B (**463**) was active against *S. aureus* (several strains) and *N. gonorrhoeae* but not active against *E. coli* and *P. aeruginosa* (*221*). Physalin D (**313**), but not physalin B, was moderately active against *Mycobacterium tuberculosis* (*222*).

7.5. Anti-inflammatory and Glucocorticoid Related Activities

7.5.1. Anti-inflammatory Activity

The antiinflammatory properties of several withanolides are well known (*3*). *Souza* and coworkers have shown that in the case of physalins B (**463**) and F (**450**), the anti-inflammatory activity parallels that of the synthetic glucocorticoid dexamethasone in preventing inflammatory injury and lethality after intestinal ischemia and reperfusion in mice (*223*). Furthermore, they found that the effect could be reversed by pretreatment with the glucocorticoid antagonist RU-486, indicating that the *in vivo* activity displayed by the physalins is mostly due to activation of the glucocorticoid receptor. As observed with dexamethasone, physalins also decreased TNF-α concentration and enhanced the anti-inflammatory interleukin IL-10 concentration in tissues. Physalins B (**463**) and F (**450**) have been shown to inhibit TNF-α induced activation of NF-κB and either a 5,6 double bond or a 5*β*,6*β*-epoxide are required for activity; the 5*α*,6*β*-diol physalin D (**313**) is inactive (*138*).

An ethanol extract of *W. somnifera* significantly suppressed lipopolysaccharide (LPS)-induced production of the pro-inflammatory cytokines TNF-α, IL-1β, and IL-12p40 in peripheral blood mononuclear cells of normal individuals and reumathoid arthritis patients, and inhibited nuclear translocation of the transcription factors NF-κB and AP-1 and phosphorylation of IκBα. The major component, withaferin A (**1**), inhibited NF-κB translocation and was associated with these effects (*224*). Withaferin A inhibited NF-κB activation by preventing the TNF-induced activation of IκB kinase via a thioalkylation-sensitive redox mechanism

(*225*). 12-Deoxywithastramonolide (**503**) and withanolide A (**504**) were much less effective. A more detailed study on the inhibition of NF-κB activation by various agents was conducted by *Ichikawa et al.* (*226*). The authors isolated a series of withanolides from a *W. somnifera* leaf extract, which included withaferin A (**1**), viscosalactone B (**505**), withanosides IV (**31**) and X (**37**), and related compounds. The 1α,3β-dihydroxy-Δ^5-withanolides and their glycosides did not inhibit NF-κB activation while withaferin A and its diacetate were strong inhibitors. Reduction of the 2,3-double bond of withaferin A rendered the compound inactive but addition of a 3-hydroxy substituent (as in viscosalactone B) restored activity. Acetylation of the hydroxy groups of the latter compound did not affect activity, but glycosylation at C-27 gave an inactive compound.

Nair and coworkers reported the selective cyclooxygenase-2 (COX-2) inhibitory activity of leaf extracts of *W. somnifera* and related such information to the use of this plant as an antiinflammatory. From the methanolic extract they isolated 12 withanolides and evaluated their abilities to inhibit COX-1 and -2. None of the withanolides inhibited COX-1 even at high doses but most of them exhibited some inhibitory activity on COX-2, with withaferin A (**1**) and viscosalactone B (**505**) being the most active (*15*). Molecular docking studies showed that most of the withanolides had more favorable binding to COX-2 than to COX-1 (*227*).

The 12-oxygenated withanolide **154** from *Discopodium penninervium* (see Sect. 4.2.3.) was a selective inhibitor of cyclooxygenase-2 and also of leukotriene formation; as both these pathways are involved in cell proliferation and angioneogenesis, the dual inhibition of COX-2 and leukotriene formation by **154** has been proposed as a starting point for the development of anti-inflammatory and cancer chemopreventive agents (*82*).

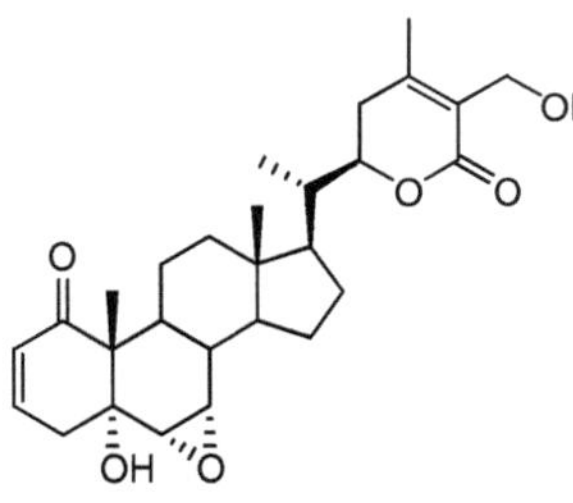

503 (12-deoxywithastramonolide)

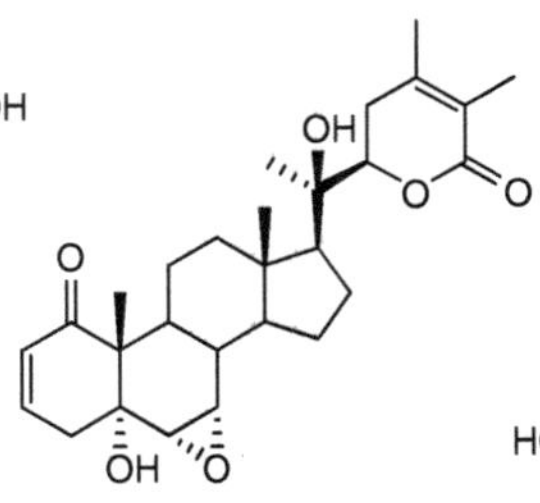

504 (withanolide A)

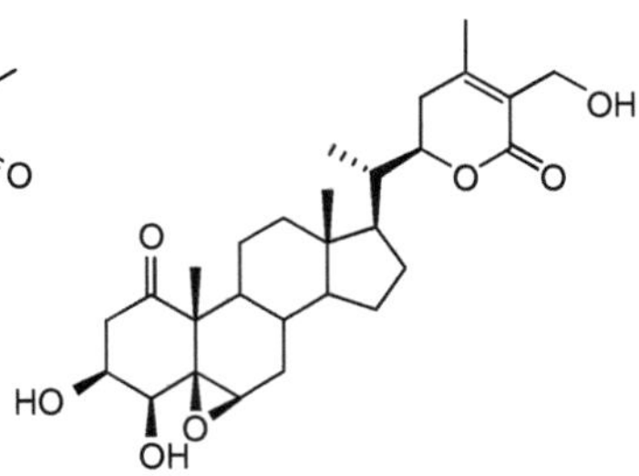

505 (viscosalactone B)

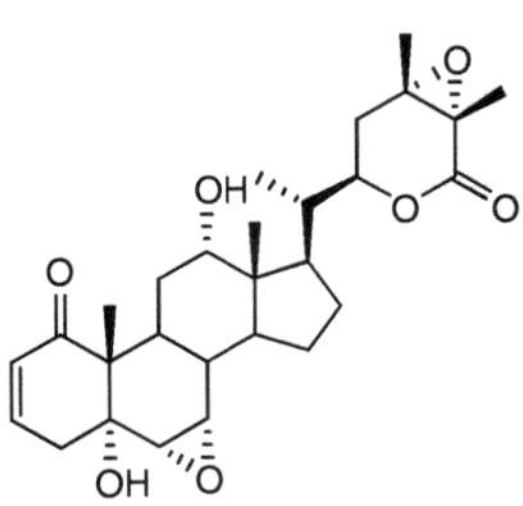

506 (daturalactone 1)

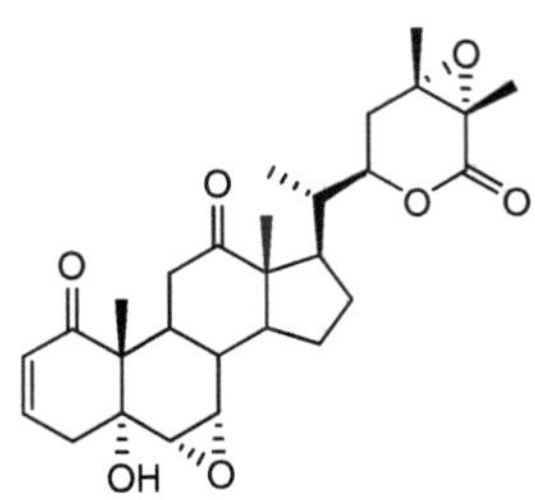

507 (daturalactone 2)

7.5.2. Antistress Activity

Withafastuosin D (**168**), the major withanolide in *Datura fastuosa* leaves, maintained corticosterone levels in male albino rats during experimental stress and exhibited antistress activity evidenced as an anxiolytic effect (*228*). Withafastuosin E (**169**) has also been found to increase the release of prostaglandins, which play an important role in the resistance to gastroduodenal mucosa to ulceration (*229*, *230*). On the other hand, the antistress activity of *W. somnifera* glycowithanolides (*e.g.* sitoindosides) has been linked to their antioxidant activity (*231*, *232*).

7.5.3. Immunosuppressive and Immunomodulatory Activity

The 16-oxygenated withanolides **188–190** isolated from *Discopodium penninervium* (see Sect. 4.2.4.) exhibited potent immunosuppresive activity; they inhibited the incorporation of [^{3}H]-thymidine in cultured rat spleen cells without being overtly toxic to the cells (*233*). Acnistins A (**259**), B (**261**), and E (**260**) from *Dunalia solanacea* significantly inhibited the incorporation of [^{3}H]-thymidine in human lymphocytes at doses as low as 0.1 μg/cm^3 (*124*). Coagulin H (**7**) isolated from *W. coagulans* also was a strong inhibitor of incorporation of [^{3}H]-thymidine in stimulated human mononuclear cells. It inhibited T-cell proliferation with an activity similar to that of the synthetic glucocorticoid, prednisolone. Like prednisolone, the T-cell suppression effect was correlated with a decrease in production of the cytokine IL-2. However, at variance with the glucocorticoid, coagulin H did not have any damaging effects on the cells (*234*).

Immunomodulatory effects of three daturalactones from *Datura quercifolia* were evaluated by observing their effects on antibody production, T-cell and B-cell activation, and cytokine production from splenocytes. The 1β-alcohol **153** was immunosuppressive at lower doses while daturalactones 1 (**506**) and 2 (**507**) were immunostimulators (*81*).

Physalins B (**463**), F (**450**), and G (**314**), but not D (**313**), inhibited nitric oxide production by activated macrophages. Addition of physalin B to lipopolysaccaride-stimulated peritoneal macrophage cultures induced decreases of TNF-α IL-6, and IL-12 production. Physalins B, F, and G also protected mice against administration of a lethal dose of lipopolysaccaride (*235*). The effects of the above-mentioned physalins were not blocked by the antiglucocorticoid, RU-486, suggesting that they act by a mechanism different from that of the glucocorticoids. Physalins B, F, and G also have potent suppressive activities *in vitro* on splenocyte cultures and *in vivo* on allogeneic transplants (*236*). The effects of physalins on transplant rejection could be explained by a direct effect of these withanolides on lymphocytes.

Several physalins isolated from *Physalis alkekengi* were found to be strong inhibitors of nitric oxide production induced by lipopolysaccaride, including the above-mentioned physalins B and F and also physalins A (**468**) and O (**469**). Compounds of the neophysalins class (see Sect. 5.2.2.) were inactive (*139*).

7.6. Cancer-Related Activities

Many studies have dealt with the potential antitumor activity of withanolides, and large amounts of data are available, mostly their cytotoxicity to cancer cell lines. However, other cancer-related activities specific to certain withanolide structural types have also been investigated in the last decade. Particularly interesting are those related to cancer chemoprevention, inhibition of angiogenesis, and microtubule stabilization.

7.6.1. Cytotoxicity

Most withanolides exhibit some level of cytotoxicity against different tumor cell lines, but usually this is non-selective and in the micromolar concentration range. Withaferin A (**1**) has been investigated extensively in this respect, with several recent studies also addressing its mechanism of action (*191*, *237–239*).

Nair and coworkers assayed several withanolides from *W. somnifera* against human lung, breast, CNS, and colon cancer cell lines. Withaferin A (**1**), its diacetate, viscosalactone B (**505**), compound **6**, and ashwagandhanolide (**78**) were the most active (IC_{50} range 0.5–1 μM) and comparable in potency to a reference compound, adriamycin (*47*, *240*). Compound **78** also inhibited lipid peroxidation and the activity of the enzyme cyclooxygenase-2 *in vitro*. Several withanolides from *Acnistus arborescens* were evaluated against panels of human cancer lines, including some 16-acetoxywithanolides and withaphysalins F (**316**), M (**323**), N (**324**), and O (**325**), with IC_{50} values in the 0.2–2 μM range (*98*, *152*, *153*, *241*). The 16-oxygenated withanolides **188–190** isolated from *Discopodium penninervium* (see Sect. 4.2.4.) exhibited significant cytotoxicity only against murine RAW 264.7 carcinoma cell lines (*233*). The 12β-acetoxywithanolide **156** did not show any cytotoxic activity up to 10 μM (*83*).

Bioassay of seventeen withanolides from *Tubocapsicum anomalum* showed significant cytotoxic activity (comparable to that of doxorubicin) against a panel of human cancer cell lines for eight of the compounds (*107*). From the 18-nor withanolides (see Sect. 5.5.3.), only tubocapsenolide A (**441**) exhibited potent cytotoxicity; this compound was shown to inhibit proliferation and induce apoptosis in MDA-MB-231 cells by thiol oxidation of heat shock proteins (*242*). Potent cytotoxicity was also exhibited by tubocapsanolide A (**201**) and its 20-hydroxy (**204**) and 23-hydroxy (**205**) derivatives, tubocapsanolide F (**203**), and anomanolide B (**269**) (*107*, *243*). The other active withanolides isolated were the known withanolide D (**508**) and its 17α-hydroxy derivative (**509**). Withaferin A (**1**), also used as a reference compound, was *ca.* 10 times more active.

Withametelins I (**235**), K (**237**), L (**238**), and N (**240**), isolated from *Datura metel*, exhibited cytotoxicity against selected human cancer cell lines, namely, A549 (lung), BGC-823 (gastric), and K562 (leukemia), with IC_{50} values of 0.05–3.5 μ*M*. Withametelin J (**236**) was only moderately active for the last two cell lines (*92*).

Withanolides from *Physalis* species have also been studied extensively in terms of their cytotoxicity against human and murine cancer cells. Among the physalins, physalins B (**463**), D (**313**), F (**450**), and H (**312**) showed strong cytotoxicity against multiple tumor cell lines, while physalins G (**314**), I (**510**), and physanolide A (**367**) were inactive. Physalins J (**457**) and U (**304**) were marginally active (*55*, *134*, *146*, *244*, *245*). The 18-*nor*-physalin, withaphysanolide A (**438**), also was weakly active (*186*). For physalins B and D, the antineoplastic activity was confirmed *in vivo* by inhibition of tumor proliferation in mice bearing sarcoma 180 tumor cells (*245*). Both physalins B and F have been shown to inhibit hedgehog (Hh)/GLI-mediated transcriptional activation, which is involved directly in tumor formation and progression (*246*). Recently, *Magalhães et al.* reported different activities for physalins D and E isolated from *P. angulata*. However, as already mentioned in Sect. 5.2.3., the structure of physalin E was incorrectly assigned and later shown to be identical to physalin D (**313**) (*244*). Accordingly, these bioactivity results should be considered with caution. Among the *Physalis* withanolides with an unmodified skeleton, potent cytotoxic activity has been reported for withangulatins A (**511**), B (**94**), and I (**102**), physangulin B (**512**), phyperunolide A (**101**), withaphysacarpin (**195**), philadelphicalactones A (**197**) and B (**198**), 18-hydroxywithanolide D (**128**), and withanone (**513**) (*50*, *55*, *56*, *58*, *104*, *247*).

Ixocarpolactones A (**360**) and B (**361**), isolated from *P. philadelphica*, exhibited significant inhibition of murine epidermal JB6 cell transformation (*104*) and ixocarpolactone A (**360**) present in the edible fruit of the plant showed potent antiproliferative and apoptotic activity in SW480 human colon cancer cells (*247*). Evaluation of 12 withaphysalins from *P. minima* and *P. divaricata*, including **331–334**, showed only moderate cytotoxicity against the HCT-116 and H460 human cancer cell lines (*154*, *186*).

The potential antitumor activities of an alcoholic extract of *W. somnifera* roots and of withaferin A (**1**) have also been studied in conjunction with radiation therapy in experimental tumors *in vivo*. Both exhibited significant antitumor and radiosensitizing effects without systemic toxicity (*248–250*).

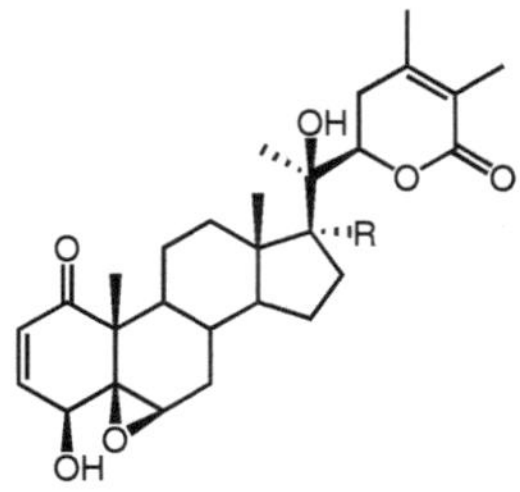

508 R = H (withanolide D)
509 R = OH

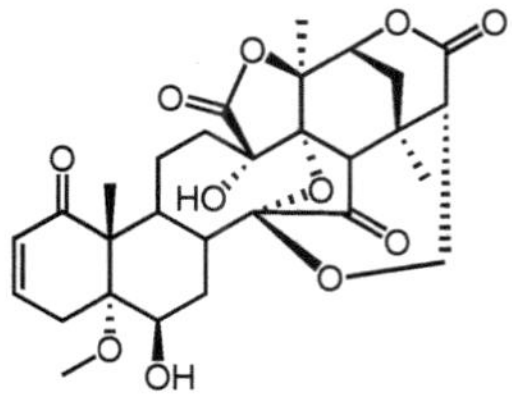

510 (physalin I)

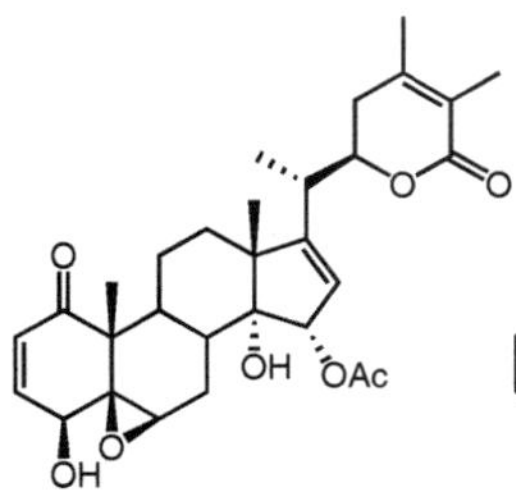

511 (withangulatin A)

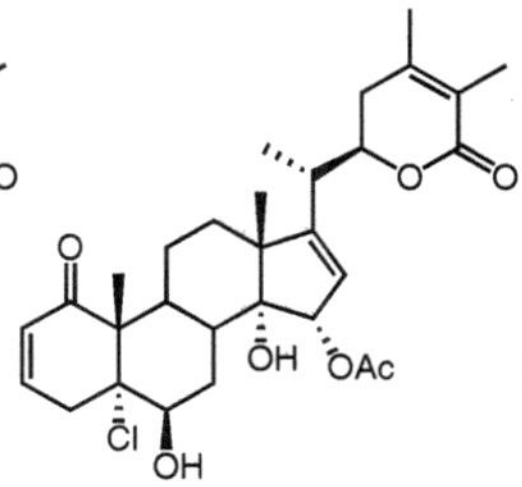

512 (physagulin B)

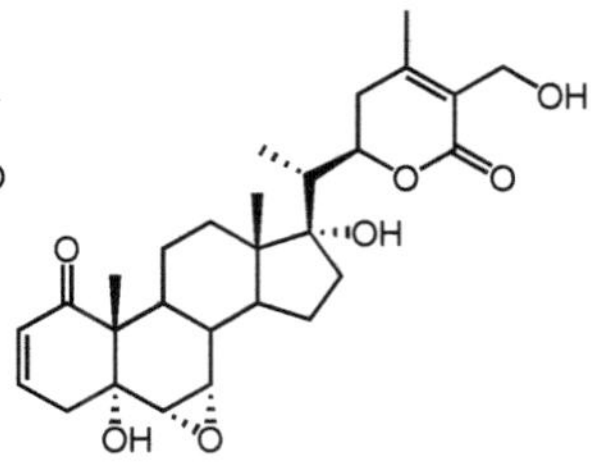

513 (withanone)

7.6.2. Cancer Chemopreventive Activity

Cancer chemoprevention by the ingestion of modulators of carcinogenesis from synthetic or natural origin has been proposed as a strategy to reduce cancer mortality in apparently healthy persons at risk of cancer (*6, 251*). *Pezzuto, Kinghorn* and collaborators have used a battery of *in vitro* bioassays to monitor inhibition of tumorigenesis at the stages of initiation, promotion, and progression by plant extracts and pure compounds. One key bioassay used, is based on the induction of the phase II drug-metabolizing enzyme NAD(P)H:quinone reductase (QR), on Hepa 1c1c7 hepatoma cells; QR induction is indicative of potential cancer prevention at the initiation phase (*252*). A large number of compounds of plant origin have been evaluated as QR inducers, including many withanolides (*6*). Of particular interest are QR inducers that exhibit low cytotoxicity and thus a high selectivity measured as the chemopreventive index, the ratio between the concentration needed to double QR activity (CD) and the concentration that inhibits cell growth by 50% (IC_{50}).

Kinghorn and coworkers investigated QR induction by the withanolides in fruits and aerial parts of *Physalis philadelphica* (see Sect. 4.2.4.) (*103–105*). With the exception of withanone (**513**), all the withanolides were potent QR inducers with activity comparable to or better than the reference compound, sulforaphane, a known chemopreventive agent. Ixocarpalactone A (**360**), philadelphicalactone A (**197**), withaphysacarpin (**195**), and 4*β*,7*β*,(20*R*)-trihydroxy-1-oxowitha-2,5-dien-22,26-olide (**199**) exhibited the highest chemopreventive indexes. The fact that ixocarpolactone A (**360**), is present in the edible fresh fruit of *P. philadelphica* (tomatillo) at a concentration level of 143 ppb (*105*), make these findings especially important.

In a broader study comprising withanolides from 13 Solanaceae species, 37 withanolides representative of a variety of structural types were evaluated for their potential to induce quinone reductase (*253*). Jaborosalactone 1 (**397**), jaborosalactone O (**353**), jaborosalactone P (**389**), trechonolide A (**368**), and withaphysalin J (**320**), were significant QR inducers with CD values in the range of 0.27–1.52 μ*M*. Results indicated that a functionalized methyl-18 plays an important role in improving QR induction while the presence of 5α-substituents is deleterious for the activity. Overall, spiranoid- and trechonolide-type withanolides exhibited good QR induction. Some of the active withanolides had low cytotoxicity, with chemopreventive indexes that compared favorably with the reference compound sulforaphane. One such compound, the spiranoid, jaborosalactone P (**389**), was selected to test its capacity to induce steady-state levels of quinone reductase in multiple organ sites of BALB/c mice. With jaborosalactone P-treated mice, a significant induction was observed in liver and colon, but not in lung, stomach, or mammary gland (*253*). This *in vivo* study confirmed the *in vitro* results, indicating that withanolides may function as potent phase II enzyme inducers.

Withanolides **184** and **185** from *A. arborescens*, bearing acetates at positions 7*β* and 16α, were also very potent QR inducers, but exhibited high cytotoxicity resulting in poor selectivity (*98*). From *Larnax subtriflora*, subtrifloralactones A (**431**), C (**433**), D (**425**), F (**436**), I (**428**), and J (**429**) containing an α,β-unsaturated ketone functionality in ring A (see Sect. 5.5.) had significant QR induction activity (*184*). Both δ- and γ-lactones were active, indicating that this part of the molecule was not critical for activity. Subtrifloralactone D was the most active, but subtrifloralactones A and F had the highest chemopreventive index (*ca.* 3 times that of sulforaphane). Subtrifloralactone L (**435**) with a doubly unsaturated ring A ketone, was inactive in the QR assay, while subtrifloralactone K (**434**) with the less common 3,6-epoxy arrangement, was active (*185*). The above data although limited, suggest that even though the presence of an α,β-unsaturated ketone unit in ring A of withanolides appears to be important for inducing QR, other structural features may compensate the lack of this functionality or block its beneficial effects.

Panjamurthy et al. have reported that pretreatment with withaferin A (**1**) significantly reduced *in vivo* 7,12-dimethylbenz[*a*]anthracene (DMBA)-induced genotoxicity, in the bone marrow of golden Syrian hamsters (*254*). This effect could also be related to the induction of phase II detoxifying enzymes.

7.6.3. Antiangiogenic Activity

Formation of new blood vessels from existing vasculature or "angiogenesis" is characteristic of all solid tumors allowing for nutrition, oxygenation, and metastasis. Thus, angiogenesis inhibitors provide an alternative way of controlling the growth of tumor cells in both pre-invasive and invasive stages. Angiogenesis is also involved in the pathogenesis of several non-malignant inflammatory diseases (age-related macular degeneration, arthritis, endometriosis, *etc.*). *Mohan et al.* reported the antiangiogenic activity of extracts of *W. somnifera* containing non-cytotoxic levels of withanolides and also of withaferin A (**1**) (*255*). Thus, withaferin A inhibited cell proliferation in human umbilical vein endothelial cells (HUVECs) (IC_{50} 12 n*M*) through a process associated with inhibition of cyclin D1 expression. A potent antiangiogenic effect was also observed *in vivo*, at doses that are 500-fold lower than those previously reported to exert antitumor activity. The authors proposed that the inhibition of NF-κB by withaferin A in HUVECs occurs by interference with the ubiquitin-mediated proteasome pathway as suggested by the increased levels of poly-ubiquitinated proteins. The biotinylated derivative **459** was used to identify protein targets of withaferin A in HUVECs (see Sect. 6.1.). Preliminary results showed that withaferin A binds irreversibly with a 56 kDa protein target and a less abundant 180 kDa species but interacts reversibly with a 70 kDa protein species (*192*). Withaferin A and withanolide D (**508**) also displayed antiangiogenic activity in human choroidal endothelial cells (HCECs). These results may lead to novel treatments of choroidal neovascularization, the major contributor to age-related macular degeneration and one of the leading causes of irreversible blindness in the Western hemisphere (*256*). Physalin B (**463**) has been identified recently as an inhibitor of the ubiquitin-mediated proteasome pathway in the DLD-1 human colon cancer cell line, producing an accumulation of ubiquitinated proteins and inhibiting TNFα-induced NF-κB activation (*257*). The antiangiogenic activity of this compound remains to be tested.

7.6.4. Microtubule Stabilizing Activity

The microtubule stabilizing activity of taccalonolides A (**405**) and E (**420**) was reported by *Mooberry* and coworkers after bioassay-directed fractionation of an extract of *Tacca chantrieri* (*258*). Microtubule stabilizers are highly effective drugs used in the treatment of many types of cancers. The taccalonolides are particularly interesting as they appear to have a unique mechanism of action, which does not involve direct binding to tubulin. Special attention has been drawn to these compounds and also to their closely related analogues, taccalonolides B (**419**) and N (**421**), as they retain efficacy in taxane- and epothilone-resistant models (*259*). Although these taccalanolides are less potent than other microtubule stabilizers in drug-sensitive cell lines, they are effective at similar concentrations against taxane-resistant cell lines. The microtubule stabilizing activity of taccalonolides has been reviewed recently by *Risinger* and *Mooberry* (*260*).

7.7. CNS-Related Activities

Reports on activities affecting the CNS by withanolides are mostly based on properties attributed to "Ashwagandha" in ayurvedic medicine and are restricted to withanolides from *W. somnifera.*

7.7.1. Synaptogenesis and Neuritic Outgrowth

Facilitating synaptogenesis and reconstructing neuronal networks in the damaged brain is required for the therapeutic treatment of neurodegenerative diseases that produce neuronal degeneration and atrophy. Several withanolides isolated from the methanolic extract of *W. somnifera* roots as well as the extract itself, have been shown to possess neurite outgrowth activity (*21*, *261*). Withanoside IV (**31**) and VI (**32**) predominantly induced dendritic outgrowth in normal cortical neurons, while withanolide A (**504**) predominantly induced axon outgrowth (*262*). These withanolides also showed neuritic regeneration and synaptic reconstruction of damaged cortical neurons and prevented both dendritic and axonal atrophy induced by Aβ(25–35) (*261*, *263*). Oral administration of withanoside IV significantly improved memory deficit in Aβ(25–35) treated mice, and sominone, the aglycone of withanoside IV, was shown to be the major metabolite after administration. Sominone was more active than withanoside IV (*264*).

7.7.2. Cholinesterase Inhibition

Natural cholinesterase inhibitors are of special interest in drug development due to the involvement of cholinesterases in *Alzheimer*'s disease and other related dementias. Acetylcholinesterase (AChE) inhibitors activate central cholinergic function by increasing the acetylcholine levels in the brain. Bracteosins A (**206**), B (**207**), and C (**208**), isolated from *Ajuga bracteosa* (*108*), and withanolide A (**504**) and withaferin A (**1**) from *W. somnifera*, were moderate inhibitors of AChE and BChE (butyrylcholinesterase) (*11*). Molecular docking studies indicated that all compounds are imbedded in the aromatic gorge of AChE. All these withanolides also showed dose-dependent spasmolytic and Ca^{2+} antagonistic activities that may help in prolonging neuron survival and function (*265*). Withanoside VI (**32**) and withaferin A attenuated the desensitization to clonidine of smooth muscle, and this effect was related to the effect of *W. somnifera* on morphine tolerance and dependence (*27*). It should be noted that clonidine inhibits the release of acetylcholine by acting on α_2-adrenoceptors in these tissues.

8. Chemotaxonomic Considerations

As already mentioned, *ca.* 650 withanolides have been described, most of them from genera of the Solanaceae, subfamily Solanoideae. The absence so far of withanolides in members of the other subfamilies is curious. In Table 3, the genera of subfamily Solanoideae containing withanolides are arranged according to the most recent phylogenetic classification (*10*). As different groups of withanolides have been reported in 23 genera and *ca.* 70 species of the Solanoideae, some chemotaxonomic considerations can be made.

Withanolides with an unmodified skeleton (Fig. 2) are the most common, occurring in 14 genera of the different tribes (Physaleae, Datureae, Hyoscyameae, Lycieae, and Solaneae). Among the six genera still not assigned phylogenetically to

Table 3. Genera of the subfamily Solanoideae containing withanolides arranged according to a established phylogenetic system

Tribe (subtribe)	Genus (section)	Main withanolide types
Physaleae (Physalinae)	*Physalis*	Physalins, neophysalins, cyclophysalins, withaphysalins, unmodified skeleton
	Margaranthus	Physalins, unmodified skeleton
	Witheringia	Physalins
	Brachistus	Physalins
Physaleae (Iochrominae)	*Acnistus*	Acnistins, withaphysalins, unmodified skeleton
	Iochroma	Unmodified skeleton
	Eriolarynx	Withaphysalins, unmodified skeleton
	Vassobia	Unmodified skeleton
	Dunalia	Acnistins, withaphysalins, unmodified skeleton
Physaleae (Withaninae)	*Withania*	Unmodified skeleton
	Tubocapsicum	Acnistins
	Discopodium	Acnistins, unmodified skeleton
Physaleae[a]	*Larnax*	Subtrifloralactones, unmodified skeleton
Datureae	*Datura (Datura)*	Unmodified skeleton
	Datura (Dutra)	Withametelins, unmodified skeleton
Hyoscyameae	*Hyoscyamus*	Unmodified skeleton
Lycieae	*Lycium*	Unmodified skeleton
Solaneae	*Solanum*	Unmodified skeleton
Genera not assigned to a more inclusive clade	*Deprea*	Withajardins
	Exodeconus	Unmodified skeleton
	Jaborosa (Jaborosa)	Aromatic ring A, spiranoid-γ-lactones, unmodified skeleton
	Jaborosa (Lonchestigma)	15,21-cyclowithanolides, sativolides, spiranoid-γ-lactones, spiranoid at C-22, trechonolides, unmodified skeleton
	Nicandra	Aromatic ring D, unmodified skeleton
	Salpichroa	Aromatic ring D, unmodified skeleton
	Schraderanthus	Physalins

[a]Genus not assigned to a subtribe

a more inclusive clade but within the Solanoideae (Table 3), withanolides with an unmodified skeleton have now been reported only in the *Larnax* (*Deprea*) and *Schraderanthus* genera.

8.1. Tribe Physaleae

Different withanolides with modified skeletons are present within this tribe (Plates 1 and 2). Physalins and related withanolides (Fig. 2) are frequent in the subtribe Physalinae (Plate 1, a–f), particularly in *Physalis*, *Margaranthus*, *Witheringia*, and *Brachistus*. The occurrence of physalins and the non-occurrence of the unmodified skeletons in *Witheringia* and *Brachistus* support the close phylogenetic relationship between both genera, which appear together in a small clade and are sisters to one another (*10*). In addition, the presence of physalins in *Schraderanthus* is noteworthy, which is a monotypic genus recently segregated from *Leucophysalis* (tribe Physaleae, subtribe Physalinae). *Schraderanthus* still remains unassigned to any tribe due to the lack of molecular analysis; the chemical information would support the inclusion of *Schraderanthus* in the subtribe Physalinae, the only subtribe where physalins have been found.

In the subtribe Physalinae, withaphysalins (Fig. 2), ixocarpalactones, and perulactones (Fig. 3) have been reported in certain *Physalis* species (*P. minima*, *P. philadelphica*, and *P. peruviana*); since this genus comprises *ca.* 90 species, these metabolites may probably appear in other species. As more information emerges from research conducted in other *Physalis* species, it will be able to evaluate the potential chemotaxonomical value of these compounds, either as chemical generic markers or as an exception in the genus.

Withaphysalins are more frequent in the subtribe Iochrominae (Plate 2), a well-supported clade but with the relationships within this clade poorly resolved (*10*). The chemical data presented in this chapter, support some taxonomic changes proposed by morphological data (*9*) and confirmed by molecular evidence (*10*), as is the case for the segregation of *Eriolarynx* containing withaphysalins F-L (see Sect. 5.2.4.) from *Vassobia* (no withaphysalins).

Acnistins (Fig. 2) appear in the subtribes Iochrominae and Withaninae and are absent in all the remaining genera investigated so far. Within the latter subtribe, three small clades have been identified with unresolved relationships between them (*10*). One clade includes *Aureliana* and *Athenaea,* two South American genera not investigated phytochemically up to now. The second clade comprises a monotypic Asiatic genus (*Mellisia*) and *Withania*, the "parent genus" of the withanolides (Plate 1, g–j); finally, *Nothocestrum*, *Tubocapsicum* and *Discopodium* are included in the third clade. The chemical evidence is consistent with this proposal since *Withania* is characterized by the presence of a large number of withanolides with the unmodified parent skeleton of withaferin A (**1**). In contrast, *Discopodium* and *Tubocapsicum* produce withanolides with structural variations. The close affinity

Plate 1. Solanaceae species of Tribe *Physaleae*. Subtribe *Physalinae*: (a) *Physalis viscosa*, branch with fruit; (b) *P. viscosa*, flower (photographs: J. Toledo and M.T. Cosa). (c) *P. pubescens*, plant with flower and fruits (photograph: G. Barboza); (d) *P. peruviana*, detail of corolla (photograph: J. Toledo); (e) *Witheringia solanacea*, stem with fruits; (f) *W. solanacea*, flower and immature fruits (photograph: G. Beltrán). Subtribe *Withaninae* (photographs with permission of the Experimental Garden and Genebank of the Radboud University of Nijmegen, The Netherlands): (g) *Withania somnifera*, flowering branch; (h) *W. somnifera*, fruits; (i) *W. adpressa*, flowers; (j) *W. aristata*, flowering branch. Genus not assigned to a subtribe: (k) *Larnax* sp., branch with flowers and fruits; (l) *Larnax* sp., flowering branch (photographs: S. Leiva)

Plate 2. Solanaceae species of Tribe *Physaleae* subtribe *Iochrominae*. (a) *Eriolarynx lorentzii*, flower; (b) *Vassobia breviflora*, branch with flowers and fruits; (c): *Acnistus arborescens*, flowering branch (photographs: F. Chiariní); (d) *Iochroma australe*, flowering branch; (e) *I. australe*, fruit (photographs: M. T. Cosa and G. Barboza); (f) *I. fuchsioides*, flowering branch and (g) *I. gesneroides*, flowering branch (photographs with permission of Experimental Garden and Genebank of the Radboud University of Nijmegen, The Netherlands)

between the latter two genera is evidenced by them both producing the 17-epiacnistins, which are absent in *Withania*.

8.2. Tribes Hyoscyameae, Lycieae, and Solaneae

In these tribes (Plate 3, d–i), only five species have been investigated (Tables 1 and 3), thus, only preliminary chemotaxonomic suggestions may be proposed. These species are consistent in always containing withanolides with an unmodified skeleton. Worth mentioning is the fact that in the largest genus of the family, *Solanum* (with *ca.* 1,500 species), withanolides have been reported in only two species (*S. ciliatum* and *S. sisymbriifolium*) (See Sect. 4.2.5.).

8.3. Tribe Datureae

The tribe comprises two genera, *Brugmansia* and *Datura*, but, however, no information is available on *Brugmansia*. In the conventional classification, the 14 species of *Datura* are included in three sections: sect. *Datura* (*D. quercifolia* and *D. stramonium*), sect. *Ceratocaula* (*D. ceratocaula*), and sect. *Dutra* (the remaining species) (Plate 3, a–c). Withanolides from seven taxa have been investigated extensively, and more than 70 different compounds have been reported. The withametelin skeleton (Fig. 2), occurring in *ca.* 30 compounds, is exclusive to *Datura*. Withametelins were reported in *D. inoxia* and *D. metel*. Genetic similarity and phylogenetic analysis both suggest that *D. metel* is related more closely to *D. inoxia* than to the other taxa of section *Dutra*, based upon the small genetic distance between them (*266*, *267*), which is supported by the available chemical information. Other chemical coincidence with conventional and phylogenetic proposals is represented by the close relationship between the two members of the section *Datura* (*266*), *i.e. D. quercifolia* and *D. stramonium*, which share the occurrence of 12α-hydroxylated unmodified withanolides.

8.4. Genera with Uncertain Positions in the Solanaceae Taxonomic System

Some genera (*Jaborosa*, *Nicandra*, *Salpichroa*) contain withanolides with exclusive arrangements, which can be considered as chemotaxonomic markers at the generic level. *Jaborosa* is a good example since more than 50% of its species have been studied (Plate 4). In this genus, several peculiar modified skeletons of the

Plate 3. Solanaceae species of Tribe *Datureae*: (a) *Datura ferox*, plant; (b) *D. ferox*, fruit (photograph: F. Chiarini); (c) *D. inoxia*, plant (photograph: M.T. Cosa). Species of Tribe *Lycieae*: (d) *Lycium chinense*, flower; (e) *L. chinense*, fruits (photograph: B. Liu); (f) *L. barbarum*, branches with fruits (photograph: M. Li). Species of Tribe *Solaneae*: (g) *Solanum sisymbriifolium*, fruits; (h) *S. sisymbriifolium*, flowers (photographs: M.T. Cosa). Species of Tribe *Hyoscyameae*: (i) *Hyoscyamus niger*, flowering branch (photograph: B. Liu). Species not assigned to a tribe: (j) *Deprea* sp., Flower (photograph: S. Leiva); (k) *Exodeconus maritimus*, plant (photograph: E. Rodríguez R.); (l) *Salpichroa origanifolia*, flowering branch (photograph: G. Barboza)

Plate 4. *Jaborosa* spp.: (a) *J. rotacea*, plant; (b) *J. rotacea*, flower (photographs: F. Chiarini); (c) *J. leucotricha*, plant; (d) *J. leucotricha*, flowers (photographs: G. Barboza). (e) *J. sativa*, flowering branch (photograph: F. Chiarini); (f) *J. laciniata*, plant and flowers in detail; (g) *J. odonelliana*, flower (photographs: G. Barboza); (h) *J. caulescens*, flower (photograph: A.A. Cocucci); (i) *J. kurtzii*, flowers (photograph: G. Barboza); (j) *J. reflexa*, plant; (k) *J. reflexa*, flowers (photographs: G. Barboza); (l) *J. bergii*, fruits; (m) *J. bergii*, flower (photographs: A.A. Cocucci); (n) *J. magellanica*, plant (photograph: G. Barboza)

withanolides appear repeatedly in different species, such as the sativolides, trechonolides, spiranoid withanolides at C-22, and spiranoid-γ-lactones (Figs. 2 and 3).

The trechonolides, sativolides, and spiranoid withanolides at C-22 are exclusive to species in *Jaborosa* section *Lonchestigma*. The trechonolides are the most widespread in the genus, being present in nine species (*J. araucana*, *J. caulescens*, *J. laciniata*, *J. lanigera*, *J. leucotricha*, *J. magellanica*, *J. parviflora*, *J. rotacea*, and *J. sativa*); sativolides appear in *J. caulescens*, *J. rotacea*, and *J. sativa*, while the spiranoid withanolides at C-22 are only present in *J. kurtzii* and *J. rotacea*.

Spiranoid-γ-lactones have been isolated from *J. odonelliana*, *J. runcinata* and *J.integrifolia* (unpublished data), all belonging to *Jaborosa* sect. *Jaborosa*. Only one spiranoid-γ-lactone, jaborosalactone 2, was obtained from *J. araucana* (section *Lonchestigma*). As was stated in Sect. 5.4.3., the structural similarity of jaborosalactone 2 with trechonolide A, both present in *J. araucana*, suggests that these compounds may have a common biosynthetic precursor, and probably this species could be a link between section *Jaborosa* and section *Lonchestigma*.

Other distinctive variations with modified skeletons are the withanolides and related steroids with an aromatic ring D, the nicandrenoids isolated from *Nicandra*, and the salpichrolides from *Salpichroa* (Plate 3, l). Moreover, both genera exhibit the same δ-lactol side chain at C-17. In spite of this chemical similarity, *Nicandra* and *Salpichroa* are not phylogenetically close (*10*).

Nicandra has been suggested as being close to *Exodeconus* (*10*) (Plate 3, k). Although the withanolides isolated from *Exodeconus* have an unmodified skeleton and the major components of *Nicandra* have an aromatic ring D, all the withanolides found in both genera share the same 1-oxo-2-ene-5α-hydroxy-6α,7α-epoxy substitution pattern in rings A and B. In this case, the phytochemical evidence coincides with the phylogenetic proposal. *Salpichroa* and *Jaborosa* have traditionally been placed in tribe Jaboroseae (*9*). However, the phylogenetic evidence proposes *Salpichroa* in a clade informally named "Salpichroina", from which *Jaborosa* is excluded. Again, chemical information supports this proposal.

Despite the fragmentary chemical contribution described in this section of the chapter, it is evident that a large and species-rich study in Solanaceae in both molecular and phytochemical aspects would provide a better comprehension of the relationships among the species contained therein.

Acknowledgments We thank CONICET, Universidad de Buenos Aires and Universidad de Córdoba for financial support.

References

1. Lavie D, Glotter E, Shvo Y (1965) Constituents of *Withania somnifera* Dun. The Structure of Withaferin A. J Chem Soc: 7517

2. Kupchan SM, Doskotch RW, Bollinger P, MacPhail AT, Sim GA, Renauld JAS (1965) The Isolation and Structural Elucidation of a Novel Steroidal Tumor Inhibitor from *Acnistus arborescens*. J Am Chem Soc **87**: 5805

3. Ray AB, Gupta M (1994) Withasteroids, a Growing Group of Naturally Occurring Steroidal Lactones. Prog Chem Org Nat Prod **63**: 1
4. Anjaneyulu ASR, Rao DS, Lequesne PW (1998) Withanolides, Biologically Active Natural Steroidal Lactones: A Review. Stud Nat Prod Chem **20**: 135
5. Veleiro AS, Oberti JC, Burton G (2005) Chemistry and Bioactivity of Withanolides from South American Solanaceae. Stud Nat Prod Chem **32**: 1019
6. Su B-N, Gu J-Q, Kang Y-H, Park E-J, Pezzuto JM, Kinghorn AD (2004) Induction of the Phase II Enzyme, Quinone Reductase, by Withanolides and Norwithanolides from Solanaceous Species. Mini-Rev Org Chem **1**: 115
7. Budhiraja RD, Krishan P, Sudhir S (2000) Biological Activity of Withanolides. J Sci Ind Res **59**: 904
8. D'Arcy WG (1991) Solanaceae III. Taxonomy, Chemistry, Evolution. Royal Botanic Gardens, Kew. In: Hawkes JG, Lester RN, Nee M, Estrada N (eds): The Solanaceae since 1976, with a Review of Its Biogeography. Royal Botanic Gardens, Kew, U.K., p 75
9. Hunziker AT (2001) Genera Solanacearum. The Genera of Solanaceae Illustrated, Arranged According to a New System, Koeltz Scientific Books, Konigstein, Germany
10. Olmstead RG, Bohs L, Abdel Migid H, Santiago-Valentín E, García VF, Collier SM (2008) A Molecular Phylogeny of the Solanaceae. Taxon **57**: 1159
11. Choudhary MI, Yousuf S, Nawaz SA, Ahmed S, Atta-ur-Rahman (2004) Cholinesterase Inhibiting Withanolides from *Withania somnifera*. Chem Pharm Bull **52**: 1358
12. Misra L, Lal P, Sangwan RS, Sangwan NS, Uniyal GC, Tuli R (2005) Unusually Sulfated and Oxygenated Steroids from *Withania somnifera*. Phytochemistry **66**: 2702
13. Benjumea D, Martin-Herrera D, Abdala S, Gutierrez-Luis J, Quinones W, Cardona D, Torres F, Echeverri F (2009) Withanolides from *Withania aristata* and Their Diuretic Activity. J Ethnopharmacol **123**: 351
14. Choudhary MI, Abbas S, Jamal SA, Atta-ur-Rahman (1996) *Withania somnifera* - a Source of Exotic Withanolides. Heterocycles **42**: 555
15. Jayaprakasam B, Nair MG (2003) Cyclooxygenase-2 Enzyme Inhibitory Withanolides from *Withania somnifera* Leaves. Tetrahedron **59**: 841
16. March J (1992): Advanced Organic Chemistry Reactions, Mechanisms, and Structure. John Wiley & Sons, New York, p 755
17. Atta-ur-Rahman, Yousaf M, Gul W, Qureshi S, Choudhary MI, Voelter W, Hoff A, Jens F, Naz A (1998) Five New Withanolides from *Withania coagulans*. Heterocycles **48**: 1801
18. Nur-e-Alam M, Yousaf M, Qureshi S, Baig I, Nasim S, Atta-ur-Rahman, Choudhary MI (2003) A Novel Dimeric Podophyllotoxin-Type Lignan and a New Withanolide from *Withania coagulans*. Helv Chim Acta **86**: 607
19. Kuroyanagi M, Shibata K, Umehara K (1999) Cell Differentiation Inducing Steroids from *Withania somnifera* L. (Dun.). Chem Pharm Bull **47**: 1646
20. Abou-Douh AM (2002) New Withanolides and Other Constituents from the Fruit of *Withania somnifera*. Arch Pharm **335**: 267
21. Zhao J, Nakamura N, Hattori M, Kuboyama T, Tohda C, Komatsu K (2002) Withanolide Derivatives from the Roots of *Withania somnifera* and Their Neurite Outgrowth Activities. Chem Pharm Bull **50**: 760
22. Sahai M, Ali A, Ray AB, Slatkin DJ, Kirson I (1983) Withaperuvin D, a Novel Withanolide from *Physalis peruviana*. J Chem Res (S): 152
23. Neogi P, Sahai M, Ray AB (1987) Withaperuvins F and G, Two Withanolides of *Physalis peruviana* Roots. Phytochemistry **26**: 243
24. Anjaneyulu ASR, Rao DS (1997) New Withanolides from the Roots of *Withania somnifera*. Indian J Chem B **36**: 424
25. Lal P, Misra L, Sangwan RS, Tuli R (2006) New Withanolides from Fresh Berries of *Withania somnifera*. Z Naturforsch, B: Chem Sci **61**: 1143
26. Misra L, Mishra P, Pandey A, Sangwan RS, Sangwan NS, Tuli R (2008) Withanolides from *Withania somnifera* Roots. Phytochemistry **69**: 1000

27. Matsuda H, Murakami T, Kishi A, Yoshikawa M (2001) Structures of Withanosides I, II, III, IV, V, VI, and VII, New Withanolide Glycosides, from the Roots of Indian *Withania somnifera* Dunal. and Inhibitory Activity for Tachyphylaxis to Clonidine in Isolated Guinea-Pig Ileum. Bioorg Med Chem **9**: 1499
28. Pramanick S, Roy A, Ghosh S, Majumder HK, Mukhopadhyay S (2008) Withanolide Z, a New Chlorinated Withanolide from *Withania somnifera*. Planta Med **74**: 1745
29. Jayaprakasam B, Strasburg GA, Nair MG (2004) Potent Lipid Peroxidation Inhibitors from *Withania somnifera* Fruits. Tetrahedron **60**: 3109
30. Atta-ur-Rahman, Shabbir M, Yousaf M, Qureshi S, Dur-e-Shahwar, Naz A, Choudhary MI (1999) Three Withanolides from *Withania coagulans*. Phytochemistry **52**: 1361
31. Abdeljebbar LH, Humam M, Christen P, Jeannerat D, Vitorge B, Amzazi S, Benjouad A, Hostettmann K, Bekkouche K (2007) Withanolides from *Withania adpressa*. Helv Chim Acta **90**: 346
32. Maurya R, Akanksha, Jayendra, Singh AB, Srivastava AK (2008) Coagulanolide, a Withanolide from *Withania coagulans* Fruits and Antihyperglycemic Activity. Bioorg Med Chem Lett **18**: 6534
33. Vande Velde V, Lavie D (1981) New Withanolides of Biogenetic Interest from *Withania somnifera*. Phytochemistry **20**: 1359
34. Atta-ur-Rahman, Dur-e-Shahwar, Naz A, Choudhary MI (2003) Withanolides from *Withania coagulans*. Phytochemistry **63**: 387
35. Huang C-F, Ma L, Sun L-J, Ali M, Arfan M, Liu J-W, Hu L-H (2009) Immunosuppressive Withanolides from *Withania coagulans*. Chem Biodivers **6**: 1415
36. Choudhary MI, Dur-e-Shahwar, Parveen Z, Jabbar A, Ali I (1995) Antifungal Steroidal Lactones from *Withania coagulance*. Phytochemistry **40**: 1243
37. Jahan E, Perveen S, Fatima I, Malik A (2010) Coagulansins A and B, New Withanolides from *Withania coagulans* Dunal. Helv Chim Acta **93**: 530
38. Atta-ur-Rahman, Choudhary MI, Yousaf M, Gul W, Qureshi S (1998) New Withanolides from *Withania coagulans*. Chem Pharm Bull **46**: 1853
39. Kumar A, Ali M, Mir SR (2004) A New Withanolide from the Roots of *Withania somnifera*. Indian J Chem, Sect B **43B**: 2001
40. Atta-ur-Rahman, Abbas S, Dur-e-Shahwar, Jamal SA, Choudhary MI (1993) New Withanolides from *Withania* sp. J Nat Prod **56**: 1000
41. Atta-ur-Rahman,, Shabbir M, Dur-e-Shahwar, Choudhary MI, Voelter W, Hohnholz D (1998) New Steroidal Lactones from *Withania coagulance*. Heterocycles **47**: 1005
42. Atta-ur-Rahman, Choudhary MI, Qureshi S, Gul W, Yousaf M (1998) Two New Ergostane-Type Steroidal Lactones from *Withania coagulans*. J Nat Prod **61**: 812
43. Ahmad S, Malik A, Muhammad P, Gul W, Yasmin R, Afza N (1998) A New Withanolide from *Physalis peruviana*. Fitoterapia **69**: 433
44. Ahmad S, Malik A, Afza N, Yasmin R (1999) A New Withanolide Glycoside from *Physalis peruviana*. J Nat Prod **62**: 493
45. Khan PM, Ahmad S, Rubnawaz H, Malik A (1999) The First Report of a Withanolide from the Family Labiatae. Phytochemistry **51**: 669
46. Ali M, Shuaib M, Ansari SH (1997) Withanolides from the Stem Bark of *Withania somnifera*. Phytochemistry **44**: 1163
47. Subbaraju GV, Vanisree M, Rao CV, Sivaramakrishna C, Sridhar P, Jayaprakasam B, Nair MG (2006) Ashwagandhanolide, a Bioactive Dimeric Thiowithanolide Isolated from the Roots of *Withania somnifera*. J Nat Prod **69**: 1790
48. Mulabagal V, Subbaraju GV, Rao CV, Sivaramakrishna C, DeWitt DL, Holmes D, Sung B, Aggarwal BB, Tsay H-S, Nair M (2009) Withanolide Sulfoxide from Ahswagandha Roots Inhibits Nuclear Transcription Factor-Kappa-B, Cyclooxygenase and Tumor Cell Proliferation. Phytother Res **23**: 987
49. Xu Y-m, Marron MT, Seddon E, McLaughlin SP, Ray DT, Whitesell L, Gunatilaka AAL (2009) 2,3-Dihydrowithaferin A-3β-*O*-Sulfate, a New Potential Prodrug of Withaferin A from Aeroponically Grown *Withania somnifera*. Bioorg Med Chem **17**: 2210

50. Lan Y-H, Chang F-R, Pan M-J, Wu C-C, Wu S-J, Chen S-L, Wang S-S, Wu M-J, Wu Y-C (2009) New Cytotoxic Withanolides from *Physalis peruviana*. Food Chem **116**: 462
51. Dinan LN, Sarker SD, Sik V (1997) 28-Hydroxywithanolide E from *Physalis peruviana*. Phytochemistry **44**: 509
52. Ahmad S, Malik A, Yasmin R, Ullah N, Gul W, Khan PM, Nawaz HR, Afza N (1999) Withanolides from *Physalis peruviana*. Phytochemistry **50**: 647
53. Ahmad S, Yasmin R, Malik A (1999) New Withanolide Glycosides from *Physalis peruviana* L. Chem Pharm Bull **47**: 477
54. Maldonado E, Alvarado VE, Torres FR, Martinez M, Perez-Castorena AL (2005) Androstane and Withanolides from *Physalis cinerascens*. Planta Med **71**: 548
55. Damu AG, Kuo P-C, Su C-R, Kuo T-H, Chen T-H, Bastow KF, Lee K-H, Wu T-S (2007) Isolation, Structures, and Structure-Cytotoxic Activity Relationships of Withanolides and Physalins from *Physalis angulata*. J Nat Prod **70**: 1146
56. Lee S-W, Pan M-H, Chen C-M, Chen Z-T (2008) Withangulatin I, a New Cytotoxic Withanolide from *Physalis angulata*. Chem Pharm Bull **56**: 234
57. Abe F, Nagafuji S, Okawa M, Kinjo J (2006) Trypanocidal Constituents in Plants 6. Minor. Withanolides from the Aerial Parts of *Physalis angulata*. Chem Pharm Bull **54**: 1226
58. He Q-P, Ma L, Luo J-Y, He F-Y, Lou L-G, Hu L-H (2007) Cytotoxic Withanolides from *Physalis angulata* L. Chem Biodivers **4**: 443
59. Glotter E (1991) Withanolides and Related Ergostane-Type Steroids. Nat Prod Rep **8**: 415
60. Choudhary MI, Yousaf S, Ahmed S, Samreen, Yasmeen K, Atta-ur-Rahman (2005) Antileishmanial Physalins from *Physalis minima*. Chem Biodivers **2**: 1164
61. Nawaz HR, Riaz M, Malik A, Khan PM, Ullah N (2000) Withanolides and Alkaloid from *Ajuga parviflora*. J Chem Soc Pak **22**: 138
62. Khan PM, Malik A, Ahmad S, Nawaz HR (1999) Withanolides from *Ajuga parviflora*. J Nat Prod **62**: 1290
63. Khan PM, Nawaz HR, Ahmad S, Malik A (1999) Ajugins C and D, New Withanolides from *Ajuga parviflora*. Helv Chim Acta **82**: 1423
64. Nawaz HR, Malik A, Khan PM, Ahmed S (1999) Ajugins E and F: Two Withanolides from *Ajuga parviflora*. Phytochemistry **52**: 1357
65. Nawaz HR, Malik A, Muhammad P, Ahmed S, Riaz M (2000) Chemical Constituents of *Ajuga parviflora*. Z Naturforsch, B: Chem Sci **55**: 100
66. Nagafuji S, Okabe H, Akahane H, Abe F (2004) Trypanocidal Constituents in Plants 4. Withanolides from the Aerial Parts of *Physalis angulata*. Biol Pharm Bull **27**: 193
67. Li Y-Z, Pan Y-M, Huang X-Y, Wang H-S (2008) Withanolides from *Physalis alkekengi* var. *francheti*. Helv Chim Acta **91**: 2284
68. Misico RI, Veleiro AS, Burton G, Oberti JC (1997) Withanolides from *Jaborosa leucotricha*. Phytochemistry **45**: 1045
69. Nicotra VE, Gil RR, Vaccarini C, Oberti JC, Burton G (2003) 15,21-Cyclowithanolides from *Jaborosa bergii*. J Nat Prod **66**: 1471
70. Misico RI, Gil RR, Oberti JC, Veleiro AS, Burton G (2000) Withanolides from *Vassobia lorentzii*. J Nat Prod **63**: 1329
71. Silva GL, Burton G, Oberti JC (1999) 18,20-Hemiacetal-Type and Other Withanolides from *Dunalia brachyacantha*. J Nat Prod **62**: 949
72. Alfonso D, Kapetanidis I (1994) Withanolides from *Iochroma gesnerioides*. Phytochemistry **36**: 179
73. Maldonado E, Torres FR, Martínez M, Pérez-Castorena AL (2004) 18-Acetoxywithanolides from *Physalis chenopodifolia*. Planta Med **70**: 59
74. Perez-Castorena A-L, Oropeza RF, Vazquez AR, Martinez M, Maldonado E (2006) Labdanes and Withanolides from *Physalis coztomatl*. J Nat Prod **69**: 1029
75. Carcamo C, Fajardo V (1993) (−)-Jaboromagellonine: New Withanolide from Seeds of *Jaborosa magellanica*. Heterocycles **36**: 1771
76. Ramacciotti NS, Nicotra VE (2007) Withanolides from *Jaborosa kurtzii*. J Nat Prod **70**: 1513

77. Cirigliano A, Veleiro AS, Oberti JC, Burton G (1995) A 15β-Hydroxywithanolide from *Datura ferox*. Phytochemistry **40**: 611
78. Veleiro AS, Cirigliano AM, Oberti JC, Burton G (1999) 7-Hydroxywithanolides from *Datura ferox*. J Nat Prod **62**: 1010
79. Yang B, Wang Q, Xia Y, Feng W, Kuang H (2007) Withanolide Compounds from the Flower of *Datura metel* L. Helv Chim Acta **90**: 1522
80. Kuang H, Yang B, Tang L, Xia Y, Dou D (2009) Baimantuoluosides A - C, Three New Withanolide Glucosides from the Flower of *Datura metel* L. Helv Chim Acta **92**: 1315
81. Bhat BA, Dhar KL, Puri SC, Qurishi MA, Khajuria A, Gupta A, Qazi GN (2005) Isolation, Characterization and Biological Evaluation of *Datura* Lactones as Potential Immunomodulators. Bioorg Med Chem **13**: 6672
82. Wube AA, Wenzig E-M, Gibbons S, Asres K, Bauer R, Bucar F (2008) Constituents of the Stem Bark of *Discopodium penninervium* and Their LTB_4 and COX-1 and −2 Inhibitory Activities. Phytochemistry **69**: 982
83. Cordero CP, Morantes SJ, Páez A, Rincón J, Aristizábal FA (2009) Cytotoxicity of Withanolides Isolated from *Acnistus arborescens*. Fitoterapia **80**: 364
84. Bravo B JA, Sauvain M, Gimenez T A, Balanza E, Serani L, Laprevote O, Massiot G, Lavaud C (2001) Trypanocidal Withanolides and Withanolide Glycosides from *Dunalia brachyacantha*. J Nat Prod **64**: 720
85. Lischewski M, Nguyen Thi Bich H, Porzel A, Adam G, Massiot G, Lavaud C (1992) Withanolide Glycosides from *Dunalia australis*. Phytochemistry **31**: 939
86. Yokosuka A, Mimaki Y, Sashida Y (2003) Chantriolides A and B, Two New Withanolide Glucosides from the Rhizomes of *Tacca chantrieri*. J Nat Prod **66**: 876
87. Liu H-Y, Ni W, Xie B-B, Zhou L-Y, Hao X-J, Wang X, Chen C-X (2006) Five New Withanolides from *Tacca plantaginea*. Chem Pharm Bull **54**: 992
88. Yokosuka A, Mimaki Y (2007) New Glycosides from the Rhizomes of *Tacca chantrieri*. Chem Pharm Bull **55**: 273
89. Manickam M, Awasthi SB, Oshima Y, Hisamichi K, Takeshita M, Sahai M, Ray AB (1994) Additional C-21-Oxygenated Withanolides from *Datura fastuosa*. J Chem Res (S): 306
90. Manickam M, Srivastava A, Ray AB (1998) Withanolides from the Flowers of *Datura fastuosa*. Phytochemistry **47**: 1427
91. Manickam M, Kumar S, Sinha-Bagchi A, Sinha SC, Ray AB (1994) Withametelin H and Withafastuosin C, Two New Withanolides from the Leaves of *Datura* Species. J Indian Chem Soc **71**: 393
92. Pan Y, Wang X, Hu X (2007) Cytotoxic Withanolides from the Flowers of *Datura metel*. J Nat Prod **70**: 1127
93. Yang B, Wang Q, Xia Y, Feng W, Kuang H (2008) Baimantuoluolines D-F, Three New Withanolides from the Flower of *Datura metel* L. Helv Chim Acta **91**: 964
94. Manickam M, Awasthi SB, Sinha-Bagchi A, Sinha SC, Ray AB (1996) Withanolides from *Datura tatula*. Phytochemistry **41**: 981
95. Srivastava A, Manickam M, Sinha-Bagchia A, Sinhaa SC, Ray AB (1996) Withasteroids. 28. Novel Withanolides from the Flowers of *Datura tatula*. Nat Prod Sci **2**: 9
96. Manickam M, Ray AB (1996) Structure of Withatatulin E, a Minor Withanolide of *Datura tatula*. Indian J Chem, Sect B **35B**: 1311
97. Gil RR, Misico RI, Sotes IR, Oberti JC, Veleiro AS, Burton G (1997) 16-Hydroxylated Withanolides from *Exodeconus maritimus*. J Nat Prod **60**: 568
98. Minguzzi S, Barata LES, Shin YG, Jonas PF, Chai H-B, Park EJ, Pezzuto JM, Cordell GA (2002) Cytotoxic Withanolides from *Acnistus arborescens*. Phytochemistry **59**: 635
99. Ma C-Y, Williams ID, Che C-T (1999) Withanolides from *Hyoscyamus niger* Seeds. J Nat Prod **62**: 1445
100. Habtemariam S, Gray AI, Waterman PG (1993) 16-Oxygenated Withanolides from the Leaves of *Discopodium penninervium*. Phytochemistry **34**: 807
101. Maldonado E, Amador S, Martinez M, Perez-Castorena AL (2010) Virginols A-C, Three New Withanolides from *Physalis virginiana*. Steroids **75**: 346

102. Habtemariam S, Gray AI (1998) Withanolides from the Roots of *Discopodium penninervium*. Planta Med **64**: 275

103. Kennelly EJ, Gerhaeuser C, Song LL, Graham JG, Beecher CWW, Pezzuto JM, Kinghorn AD (1997) Induction of Quinone Reductase by Withanolides Isolated from *Physalis philadelphica* (Tomatillos). J Agric Food Chem **45**: 3771

104. Su B-N, Misico RI, Park EJ, Santarsiero BD, Mesecar AD, Fong HHS, Pezzuto JM, Kinghorn AD (2002) Isolation and Characterization of Bioactive Principles of the Leaves and Stems of *Physalis philadelphica*. Tetrahedron **58**: 3453

105. Gu J-Q, Li W, Kang Y-H, Su B-N, Fong HHS, van Breemen RB, Pezzuto JM, Kinghorn AD (2003) Minor Withanolides from *Physalis philadelphica*: Structures, Quinone Reductase Induction Activities, and Liquid Chromatography (LC)-MS-MS Investigation as Artifacts. Chem Pharm Bull **51**: 530

106. Cardona D, Quinones W, Torres F, Velez ID, Orozco-P CI, Garzon J, Echeverri F (2005) Structure and Leishmanicidal Activity of Larnaxolide A and B, New Withanolides from *Larnax glabra* (Standl.) Sawyer. Actual Biol (Medellin, Colomb) **27**: 81

107. Hsieh P-W, Huang Z-Y, Chen J-H, Chang F-R, Wu C-C, Yang Y-L, Chiang MY, Yen M-H, Chen S-L, Yen H-F, Luebken T, Hung W-C, Wu Y-C (2007) Cytotoxic Withanolides from *Tubocapsicum anomalum*. J Nat Prod **70**: 747

108. Riaz N, Malik A, Nawaz SA, Muhammad P, Choudhary MI (2004) Cholinesterase-Inhibiting Withanolides from *Ajuga bracteosa*. Chem Biodivers **1**: 1289

109. Ma L, Xie C-M, Li J, Lou F-C, Hu L-H (2006) Daturametelins H, I, and J: 3 New Withanolide Glycosides from *Datura metel* L. Chem Biodivers **3**: 180

110. Zhu XH, Takagi M, Ikeda T, Midzuki K, Nohara T (2001) Withanolide-Type Steroids from *Solanum cilistum*. Phytochemistry **56**: 741

111. Zhu X-H, Ando J, Takagi M, Ikeda T, Nohara T (2001) Six New Withanolide-Type Steroids from the Leaves of *Solanum cilistum*. Chem Pharm Bull **49**: 161

112. Zhu X-H, Ando J, Takagi M, Ikeda T, Yoshimitsu A, Nohara T (2001) Four Novel Withanolide-Type Steroids from the Leaves of *Solanum cilistum*. Chem Pharm Bull **49**: 1440

113. Niero R, Da Silva IT, Tonial GC, Camacho BDS, Gacs-Baitz E, Delle Monache G, Delle Monache F (2006) Cilistepoxide and Cilistadiol, 2 New Withanolides from *Solanum sisymbiifolium*. Nat Prod Res Part A **20**: 1164

114. Vankar PS, Srivastava J, Molcanov K, Kojic-Prodic B (2009) Withanolide A Series Steroidal Lactones from *Eucalyptus globulus* Bark. Phytochemistry Lett **2**: 67

115. Siddiqui BS, Afreen S, Begum S (1999) Two New Withanolides from the Aerial Parts of *Datura innoxia*. Aust J Chem **52**: 905

116. Siddiqui BS, Hashmi IA, Begum S (2002) Two New Withanolides from the Aerial Parts of *Datura innoxia*. Heterocycles **57**: 715

117. Siddiqui BS, Arfeen S, Afshan F, Begum S (2005) Withanolides from *Datura innoxia*. Heterocycles **65**: 857

118. Siddiqui BS, Arfeen S, Begum S, Sattar FA (2005) Daturacin, a New Withanolide from *Datura innoxia*. Nat Prod Res **19**: 619

119. Bonetto GM, Gil RR, Oberti JC, Veleiro AS, Burton G (1995) Novel Withanolides from *Jaborosa sativa*. J Nat Prod **58**: 705

120. Nicotra VE, Ramacciotti NS, Gil RR, Oberti JC, Feresin GE, Guerrero CA, Baggio RF, Garland MT, Burton G (2006) Phytotoxic Withanolides from *Jaborosa rotacea*. J Nat Prod **69**: 783

121. Nicotra VE, Gil RR, Oberti JC, Burton G (2007) Withanolides with Phytotoxic Activity from *Jaborosa caulescens* var. *caulescens* and *J. caulescens* var. *bipinnatifida*. J Nat Prod **70**: 808

122. Usubillaga A, de Castellano G, Zabel V, Watson WH (1980) Acnistins, a New Class of Steroidal Lactones from *Acnistus ramiflorum* Miers, X-Ray Structure of Acnistin E. J Chem Soc, Chem Commun: 854

123. Luis JG, Echeverri F, Quiñones W, González AG, Torres F, Cardona G, Archbold R, Rojas M, Perales A (1994) Unambiguous ^{13}C NMR Assignment of Acnistins and Absolute Configuration of Acnistin A. Steroids **59**: 299
124. Luis JG, Echeverri F, García F, Rojas M (1994) The Structure of Acnistin B and the Immunosuppressive Effects of Acnistins A, B, and E. Planta Med **60**: 348
125. Luis JG, Echeverri F, Gonzalez AG (1994) Acnistins C and D, Withanolides from *Dunalia solanacea*. Phytochemistry **36**: 1297
126. Luis JG, Echeverri F, Gonzalez AG (1994) Acnistins F-H, Withanolides from *Dunalia solanacea*. Phytochemistry **36**: 769
127. Habtemariam S, Skelton BW, Waterman PG, White AH (2000) 17-Epiacnistin-A, a Further Withanolide from the Leaves of *Discopodium penninervium*. J Nat Prod **63**: 512
128. Kiyota N, Shingu K, Yamaguchi K, Yoshitake Y, Harano K, Yoshimitsu H, Ikeda T, Nohara T (2007) New C28 Steroidal Glycosides from *Tubocapsicum anomalum*. Chem Pharm Bull **55**: 34
129. Luis JG, Echeverri F, Quinones W, Gonzalez AG, Torres F, Cardona G, Archbold R, Perales A (1994) Withajardins, Withanolides with a New Type of Skeleton Structure of Withajardins A, B, C and D. Absolute Configuration of Withajardin C. Tetrahedron **50**: 1217
130. Echeverri F, Quinones W, Forres F, Cardona G, Archbold R, Luis JG, Gonzalez AG (1995) Withajardin E, a Withanolide from *Deprea orinocensis*. Phytochemistry **40**: 923
131. Kiyota N, Shingu K, Yamaguchi K, Yoshitake Y, Harano K, Yoshimitsu H, Miyashita H, Ikeda T, Tagawa C, Nohara T (2008) New C28 Steroidal Glycosides from *Tubocapsicum anomalum*. Chem Pharm Bull **56**: 1038
132. Glotter E, Abraham A, Guenzberg G, Kirson I (1977) Naturally Occurring Steroidal Lactones with a 17α-Oriented Side Chain. Structure of Withanolide E and Related Compounds. J Chem Soc, Perkin Trans 1: 341
133. Tomassini TCB, Barbi NS, Ribeiro IM, Xavier DCD (2000) Genero *Physalis* - Uma Revisao Sobre Vitaesteroides. Quim Nova **23**: 47
134. Fang L, Chai H-B, Castillo JJ, Soejarto DD, Farnsworth NR, Cordell GA, Pezzuto JM, Kinghorn AD (2003) Cytotoxic Constituents of *Brachistus stramoniifolius*. Phytother Res **17**: 520
135. Karikas GA, Gupta MP, Ravelo AG, Gonzales AG (1998) Physalin B from *Witheringia hunzikeri*. Fitoterapia **69**: 468
136. Pérez-Castorena AL, García M, Martínez M, Maldonado E (2004) Physalins from *Physalis solanaceus*. Biochem Syst Ecol **32**: 1231
137. Ripperger H, Kamperdick C (1998) First Isolation of Physalins from the Genus *Saracha* of Solanaceae. Pharmazie **53**: 144
138. Jacobo-Herrera NJ, Bremner P, Marquez N, Gupta MP, Gibbons S, Munoz E, Heinrich M (2006) Physalins from *Witheringia solanacea* as Modulators of the NF-κB Cascade. J Nat Prod **69**: 328
139. Qiu L, Zhao F, Jiang Z-H, Chen L-X, Zhao Q, Liu H-X, Yao X-S, Qiu F (2008) Steroids and Flavonoids from *Physalis alkekengi* var. *franchetii* and Their Inhibitory Effects on Nitric Oxide Production. J Nat Prod **71**: 642
140. Sunayama R, Kuroyanagi M, Umehara K, Ueno A (1993) Physalin and Neophysalins from *Physalis alkekengi* var. *francheti* and Their Differentiation Inducing Activity. Phytochemistry **34**: 529
141. Qiu L, Jiang Z-H, Liu H-X, Chen L-X, Yao X-S, Qiu F (2008) A Pair of 3-Epimeric Physalins from *Physalis alkekengi* L. var. *franchetii*. J Asian Nat Prod Res **10**: 881
142. Makino B, Kawai M, Kito K, Yamamura H, Butsugan Y (1995) New Physalins Possessing an Additional Carbon-Carbon Bond from *Physalis alkekengi* var. *francheti*. Tetrahedron **51**: 12529
143. Kawai M, Yamamoto T, Makino B, Yamamura H, Araki S, Butsugan Y, Saito K (2001) The Structure of Physalin T from *Physalis alkekengi* var. *francheti*. J Asian Nat Prod Res **3**: 199

144. Makino B, Kawai M, Iwata Y, Yamamura H, Butsugan Y, Ogawa K, Hayashi M (1995) Physalins Possessing an Endoperoxy Structure from *Physalis alkekengi* var. *francheti*. Structural Revision of Physalin K. Bull Chem Soc Jpn **68**: 219

145. Choudhary MI, Yousuf S, Samreen, Ahmed S, Atta-ur-Rahman (2007) New Leishmanicidal Physalins from *Physalis minima*. Nat Prod Res **21**: 877

146. Kuo P-C, Kuo T-H, Damu AG, Su C-R, Lee E-J, Wu T-S, Shu R, Chen C-M, Bastow KF, Chen T-H, Lee K-H (2006) Physanolide A, a Novel Skeleton Steroid, and Other Cytotoxic Principles from *Physalis angulata*. Org Lett **8**: 2953

147. Chen R, Liang J-Y, Liu R (2007) Two Novel Neophysalins from *Physalis alkekengi* L. var. *franchetii*. Helv Chim Acta **90**: 963

148. Makino B, Kawai M, Ogura T, Nakanishi M, Yamamura H, Butsugan Y (1995) Structural Revision of Physalin H Isolated from *Physalis angulata*. J Nat Prod **58**: 1668

149. Chen R, Liang JY, Yang Y, Liu R (2007) Chemical Constituents from *Physalis alkekengi* and Structural Revision of Physalin G. Chin J Nat Med **5**: 186

150. Sen G, Pathak HD (1995) Physalin L, a 13,14-Seco-16,24 Cyclosteroid from *Physalis minima*. Phytochemistry **39**: 1245

151. Kawai M, Makino B, Yamamura H, Butsugan Y (1996) Upon "Physalin L" Isolated from *Physalis minima*. Phytochemistry **43**: 661

152. Veras ML, Bezerra MZB, Braz-Filho R, Pessoa ODL, Montenegro RC, Do Ó Pessoa C, De Moraes MO, Costa-Lotufo LV (2004) Cytotoxic Epimeric Withaphysalins from Leaves of *Acnistus arborescens*. Planta Med **70**: 551

153. Veras ML, Bezerra MZB, Lemos TLG, Uchoa DEDA, Braz-Filho R, Chai HB, Cordell GA, Pessoa ODL (2004) Cytotoxic Withaphysalins from the Leaves of *Acnistus arborescens*. J Nat Prod **67**: 710

154. Ma L, Gan X-W, He Q-P, Bai H-Y, Arfan M, Lou F-C, Hu L-H (2007) Cytotoxic Withaphysalins from *Physalis minima*. Helv Chim Acta **90**: 1406

155. Sahai M, Kirson I (1984) Withaphysalin D, a New Withaphysalin from *Physalis minima* Linn. var. *indica*. J Nat Prod **47**: 527

156. Veleiro AS, Oberti JC, Burton G (1992) A Ring D Aromatic Withanolide from *Salpichroa origanifolia*. Phytochemistry **31**: 935

157. Veleiro AS, Burton G, Bonetto GM, Gil RR, Oberti JC (1994) New Withanolides from *Salpichroa origanifolia*. J Nat Prod **57**: 1741

158. Tettamanzi MC, Veleiro AS, Oberti JC, Burton G (1998) New Hydroxylated Withanolides from *Salpichroa origanifolia*. J Nat Prod **61**: 338

159. Tettamanzi MC, Veleiro AS, de la Fuente JR, Burton G (2001) Withanolides from *Salpichroa origanifolia*. J Nat Prod **64**: 783

160. Tettamanzi MC, Veleiro AS, Oberti JC, Burton G (1996) Ring D Aromatic Ergostane Derivatives from *Salpichroa origanifolia*. Phytochemistry **43**: 461

161. Gill HK, Smith RW, Whiting DA (1990) Ring D Expansion and Aromatisation in the Biosynthesis of Nic-1, an Antifeedant Steroid from *Nicandra physaloides*. J Chem Soc, Perkin Trans 1: 2989

162. Monteagudo ES, Burton G, Gros EG, Gonzalez CM, Oberti JC (1989) A 19-Hydroxywithanolide from *Jaborosa leucotricha*. Phytochemistry **28**: 1514

163. Misico RI, Oberti JC, Veleiro AS, Burton G (1996) New 19-Hydroxywithanolides from *Jaborosa leucotricha*. J Nat Prod **59**: 66

164. Veleiro AS, Trocca CE, Burton G, Oberti JC (1992) A Phenolic Withanolide from *Jaborosa leucotricha*. Phytochemistry **31**: 2550

165. Kirson IA, Cohen M, Greenberg M, Gottlieb HE, Varenne P, Abraham A (1979) Ixocarpalactone A and B. Two Unusual Naturally Occurring Steroids of the Ergostane Type. J Chem Res (S): 103

166. Huang Y, Müehlbauer A, Henkel T, Liu J-K (2003) Two New Taccalonolides from Tropic Plant *Tacca subflaellata*. Chin Chem Lett **14**: 68

167. Huang Y, Liu J-K, Mühlbauer A, Henkel T (2002) Three Novel Taccalonolides from the Tropical Plant *Tacca subflaellata*. Helv Chim Acta **85**: 2553

168. Fang S-T, Li B, Liu J-K (2009) Two New Withanolides from *Physalis peruviana*. Helv Chim Acta **92**: 1304

169. Lavie D, Besalle R, Pestchanker MJ, Gottlieb HE, Frolow F, Giordano OS (1987) Trechonolide A, a New Withanolide Type from *Trechonaetes laciniata*. Phytochemistry **26**: 1791

170. Parvez M, Fajardo V, Shamma M (1988) (+)-Jaborosalactone M, a Hemiketal Withanolide from *Jaborosa magellanica*. Acta Crystallogr, Sect C: Cryst Struct Commun **44**: 553

171. Cirigliano AM, Veleiro AS, Misico RI, Tettamanzi MC, Oberti JC, Burton G (2007) Withanolides from *Jaborosa laciniata*. J Nat Prod **70**: 1644

172. Pecci A, Alvarez LD, Veleiro AS, Ceballos NR, Lantos CP, Burton G (2009) New Lead Compounds in the Search for Pure Antiglucocorticoids and the Dissociation of Antiglucocorticoid Effects. J Steroid Biochem Mol Biol **113**: 155

173. Baggio RF (2005). Personal communication

174. Garcia ME, S. P, Navarro-Vázquez A, Phillips DD, Gayathri C, Krakauer H, Stephens PW, Nicotra VE, Gil RR (2009) Stereochemistry Determination by Powder X-Ray Diffraction Analysis and NMR Spectroscopy Residual Dipolar Couplings. Angew Chem Int Ed **48**: 5670

175. Tettamanzi MC, Biurrun FN, Cirigliano AM (2007) A New Antifeedant Withanolide from *Jaborosa lanigera*. Z Naturforsch, B: Chem Sci **62**: 573

176. Monteagudo ES, Oberti JC, Gros EG, Burton G (1990) A Spiranic Withanolide from *Jaborosa odonelliana*. Phytochemistry **29**: 933

177. Cirigliano AM, Veleiro AS, Oberti JC, Burton G (2002) Spiranoid Withanolides from *Jaborosa odonelliana*. J Nat Prod **65**: 1049

178. Cirigliano AM, Misico RI (2005) Spiranoid Withanolides from *Jaborosa odonelliana* and *Jaborosa runcinata*. Z Naturforsch, B: Chem Sci **60**: 867

179. Cirigliano AM, Veleiro AS, Bonetto GM, Oberti JC, Burton G (1996) Spiranoid Withanolides from *Jaborosa runcinata* and *Jaborosa araucana*. J Nat Prod **59**: 717

180. Shen J, Chen Z, Gao Y (1996) Taccalonolides from *Tacca plantaginea*. Phytochemistry **42**: 891

181. Chen Z-l, Shen J-h, Gao Y-s, Wichtl M (1997) Five Taccalonolides from *Tacca plantaginea*. Planta Med **63**: 40

182. Yang J-Y, Zhao R-H, Chen C-X, Ni W, Teng F, Hao X-J, Liu H-Y (2008) Taccalonolides W - Y, Three New Pentacyclic Steroids from *Tacca plantaginea*. Helv Chim Acta **91**: 1077

183. Mühlbauer A, Seip S, Nowak A, Tran VS (2003) Five Novel Taccalonolides from the Roots of the Vietnamese Plant *Tacca paxiana*. Helv Chim Acta **86**: 2065

184. Su B-N, Park EJ, Nikolic D, Santarsiero BD, Mesecar AD, Vigo JS, Graham JG, Cabieses F, van Breemen RB, Fong HHS, Farnsworth NR, Pezzuto JM, Kinghorn AD (2003) Activity-Guided Isolation of Novel Norwithanolides from *Deprea subtriflora* with Potential Cancer Chemopreventive Activity. J Org Chem **68**: 2350

185. Su B-N, Park EJ, Nikolic D, Vigo JS, Graham JG, Cabieses F, Van Breemen RB, Fong HHS, Farnsworth NR, Pezzuto JM, Kinghorn AD (2003) Isolation and Characterization of Miscellaneous Secondary Metabolites of *Deprea subtriflora*. J Nat Prod **66**: 1089

186. Ma L, Ali M, Arfan M, Lou L-G, Hu L-H (2007) Withaphysanolide A, a Novel C-27 Norwithanolide Skeleton, and Other Cytotoxic Compounds from *Physalis divericata*. Tetrahedron Lett **48**: 449

187. Shingu K, Marubayashi N, Ueda I, Yahara S, Nohara T (1990) Two New Ergostane Derivatives from *Tubocapsicum anomalum* (Solanaceae). Chem Pharm Bull **38**: 1107

188. Kovganko NV, Kashkan ZhN (1997) Advances in the Chemical Synthesis of Withanolides. Chem Nat Compd **33**: 133

189. Mareggiani G, Picollo MI, Zerba E, Burton G, Tettamanzi MC, Benedetti-Doctorovich MOV, Veleiro AS (2000) Antifeedant Activity of Withanolides from *Salpichroa origanifolia* on *Musca domestica*. J Nat Prod **63**: 1113

190. Bado S, Mareggiani G, Amiano N, Burton G, Veleiro AS (2004) Lethal and Sublethal Effects of Withanolides from *Salpichroa origanifolia* and Analogues on *Ceratitis capitata*. J Agric Food Chem **52**: 2875

191. Misra L, Lal P, Chaurasia ND, Sangwan RS, Sinha S, Tuli R (2008) Selective Reactivity of 2-Mercaptoethanol with 5β,6β-Epoxide in Steroids from *Withania somnifera*. Steroids **73**: 245

192. Yokota Y, Bargagna-Mohan P, Ravindranath PP, Kim KB, Mohan R (2006) Development of Withaferin A Analogs as Probes of Angiogenesis. Bioorg Med Chem Lett **16**: 2603

193. Choudhary MI, Yousuf S, Samreen, Shah SAA, Ahmed S, Atta-ur-Rahman (2006) Biotransformation of Physalin H and Leishmanicidal Activity of its Transformed Products. Chem Pharm Bull **54**: 927

194. Wijeratne EMK, Xu Y, Marron MT, Whitesell L, Gunatilaka AAL (2008) Isolation and Synthesis of Analogs of the Anticancer Natural Product Withaferin A for Structure-Activity Relationship Studies. Abstracts, 42nd Western Regional Meeting of the American Chemical Society, Las Vegas, NV, United States, September 23–27.

195. Jha S, Bandyopadhyay M, Chaudhuri KN, Ghosh S, Ghosh B (2005) Biotechnological Approaches for the Production of Forskolin, Withanolides, Colchicine and Tylophorine. Plant Genet Resour **3**: 101

196. Ray S, Jha S (2001) Production of Withaferin A in Shoot Cultures of *Withania somnifera*. Planta Med **67**: 432

197. Sangwan RS, Das Chaurasiya N, Lal P, Misra L, Uniyal GC, Tuli R, Sangwan NS (2007) Withanolide a Biogeneration in *in Vitro* Shoot Cultures of Ashwagandha (*Withania somnifera* Dunal), a Main Medicinal Plant in Ayurveda. Chem Pharm Bull **55**: 1371

198. Sharada M, Ahuja A, Suri KA, Vij SP, Khajuria RK, Verma V, Kumar A (2007) Withanolide Production by *in Vitro* Cultures of *Withania somnifera* and its Association with Differentiation. Biologia Plantarum **51**: 161

199. Madina BR, Sharma LK, Chaturvedi P, Sangwan RS, Tuli R (2007) Purification and Physico-Kinetic Characterization of 3β-Hydroxy Specific Sterol Glucosyltransferase from *Withania somnifera* (L) and its Stress Response. Biochim Biophys Acta **1774**: 392

200. Madina BR, Sharma LK, Chaturvedi P, Sangwan RS, Tuli R (2007) Purification and Characterization of a Novel Glucosyltransferase Specific to 27*β*-Hydroxy Steroidal Lactones from *Withania somnifera* and its Role in Stress Responses. Biochim Biophys Acta **1774**: 1199

201. Elliger CA, Waiss AC, Jr. (1989) Insect Growth Inhibitors from *Petunia* and Other Solanaceous Plants. In: Arnason JT, Philogene BJR, Morand P (eds): Pesticides of Plant Origin, ACS Symposium Series No. 387. American Chemical Society, Washington DC, p 188

202. Waiss AC, Elliger CA, Haddon WF, Benson M (1993) Insect Inhibitory Steroidal Saccharide Esters from *Physalis peruviana*. J Nat Prod **56**: 1365

203. Elliger CA, Haddon WF, Harden L, Waiss AC, Wong RY (1994) Insect Inhibitory Steroidal Saccharide Esters from *Physalis peruviana*. J Nat Prod **57**: 348

204. Baumann TW, Meier CM (1993) Chemical Defense by Withanolides During Fruit Development in *Physalis peruviana*. Phytochemistry **33**: 317

205. Mareggiani G, Picollo MI, Veleiro AS, Tettamanzi MC, Benedetti-Doctorovich MOV, Burton G, Zerba E (2002) Response of *Tribolium castaneum* (Coleoptera, Tenebrionidae) to *Salpichroa origanifolia* Withanolides. J Agric Food Chem **50**: 104

206. Vaccarini CE, Bonetto GM (2000) Antifeedant Activity Evaluation of Withanolides from *Jaborosa integrifolia*. Molecules **5**: 422

207. Enriz RD, Baldoni HA, Zamora MA, Jauregui EA, Sosa ME, Tonn CE, Luco JM, Gordaliza M (2000) Structure-Antifeedant Activity Relationship of Clerodane Diterpenoids. Comparative Study with Withanolides and Azadirachtin. J Agric Food Chem **48**: 1384

208. Garcia ES, Castro DP, Ribeiro IM, Tomassini TCB, Azambuja P (2006) *Trypanosoma rangeli*: Effects of Physalin B on the Immune Reactions of the Infected Larvae of *Rhodnius prolixus*. Exp Parasitol **112**: 37

209. Castro DP, Figueiredo MB, Ribeiro IM, Tomassini TCB, Azambuja P, Garcia ES (2008) Immune Depression in *Rhodnius prolixus* by Seco-Steroids, Physalins. J Insect Physiol **54**: 555

210. Castro DP, Figueiredo MB, Genta FA, Ribeiro IM, Tomassini TCB, Azambuja P, Garcia ES (2009) Physalin B Inhibits *Rhodnius prolixus* Hemocyte Phagocytosis and Microaggregation by the Activation of Endogenous PAF-Acetyl Hydrolase Activities. J Insect Physiol **55**: 532

211. Dinan L, Whiting P, Alfonso D, Kapetanidis I (1996) Certain Withanolides from *Iochroma gesnerioides* Antagonize Ecdysteroid Actin in a *Drosophila melanogaster* Cell Line. Entomol Exp Appl **80**: 415

212. Savchenko T, Whiting P, Germade A, Dinan L (2000) Ecdysteroid Agonist and Antagonist Activities in Species of the Solanaceae. Biochem Syst Ecol **28**: 403

213. Dinan L, Savchenko T, Whiting P, Sarker SD (1999) Plant Natural Products as Insect Steroid Receptor Agonists and Antagonists. Pestic Sci **55**: 331

214. Dinan L, Bourne PC, Meng Y, Sarker SD, Tolentino RB, Whiting P (2001) Assessment of Natural Products in the *Drosophila melanogaster* BII Cell Bioassay for Ecdysteroid Agonist and Antagonist Activities. Cell Mol Life Sci **58**: 321

215. Vaccarini CE, Bonetto GM (2000) Selective Phytotoxic Activity of Withanolides from *Iochroma australe* to Crop and Weed Species. J Chem Ecol **26**: 2187

216. Vieira NC, Espíndola LS, Santana JM, Veras ML, Pessoa ODL, Pinheiro SM, de Araújo RM, Lima MAS, Silveira ER (2008) Trypanocidal Activity of a New Pterocarpan and Other Secondary Metabolites of Plants from Northeastern Brazil Flora. Bioorg Med Chem **16**: 1676

217. Atta-ur-Rahman, Samreen, Atia-tul-Wahab, Choudhary MI (2008) Discovery of Leishmanicidal Agents from Medicinal Plants. Pure Appl Chem **80**: 1783

218. Cardona D, Quiñones W, Torres F, Robledo S, Vélez ID, Cruz V, Notario R, Echeverri F (2006) Leishmanicidal Activity of Withajardins and Acnistins. An Experimental and Computational Study. Tetrahedron **62**: 6822

219. Sen N, Banerjee B, Das BB, Ganguly A, Sen T, Pramanik S, Mukhopadhyay S, Majumder HK (2006) Apoptosis is Induced in Leishmanial Cells by a Novel Protein Kinase Inhibitor Withaferin A and is Facilitated by Apoptotic Topoisomerase I-DNA Complex. Cell Death Differ **14**: 358

220. Kannan ND, Kulandaivelu G (2007) Novel Method to Isolate Withaferin A from *Withania somnifera* Roots and Its Bioactivity. Allelopathy J **20**: 213

221. Silva MTG, Simas SM, Batista TG, Cardarelli P, Tomassini TCB (2005) Studies on Antimicrobial Activity, in Vitro, of *Physalis angulata* L. (Solanaceae) Fraction and Physalin B Bringing out the Importance of Assay Determination. Mem Inst Oswaldo Cruz **100**: 779

222. Januário AH, Rodrigues Filho E, Pietro RCLR, Kashima S, Sato DN, França SC (2002) Antimycobacterial Physalins from *Physalis angulata* L. (Solanaceae). Phytother Res **16**: 445

223. Vieira AT, Pinho V, Lepsch LB, Scavone C, Ribeiro IM, Tomassini T, Ribeiro-dos-Santos R, Soares MBP, Teixeira MM, Souza DG (2005) Mechanisms of the Anti-Inflammatory Effects of the Natural Secosteroids Physalins in a Model of Intestinal Ischaemia and Reperfusion Injury. Br J Pharmacol **146**: 244

224. Singh D, Aggarwal A, Maurya R, Naik S (2007) *Withania somnifera* Inhibits NF-κB and AP-1 Transcription Factors in Human Peripheral Blood and Synovial Fluid Mononuclear Cells. Phytother Res **21**: 905

225. Kaileh M, Vanden Berghe W, Heyerick A, Horion J, Piette J, Libert C, De Keukeleire D, Essawi T, Haegeman G (2007) Withaferin A Strongly Elicits IκB Kinase Hyper-phosphorylation Concomitant with Potent Inhibition of Its Kinase Activity. J Biol Chem **282**: 4253

226. Ichikawa H, Takada Y, Shishodia S, Jayaprakasam B, Nair MG, Aggarwal BB (2006) Withanolides Potentiate Apoptosis, Inhibit Invasion, and Abolish Osteoclastogenesis through Suppression of Nuclear Factor-κB (Nf-κB) Activation and NF-κB-Regulated Gene Expression. Mol Cancer Ther **5**: 1434

227. Punetha H, Gaur AK, Pant AK, Mishra DP (2006) Cyclooxygenase (Cox) Inhibitory Activity of Withanolides from *Withania somnifera* (L.) Dunal as Revealed by Molecular Docking. Bioinf Trends **1**: 41
228. Manickam M, Padma P, Chansouria JPN, Ray AB (1997) Evaluation of Antistress Activity of Withafastuosin D, a Withanolide of *Datura fastuosa*. Phytother Res **11**: 384
229. Goel RK, Maiti RN, Manickam M, Ray AB (1997) Effect of Withafastuosin E on Prostanoid Synthesis by Human Intestinal Mucosa 'in Vitro'. Fitoterapia **68**: 345
230. Maiti RN, Manickam M, Ray AB, Goel RK (1997) Effect of Withafastuosin E on Gastric Mucosal Offensive and Defensive Factors in Rats. Indian J Exp Biol **35**: 751
231. Bhattacharya A, Ghosal S, Bhattacharya SK (2001) Anti-Oxidant Effect of *Withania somnifera* Glycowithanolides in Chronic Footshock Stress-Induced Perturbations of Oxidative Free Radical Scavenging Enzymes and Lipid Peroxidation in Rat Frontal Cortex and Striatum. J Ethnopharmacol **74**: 1
232. Bhattacharya SK, Satyan KS, Ghosal S (1997) Antioxidant Activity of Glycowithanolides from *Withania somnifera*. Indian J Exp Biol **35**: 236
233. Habtemariam S (1997) Cytotoxicity and Immunosuppressive Activity of Withanolides from *Discopodium penninervium*. Planta Med **63**: 15
234. Mesaik MA, Zaheer Ul H, Murad S, Ismail Z, Abdullah NR, Gill HK, Atta-ur-Rahman, Yousaf M, Siddiqui RA, Ahmad A, Choudhary MI (2006) Biological and Molecular Docking Studies on Coagulin-H: Human Il-2 Novel Natural Inhibitor. Mol Immunol **43**: 1855
235. Soares MBP, Bellintani MC, Ribeiro IM, Tomassini TCB, Ribeiro dos Santos R (2003) Inhibition of Macrophage Activation and Lipopolysaccaride-Induced Death by Seco-Steroids Purified from *Physalis angulata* L. Eur J Pharmacol **459**: 107
236. Soares MBP, Brustolim D, Santos LA, Bellintani MC, Paiva FP, Ribeiro YM, Tomassini TCB, Ribeiro dos Santos R (2006) Physalins B, F and G, Seco-Steroids Purified from *Physalis angulata* L., Inhibit Lymphocyte Function and Allogeneic Transplant Rejection. Int Immunopharm **6**: 408
237. Yang H, Shi G, Dou QP (2007) The Tumor Proteasome is a Primary Target for the Natural Anticancer Compound Withaferin A Isolated from "Indian Winter Cherry". Mol Pharmacol **71**: 426
238. Mandal C, Dutta A, Mallick A, Chandra S, Misra L, Sangwan RS (2008) Withaferin A Induces Apoptosis by Activating P38 Mitogen-Activated Protein Kinase Signaling Cascade in Leukemic Cells of Lymphoid and Myeloid Origin through Mitochondrial Death Cascade. Apoptosis **13**: 1450
239. Stan SD, Hahm E-R, Warin R, Singh SV (2008) Withaferin A Causes FOXO3a- and Bim-Dependent Apoptosis and Inhibits Growth of Human Breast Cancer Cells *In Vivo*. Cancer Res **68**: 7661
240. Jayaprakasam B, Zhang Y, Seeram NP, Nair MG (2003) Growth Inhibition of Human Tumor Cell Lines by Withanolides from *Withania somnifera* Leaves. Life Sci **74**: 125
241. Damasceno Rocha D, Gadelha Militão GC, Veras ML, Deusdênia Loiola Pessoa O, Rocha Silveira E, Negreiros Nunes Alves AP, Odorico de Moraes M, Pessoa C, Veras Costa-Lotufo L (2006) Selective Cytotoxicity of Withaphysalins in Myeloid Leukemia Cell Lines Versus Peripheral Blood Mononuclear Cells. Life Sci **79**: 1692
242. Chen W-Y, Chang F-R, Huang Z-Y, Chen J-H, Wu Y-C, Wu C-C (2008) Tubocapsenolide A, a Novel Withanolide, Inhibits Proliferation and Induces Apoptosis in MDA-MB-231 Cells by Thiol Oxidation of Heat Shock Proteins. J Biol Chem **283**: 17184
243. Chang H-C, Chang F-R, Wang Y-C, Pan M-R, Hung W-C, Wu Y-C (2007) A Bioactive Withanolide Tubocapsanolide A Inhibits Proliferation of Human Lung Cancer Cells Via Repressing Skp2 Expression. Mol Cancer Ther **6**: 1572
244. Magalhães HIF, Veras ML, Pessoa ODL, Silveira ER, Moraes MO, Pessoa C, Costa-Lotufo LV (2006) Preliminary Investigation of Structure-Activity Relationship of Cytotoxic Physalins. Lett Drug Design Discovery **3**: 9

245. Magalhães HIF, Veras ML, Torres M, Rocha, Alves APNN, Pessoa ODL, Silveira ER, Costa-Lotufo LV, de Moraes MO, Pessoa C (2006) In-Vitro and in-Vivo Antitumour Activity of Physalins B and D from *Physalis angulata*. J Pharm Pharmacol **58**: 235
246. Hosoya T, Arai MA, Koyano T, Kowithayakorn T, Ishibashi M (2008) Naturally Occurring Small-Molecule Inhibitors of Hedgehog/Gli-Mediated Transcription. ChemBioChem **9**: 1082
247. Choi JK, Murillo G, Su B-N, Pezzuto JM, Kinghorn AD, Mehta RG (2006) Ixocarpalactone A Isolated from the Mexican Tomatillo Shows Potent Antiproliferative and Apoptotic Activity in Colon Cancer Cells. FEBS J **273**: 5714
248. Uma Devi P (1996) *Withania somnifera* Dunal (Ashwagandha): Potential Plant Source of a Promising Drug for Cancer Chemotherapy and Radiosensitization. Indian J Exp Biol **34**: 927
249. Sharada AC, Solomon FE, Devi PU, Udupa N, Srinivasan KK (1996) Antitumor and Radiosensitizing Effects of Withaferin A on Mouse Ehrlich Ascites Carcinoma in Vivo. Acta Oncol **35**: 95
250. Uma Devi P, Kamath R (2003) Radiosensitizing Effect of Withaferin A Combined with Hyperthermia on Mouse Fibrosarcoma and Melanoma. J Radiat Res **44**: 1
251. Kinghorn AD, Su BN, Lee D, Gu JQ, Pezzuto JM (2003) Cancer Chemopreventive Agents Discovered by Activity-Guided Fractionation: An Update. Curr Org Chem **7**: 213
252. Kang Y-H, Pezzuto JM, Helmut S, Lester P (2004) Induction of Quinone Reductase as a Primary Screen for Natural Product Anticarcinogens: Methods Enzymol, vol. 382, p 380. Elsevier, Inc., San Diego.
253. Misico RI, Song LL, Veleiro AS, Cirigliano AM, Tettamanzi MC, Burton G, Bonetto GM, Nicotra VE, Silva GL, Gil RR, Oberti JC, Kinghorn AD, Pezzuto JM (2002) Induction of Quinone Reductase by Withanolides. J Nat Prod **65**: 677
254. Panjamurthy K, Manoharan S, Menon VP, Nirmal MR, Senthil N (2008) Protective Role of Withaferin-A on 7,12-Dimethylbenz(a)Anthracene-Induced Genotoxicity in Bone Marrow of Syrian Golden Hamsters. J Biochem Mol Toxic **22**: 251
255. Mohan R, Hammers H, Bargagna-Mohan P, Zhan X, Herbstritt C, Ruiz A, Zhang L, Hanson A, Conner B, Rougas J, Pribluda V (2004) Withaferin A Is a Potent Inhibitor of Angiogenesis. Angiogenesis **7**: 115
256. Bargagna-Mohan P, Ravindranath Padma P, Mohan R (2006) Small Molecule Anti-Angiogenic Probes of the Ubiquitin Proteasome Pathway: Potential Application to Choroidal Neovascularization. Invest Ophthalmol Vis Sci **47**: 4138
257. Vandenberghe I, Créancier L, Vispé S, Annereau J-P, Barret J-M, Pouny I, Samson A, Aussagues Y, Massiot G, Ausseil F, Bailly C, Kruczynski A (2008) Physalin B, a Novel Inhibitor of the Ubiquitin-Proteasome Pathway, Triggers NOXA-Associated Apoptosis. Biochem Pharmacol **76**: 453
258. Tinley TL, Randall-Hlubek DA, Leal RM, Jackson EM, Cessac JW, Quada JC, Jr., Hemscheidt TK, Mooberry SL (2003) Taccalonolides E and A: Plant-Derived Steroids with Microtubule-Stabilizing Activity. Cancer Res **63**: 3211
259. Risinger AL, Jackson EM, Polin LA, Helms GL, LeBoeuf DA, Joe PA, Hopper-Borge E, Luduena RF, Kruh GD, Mooberry SL (2008) The Taccalonolides: Microtubule Stabilizers That Circumvent Clinically Relevant Taxane Resistance Mechanisms. Cancer Res **68**: 8881
260. Risinger AL, Mooberry SL (2010) Taccalonolides: Novel Microtubule Stabilizers with Clinical Potential. Cancer Lett **291**: 14
261. Tohda C, Kuboyama T, Komatsu K (2005) Search for Natural Products Related to Regeneration of the Neuronal Network. NeuroSignals **14**: 34
262. Kuboyama T, Tohda C, Zhao J, Nakamura N, Hattori M, Komatsu K (2002) Axon- or Dendrite-Predominant Outgrowth Induced by Constituents from Ashwagandha. NeuroReport **13**: 1715
263. Kuboyama T, Tohda C, Komatsu K (2005) Neuritic Regeneration and Synaptic Reconstruction Induced by Withanolide A. Br J Pharmacol **144**: 961

264. Kuboyama T, Tohda C, Komatsu K (2006) Withanoside IV and Its Active Metabolite, Sominone, Attenuate Aβ(25–35)-Induced Neurodegeneration. Eur J Neurosci **23**: 1417
265. Choudhary MI, Nawaz SA, Zaheer-ur-Haq, Lodhi MA, Ghayur MN, Jalil S, Riaz N, Yousuf S, Malik A, Gilani AH, Atta-ur-Rahman (2005) Withanolides, a New Class of Natural Cholinesterase Inhibitors with Calcium Antagonistic Properties. Biochem Biophys Res Commun **334**: 276
266. Jiao M, Luna-Cavazos M, Bye R (2002) Allozyme Variation in Mexican Species and Classification of *Datura* (Solanaceae). Plant Syst Evol **232**: 155
267. Luna-Cavazos M, Bye R, Jiao M (2009) The Origin of *Datura metel* (Solanaceae): Genetic and Phylogenetic Evidence. Gen Resour Crop Evol **56**: 263

Author Index

A.D. Kinghorn, H. Falk, J. Kobayashi (eds.), *Progress in the Chemistry of Organic Natural Products, Vol. 94*, DOI 10.1007/978-3-7091-0748-5,

B

C

D

I

R

S

Subject Index

MIX
Papier aus verantwortungsvollen Quellen
Paper from responsible sources
FSC® C105338

If you have any concerns about our products,
you can contact us on
ProductSafety@springernature.com

In case Publisher is established outside the EU,
the EU authorized representative is:
Springer Nature Customer Service Center GmbH
Europaplatz 3, 69115 Heidelberg, Germany

Printed by Libri Plureos GmbH
in Hamburg, Germany

Springer Proceedings in Mathematics & Statistics

Volume 88

Springer Proceedings in Mathematics & Statistics

This book series features volumes composed of selected contributions from workshops and conferences in all areas of current research in mathematics and statistics, including OR and optimization. In addition to an overall evaluation of the interest, scientific quality, and timeliness of each proposal at the hands of the publisher, individual contributions are all refereed to the high quality standards of leading journals in the field. Thus, this series provides the research community with well-edited, authoritative reports on developments in the most exciting areas of mathematical and statistical research today.

More information about this series at http://www.springer.com/series/10533

De-Jun Feng · Ka-Sing Lau
Editors

Geometry and Analysis of Fractals

Hong Kong, December 2012

Springer

Editors
De-Jun Feng
Ka-Sing Lau
Department of Mathematics
The Chinese University of Hong Kong
Hong Kong

ISSN 2194-1009 ISSN 2194-1017 (electronic)
ISBN 978-3-662-51376-7 ISBN 978-3-662-43920-3 (eBook)
DOI 10.1007/978-3-662-43920-3

Mathematics Subject Classification: 28-XX, 31-XX, 37-XX, 60-XX

Springer Heidelberg New York Dordrecht London

Printed on acid-free paper

Springer is part of Springer Science+Business Media (www.springer.com)

Foreword

It was about half a century ago when Mandelbrot first formulated a revolutionary idea in the notion of fractals, and used it to describe irregular geometric objects and physical phenomena in nature. The importance of this concept was immediately recognized in many disciplines such as physics, geosciences, and biological sciences. In mathematics, it emerged as a new area named fractal geometry. Over the years, rather complete theories on fractal dimensions, multifractals, and random fractals have been developed; their connections with dynamical systems, ergodic theory, and probability theory are apparent. In recent developments, the close interplay of fractals with harmonic analysis, complex analysis, geometric measure theory, graph theory, and algebra also enriches the area greatly.

In China, we share with the world the great enthusiasm in the development of the area. In 2011–2012, we were grateful to have support from the Focused Investment Scheme of the Chinese University of Hong Kong on a two-year special program, *Advances on Fractals*. The program was intended to bring together interested researchers from China and abroad to discuss and carry out fractal research. There were seminars and lectures throughout, which culminated in the *International Conference on Advances of Fractals and Related Topics* on December 10–14, 2012. The conference was attended by 150 participants from all over the world and 90 lectures were presented.

This monograph is the fruit of the conference. It contains 13 polished versions of lectures from the conference, covering a wide range of current research directions:

- dimension theory—multifractal analysis—geometric measure theory—harmonic analysis and spectral analysis—analysis on metric spaces and heat kernels.

They are in the form of a survey with sufficient introduction to the topics and with selected proofs. We regret that we are not able to include many of the interesting lectures because of the limited size of the monograph.

Dimension theory is classical in fractal geometry. We have four articles on this topic. The Marstrand's projection theorem in the setup of Heisenberg groups is being surveyed by Mattila. Another consideration is due to Rams and Simon on fractal percolation and random self-similar sets. For self-similar sets of overlapping IFS, Hochman provides a heuristic argument to illustrate his in-depth result on the comparison of Hausdorff dimension and self-similar dimension. Also, Shmerkin outlines his joint work with Feng on the self-affinity dimension and its continuity on self-affine sets.

On multifractal analysis, Falconer extends a potential theoretic method and uses it to estimate the L^p-dimensions of self-affine measures, images of measures under Brownian-type processes, and moments of random cascade measures. The random cascade, introduced by Mandelbrot, is intended to give a simple tractable model to turbulence and is an origin of multifractals. In his survey, Barral gives an exposition of this theory as well as the recent development on branching random walks and directed polymers on disordered trees. Fan, on the other hand, provides a study of the multifractals from the point of view of ergodic average, especially on his recent work on multiple ergodic averages.

On the geometric measure theory, Zähle reviews the fractal extension of the curvature measures in classical convex geometry and differential geometry, and investigates the stability problem on different approximations.

Self-similarity plays a special role in harmonic analysis. Dutkay and Jorgensen give an operator theoretic approach to consider the affine function systems, and build a Cantor-wavelet analysis as an application. In another direction, Dai, He, and Lai review some of the recent development concerning the spectral problem of self-similar measures, namely the existence of exponential orthonormal bases and Fourier frames.

Analysis on metric spaces generalizes the analysis on manifolds, graphs, as well as fractals. In this exciting topic, Kigami brings in the quasisymmetric classification of metrics on self-similar sets, and applies this to estimate the quasiconformal dimension of the Sierpinski carpet. Grigor'yan, Hu, and Lau give an expository survey on the heat kernel estimates on metric measure spaces, and the new development on nonlocal Dirichlet forms. Furthermore, Grigor'yan and Huang give a detailed discussion of a stochastic completeness criterion of a nonlocal Dirichlet form based on volume growth.

We are grateful to Profs. Chi-Wai Leung, Sze-Man Ngai, Yan-Hui Qu, and Ms. Annie Wong for their help in preparing the conference, and to the Department of Mathematics of CUHK for providing the facilities and secretarial support. Finally, we would like to thank the following organizations for their generous financial support to the conference: CRS 701 of the German Research Council, Hong Kong Pei Hua Education Foundation, K. C. Wong Education Foundation, and the Hong Kong Mathematical Society.

De-Jun Feng
Ka-Sing Lau

Contents

Mandelbrot Cascades and Related Topics

Julien Barral

Abstract This article is an extended version of the talk given by the author at the conference "Advances in fractals and related topics", in December 2012 at the Chinese Hong-Kong University. It gathers recent advances in Mandelbrot cascades theory and related topics, namely branching random walks, directed polymers on disordered trees, multifractal analysis, and dynamical systems.

1 Introduction

In the late sixties, motivated by Kolmogorov's work [Kol62] on turbulence in which the "lognormal hypothesis" appeared, Mandelbrot introduced lognormal multiplicative processes (see [Man72]) to build random measures obtained as limit of martingales, and describing the distribution of the energy dissipation in intermittent turbulence. As his model turned out to be too difficult to found and study in complete rigor, Mandelbrot defined (not necessarily log-normal or conservative) multiplicative cascades on homogeneous trees [Man74a, Man74b], now called Mandelbrot cascades, to provide a mathematically easier to define and, a priori, tractable model of turbulence. Some of his conjectures on the model behavior were then proved by Kahane and Peyrière in [Kah74, Pey74, KP76], and some questions remained open for a long time. In the eighties, Kahane developed multiplicative chaos theory [Kah85, Kah87a] to give a completely rigorous framework to Mandelbrot original lognormal multiplicative processes, and go beyond. This theory (which covers the case of Mandelbrot cascades) and its applications have been particularly enriched by the regular introduction of new models of multiplicative cascades [BF05, BJM10, BJM10a, BM02, BM03, BM04, BM04a, BM09, CRV13, Fan89, Fan97, LRV00, Pey77, Pey79, RV10, WW69], and during the last five years by the rigorous

The author thanks Dr Xiong Jin for his kind help in the figures elaboration.

J. Barral (✉)
LAGA (UMR 7539), Département de Mathématiques, Institut Galilée, Université Paris 13, Paris-Sorbonne Cité, 99 avenue Jean-Baptiste Clément, 93430 Villetaneuse, France
e-mail: barral@math.univ-paris13.fr

D.-J. Feng and K.-S. Lau (eds.), *Geometry and Analysis of Fractals*,
Springer Proceedings in Mathematics & Statistics 88,
DOI 10.1007/978-3-662-43920-3_1

connexion between lognormal multiplicative chaos and KPZ relations in quantum gravity [BJRV13, BS09, DS11, KPZ98, RV08] and the connexion between this chaos and SLE curves [AJKS11, She00], as well as by fine renormalization results for degenerate multiplicative chaos [AS14, BRV12, DRSV00a, DRSV00, JW11, Mad00, Web11] (the interested reader should consult the recent survey [RV13] on lognormal multiplicative chaos for more details).

This paper focuses on Mandelbrot cascades, and aims at gathering recent advances in its theory and related topics, namely branching random walks, directed polymers on disordered trees, multifractal analysis, and dynamical systems (other surveys dealing with Mandelbrot cascades, multiplicative chaos and applications are [BFP10, Fan04, Kah91, Pey00]).

2 Mandelbrot Cascades

Consider the set $\mathscr{A} = \{0, \dots, b-1\}$, where $b \geq 2$. Set $\mathscr{A}^* = \bigcup_{n\geq 0} \mathscr{A}^n$, where, by convention, $\mathscr{A}^0$ is the singleton $\{\epsilon\}$ whose the only element is the empty word ϵ. If $w \in \mathscr{A}^*$, we denote by $|w|$ the integer such that $w \in \mathscr{A}^{|w|}$. If $n \geq 1$ and $w = w_1 \cdots w_n \in \mathscr{A}^n$ then for $1 \leq k \leq n$ the word $w_1 \cdots w_k$ is denoted by $w|k$. By convention, $w|0 = \epsilon$.

Given v and w in $\mathscr{A}^n$, $v \wedge w$ is defined to be the longest prefix common to both v and w, i.e., $v|n_0$, where $n_0 = \sup\{0 \leq k \leq n \,:\, v|k = w|k\}$.

Let $\mathscr{A}^\omega$ stand for the set of infinite sequences $w = w_1 w_2 \cdots$ of elements of $\mathscr{A}$. Also, for $x \in \mathscr{A}^\omega$ and $n \geq 0$, let $x|n$ stand for the projection of x on $\mathscr{A}^n$.

If $w \in \mathscr{A}^*$, we consider the cylinder $[w]$ consisting of infinite words in $\mathscr{A}^\omega$ whose w is a prefix.

We index the closed b-adic subintervals of $[0, 1]$ by $\mathscr{A}^*$: for $w \in \mathscr{A}^*$, we set

$$I_w = \left[\sum_{1\leq k\leq |w|} w_k b^{-k}, \sum_{1\leq k\leq |w|} w_k b^{-k} + b^{-|w|} \right].$$

Let W be a non negative random variable such that $\mathbb{E}(W) = 1$ and $\big(W(w)\big)_{w\in\mathscr{A}^*}$ a family of independent random variables, identically distributed with W. Denote by $(\Omega, \mathcal{A}, \mathbb{P})$ the probability space over which these random variables are defined.

Then the non negative martingale

$$Y_n = b^{-n} \sum_{w\in\mathscr{A}^n} W(w|1)W(w|2)\cdots W(w|n) \tag{2.1}$$

converges to a non negative random variable Y, and the sequence of Borel positive measures $(\mu_n)_{n\geq 1}$ defined on $[0, 1]$ by

$$\mu_n(\mathrm{d}x) = W(w|1)W(w|2)\cdots W(w|n)\,\mathrm{d}x \quad \text{if } x \in I_w,\ w \in \mathscr{A}^n, \tag{2.2}$$

weakly converges to a non negative measure $\mu = \mu_W$ with total mass $\|\mu\| = Y$ almost surely. In the literature, the terminology *Mandelbrot cascade*, or *Mandelbrot martingale* denotes the martingale $(Y_n)_{n\geq 1}$ or the measure-valued sequence $(\mu_n)_{n\geq 1}$, while μ is called a *Mandelbrot measure*.

For each $0 \leq k \leq b-1$, substituting $\big(W(kw)\big)_{w\in\mathscr{A}^*}$ to $\big(W(w)\big)_{w\in\mathscr{A}^*}$ yields a copy $Y(k)$ of Y, so that the $Y(k)$ are independent, and independent of $(W(0), \dots, W(b-1))$. The statistical self-similarity of the construction is summarized in the fundamental almost sure equation

$$Y = b^{-1}\sum_{k=0}^{b-1} W(k)Y(k). \tag{2.3}$$

Moreover, defining more generally

$$Y_n(w) = b^{-n}\sum_{v\in\mathscr{A}^n} W(w\cdot v|1)W(w\cdot v|2)\cdots W(w\cdot v|n), \quad (w \in \mathscr{A}^*),$$

and $Y(w) = \lim_{n\to\infty} Y_n(w)$, $Y(w)$ is a copy of $\|\mu\|$ and one gets the following multiplicative structure for the the mass of b-adic intervals:

$$\mu_{|w|+n}(I_w) = b^{-|w|}Y_n(w)\prod_{1\leq j\leq |w|} W(w|j),$$

which leads

$$\mu(I_w) = b^{-|w|}Y(w)\prod_{1\leq j\leq |w|} W(w|j), \tag{2.4}$$

since b-adic numbers cannot be atoms of μ (indeed, fixing $x_0 \in [0,1]$, and, for $\epsilon > 0$, f_ϵ a non negative function bounded by 1, with compact support in $[x_0-\epsilon, x_0+\epsilon]$ and taking the value 1 over $[x_0-\epsilon/2, x_0+\epsilon/2]$, we have $\mathbb{E}(\mu(\{x_0\})) \leq \mathbb{E}(\mu(f_\epsilon)) \leq \liminf_{n\to\infty} \mathbb{E}(\mu_n(f_\epsilon)) \leq \mathrm{Leb}(f_\epsilon) \leq 2\epsilon$, where we used the martingale property of $(\mu_n)_{n\geq 1}$).

Finally, μ possesses the statistical self-similar structure:

$$\mu_{|I_w} = \Big(b^{-|w|}\prod_{1\leq j\leq |w|} W(w|j)\Big)\mu^{(w)} \circ f_w^{-1} \quad (w \in \mathscr{A}^*), \tag{2.5}$$

where f_w is the direct similitude mapping $[0,1]$ onto I_w and $\mu^{(w)}$ is the copy of μ built from $(W(w\cdot v))_{v\in\mathscr{A}^*}$.

It turns out that μ may vanish almost surely. This had been observed by Mandelbrot, who considered the function

$$\varphi(q) = \varphi_W(q) = q - 1 - \log_b \mathbb{E}(W^q) \quad (q \in \mathbb{R}_+), \tag{2.6}$$

and conjectured that $\mathbb{P}(\mu \neq 0) > 0$ if and only if $\varphi'(1^-) > 0$. It was Kahane who proved this result in [Kah74, KP76] (this was also proved independently and a bit later by Biggins [Big77] in the more general context of the branching random walk, with a different approach; another approach is presented in [WW69]).

Theorem 1 *One has* $\mathbb{P}(\mu \neq 0) > 0$ *if and only if* $\mathbb{E}(W \log W) < \log(b)$, *i.e.* $\varphi'_W(1^-) > 0$. *Moreover, in this case,* $(Y_n)_{n\geq 1}$ *is uniformly integrable*: $\mathbb{E}(Y) = 1$.

The necessity of $\varphi'(1^-) \geq 0$ is easy to see, for otherwise since $\varphi(1) = 0$ one has $\varphi(h) > 0$ for h close to 1^-, so that

$$\mathbb{E}(Y_n^h) \leq b^{-nh} \mathbb{E} \sum_{w \in \mathscr{A}^n} W(w|1)^h W(w|2)^h \cdots W(w|n)^h = b^{-n\varphi(h)}$$

tends to 0, and $Y = 0$ almost surely by applying Fatou's lemma. The sufficiency of $\varphi'(1^-) > 0$ is also quite direct if one assumes $\varphi(h) > -\infty$ for some $h > 1$. Indeed, in this case choosing h close enough to 1^+ yields $\varphi(h) < 0$. Then, following Kahane in [KP76], if we write $Y_n = b^{-1} \sum_{k=0}^{b-1} W(k) Y_{n-1}(k)$ and take $h \in (1, 2]$, then $Y_n^h = ((b^{-1} \sum_{k=0}^{b-1} W(k) Y_{n-1}(k))^{h/2})^2 \leq b^{-h} (\sum_{k=0}^{b-1} W(k)^{h/2} Y_{n-1}(k)^{h/2})^2$, so $\mathbb{E}(Y_n^h) \leq b^{-h} b \mathbb{E}(W^h) \mathbb{E}(Y_{n-1}^h) + b^{-h} b(b-1) \mathbb{E}(W^{h/2})^2 \mathbb{E}(Y_{n-1}^{h/2})^2 \leq b^{\varphi(h)} \mathbb{E}(Y_{n-1}^h) + b^{-h} b(b-1)$ since $\mathbb{E}(W) = \mathbb{E}(Y_{n-1}) = 1$, and finally $\mathbb{E}(Y_n^h) \leq (1 - b^{\varphi(h)})^{-1} b^{-h} b\,(b-1)$ after noting that $\mathbb{E}(Y_{n-1}^h) \leq \mathbb{E}(Y_n^h)$ by Jensen's inequality. Thus $(Y_n)_{n\geq 1}$ is bounded in L^h, hence uniformly integrable. The sharp result is much more delicate.

When $\mathbb{E}(W \log W) \geq \log(b)$, i.e. $\varphi'(1^-) \leq 0$, Mandelbrot naturally asked in [Man74a] for the existence of a normalizing positive sequence $(A_n)_{n\geq 1}$ so that $(Y_n/A_n)_{n\geq 1}$ converges, at least in law, to some non degenerate random variable. He observed that the convergence in law of Y_n/A_n to some random variable Z imposes that A_{n+1}/A_n converges to a positive constant c, so that Z must satisfy the equation

$$Z \stackrel{\text{dist}}{=} \sum_{k=0}^{b-1} \frac{W(k)}{bc} Z(k), \tag{2.7}$$

where $\stackrel{\text{dist}}{=}$ means equality in distribution, the $Z(k)$ are independent, identically distributed with Z, and independent of $(W(0), \ldots, W(b-1))$.

The non trivial solutions of this equation and its generalization to the branching random walk context, also called fixed points of the smoothing transformation, have been studied intensively, starting with the fundamental paper [DL83] by Durrett and Liggett (their motivation came from interacting particle systems) and followed by regular notable advances. We will see (Sect. 3.1) that its general solutions are natural combinations of stable laws and the laws of random variables of the same nature as $\|\mu\|$, or combinations of stable laws and the laws of the total mass of *critical* Mandelbrot measures. These solutions have counterparts in terms of statistically self-similar measures, namely derivatives of stable Lévy subordinators in Mandelbrot time, or critical Mandelbrot time. These statistically self-similar measures, as well as Mandelbrot and critical Mandelbrot measures provide fundamental illustrations of the multifractal formalism, a notion that was pointed out by Frisch and Parisi to provide a fine geometric description of the energy dissipation in intermittent turbulence, with Mandelbrot measures as main illustration [FP88]. This will be explained in Sect. 5, where we present recent progress in the multifractal analysis of these objects, as well as in the control of the modulus of continuity of Mandelbrot and critical Mandelbrot measures. These moduli of continuity controls are partly based on the remarkable advances achieved in the solution to the original normalization question (see Sect. 3.2). It turns out that combining these renormalization results with fixed points of the smoothing transformation theory yields, in the case of a second order phase transition, a precise description of the asymptotic behavior of the partition functions and the limiting laws of Gibbs measures associated with polymers on disordered trees expressed in terms of the statistically self-similar measures mentioned above (Sect. 4).

In fact, in his notes on multiplicative cascades [Man74a], Mandelbrot starts by raising the following general problem: assume only that W is a random variable taking values in $\mathbb{R}$, and consider the sequence of functions

$$F_n(x) = F_{W,n}(x) = \int_0^x Q_n(u)\,\mathrm{d}u, \tag{2.8}$$

where $Q_n(x) = W(w|1)W(w|2)\cdots W(w|n)\,\mathrm{d}x$ if $x \in I_w,\ w \in \mathscr{A}^n$; under which condition does there exist a normalizing sequence $(A_n)_{n\geq 1}$ such that $F_n(x)/A_n$ converges in law, or in a stronger sense?

We will give some results in this direction in Sect. 6. In Sect. 7 we explain how Mandelbrot cascades define a natural dynamical system on fixed points of the smoothing transformation with finite expectation, to which is associated a functional CTL whose limit process is obtained as the limit of an *additive cascade* on $\mathscr{A}^*$.

Finally, let us mention that a lot of results presented in this paper are not specific to multiplicative cascades on regular trees, and have extensions in the context of branching random walks on Galton-Watson trees. Also, while we consider measures on the interval $[0, 1]$, the results of Sect. 5 can be extended to Mandelbrot measures on $[0, 1]^d$ ($d \geq 2$). The only difference in that one must use a different way to define the Lévy-Mandelbrot measures defined in Sect. 3.1.2; the procedure is explained in [BJRV13, BRV12].

In Sects. 3–5, to simplify the discussion we assume that $W > 0$ and the probability distribution of $\log(W)$ is non lattice.

3 Fixed Points of the Smoothing Transformation and Associated Statistically Self-similar Measures; Renormalization of Mandelbrot Cascades

This section gathers old and new information related to the solutions of the functional equation pointed out by Mandelbrot in connexion with the renormalization of the martingale $(Y_n)_{n\geq 1}$ when $\varphi'(1) \leq 0$.

3.1 Fixed Points of the Smoothing Transformation

If Z is a non negative random variable, denote by ϕ_Z the Laplace transform of its probability distribution.

Consider the positive random variable W and do not assume a priori that $\mathbb{E}(W) = 1$. Then consider the equation

$$Z \stackrel{\text{dist}}{=} b^{-1} \sum_{k=0}^{b-1} W(k)Z(k), \tag{3.1}$$

where the random variables $W(k)$, $0 \leq k \leq b-1$, are independent copies of W, the random variables $Z(k)$ are independent copies of the non negative random variable Z, and $(W(0), \dots, W(b-1))$ and $(Z(0), \dots, Z(b-1))$ are independent. If Z is not identically equal to 0, then it is easily deduced from (3.1) that in fact Z is positive almost surely; moreover any positive multiple of Z satisfies (3.1).

In terms of Laplace transform, (3.1) means

$$\phi_Z = \left(\mathbb{E}(\phi_Z(b^{-1}tW))\right)^b, \tag{3.2}$$

so that ϕ_Z is a fixed point of the *smoothing transformation* T_W defined on the space of Laplace transforms of probability distributions on $\mathbb{R}_+$ as

$$T_W(\phi) = (\mathbb{E}(\phi(b^{-1}tW)))^b.$$

As indicated in the previous section, non trivial solutions of Eq. (3.1) have been studied intensively [DL83, Gui90, BK97, Liu98, Liu00, Liu01, BK04, BK05, AM12, ABM12] (see also [AM13] for the study of non necessarily positive solutions, still with $W > 0$).

Let φ be defined as in Sect. 5.1.3.

Theorem 2 *Equation* (3.1) (*or equivalently* (3.2)) *has a non trivial solution if and only if there exists* $\alpha \in (0, 1]$ *such that* $\varphi_W(\alpha) = 0$ *and* $\varphi'_W(\alpha^-) \geq 0$. *Moreover, if* Z_1 *and* Z_1 *are two such solutions, there exists* $c > 0$ *such that* $Z_1 \stackrel{\text{dist}}{=} cZ_2$.

This was proved by Durrett and Liggett under the assumption $\varphi(1+\epsilon) > -\infty$ for some $\epsilon > 0$. This assumption has been removed thanks to recent advances due to Alsmeyer, Biggins, Kyprianou and Meiners works in [ABM12, AM12, BK04, BK05]. We do not try to outline the extremely involved proof of this result; we just mention that renewal theory and fluctuation theory of random walks on $\mathbb{R}$ play a central role here.

It turns out that assuming that $\varphi''(\alpha^-) > -\infty$ whenever $\varphi'(\alpha^-) = 0$ simplifies the description of the solutions. This is what we will do in the next subsections.

3.1.1 Special Solutions from Mandelbrot and Critical Mandelbrot Cascades When $\alpha = 1$

For $q \in \mathbb{R}_+$ and $n \geq 1$ set

$$Y_{q,n} = b^{n(\varphi(q)-q)} \sum_{w \in \mathscr{A}^n} W(w|1)^q W(w|2) \cdots W(w|n)^q,$$

and

$$\widetilde{Y}_n = -\frac{\mathrm{d}Y_{q,n}}{\mathrm{d}q}\Big|_{q=1}.$$

Notice that if $\varphi(1) = 0$, i.e. $\mathbb{E}(W) = 1$, then $Y_{1,n} = Y_n$ (see (2.1)). If, moreover, $\varphi'(1) = 0$, we are in the critical case of degeneracy of Mandelbrot cascades, and $\widetilde{Y}_n$ takes the form

$$\widetilde{Y}_n = -\sum_{w \in \mathscr{A}^n} (b^{-n} W(w|1)W(w|2) \cdots W(w|n)) \log(b^{-n} W(w|1)W(w|2) \cdots W(w|n));$$

in this case it is a martingale with respect to $(\sigma(W(w) : w \in \mathscr{A}^k,\ 1 \leq k \leq n))_{n \geq 1}$ and it is called *derivative martingale.*

The following result gathers information about the non trivial solutions of (3.1) when $\varphi(1) = 0$ and $\varphi'(1) \leq 0$.

Theorem 3 *Assume that* $\varphi(1) = 0$.

(1) *Suppose that* $\varphi'(1^-) > 0$. *Let* Y *be the almost sure limit of* $(Y_n)_{n \geq 1}$ *given by Theorem 1.*

   (a) *For* $q > 1$ *one has* $\mathbb{E}(Y^q) < \infty$ *if and only if* $\varphi(q) > 0$;
   (b) *If* $\varphi(q_0) = 0$ *and* $\varphi'(q_0) > -\infty$ *for some* $q_0 > 1$ *(such a* q_0 *is necessarily unique due to the concavity of* φ*), one has* $\mathbb{P}(Y > x) \sim_\infty Ax^{-q_0}$ *for some* $A > 0$. *(c) For* $q < 0$, $\varphi(q/b) > -\infty$ *implies* $E(Y^{q'}) < \infty$ *for all* $q' \in (q, 0)$, *and* $E(Y^q) < \infty$ *implies* $\varphi(q/b) > -\infty$.

Moreover, any non trivial solution Z *of* (3.1) *satisfies* $Z \overset{dist}{=} cY$ *for some* $c > 0$.

(2) *Suppose that* $\varphi'(1^-) = 0$ *and* $\varphi''(1^-) > -\infty$. *Then* $(\widetilde{Y}_n)_{n\geq 1}$ *is a martingale which converges almost surely to a positive random variable* $\widetilde{Y}$ *satisfying the following properties*:

   (a) *for* $q > 0$ *one has* $\mathbb{E}(Y)^q < \infty$ *if and only if* $q < 1$;
   (b) $\phi_{\widetilde{Y}}(t) \sim_{0^+} 1 - ct\log(1/t)$ *for some* $c > 0$;
   (c) *if* $\varphi(1+\epsilon) > -\infty$ *for some* $\epsilon > 0$, $\mathbb{P}(\widetilde{Y} > x) \sim_\infty Ax^{-1}$ *for some* $A > 0$.
   (d) *For* $q < 0$, $\varphi(q/b) > -\infty$ *implies* $E(Y^{q'}) < \infty$ *for all* $q' \in (q, 0)$, *and* $E(Y^q) < \infty$ *implies* $\varphi(q/b) > -\infty$.

Moreover, any non trivial solution Z *of* (3.1) *satisfies* $Z \stackrel{\text{dist}}{=} c\widetilde{Y}$ *for some* $c > 0$.

Under the assumption that $\varphi(1+\epsilon) > -\infty$ for some $\epsilon > 0$, Durrett and Liggett showed that when $\varphi(1) = 0$ and $\varphi'(1^-) = 0$, Eq. (3.2) possesses non trivial solutions satisfying $\phi(t) \sim_{0^+} 1 - ct\log(1/t)$ for some $c > 0$, and unique up to a scaling factor in their argument. Then Liu [Liu98] used this behavior of ϕ at 0^+ to identify the solutions as the Laplace transform of $\widetilde{Y}$ by proving that $\widetilde{Y}_n$ converges to a non trivial limit (the point is that for any fixed $t > 0$, $\prod_{w\in\mathscr{A}^n}\phi(tW(w|1)W(w|2)\cdots W(w|n))$ is a bounded non negative martingale with expectation in $(0, 1)$, which is asymptotically equivalent to $e^{-tc\widetilde{Y}_n}$). The weaker condition used in Theorem 3 is established in [BK04, BK05].

The necessary and sufficient condition for the finitness of moments of order greater than 1 when $\varphi'(1) > 0$ (conjectured in [Man74a]) was established by Kahane in [Kah74, KP76] by generalizing the argument presented just after Theorem 1. The right tail behaviors of solutions when $\varphi'(1) > 0$ and $\varphi'(1) = 0$ are due Guivarc'h [Gui90] (conjectured in [Man74a]) and Buraczewski [Bur09] respectively. The proofs strongly rely on random difference equations and renewal theories.

The result on moments of negative orders is first established in [Kah91] (see also [Liu01, Mol96]). Its proof consists in estimating from (3.2) the asymptotic behavior of ϕ_Y or $\phi_{\widetilde{Y}}$ at ∞.

Associated statistically self-similar measures Under the assumptions of Theorem 3(1), we already now the Mandelbrot measure μ.

Under the assumptions of Theorem 3(2), defining

$$\widetilde{Y}_n(w) = -\sum_{v\in\mathscr{A}^n} \frac{W(w\cdot v|1)\cdots W(w\cdot v|n)}{b^n}\log\frac{W(w\cdot v|1)\cdots W(w\cdot v|n)}{b^n} \quad (w\in\mathscr{A}^*)$$

and $\widetilde{Y}(w) = \lim_{n\to\infty}\widetilde{Y}_n(w)$, $\widetilde{Y}(w)$ is a copy of $\|\widetilde{\mu}\|$ and

$$\widetilde{\nu}([w]) = b^{-|w|}\widetilde{Y}(w)\prod_{1\leq j\leq|w|} W(w|j), \tag{3.3}$$

defines almost surely a measure on $\mathscr{A}^\omega$. One can show (see [Bar00]) $\widetilde{\nu}$-amost every $t\in\mathscr{A}^\omega$ is normal, so that $\widetilde{\nu}$ has no atom in the set of infinite branches of $\mathscr{A}^\omega$

encoding b-adic numbers of [0, 1]. Thus $\widetilde{\nu}$ naturally projects on [0, 1] to a measure $\widetilde{\mu} = \widetilde{\mu}_W$, called *critical Mandelbrot measure*, such that for any $w \in \mathscr{A}^*$,

$$\widetilde{\mu}_{|I_w} = \Big(b^{-|w|} \prod_{1\le j\le |w|} W(w|j)\Big)\widetilde{\mu}^{(w)} \circ f_w^{-1} \quad (w \in \mathscr{A}^*), \tag{3.4}$$

where $\widetilde{\mu}^{(w)}$ is the copy of $\widetilde{\mu}$ built from $(W(w\cdot v))_{v\in\mathscr{A}^*}$.

3.1.2 General Form of the Solutions

Theorem 4 *Assume that $\varphi(\alpha) = 0$ and $\varphi'(\alpha^-) \ge 0$ for some $\alpha \in (0, 1]$ (this α is then unique by concavity of φ). Let L_α be a stable Lévy subordinator of index α if $\alpha \in (0, 1)$ and the identity map of $\mathbb{R}_+$ if $\alpha = 1$. Assume that L_α is independent of $(W(w))_{w\in\mathscr{A}^*}$. Let $W_\alpha = W^\alpha/\mathbb{E}(W^\alpha)$.*

(1) *Suppose that $\varphi'(\alpha^-) > 0$. Then W_α satisfies the assumptions of Theorem* 3(1). *Let Y_α be the limit of the associated Mandelbrot cascade built from $(W_\alpha(w))_{w\in\mathscr{A}^*}$. Any non trivial solution Z of* (3.1) *satisfies $Z \overset{dist}{=} cL_\alpha(Y_\alpha)$ for some $c > 0$. Moreover, $\phi_Z(t) \sim_{0^+} 1 - c't^\alpha$ for some $c' > 0$.*
(2) *Suppose that $\varphi'(\alpha^-) = 0$ and $\varphi''(\alpha^-) > -\infty$. Then W_α satisfies the assumptions of Theorem* 3(2). *Let $\widetilde{Y}_\alpha$ be the limit of the associated derivative martingale built from $(W_\alpha(w))_{w\in\mathscr{A}^*}$. Any non trivial solution Z of* (3.1) *satisfies $Z \overset{dist}{=} cL_\alpha(\widetilde{Y}_\alpha)$ for some $c > 0$. Moreover, $\phi_Z(t) \sim_{0^+} 1 - c't^\alpha \log(1/t)$ for some $c' > 0$.*

Durrett and Liggett obtained the same sets of solutions under the stronger assumption mentioned above, and the sharp result, without the assumption $\varphi''(\alpha^-) > -\infty$ when $\varphi'(\alpha^-) = 0$, is obtained in [ABM12]. Notice that checking that $cL_\alpha(Y_\alpha)$ is solution of (3.1) in case (1) and $cL_\alpha(\widetilde{Y}_\alpha)$ is solution of (3.1) in case (2) is a simple exercise using that $\phi_{L_\alpha}(t) = e^{-\gamma t^\alpha}$ for some $\gamma > 0$.

Associated statistically self-similar measures Denote by μ and $\widetilde{\mu}$ the Mandelbrot measure and critical Mandelbrot measure considered at the end of Sect. 3.1.1. Let $\alpha \in (0, 1)$ and L_α an α-stable Lévy subordinator independent of $(W(w))_{w\in\mathscr{A}^*}$. Then let μ^α (resp. $\widetilde{\mu}^\alpha$) be the positive measure obtained as the derivative of $L_\alpha(\mu([0, \cdot]))$ (resp. $L_\alpha(\widetilde{\mu}([0, \cdot]))$). Let us also call μ^α a *Lévy-Mandelbrot measure* and $\widetilde{\mu}^\alpha$ a *critical Lévy-Mandelbrot measure*. These measures are statistically self-similar in the sense that we have for any $w \in \mathscr{A}^*$

$$(\mu^\alpha)_{|I_w} \overset{\text{dist}}{=} \Big(b^{-|w|} \prod_{1\le j\le |w|} W(w|j)\Big)^{1/\alpha} \mu^{\alpha,(w)} \circ f_w^{-1} \quad (w \in \mathscr{A}^*), \tag{3.5}$$

where $\mu^{\alpha,(w)}$ is a copy of μ^α independent of $\prod_{1\le j\le |w|} W(w|j)$, and the same holds for $\widetilde{\mu}_\alpha$.

In the context of Theorem 4, (3.5) reads

$$(\mu^{\alpha}_{W_\alpha})_{|I_w} \overset{\text{dist}}{=} \Big(b^{-|w|} \prod_{1\le j\le |w|} W(w|j)\Big)\mu^{\alpha,(w)}_{W_\alpha} \circ f_w^{-1} \quad (w \in \mathscr{A}^*),$$

which is similar to (2.5) and (3.4).

3.2 Renormalization of Mandelbrot Cascades

We now suppose that $\varphi(1) = 0$ and $\varphi'(1^-) \le 0$ and come to the existence of a sequence $(A_n)_{n\ge 1}$ such that Y_n/A_n converges in law to a non trivial limit. When $\log W$ is Gaussian, the solution to this question has been conjectured by Derrida and Spohn [DS88] in the context of directed polymers on disordered trees and rigorously established recently by Webb [Web11], while in the general result presented below, the first part has been obtained by Aidekon and Shi [AS14], and the second one combines a convergence in law result obtained by Madaule [Mad00] with an identification of the limiting law using Theorem 4.

Theorem 5

(1) *Suppose that* $\varphi(1) = 0$, $\varphi'(1^-) = 0$ *and* $\varphi''(1^-) > -\infty$. *Then* $(\sqrt{n}\, Y_n)_{n\ge 1}$ *converges in probability to* $\sqrt{\dfrac{-2}{\pi\varphi''(0)}}\,\widetilde{Y}$.

(2) *Suppose that* $\varphi(1) = 0$ *and* $\varphi'(1^-) < 0$. *Let* α *be the unique solution of* $\varphi'(\alpha) = \varphi(\alpha)/\alpha$ *in* $(0,1)$. *The random variable* $W_\alpha = W^\alpha/\mathbb{E}(W^\alpha)$ *satisfies the assumptions of Theorem* 4(2). *Let* $\widetilde{Y}_\alpha$ *be the associated limit of the derivative martingale built from* $(W_\alpha(w))_{w\in\mathscr{A}^*}$.

The sequence $((n^{3/2}b^{n(\varphi(\alpha)-\alpha)})^{1/\alpha}\, Y_n)_{n\ge 1}$ *converges in law to* $L_\alpha(\widetilde{Y}_\alpha)$, *where* L_α *is a stable Lévy subordinator of index* α *independent of* $\widetilde{Y}_\alpha$.

Theorem 5(2) is a special case of a more general renormalization results for Y_n when one assumes only that $W > 0$ and there exists $\alpha \in (0,1)$ such that $\varphi'(\alpha) = \varphi(\alpha)/\alpha$ (see Sect. 4.2).

The limiting law of μ_n/A_n is described in the next section.

4 Directed Polymers on $\mathscr{A}^*$: Partition Functions, Free Energies and Gibbs Measures

Here we consider the random variables $(W(w))_{w\in\mathscr{A}^*}$ of the previous sections and we only assume that $W > 0$. If we define the potential $V = \log(b) - \log W$ and set $V(w) = \log(b) - \log W(w)$ $(w \in \mathscr{A}^*)$, in the setting of [DS88] the branching random walk $H(w) = V(w|1) + \cdots + V(w|n)$, $w \in \mathscr{A}^n$, defines for each $n \geq 1$ a polymer on the tree $\mathscr{A}^n$ in the random medium $(V(w))_{w \in \bigcup_{1 \leq k \leq n} \mathscr{A}^n}$ by associating the energy $H(w)$ to each path of length n; moreover, this model possesses *logarithmic correlations*.

Then, of first importance is the asymptotic behavior, as n tends to ∞, of thermodynamical objects such as the partition function

$$Z_n(\beta) = \sum_{w \in \mathscr{A}^n} e^{-\beta H(w)} \quad (\beta \geq 0),$$

the free energy

$$P_n(\beta) = \frac{\log Z_n(\beta)}{n \log(b)},$$

and the Gibbs measures defined on [0, 1] by

$$\nu_{\beta,n}(\mathrm{d}x) = b^{-n} \frac{e^{-\beta H(w)}}{Z_n(\beta)} \mathrm{d}x \quad \text{if } x \in I_w \text{ and } w \in \mathscr{A}^n,$$

where β stands for the inverse of the temperature.

Writing $e^{-\beta H(w)} = (b^{-n} W(w|1) \cdots W(w|n))^\beta$ shows the direct connexion with Mandelbrot cascades.

The understanding of these asymptotic behaviors has made enormous progress in the recent years. We still define $\varphi(\beta)$ as in Sect. 5.1.3.

Four different situations can occur; they are described in the Sects. 4.1–4.4 and depend on the behavior of φ at

$$\beta_c = \sup\{\beta > 0 : \varphi'(\beta^-)\beta - \varphi(\beta) > 0\},$$

with the convention $\sup(\emptyset) = 0$.

For $\beta \geq 0$ and $n \geq 1$, we denote by $\mu_{\beta,n}$ the measure defined by (2.2) when the weight W is taken equal to

$$W_\beta = \frac{W^\beta}{\mathbb{E}(W^\beta)}. \tag{4.1}$$

Notice that by construction we have

$$\|\mu_{\beta,n}\| = \frac{Z_n(\beta)}{\mathbb{E}(Z_n(\beta))} \quad \text{and} \quad \varphi'_{W_\beta}(1^-) = \varphi'(\beta^-)\beta - \varphi(\beta).$$

Section 4.5 presents a unified result for the free energy behavior.

4.1 No Phase Transition: $\beta_c = +\infty$

Theorem 6 *With probability 1, for all $\beta \in [0, \beta_c)$, $(\mu_{\beta,n})_{n\geq 1}$ weakly converges to a non degenerate Mandelbrot measure μ_β.*

Consequently, $\left(\frac{Z_n(\beta)}{\mathbb{E}(Z_n(\beta))}\right)_{n\geq 1}$ converges to $\|\mu_\beta\|$, $P_n(\cdot)$ converges to the analytic function $-\varphi(\cdot)$ uniformly on compact subintervals of $[0, \beta_c)$, and $(\nu_{\beta,n})_{n\geq 1}$ weakly converges to $\frac{\mu_\beta}{\|\mu_\beta\|}$, as $n \to \infty$.

That the result holds for each fixed β almost surely essentially follows from Theorem 1 applied when W is taken equal to W_β. The uniform version of this result essentially follows from [Big92] and [Bar00]. Biggins considers in [Big92] the analytic extensions of the mappings $\mathscr{Y}_n : \beta \mapsto \frac{Z_n(\beta)}{\mathbb{E}(Z_n(\beta))}$ in a complex neighborhood of $\{\beta : \varphi'(\beta^-)\beta - \varphi(\beta) > 0\}$. He proves the almost sure uniform convergence of these extensions on a common domain U of $\mathbb{C}$ by combining Cauchy's formula with the fact that, for each compact subset of U, $\mathscr{Y}_n(z) - \mathscr{Y}_{n-1}(z)$ converges uniformly exponentially to 0 in L^p for some $p > 1$. It remains to prove that almost surely, for all β such that $\varphi'(\beta^-)\beta - \varphi(\beta) > 0$, we have $\lim_{n\to\infty} \mathscr{Y}_n(\beta) > 0$; this is done in [Bar00] (Fig. 1).

4.2 Second Order Phase Transition: $\beta_c \in (0, \infty)$ and $\varphi'_{W_{\beta_c}}(1^-) = \varphi'(\beta_c^-)\beta_c - \varphi(\beta_c) = 0$

This situation is illustrated by the case where the potential V is Gaussian (Fig. 2).

Theorem 7

(1) *The same conclusions as in Theorem 6 hold over $[0, \beta_c)$.*
(2) *Suppose that $\mathbb{E}(W^{\beta_c}|\log(W)|^2) < \infty$, i.e. $\varphi''_{W_{\beta_c}}(1^-) > -\infty$. Let $\widetilde{\mu}_{\beta_c}$ be the critical Mandelbrot measure built from $(W_{\beta_c}(w))_{w\in\mathscr{A}^*}$ in Sect. 3.1.1. Then $\left(\sqrt{n}\frac{Z_n(\beta_c)}{\mathbb{E}(Z_n(\beta_c))}\right)_{n\geq 1}$ converges in probability to $c\|\widetilde{\mu}_{\beta_c}\|$ for some explicit $c > 0$, and $(\nu_{\beta_c,n})_{n\geq 1}$ weakly converges in probability to $\frac{\widetilde{\mu}_{\beta_c}}{\|\widetilde{\mu}_{\beta_c}\|}$ as $n \to \infty$.*
(3) *Suppose that $\mathbb{E}(W^{\beta_c}|\log(W)|^3) < \infty$. For $\beta > \beta_c$, let $\widetilde{\mu}_{\beta_c}^{\frac{\beta_c}{\beta}}$ be the critical Lévy-Mandelbrot measure built from $\widetilde{\mu}_{\beta_c}$ and a stable Lévy subordinator of index β_c/β in Sect. 3.1.2.*

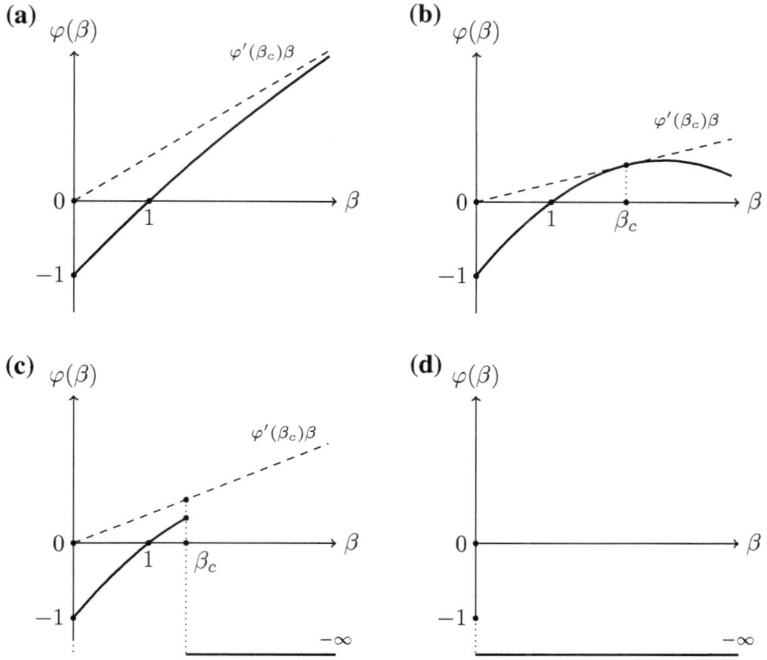

Fig. 1 The four possible behaviors of φ (by design, when $\beta_c > 0$, we make the pictures with $\varphi(1) = 0$ and $\varphi'(1) > 0$ to link these behaviors with multifractal analysis of Mandelbrot measures in Sect. 5). **a** No phase transition. **b** Second order phase transition. **c** First order phase transition. **d** The degenerate case

Then, for all $\beta > \beta_c$, $\left(n^{\frac{3}{2}\frac{\beta}{\beta_c}} \dfrac{Z_n(\beta)}{(\mathbb{E}(Z_n(\beta_c)))^{\frac{\beta}{\beta_c}}}\right)_{n\geq 1}$ *weakly converges in distribution to* $c\|\widetilde{\mu}_{\beta_c}^{\frac{\beta_c}{\beta}}\|$ *for some* $c > 0$, *and* $(\nu_{\beta,n})_{n\geq 1}$ *weakly converges in distribution to* $\widetilde{\mu}_{\beta_c}^{\frac{\beta_c}{\beta}}/\|\widetilde{\mu}_{\beta_c}^{\frac{\beta_c}{\beta}}\|$ *as* $n \to \infty$.

(4) $P_n(\cdot)$ *converges almost surely, uniformly over the compact subsets of* $\mathbb{R}_+$ *to*

$$P(\beta) = \begin{cases} -\varphi(\beta) & \text{if } 0 \leq \beta \leq \beta_c, \\ \dfrac{-\varphi(\beta_c)}{\beta_c}\beta = -\varphi'(\beta_c^-)\beta & \text{if } \beta > \beta_c \end{cases}.$$

Thus P *is analytic except at* β_c *where it is differentiable with a discontinuity of its second derivative, hence it presents a second order phase transition at* β_c.

Part (1) is obtained in a similar way as in the previous case.

For parts (2) and (3), the convergences of the renormalized partition functions are established in [Web11] for the log-normal case (with convergence in law for the

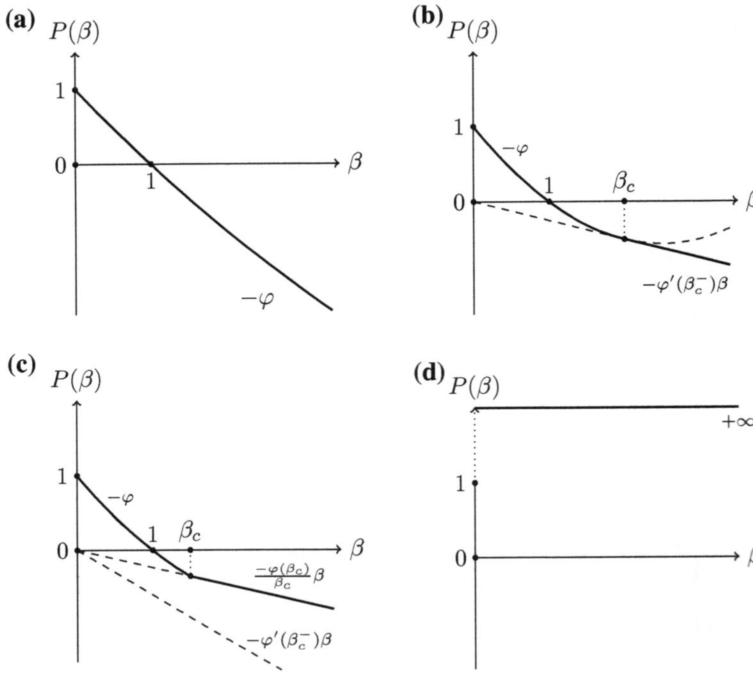

Fig. 2 The corresponding pressure function. **a** No phase transition. **b** Second order phase transition. **c** First order phase transition. **d** The degenerate case

part (2)), and [AS14] and [Mad00] for part (2) and (3) respectively. The identification of the limit in case (3) follows easily from (2.7) and Theorem 4.

In part (3), the exponent $\frac{3}{2}$ is reminiscent from the asymptotic behavior of $M_n = \min\{\beta_c H(w) - n\varphi(\beta_c) : w \in \mathscr{A}^n\}$, for which one has, due to [Web11] for the log-normal case, and [Aid13] for the general case.

Theorem 8 *Suppose that* $\mathbb{E}(W^{\beta_c}|\log(W)|^2) < \infty$. *Then, there exists* $c > 0$ *such that for all* $x \in \mathbb{R}$,

$$\lim_{n\to\infty} \mathbb{P}(M_n \geq \frac{3}{2}n + x) = \mathbb{E}(e^{-ce^x\|\widetilde{\mu}_{\beta_c}\|}).$$

The asymptotic behavior of the Gibbs measures reflects the fact that when the temperature $1/\beta$ becomes lower than $1/\beta_c$, the systems is in its glassy phase: asymptotically with n, the energy concentrates around a few local minima of $H(x|n)$, a phenomenon amplified as β tends to ∞, and also called freezing transition (see [CD01]).

It is out of our scope to describe the techniques developed in [Aid13, AS14, Mad00, Web11]. The convergence of the Gibbs measures in case (2) and (3) requires additional arguments provided in [JW11] for the case $\beta = \beta_c$, and in [BRV12] for $\beta \geq \beta_c$.

Regarding point (4), this result or its analogue for Mandelbrot measures, has been obtained by several authors [CK92, Fra95, HW92, Mol96, OW00, WW13]. Due to (1), only the case $\beta \geq \beta_c$ must be considered. Due to the convexity of the functions P_n, it is quite direct that $\liminf_{n\to\infty} P_n(\beta) \geq -\varphi'(\beta_c^-)\beta$ for $\beta \geq \beta_c$. Moreover, for $\beta \geq \beta_c$, if $0 < \beta' < \beta_c$, we have $Z_n(\beta) = \sum_{w\in\mathscr{A}^n} e^{-\beta H(w)} = \sum_{w\in\mathscr{A}^n} (e^{-\beta' H(w)})^{\beta/\beta'} \leq Z_n(\beta')^{\beta/\beta'}$. Consequently, since $P_n(\beta')$ converges to $-\varphi(\beta')$, we get $\limsup_{n\to\infty} P_n(\beta) \leq \frac{\beta}{\beta'}(-\varphi(\beta'))$. Letting β' tend to β_c yields the result.

4.3 First Order Phase Transition: $\beta_c \in (0, \infty)$ and $\varphi'_{W_{\beta_c}}(1^-) = \varphi'(\beta_c^-)\beta_c - \varphi(\beta_c) > 0$

Notice that in this case one necessarily has $\varphi(\beta) = -\infty$ for $\beta > \beta_c$. Also, the measures $\mu_{\beta_c,n}$ weakly converge almost surely to the non degenerate Mandelbrot measure μ_{β_c}, for which Theorem 1 is optimal in the sense that $\varphi_{W_{\beta_c}}(\beta) = -\infty$ for $\beta > 1$, hence in this case $(\|\mu_{\beta_c,n}\|)_{n\geq 1}$ is not bounded in any L^q, $q > 1$.

We have the following result.

Theorem 9

(1) *The same conclusions as in Theorem* 6 *hold over* $[0, \beta_c]$.
(2) *With probability* 1, $P_n(\cdot)$ *converges almost surely, uniformly over the compact subsets of* $\mathbb{R}_+$ *to*

$$P(\beta) = \begin{cases} -\varphi(\beta) & \text{if } 0 \leq \beta \leq \beta_c, \\ \dfrac{-\varphi(\beta_c)}{\beta_c}\beta > -\varphi'(\beta_c^-)\beta & \text{if } \beta > \beta_c \end{cases}.$$

Thus P is analytic except at β_c where it is continuous and not differentiable, hence it presents a first order phase transition at β_c.

Remark 1 The expression of P shows that the concave Legendre-Fenchel transform of $-P$, i.e. the mapping $\alpha \in \mathbb{R} \mapsto \inf\{\alpha\beta + P(\beta) : \beta \geq 0\}$ is non negative at $\alpha = \varphi(\beta_c)/\beta_c$. This proves that $|\beta_c H(w) - n\varphi(\beta_c)|$ behaves sub-linearly at some $w \in \mathscr{A}^n$, like in the case of the previous section.

Consequently, this raises the following questions: does $\min\{\beta_c H(w) - n\varphi(\beta_c) - s_n : w \in \mathscr{A}^n\}$ converge in law for some sub-linear sequence $(s_n)_{n\geq 1}$? If so, is there a freezing phenomenon like in the previous case?

Then, one can wondering if for $\beta > \beta_c$, $\left(e^{\frac{\beta}{\beta_c}s_n}\dfrac{Z_n(\beta)}{(\mathbb{E}(Z_n(\beta_c)))^{\frac{\beta}{\beta_c}}}\right)_{n\geq 1}$ weakly converges in distribution to $c\|\mu_{\beta_c}^{\frac{\beta_c}{\beta}}\|$ for some $c > 0$, and $(\nu_{\beta,n})_{n\geq 1}$ weakly converges in distribution to $\mu_{\beta_c}^{\frac{\beta_c}{\beta}}/\|\mu_{\beta_c}^{\frac{\beta_c}{\beta}}\|$ as $n \to \infty$, where $\mu_{\beta_c}^{\frac{\beta_c}{\beta}}$ is the Lévy-Mandelbrot measure built from μ_{β_c} and a stable Lévy subordinator of index β_c/β in Sect. 3.1.2.

The upper bound for $\limsup_{n\to\infty} P_n(\beta)$ over (β_c, ∞) is obtained like in the previous case. The lower bound for $\liminf_{n\to\infty} P_n(\beta)$ is more involved. It was obtained by Molchan in [Mol96] only along a deterministic subsequence. However, we believe that Molchan's approach can give the convergence along the whole sequence. Anyway, in [AB14], an alternative approach gives the result in connection with the multifractal analysis of $H(w)$ (see Sect. 5.1.3).

4.4 The Degenerate Case: $\beta_c = 0$, i.e. $\varphi = -\infty$ Over $\mathbb{R}_+^*$

In this case, we have:

Theorem 10 *With probability* 1, P_n *converges pointwise to* $-\varphi$ *as* $n \to \infty$.

This convergence result is shown in [Mol96] along a deterministic subsequence, and in this form in [AB14].

Remark 2 One can wonder if there is a precise superexponential speed of divergence of $Z_n(\beta)$ to ∞ when $\beta > 0$.

4.5 A Uniform Point of View for the Free Energy

We can deduce from the previous results the following synthetic presentation:

Theorem 11 *With probability* 1, $\lim_{n\to\infty} P_n(\beta) = \beta \inf\{-\varphi(\beta')/\beta' : \beta' \in (0, \beta]\} = \inf\{-\varphi(\theta\beta)/\theta : \theta \in (0, 1]\}$ *for all* $\beta > 0$.

5 Fine Geometric Properties of Statistically Self-similar Measures

This section presents recent results about the modulus of continuity of Mandelbrot and critical Mandelbrot measures, as well as recent progress in their multifractal analysis, with consequences for the multifractal analysis of their discrete companions, i.e. Lévy-Mandelbrot and critical Lévy-Mandelbrot measures. We also give results about the dimension of these measures. Finally, we say a word about KPZ formula.

5.1 *Dimension, Modulus of Continuity, and Multifractal Analysis of Mandelbrot and Critical Mandelbrot Measures*

Here we suppose that $\mathbb{E}(W) = 1$.

5.1.1 Dimension

Recall that a positive and finite Borel measure ν on $[0, 1]$ is said to be exact dimensional with dimension D if $\lim_{r\to 0^+} \frac{\log(\mu(B(x,r)))}{\log(r)} = D$, ν-almost everywhere, or equivalently $\lim_{n\to\infty} \frac{\log(\mu(I_n(x)))}{-n\log(b)} = D$, ν-almost everywhere, where $I_n(x)$ stands for the closure of the semi-open to the right b-adic interval of generation$x$ which contains x when $x \in [0, 1)$ and $[1 - b^{-n}, 1]$ if $x = 1$.

Theorem 12

(1) *Suppose that* $\varphi'(1^-) > 0$. *Then the Mandelbrot measure* μ *is exact dimentional with dimension* $\varphi'(1^-)$*; in particular it is continuous.*
If, moreover, $\mathbb{E}(W|\log W|^3) < \infty$ *and* $-\varphi''(1^-) > 0$ *(i.e.* $W \overset{dist}{\not\equiv} 1$*), then for* μ*-almost every* x *we have*

$$\liminf_{n\to\infty}(\mathit{resp.}\ \limsup_{n\to\infty})\frac{\log(\mu(I_n(x))) + n\log(b)\varphi'(1^-)}{\sqrt{2\log(b)(-\varphi''(1^-))n\log\log(n)}} = -1(\mathit{resp.}\ -1).$$

(2) *Suppose that* $\varphi'(1^-) = 0$ *and* $\varphi''(1^-) > -\infty$. *Then the critical Mandelbrot measure* $\widetilde{\mu}$ *is exact dimensional with dimension 0. If, moreover,* $\varphi(1+\epsilon) > -\infty$ *for some* $\epsilon > 0$, *then* $\widetilde{\mu}$ *is continuous.*
Also, if $\log(W)$ *is Gaussian, for* $\widetilde{\mu}$*-almost every* x*, for all* $\alpha > 1/3$ *and* $k \in \mathbb{N}$*, for* n *large enough, we have*

$$\exp\left(-\sqrt{6\log(b)}\sqrt{n}\,(\log n + \alpha\log\log n)\right) \leq \widetilde{\mu}(I_n(x)) \leq n^{-k}. \qquad (5.1)$$

For part (1), the fact that μ is exact dimensional, conjectured in [Man74b], was proved by Peyrière in [KP76, Pey74] under the assumption $\mathbb{E}(Y\log^+(Y)) < \infty$, which is shown to be equivalent to $\varphi''(1) > -\infty$ in [Big79]. In [Kah87], Kahane used a powerful percolation argument combined with Theorem 1 to eliminate the assumption $\mathbb{E}(Y\log^+(Y)) < \infty$.

The law of the iterated logarithm is stated in [Liu00] assuming $\mathbb{E}(W|\log W|^2) < \infty$. To get such a law, consider the now called "Peyrière measure" $\mathbb{Q}$ introduced in [Pey74] and defined on $(\Omega \times [0, 1], \boldsymbol{\mathcal{A}} \otimes \boldsymbol{\mathcal{B}}([0, 1]))$ by

$$\mathbb{Q}(A) = \mathbb{E}\left(\int_{[0,1]} \mathbf{1}_A(\omega, x)\, \mu(\mathrm{d}x)\right),$$

so that "$\mathbb{Q}$-almost everywhere" means "almost surely, μ-almost everywhere". Its proof first uses the standard law of the iterated logarithm applied to the sequence of centered i.i.d. random variables $\log(W(x|k)) - \log(b) + \log(b)\varphi'(1^-)$ with variance $-\log(b)\varphi'(1^-)$ with respect to $\mathbb{Q}$, to control $\frac{\sum_{k=1}^n \log(W(x|k)) - \log(b) + \log(b)\varphi''(1^-)}{\sqrt{2\log(b)(-\varphi''(1^-))n\log\log(n)}}$. However the control of $Y(x|n)$ as $o(\sqrt{n\log\log(n)})$ has a gap in [Liu00]. Assuming $\mathbb{E}(W|\log W|^3) < \infty$, we know from [Big79] that $\mathbb{E}(Y\log(Y)^2) < \infty$. Then, for $\epsilon > 0$, a calculation yields

$$\begin{aligned}
\mathbb{Q}\left(\{|\log Y(x|n)| \geq \sqrt{n}\epsilon\}\right) &= \mathbb{Q}\left(\{Y(x|n) \geq e^{\sqrt{n}\epsilon}\}\right) + \mathbb{Q}\left(\{Y(x|n) < e^{-\sqrt{n}\epsilon}\}\right) \\
&= \mathbb{E}\left(Y \cdot \mathbf{1}_{\{Y \geq e^{\sqrt{n}\epsilon}\}}\right) + \mathbb{E}\left(Y \cdot \mathbf{1}_{\{Y < e^{-\sqrt{n}\epsilon}\}}\right) \\
&\leq \mathbb{E}\left(Y \cdot \mathbf{1}_{\{Y \geq e^{\sqrt{n}\epsilon}\}}\right) + e^{-\sqrt{n}\epsilon}.
\end{aligned}$$

Applying the elementary inequality $\sum_{n\geq 1} \mathbf{1}_{\{X \geq \sqrt{n}\}} \leq X^2$ we get

$$\begin{aligned}
\sum_{n\geq 1} \mathbb{Q}\left(\{|\log Y(x|n)| \geq \sqrt{n}\epsilon\}\right) &\leq \sum_{n\geq 1} \mathbb{E}\left(Z \cdot \mathbf{1}_{\{Z \geq e^{\sqrt{n}\epsilon}\}}\right) + \sum_{n\geq 1} e^{-\sqrt{n}\epsilon} \\
&= \mathbb{E}\left(Z \cdot \sum_{n\geq 1} \mathbf{1}_{\left\{\frac{\log Z}{\epsilon} \geq \sqrt{n}\right\}}\right) + \sum_{n\geq 1} e^{-\sqrt{n}\epsilon} \\
&\leq \epsilon^{-2}\mathbb{E}(Z(\log Z)^2) + \sum_{n\geq 1} e^{-\sqrt{n}\epsilon} < \infty.
\end{aligned}$$

Then the Borel-Cantelli lemma yields $|\log Y(x|n)| = o(\sqrt{n})$, μ-almost everywhere.

For part (2), the fact that $\widetilde{\mu}$ is exact dimensional with dimension 0 is established in [Bar00] using large deviations estimates to prove that $\sum_{n\geq 1} \widetilde{\mu}(\{x : \widetilde{\mu}(I_n(x)) \leq b^{-n\epsilon}\}) < \infty$ for all $\epsilon > 0$. This is refined in [BKNSW14a], using Theorem 3(2, 3), to get the lower bound in (5.1) (this bound is easy to extend to general distributions for W). The fact that $\widetilde{\mu}$ is atomless under the assumption $\varphi(1+\epsilon) > -\infty$ for some $\epsilon > 0$ is also established in [BKNSW14a]; this exploits Theorems 3(2, 3) and 5(1) to prove that $n^\gamma \max_{w\in\mathscr{A}^n} \mu(I_w)$ converge to 0 in probability as $n \to \infty$ for all $\gamma \in [0, 1/2)$. The upper bound in (5.1) is more involved than the lower one; we refer to [BKNSW14a] for the details of (5.1).

Remark 3 The bounds (5.1) are not completely satisfactory since we do not know whether they are sharp or not; in fact, we believe that at least the second one is not sharp, and also that the order of magnitude of the sharp upper bound should differ from that of the lower bound and reflect the fluctuations of random walks conditioned to stay positive.

5.1.2 Modulus of Continuity

At first we notice that if μ is a non degenerate Mandelbrot measure then $\beta_c \geq 1$ and $\beta_c > 1$ if $\varphi'_{W_{\beta_c}}(1^-) = 0$, and if $\widetilde{\mu}$ is a critical Mandelbrot measure, then $\beta_c = 1$, where β_c is defined in Sect. 4.

Also, it is important to have in mind that for either of these measures, the smallest pointwise Hölder exponent is $\varphi(\beta_c)/\beta_c$, which equals $\varphi'(\beta_c^-)$ if $\varphi'_{W_{\beta_c}}(1^-) = 0$ (see Sect. 5.1.3). It is natural to complete this information by estimating the modulus of continuity of these measures.

Theorem 13

(1) *Assume that* $\varphi'(1) > 0$, $\varphi'_{W_{\beta_c}}(1^-) = 0$, *and there exists* $\beta > \beta_c$ *such that* $\varphi(\beta) > -\infty$.
With probability 1, *for all* $\gamma \in (0, 1/2)$, *there exists* $C > 0$ *such that for all subintervals* $I \subset [0, 1]$, *the Mandelbrot measure* μ *satisfies*

$$\mu(I) \leq C|I|^{\varphi(\beta_c)/\beta_c}\left(\log\left(1+\frac{1}{|I|}\right)\right)^{-\gamma/\beta_c}. \tag{5.2}$$

(2) *Assume that* $\varphi(-\beta) > -\infty$ *and* $\varphi(1+\beta) > -\infty$ *for some* $\beta > 0$, *as well as* $\varphi'(1) = 0$.
With probability 1, *for all* $\gamma \in (0, 1/2)$, *there exists* $C > 0$ *such that for all subintervals* $I \subset [0, 1]$, *the critical Mandelbrot measure* $\widetilde{\mu}$ *satisfies*

$$\widetilde{\mu}(I) \leq C\left(\log\left(1+\frac{1}{|I|}\right)\right)^{-\gamma}. \tag{5.3}$$

Moreover, one cannot take $\gamma > 1/2$ *in the above statement.*

This result is proved in [BKNSW14a].

For (1), one needs that $\mathbb{E}(Y^\beta) < \infty$ for some $\beta > \beta_c$ to use the fact that $h(x) = \mathbb{P}(Y \geq x) \leq cx^{-\beta}$ for such a β; this is the case by Theorem 3(1) since $\varphi(\beta_c) > 0$ and our assumption imply $\varphi(\beta) > 0$ near β_c^+. This is combined in a non obvious way with the upper bound for $\mathbb{E}\Big(\Big(n^{\frac{3}{2}\frac{\beta}{\beta_c}}\frac{Z_n(\beta)}{(\mathbb{E}(Z_n(\beta_c)))^{\frac{\beta}{\beta_c}}}\Big)^\theta\Big)$ provided for $\theta \in (0, 1)$ in [Mad00] (and used here with $\theta = \beta_c/\beta$) to prove that

$$\begin{aligned}\mathbb{P}\big(\max_{w\in\mathscr{A}^n}\{\mu(I_w)\} \geq 2^{-n\frac{\varphi(\beta_c)}{\beta_c}}n^{-\frac{\gamma}{\beta_c}}\big) &= \mathbb{E}\Big(\prod_{w\in\mathscr{A}^n}(1-h(e^{\beta_c H(w)-n\varphi(\beta_c)}n^{-\gamma}))\Big)\\ &\leq C_\epsilon n^{(1-\epsilon)(\gamma-3/2)}\end{aligned}$$

for all $\epsilon > 0$ and $\gamma \in (0, 3/2)$. This yields (5.2).

Similarly, the upper bound for $\mathbb{E}\big((n^{\frac{1}{2}} Z_n(1))^\theta\big)$ provided for $\theta \in (0, 1)$ in [HS09] is combined with the information $\mathbb{P}(\widetilde{Y} \geq x) \leq cx^{-1}$ provided by Theorem 3(2, 3) to prove that $\mathbb{P}(\max_{w \in \mathscr{A}^n} \mu(I_w) \geq n^{-\gamma}) \leq C_\epsilon n^{(1-\epsilon)(\gamma-1/2)}$ for all $\epsilon > 0$ and $\gamma \in (0, 1/2)$. This yields (5.3). The fact that one cannot take $\gamma > 1/2$ in (5.3) is proved by using Theorems 5(1) and 8, as well as the information $\mathbb{P}(\widetilde{Y} \geq x) \geq c'x^{-1}$ $(x \geq 1)$ provided by Theorem 3(2, 3) to obtain that $\mathbb{P}(\max_{w \in \mathscr{A}^n} \mu(I_w) < n^{-\gamma})$ tends to 0 for all $\gamma > 1/2$.

Remark 4

(1) It is not known if the choice $\gamma < 1/2$ in (5.2) can be improved.
(2) Property (5.3) is extended to critical lognormal multiplicative chaos measures in [BKNSW14+].
(3) Theorem 13 only deals with the second order phase transition case (in the frame of Sect. 4). To get similar results in the first order phase transition case is desirable. This is related to Remark 1.

5.1.3 Multifractal Analysis

Recall that given a positive Borel measure ν supported on [0, 1], its multifractal analysis consists in computing the Hausdorff dimension, denoted dim, of the level sets of the pointwise Hölder exponent of ν, namely the sets

$$E_\nu(\gamma) = \Big\{x \in [0, 1] : \gamma(\nu, x) = \gamma\Big\} \quad (\gamma \geq 0),$$

where

$$\gamma(\nu, x) = \liminf_{r \to 0^+} \frac{\log \nu(B(x, r))}{\log(r)}.$$

The L^q-spectrum of ν is defined as

$$\tau_\nu : q \in \mathbb{R} \mapsto \liminf_{r \to 0^+} \frac{\log \sup \Big\{ \sum_i \mu(B(x_i, r))^q \Big\}}{\log(r)}, \tag{5.4}$$

where the supremum is taken over all the centered packing of [0, 1] by closed balls of radius r. Define also

$$\widetilde{\tau}_\nu : q \in \mathbb{R} \mapsto \liminf_{n \to \infty} \frac{-1}{n \log(b)} \log \sum_{w \in \mathscr{A}^n} \mu(I_w))^q, \tag{5.5}$$

Throughout, we adopt the convention that a set has a negative dimension if and only if it is empty.

One always has

$$\dim E_\nu(\gamma) \le \tau_\nu^*(\gamma),$$

and one says that ν obeys the multifractal formalism at γ if $\dim E_\nu(\gamma) = \tau_\nu^*(\gamma)$ (see [Ols95]).

Let us naturally extend φ to the real line as

$$\varphi(q) = \varphi_W(q) = q - 1 - \log_b \mathbb{E}(W^q) \quad (q \in \mathbb{R}),$$

Given a function $\psi : \mathbb{R} \to \mathbb{R} \cup \{-\infty\}$, define its concave Legendre-Fenchel conjugate as

$$\psi^* : \gamma \in \mathbb{R} \mapsto \inf\{\gamma q - \psi(q) : q \in \mathbb{R}\}.$$

Let

$$f : \gamma \mapsto \begin{cases} \varphi^*(\gamma) & \text{if } \varphi^*(\gamma) \ge 0 \\ -\infty & \text{otherwise} \end{cases}.$$

For non degenerate Mandelbrot measures and critical Mandelbrot measures, one has the following result (for point (3) we refer the reader to appendix for the definition of Hausdorff measures).

Theorem 14 *Let ν be the non degenerate Mandelbrot measure μ if $\varphi'(1^-) > 0$, or the critical Mandelbrot measure $\widetilde{\mu}$ if $\varphi'(1^-) = 0$ and $\varphi''(1^-) > -\infty$.*

Suppose that $\varphi(-\epsilon) > -\infty$ for some $\epsilon > 0$. Then, with probability 1,

(1) $\tau_\nu(0) = -1$ *and for all* $q \in \mathbb{R}^*$, $\tau_\nu(q) = \sup\{\varphi(\theta q)/\theta : \theta \in (0, 1]\}$. *Moreover,* $\tau_\nu = f^*$, $f = \tau_\nu^*$, *and in* (5.4), lim inf *can be replaced by* lim.
(2) *For all* $\gamma \ge 0$, $\dim E_\nu(\gamma) = \tau_\nu^*(\gamma) = f(\gamma)$.
(3) $E_\nu(\tau_\nu'(0))$ *is of full Lebesgue measure;*
(0–∞ *law*) *for all* $\gamma \ge 0$ *such that* $\dim 0 < \tau_\nu^*(\gamma) < 1$, *for all gauge functions* g, *we have* $\mathcal{H}^g(E_\nu(\gamma)) = \infty$ *if* $\limsup_{t\to 0^+} \log(g(t))/\log(t) \le \tau_\nu^*(\gamma)$ *and* $\mathcal{H}^g(E_\nu(\gamma)) = 0$ *otherwise.*
(4) *For* $n \ge 1$, $\gamma \in \mathbb{R}_+$ *and* $\epsilon > 0$ *let*

$$f(n, \gamma, \epsilon) = n^{-1} \log_b \#\Big\{w \in \mathscr{A}^n : \frac{\log \mu(I_w)}{-n \log(b)} \in [\gamma - \epsilon, \gamma + \epsilon]\Big\}.$$

We have

$$\text{for all } \gamma \ge 0, \quad \lim_{\epsilon\to 0^+} \liminf_{n\to\infty} f(n, \gamma, \epsilon) = \lim_{\epsilon\to 0^+} \limsup_{n\to\infty} f(n, \gamma, \epsilon) = f(\gamma).$$

Let us sketch the ideas of the proof of this result, and start with a remark.

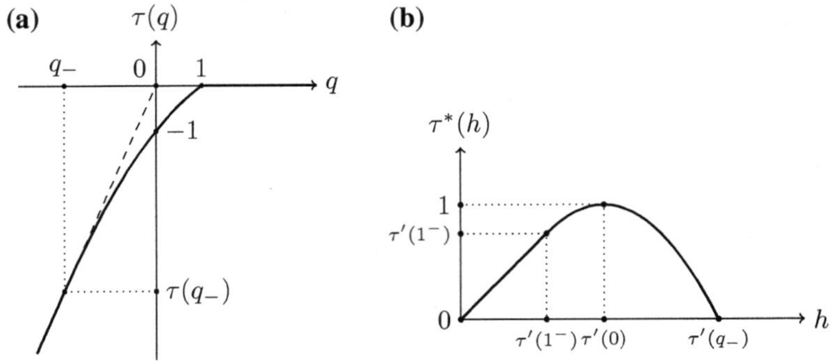

Fig. 3 Multifractal nature of a Mandelbrot measure with a second order phase transition at q_- and a first order phase transition at $q_+ = 1$, i.e. in the case where Theorem 1 is sharp. **a** The L^q spectrum of μ. **b** Its Legendre transform

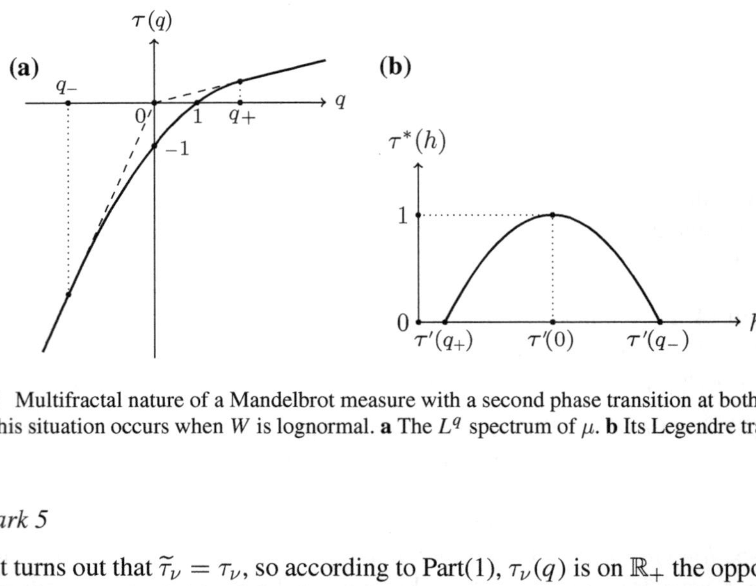

Fig. 4 Multifractal nature of a Mandelbrot measure with a second phase transition at both q_- and q_+. This situation occurs when *W* is lognormal. **a** The L^q spectrum of μ. **b** Its Legendre transform

Remark 5

(1) It turns out that $\widetilde{\tau}_\nu = \tau_\nu$, so according to Part(1), $\tau_\nu(q)$ is on $\mathbb{R}_+$ the opposite of the free energy $P(q)$ of the directed polymer associated with $H(w)$ defined in Sect. 4, and on $\mathbb{R}_-$, $\tau_\nu(q)$ is the opposite of $P(-q)$, where P is the free energy of the directed polymer associated with $-H(w)$. Thus, there may be two phase transitions for τ_ν: one on $\mathbb{R}_+$ according to $q_+ = \sup\{q > 0 : \varphi'(q)q - \varphi(q) > 0\}$ is finite or not, and one on $\mathbb{R}_-$ according to $q_- = \inf\{q < 0 : \varphi'(q)q - \varphi(q) > 0\}$ is finite or not. This yields nine possible situations under our assumptions in the case of Mandelbrot measures, and three in the case of critical Mandelbrot measures, since on $\mathbb{R}_+$ there is automatically a second order phase transition at 1. Some of these possibilities are illustrated in Figs. 3, 4, 5 and 6.

(2) Due to (2.4) and (3.3), and Parts (2) and (4) of Theorem 14 are geometric and statistical counterparts in $\mathscr{A}^\omega$ and $\mathscr{A}^*$ of Cramer's theorem and its extension by Bahadur and Zabell (see [DZ98]), which ensures that for all $\gamma \in \mathbb{R}$, over any fixed

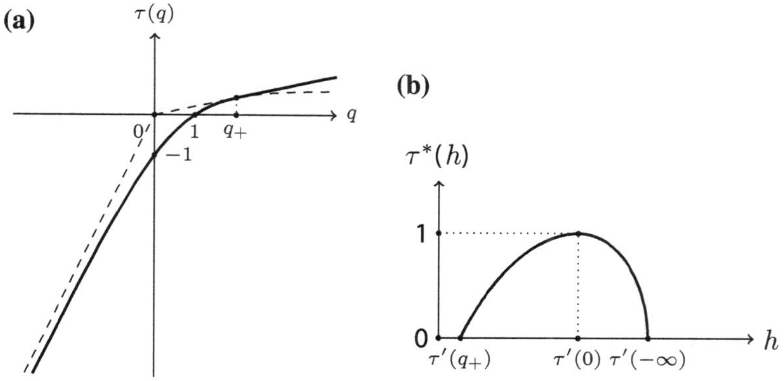

Fig. 5 Multifractal nature of a Mandelbrot measure with $q_- = -\infty$ and a second order phase transition at q_+. **a** The L^q spectrum of μ. **b** Its Legendre transform

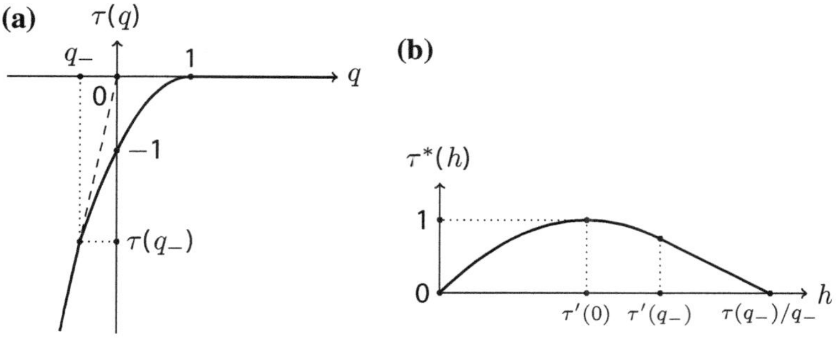

Fig. 6 Multifractal nature of a critical Mandelbrot measure (one always has $q_+ = 1$, where a second order phase transition occurs), with a first order phase transition at q_-. **a** The L^q spectrum of $\widetilde{\mu}$. **b** Its Legendre transform

infinite branch $w_1 \cdots w_n \cdots$ of $\mathscr{A}^\omega$, after setting $L(q) = \log_b \mathbb{E}((bW^{-1})^q)$, one has

$$\begin{aligned} r(\gamma) &:= \inf_{q\in\mathbb{R}} L(q) - \gamma q \\ &= \lim_{\epsilon\to 0} \liminf_{n\to\infty} n^{-1} \log_b \mathbb{P}(|\gamma - (n\log b)^{-1} \sum_{k=1}^{n} \log(b) - \log(W(w_1\cdots w_n)| \le \epsilon) \\ &= \lim_{\epsilon\to 0} \limsup_{n\to\infty} n^{-1} \log_b \mathbb{P}(|\gamma - (n\log b)^{-1} \sum_{k=1}^{n} \log(b) - \log(W(w_1\cdots w_n)| \le \epsilon). \end{aligned}$$

Indeed, heuristically, due to the rate of growth of the trees $\{\mathscr{A}\}^n$, $n \ge 1$, almost surely, for all $\gamma \in \mathbb{R}$ such that $1 + r(\gamma) \ge 0$, one should have the "logarithmic frequency"

$$\begin{aligned}\lim_{\epsilon\to 0^+}\liminf_{n\to\infty} f(n,\gamma,\epsilon) &= \lim_{\epsilon\to 0^+}\limsup_{n\to\infty} f(n,\gamma,\epsilon)\\ &= \lim_{\epsilon\to 0}\liminf_{n\to\infty} n^{-1}\log_b\Big(\#(\mathscr{A}^n)\cdot\mathbb{P}(|\gamma-(n\log b)^{-1}\sum_{k=1}^{n}\log(b)\\ &\qquad -\log(W(w_1\cdots w_n)|\le\epsilon)\Big)\\ &= 1+r(\gamma)=\varphi^*(\gamma).\end{aligned}$$

This observation about point (4) was made by Mandelbrot in [Man74a].

We notice that the formula given by Theorem 14 for τ_ν simplifies to $\tau_\nu=\varphi$ over (q_-,q_+), and also $\tau_\nu^*(\gamma)=\varphi^*(\gamma)$ for $\gamma=\varphi'(q)$, $q\in(q_-,q_+)$.

The lower bound $\tau_\nu(q)\ge T(q)=\sup\{\varphi(\theta q)/\theta:\theta\in(0,1]\}$ is quite easy. At first the problem reduces to packings by b-adic intervals, since $\tau_\nu=\widetilde{\tau}_\nu$. Then, notice that for $q\in(q_-,q_+)$, after Theorem 3 we have $\mathbb{E}(\|\nu\|^q)<\infty$, so (using (2.5) or (3.4))

$$\sum_{n\ge 1}\mathbb{E}\Big(\sum_{w\in\mathscr{A}^n}\nu(I_w)^q b^{n(\varphi(q)-\epsilon)}\Big)=\mathbb{E}(\|\nu\|^q)\sum_{n\ge 1} b^{-n\epsilon}<\infty$$

for all $\epsilon>0$. This yields $\widetilde{\tau}_\nu(q)\ge\varphi(q)$, but $\varphi(q)=\sup\{\varphi(\theta q)/\theta:\theta\in(0,1]\}$ if $q\in(q_-,q_+)$. For the other values of q, the argument giving the lower bound for $\widetilde{\tau}_\nu$ is similar to that used to obtain the upper bound for $\limsup_{n\to\infty}P_n$ in Sect. 4.2 when $\beta\ge\beta_c$. Moreover, one can show that $T^*=f$ and $f=T^*$ (see [AB14] for the details).

Thus, for all $\gamma\ge 0$, $\dim E_\nu(\gamma)\le\tau_\nu^*(\gamma)\le f(\gamma)\le\varphi^*(\gamma)$.

Let $\gamma\ge 0$ such that $\varphi^*(\gamma)\ge 0$. When $\nu=\mu$, the sharp lower bound $\varphi^*(\gamma)$ for $\dim E_\nu(\gamma)$ is determined in several papers (under different kind of assumptions, and sometimes for the level sets associated with $\widetilde{\gamma}(\nu,x)$ rather than $\gamma(\nu,x)$) in the case that $\gamma=\varphi'(q)$ with $q\in(q_-,q_+)$ [Bar00, BBP03, BHJ11, BJ10, Fal94, HW92, Kah91, Mol96, Ols94] or $q\in\{q_-,q_+\}$ [Bar00, BJ10]. This fully describes the range of γ for which $E_\nu(\gamma)\neq\emptyset$ in the cases for which q_- and q_+ correspond to no phase transition or a second order phase transition: one can prove that, with probability 1, for all $q\in(q_-,q_+)$, the Mandelbrot measure μ_q associated with the weights $W_q(w)$ (see (4.1)) is exact dimensional with a Hausdorff dimension equal to $\varphi^*(\varphi'(q))=\varphi'(q)q-\varphi(q)$ and is carried by $E_\nu(\alpha)$ [Bar00] (in [BHJ11, Fal94, HW92, Kah91, Mol96, Ols94], one finds the weaker version: for each fixed $q\in(q_-,q_+)$, $\dim E_\nu(\varphi'(q))\ge\varphi^*(\varphi'(q))$ almost surely). For $\gamma=\varphi'(q)$ with $q\in\{q_-,q_+\}$, if q is finite then one shows that the critical Mandelbrot measure $\widetilde{\mu}_q$ is carried by $E_\nu(\varphi'(q))$ [Bar00], otherwise one builds an inhomogeneous Mandelbrot measure carried by $E_\nu(\varphi'(q))$ and whose Hausdorff dimension is $\varphi^*(\varphi'(q))$ [BJ10].

The same approach is used in [Bar00] to achieve the multifractal analysis of $\widetilde{\mu}$ in absence of first order phase transition on $\mathbb{R}_-$.

However, in case of first order phase transition at $q\in\{q_-,q_+\}$, the previous method does not make it possible to study $E_\nu(\gamma)$ for the exponents γ in

$[\varphi(q)/q, \varphi'(q^-))$ if $q = q_+$ or in $(\varphi'(q^+), \varphi(q)/q]$ if $q = q_-$. Moreover, in any situation, they cannot provide the 0–∞ law obtained in point (3).

In [AB14], a new approach is developed to treat all the possible types of phase transition in the multifractal analysis of branching random walks. Moreover, following the approach developed in [BBP03] or [BJ10], one can reduce the study of the level sets of $\gamma(\nu, x)$ to that of the level sets of $\liminf_{n\to\infty} \frac{\log \nu(I_n(x))}{-n\log(b)}$. Also, the assumption $\varphi(-\epsilon) > -\infty$ for some $\epsilon > 0$ assures that $\mathbb{E}(\|\nu\|^{-\epsilon}) < \infty$, and $\mathbb{E}(\|\nu\|^{\epsilon}) < \infty$ for $\epsilon \in (0,1)$, so that the study of the level sets of $\liminf_{n\to\infty} \frac{\log \nu(I_n(x))}{-n\log(b)}$ reduces to the level sets of $\liminf_{n\to\infty} \frac{H(x|n)}{n\log(b)}$, by using (2.4).

One can also assume that $W \overset{\text{dist}}{\neq} 1$ without loss of generality. Then (we still use the notations of Sect. 4), there exists $A_0 > 0$ such that for $A \geq A_0$, $\mathbb{P}(|V| \leq A) > 0$. One considers the random weights

$$W_{q,A}(w) = \frac{\mathbf{1}_{\{|V(w)|\leq A\}} W^q(w)}{\mathbb{E}(\mathbf{1}_{\{|V(w)|\leq A\}} W^q(w))}$$

and the functions

$$\varphi_A(q) = -1 + q - \log_b \mathbb{E}(\mathbf{1}_{\{|V(w)|\leq A\}} W^q).$$

One has $\varphi_A \searrow \varphi$ pointwise as $A \to \infty$, and it can be shown that this implies that $\varphi_A^* \nearrow \varphi^*$ pointwise on the interior of $\{\gamma : \varphi^*(\gamma) > -\infty\}$ as $A \to \infty$. It follows that one can find an increasing sequence $(A_k)_{k\geq 1}$ converging to ∞ and for each $k \geq 1$ a finite set $D_k \subset \{q : \varphi_{A_k}^*(\varphi'_{A_k}(q)) > 0\}$ such that for each γ such that $\varphi^*(\gamma) \geq 0$, there exists $(q_k)_{k\geq 1} \in \prod_{k\geq 1} D_k$ such that $\lim_{k\to\infty} \varphi'_{A_k}(q_k) = \gamma$ and $\lim_{k\to\infty} \varphi_{A_k}^*(\varphi'_{A_k}(q_k)) = \varphi^*(\gamma)$.

Instead of considering Mandelbrot measures like the measures μ_q, $q \in (q_-, q_+)$, one considers the family of inhomogeneous Mandelbrot measures obtained as follows.

Fix an increasing sequence of integers $(N_k)_{k\geq 0}$ with $N_0 = 0$. Set $M_k = \sum_{i=1}^k N_i$ and for $n \geq 1$ define k_n so that

$$M_{k_n} + 1 = 1 + \sum_{k=1}^{k_n} N_k \leq n \leq \sum_{k=1}^{k_n+1} N_k.$$

For every sequence $B = (q_k)_{k\geq 1} \in \prod_{k=1}^{\infty} D_k$, consider the inhomogeneous branching random walk

$$H_B(w) = n\log(b) - \sum_{k=1}^{k_n}\sum_{i=1}^{N_k} \log W_{q_k,A_k}(w|M_{k-1}+i) - \sum_{i=1}^{n-M_{k_n}} \log W_{q_k,A_k}(w|M_{k_n}+i)$$

defined on $E_{B,n} = \{w \in \mathscr{A}^n : |V(w|M_k + i)| \leq A_k,\ \forall k, i \text{ such that } M_k + i \leq n\}$. Then define the random measures

$$\mu_{B,n}(\mathrm{d}x) = \mathbf{1}_{E_{B,n}}(w)e^{-H_B(w)}\, b^n \mathrm{d}x, \text{ if } x \in I_w.$$

They form a martingale which converges almost surely to a random measure μ_B supported on $E_B = \bigcap_{n\geq 1}\bigcup_{w\in E_{B,n}} I_w$. Moreover, it is possible to choose $(N_k)_{k\geq 1}$ suitably so that all the measures μ_B are simultaneously defined and non degenerate conditionally on $E_B \neq \emptyset$. The sequence $(N_k)_{k\geq 1}$ can also be fixed so that for all γ such that $\varphi^*(\gamma) \geq 0$, each time $B = (q_k)_{k\geq 1}$ is such that $\lim_{k\to\infty} \varphi'_{A_k}(q_k) = \gamma$ and $\lim_{k\to\infty} \varphi^*_{A_k}(\varphi'_{A_k}(q_k)) = \varphi^*(\gamma)$, then, conditionally on $\{E_B \neq \emptyset\}$, μ_B is exact dimensional with dimension $\varphi^*(\gamma)$ and carried both by the set $E_H(\gamma) = \{x : \lim_{n\to\infty} \frac{H(x|n)}{n\log(b)} = \gamma\}$ and the set $\widetilde{E}_\nu(\gamma) = \Big\{x \in [0,1] : \lim_{n\to\infty} \frac{\log\nu(I_n(x))}{-n\log(b)} = \gamma\Big\} \cap \Big\{x \in [0,1] : \lim_{r\to 0^+} \frac{\log\nu(B(x,r))}{\log(r)} = \gamma\Big\}$. Moreover, there are uncountably many such measures for a given γ, and two such measures μ_B and $\mu_{B'}$ are mututally singular if B and B' are not ultimately equal. Since, moreover, $\mathbb{P}(\{E_B \neq \emptyset\})$ tends to 1 as the first terms A_1 of $(A_k)_{k\geq 1}$ tends to ∞, the measures μ_B are the main tool to get parts (2) of the theorem. Part (3) requires additional work, and uses in an essential way the reach family exhibited above of inhomogeneous measures of dimension $\varphi^*(\gamma)$ supported by $E_\nu(\gamma)$.

To get what remains of part (1), it is first not hard to deduce point (4) from the previous estimations for the Hausdorff dimensions: we have

$$\text{a.s., for all } \gamma \geq 0, \quad \lim_{\epsilon\to 0^+}\liminf_{n\to\infty} f(n,\gamma,\epsilon) = \lim_{\epsilon\to 0^+}\limsup_{n\to\infty} f(n,\gamma,\epsilon) = f(\gamma).$$

Then the conclusion follows from Varadhan's integral Lemma [DZ98, Theorem 4.3.1].

Remark 6

(1) The measures of type μ_B can be used to control the Hausdorff and packing dimensions of the wider family of sets

$$\mathbb{E}_\nu(\gamma;\gamma') = \Big\{x \in [0,1] : \liminf_{n\to\infty} \frac{\log\nu(I_n(x))}{-n\log(b)} = \gamma,\ \limsup_{n\to\infty} \frac{\log\nu(I_n(x))}{-n\log(b)} = \gamma'\Big\},$$

$0 \leq \gamma \leq \gamma'$, (see [AB14] for the details).

(2) Consider the branching random walk $H(w)$, $w \in \mathscr{A}^*$, for itself, and do not assume any integrability properties for H (or W). Set

$$\widetilde{f}(n,\gamma,\epsilon) = n^{-1}\log_b \#\Big\{w \in \mathscr{A}^n : \frac{H_n(w)}{n\log(b)} \in [\gamma-\epsilon, \gamma+\epsilon]\Big\}.$$

The same approach as above yields that with probability 1,

$$\text{for all } \gamma \in \mathbb{R}, \quad \dim E_H(\gamma) = \lim_{\epsilon \to 0^+} \liminf_{n \to \infty} \widetilde{f}(n, \gamma, \epsilon)$$
$$= \lim_{\epsilon \to 0^+} \limsup_{n \to \infty} \widetilde{f}(n, \gamma, \epsilon) = f(\gamma).$$

(see [AB14]).

This includes in particular the both sided degenerate case $\varphi(q) = -\infty$ for all $q \neq 0$ (see Sect. 4.4 for the definiton of the degenerate case), and in this case one has $\dim E_H(\gamma) = 1$ for all $\gamma \in \mathbb{R}$.

5.2 Multifractal Analysis of Lévy-Mandelbrot and Critical Lévy-Mandelbrot Measures

We now describe the multifractal nature of the discrete statistically self-similar measures defined in Sect. 3.1.2 (notice that by construction these measures are exact dimensional with dimension 0) (Fig. 7).

Theorem 15 *Let $\alpha \in (0, 1)$ and ν the non degenerate Mandelbrot measure μ if $\varphi'(1^-) > 0$, or the critical Mandelbrot measure $\widetilde{\mu}$ if $\varphi'(1^-) = 0$ and $\varphi''(1^-) > -\infty$. Let ν^α be the associated Lévy-Mandelbrot or critical Lévy-Mandelbrot measure in Sect.* 3.1.2.

Suppose that $\varphi(-\epsilon) > -\infty$ for some $\epsilon > 0$. With probability 1, *we have*

(1) *for all $q \in \mathbb{R}$, $\tau_{\nu^\alpha}(q) = \min(\tau_\nu(q/\alpha), 0)$;*
(2) *for all $\gamma \in \mathbb{R}_+$, $\dim E_{\nu^\alpha}(\gamma) = \tau^*_{\nu^\alpha}(\gamma)$.*

Remark 7 There is a phase transition at $q_+ = \alpha$; it is of first order when $\nu = \mu$ and of second order when $\nu = \widetilde{\mu}$. On $\mathbb{R}_-$, we have the same three possibilities as for ν.

We are going to see that when $\nu = \mu$, the first order phase transition at $q = \alpha$ also corresponds to a transition in the geometric properties responsible for the Hausdorff

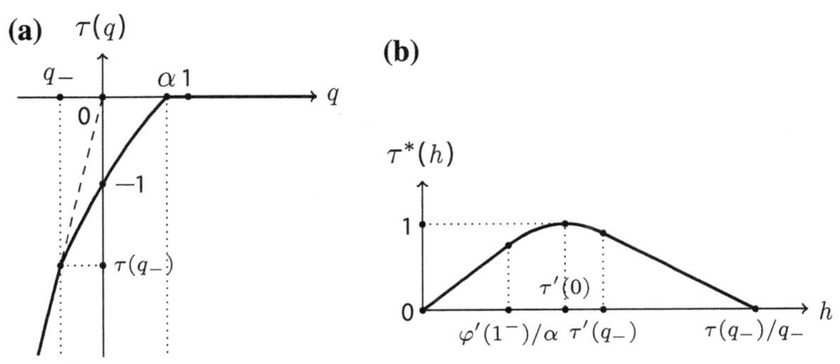

Fig. 7 Multifractal nature of a Lévy-Mandelbrot measure, with the necessary first order phase transition at $\alpha = q_+$, and another first order phase transition at q_-. **a** The L^q spectrum of μ^α. **b** Its Legendre transform

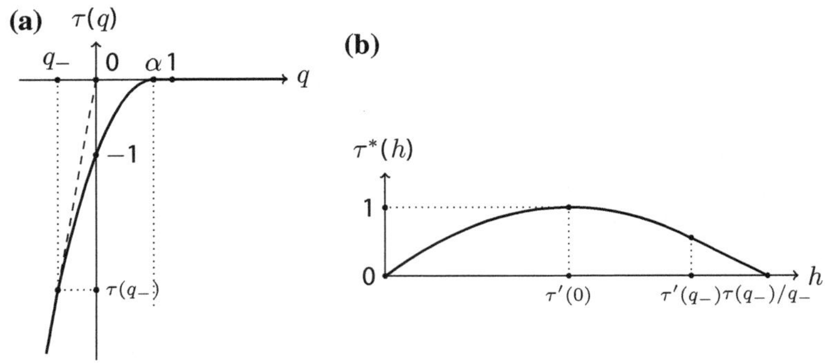

Fig. 8 Multifractal nature of a critical Lévy-Mandelbrot measure, with the necessary second order phase transition at $\alpha = q_+$, and a first order phase transition at q_-. **a** The L^q spectrum of $\widetilde{\mu}^\alpha$. **b** Its Legendre transform

dimensions of the sets $E_{\nu^\alpha}(\gamma)$ for $\gamma \in [0, \varphi'(1^-)/\alpha]$: for these exponents the rate of approximations of the elements of $E_{\nu^\alpha}(\gamma)$ by the atoms of ν^α come into play, while for the other exponents, and the same holds when $\nu = \widetilde{\mu}$, the Hausdorff dimension of $E_{\nu^\alpha}(\gamma)$ can be captured essentially in the same way as for the sets $E_\nu(\alpha\gamma)$ (Fig. 8).

We can reduce the study of the L^q-spectrum to packings by b-adic intervals. For $\alpha q_- < q < \alpha$ we have, using (3.5)

$$\sum_{n\geq 1} \mathbb{E}\Big(\sum_{w\in\mathscr{A}^n} \nu^\alpha(I_w)^q b^{n(\varphi(q/\alpha)-\epsilon)}\Big) = \mathbb{E}(\|\nu^\alpha\|^q) \sum_{n\geq 1} b^{-n\epsilon} < \infty$$

for all $\epsilon > 0$. This yields $\tau_\nu(q) \geq \varphi(q/\alpha)$, and with manipulations similar to those use in the sketch of proof of Theorem 14, we get the lower bound $\tau_{\nu^\alpha}(q) \geq \min(\tau_\nu(q/\alpha), 0)$ almost surely. In particular, $\tau_{\nu^\alpha}(\alpha) \geq 0$, so that since $\tau_{\nu^\alpha}(1) = 0$ and τ_{ν^α} is non decreasing and concave, we must have $\tau_{\nu^\alpha} = 0$ over $[\alpha, \infty)$. The upper bound for τ_{ν^α} is a consequence of the lower bound for the Hausdorff dimension in part (2) of the theorem, as well as the inverse Legendre transform for φ^*.

Part (2) is proven in [Jaff99] when ν is the Lebesgue measure restricted to [0, 1] ($W = 1$ almost surely), i.e. ν^α is just the derivative of a Lévy subordinator, in the following form: $\dim E_{\nu^\alpha}(\gamma) = \alpha\gamma$ if $\gamma \in [0, 1/\alpha]$, and $\dim E_{\nu^\alpha}(\gamma) = -\infty$ otherwise.

Fix $T > 0$ and let $\mathcal{P} = \{(y_n, r_n) : n \geq 1\}$ be a Poisson point process with intensity $\mathrm{d}y \otimes \frac{\mathrm{d}r}{r^2}$ in $[0, T] \times \mathbb{R}_+^*$, so that the sequence $(r_n)_{n\geq 1}$ tends to 0 as $n \to \infty$. Then take for L_α the α-stable Lévy subordinator $L_\alpha(y) = \sum_{n: y_n \leq y} r_n^{1/\alpha}$, so that $\nu^\alpha = \sum_{n\geq 1} r_n^{1/\alpha} \delta_{y_n} := \rho$ when ν is the Lebesgue measure. The multifractal analysis in this case uses the following facts. At first, it results from quite direct estimates that for any $y \in \mathbb{R}_+ \setminus \{y_n : n \geq 1\}$, the pointwise exponent $\gamma(\rho, y)$ equals $1/(\alpha s(y))$, where $s(y)$ is the rate of approximation of y by the family $\{(y_n, r_n) : n \geq 1\}$, defined as

$$s(y) = \limsup_{n\to\infty} \frac{\log(|y - y_n|)}{\log(r_n)}.$$

By a theorem of Shepp [She72], one has $s(y) \geq 1$ for all $y \in [0, T] \setminus \{y_n : n \geq 1\}$, hence $\gamma(\rho, y) \leq 1/\alpha$, so $E_\rho(\gamma) = \emptyset$ if $\gamma > 1/\alpha$. Moreover, as a consequence of a theorem on the "ubiquitous systems" established independently in [Jaf00a] and [DMPV95] and applied to the family $\{(y_n, r_n) : n \geq 1\}$, one has $\dim\{y : s(y) : s\} = 1/s$ for all $s \geq 1$, hence the Hausdorff dimension of $E_\rho(\gamma) = \{y \in [0, T] : s(y) : 1/(\alpha\gamma)\} = \alpha\gamma$.

The general case is treated in [BS07] under stronger assumptions on φ that only allow no phase transition or a second order phase transition on $\mathbb{R}_-$. Conditionally on ν, the Lévy subirdinator L_α is considered over $[0, T]$, with $T = \|\nu\|$. The composition by the indefinite integral of ν when $W \overset{\text{dist}}{\neq} 1$ induces distorsions with respect to the situation explained above. For $\gamma \geq \varphi'(1^-)/\alpha$ such that $\varphi^*(\alpha\gamma) \geq 0$, one takes a measure μ_B of Hausdorff dimension $\varphi^*(\alpha\gamma) \geq 0$ carried by $E_\nu(\alpha\gamma)$ as in the proof of Theorem 14 and proves that for μ_B-almost very x, noting $F(x) = \nu([0, x])$, we have $s(F(x)) = 1$, which combined with $\lim_{r\to 0^+} \frac{\log(\nu(B(x,r)))}{\log(r)} = \alpha\gamma$ and $\gamma(\varrho, F(x)) = 1/(\alpha s(F(x)))$ yields $\gamma(\nu^\alpha, x) = \gamma$. Hence $\dim E_{\nu^\alpha}(\gamma) \geq \varphi^*(\alpha\gamma) = (\min(\tau_\nu(q/\alpha), 0))^*(\gamma)$.

When $\nu = \widetilde{\mu}$, since $\varphi'(1^-) = 0$, this yields the whole spectrum for $\gamma > 0$. Since $E_{\nu^\alpha}(0)$ contains the atoms of ν^α, i.e. $\{F^{-1}(\{y_n : n \geq 1\})$, we also have $\dim E_{\nu^\alpha}(0) \geq 0 = (\min(\tau_\nu(q/\alpha), 0))^*(0)$, and this is also valid when $\nu = \mu$.

The most delicate sets are the $E_{\nu^\alpha}(\gamma)$ for $\gamma \in (0, \varphi'(1^-)/\alpha)$ when $\nu = \mu$, for which we must prove $\dim E_{\nu^\alpha}(\gamma) \geq \alpha\gamma$. This requires a non trivial extension of classical "ubiquitous systems" (which deserve to be called homogeneous) to "heterogeneous ubiquitous systems". This is achieved in [BS07], and applied to the present situation in [BS07]. The main tool provided by these papers is, for each $s > 1$, a Borel probability measure ρ_s on $[0, 1]$ such that for ρ_s-almost every x, $\liminf_{r\to 0^+} \frac{\log(\rho_s(B(x,r)))}{\log(r)} \geq \frac{\varphi'(1^-)}{s}$ (which implies that for any Borel set E, one has $\rho_s(E) = 0$ if $\dim E < \frac{\varphi'(1^-)}{s}$) and there exists a decreasing sequence $(\epsilon_j)_{j\geq 1}$ and a subsequence $(n_j)_{j\geq 1}$ such that for each $j \geq 1$ one has $F^{-1}(y_{n_j}) \in B(x, r_{n_j}^{s/\varphi'(1^-)-\epsilon_j})$. This second property implies that $\nu^\alpha(B(x, r_{n_j}^{s/\varphi'(1^-)-\epsilon_j})) \geq r_{n_j}^{1/\alpha}$, hence $\gamma(\nu^\alpha, x) \leq \varphi'(1^-)/\alpha s$. In particular for $\gamma = \varphi'(1^-)/(\alpha s)$, $\rho_s(F_{\nu^\alpha}(\gamma)) = 1$, where $F_{\nu^\alpha}(\gamma) = \{x \in [0, 1] : \gamma(\nu^\alpha, x) \leq \gamma\}$.

Now, we notice that for $\gamma \in (0, \varphi'(1^-)/\alpha)$, we have $(\min(\tau_\nu(q/\alpha), 0))^*(\gamma) = \alpha\gamma$. Moreover, it also comes from the multifractal formalism [Ols95] that $\dim F_{\nu^\alpha}(\gamma') \leq \tau^*_{\nu^\alpha}(\gamma')$ for $\gamma' \in [0, \tau'_{\nu^\alpha}(0^-)]$. Since $\tau_{\nu^\alpha}(q) \geq \min(\tau_\nu(q/\alpha), 0)$ almost surely for all q, we get $\dim F_{\nu^\alpha}(\gamma') \leq \tau^*_{\nu^\alpha}(\gamma') \leq \alpha\gamma'$ for all $0 \leq \gamma' < \gamma < \varphi'(1^-)/\alpha)$. Now, setting $\gamma = \varphi'(1^-)/\alpha s$, since $\liminf_{r\to 0^+} \frac{\log(\rho_s(B(x,r)))}{\log(r)} \geq \frac{\varphi'(1^-)}{s} = \alpha\gamma$, we get $\rho_s(F_{\nu^\alpha}(\gamma')) = 0$ for all $0 \leq \gamma' < \gamma$, hence $\rho_s\Big(\bigcup_{0\leq\gamma'<\gamma} F_{\nu^\alpha}(\gamma')\Big) = 0$

noting that the sets $F_{\nu^\alpha}(\gamma')$ are non decreasing. Finally, writing $E_{\nu^\alpha}(\gamma) = F_{\nu^\alpha}(\gamma) \setminus \bigcup_{0\le\gamma'<\gamma} F_{\nu^\alpha}(\gamma')$, we obtain $\rho_s(E_{\nu^\alpha}(\gamma)) > 0$ hence $\dim E_{\nu^\alpha}(\gamma)) \ge \alpha\gamma$.

The construction of the measures ρ_s is rather involved, and uses a combination of Shepp's theorem and the statistical self-similarity and exact dimensionality properties of μ.

5.3 KPZ Formula

Here we give, in the Mandelbrot cascade context, results related to KPZ formula [KPZ98] of two dimensional quantum gravity. The KPZ formula was reformulated and proved by Duplantier and Sheffield [DS09, DS11] as a relation between the box counting dimensions of sets in the Euclidean geometry and the expecting box counting dimensions of sets computed using a random measure given by exponential of the Gaussian Free Field (a fundamental example of log-Gaussian multiplicative chaos), and reformulated by Benjamini and Schramm [BS09] and Rhodes and Vargas [RV08] as a relation between Hausdorff dimensions of sets computed using the Lebesgue measure to measure the size of balls, and the Hausdorff dimensions of sets when the Lebesgue measure is replaced by a non degenerate Mandelbrot measure or the limit of a non degenerate log-infinitely divisible cascade. In dimension 1, this can be directly interpreted in terms of a change of metric. Then, a rigorous proof of the "dual" KPZ formula was given in [BJRV13], using discrete random measures which are the analogue of the measure μ^α of Sect. 3.1.2 in the log-normal multiplicative chaos framework (see [RV13] for a review of this). This has been extended to the cascade case in [BKNSW14a], with similar formulas for critical Mandelbrot measures $\widetilde{\mu}$ and the associated discrete measures $\widetilde{\mu}^\alpha$.

Since we work in dimension 1, it is convenient to present KPZ formulas as a relation between the Hausdorff dimension of a set and the Hausdorff dimension of its image by the indefinite integral of a given statistically self-similar measure:

Theorem 16 *Let ν be the non degenerate Mandelbrot measure μ if $\varphi'(1^-) > 0$, or the critical Mandelbrot measure $\widetilde{\mu}$ if $\varphi'(1^-) = 0$ and $\varphi''(1^-) > -\infty$. Denote also ν by ν^1. Suppose that $\varphi(-q) < -\infty$ for all $q \in (0, 1/b)$.*

Let $\alpha \in (0, 1]$ and define $F_\alpha(t) = \nu^\alpha([0, t])$ for $t \in [0, 1]$ (i.e. F_α is the α-stable Lévy process L_α in independent multifractal time $F : t \mapsto \nu([0, t])$). If E is a Borel subset of $[0, 1]$ of Hausdorff dimension ξ_0, then, with probability 1, the Hausdorff dimension of $F_\alpha(E)$ is the unique solution ξ of the equation

$$1 + \varphi(\xi/\alpha) = \xi_0.$$

Notice that when $\alpha = 1$, $F_1 = F$. In this case, when $\nu = \mu$, the result is proved in [BS09], since the authors prove that the dimension ξ of E under the metric $d(x, y) = \mu([x, y])$ is given by the above formula. The formula is quite easy to guess using a natural covering argument and the fact that for all $x, x' \in [0, 1]$,

$\mathbb{E}(d(x,y)^s) \le 8|x-y|^{1+\varphi(s)}$ for all $s \in [0,1]$ to find an upper bound of dim E under the metric d. For the lower bound, one first reduces the situation to E being compact. If $\xi_0 = 0$ there is nothing to prove. Otherwise, for each $t \in [0,\xi_0)$ one fixes a Frostman Borel measure ρ such that the energy $\int\int \frac{\rho(\mathrm{d}x)\rho(\mathrm{d}y)}{|y-x|^t}$ is finite, sets s the solution of $1+\varphi(s) = t$, and then shows that the measures $\rho_n(\mathrm{d}x) = W_s(w|1)W_s(w|2)\cdots W_s(w|n)\,\rho(\mathrm{d}x)$ if $x \in I_w$, $w \in \mathscr{A}^n$ weakly converge to a non degenerate measure ρ_s supported on E almost surely and such that $\int\int \frac{\rho_s(\mathrm{d}x)\rho_s(\mathrm{d}y)}{d(x,y)^s} < \infty$ almost surely, which implies that the Hausdorff dimension of E under d is at least s. Since s tends to ξ as t tends to ξ_0, this is enough to conclude.

This approach can be adapted to get the result when $\alpha = 1$ and $\nu = \widetilde{\mu}$ (see [BKNSW14a]). The general case then follows from the fact that a.s. $L_\alpha(A) = \alpha \dim A$ for all subsets A of $[0,1]$ [Ber96, III.5].

6 On Signed and Complex Multiplicative Cascades

We give some convergence and renormalization results proved in [BM09, BJM10, BJM10a] for the continuous function-valued sequence $(F_n = F_{W,n})_{n\ge1}$ defined in (2.8). The asymptotic behavior of $(F_n)_{n\ge1}$ is far from being completely understood in general, and deserves to be further explored. An interesting related and earlier work is [DES93] about the mean field theory of directed polymers with random complex weights whose modulus is independent of the argument; under this kind of assumptions, renormalization results in the space of distributions have been obtained very recently in [LRV00] for complex Gaussian multiplicative chaos on $\mathbb{R}^d$ when the modulus and argument which are independent.

When $(F_n)_{n\ge1}$ has a non degenerate limit F, the multifractal analysis of F is similar to that of Mandelbrot measures, but new multifractal phenomena can emerge when working with conservative complex cascades (see [BJ10]), a situation excluded in the present setting. This classical multifractal analysis can be completed by the natural notion of multifractal analysis of the graph roughness, a notion explored for non degenerate limits of $(F_n)_{n\ge1}$ in [Jin11].

For simplicity we suppose that $|W| > 0$ almost surely. Also, we assume that $\mathbb{P}(W \in \mathbb{C}\setminus\mathbb{R}_+) > 0$ and the normalization $E(W) = 1$ holds, so that $(F_n)_{n\ge1}$ is a martingale. Then we set

$$\varphi(q) = \varphi_W(q) = q - 1 - \log_b \mathbb{E}(|W|^q).$$

6.1 Some Convergence Theorems

6.1.1 Strong Convergence of Complex Cascades

There are conditions sufficient to imply the convergence of $F_n = F_{W,n}$, which in view of the positive case (Theorems 1 and 3(b)) seem optimal. The following result is proved in [BJM10a] (see [BJM10] for an extended version to general complex multiplicative chaos).

Theorem 17 *Assume that there exists $q > 1$ such that $\varphi_W(q) > 0$. Also suppose $q \in (1, 2]$ or $\varphi_W(2) > 0$.*

(1) *$(F_n)_{n\geq 1}$ uniformly converges, with probability 1, and in L^q norm, when n goes to ∞, towards a Hölder function $F = F_W$.*
(2) $F = \sum_{i=0}^{b-1} \mathbf{1}_{[i/b,(i+1)/b]}\Big(F(i/b)+W(i)\, F_i \circ f_i^{-1}\Big)$, *where $(W(0), \dots, W(b-1))$, $F_0, \dots, F_{b-1}$ are independent, F_i is equidistributed with F, and the equality holds with probability* 1.

With respect to the convergence of $(\mu_n)_{n\geq 1}$ when $W > 0$, the proof necessitates an additional compactness argument.

Remark 8

(1) The statistical self-affinity expressed by Theorem 17(2) implies, setting $Z = F(1) - F(0)$ and $Z(i) = F_i(1) - F_i(0)$:

$$Z = b^{-1} \sum_{k=0}^{b-1} W(k)\, Z(k),$$

which could be considered as an extension of (2.3) to the case of complex weights $W(k)$.
(2) When the weight W is real and such that $|W| = b^{1-H}$, Theorem 17 yields a limit function F which is a monofractal object obtained by a multiplicative cascade, which shares lots of properties with the fractional brownian motion of index H, but its construction is more straightforward. This remarkable fact was one of Mandelbrot's first motivations to consider the signed cascades. Nevertheless there is a constraint on the exponent H: it should lie within the interval $(1/2, 1]$ (see [BM09] for a specific study of the monofractal case).
For a pair $B_H = (F_1, F_2)$ of two independent copies of such a monofractal process, Jin has proved in [Jin14] an analogue of Kaufman's theorem about the Hausdorff dimension of the image of Borel subsets of $\mathbb{R}_+$, namely $\mathbb{P}(\forall\, E \in \mathcal{B}([0, 1]), \dim B_H(E) = H^{-1} \dim E) = 1$.
(3) In [BJM10a], when $|W|$ is not constant, the natural question of deciding whether the limit F can be decomposed as a monofractal process B composed with

the indefinite integral of a Mandelbrot measure μ is discussed. Under stronger assumptions than that of Theorem 17, denoting by β the unique solution of $\varphi(q) = 0$ in $[1, 2]$, and setting $W_\beta = |W|^\beta/\mathbb{E}(|W|^\beta)$, such a decomposition exists, B being of index $H = 1/\beta$, and $\mu = \mu_{W_\beta}$.

6.1.2 Weak Convergence Towards a Brownian Motion in Multifractal Time When W Is Real Valued

When one cannot use Theorem 17, it is natural to seek for a suitable normalisation of the process $F_{W,n}$ in order it be convergent in distribution. We present one result in this direction.

Assume

(1) $W \in \mathbb{R}$ almost surely.
(2) $\varphi_W > -\infty$ on $\mathbb{R}_+$.
(3) $\varphi_W(2) \leq 0$ and φ_W is non-decreasing.

It follows from these hypotheses that $|W| \leq b$ a.s., and the hypotheses of Theorem 17 are not fulfilled. Also, a direct computation shows that the martingale $F_n(1)$ is not bounded in L^2. More precisely

$$\sigma_n^2 = \mathbb{E}(|F_n(1)|^2) \sim \begin{cases} \sigma^2 b^{-n\varphi_W(2)} \text{ with } \sigma^2 = \dfrac{\mathbb{E}(|\sum_{i=0}^{b-1} W(i)|^2) - b^2}{\mathbb{E}(\sum_{i=0}^{b-1} |W(i)|^2) - b^2} & \text{if } \varphi_W(2) < 0 \\ \sigma^2 n \text{ with } \sigma^2 = b^{-2} \sum_{i \neq j} \mathbb{E}(W(i)W(j)) & \text{if } \varphi_W(2) = 0 \end{cases}.$$

Moreover, (3) implies $\varphi_{W_2}(q) > 0$ for all $q > 1$ (where $W_2 = W^2/\mathbb{E}(W^2)$. So, the non decreasing function F_{W_2}, which is nothing but the indefinite integral of the Mandelbrot measure μ_{W_2}, is non degenerate, and $F_{W_2}(1)$ is bounded in L^q for all $q > 1$ after Theorem 3(1).

Theorem 18 *Under the above assumptions, the random continuous function F_n/σ_n converges in distribution towards $B \circ F_{W^{(2)}}$, where B is a standard brownian motion independent of $F_{W^{(2)}}$.*

This is established in [BJM10a]. The assumption (3) is used to identify the law of the limit process thanks to the moments method and the equation $F_n(1) = \sum_{k=0}^{b-1} b^{-1} W(k) F_{n-1,k}$, after proving the tightness of the normalized sequence. However, it is desirable to relax this assumption.

Remark 9 A few additional information about the asymptotic behavior of F_n when the assumptions of Theorems 17 and 18 fail can be found in [BJM10a]. When $W/|W|$ and $|W|$ are independent, precise information about the convergence in probability of the free energy $n^{-1} \log |F_n(1)|$ can be found in [DES93], with three main possible phases. The recent related work achieved in [LRV00] leads to believe that in general, possible limit of F_n/A_n when the conditions of Theorem 17 fail are Brownian motion in critical Mandelbrot or critical Lévy-Mandelbrot time when φ presents a second order phase transition at β_c.

6.2 *Multifractal Analysis of Roughness in the Graph of F*

The multifractal analysis of the limit function F in Theorem 17 is a refinement of that of the Mandelbrot measure μ: one considers the Hölder exponent associated with the oscillations of order 1 of F,

$$h_F(t) = \liminf_{r\to 0^+} \frac{\log \mathrm{Osc}_F([t-r, r+r])}{\log r}$$

and the corresponding level sets

$$E_F(h) = \{t \in [0,1] : h_F(t) = h\} \quad (h \geq 0).$$

The exponent $h_F(\cdot)$ should be considered as a measure of the aspect more or less rough of the graph of F at each point: the smaller $h_F(t)$, the larger roughness at t is.

Let

$$\tau_F(q) = \liminf_{r\to 0} \frac{\log \sup\left\{\sum_i \mathrm{Osc}_F(B_i)^q\right\}}{\log r},$$

where the supremum is taken over all families of pairwise disjoint closed intervals B_i, of diameter $2r$, and centered in $\mathrm{supp}(F')$, and F' is the derivative of F in the distribution sense. The following statement is similar to Theorem 14. It is originally proved in [BJ10] under much stronger assumptions (also, in [BJ10] a finer multifractal analysis is achieved; also see [BFP10, BS05, Jaf98, Jaf00, Jaf04] to have a substantial overview of the multifractal formalism for functions), but the techniques introduced in [AB14] makes it possible to relax them, and now cover cases presenting first order phase transition.

Theorem 19 *Under the assumptions of Theorem* 17, *if* $\varphi(-\epsilon) > -\infty$ *for some* $\epsilon > 0$, *then with probability* 1, *all the conclusions of Theorem* 14 *hold with* ν *replaced by* F.

The sets $E_F(h)$ are only indirectly linked to the apparent roughness of the graph because they lie in the support of F. This leads to consider the *roughness spectrum* which gives the Hausdorff dimension of the sets obtained by lifting on the graph of F the sets $E_F(h)$. In other terms, this spectrum is the Hausdorff dimensions of the sets

$$G_F(h) = \{(t, F(t)) : t \in E_F(h)\} \quad (h \geq 0).$$

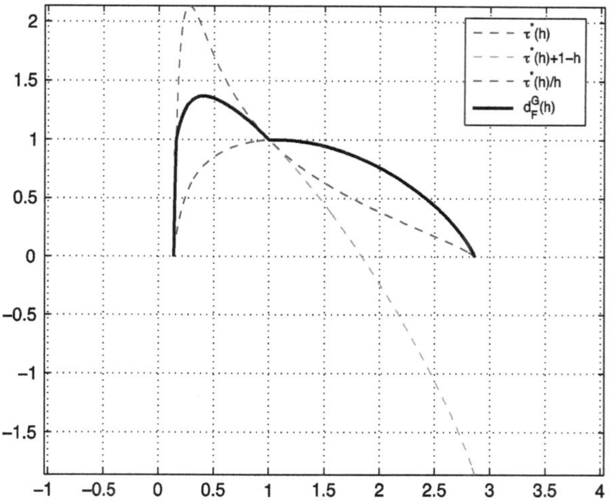

Fig. 9 Example of roughness spectrum $d_F^G : h \mapsto \dim G_F(h)$

The following result [Jin11] is the first of this kind for multifractal stochastic processes. We present here a stronger version using the assumptions of Theorem 19.

Theorem 20 *Under the assumptions of Theorem* 19, *if W is real valued, with probability* 1, *one has*

$$\dim \mathrm{Graph}(F) = 1 - \tau(1),$$

and, for all $h \geq 0$ *such that* $E_F(h) \neq \emptyset$, *one has*

$$\dim E_F(h) = \tau_F^*(h)$$

and

$$\dim G_F(h) = \Big(\frac{\dim E_F(h)}{h} \wedge \big(\dim E_F(h) + 1 - h\big)\Big) \vee \dim E_F(h).$$

The upper bounds for Hausdorff dimensions follows from quite standard arguments and is true in general. For the lower bound, the proof uses the auxiliary measures introduced to find a lower bound for $\dim E_F(h)$. These measures are lifted to the graph of F, and the lifted measures do have the desired lower Hausdorff dimension. This dimension is estimated by showing that they have a finite energy with respect to suitable Riesz kernels. However, while for the calculation of the dimension of the graph this method works quite straightforward, for the sets $G_F(h)$ in general the study is very delicate (see [Jin11] for details) (Fig. 9).

7 Mandelbrot Cascades as a Dynamical System

We present the results obtained in [BPW09], showing that Mandelbrot cascades define a dynamical system on a subset of the fixed points of the smoothing transformation. The asymptotic behavior of this dynamical systems exhibits a functional central limit theorem whose Gaussian limit process is, unexpectedly, the limit of an additive cascade on a tree. Fine properties of this process are also detailed.

7.1 A Dynamical System

Let $\mathcal{P}$ the set of Borel probability measures on $\mathbb{R}_+$. If $\mu \in \mathcal{P}$ and $p > 0$, we denote by $\mathbf{m}_p(\mu)$ the moment of order p of μ, i.e.,

$$\mathbf{m}_p(\mu) = \int_{\mathbb{R}_+} x^p \, \mu(\mathrm{d}x).$$

Then let $\mathcal{P}_1$ be the set of elements of $\mathcal{P}$ whose first moment equals 1:

$$\mathcal{P}_1 = \{\rho \in \mathcal{P} \ : \ \mathbf{m}_1(\rho) = 1\}.$$

Using the characterizations of the elements of $\mathcal{P}$ by their Laplace transform, the smoothing transformation S_ρ associated with $\rho \in \mathcal{P}$ considered in Sect. 3.1 is nothing but the mapping from $\mathcal{P}$ to itself so defined: If $\nu \in \mathcal{P}$, one considers $2b$ independent random variables, $Z(0), Z(1), \dots, Z(b-1)$, whose common probability distribution is ν, and $W(0), W(1), \dots, W(b-1)$ whose common probability distribution is ρ; then $\mathsf{S}_\rho \nu$ is the probability distribution of the random variable $b^{-1} \sum_{0 \le j < b} W(j)\, Z(j)$.

Since the measure ρ is in $\mathcal{P}_1$, S_ρ maps $\mathcal{P}_1$ into itself. We have seen (Theorem 1) that the condition $\int x \log(x)\, \rho(\mathrm{d}x) < \log b$ is necessary and sufficient for the weak convergence of the sequence $\mathsf{S}_\rho^n \delta_1$ (where δ_1 stands for the Dirac mass at point 1) towards a probability measure ν, which therefore is a fixed point of S_ρ. Indeed, under this assumption $\mathsf{S}_\rho^n \delta_1$ is the probability distribution of the uniformly integrable non-negative martingale $(Y_n)_{n \ge 1}$ whose limit Y has probability distribution ν belonging to $\mathcal{P}_1$ and satisfies $\mathsf{S}_\rho \nu = \nu$ due to (2.3). In this case, we denote the measure ν by $\mathsf{T}\rho$. It is natural to try and iterate T. But, in general this is not possible because $\nu = \mathsf{T}\rho$ may not inherit the property $\int x \log(x)\, \nu(\mathrm{d}x) < \log b$. So, we have to find a domain stable under the action of T. This is done by imposing conditions on moments.

Indeed, it is easily seen that the sequence $(Y_n)_{n \ge 1}$ defined by (2.1) remains bounded in L^2 norm if and only if $\mathbb{E}(W^2) = \mathbf{m}_2(\rho) < b$, and that in this case Formula (2.3) yields

$$\mathbb{E}Y^2 = \frac{b-1}{b-\mathbb{E}W^2} \tag{7.1}$$

(since the random variables $W(j)$ and $Y(j)$ are independent and of expectation 1, squaring both sides of Formula (2.3) yields $b^2\mathbb{E}Y^2 = b\mathbb{E}W^2\mathbb{E}Y^2 + b(b-1)$). It follows that if $b \geq 3$ and $1 \leq \mathbb{E}W^2 < b-1$, we have $\mathbb{E}Y^2 \leq \mathbb{E}W^2$ (the equality holding only if $W = 1$). Therefore, since the condition $\mathbb{E}W^2 < b$ is stronger than $\mathbb{E}(W \log W) < \log b$ when $\mathbb{E}W = 1$ (since the function $t \mapsto \log \mathbb{E}W^t$ is convex), T is a transformation on the subset of $\mathcal{P}_1$ defined by

$$\mathscr{P}_b = \{\rho \in \mathcal{P}_1 \ : \ 1 < \mathbf{m}_2(\rho) < b-1\}.$$

If $\rho \in \mathscr{P}_b$, due to (2.3), we can associate with each $n \geq 0$ a random variable W_{n+1} as well as $2b$ independent random variables $W_n(0), \ldots, W_n(b-1)$ and $W_{n+1}(0), \ldots, W_{n+1}(b-1)$ such that

$$W_{n+1} = \frac{1}{b}\sum_{k=0}^{b-1} W_n(k)W_{n+1}(k), \tag{7.2}$$

$\mathsf{T}^n\rho$ is the probability distribution of $W_n(k)$ for every k such that $0 \leq k \leq b-1$, and $\mathsf{T}^{n+1}\rho$ is the probability distribution of W_{n+1} and $W_{n+1}(k)$ for every $0 \leq k \leq b-1$. We advise the reader that if one writes Formula (7.2) with $n-1$ instead of n, the variables $W_n(k)$ which then appear are different from the previous ones.

We have seen in Sect. 2 that the random variable Y represents the increment between 0 and 1 of the non-decreasing continuous function F on $[0, 1]$ obtained as the almost sure uniform limit of the sequence of non-decreasing continuous functions F_n defined in (2.8), and F is nothing but the indefinite integral of the Mandelbrot measure μ. Thus, (2.4) rewrites, for $w \in \mathscr{A}^*$,

$$\Delta(F, I_w) = b^{-|w|}Y(w) \prod_{1\leq j\leq |w|} W(w|j), \tag{7.3}$$

where for any bounded $f : [0, 1] \mapsto \mathbb{R}$, for every sub-interval $I = [\alpha, \beta]$ of $[0, 1]$, we denote by $\Delta(f, I)$ the increment $f(\beta) - f(\alpha)$ of f over the interval I.

Let us denote by $\Phi(\rho)$ the probability distribution of the limit F, considered as a random continuous function.

We can describe the asymptotic behavior of the dynamical system $(\mathscr{P}_b, \mathsf{T})$, as well as the asymptotic behavior of $\left(\mathsf{T}^n\rho, \Phi(\mathsf{T}^{n-1}\rho)\right)_{n\geq 1}$ as n goes to ∞.

We need some more definitions. For $b \geq 3$, set

$$w_2(b) = \min\left(b-1, \ b\frac{b^4 - 4b^2 + 12b - 8}{b^4 + 8b^2 - 12b + 4}\right)$$

and, for t such that $1 < t < w_2(b)$,

$$w_3(b,t) = \frac{b^2}{2} + \frac{1}{2}\sqrt{\frac{b(b^4-4b^2+12b-8) - t\big(b^4+8b^2-12b+4\big)}{b-t}}.$$

One always has $w_3(b,t) < b^2 - 1$.

Also set

$$\mathscr{D}_b = \Big\{\rho \in \mathcal{P} \,:\, \mathbf{m}_1(\rho) = 1,\ 1 < \mathbf{m}_2(\rho) < w_2(b),\ \text{ and } \mathbf{m}_3(\rho) < w_3\big(b, \mathbf{m}_2(\rho)\big)\Big\}.$$

Theorem 21 *Suppose $b \geq 3$. Let $\rho \in \mathscr{P}_b$, and, for $n \geq 0$, define $\sigma_n = \left(\int (x-1)^2\, \mathsf{T}^n\mu(dx)\right)^{1/2}$. Then*

(1) *The limit of $(b-1)^{n/2}\sigma_n$ exists and is positive; so $\lim_{n\to\infty} \mathsf{T}^n\mu = \delta_1$.*
More generally, the probability distributions $\Phi(\mathsf{T}^n\rho)$ weakly converges towards δ_{Id}.

(2) *Suppose that ρ lies in the domain $\mathscr{D}_b \subset \mathscr{P}_b$. Then, if W_n is a variable whose distribution is $\mathsf{T}^n\mu$, $\dfrac{W_n - 1}{\sigma_n}$ converges in distribution towards $\mathcal{N}(0,1)$.*
More generally, if h_n is a random function distributed according to $\Phi(\mathsf{T}^{n-1}\rho)$, the distribution of $\dfrac{h_n - Id}{\sigma_n}$ weakly converges towards the distribution of the unique continuous Gaussian process $(X_t)_{t\in[0,1]}$, such that $X(0) = 0$ and, for all $j \geq 1$, the covariance matrix M_j of the vector $\big(\Delta(X, I_w)\big)_{w\in\mathscr{A}^j}$ is given by

$$\mathsf{M}_j(w,w') = \begin{cases} b^{-2j}\big(1 + (b-1)|w|\big) & \text{if } w = w', \\ b^{-2j}(b-1)|w \wedge w'| & \text{otherwise.}\end{cases}$$

Part (1) of the theorem follows from an easy calculation. For part (2), setting $Z_n = \frac{W_n - 1}{\sigma_n}$ one first exploits (7.2) to prove that $(|W_n|)_{n\geq 1}$, and then $(|Z_n|)_{n\geq 1}$, is bounded in L^3. This uses a recursion in which the domain of attraction $\mathscr{D}_b$ is introduced. Then, using (7.2) again one gets, after setting,

$$R_n = \frac{1}{b}\sum_{j=0}^{b-1} Z_n(j) Z_{n-1}(j)\sigma_{n-1} + \frac{1}{b}\left(\frac{\sigma_{n-1}}{\sigma_n} - \sqrt{b-1}\right)\sum_{j=0}^{b-1} Z_{n-1}(j),$$

$$Z_n = R_n + \frac{\sqrt{b-1}}{b}\sum_{k=0}^{b-1} Z_{n-1}(k) + \frac{1}{b}\sum_{k=0}^{b-1} Z_n(k). \tag{7.4}$$

Due to the equivalence $\sigma_n \sim c(b-1)^{-n/2}$, the situation is essentially reducible to the relation $Z_n = \frac{\sqrt{b-1}}{b}\sum_{k=0}^{b-1} Z_{n-1}(k) + \frac{1}{b}\sum_{k=0}^{b-1} Z_n(k)$. Using the relation (7.4) recursively n times yields a relation of the form $Z_n \stackrel{\text{dist}}{=} o(1) + \sum_{k=1}^{N_n} a_{n,k} Z_{n,k}$, where the $Z_{n,k}$ are independent centered variables, each of which is distributed like one of the Z_k, $1 \le k \le n$, $\lim_{n\to\infty} \sup\{a_{n,k} = 1 \le k \le N_n\} = 0$ and $\sum_{k=1}^{N_n} a_{n,k}^2 = 1$. An application of Lindeberg theorem (using the boundedness of $(|Z_k|)_{k\ge1}$ in L^3) yields the convergence in law of $(Z_n)_{n\ge1}$ to $\mathcal{N}(0, 1)$.

For the functional central limit theorem, set $\mathcal{Z}_n = \dfrac{h_n - \mathrm{Id}}{\sigma_n}$. From (7.3) one deduces

$$\begin{aligned}\Delta(\mathcal{Z}_n, I_w) &= b^{-j}\sigma_n^{-1}\left[W_n(w)\prod_{k=1}^{j} W_{n-1}(w|k) - 1\right]\\ &= b^{-j} Z_n(w)\prod_{k=1}^{j} W_{n-1}(w|k)\\ &\quad + b^{-j}\sum_{l=1}^{j}\frac{\sigma_{n-1}}{\sigma_n} Z_{n-1}(w|l)\prod_{k=1}^{l-1} W_{n-1}(w|k).\end{aligned} \tag{7.5}$$

From this relation one can both show the tightness of the distributions of the processes $\mathcal{Z}_n$, $n \ge 1$, and the convergence in distribution of the increments by simply passing to the limit and using the independences between the $W_{n-1}(w|k)$ and $W_n(w)$: there exist $\big(\mathcal{N}(v)\big)_{v\in\bigcup_{k=1}^{j}\mathscr{A}^k}$ and $\big(\widetilde{\mathcal{N}}(w)\big)_{w\in\mathscr{A}^j}$ two families of $\mathcal{N}(0, 1)$ random variables so that all the random variables involved in these families are independent, and

$$\lim_{n\to\infty}\big(\Delta(\mathcal{Z}_n, I_w)\big)_{w\in\mathscr{A}^j} \overset{\text{dist}}{=} b^{-j}\left(\widetilde{\mathcal{N}}(w) + \sqrt{b-1}\sum_{k=1}^{j}\mathcal{N}(w|k)\right)_{w\in\mathscr{A}^j}. \tag{7.6}$$

This is enough to derive the convergence in law of the $\mathcal{Z}_n$ to the Gaussian process X of the statement.

Remark 10 It is natural to seek for a larger domain of attraction than $\mathscr{D}_b$. This requires to be able to keep controls similar to the previous ones in $L^{2+\epsilon}$ (if not in L^2) rather than in L^3.

7.2 The Limit Process X as the Limit of an Additive Cascade

Recall that, if $v \in \mathscr{A}^*$, $[v]$ stands for the cylinder in $\mathscr{A}^\omega$ consisting of sequences beginning by v.

Let $(\xi(w))_{w\in\mathscr{A}^*}$ be a sequence of independent $\mathcal{N}(0, 1)$ random variables. For each $w \in \mathscr{A}^*$,

$$\zeta_n(w) = \sqrt{b-1}\sum_{j=1}^{n} b^{-j}\sum_{v\in\mathscr{A}^j}\xi(wv),$$

is a martingale bounded in L^2 norm, whose limit $\zeta(w)$ is a $\mathcal{N}(0,1)$ random variable, independent of the $\xi(v)$, $|v| \leq |w|$. Moreover, all the $\zeta(w)$, $w \in \mathscr{A}^n$ are independent.

One can then check that

$$M([w]) = b^{-|w|}\left(\zeta(w) + \sqrt{b-1}\sum_{1\leq k\leq |w|}\xi(w|k)\right) \tag{7.7}$$

defines a finitely additive Gaussian random measure defined on the cylinders of $\mathscr{A}^{\omega}$ Then, the limit process X of the previous sections can be seen as the primitive of the projection of M on $[0,1]$, and the structure of the increment $\Delta(X, I_w)$ given by (7.7) is an additive counterpart to the multiplicative structure of the increment of F given by (7.3).

Of course (7.7) makes sense even for $b = 2$.

7.3 Fine Properties of X

Due to the structure of the increments of X, it is natural to consider for all $\alpha \in \mathbb{R}$ the sets

$$\overline{E}_\alpha = \left\{t \in [0,1) : \limsup_{n\to\infty} \frac{\Delta(X, I_n(t))}{nb^{-n}} = \alpha\sqrt{b-1}\right\},$$
$$\underline{E}_\alpha = \left\{t \in [0,1) : \liminf_{n\to\infty} \frac{\Delta(X, I_n(t))}{nb^{-n}} = \alpha\sqrt{b-1}\right\},$$

and

$$E_\alpha = \underline{E}_\alpha \bigcap \overline{E}_\alpha,$$

where $I_n(t)$ stands for the semi-open to the right b-adic interval of generation n containing t. One has

Theorem 22 *With probability 1,*

(1) *the modulus of continuity of X is a $O\big(\delta\log(1/\delta)\big)$,*
(2) *X does not belong to the Zygmund class,*
(3) *the set E_0 contains a set of full Lebesgue measure at each point of which X is not differentiable,*
(4) $\dim E_\alpha = \dim \underline{E}_\alpha = \dim \overline{E}_\alpha = 1 - \dfrac{\alpha^2}{2\log b}$ *if* $|\alpha| \leq \sqrt{2\log b}$, *and* $E_\alpha = \emptyset$ *if* $|\alpha| > \sqrt{2\log b}$.

The multifractal analysis part (4) is directly deduced from the multifractal analysis of the branching random walk $S(w) = \sum_{1\leq j\leq |w|}\xi(w|j)$ which follows from the approach explained in Sect. 5.1.3. Part (3) is a consequence of the law of the iterated logarithm with respect to $\mathbb{P} \otimes$ Leb. Points (1) and (2) follow from quite direct calculations.

It would be good to decide whether or not X is nowhere differentiable.

Appendix: Hausdorff Measures and Dimension

Given $g : \mathbb{R}_+ \to \mathbb{R}_+$ a continuous non-decreasing function near 0 and such that $g(0) = 0$, and E a subset of $[0, 1]$, the Hausdorff measure of E with respect to the gauge function g is defined as

$$\mathcal{H}^g(E) = \lim_{\delta \to 0^+} \inf \Big\{ \sum_{i \in \mathbb{N}} g(\mathrm{diam}(U_i)) \Big\},$$

the infimum being taken over all the countable coverings $(U_i)_{i \in \mathbb{N}}$ of E by subsets of K of diameters less than or equal to δ.

If $s \in \mathbb{R}_+^*$ and $g(u) = u^s$, then $\mathcal{H}^g(E)$ is also denoted $\mathcal{H}^s(E)$ and called the s-dimensional Hausdorff measure of E. Then, the Hausdorff dimension of E is defined as

$$\dim E = \sup\{s > 0 : \mathcal{H}^s(E) = \infty\} = \inf\{s > 0 : \mathcal{H}^s(E) = 0\},$$

with the convention $\sup \emptyset = 0$ and $\inf \emptyset = \infty$.

For more information the reader is referred to [Fal03, Mat95].

References

[Aid13] Aidekon, E.: Convergence in law of the minimum of a branching random walk. Ann. Probab. **41**, 1362–1426 (2013)

[AS14] Aidekon, E., Shi, Z.: The Seneta-Heyde scaling for the branching random walk. Ann. Probab. **42**, 959–993 (2014)

[ABM12] Alsmeyer, G., Biggins, J.-D., Meiners, M.: The functional equation of the smoothing transformation. Ann. Probab. **140**, 2069–2105 (2012)

[AM12] Alsmeyer, G., Meiners, M.: Fixed points of inhomogeneous smoothing transforms. J. Differ. Equ. Appl. **18**, 1287–1304 (2012)

[AM13] Alsmeyer, G., Meiners, M.: Fixed points of the smoothing transformation: two-sided case. Probab. Theory Relat. Fields **155**, 165–199 (2013)

[AJKS11] Astala, K., Jones, P., Kupiainen, A., Saksman, E.: Random conformal weldings. Acta Math. **207**(2), 203–254 (2011)

[AB14] Attia, N., Barral, J.: Hausdorff and packing spectra, large deviations, and free energy for branching random walks. in $\mathbb{R}^d$. To appear in Comm. Math. Phys (2014). arXiv:1305.2034

[BM03] Bacry, E., Muzy, J.F.: Log-infinitely divisible multifractal processes. Commun. Math. Phys. **236**, 449–475 (2003)

[Bar00] Barral, J.: Continuity of the multifractal spectrum of a random statistically self-similar measure. J. Theor. Probab. **13**, 1027–1060 (2000)

[BBP03] Barral, J., Ben Nasr, F., Peyrière, J.: Comparing multifractal formalisms: the neighboring boxes conditions. Asian J. Math. **7**, 149–166 (2003)

[BF05] Barral, J., Fan, A.H.: Covering numbers of different points in the Dvoretzky covering. Bull. Sci. Math. **129**, 275–317 (2005)

[BFP10] Barral, J., Fan, A.-H., Peyrière, J.: Mesures engendrées par multiplications. In: Quelques Interactions Entre Analyse, Probabilités et Fractals, Panoramas et Synthéses, vol. 32, Soc. Math., France (2010)

[BJ10] Barral, J., Jin, X.: Multifractal analysis of complex random cascades. Commun. Math. Phys. **219**, 129–168 (2010)

[BJM10] Barral, J., Jin, X., Mandelbrot, B.B.: Uniform convergence for complex [0, 1]-martingales. Ann. Appl. Probab. **20**, 1205–1218 (2010)

[BJM10a] Barral, J., Jin, X., Mandelbrot, B.B.: Convergence of signed multiplicative cascades. Ann. Appl. Probab. **20**, 1219–1252 (2010)

[BJRV13] Barral, J., Jin, X., Rhodes, R., Vargas, V.: Gaussian multiplicative chaos and KPZ duality. Commun. Math. Phys. **323**, 451–485 (2013)

[BKNSW14a] Barral, J., Kupiainen, A., Nikula, M., Saksman, E., Webb, C.: Critical Mandelbrot cascades. Commun. Math. Phys. **325**, 685–711 (2014)

[BKNSW14+] Barral, J., Kupiainen, A., Nikula, M., Saksman, E., Webb, C.: Basic properties of critical lognormal multiplicative chaos. To appear in Ann. Probab. arXiv:1303.4548

[BM02] Barral, J., Mandelbrot, B.: Multifractal products of cylindrical pulses. Probab. Theory Relat. Fields **124**, 409–430 (2002)

[BM04] Barral, J., Mandelbrot, B.: Introduction to infinite products of random independent functions (Random multiplicative multifractal measures, Part I). In: Lapidus, M.L., van Frankenhuysen, M. (eds.) Fractal Geometry and Applications: A Jubilee of Benot Mandelbrot. Proceedings of Symposia in Pure Mathematics, vol. 72, Part 2, pp. 3–16. AMS, Providence, RI (2004)

[BM04a] Barral, J., Mandelbrot, B.: Non-degeneracy, moments, dimension, and multifractal analysis for random multiplicative measures (Random multiplicative multifractal measures, Part II), pp. 17–52. In: Proceedings of Symposia in Pure Mathematics, vol. 72, Part 2. AMS, Providence, RI (2004)

[BM09] Barral, J., Mandelbrot, B.B.: Fractional multiplicative processes. Ann. Inst. Henry Poincaré Probab. Stat. **45**, 1116–1129 (2009)

[BPW09] Barral, J., Peyrière, J., Wen, Z.: Dynamics of Mandelbrot cascades. Probab. Theory Relat. Fields **144**, 615–631 (2009)

[BRV12] Barral, J., Rhodes, R., Vargas, V.: Limiting laws of supercritical branching random walks, C. R. Acad. Sci. Paris Ser. I **350**, 535–538 (2012)

[BS05] Barral, J., Seuret, S.: From multifractal measures to multifractal wavelet series. J. Fourier Anal. Appl. **11**, 589–614 (2005)

[BS07] Barral, J., Seuret, S.: Heterogeneous ubiquitous systems in $\mathbb{R}^d$ and Hausdorff dimension. Bull. Braz. Math. Soc. **38**, 467–515 (2007)

[BS07] Barral, J., Seuret, S.: The singularity spectrum of Lévy processes in multifractal time. Adv. Math. **214**, 437–468 (2007)

[BS09] Benjamini, I., Schramm, O.: KPZ in one dimensional random geometry of multiplicative cascades. Commun. Math. Phys. **289**, 653–662 (2009)

[Ber96] Bertoin, J.: Lévy Processes. Cambridge University Press, Cambridge (1996)

[Big77] Biggins, J.D.: Martingale convergence in the branching random walk. J. Appl. Probab. **14**, 25–37 (1977)

[Big79] Biggins, J.D.: Growth rates in the branching random walk. Z. Wahrsch. Verw. Geb. **48**(48), 17–34 (1979)

[Big92] Biggins, J.D.: Uniform convergence of martingales in the branching random walk. Ann. Probab. **20**, 137–151 (1992)

[BK97] Biggins, J.D., Kyprianou, A.: Seneta-Heyde norming in the branching random walk. Ann. Probab. **25**, 337–360 (1997)

[BK04] Biggins, J.D., Kyprianou, A.: Measure change in multitype branching. Adv. Appl. Probab. **36**, 544–581 (2004)

[BK05] Biggins, J.D., Kyprianou, A.: The smoothing transform: the boundary case. Electron. J. Probab. **10**, 609–631 (2005)

[BHJ11] Biggins, J.D., Hambly, B.M., Jones, O.D.: Multifractal spectra for random self-similar measures via branching processes. Adv. Appl. Probab. **43**, 1–39 (2011)

[Bov06] Bovier, A.: Statistical mechanics of disordered system. A mathematical perspective. Cambridge Series in Statistical and Probabilistic Mathematics, vol. 18. Cambridge University Press, Cambridge (2006)

[Bra83] Bramson, M.: Convergence of solutions of the Kolmogorov equation to travelling waves. Mem. Am. Math. Soc. **44**(285), iv+190 pp. (1983)

[Bur09] Buraczewski, D.: On tails of fixed points of the smoothing transform in the boundary case. Stoch. Process. Appl. **119**(11), 3955–3961 (2009)

[CD01] Carpentier, D., Le Doussal, P.: Glass transition of a particle in a random potential, front selection in nonlinear RG and entropic phenomena in Liouville and Sinh-Gordon models. Phys. Rev. E **63**, 026110 (2001)

[CRV13] Chevillard, L., Rhodes, R., Vargas, V.: Gaussian multiplicative chaos for symmetric isotropic matrices. J. Stats. Phys. **150**, 678–703 (2013)

[CK92] Collet, P., Koukiou, F.: Large deviations for multiplicative chaos. Commun. Math. Phys. **147**, 329–342 (1992)

[DZ98] Dembo, A., Zeitouni, O.: Large deviations techniques and applications. In: Applications of Mathematics, vol. 38, 2nd edn. Springer, New York (1998)

[DS88] Derrida, B., Spohn, H.: Polymers on disordered trees, spin glasses and traveling waves. J. Stat. Phys. **51**, 817 (1988)

[DES93] Derrida, B., Evans, M.R., Speer, E.R.: Mean field theory of directed polymers with random complex weights. Commun. Math. Phys. **156**(2), 221–244 (1993)

[DMPV95] Dodson, M.M., Melián, M.V., Pestana, D., Vélani, S.L.: Patterson measure and ubiquity. Ann. Acad. Sci. Fenn. Ser. A I Math. **20** 37–60 (1995)

[DS09] Duplantier, B., Sheffield, S.: Duality and KPZ in Liouville quantum gravity. Phys. Rev. Lett. **102**, 150603 (2009)

[DS11] Duplantier, B., Sheffield, S.: Liouville quantum gravity and KPZ. Invent. Math. **185**(2), 333–393 (2011)

[DRSV00a] Duplantier, B., Rhodes, R., Sheffield, S., Vargas, V.: Critical Gaussian multiplicative chaos: convergence of the derivative martingale. arXiv:1206.1671

[DRSV00] Duplantier, B., Rhodes, R., Sheffield, S., Vargas, V.: Renormalization of critical Gaussian multiplicative chaos and KPZ formula. arXiv:1212.0529

[DL83] Durrett, R., Liggett, T.: Fixed points of the smoothing transformation. Z. Wahrsch. Verw. Geb. **64**, 275–301 (1983)

[Fal94] Falconer, K.J.: The multifractal spectrum of statistically self-similar measures. J. Theor. Probab. **7**(3), 681–702 (1994)

[Fal03] Falconer, K.J.: Fractal Geometry. Mathematical Foundations and Applications, 2nd edn. Wiley, New York (2003)

[Fan89] Fan, A.H.: Chaos additifs et chaos multiplicatifs de Lévy. C. R. Acad. Sci. Paris Sér. I **308**, 151–154 (1989)

[Fan97] Fan, A.H.: Sur le chaos de Lévy d'indice $0 < \alpha < 1$. Ann. Sci. Math. Que. **21**, 53–66 (1997)

[Fan04] Fan, A.-H.: Some topics in the theory of multiplicative chaos. In: Bandt, Ch., Zaehle, M., Mosco, U. (eds.) Progress in Probability, vol. 57, pp. 119–134. Birkhäuser, Basel (2004)

[Fra95] Franchi, J.: Chaos multiplicatif: un traitement simple et complet de la fonction de partition. Séminaire Probab. Strasbourg **29**, 194–201 (1995)

[FP88] Frisch, U., Parisi, G.: Fully developed turbulence and intermittency in turbulence. Proceedings of the International School of Physics "Enrico Fermi". In: Ghil, M. (ed.) Turbulence and Predictability in Geophysical Fluid Dynamics and Climate Dynamics, course 88, pp. 84–88. North Holland, Amsterdam (1985)

[Gui90] Guivarch, Y.: Sur une extension de la notion de loi semi-stable. Ann. Inst. Henry Poincaré Probab. Stat. **26**, 261–285 (1990)

[HW92] Holley, R., Waymire, E.C.: Multifractal dimensions and scaling exponents for strongly bounded random fractals. Ann. Appl. Probab. **2**, 819–845 (1992)

[HS09] Hu, Y., Shi, Z.: Minimal position and critical martingale convergence in branching random walks, and directed polymers on disordered trees. Ann. Probab. **37**(2), 742–789 (2009)

[Jaf98] Jaffard, S.: Oscillations spaces: properties and applications to fractal and multifractal functions. J. Math. Phys. **39**(8), 4129–4141 (1998)

[Jaff99] Jaffard, S.: The multifractal nature of Lévy processes. Probab. Theory Relat. Fields **114**, 207–227 (1999)

[Jaf00a] Jaffard, S.: On lacunary wavelet series. Ann. Appl. Probab. **10**, 313–329 (2000)

[Jaf00] Jaffard, S.: On the Frisch-Parisi conjecture. J. Math. Pures Appl. **79**(6), 525–552 (2000)

[Jaf04] Jaffard, S.: Wavelets techniques in multifractal analysis. In: Lapidus, M., Frankenhuijsen, M.V. (eds.) Fractal Geometry and Applications: A Jubilee of Benoît Mandelbrot. Proceedings of Symposia in Pure Mathematics, vol. 72(2), pp. 91–151 (2004)

[Jin11] Jin, X.: The graph and range singularity spectra of b-adic independent cascade functions. Adv. Math. **226**, 4987–5017 (2011)

[Jin12] Jin, X.: Dimension result and KPZ formula for two-dimensional multiplicative cascade processes. Ann. Probab. **40**, 1–18 (2012)

[Jin14] Jin, X.: A uniform result for two dimensional fractional multiplicative processes. Ann. Inst. H. Poincaré Probab. Stat. **50**(2), 512–523 (2014)

[JW11] Johnson, T., Waymire, E.: Tree polymers in the infinite volume limit at critical strong disorder. J. Appl. Probab. **48**, 885–891 (2011)

[Kah74] Kahane, J.-P.: Sur le modele de turbulence de Benoit Mandelbrot. C. R. Acad. Sci. Paris, **278**, 567–569 (1974)

[Kah85] Kahane, J.-P.: Sur le chaos multiplicatif. Ann. Sci. Math. Que. **9**, 105–150 (1985)

[Kah87a] Kahane, J.-P.: Positive martingales and random measures. Chi. Ann. Math. **8B1**, 1–12 (1987)

[KP76] Kahane, J.-P., Peyrière, J.: Sur certaines martingales de B. Mandelbrot Adv. Math. **22**, 131–145 (1976)

[Kah87] Kahane, J.-P.: Multiplications aléatoires et dimensions de Hausdorff. Ann. Inst. Henri Poincaré Probab. Stat. **23**, 289–296 (1987)

[Kah91] Kahane, J.-P.: Produits de poids aléatoires indépendants et applications. In: Fractal Geometry and Analysis (Montreal, PQ, 1989), pp. 277–324. NATO Adv. Sci. Inst. Ser. C Math. Phys. Sci., vol. 346. Kluwer, Dordrecht (1991)

[KPZ98] Knizhnik, V.G., Polyakov, A.M., Zamolodchikov, A.B.: Fractal structure of 2D-quantum gravity. Mod. Phys. Lett A **3**(8), 819–826 (1988)

[Kol62] Kolmogorov, A.N.: Précisions sur la structure locale de la turbulence dans un fluide visqueux aux nombres de Reynolds élevés, Mécanique de la turbulence, Colloq. Intern. CNRS, Marseille (1961) [Editions CNRS, pp. 447–451 (1962)]

[LRV00] Lacoin, H., Rhodes, R., Vargas, V.: Complex Gaussian multiplicative chaos. arXiv:1307.6117v1

[Liu98] Liu, Q.: Fixed points of a generalized smoothing transformation and applications to the branching random walk. Adv. Appl. Probab. **30**, 85–112 (1998)

[Liu00] Liu, Q.: On generalized multiplicative cascades. Stoch. Process. Appl. **86**(2), 263–286 (2000)

[Liu01] Liu, Q.: Asymptotic properties and absolute continuity of laws stable by random weightedmean. Stoch. Proc. Appl. **95**, 83–107 (2001)

[Mad00] Madaule, T.: Convergence in law for the branching random walk seen from its tip. arXiv:1107.2543

[Man72] Mandelbrot, B.B.: Possible refinement of the lognormal hypothesis concerning the distribution of energy in intermittent turbulence. Statistical models and turbulence. In: Rosenblatt, M., Atta, C.V. (eds.) Lectures Notes in Physics, vol. 12, 333–351. Springer, New York (1972)

[Man74a] Mandelbrot, B.B.: Multiplications aléatoires itérées et distributions invariantes par moyennes pondérées. C. R. Acad. Sci. Paris **278**, 289–292 and 355–358 (1974)

[Man74b] Mandelbrot, B.B.: Intermittent turbulence in self-similar cascades, divergence of high moments and dimension of the carrier. J. Fluid Mech. **62**, 331–358 (1974)

[Mat95] Mattila, P.: Geometry of sets and measures in Euclidean spaces, fractals and rectifiability. In: Cambridges Studies in Advanced Mathematics, vol. 44. Cambridge University Press, Cambridge (1995)

[Mol96] Molchan, G.M.: Scaling exponents and multifractal dimensions for independent random cascades. Commun. Math. Phys. **179**(3), 681–702 (1996)

[Ols94] Olsen, L.: Random geometrically graph directed self-similar multifractals, Pitman Res. Notes Math. Ser. **307** (1994)

[Ols95] Olsen, L.: A multifractal formalism. Adv. Math. **116**, 92–195 (1995)

[OW00] Ossiander M., Waymire, E.C.: Statistical estimation for multiplicative cascades. Ann. Stat. **28**, 1–29 (2000)

[Pey74] Peyrière, J.: Turbulence et dimension de Hausdorff. C. R. Acad. Sci. Paris **278**, 567–569 (1974)

[Pey77] Peyrière, J.: Calculs de dimensions de Hausdorff. Duke Math. J. **44**, 591–601 (1977)

[Pey79] Peyrière, J.: A singular random measure generated by spliting [0, 1]. Z. Wahrsch. Verw. Geb. **47**, 289–297 (1979)

[Pey00] Peyrière, J.: Recent results on Mandelbrot multiplicative cascades. In: Bandt, Ch., Graf, S., Zähle, M. (eds.) Fractal Geometry and Stochastics II, pp. 147–159. Progress in Probability, vol. 46. Birkhäuser, Basel (2000)

[RV10] Robert, R., Vargas, V.: Gaussian multiplicative chaos revisited. Ann. Probab. **38**(2), 605–631 (2010)

[RV08] Rhodes, R., Vargas, V.: KPZ formula for log-infinitely divisible multifractal random measures. arXiv:0807.1036

[RV13] Rhodes, R., Vargas, V.: Gaussian multiplicative chaos and applications: a review. arXiv:1305.6221

[She00] Sheffield, S.: Conformal weldings of random surfaces: SLE and the quantum gravity zipper. arXiv:1012.4797

[She72] Shepp, L.: Covering the line with random arcs. Z. Wahrsch. Verw. Geb. **23**, 163–170 (1972)

[WW69] Waymire, E.C., Williams, S.C.: A cascade decomposition theory with applications to Markov and exchangeable cascades. Trans. Am. Math. Soc. **348**, 585–632 (1996)

[WW13] Waymire, E.C., Williams, S.C.: T-martingales, size-biasing and tree polymer cascades. In: Barral, J., Seuret, S. (eds.) Further Developments in Fractals and Related Fields. Birkhäuser, Basel (2013)

[Web11] Webb, C.: Exact asymptotics of the freezing transitions of a logarithmically correlated random energy model. J. Stat. Phys. **145**, 1595–1619 (2011)

Law of Pure Types and Some Exotic Spectra of Fractal Spectral Measures

Xin-Rong Dai, Xing-Gang He and Chun-Kit Lai

Abstract Let μ be a Borel probability measure with compact support in $\mathbb{R}^d$ and let $E(\Lambda) = \{e^{-2\pi\lambda\cdot x} : \lambda \in \Lambda\}$. We make a review on some recent progress about spectral measures. We first show that the law of pure types holds for spectral measures, i.e. if $E(\Lambda)$ is a frame for $L^2(\mu)$, then μ is discrete or absolutely continuous or singular continuous with respect to Lebesgue measure (see [HLL13]). And we discuss the spectral properties of Cantor measures (see [DaHL13]), where we focus on some exotic properties of the spectra of some Cantor measures.

Keywords Cantor measures fourier frames · Law of pure types · Spectral measures

2010 Mathematics subject classification 28A80 · 42C05

The research is partially supported by the RGC grant of Hong Kong and the Focused Investment Scheme of CUHK; The first two authors are also supported by the National Natural Science Foundation of 10871180 and 11271148, and Guangdong Province Key Laboratory of Computational Science at the Sun Yat-sen University.

X.-R. Dai
School of Mathematics and Computational Science, Sun Yat-Sen University,
Guangzhou 510275, P. R. China
e-mail: daixr@mail.sysu.edu.cn

X.-G. He (✉)
School of Mathematics and Statistics, Central China Normal University,
Wuhan 430079, P. R. China
e-mail: xingganghe@sina.com

C.-K. Lai
Department of Mathematics and Statistics, McMaster University, Hamilton,
ON L8S 4K1, Canada
e-mail: cklai@math.mcmaster.ca

D.-J. Feng and K.-S. Lau (eds.), *Geometry and Analysis of Fractals*,
Springer Proceedings in Mathematics & Statistics 88,
DOI 10.1007/978-3-662-43920-3_2

1 Introduction to the General Spectral Measures

Let μ be a Borel probability measure on $\mathbb{R}^d$ with compact support. We call a family $E(\Lambda) = \{e_\lambda := e^{-2\pi i\lambda\cdot x} : \lambda \in \Lambda\}$ (Λ is a countable set) a *Fourier frame* for the Hilbert space $L^2(\mu)$ if there exist $A, B > 0$ such that

$$A\|f\|^2 \leq \sum_{\lambda\in\Lambda} |\langle f, e_\lambda\rangle_\mu|^2 \leq B\|f\|^2, \quad \forall\, f \in L^2(\mu). \tag{1.1}$$

Here the inner product is defined as usual, $x \cdot y = \sum_{i=1}^{d} x_i y_i$ for $x, y \in \mathbb{R}^d$ and

$$\langle f, e_\lambda\rangle_\mu = \int_{\mathbb{R}^d} f(x)e_{-\lambda}d\mu(x).$$

$E(\Lambda)$ is called an *(exponential) Riesz basis* if it is both a basis and a frame for $L^2(\mu)$. Fourier frames and exponential Riesz bases are natural generalizations of exponential orthonromal bases in $L^2(\mu)$. They have fundamental importance in non-harmonic Fourier analysis and wavelet. When (1.1) is satisfied, $f \in L^2(\mu)$ can be expressed as $f(x) = \sum_{\lambda\in\Lambda} c_\lambda e^{2\pi i\lambda x}$, and the expression is unique if it is a Riesz basis.

When $E(\Lambda)$ is an orthonormal basis (Riesz basis, or frame) for $L^2(\mu)$, we say that μ is a *spectral measure* (*R-spectral measure, or F-spectral measure respectively*) and Λ is called a *spectrum* (*R-spectrum, or F-spectrum respectively*) of μ. We will also use the term *orthonormal spectrum* instead of spectrum when we need to emphasize the orthonormal property. If $E(\Lambda)$ only satisfies the upper bound condition in (1.1), then it is called a *Bessel sequence*; for convenience, we also call Λ a Bessel sequence of $L^2(\mu)$.

Since Fuglede proposed the spectral set conjecture [Fug74] and Jorgensen and Pedersen [JP98] discovered the first singular fractal spectral measure, there has been a lot of interest in understanding which kind of measures are spectral and its delicate connection with translational tiling. In this short note, we aim at giving a systematic survey on the recent progress in this line of research and some more detailed explanations about our discovery in [HLL13, DaHL13] will be given.

The first fundamental result is about the law of pure type. It was proved by He et al. [HLL13], which generalized the early investigation of spectral measures by Łaba and Wang [LaW06]. Recall that a σ-finite Borel measure μ on $\mathbb{R}^d$ can be decomposed uniquely as *discrete*, *singularly continuous* and *absolutely continuous* measures, i.e., $\mu = \mu_d + \mu_s + \mu_a$. The measure μ is said to be of *pure types* if μ equals only one of the three components.

Theorem 1.1 *Let μ be an F-spectral measure on $\mathbb{R}^d$. Then it must be one of the three pure types: discrete (and finite), singularly continuous or absolutely continuous.*

By the law of pure types, we can study spectral measures according to its type. When μ is a discrete counting measure with finite support, it is an R-spectral measure

[HLL13]; When μ is absolutely continuous, Lai proved that μ is an F-spectral measure if and only if its density function is bounded above and bounded away from 0 almost everywhere on its support [Lai11]. Furthermore, Dutkay and Lai proved that if μ is a spectral measure, then its density function is a constant on its support, that is, μ is essentially the Lebesgue measure restricted on its support [DL00]. However, classification of spectral measures is far from complete. For the study of R-spectral absolutely continuous measures, one can refer to the recent work of Lev et al. with their emphasis on the use of quasicrystals [KN00, Lev12, GL14].

From now on, we concentrate on orthogonally spectral measures. We call Λ an *orthogonal set* if $E(\Lambda)$ is a mutually orthogonal sequence for $L^2(\mu)$. Define

$$Q_\Lambda(\xi) = \sum_{\lambda\in\Lambda} |\widehat{\mu}(\xi+\lambda)|^2,$$

where the Fourier transform of μ is define as usually by

$$\widehat{\mu}(\xi) = \int_{\mathbb{R}^d} e^{-2\pi\xi\cdot x} d\mu(x).$$

Q_Λ is crucial in determining whether $E(\Lambda)$ is complete. It is well-known that an orthogonal sequence $E(\Lambda)$ is complete in $L^2(\mu)$ if and only if $Q_\Lambda \equiv 1$ [JP98]. Here, we give a slight generalization of this result and also exploit the analytic property of Q_Λ.

Theorem 1.2 *Let μ be a compactly supported Borel probability measure with compact support in $\mathbb{R}^d$.*

(i) Suppose that $\overline{\text{span}}E(\Gamma) = L^2(\mu)$ and $E(\Lambda)$ is an orthogonal set for $L^2(\mu)$. Then Λ is a spectrum of μ if and only if

$$Q_\Lambda(\gamma) = 1, \quad \text{for } \gamma \in \Gamma.$$

(ii) Suppose that $E(\Lambda)$ is a Bessel sequence for $L^2(\mu)$. Then $Q_\Lambda(\cdot)$ is an entire function in $\mathbb{C}^d$.

The entire property is a simple extension of [JP98, Lemma 4.3]. In our proofs, this property helps us establish the completeness by allowing us to focus on small values of ξ.

Our main interest on the spectral measures is when μ is singularly continuous. The one-fourth contraction Cantor measure was the first example of such spectral measures, which was found by Jorgensen and Pedersen [JP98] in 1998. From that time on, various properties of singular spectral measures are studied extensively [Dai12, DHJ09, DHS09, DHSW11, LaW02, LaW06]. In particular, many exotic spectra were discovered and they do not appear in their absolutely continuous counterpart. Here, we list some of the interesting ones.

(1) There exists a spectrum Λ of a singularly continuous measure μ such that $k\Lambda$ is also a spectrum of μ for some $k \neq 1$;
(2) There exists a Λ so that $E(\Lambda)$ is a maximal orthogonal collection of exponentials for $L^2(\mu)$, but not a basis;
(3) There exists a spectrum Λ of a singularly continuous measure so that its Beurling dimension is zero.

Property (1) means that we can sort of dilate a spectrum but preserve its completeness. It was first given by Łaba and Wang [LaW02] and some studies are given in [DJ12].

Property (2) has two types of variants. First, some measures have maximal orthogonal collections of infinite cardinality without being spectral [HuL08, Dai12]. Second, even though the measure is spectral, there still exists some incomplete maximal orthogonal collections. In [DHS09], Dutkay et al. tried to give a classification on maximal orthogonal collection for one-fourth Cantor measures and tried to study which of them are complete. This investigation was generalized and improved in [DaHL13]. Furthermore, we can demonstrate the existence of spectrum satisfying property (3). Beurling dimension is a concept defined in [DHSW11], who tried to generalize Beurling density and the elegant result of Landau [Lan67] on Fourier frame spectra to fractal setting. Their work gave some partial positive results, letting alone a technical assumption on the spectra. In person communication with Wang in 2011, we were told that he can construct an example such that a spectral measure can have a spectrum with zero Beurling dimension. However, he cannot explain why there can be such phenomenon. Our construction gave a better picture of it.

For the rest of our paper, we will prove Theorem 1.1 and 1.2 in Sect. 2. In Sect. 3, we will present a simplified content of [DaHL13] and the examples of zero Beurling dimension spectra will be given. For more results on this issue, reader may refer to [DaHL13, DHS09].

2 Law of Pure Types

In this section, we will present a self-contained proof for the law of pure types of F-spectral measures. First, we need the following proposition, which was proved in [DHSW11]. This can be viewed as the stability of Bessel sequence under a constant perturbation of a Bessel sequence. It has its origin in the paper of Duffin and Schaeffer [DS52].

Proposition 2.1 *Let* $\{\lambda_n\}_{n=0}^{\infty}$ *be a Bessel sequence of* μ*. If there exists* C *such that* $|\lambda_n - \gamma_n| \le C$ *for* $n \ge 0$*, then* $\{\gamma_n\}_{n=0}^{\infty}$ *is also a Bessel sequence of* μ*.*

Proof It is sufficient to show that all $\gamma_n = (\gamma_1^{(n)}, \dots, \gamma_d^{(n)})$ differs $\lambda_n = (\lambda_1^{(n)}, \dots, \lambda_d^{(n)})$ only on the first component, and the statement follows by induction on the number of components.

Let $\text{supp}\mu \subseteq [-P, P]^d$ for some $P > 0$. We have that

$$\begin{aligned}
\sum_{n=0}^{\infty} \left| < f(x), e^{-2\pi i \gamma_n \cdot x} > \right|^2 &= \sum_{n=0}^{\infty} \left| < f(x) e^{2\pi i (\gamma_n - \lambda_n)\cdot x}, e^{-2\pi i \lambda_n \cdot x} > \right|^2 \\
&= \sum_{n=0}^{\infty} \left| < f(x) e^{2\pi i (\gamma_1^{(n)} - \lambda_1^{(n)}) x_1}, e^{-2\pi i \lambda_n \cdot x} > \right|^2 \\
&= \sum_{n=0}^{\infty} \left| \sum_{k=0}^{\infty} \frac{(2\pi i (\gamma_1^{(n)} - \lambda_1^{(n)}))^k}{k!} < f(x) x_1^k, e^{-2\pi i \lambda_n \cdot x} > \right|^2 \\
&\le \sum_{n=0}^{\infty} \sum_{k=0}^{\infty} \frac{(2\pi C)^{2k}}{k!} \sum_{k=0}^{\infty} \frac{| < f(x) x_1^k, e^{-2\pi i \lambda_n \cdot x} > |^2}{k!} \\
&\le e^{(2\pi C)^2} \sum_{k=0}^{\infty} \frac{B \| f(x) x_1^k \|^2}{k!} \\
&\le B e^{(2\pi C)^2 + P^2} \| f \|^2.
\end{aligned}$$

Note that the fourth line above uses Cauchy-Schwarz inequality. Hence, the assertion follows. □

In the proof of the pure type property of the F-spectral measures, we need to use the *lower Beurling density* of an infinite discrete set $\Lambda \subset \mathbb{R}^d$:

$$D^{-}\Lambda := \liminf_{h\to\infty} \inf_{x\in\mathbb{R}^d} \frac{\#(\Lambda \cap Q_h(x))}{h^d},$$

where $Q_h(x)$ is the standard cube of side length h centered at x. Intuitively Λ is distributed like a lattice if $D^{-}\Lambda$ is positive. In the seminal paper [Lan67], Landau gave an elegant and useful necessary condition for Λ to be an F-spectrum on $L^2(\Omega)$: $D^{-}\Lambda \ge \mathcal{L}(\Omega)$, where $\mathcal{L}$ is the Lebesgue measure. The following proposition provides some relationships between the lower Beurling density and the types of the measures.

Proposition 2.2 *Let μ be a compactly supported probability measure on $\mathbb{R}^d$ and let Λ be an F-spectrum of μ. We have*

(i) *If $\mu = \sum_{c\in C} p_c\delta_c$ is discrete, then $\#\Lambda < \infty$ and $\#C < \infty$;*
(ii) *If μ is singularly continuous, then $D^{-}\Lambda = 0$;*
(iii) *If μ is absolutely continuous, then $D^{-}\Lambda > 0$.*

Proof (i) By the definition of Fourier frame, we have for all $f \in L^2(\mu)$,

$$\sum_{\lambda\in\Lambda} |\sum_{c\in C} f(c) e^{2\pi i \langle\lambda, c\rangle} p_c|^2 \le B \sum_{c\in C} |f(c)|^2 p_c.$$

Taking $f = \chi_{c_0}$, where χ_{c_0} is the indictor function of the set $\{c_0\}$ and $p_{c_0} > 0$, we have $(\#\Lambda) \cdot p_{c_0}^2 \le B p_{c_0}$. Hence $\#\Lambda \le B/p_{c_0} < \infty$. This implies $\#\mathcal{C} < \infty$ by the completeness of Fourier frame.
(ii) Suppose on the contrary that $D^-\Lambda \ge c > 0$. We claim that $\mathbb{Z}^d$ is a Bessel sequence of $L^2(\mu)$. By the definition of $D^-\Lambda$, we can choose a large $h \in \mathbb{N}$ such that

$$\inf_{x\in\mathbb{R}^d} (\#(\Lambda \cap Q_h(x))) \ge ch^d > 1.$$

Taking $x = h\mathbf{n}$, where $\mathbf{n} \in \mathbb{Z}^d$, we see that all cubes of the form $h\mathbf{n} + [-h/2, h/2)^d$ contains at least one points of Λ, say $\lambda_{\mathbf{n}}$. Since Λ is an F-spectrum, $\{\lambda_{\mathbf{n}}\}_{\mathbf{n}\in\mathbb{Z}^d}$ is a Bessel sequence. Observing that

$$|\lambda_{\mathbf{n}} - h\mathbf{n}| \le \operatorname{diam}([-h/2, h/2)^d) = \sqrt{d}\, h,$$

then $h\mathbb{Z}^d$ is also a Bessel sequence of $L^2(\mu)$ by Proposition 2.1. As a Bessel sequence is invariant under translation, we see that the finite union $\mathbb{Z}^d = \bigcup_{\mathbf{k}\in\{0,\dots,h-1\}^d} (h\mathbb{Z}^d + \mathbf{k})$ is again a Bessel sequence of $L^2(\mu)$, which proves the claim.

Now consider

$$G(x) := \sum_{\mathbf{n}\in\mathbb{Z}^d} |\widehat{\mu}(x+\mathbf{n})|^2.$$

G is a periodic function (mod $\mathbb{Z}^d$). As $\mathbb{Z}^d$ is a Bessel sequence, applying the definition to e_{-x}, we see that $G(x) \le B < \infty$. Hence $G \in L^1([0,1)^d)$ and

$$\int_{\mathbb{R}^d} |\widehat{\mu}(x)|^2\, dx = \sum_{\mathbf{n}\in\mathbb{Z}^d} \int_{[0,1)^d} |\widehat{\mu}(x+\mathbf{n})|^2 dx = \int_{[0,1)^d} |G(x)| dx < \infty.$$

This means that $\widehat{\mu} \in L^2(\mathbb{R}^d)$, which implies that μ must be absolutely continuous. This is a contradiction.
(iii) If μ is absolutely continuous, we write $d\mu(x) = \varphi(x)dx$, for some L^1 function φ and denote by Ω the support of μ. Let

$$E_N = \left\{x \in \Omega : \frac{1}{N} \le \varphi(x) \le N\right\}.$$

Since μ is absolutely continuous, the support Ω must have positive Lebesgue measure and E_N also has positive Lebesgue measure for N large, which we may assume it holds for all E_N. Now, we claim that Λ is an F-spectrum of $L^2(E_N)$. To see this, let $f \in L^2(E_N)$, then we have $\int_{E_N} |\frac{f(x)}{\varphi(x)}|^2 \varphi(x) dx \le N \int_{E_N} |f|^2 < \infty$. Hence,

$$\sum_{\lambda\in\Lambda}\left|\int_{E_N} f(x)e^{2\pi i\lambda x}dx\right|^2 = \sum_{\lambda\in\Lambda}\left|\int_{E_N}\frac{f(x)}{\varphi(x)}e^{2\pi i\lambda x}\varphi(x)dx\right|^2$$
$$\leq B\int_{E_N}\left|\frac{f(x)}{\varphi(x)}\right|^2\varphi(x)dx \leq BN\int_{E_N}|f(x)|^2dx.$$

This establishes the upper frame bound. The lower bound can also be established analogously. This justify the claim. By the Landau's density theorem, we have $D^-\Lambda \geq \mathcal{L}(E_N)$. As E_N are increasing sequence of sets and $\bigcup_N E_N = \Omega$ up to a Lebesgue measure zero set, we have

$$D^-\Lambda \geq \mathcal{L}(\Omega) > 0. \qquad \square$$

Now it is easy to conclude that an F-spectral measure is of pure type.

Proof of Theorem 1.1 First let us assume that if μ is decomposed into non-trivial discrete and continuous parts, $\mu = \mu_d + \mu_c$. Let Λ be an F-spectrum of μ. As $L^2(\mu_d)$ and $L^2(\mu_s)$ are non-trivial subspaces of $L^2(\mu)$, it is easy to see that Λ is also an F-spectrum of both $L^2(\mu_d)$ and $L^2(\mu_c)$. Then $\#\Lambda < \infty$ by Proposition 2.2(i); but $\#\Lambda = \infty$ since $L^2(\mu_c)$ is an infinite dimensional Hilbert space. This contradiction shows that μ is either discrete or purely continuous.

Suppose μ is continuous and has non-trivial singular part μ_s and absolutely continuous part μ_a. By applying the same argument as the above, Λ is an F-spectrum of $L^2(\mu_s)$ and $L^2(\mu_a)$. This is impossible in view of the Beurling density of Λ in Proposition 2.2(ii) and (iii). $\square$

The following corollary is immediate from Theorem 1.1.

Corollary 2.3 *A spectral measure or an R-spectral measure must be of pure type.*

In the rest of this section, we will prove Theorem 1.2 and it will be needed in the next section.

Proof of Theorem 1.2 (i) It is easy to see that the necessity follows by applying Parseval's identity to e_γ for $\gamma \in \Gamma$. Now we show the sufficiency. By the hypotheses, it is sufficient to show that $e_\gamma \in \overline{\text{span}}E(\Lambda)$ for each $\gamma \in \Gamma$. Let Π be the projection from $L^2(\mu)$ to $\overline{\text{span}}E(\Lambda)$. Then $e_\gamma = \Pi(e_\gamma) + (Id - \Pi)(e_\gamma)$ and thus $1 = \|\Pi(e_\gamma)\|^2 + \|(Id-\Pi)(e_\gamma)\|^2$. Note that

$$\|\Pi(e_\gamma)\|^2 = \sum_{\lambda\in\Lambda}|\langle\Pi(e_\gamma), e_\lambda\rangle|^2 = \sum_{\lambda\in\Lambda}|\langle e_\gamma, e_\lambda\rangle|^2 = \|e_\gamma\|^2 = 1.$$

Then $(Id-\Pi)(e_\gamma) = 0$ and thus $e_\gamma \in \overline{\text{span}}E(\Lambda)$.

(ii) Let $M > 0$ so that $\text{supp}\mu \subseteq B(0, M)$, where $B(0, M)$ is the ball with center at 0 and radius M. Denote $\Lambda = \{\lambda_n\}_{n=0}^\infty$ and

$$Q_N(w) = \sum_{n=0}^{N}|\widehat{\mu}(w+\lambda_n)|^2, \quad \forall w \in \mathbb{C}^d.$$

Let B be the upper bound of $E(\Lambda)$. Note that

$$Q_N(w) = \sum_{n=0}^{N} |< e_{-w}, e_{\lambda_n} >|^2 \leq B\|e_{-w}\|^2 \leq Be^{4\pi M|\mathcal{I}m(w)|},$$

where $\mathcal{I}m(w)$ is the imaginary part of w. This implies that the sequence $\{Q_N(w)\}_{N=1}^{\infty}$ is uniformly bounded on each compact set of $\mathbb{C}^d$. By Montel theorem (see, e.g., [Gun90] p. 54), we have Q_Λ is an entire function on $\mathbb{C}^d$ and

$$|Q_\Lambda(w)| \leq Be^{4\pi M|\mathcal{I}m(w)|}, \quad \forall w \in \mathbb{C}^d.$$

□

Now the standard Jorgensen-Pedersen Lemma follows as a corollary.

Corollary 2.4 *An orthogonal sequence $E(\Lambda)$ is complete in $L^2(\mu)$ if and only if $Q_\Lambda \equiv 1$.*

Proof We only need to show that span$E(\mathbb{R}^d)$ is dense in $L^2(\mu)$ by Theorem 1.2. Let $K = \text{supp}\,\mu$. Since span$E(\mathbb{R}^d)$ is a subalgebra of Banach algebra $C(K)$, the space of all continuous function on K, and it separates points K. By Stone-Weierstrass theorem, we have that span$E(\mathbb{R}^d)$ is dense in the space $C(K)$. According to Lusin theorem, $C(K)$ is dense in $L^2(\mu)$. This implies the assertion. □

3 Spectral Properties of Cantor Measures on $\mathbb{R}$

This section is devoted to a simplified content of [DaHL13]. Our aim is to show the existence of spectra with zero Beurling dimension (Theorem 3.5) when the measures are the Cantor measure with consecutive digits. Let b, q be two integers > 1 with $b > q$ and $q \mid b$. Then there exists unique Borel probability measure, denoted by $\mu_{b,q}$, satisfying

$$\mu_{b,q}(\cdot) = \frac{1}{q}\sum_{i=0}^{q-1} \mu_{b,q}(q \cdot -i). \tag{3.1}$$

$\mu_{b,q}$ is called a *Cantor measure* (with consecutive digit). It is well-known that the Hausdorff dimension of the support of $\mu_{b,q}$ is $\ln q/\ln b < 1$ and thus $\mu_{b,q}$ is singularly continuous with respect to Lebesgue measure. We will construct a class of orthogonal set of $\mu_{b,q}$.

Denote $\Sigma_q = \{0, \cdots, q-1\}$, $\Sigma_q^0 = \{\vartheta\}$ and $\Sigma_q^n = \underbrace{\Sigma_q \times \cdots \times \Sigma_q}_{n}$. Let $\Sigma_q^* = \bigcup_{n=0}^{\infty} \Sigma_q^n$ be the set of all finite words. Given $\sigma = \sigma_1\sigma_2\cdots \in \Sigma^*$, we define $\vartheta\sigma = \sigma$, $\sigma|_k = \sigma_1\cdots\sigma_k$ for $k \geq 0$ where $\sigma|_0 = \vartheta$ for any σ and adopt the notation $0^k = \underbrace{0\cdots 0}_{k}$ and $\sigma\sigma'$ is the concatenation of σ and σ'. We start with two definitions.

Definition 3.1 Let Σ_q^* be all the finite words defined as above. We say it is a $q-$*adic tree* if we set naturally the root is ϑ, all the k-th level nodes are Σ_q^k for $k \geq 1$ and all the offsprings of $\sigma \in \Sigma_q^*$ are σi for $i = 0, 1, \ldots, q-1$.

Definition 3.2 Let Σ_q^* be a $q-$adic tree, τ is called a *regular mapping* from Σ_q^* to $\{-1, 0, ..., b-2\}$ if it satisfies

(i) $\tau(\vartheta) = \tau(0^n) = 0$ for all $n \geq 1$.
(ii) For $\sigma_1 \cdots \sigma_k \in \Sigma_q^k$, $\tau(\sigma_1 \cdots \sigma_k) \in (\sigma_k + q\mathbb{Z}) \cap \{-1, 0, ..., b-2\}$.
(iii) For any $\sigma \in \Sigma_q^*$, $\tau(\sigma 0^\ell) = 0$ for ℓ large enough.

Let τ be a regular mapping from Σ_q^* to $\{-1, 0, ..., b-2\}$. For any $n \in \mathbb{N}$ with $q^{N-1} \leq n < q^N$, there exists unique $\sigma = \sigma_1 \cdots \sigma_N \in \Sigma_q^N$ such that $\sigma_N \neq 0$ and

$$n = \sigma_1 + \sigma_2 q + \cdots + \sigma_N q^{N-1}.$$

Associated to τ, we define a sequence of integers by $\lambda_0 = 0$ and

$$\lambda_n = \tau(\sigma|_1) + \tau(\sigma|_2)b + \cdots + \tau(\sigma|_N)b^{N-1} + \sum_{k=N}^{\infty} \tau(\sigma 0^{k-N+1})b^k.$$

Note that λ_n is uniquely determined by $\tau(\sigma|_1), \tau(\sigma|_2), \ldots, \tau(\sigma|_N) = \tau(\sigma)$. We call $\Lambda = \{\lambda_n\}_{n=0}^{\infty}$ a τ-*sequence*. Let ℓ_n be the number of nonzero terms in the sum $\sum_{k=N}^{\infty} \tau(\sigma 0^{k-N+1})b^k$, that is

$$\ell_n = \#\{k : \tau(\sigma 0^k) \neq 0 \text{ for } k \geq 1\}. \tag{3.2}$$

We assume that $b, q, r = b/q$ are integers with $b > q$. The following are our main theorems.

Theorem 3.3 *Let τ be a regular mapping from Σ_q^* to $\{-1, 0, ..., b-2\}$ and let $\Lambda = \{\lambda_n\}_{n=0}^{\infty}$ be the τ-sequence. Then $E(r\Lambda)$ is a maximal orthogonal collection of exponentials for $L^2(\mu_{b,q})$.*

Theorem 3.4 *Let τ be a regular mapping and let $\Lambda = \{\lambda_n\}_{n=0}^{\infty}$ be the τ-sequence. We have the following:*

(i) *If $\max_{n\geq 1}\{\ell_n\} < \infty$, then $r\Lambda$ is a spectrum of $\mu_{b,q}$;*
(ii) *If $\ell_n \geq \log_q n$ for sufficient large n, then $r\Lambda$ is not a spectrum of $\mu_{b,q}$.*

Theorem 3.5 *Let $g(x)$ be an increasing non-negative function on $[0, \infty)$. Then there exists a spectrum Λ of $L^2(\mu_{b,q})$ such that*

$$\lim_{R\to\infty} \sup_{x\in\mathbb{R}} \frac{\#(\Lambda \cap (x-R, x+R))}{g(R)} = 0. \tag{3.3}$$

Let $\mu_{b,q}$ be the Cantor measure given by (3.1) and let

$$M(\xi) = \frac{1}{q}(1 + e^{2\pi i\xi} + \cdots + e^{2\pi i(q-1)\xi}).$$

Then it is easy to obtain that

$$\widehat{\mu}_{b,q}(\xi) = M(b^{-1}\xi)\widehat{\mu}_{b,q}(b^{-1}\xi) = \prod_{k=1}^{\infty} M(b^{-k}\xi). \tag{3.4}$$

Note that $|M(\xi)| = |\sin q\pi\xi|/q|\sin \pi\xi|$. Then

$$\mathcal{Z}_M := \{\xi : M(\xi) = 0\} = \frac{1}{q}(\mathbb{Z} \setminus q\mathbb{Z})$$

and

$$\mathcal{Z}_\mu := \{\xi : \widehat{\mu}_{b,q}(\xi) = 0\} = r\{b^k a : k \geq 0, a \in \mathbb{Z} \setminus q\mathbb{Z}\}.$$

Clearly, Θ is an orthogonal set of $\mu_{b,q}$ if and only if

$$\Theta - \Theta \subseteq \mathcal{Z}_\mu \cup \{0\}. \tag{3.5}$$

Proof of Theorem 3.3 We first prove the orthogonal property of $E(r\Lambda)$. Denote $\lambda_{n'} = \tau(\sigma'|_1) + \tau(\sigma'|_2)b + \cdots + \tau(\sigma'|_{N'})b^{N'-1} + \sum_{k=N'}^{\infty} \tau(\sigma 0^{k-N'+1})b^k$ and $n \neq n'$. Let s be the smallest index such that $\tau(\sigma' 0^{|\sigma|}|_s) \neq \tau(\sigma 0^{|\sigma'|}|_s)$, where $|\sigma|$ is the length of σ. Then

$$\lambda_{n'} - \lambda_n = (\tau(\sigma'|_s) - \tau(\sigma|_s))b^s + b^{s+1}M$$

for some $M \in \mathbb{Z}$. Then $r(\lambda_{n'} - \lambda_n)$ is the zero point of $M(b^{-s+1}\xi)$ by the definition of τ and thus is a zero point of $\widehat{\mu}_{b,q}$ by (3.4). This implies that $r\Lambda$ is an orthogonal set of $\mu_{b,q}$.

Now we show the maximal property of $E(r\Lambda)$. Suppose that $r\Lambda \cup \{\gamma\}$ is an orthogonal set of $\mu_{b,q}$ with $\gamma \notin r\Lambda$. By (3.5) and $0 \in r\Lambda$, we have $\gamma = rb^k a$ for some $k \geq 0$ and $a \in \mathbb{Z} \setminus q\mathbb{Z}$. Since a can be expressed uniquely as

$$a = a_0 + a_1 b + \cdots + a_m b^m,$$

where all $a_i \in \{-1, 0, 1, \ldots, b-1\}$, $a_m \neq 0$ and $a_0 \in \mathbb{Z} \setminus q\mathbb{Z}$, there exists unique $i_0 \in \{1, 2, \ldots, q-1\}$ such that $a_0 - \tau(0^k i_0) \in q\mathbb{Z}$. By the assumption we have $ab^k - \lambda_{i_0 q^k} \in \mathcal{Z}_\mu$, that is,

$$\frac{ab^k - \lambda_{i_0 q^k}}{b^k} = a_0 - \tau(0^k i_0) + \sum_{s=1}^{m}(a_i - \tau(0^k i_0 0^s))b^s - \sum_{s=m+1}^{\infty} \tau(0^k i_0 0^s)b^s \in \mathcal{Z}_\mu.$$

Since $q \mid (a_0 - \tau(0^k i_0))$, but $b \nmid (a_0 - \tau(0^k i_0))$ if $a_0 \neq \tau(0^k i_0)$, one has $a_0 = \tau(0^k i_0)$. Similarly, there exists unique $i_1 \in \{0, 1, \ldots, q-1\}$ such that $a_1 - \tau(0^k i_0 i_1) \in q\mathbb{Z}$. From $ab^k - \lambda_{i_0 q^k + i_1 q^{k+1}} \in \mathcal{Z}_\mu$, one has $a_1 = \tau(0^k i_0 i_1)$. By m-steps one has $a_s = \tau(0^k i_0 \cdots i_s)$ for $0 \leq s \leq m$.

Let $p = \sum_{s=0}^{m} i_s q^{k+s}$. We claim that $\gamma = r\lambda_p$ and the result follows if the claim holds. In fact,

$$\begin{aligned}\frac{ab^k - \lambda_p}{b^k} &= \sum_{s=0}^{m}(a_s - \tau(0^k i_0 \cdots i_s))b^s - \sum_{s=m+1}^{\infty} \tau(0^k i_0 \cdots i_m 0^{s-m})b^s \\ &= -\sum_{s=m+1}^{\infty} \tau(0^k i_0 \cdots i_m 0^{s-m})b^s.\end{aligned}$$

If $ab^k \neq \lambda_p$, the above implies that $ab^k - \lambda_p \notin \mathcal{Z}_\mu$, which contradicts to the assumption. Hence the claim follows. □

Let δ_a be the Dirac measure with center a. We define

$$\delta_{\mathcal{E}} = \frac{1}{\#\mathcal{E}} \sum_{e \in \mathcal{E}} \delta_e$$

for any finite set $\mathcal{E}$, where $\#\mathcal{E}$ is the cardinality of $\mathcal{E}$. Write $\mathcal{D} = \{0, 1, \ldots, q-1\}$ and $D_N = \frac{1}{b}\mathcal{D} + \cdots + \frac{1}{b^N}\mathcal{D}$ for $N \geq 1$. Let $\mu_N = \delta_{\mathcal{D}_N}$. Then

$$\widehat{\mu_N}(\xi) = \prod_{j=1}^{N} M(b^{-j}\xi).$$

By (3.4) we have

$$\widehat{\mu}_{b,q}(\xi) = \widehat{\mu_N}(\xi)\widehat{\mu}_{b,q}(\frac{\xi}{b^N}). \tag{3.6}$$

Lemma 3.6 *Let τ be a regular mapping and let $\{\lambda_n\}_{n=0}^{\infty}$ be the τ-sequence. Then for all $N \geq 1$,*

$$\sum_{n=0}^{q^N-1} |\widehat{\mu_N}(\xi + r\lambda_n)|^2 \equiv 1. \tag{3.7}$$

Proof Since the dimension of $L^2(\mu_N)$ is q^N, the assertion follows by Corollary 2.4 if $\{r\lambda_n\}_{n=0}^{q^N-1}$ is an orthogonal set of μ_N, which can be proved by the same proof of Theorem 3.3. □

For $m \geq 1$, let

$$Q_m(\xi) = \sum_{n=0}^{q^m-1} \left|\widehat{\mu}_{b,q}(\xi + r\lambda_n)\right|^2 \text{ and } Q(\xi) = \sum_{n=0}^{\infty} \left|\widehat{\mu}_{b,q}(\xi + r\lambda_n)\right|^2.$$

Let $\mu = \mu_{b,q}$. For any $m, p > 0$, we have the following identity:

$$\begin{aligned} Q_{m+p}(\xi) &= Q_m(\xi) + \sum_{n=q^m}^{q^{m+p}-1} |\widehat{\mu}(\xi + r\lambda_n)|^2 \\ &= Q_m(\xi) + \sum_{n=q^m}^{q^{m+p}-1} \left|\widehat{\mu_{m+p}}(\xi + r\lambda_n)\right|^2 \left|\widehat{\mu}(\frac{\xi + r\lambda_n}{b^{m+p}})\right|^2. \end{aligned} \tag{3.8}$$

Our goal is see whether $Q(\xi) \equiv 1$. Then by invoking Corollary 2.4, we can determine whether we have a spectrum. As Q is an entire function by Theorem 1.2(ii), we just need to see the value of $Q(\xi)$ for some small values of ξ. To do this, we need to make a fine estimation of the terms $\left|\widehat{\mu}(\frac{\xi+r\lambda_n}{b^{m+p}})\right|^2$ in the above. Write

$$\alpha = \min\left\{|M(\xi)\widehat{\mu}(\xi)|^2 : |\xi| \leq \frac{b-1}{qb}\right\} > 0$$

and

$$\beta = \max\left\{|M(\xi)|^2 : \frac{1}{b^2} \leq |\xi| \leq \frac{b-1}{qb}\right\} < 1.$$

where $|M(\xi)| = \frac{|\sin \pi q\xi|}{q|\sin \pi\xi|}$.

Proposition 3.7 *Let* $|\xi| \leq \frac{r(b-2)}{b-1}$ *and let* $t = \xi + \sum_{k=1}^{N} d_i b^{n_k}$, *where* $d_i \in \{1, 2, \cdots r-1\}$ *and* $1 \leq n_1 < \cdots < n_N$. *Then*

$$\alpha^{N+1} \leq |\widehat{\mu}(t)|^2 \leq \beta^N. \tag{3.9}$$

Proof First it is easy to check that, for $|\xi| \leq \frac{r(b-2)}{b-1}$ and all $d_k \in \{0, 1, 2, \ldots, r-1\}$, we have

$$\left|\frac{\xi+\sum_{k=1}^{n} d_k b^k}{b^{n+1}}\right| \le \frac{1}{b^{n+1}}\left(\frac{r(b-2)}{b-1}+(r-1)(b+b^2+\cdots+b^n)\right)$$
$$=\frac{r(b-2)+(r-1)(b^{n+1}-b)}{b^{n+1}(b-1)}$$
$$\le \frac{b-1}{qb} \tag{3.10}$$

for $n \ge 1$. The inequality in the last line follows from a direct comparison of the difference and $q \ge 2$. To simplify notations, we let $n_0 = 0$ and $n_{N+1} = \infty$. Then $|\widehat{\mu}(t)|^2$ equals

$$\prod_{j=1}^{\infty}\left|M\left(b^{-j}t\right)\right|^2=\prod_{i=0}^{N}\prod_{j=n_i+1}^{n_{i+1}}\left|M\left(b^{-j}t\right)\right|^2. \tag{3.11}$$

We now estimate the products one by one. By (3.10), we have

$$\left|\frac{\xi+\sum_{k=1}^{i} d_k b^{n_k}}{b^{n_i+1}}\right| \le \frac{b-1}{qb}.$$

Hence, together with the integral periodicity of $M(\xi)$ and the definition of α, we have for all $i > 0$,

$$\prod_{j=n_i+1}^{n_{i+1}}\left|M(b^{-j}t)\right|^2=\prod_{j=n_i+1}^{n_{i+1}}\left|M\left(b^{-j}(\xi+\sum_{k=1}^{i} d_k b^{n_k})\right)\right|^2$$
$$\ge\prod_{j=0}^{\infty}\left|M\left(b^{-j}\left(\frac{\xi+\sum_{k=1}^{i} d_k b^{n_k}}{b^{n_i+1}}\right)\right)\right|^2 \ge \alpha. \tag{3.12}$$

For the case $i = 0$, it is easy to see that $\left|\frac{\xi}{b}\right| \le \frac{b-2}{q(b-1)} < \frac{b-1}{qb}$. Hence, $\prod_{j=n_0+1}^{n_1}\left|M(b^{-j}t)\right|^2 \ge \prod_{j=0}^{\infty}\left|M\left(b^{-j}(\xi/b)\right)\right|^2 \ge \alpha$. Putting this fact and (3.12) into (3.11), we have $|\widehat{\mu}_{b,q}(t)|^2 \ge \alpha^{N+1}$.

We next prove the upper bound. From $|M(\xi)| \le 1$, (3.11) and the integral periodicity of $M(\xi)$,

$$|\widehat{\mu}(t)|^2 \le \prod_{i=1}^{N}\left|M\left(b^{-(n_i+1)}t\right)\right|^2=\prod_{i=1}^{N}\left|M\left(b^{-(n_i+1)}(\xi+\sum_{k=1}^{i} d_k b^{n_k})\right)\right|^2. \tag{3.13}$$

By (3.10) we have

$$|\xi + \sum_{k=1}^{i} d_k b^{n_k}| \geq b^{n_i} - |\xi + \sum_{k=1}^{i-1} d_k b^{n_k}| \geq b^{n_i} - \frac{b^{n_{i-1}}(b-1)}{q} \geq b^{n_i - 1}.$$

By (3.10), (3.13), the above and the definition of β, we obtain that $|\widehat{\mu}(t)|^2 \leq \beta^N$. □

Proof of Theorem 3.4 (i) Without loss generality we assume that $|\xi| \leq \frac{r(b-2)}{b-1}$. Recall that

$$Q_{m+p}(\xi) = Q_m(\xi) + \sum_{n=q^m}^{q^{m+p}-1} \left|\widehat{\mu}_{m+p}(\xi + r\lambda_n)\right|^2 \left|\widehat{\mu}\left(\frac{\xi + r\lambda_n}{q^{m+p}}\right)\right|^2. \tag{3.14}$$

Let also $L = \max_{n\geq 1} \ell_n$ ($< \infty$ by assumption). For $q^m \leq n < q^{m+p}$, there exists unique N, $m < N \leq m+p$, such that $q^{N-1} \leq n < q^N$. By the definition of τ, we have $\tau(\sigma 0^k) \in \{0, q, 2q, \ldots, (r-1)q\}$ for $k \geq 1$. We therefore have

$$\begin{aligned}
\xi + r\lambda_n &= \xi + r\tau(\sigma|_1) + r\tau(\sigma|_2)b + \cdots + r\tau(\sigma|_N)b^{N-1} + \sum_{s=N}^{\infty} r\tau(\sigma 0^{s-N+1})b^s \\
&= \xi + r\sum_{i=1}^{N} \tau(\sigma|_i)b^{i-1} + r\tau(\sigma 0)b^N + \cdots + r\tau(\sigma 0^{m+p-N})b^{m+p-1} \\
&\quad + \sum_{s=m+p}^{\infty} d_s b^{s+1}
\end{aligned}$$

Hence,

$$\begin{aligned}
\frac{\xi + r\lambda_n}{b^{m+p}} &= \frac{1}{b^{m+p}}\left(\xi + r\sum_{i=1}^{N} \tau(\sigma|_i)b^{i-1} + r\tau(\sigma 0)b^N + \cdots + r\tau(\sigma 0^{m+p-N})b^{m+p-1}\right) \\
&\quad + \sum_{s=m+p}^{\infty} d_s b^{s+1-(m+p)} := t + \sum_{s=m+p}^{\infty} d_s b^{s+1-(m+p)}.
\end{aligned}$$

Note that, from $|\tau(\sigma)| \leq b-2$ for any multi-indices σ,

$$|t| \leq \frac{1}{b^{m+p}}\left(|\xi| + r(b-2)(1 + b + b^2 + \cdots + b^{m+p-1})\right) \leq \frac{r(b-2)}{b-1}.$$

Also, $d_s \in \{0, 1, \ldots, r-1\}$ and there are at most L non-zero terms. By Proposition 3.7, we conclude that $\left|\widehat{\mu}\left(\frac{\xi + r\lambda_n}{q^{m+p}}\right)\right|^2 \geq \alpha^{L+1}$. Using (3.14) and Lemma 3.6. we obtain

$$\begin{aligned} Q_{m+p}(\xi) &\geq Q_m(\xi) + \alpha^{L+1} \sum_{n=q^m}^{q^{m+p}-1} \left|\widehat{\mu}_{m+p}(\xi + r\lambda_n)\right|^2 \\ &= Q_m(\xi) + \alpha^{L+1} \left(1 - \sum_{n=0}^{q^m-1} \left|\widehat{\mu}_{m+p}(\xi + r\lambda_n)\right|^2\right). \end{aligned}$$

Fixing m, we first let p approaches infinity and obtain

$$Q(\xi) \geq Q_m(\xi) + \alpha^{L+1} \left(1 - \sum_{n=0}^{q^m-1} |\widehat{\mu}(\xi + r\lambda_n)|^2\right).$$

We then finally let m goes to infinity.

$$\alpha^{L+1}\left(1 - \sum_{n=0}^{\infty} |\widehat{\mu}(\xi + r\lambda_n)|^2\right) \leq 0.$$

This means that $Q(\xi) \geq 1$ for $|\xi| \leq r(b-2)/(b-1)$. As $Q(\xi) \leq 1$ for mutually orthogonal sets and by the entire function property of Q on $\mathbb{C}$, we must have $Q(\xi) \equiv 1$ and hence Λ is a spectrum for μ.

(ii) With loss of generality we assume that $\ell_n \geq \log_q n$ for $n \geq 1$. Again we begin with

$$Q_m(\xi) = Q_{m-1}(\xi) + \sum_{n=q^{m-1}}^{q^m-1} |\widehat{\mu}_m(\xi + r\lambda_n)|^2 \left|\widehat{\mu}(\frac{\xi + r\lambda_n}{q^m})\right|^2.$$

Note that for $q^{m-1} \leq n < q^m$, $\ell_n \geq \log_q n \geq m-1$. Using it and the same estimate as in (i) so as to apply Proposition 3.7, we have

$$\begin{aligned} Q_m(\xi) &\leq Q_{m-1}(\xi) + \sum_{n=q^{m-1}}^{q^m-1} |\widehat{\mu}_m(\xi + r\lambda_n)|^2 \, \beta^{\ell_n} \\ &\leq Q_{m-1}(\xi) + \beta^{m-1} \sum_{n=q^{m-1}}^{q^m-1} |\widehat{\mu}_m(\xi + r\lambda_n)|^2 \\ &= Q_{m-1}(\xi) + \beta^{m-1}(1 - \sum_{n=0}^{q^{m-1}-1} |\widehat{\mu}_m(\xi + r\lambda_n)|^2) \\ &\leq Q_{m-1}(\xi) + \beta^{m-1}(1 - Q_{m-1}(\xi)). \end{aligned}$$

Consequently,

$$1 - Q_m(\xi) \geq (1 - Q_{m-1}(\xi))(1 - \beta^{m-1}) \geq (1 - Q_1(\xi)) \prod_{k=1}^{m-1} (1 - \beta^k).$$

By letting m to infinity, we have

$$1 - Q(\xi) \geq (1 - Q_1(\xi)) \prod_{k=1}^{\infty} (1 - \beta^k).$$

Since $Q_1(\xi) < 1$ for almost all $\xi \in \mathbb{R}$, the second assertion follows by Corollary 2.4. □

Proof of Theorem 3.5 Let $\{m_k\}_{k=1}^{\infty}$ be a strictly increasing sequence of positive integers with $m_1 \geq 2$. Then $m_k > k$ for $k \geq 1$. We now define a regular mapping in terms of this sequence by induction. Let $\tau(\vartheta) = \tau(0^k) = 0$ for $k \geq 1$. For $\sigma \in \{1, 2, \ldots, q-1\} \subset \Sigma_q^1$, we define $\tau(\sigma) = \sigma$ and $\tau(\sigma 0^l) = 0$ or q according to $l \neq m_\sigma$ or $l = m_\sigma$, respectively. Suppose we have defined all $\tau(\sigma)$, $\sigma = \sigma_1 \cdots \sigma_s$ with $s \leq k$ and $\sigma_s \neq 0$, and $\tau(\sigma 0^l)$ for $l \geq 1$. For $\sigma = \sigma_1 \cdots \sigma_{k+1} \in \Sigma_q^{k+1}$ with $\sigma_{k+1} \neq 0$, we define $\tau(\sigma) = \sigma_{k+1}$ and $\tau(\sigma 0^l) = 0$ or q according to $l \neq m_{p_\sigma}$ or $l = m_{p_\sigma}$, respectively, where $p_\sigma = \sum_{i=1}^{k+1} \sigma_i q^{i-1}$. By induction we have well-defined a regular mapping from the q-adic tree to $\{-1, 0, 1, \ldots, b-1\}$.

For any $n \in \mathbb{N}$, there exists unique $k \geq 1$ such that $q^{k-1} \leq n < q^k$. Then n can be expressed by

$$n = \sum_{j=1}^{k} \sigma_j q^{j-1}, \tag{3.15}$$

where all $\sigma_j \in \{0, 1, \ldots, q-1\}$ and $\sigma_k \neq 0$. By the definition of τ-sequence, we have $\lambda_0 = 0$ and

$$\lambda_n = \sum_{j=1}^{k} \tau(\sigma_1 \cdots \sigma_j) b^{j-1} + q b^{m_n},$$

consequently, $\ell_n = 1$ and by Theorem 3.4(i), $\Lambda = \{\lambda_n\}_{n=0}^{\infty}$ is a spectrum of $\mu_{b,q}$.

We now find Λ satisfying (3.9) by choosing m_n. To do this, we first note that there exists a strictly increasing continuous function $h(t)$ from $[0, \infty)$ onto itself such that $h(t) \leq g(t)$ for $t \geq 0$ and it is sufficient to replace $g(t)$ by $h(t)$ in the proof. In this way, the inverse of $h(t)$ exists, and we denote it by $h^{-1}(t)$.

Now, note that

$$\lambda_n \leq q \frac{b^k - 1}{b - 1} + q b^{m_n} \leq (q+1) b^{m_n}.$$

Hence,

$$\lambda_{n+1} - \lambda_n \geq q b^{m_{n+1}} - (q+1) b^{m_n} \geq b^{m_n + 1}. \tag{3.16}$$

Therefore, we choose m_n so that $b^{m_n} \geq 2h^{-1}(b^{n+1})$ for all $n \geq 1$. For any $h(R) \geq 1$, there exists unique $s \in \mathbb{N}$ such that $b^{s-1} \leq h(R) < b^s$. Then

$$\frac{\sup_{x\in\mathbb{R}} \#(\Lambda \cap (x - R, x + R))}{h(R)} \leq \frac{\sup_{x\in\mathbb{R}} \#(\Lambda \cap (x - h^{-1}(b^s), x + h^{-1}(b^s)))}{b^{s-1}}. \tag{3.17}$$

Note from (3.16) that the length of the open intervals $(x - h^{-1}(b^s), x + h^{-1}(b^s))$ is less than $\lambda_{n+1} - \lambda_n$ whenever $n \geq s$. This implies that the set $\Lambda \cap (x - h^{-1}(b^s), x + h^{-1}(b^s))$ contains at most one λ_n where $n \geq s$. We therefore have

$$\sup_{x\in\mathbb{R}} \#(\Lambda \cap (x - h^{-1}(b^s), x + h^{-1}(b^s))) \leq s + 1.$$

Thus the result follows by taking limit in (3.17). □

We conclude the paper with some remarks.

Remark (1) When observing the proofs of theorems, the main crux of the proof to spectra of zero Beurling dimension is in Proposition 3.7. The uniform control on the Fourier transform depends only on the number of non-zero digits in the b-adic expansion rather than the size of the frequencies.

(2) Indeed, all maximal orthogonal exponentials for $\mu_{b,q}$ can be classified through either *regular* or *irregular* mappings. This note discusses only the regular mappings. For irregular mappings, we can discuss its spectral properties if the number of irregular paths is finite. One can refer the details to [DaHL13].

(3) Much less is known about dilating a spectrum of a spectral measure. A standard example is that if $\Lambda = \{0, 1\} \oplus 4\{0, 1\} \oplus ....$, then 5Λ is also a spectrum for the standard one-fourth Cantor measure (i.e. $q = 2, b = 4$) [DHSW11]. However, one can prove that the tree mapping corresponding to 5Λ is irregular with infinitely many irregular paths. To see this, we re-write the following elements 5Λ into our standard 4-adic expansions.

$$5 \cdot 4^n + 5 \cdot 4^{n+1} + \cdots + 5 \cdot 4^m = 4^n + 2 \cdot 4^{n+1} + 2 \cdot 4^{n+2} + ... + 2 \cdot 4^m + 4^{m+1}.$$

This means the paths $0^{n-1}10^{\infty}$ are irregular paths. Hence, there are infinitely many such paths. This example of spectra cannot be covered by our theory and is also the first example of spectra with infinitely many irregular paths .

References

[Dai12] Dai, X.-R.: When does a Bernoulli convolution admit a spectrum? Adv. Math. **231**(3–4), 1681–1693 (2012)

[DaHL13] Dai, X.-R., He, X.-G., Lai, C.-K.: Spectral property of Cantor measures with consecutive digits. Adv. Math. **242**, 187–208 (2013)

[DHJ09] Dutkay, D., Han, D., Jorgensen, P.: Orthogonal exponentials, translations and Bohr completions. J. Funct. Anal. **257**, 2999–3019 (2009)
[DHS09] Dutkay, D., Han, D., Sun, Q.: On spectra of a Cantor measure. Adv. Math. **221**, 251–276 (2009)
[DHSW11] Dutkay, D., Han, D., Sun, Q., Weber, E.: On the Beurling dimension of exponential frames. Adv. Math. **226**, 285–297 (2011)
[DJ12] Dutkay, D., Jorgensen, P.: Fourier duality for fractal measures with affine scales. Math. Comp. **81**, 2253–2273 (2012)
[DL00] Dutkay, D., Lai, C.-K.: Uniformity of Measures with Fourier Frames . Adv. Math. **252**, 684–707 (2014)
[DS52] Duffin, R., Schaeffer, A.: A class of nonharmonic Fourier series. Tran. Amer. Math. Soc. **72**, 341–366 (1952)
[Fal90] Falconer, K.J.: Fractal Geometry. Mathematical Foundations and Applications. Wiley, New York (1990)
[Fug74] Fuglede, B.: Commuting self-adjoint partial differential operators and a group theoretic problem. J. Funct. Anal. **16**, 101–121 (1974)
[JP98] Jorgensen, P., Pedersen, S.: Dense analytic subspaces in fractal L^2 spaces. J. Anal. Math. **75**, 185–228 (1998)
[GL14] Grepstad, S., Lev, N.: Multi-tiling and Riesz bases. Adv. Math. **252**, 1–6 (2014)
[Gun90] Gunning, R.: intruduction to Holomorphic Functions of Several Variables Volume I, Function Theory. Wadsworth and Brooks , Pacific Grove, Calif (1990)
[HLL13] He, X.-G., Lai, C.-K., Lau, K.-S.: Exponential spectra in $L^2(\mu)$. Appl. Comput. Harmon. Anal. **34**, 327–338 (2013)
[HuL08] Hu, T.-Y., Lau, K.-S.: Spectral property of the Bernoulli convolutions. Adv. Math. **219**, 554–567 (2008)
[KN00] Kozma, G., Nitzan, S.: Combining Riesz bases. arXiv:1210.6383
[Lab01] Łaba, I.: The spectral set conjecture and multiplicative properties of roots of polynomials. J. London Math. Soc. **65**, 661–671 (2001)
[LaW02] Łaba, I., Wang, Y.: On spectral Cantor measures. J. Funct. Anal. **193**, 409–420 (2002)
[LaW06] Łaba, I., Wang, Y.: Some properties of spectral measures. Appl. Comput. Harmon. Anal. **20**, 149–157 (2006)
[Lai11] Lai, C.-K.: On Fourier frame of absolutely continuous measures. J. Funct. Anal. **261**, 2877–2889 (2011)
[Lan67] Landau, H.: Necessary density conditions for sampling and interpolation of certain entire functions. Acta Math. **117**, 37–52 (1967)
[Lev12] Lev, N.: Riesz bases of exponentials on multiband spectra. Proc. Amer. Math. Soc. **140**, 3127–3132 (2012)

The Role of Transfer Operators and Shifts in the Study of Fractals: Encoding-Models, Analysis and Geometry, Commutative and Non-commutative

Dorin Ervin Dutkay and Palle E.T. Jorgensen

Abstract We study a class of dynamical systems in L^2 spaces of infinite products X. Fix a compact Hausdorff space B. Our setting encompasses such cases when the dynamics on $X = B^{\mathbb{N}}$ is determined by the one-sided shift in X, and by a given transition-operator R. Our results apply to any positive operator R in $C(B)$ such that $R1 = 1$. From this we obtain induced measures Σ on X, and we study spectral theory in the associated $L^2(X, \Sigma)$. For the second class of dynamics, we introduce a fixed endomorphism r in the base space B, and specialize to the induced solenoid $\mathrm{Sol}(r)$. The solenoid $\mathrm{Sol}(r)$ is then naturally embedded in $X = B^{\mathbb{N}}$, and r induces an automorphism in $\mathrm{Sol}(r)$. The induced systems will then live in $L^2(\mathrm{Sol}(r), \Sigma)$. The applications include wavelet analysis, both in the classical setting of $\mathbb{R}^n$, and Cantor-wavelets in the setting of fractals induced by affine iterated function systems (IFS). But our solenoid analysis includes such hyperbolic systems as the Smale-Williams attractor, with the endomorphism r there prescribed to preserve a foliation by meridional disks. And our setting includes the study of Julia set-attractors in complex dynamics.

Keywords Fractal · Solenoid · Shift-spaces · Transition operators · Ruelle-operators · Attractors · Cantor · Infinite-product measures · Wavelets · Wavelet representations

2010 Mathematics Subject Classification 28A80 · 37A30 · 37C30 · 37C85 · 46G25 · 47B65

D.E. Dutkay
Department of Mathematics, University of Central Florida, 4000 Central Florida Blvd., P.O. Box 161364, Orlando, FL 32816-1364, USA
e-mail: Dorin.Dutkay@ucf.edu

P.E.T. Jorgensen (✉)
Department of Mathematics, University of Iowa, 14 MacLean Hall, Iowa City, IA 52242-1419, USA
e-mail: jorgen@math.uiowa.edu

D.-J. Feng and K.-S. Lau (eds.), *Geometry and Analysis of Fractals*,
Springer Proceedings in Mathematics & Statistics 88,
DOI 10.1007/978-3-662-43920-3_3

1 Introduction

The purpose of this paper is to offer a general framework for geometry and analysis of iteration systems. We offer a setting encompassing the kind of infinite product, or solenoid constructions arising in the study of iterated function systems (IFSs). Our aim is to give an operator theoretic construction of infinite product measures in a general setting that includes wavelet analysis of IFSs. To motivate this, recall, that to every affine function system S with fixed scaling matrix, and a fixed set of translation points in $\mathbb{R}^n$, we may associate to S a solenoid. By this we mean a measure space whose L^2 space includes $L^2(\mathbb{R}^n)$ in such a way that $\mathbb{R}^n$ embeds densely in the solenoid. (In the more familiar case of $n = 1$, we speak of a dense curve in an infinite-dimensional "torus". The latter being a geometric model of the solenoid.

The need for this generality arose in our earlier investigations, for example in the building of wavelet systems on Cantor systems, of which the affine IFSs are special cases. In these cases (see Theorem 2.6 and Corollary 4.5 below) we found that one must pass to a suitable L^2 space of a solenoid. Indeed, we showed that such wavelet bases fail to exist in the usual receptor Hilbert space $L^2(\mathbb{R}^n)$ from wavelet theory.

For reference to earlier papers dealing with measures on infinite products, and their use in harmonic analysis and wavelet theory on fractals; see e.g., [DJ12, DJS12, DLS11, DJ11a, DS11, DJ11b, DHSW11, DJ10, DJP09, LN12, DL10, LW09].

The paper is organized as follows: Starting with a compact Hausdorff space B, and a positive operator R in $C(B)$, we pass to a family of induced probability measures Σ (depending on R) on the infinite product $\Omega = B^{\mathbb{N}}$. Among all probability measures on Ω, we characterize those which are induced. In Sect. 2, we prove a number of theorems about R-induced measures on Ω, and we include applications to random walks, and to fractal analysis. In Sect. 3, we then introduce an additional structure: a prescribed endomorphism r in the base space B, and we study the corresponding solenoid $\mathrm{Sol}(r)$, contained in Ω, and its harmonic analysis, including applications to generalized wavelets. The latter are studied in detail in Sect. 4 where we introduce wavelet-filters, in the form of certain functions m on B.

2 Analysis of Infinite Products

Definition 2.1 Let B be some compact Hausdorff space. $\mathcal{B}$ refes to a σ-algebra, usually generated by the open sets, so Borel. We will denote by $\mathcal{C}$ the cylinder sets, see below. We denote by $\mathcal{M}(B)$ the set of positive Borel masures on B, and by $\mathcal{M}_1(B)$ those that have $\mu(B) = 1$. Let V be some set.

$$B^V = \prod_V B = \text{all functions from } V \text{ to } B.$$

For example $V = \mathbb{N}$ or $\mathbb{Z}$.

For $x \in B^V$ we denote by $\pi_v(x) := x_v$, $v \in V$. If $V = \mathbb{N}$, then we denote by

$$\pi_1^{-1}(x) = \{(x_1, x_2, \dots) \in B^{\mathbb{N}} : x_1 = x\}.$$

Let $r : B \to B$ be some onto mapping, and μ a Borel probability measure on B, $\mu(B) = 1$.

To begin with we do not introduce μ and r, but if r is fixed and

$$1 \leq \#r^{-1}(x) < \infty, \quad \text{for all } x \in B, \tag{2.1}$$

then we introduce two objects

(1) $R = R_W$, the Ruelle operator;
(2) $\mathrm{Sol}(r)$, the solenoid.

For (1), fix $W : B \to [0, \infty)$ such that

$$\sum_{r(y)=0x} W(y) = 1, \text{ for all } x \in B,$$

and set

$$(R_W\varphi)(x) = \sum_{r(y)=x} W(y)\varphi(y). \tag{2.2}$$

For (2),

$$\mathrm{Sol}(r) = \left\{x \in B^{\mathbb{N}} : r(x_{i+1}) = x_i, i = 1, 2, \dots\right\} \tag{2.3}$$

$$\sigma(x)_i = x_{i+1}, \quad (x \in B^{\mathbb{N}}), \qquad \widehat{r}(x) = (r(x_1), x_1, x_2, \dots). \tag{2.4}$$

More generally, consider

$$R : C(B) \to C(B) \text{ or } R : M(B) \to M(B), \tag{2.5}$$

where $M(B)$ is the set of all measurable functions on B.

Definition 2.2 We say that R is *positive* iff

$$\varphi(x) \geq 0 \text{ for all } x \in B \text{ implies } (R\varphi)(x) \geq 0, \text{ for all } x \in B. \tag{2.6}$$

We will always assume $R1 = 1$ where 1 indicates the constant function 1 on B. This is satisfied if $R = R_W$ in (2.2), but there are many other positive operators R with these properties.

While what we call "the transfer operator" or a "Ruelle operator" has a host of distinct mathematical incarnations, each dictated by a particular family of applications, they are all examples of positive operators R in the sense of our Definition 2.2. Our paper has two aims: One is to unify, and extend earlier studies; and the other is to prove a number of theorems on measures, dynamical systems, stochastic processes built from infinite products. Indeed there are many positive operators R which might not fall in the class of operators studied as "transfer operators". The earlier literature on transfer operators includes applications to physics [LR69], to the Selberg zeta function [FM12], to dynamical zeta functions [Rue02, Rue96, Nau12, MMS12]; to C^*-dynamical systems [Kwa12, ABL11]; to the study of Hausdorff dimension [Hen12]; to spectral theory [ABL12].

These applications are, in addition to the aforementioned, to analysis on fractals, and to generalized wavelets. For book treatments, we refer the reader to [Bal00], and [BJ02]. The literature on positive operators R, in the general sense, is much less extensive; but see [Arv86].

Definition 2.3 A subset S of $B^{\mathbb{N}}$ is said to be *shift-invariant* iff $\sigma(S) \subset S$, where σ is as in (2.4), $\sigma(x)_i = x_{i+1}$.

Remark 2.4 Every solenoid $\mathrm{Sol}(r)$ is shift-invariant.

Example 2.5 The solenoids introduced in connection with generalized wavelet constructions:

Let $r : B \to B$ as above and let μ be a strongly invariant measure, i.e.,

$$\int f \, d\mu = \int \frac{1}{\# r^{-1}(x)} \sum_{r(y)=x} f(y) \, d\mu(x)$$

for all $f \in C(B)$.

A *quadrature mirror filter* (*QMF*) for r is a function m_0 in $L^\infty(B, \mu)$ with the property that

$$\frac{1}{N} \sum_{r(w)=z} |m_0(w)|^2 = 1, \quad (z \in B) \tag{2.7}$$

As shown by Dutkay and Jorgensen [DJ05, DJ07], every quadrature mirror filter (QMF) gives rise to a wavelet theory. Various extra conditions on the filter m_0 will produce wavelets in $L^2(\mathbb{R})$ [Dau92], on Cantor sets [DJ06, MP11], on Sierpinski gaskets [DMP08] and many others.

Theorem 2.6 [DJ05, DJ07] *Let m_0 be a QMF for r. Then there exists a Hilbert space $\mathcal{H}$, a representation π of $L^\infty(B)$ on $\mathcal{H}$, a unitary operator U on $\mathcal{H}$ and a vector φ in $\mathcal{H}$ such that*

(i) **Covariance**

$$U\pi(f)U^* = \pi(f \circ r), \quad (f \in L^\infty(B)) \tag{2.8}$$

(ii) **Scaling equation**

$$U\varphi = \pi(m_0)\varphi \tag{2.9}$$

(iii) **Orthogonality**

$$\langle \pi(f)\varphi\,,\,\varphi\rangle = \int f\,d\mu, \quad (f \in L^\infty(B)) \tag{2.10}$$

(iv) **Density**

$$\overline{span}\left\{U^{-n}\pi(f)\varphi : f \in L^\infty(B), n \geq 0\right\} = \mathcal{H} \tag{2.11}$$

The system $(\mathcal{H}, U, \pi, \varphi)$ in Theorem 2.6 is called *the wavelet representation associated to the QMF* m_0.

While, as we mentioned before, these representations can have incarnations on the real line, or on Cantor sets, they can be also represented using certain random-walk measures on the solenoid (see [DJ05, DJ07, Dut06]).

Remark 2.7 In examples when the condition (2.1) is not satisfied, the modification of the family of relevant integral operators is as follows.

In the general case when $r : B \to B$ is given, but $\# r^{-1}(x) = \infty$, the modification of the operators R, extending those from Example 2.5, is as follows:
Consider

(i) $W : B \to [0, \infty)$ Borel
(ii) $p : B \times \mathcal{B}(B) \to [0, \infty)$ such that for all $x \in B$, $p(x, \cdot) \in \mathcal{M}(r^{-1}(x))$, so is a positive measure such that

$$\int_{r^{-1}(x)} W(y)p(x, dy) = 1, \quad (x \in B).$$

Then set

$$(R\varphi)(x) = \int_{r^{-1}(x)} \varphi(y)W(y)p(x, dy).$$

Example 2.8 $G = (V, E)$ infinite graph, V are the vertices, E are the edges.

$$i(e) = \text{initial vertex}, \quad t(e) = \text{terminal vertex}. \tag{2.12}$$

$$S^{(G)} = \text{solenoid of } G = \left\{\tilde{e} \in E^{\mathbb{N}} : t(e_j) = i(e_{j+1}) \text{ for all } j \in \mathbb{N}\right\}. \tag{2.13}$$

For example $V = \mathbb{Z}^2$ and the edges are given by $x \sim y$ iff $\|x - y\| = 1$. For details and applications, see [JP10].

Definition 2.9 Now back to $\mathcal{C}$, the cylinder sets mentioned in Definition 2.1. $C \in \mathcal{C}$ are subsets of B^V indexed by finite sytems $v_1, \ldots, v_n$, $O_1, \ldots, O_n$, $v_i \in V$, $O_i \subset B$ open subsets, $i = 1, \ldots, n$, $n \in \mathbb{N}$.

$$C_{v_i, O_i} := \left\{ \tilde{x} \in B^V : x_{v_i} \in O_i \text{ for all } i = 1, \ldots, n \right\}. \tag{2.14}$$

Notation: $\mathcal{C}$ generates the topology and the σ-algebra of subsets in B^V in the usual way, and B^V is compact by Tychonoff's theorem.

If φ is a function on B, we denote by M_φ the multiplication operator $M_\varphi f = \varphi f$, defined on functions f on B. In the applications below, we will use $C(B)$, all continuous functions from B to $\mathbb{R}$.

Lemma 2.10 *Consider $B^{\mathbb{N}}$ and the algebra generated by cylinder functions of the form $f = \varphi_1 \otimes \cdots \otimes \varphi_n$, $\varphi_i \in C(B)$, $n \in \mathbb{N}$, $1 \le i \le n$,*

$$(\varphi_1 \otimes \cdots \otimes \varphi_n)(\tilde{x}) = \varphi_1(x_1)\varphi_2(x_2)\ldots\varphi_n(x_n), \quad (\tilde{x} \in B^{\mathbb{N}}), \tag{2.15}$$

or

$$\varphi_1 \otimes \cdots \otimes \varphi_n = (\varphi_1 \circ \pi_1)(\varphi_2 \circ \pi_2)\ldots(\varphi_n \circ \pi_n). \tag{2.16}$$

Let $\mathcal{A}_{\mathcal{C}}$ be the algebra of all cylinder functions. Then $\mathcal{A}_{\mathcal{C}}$ is dense in $C(B^{\mathbb{N}})$.

Proof Easy consequence of Stone-Weierstrass. □

Theorem 2.11 *Let R be a positive operator as in* (2.5)*, with $R1 = 1$. Then for each $x \in B$ there exists a unique Borel probability measure $\mathbb{P}_x$ on $B^{\mathbb{N}}$ such that*

$$\int_{B^{\mathbb{N}}} \varphi_1 \otimes \cdots \otimes \varphi_n \, d\mathbb{P}_x = \left(M_{\varphi_1} R M_{\varphi_2} \ldots R M_{\varphi_n} 1\right)(x), \quad (\varphi_i \in C(B), n \in \mathbb{N}). \tag{2.17}$$

Proof We only need to check that the right-hand side of (2.17) for $\varphi_1 \otimes \ldots \varphi_n$ equals the right-hand side of (2.17) for $\varphi_1 \otimes \ldots \varphi_n \otimes 1$; but this is immediate from (2.17) and the fact that $R1 = 1$. The existence and uniqueness of $\mathbb{P}_x$ the follows form the inductive method of Kolmogorov. □

Corollary 2.12 *Let B and $R : C(B) \to C(B)$ be as in* Theorem 2.11*, and let $\mu \in \mathcal{M}_1(B)$ be given. Let $\Sigma = \Sigma^{(\mu)}$ be the measure on $\Omega = B^{\mathbb{N}}$ given by*

$$\int f \, d\Sigma := \int_B \int_{\pi_1^{-1}(x)} f \, d\mathbb{P}_x \, d\mu(x). \tag{2.18}$$

Then

(i) *$V_1 : L^2(B, \mu) \to L^2(\Omega, \Sigma)$ given by $V_1\varphi := \varphi \circ \pi_1$ is isometric.*

(ii) *For its adjoint operator* V_1^*, *we have* $V_1^* : L^2(\Omega, \Sigma) \to L^2(B, \mu)$ *with*

$$(V_1^* f)(x) = \int_{\pi_1^{-1}(x)} f \, d\mathbb{P}_x. \tag{2.19}$$

Proof The assertion (i) is immediate from Theorem 2.11. To prove (ii) we must show that the following formula holds:

$$\int_B \left(\int_{\pi_1^{-1}(x)} f \, d\mathbb{P}_x \right) \psi(x) \, d\mu(x) = \int_\Omega f \, \psi \circ \pi_1 \, d\Sigma \tag{2.20}$$

for all $f \in L^2(\Omega, \Sigma)$ and all $\psi \in C(B)$.

Recall that V_1^* is determined by

$$\langle V_1^* f \, , \, \psi \rangle_{L^2(\mu)} = \langle f \, , \, V_1 \psi \rangle_{L^2(\Omega, \Sigma)} \, . \tag{2.21}$$

But by Lemma 2.10 (Stone-Weierstrass), to verify (2.20), we may restrict attention to the special case when f has the form given in (2.16). Note that if $f = (\varphi_1 \circ \pi_1)(\varphi_2 \circ \pi_2) \ldots (\varphi_n \circ \pi_n)$ then

$$f(\psi \circ \pi_1) = ((\varphi_1 \psi) \circ \pi_1)(\varphi_2 \circ \pi_2) \ldots (\varphi_n \circ \pi_n),$$

and so the right-hand side of (2.20) is equal to

$$= \int_\Omega ((\varphi_1 \psi) \circ \pi_1)(\varphi_2 \circ \pi_2) \ldots (\varphi_n \circ \pi_n) \, d\Sigma$$

$$= \int_B \varphi_1(x) \psi(x) R(\varphi_2 R(\ldots \varphi_{n-1} R(\varphi_n)) \ldots)))(x) \, d\mu(x) = \int_B \psi(x) \int f \, d\mathbb{P}_x \, d\mu(x)$$

which is the left-hand side of (2.20) and (ii) follows. □

Remark 2.13 When $R : C(B) \to C(B)$ is a given positive operator, we induce measures on $\Omega = B^{\mathbb{N}}$ by the inductive procedure outlined in the proof of Theorem 2.11; but implicit in this construction is an extension of $\varphi \mapsto R(\varphi)$ from all φ continuous to all Borel measurable functions. This extension uses the Riesz theorem in the usual way as follows: Fix $x \in B$ and then apply Riesz' theorem to the positive linear functional $C(B) \ni \varphi \mapsto R(\varphi)(x)$. There is a unique regular Borel measure μ_x on B such that

$$R(\varphi)(x) = \int_B \varphi(y)\, d\mu_x(y), \quad (\varphi \in C(B)).$$

If $E \subset B$ is Borel, we define

$$\tilde{R}(E)(x) = \tilde{R}(\chi_E)(x) := \mu_x(E);$$

but we shall use this identification without overly burdening our notation with tildes.

Lemma 2.14 *Let B and R be specified as above. Given $\mu \in \mathcal{M}_1(B)$, let $\Sigma = \Sigma^{(\mu)}$ denote the corresponding measure on Ω, i.e.,*

$$\int_\Omega f\, d\Sigma = \int_B \int_{\pi_1^{-1}(x)} f\, d\mathbb{P}_x^{(R)}\, d\mu(x) \tag{2.22}$$

We shall consider $V_1 : L^2(B, \mu) \to L^2(\Omega, \Sigma)$ and its adjoint operator $V_1^* : L^2(\Omega, \Sigma) \to L^2(B, \mu)$, where $V_1\varphi = \varphi \circ \pi$, for all $\varphi \in L^2(B, \mu)$. Note that the adjoint operator V_1^* makes reference to the choice of R at the very outset. The following two hold:

(i) *R naturally extends to $L^2(B, \mu)$; and*
(ii)

$$RV_1^* f = V_1^*(f \circ \sigma), \quad (f \in L^2(\Omega, \Sigma)) \tag{2.23}$$

Remark 2.15 Given R, we say that a function $\varphi \in B$ is *harmonic* iff $R\varphi = \varphi$. It follows that harmonic functions contain the range of V_1^*, applied to $\{f : f \circ \sigma = f\}$. For a stronger conclusion, see Corollary 2.21.

Proof of Lemma 2.14 Using the Stone-Weierstrass theorem, applied to $C(\Omega)$, we note that it is enough for us to check the validity of formula (2.23) on the algebra $\mathcal{A}^{(cyl)}$ spanned by all cylinder functions

$$f = (\varphi_1 \circ \pi_1)(\varphi_2 \circ \pi_2) \dots (\varphi_n \circ \pi_n) \tag{2.24}$$

$n \in \mathbb{N}$, $\varphi_i \in C(B)$. But note that if f is as in (2.24) then

$$f \circ \sigma = (\varphi_1 \circ \pi_2)(\varphi_2 \circ \pi_3) \dots (\varphi_n \circ \pi_{n+1}) \tag{2.25}$$

Using then (2.19) in Corollary 2.12 above, we conclude that

$$(V_1^*(f \circ \sigma))(x) = R(\varphi_1 R(\varphi_2(R \dots \varphi_{n-1} R(\varphi_n)) \dots))(x) = (RV_1^* f)(x);$$

The extension from the cylinder functions $\mathcal{A}^{(cyl)}$ to all of $L^2(\Omega, \Sigma)$ now follows from the usual application of Stone-Weierstrass; recall that $C(\Omega)$ is dense in $L^2(\Omega, \Sigma)$ relative to the L^2-norm; and we have the desired conclusion. □

2.1 What Measures on $B^{\mathbb{N}}$ have a Transfer Operator?

Below we characterize, among all Borel probability measures Σ on $B^{\mathbb{N}}$, precisely those which arise from a pair μ and R with a transfer operator R and μ a measure on B. The characterization is general and involves only the one-sided shift σ on $B^{\mathbb{N}}$.

Lemma 2.16 *Let $\Sigma \in \mathcal{M}_1(B^{\mathbb{N}})$ and set $\mu := \Sigma \circ \pi_1^{-1} \in \mathcal{M}(B)$; then for μ-almost all $x \in B$ there is a field $\mathbb{P}_x \in \mathcal{M}(\pi_1^{-1}(x))$ such that*

$$d\Sigma = \int_B d\mathbb{P}_x \, d\mu(x) \tag{2.26}$$

and the following hold

(i) *The operator $V_1 : L^2(B, \mu) \to L^2(B^{\mathbb{N}}, \Sigma)$ given by*

$$V_1\varphi = \varphi \circ \pi_1 \tag{2.27}$$

is isometric.

(ii) *Its adjoint operator $V_1^* : L^2(B^{\mathbb{N}}, \Sigma) \to L^2(B, \mu)$ satisfies*

$$(V_1^* f)(x) = \int_{\pi_1^{-1}(x)} f \, d\mathbb{P}_x =: \mathbb{E}_x(f), \quad (x \in B). \tag{2.28}$$

Proof (i) For $\varphi \in C(B)$, we have

$$\begin{aligned} \|V_1\varphi\|^2_{L^2(\Sigma)} &= \int_{B^{\mathbb{N}}} |\varphi \circ \pi_1|^2 \, d\Sigma = \int_{B^{\mathbb{N}}} |\varphi|^2 \circ \pi_1 \, d\Sigma \\ &= \int_B |\varphi|^2 \, d(\Sigma \circ \pi_1^{-1}) = \int_B |\varphi|^2 \, d\mu. \end{aligned}$$

(ii) For $\varphi \in C(B)$ and $f \in L^2(B^{\mathbb{N}}, \Sigma)$, we have

$$\int_{B^{\mathbb{N}}} (V_1\varphi) f \, d\Sigma = \int_B \varphi(x)(V_1^* f)(x) \, d\mu(x), \tag{2.29}$$

where $V_1^* f \in L^2(B, \mu)$. Hence

$$\int_{B^{\mathbb{N}}} (\varphi \circ \pi_1) f \, d\Sigma = \int_B \varphi(x)(V_1^* f)(x) \, d\mu(x) \tag{2.30}$$

and $(V_1^* f)(x)$ is well defined for μ-almost all $x \in B$. Moreover, the mapping

$$C(B^{\mathbb{N}}) \ni f \mapsto (V_1^* f)(x) \tag{2.31}$$

is positive; i.e., $f \geq 0$ implies $(V_1^* f)(x) \geq 0$. This follows from (2.30). For if $E \subset B$, $\mu(E) > 0$, and $V_1^* f < 0$ on E then there exists $\varphi \in C(B)$, $\varphi > 0$ such that $\int_B \varphi(x)(V_1^* f)(x)\,d\mu(x) < 0$, which contradicts (2.30). Now the conclusion in (2.28) follows from an application of Riesz' theorem to (2.31). □

Proposition 2.17 *Let $\Sigma \in \mathcal{M}(B^{\mathbb{N}})$, then $(\mathbb{P}_x)_{x\in B}$ from* Lemma 2.16 *has the form* (2.17) *in* Theorem 2.11 *if and only if there is a positive operator R such that $R1 = 1$ and*

$$\mathbb{E}_x(f \circ \sigma) = (R(\mathbb{E} \bullet f))(x) \tag{2.32}$$

holds for all $x \in B$ and for all $f \in L^2(B^{\mathbb{N}}, \Sigma)$, where in (2.32) *we use the notation*

$$\mathbb{E}_x(\dots) = \int\limits_{\pi_1^{-1}(x)} \dots\, d\mathbb{P}_x = \mathbb{E}^{(\Sigma)}(\dots \mid \pi_1 = x) \tag{2.33}$$

for the field of conditional expectations, and $\mathbb{E} \bullet f$ denotes the map $x \mapsto \mathbb{E}_x f$.

Proof The implication (2.17) ⇒ (2.32) is already established. It is Lemma 2.14(ii). Now assume some positive operator R exists such that (2.32) holds. We will then prove that Σ is the measure determined in Theorem 2.11 from R and $\mu = \Sigma \circ \pi_1^{-1}$. It is enough to verify (2.17) on all finite tensors

$$f = (\varphi_1 \circ \pi_1)(\varphi_2 \circ \pi_2)\dots(\varphi_n \circ \pi_n) \tag{2.34}$$

as in (2.16); and we now establish (2.17) by induction, using the assumed (2.32).

The case $n = 1$ is

$$\mathbb{E}_x(\varphi \circ \pi_1) = \varphi(x), \quad (\varphi \in C(B), x \in B);$$

and this follows from Lemma 2.16.

For $n = 2$, we compute as follows

$$\mathbb{E}_x(\varphi_1 \circ \pi_1\, \varphi_2 \circ \pi_2) = \varphi_1(x)(R\varphi_2)(x). \tag{2.35}$$

To do this, we shall prove the following fact, obtained from assumption (2.32):

For $\psi \in C(B)$ and $f \in L^2(B^{\mathbb{N}}, \Sigma)$ we have

$$\mathbb{E}_x((\psi \circ \pi_1) f) = \psi(x)\mathbb{E}_x(f). \tag{2.36}$$

Using (2.28) in Lemma 2.16(ii), note that (2.36) is equivalent to

$$\int\limits_{B^{\mathbb{N}}} (\varphi \circ \pi_1)(\psi \circ \pi_1) f\, d\Sigma = \int\limits_B \varphi\psi V_1^* f\, d\mu,$$

which in turn follows from $V_1(\varphi\psi) = (V_1\varphi)(V_1\psi)$ since $\varphi \mapsto \varphi \circ \pi_1$ is multiplicative.

Returning to (2.35), we then get

$$\begin{aligned}\mathbb{E}_x((\varphi_1 \circ \pi_1)(\varphi_2 \circ \pi_2)) &= \varphi_1(x)\mathbb{E}_x(\varphi_2 \circ \pi_2) = \varphi_1(x)\mathbb{E}_x(\varphi_2 \circ \pi_1 \circ \sigma) \\ &= \varphi_1(x)R(\mathbb{E} \bullet (\varphi_2 \circ \pi_1))(x) = \varphi_1(x)(R\varphi_2)(x).\end{aligned}$$

We shall now be using $\pi_i \circ \sigma = \pi_{i+1}$.

Assume that

$$\mathbb{E}_x(f) = \varphi_1(x)R(\varphi_2 R(\dots R(\varphi_n)\dots))(x) \tag{2.37}$$

holds when f in (2.34) has length $n-1$; then we show it must hold if it has length n. We set

$$\mathbb{E}_x(f) = \mathbb{E}_x((\varphi_1 \circ \pi_1)(g \circ \sigma)),$$

where g is a tensor of length $n-1$. Hence the induction hypothesis yields

$$\mathbb{E}_x(f) = \varphi_1(x)\mathbb{E}_x(g \circ \sigma) = \varphi_1(x)R(\mathbb{E} \bullet (g))$$

which is the right-hand side of (2.37). □

2.2 *Subalgebras in $L^\infty(\Omega, \Sigma)$ and a Conditional Expectation*

Let B, $R : C(B) \to C(B)$, $\mu \in \mathcal{M}_1(B)$ and $\Sigma = \Sigma^{(\mu)}$ be as specified. The only assumptions on R are that

(i) it is linear;
(ii) it is positive and
(iii) $R1 = 1$.

We will be using Theorem 2.11 and Corollaries 2.12 and 3.14 referring to the measures

$$\{\mathbb{P}_x^{(R)} : x \in B\} \text{ on } \pi_1^{-1}(x), \quad (x \in B). \tag{2.38}$$

The theorem below is about the operators $\{V_n : n \in \mathbb{N}\}$, $V_n : L^2(B, \mu) \to L^2(\Omega, \Sigma)$ given by

$$V_n\varphi = \varphi \circ \pi_n, \quad (\varphi \in C(B), n \in \mathbb{N}).$$

Since $V_1 : L^2(B, \mu) \to L^2(\Omega, \Sigma)$ is isometric, it follows that

$$Q_1 := V_1 V_1^* \tag{2.39}$$

is a projection in each of the Hilbert spaces $L^2(\Omega, \Sigma^{(\mu)})$.

Theorem 2.18 *With $B, R, \mu, \Sigma = \Sigma^{(\mu)}$ and V_n specified as above, we have the following formulas:*

(i) $V_1^* V_{n+1} = R^n$ *on* $L^2(B, \mu)$, $n = 0, 1, 2 \ldots$;
(ii) $Q_1 := V_1 V_1^*$ *is a conditional expectation onto*

$$\mathcal{A}_1 := \{\varphi \circ \pi_1 : \varphi \in L^\infty(B, \mu)\}$$

$$Q_1((\varphi \circ \pi_1) f) = (\varphi \circ \pi_1) Q_1(f) \text{ for all } \varphi \in L^\infty(B, \mu),\ f \in L^\infty(\Omega, \Sigma). \tag{2.40}$$

(iii) $Q_1(\varphi \circ \pi_{n+1}) = (R^n \varphi) \circ \pi_1$ *for all* $\varphi \in C(B)$, $n = 0, 1, 2, \ldots$.

Proof (i) As a special case of Theorem 2.11, we see that

$$\int_{\pi_1^{-1}(x)} (\varphi \circ \pi_{n+1})\, d\mathbb{P}_x^{(R)} = (R^n \varphi)(x) \tag{2.41}$$

holds for all $\varphi \in C(B)$. We further see that (2.41) extends to both $L^\infty(B, \mu)$ and to $L^2(B, \mu)$. Hence

$$(V_1^* V_{n+1} \varphi)(x) = \int_{\pi_1^{-1}(x)} (\varphi \circ \pi_{n+1})\, d\mathbb{P}_x^{(\mathbb{R})} = (R^n \varphi)(x), \quad (x \in B). \tag{2.42}$$

(ii) By Lemma 2.10, we see that to verify (2.40), it is enough to check it for cylinder functions f, i.e.,

$$f = (\psi_1 \circ \pi_1)(\psi_2 \circ \pi_2) \ldots (\psi_n \circ \pi_n), \tag{2.43}$$

$n \in \mathbb{N}$, $\psi_i \in C(B)$. But if f is as in (2.43), then

$$(\varphi \circ \pi_1) f = ((\varphi \psi_1) \circ \pi_1)(\psi_2 \circ \pi_2) \ldots (\psi_n \circ \pi_n), \tag{2.44}$$

and the desired formula (2.40) is immediate.

(iii) Given (i), we may apply V_1 to both sides in (2.42), and the desired formula (iii) follows. □

It is important to stress that one obtains a closed-form expression for V_1^* where the operator $V_1 : \varphi \mapsto \varphi \circ \pi_1$ is introduced in Corollary 2.12. Indeed V_1^* is a conditional expectation:

$$(V_1^* f)(x) = \mathbb{E}^{(\Sigma)}(f \mid \pi_1 = x) = \mathbb{E}_x^{(\Sigma)}(f), \quad (x \in B, f \in L^2(\Omega, \Sigma)) \qquad (2.45)$$

By contrast, the situation for V_n^*, $n > 1$ is more subtle.

Proposition 2.19 *Let* B, $R : C(B) \to C(B)$ *and* $\Sigma = \Sigma^{(\mu)} \in \mathcal{M}_1(\Omega)$ *be as above, i.e.,* $\mu = \Sigma \circ \pi_1^{-1}$. *Let* R^* *be the adjoint of the operator* R *when considered as a bounded operator in* $L^2(B, \mu)$. *For* V_2^* *we have*

$$V_2^*((\varphi_1 \circ \pi_1)(\varphi_2 \circ \pi_2) \dots (\varphi_n \circ \pi_n)) = R^*(\varphi_1)\varphi_2 R(\varphi_3 \dots R(\varphi_n) \dots). \qquad (2.46)$$

Proof Let ψ be the function given on the right hand side in (2.46). The operator $V_2 : \varphi \mapsto \varphi \circ \pi_2$ maps from $L^2(B, \mu)$ into $L^2(\Omega, \Sigma)$. The assertion in (2.46) follows if we check that, for all $\xi \in C(B)$, we have the following identity:

$$\int_\Omega (\varphi_1 \circ \pi_1)((\xi\varphi_2) \circ \pi_2)(\varphi_3 \circ \pi_3) \dots (\varphi_n \circ \pi_n)\, d\Sigma = \int_B \xi\psi\, d\mu. \qquad (2.47)$$

But we may compare the left-hand side in (2.47) with the use of Theorem 2.11:

$$= \int_B \varphi_1 R((\xi\varphi_2) R(\varphi_3 R(\dots R(\varphi_n)\dots)))\, d\mu = \int_B (R^*\varphi_1)\xi\varphi_2 R(\varphi_3 R(\dots R(\varphi_n)\dots))\, d\mu,$$

which is the desired conclusion (2.46).

Recall that, by Theorem 2.18, we have $R = V_1^* V_2$, and so $R^* = V_2^* V_1$. □

The next result is an extension of Lemma 2.14(ii). Note that (2.23) is the assertion that V_1^* intertwines the two operations, R and $f \mapsto f \circ \sigma$. The next result shows that, by contrast, V_2^* acts as a multiplier.

Corollary 2.20 *Let* B, R, Σ *and* $\mu = \Sigma \circ \pi_1^{-1}$ *be as in* Proposition 2.19, *and set* $\rho := R^* 1 \in L^2(B, \mu)$; *then*

$$(V_2^*(f \circ \sigma))(x) = \rho(x)\mathbb{E}_x(f) = \rho(x)(V_1^* f)(x), \quad (x \in B, f \in L^2(\Omega, \Sigma)).$$

Proof This is immediate from Proposition 2.19, see (2.46). Recall that the span of the tensors is dense in $L^2(\Omega, \Sigma)$ and that if $f = (\varphi_1 \circ \pi_1)(\varphi_2 \circ \pi_2) \dots (\varphi_n \circ \pi_n)$, then $f \circ \sigma = (\varphi_1 \circ \pi_2)(\varphi_2 \circ \pi_3) \dots (\varphi_n \circ \pi_{n+1})$. □

In Proposition 4.6 we calculate the multiplier ρ for the special case of the wavelet representation from Example 2.5.

Corollary 2.21 *Let* B *and* R *be as in* Theorem 2.18, *and let* $\mu \in \mathcal{M}_1(B)$ *be given. The induced measure on* $\Omega = B^{\mathbb{N}}$ *is denoted* $\Sigma^{(\mu)}$ *and specified as in* (2.18). *We then have the following equivalence:*

(i) *$h \in L^2(B, \mu)$ and $Rh = h$, i.e., h is harmonic; and*
(ii) *There exists $f \in L^2(\Omega, \Sigma^{(\mu)})$ such that $f = f \circ \sigma$ and*

$$h(x) = \int_{\pi_1^{-1}(x)} f \, d\mathbb{P}_x^{(R)}. \tag{2.48}$$

Proof The implication (ii)⇒(i) follows from Lemma 2.14 and Remark 2.15. For (i)⇒(ii), let h be given and assume it satisfies (i). An application of (iii) from Theorem 2.18 now yields

$$Q_1(h \circ \pi_{n+1}) = R^n(h) \circ \pi_1 = h \circ \pi_1 = V_1 h.$$

Using (2.39), we get $V_1(h - V_1^*(h \circ \pi_{n+1})) = 0$ for all $n = 0, 1, 2, \ldots$; and therefore

$$h = V_1^*(h \circ \pi_{n+1}), \quad (n \in \mathbb{N}). \tag{2.49}$$

Recalling

$$V_1^*(h \circ \pi_{n+1})(x) = \int_{\pi_1^{-1}(x)} h \circ \pi_{n+1} \, d\mathbb{P}_x^{(R)} \tag{2.50}$$

and using Theorem 2.11, we conclude that $\{h \circ \pi_{n+1}\}_{n \in \mathbb{N}}$ is a bounded $L^2(\Omega, \Sigma^{(\mu)})$-martingale.

By Doob's theorem, there is a $f \in L^2(\Omega, \Sigma^{(\mu)})$ such that

$$\lim_{n \to \infty} \|f - h \circ \pi_{n+1}\|_{L^2(\Sigma^{(\mu)})} = 0.$$

Since $\pi_{n+1} \circ \sigma = \pi_n$, it follows that $f \circ \sigma = f$. Taking the limit in (2.49) and using that the operator norm of V_1^* is one, we get that $h = V_1^* f$ and therefore the desired formula (2.48) holds. □

2.3 A Stochastic Process Indexed by $\mathbb{N}$

Remark 2.22 In the literature one has a number of theorems dealing with the existence of measures μ satisfying the various conditions; and if $\mu \circ R = \mu$ is satsified, then the measure is called a Ruelle equilibrium measure.

Theorem 2.23 *Let B be compact Hausdorff and $R : C(B) \to C(B)$ positive, $R1 = 1$. Let $\mu \in \mathcal{M}_1(B)$ such that $\mu(B) = 1$, $\mu \circ R = \mu$. Set*

$$X_n(\varphi) = \varphi \circ \pi_n, \quad (\varphi \in C(B), n \in \mathbb{N})$$

and

$$\int f \, d\Sigma = \int_B \int_{\pi_1^{-1}(x)} f \, d\mathbb{P}_x^{(R)} \, d\mu(x) \tag{2.51}$$

Then

$$\mathbb{E}(\ldots) = \int \ldots \, d\Sigma$$

satisfies

$$\mathbb{E}(X_n(\varphi)X_{n+k}(\psi)) = \int_B \varphi(x)(R^k\psi)(x) \, d\mu, \quad (n, k \in \mathbb{N}, \varphi, \psi \in C(B)) \tag{2.52}$$

i.e., R^k is the transfer operator governing distances k. Asymptotic properties as k goes to infinity govern long-range order.

Proof From the definition of $\mathbb{P}_x^{(R)}$ we have

$$\mathbb{P}_x^{(R)}(X_n(\varphi)) = R^{n-1}(\varphi)(x), \quad \varphi \in C(B) \tag{2.53}$$

Now let n, k, φ, ψ as in the statement in (2.52). Let Σ be the measure on $B^{\mathbb{N}}$ in (2.51). Then

$$\begin{aligned}
\mathbb{E}(X_n(\varphi)X_{n+k}(\psi)) &= \int_{B^{\mathbb{N}}} (\varphi \circ \pi_n)(\psi \circ \pi_{n+k}) \, d\Sigma \\
&= \int_B R^{n-1}(\varphi R^k(\psi))(x) \, d\mu(x) \\
&= \int \varphi(x) R^k(\psi)(x) \, d\mu(x)
\end{aligned}$$

which is the desired conclusion. □

Definition 2.24 We say that $\{X_k(\varphi)\}$ is independent at ∞ if

$$\lim_{k \to \infty} \mathbb{E}(X_n(\varphi)X_{n+k}(\psi)) = \left(\int \varphi \, d\mu\right)\left(\int \psi \, d\mu\right), \quad (\varphi, \psi \in C(B), n \in \mathbb{N}). \tag{2.54}$$

Corollary 2.25 *Suppose for all φ in $C(B)$ we have*

$$\lim_{k\to\infty} R^k(\varphi) = \left(\int \varphi\, d\mu\right) 1,$$

then (2.54) *is satisfied.*

Proof We proved

$$\mathbb{E}(X_n(\varphi)X_{n+k}(\psi)) = \int \varphi R^k \psi\, d\mu.$$

Now take the limit as $k \to \infty$, the desired conclusion (2.54) follows. □

The next result answers the question: what is the distribution of the random variable $X_n(\varphi)$?

Corollary 2.26 *Assume $\mu \circ R = \mu$. The distribution of $X_n(\varphi)$ is $\mu(\{x \in B : \varphi(x) \leq t\})$ for all n.*

Proof Take φ real valued for simplicity.
For $t \in \mathbb{R}$,

$$\begin{aligned}\Sigma(\{\tilde{x} \in B^{\mathbb{N}} : \varphi \circ \pi_n(\tilde{x}) \leq t\}) &= \int_B \int_{\pi_1^{-1}(x)} \chi_{\{\varphi\leq t\}} \circ \pi_n\, d\mathbb{P}_x^{(R)}\, d\mu(x) \\ &= \int_B R^{n-1}\chi_{\{\varphi\leq t\}}\, d\mu = \int_B \chi_{\{\varphi\leq t\}}\, d\mu.\end{aligned}$$

In particular, it follows that all the random variables $X_n(\varphi)$ have the same distribution. □

2.4 Application to Random Walks

Corollary 2.27 *Let (r, W) be as in* Definition 2.1, *and let R_W be the Ruelle operator in* (2.2), *$\mathbb{P}_x^{(W)}$—the random walk measure with transition probability specified as follows*

$$\text{Prob}(x \to y) = \begin{cases} W(y), & \textit{if } r(y) = x \\ 0, & \textit{otherwise} \end{cases} \tag{2.55}$$

Then $\mathbb{P}_x$ from Theorem 2.11 *is equal to $\mathbb{P}_x^{(W)}$.*

Proof We apply Theorem 2.11 to $R = R_W$ in (2.2) and we compute the right-hand side in (2.15) with induction

$$(M_{\varphi_1} R_W \dots R_W M_{\varphi_{n+1}})(x) = \varphi_1(x) \sum_{y_1} \cdots \sum_{y_n} W(y_1)W(y_2)\dots W(y_n)\varphi_2(y_1)\dots\varphi_{n+1}(y_n),$$

where $r(y_{i+1}) = y_i$, $1 \leq i < n$, $r^{(n)}(y_n) = x$. Further,

$$= \sum \cdots \sum \mathrm{Prob}(x \to y_1)\,\mathrm{Prob}(y_1 \to y_2 \,|, y_1)\dots\mathrm{Prob}(y_{n-1} \to y_n \,|\, y_{n-1})\varphi(y_1)\dots\varphi(y_n)$$

$$= \int d\mathbb{P}_x^{(W\text{-transition } R_W\text{-measure})} \varphi_1 \otimes \cdots \otimes \varphi_n \tag{2.56}$$

□

Remark 2.28 The assertion in (2.56) applies to any random walk measure, for example, the one in Example 2.8.

Let $G = (V, E)$ be as in Example 2.8, with E un-directed edges. Let $c : E \to [0, \infty)$ be such that

$$c_{(xy)} = c_{(yx)} \text{ for all } (xy) \in E, \quad c_{(xy)} \neq 0 \text{ if } (xy) \notin E. \tag{2.57}$$

A function as in (2.57) is called conductance.

Set $p = p^c$, where

$$p_{xy} = \frac{c_{xy}}{\sum_{z, z\sim x} c_{xz}} = \frac{c_{xy}}{c(x)}, \tag{2.58}$$

where

$$c(x) = \sum_{z, z\sim x} c_{xz}, \text{ and } z \sim x \text{ means } (zx) \in E.$$

Then there is a unique $\mathbb{P}_x^{(c)}$ such that

$$\int \varphi_1 \otimes \cdots \otimes \varphi_n \, d\mathbb{P}_x^{(c)} = \sum_{y_1} \cdots \sum_{y_n} p_{xy_1} p_{y_1 y_2} \cdots p_{y_{n-1} y_n} \varphi_1(y_1) \dots \varphi_n(y_n),$$

where the sums are over all $y_1, y_2, \dots, y_n$ such that $(y_i y_{i+1}) \in E$.

Note that $\mathbb{P}_x^{(W)}$ is supported on the solenoid, and $\mathbb{P}_x^{(c)}$ is supported on $S^{(G)}$ (see (2.13)).

Remark 2.29 The last application is useful in the setting of harmonic functions on graphs $G = (V, E)$ with prescribed conductance function c as in (2.57). Set

$$(\Delta\varphi)(x) = \sum_{y\in V, y\sim x} c_{xy}(\varphi(x) - \varphi(y)) \tag{2.59}$$

the graph Laplacian with conductance c.

A function φ on V satisfies $\Delta\varphi \equiv 0$, iff

$$\varphi(x) = \sum_{y \in V, y \sim x} p_{xy}^{(c)} \varphi(y), \tag{2.60}$$

where $p_{xy}^{(c)} = \frac{c_{xy}}{c(x)}$ as in (2.58).

Application. Use $\mathbb{P}_x$ to get harmonic functions. The study of classes of harmonic functions is of interest for infinite networks (see Remarks 2.28 and 2.29), and in Corollary 2.21 is is shown that the harmonic functions h are precisely those that arise from applying $\mathbb{E}_x$ to functions f, $f \circ \sigma = f$ on $B^{\mathbb{N}}$, i.e.,

$$h(x) = \int f(x \dots)\, d\mathbb{P}_x,$$

and conversely a martingale limit constructs f from h. For more details on this construction, see Corollary 2.21.

2.5 An Application to Integral Operators

Let $K : B \times B \to [0, \infty)$ be a continuous function and let μ be a probability measure on B such that

$$\int_B K(x, y)\, d\mu(y) = 1 \text{ for all } x \in B. \tag{2.61}$$

Define

$$R_K f(x) = \int_B K(x, y) f(y)\, d\mu(y), \quad (x \in B, f \in C(B)).$$

Then $R = R_K$ defines a positive operator as in Definition 2.2, $R_K 1 = 1$ and then $\mathbb{P}_x$ in Theorem 2.11 satisfies

$$\int_{B^{\mathbb{N}}} \varphi_1 \otimes \dots \otimes \varphi_{n+1}\, d\mathbb{P}_x =$$

$$\varphi_1(x) \int \dots \int K(x, y_1) K(y_1, y_2) \dots K(y_{n-1}, y_n) \varphi_2(y_1) \dots \varphi_{n+1}(y_n)\, d\mu(y_1) \dots d\mu(y_n)$$

We get a measure Σ on $B^{\mathbb{N}}$ as follows

$$\int f\, d\Sigma = \int f\, d\mathbb{P}_x\, d\mu(x) \tag{2.62}$$

since the right-hand side in (2.62) is independent of x.

3 Positive Operators and Endomorphisms

3.1 Preliminaries About $r : B \to B$

Given an endomorphism r, we form the solenoid $\mathrm{Sol}(r) \subset B^{\mathbb{N}}$. Below we will study $\widehat{r} : \mathrm{Sol}(r) \to \mathrm{Sol}(r)$,

$$\widehat{r}(x_1 x_2 \ldots) = (r(x_1) x_1 x_2 \ldots)$$

and $\widehat{r} \in \mathrm{Aut}(\mathrm{Sol}(r))$.

Given a positive operator $R : C(B) \to C(B)$, $R1 = 1$ we then form the measure $\mathbb{P}_x^{(R)}$ in the usual way. We will prove the following property $\mathbb{P}_x^{(R)} \circ \widehat{r}^{-1} = \mathbb{P}_x^{(R)}$ on the solenoid but not on $B^{\mathbb{N}}$.

We will impose the conditon (3.10)

$$R((\varphi \circ r)\psi) = \varphi R\psi$$

as the only axiom. It may or may not be satisfied for some examples of positive operators R. But it does hold in the following two examples:

$$(R\varphi)(x) = \sum_{r(y)=x} W(y)\varphi(y) \text{ and}$$

$$(R\varphi)(x) = \frac{1}{\# r^{-1}(x)} \sum_{r(y)=x} |m(y)|^2 \varphi(y),$$

where the functions W and m are given subject to the usual conditions.

For reference to earlier papers dealing with measures on infinite products, random walk, and stochastic processes; see e.g., [JP11, JP10, AJ12].

Example 3.1 Classical wavelet theory on the real line. Let $N = 2$, $B = \mathbb{T} = \{z \in \mathbb{C} : |z| = 1\} \simeq \mathbb{R}/\mathbb{Z} \simeq (-\frac{1}{2}, \frac{1}{2}]$ via $z = e^{2\pi i\theta}$, $\theta \in \mathbb{R}/\mathbb{Z}$; $\mu = d\theta$; $L^2(B, \mu) = L^2((-\frac{1}{2}, \frac{1}{2}], d\theta)$, $r : B \to B$,

$$r(z) = z^2, \text{ or equivalently} r(\theta \mod \mathbb{Z}) = 2\theta \mod \mathbb{Z}. \tag{3.1}$$

Let

$$m_0(\theta) = \sum_{n\in\mathbb{Z}} h_n e^{2\pi i n\theta}, \text{ or equivalently } m_0(z) = \sum_{n\in\mathbb{Z}} h_n z^n, \tag{3.2}$$

where we assume

$$\sum_{n\in\mathbb{Z}} h_n = \sqrt{2}, \quad \sum_{n\in\mathbb{Z}} |h_n|^2 < \infty.$$

Lemma 3.2 *With m_0 as in* (3.2)*, the condition* (2.7) *is equivalent to*

$$\sum_{k\in\mathbb{Z}} h_k \overline{h}_{k-2n} = \frac{1}{2}\delta_{n,0}. \tag{3.3}$$

Proposition 3.3 ([BJ02, DJ05]) *Suppose that m_0 is as above, and that there is a solution $\varphi \in L^2(\mathbb{R})$ satisfying*

$$\frac{1}{\sqrt{2}}\varphi\left(\frac{x}{2}\right) = \sum_{n\in\mathbb{Z}} h_n \varphi(x-n), \tag{3.4}$$

and

$$\textit{The translates } \varphi(\cdot - n) \textit{ are orthogonal in } L^2(\mathbb{R}),\ n \in \mathbb{Z}. \tag{3.5}$$

Set $W : L^2(\mathbb{T}) \to L^2(\mathbb{R})$,

$$(W\xi)(x) = \sum_{n\in\mathbb{Z}} \widehat{\xi}(n)\varphi(x-n) =: \pi(\xi)\varphi, \tag{3.6}$$

where $\xi \in L^2(\mathbb{T})$, and $\widehat{\xi}(n) = \int_{\mathbb{T}} \overline{e}_n \xi \, d\mu$;

$$(S_0\xi)(z) = m_0(z)\xi(z^2), \quad (z \in \mathbb{T}); \tag{3.7}$$

and

$$(Uf)(x) = \frac{1}{\sqrt{2}} f\left(\frac{x}{2}\right), \quad f \in L^2(\mathbb{R}). \tag{3.8}$$

(i) *Then S_0 is isometric, and $(L^2(\mathbb{R}), \varphi, \pi, U)$ is a wavelet representation.*
(ii) *The dilation $W : L^2(\mathbb{T}) \to L^2(\mathbb{R})$ then takes the following form: W is isometric and it intertwines S_0 and the unitary operator U, i.e., we have*

$$(WS_0\xi)(x) = (UW\xi)(x) = \frac{1}{\sqrt{2}}(W\xi)\left(\frac{x}{2}\right), \quad (\xi \in L^2(\mathbb{T}), x \in \mathbb{R}). \tag{3.9}$$

Remark 3.4 With m_0 as specified in Proposition 3.3, we conclude that the wavelet representation can be realized on $L^2(\mathbb{R})$. On the other hand, we will see in Corollary 4.5 that it can be also realized on the solenoid. The two representations have to be isomorphic. The identifications can be done via the usual embedding of $\mathbb{R}$ into the solenoid $x \mapsto (e^{2\pi ix}, e^{2\pi ix/2}, e^{2\pi ix/2^2}, \ldots)$. The measure Σ in this case is supported on the image of $\mathbb{R}$ under this embedding. For details, see [Dut06].

Axioms. B compact Hausdorff space, $R : C(B) \to C(B)$ positive linear operator such that $R1 = 1$, $r : B \to B$ onto, continuous.

Assume

$$R((\varphi \circ r)\psi) = \varphi R(\psi), \quad (\varphi, \psi \in C(B)) \tag{3.10}$$

Note that (3.10) is the only property that we assume on the operator R.

Lemma 3.5 *On the solenoid*

$$\mathrm{Sol}(r) = \{(x_1, x_2, \ldots) \in B^{\mathbb{N}} : r(x_{i+1}) = x_i\},$$

$$\pi_i \circ \widehat{r} = r \circ \pi_i.$$

Proof For $\tilde{x} = (x_1, x_2, \ldots)$,

$$\widehat{r}(\tilde{x}) = (r(x_1), x_1, x_2, \ldots). \tag{3.11}$$

$$\pi_i \circ \widehat{r}(\tilde{x}) = x_{i-1} = r(x_i) = r \circ \pi_i(\tilde{x}).$$

□

Remark 3.6 Our initial setup for a given endomorphism r in our present setup is deliberately left open to a variety of possibilities. Indeed, the literature on solenoid analysis is vast, but divides naturally into cases when $r : B \to B$ has only one contractivity degree; as opposed to a mix of non-linear contractive directions. The first case is common in wavelet analysis, such as those studied in [DJ06, DJ07, DJ10, DJ12]. Examples of the second class, often called "hyperbolic" systems, includes the Smale-Williams attractor, with the endomorphism r there prescribed to preserve a foliation by meridional disks; see e.g., [Kuz10, KP07, KP07, Rue04]. Or the study of complex dynamics and Julia sets; see e.g., [BCMN04] .

Lemma 3.7 *Let $r : B \to B$ be given and let $\widehat{r} \in \mathrm{Aut}(\mathrm{Sol}(r))$ be the induced automorphism on the solenoid. Then*

$$\widehat{r}(\pi_1^{-1}(x)) = \pi_1^{-1}(r(x)) \cap \pi_2^{-1}(x), \quad (x \in B).$$

Proof Use the definition of $\widehat{r}$ in (3.11). □

Definition 3.8 Given μ and R, they generate the probability measure $\Sigma = \Sigma^{(\mu)}$ on $B^{\mathbb{N}}$. We assume $R1 = 1$ and $\mu(B) = 1$. Define

$$\mathbb{E}(f) = \int_{B^{\mathbb{N}}} f \, d\Sigma \tag{3.12}$$

$$\mathbb{E}_x(f) := \mathbb{E}(f \mid \pi_1 = x) = \int_{\pi_1^{-1}(x)} f \, d\Sigma \tag{3.13}$$

$$\mathbb{E}_{x_1,x_2}(f) := \mathbb{E}(f \mid \pi_1 = x_1, \pi_2 = x_2) = \int_{\pi_1^{-1}(x_1)\cap\pi_2^{-1}(x_2)} f \, d\Sigma. \tag{3.14}$$

for all $x_1, x_2 \in B$. As before we take

$$\mathbb{E}(f) = \int_B \int_{\pi^{-1}(x)} f \, d\mathbb{P}_x \, d\mu(x) \tag{3.15}$$

and we then get

$$\mathbb{E}_x(f) = \int_{\pi^{-1}(x)} f \, d\mathbb{P}_x \tag{3.16}$$

Lemma 3.9 *Let B, R and r be given as above.*

$$R((\varphi \circ r)\psi) = \varphi R(\psi) \tag{3.17}$$

Then the following two are equivalent for some measure μ on B:

(i) $\mu \circ R = \mu$
(ii) $\int (\varphi \circ r)\psi \, d\mu = \int \varphi R\psi \, d\mu$.

Proof (i)⇒(ii). Assume (i) and (3.17). Then

$$\int \varphi \circ r \cdot \psi \, d\mu = \int R((\varphi \circ r)\psi) \, d\mu = \int \varphi R\psi \, d\mu$$

which is condition (ii).

(ii)⇒(i). Assume (ii). Then set $\varphi = 1$ in (ii) and we get $\int \psi \, d\mu = \int R\psi \, d\mu$ which is the desired property (i). □

Lemma 3.10 *Assume the basic axiom* (3.17). *For $f \in L^1(\Sigma)$, we denote by $\mathbb{E}\bullet(f)$, the function $x \mapsto \mathbb{E}_x(f)$, $x \in B$. Then*

$$\mathbb{E}_x(f \circ \sigma) = R(\mathbb{E} \bullet (f))(x) \tag{3.18}$$

Also,

$$\mathbb{E}_x(f \circ \widehat{r}) = \mathbb{E}_{r(x),x}(f) \tag{3.19}$$

for all $x \in B$, $f \in L^1(\Sigma)$, or equivalently

$$\mathbb{E}(f \circ \widehat{r} \mid \pi_1 = x) = \mathbb{E}(f \mid \pi_1 = r(x), \pi_2 = x); \tag{3.20}$$

see the notations in Definition 3.8.

Proof Equation (3.18) is proved in (2.32). For (3.19), we use the Stone-Weierstrass approximation as before. If

$$f = (\varphi_1 \circ \pi_1)(\varphi_2 \circ \pi_2) \dots (\varphi_n \circ \pi_n),$$

then

$$f \circ \widehat{r} = (\varphi_1 \circ r \circ \pi_1)(\varphi_2 \circ r \circ \pi_2) \dots (\varphi_n \circ r \circ \pi_n),$$

and so

$$\begin{aligned}\mathbb{E}_x(f \circ \widehat{r}) &= \varphi_1(r(x)) R(\varphi_2 \circ r R(\varphi_3 \circ r \dots R(\varphi_n \circ r)) \dots)(x)\\ &= \varphi_1(r(x))\varphi_2(x) R(\varphi_3 R(\dots \varphi_{n-1} R(\varphi_n)) \dots)(x) = \mathbb{E}_{r(x),x}(f),\end{aligned}$$

or equivalentlly (3.20). □

Proposition 3.11 *Let B and $R : C(B) \to C(B)$ be as stated in* Theorem 2.11. *For every $\mu \in \mathcal{M}_1(B)$ we denote the induced measure on $\Omega = B^{\mathbb{N}}$ by $\Sigma^{(\mu)}$. If some $r : B \to B$ satisfies*

$$R((\varphi \circ r)\psi) = \varphi R(\psi), \quad (\varphi, \psi \in C(B)) \tag{3.21}$$

then every one of the induced measures $\Sigma^{(\mu)}$ has its support contained in the solenoid $\mathrm{Sol}(r)$.

Proof Using Lemma 2.14, it is enough to prove that each of the measures $\mathbb{P}_x^{(R)}$ with x fixed (from Corollary 2.12) has its support equal to

$$\pi_1^{-1}(x) \cap \mathrm{Sol}(r) \tag{3.22}$$

For every n, consider all infinite words indexed by $y \in r^{-n}(x)$ and specified on the beginning length-n segments as follows $\Omega_n(r, x)$: $(x, r^{n-1}(y), \dots, r(y), y,$ free infinite tail) and note that

$$\pi_1^{-1}(x) \cap \mathrm{Sol}(r) = \bigcap_n \Omega_n(r, x) \tag{3.23}$$

For $n = 1$, we have

$$\mathbb{P}_x^{(R)}(\Omega_1(r, x)) = R(\chi_{\{x\}} \circ r)(x) = \chi_{\{x\}}(x) R(1) = 1,$$

where we used assumption (3.21) in the last step in the computation.

The remaining reasoning in the proof is an induction. Indeed, one checks that

$$\begin{aligned}\mathbb{P}_x^{(R)}(\Omega_n(r,x)) &= R((\chi_{\{x\}}\circ r)R((\chi_{\{x\}}\circ r^2)R(\dots(\chi_{\{x\}}\circ r^{n-1})R(\chi_{\{x\}}\circ r^n)\dots)))(x)\\ &= R((\chi_{\{x\}}\circ r)R(\dots R(\chi_{\{x\}}\circ r^{n-1})\dots))(x).\end{aligned}$$

Hence the assertion for $n-1$ implies the next step n. By induction, we get

$$\mathbb{P}_x^{(R)}(\Omega_n(r,x)) = 1, \quad (n\in\mathbb{N}, x\in B).$$

Using (3.23), we get

$$\mathbb{P}_x^{(R)}(\pi_1^{-1}(x)\cap \mathrm{Sol}(r)) = \lim_{n\to\infty}\mathbb{P}_x^{(R)}(\Omega_n(r,x)) = 1.$$

As a consequence, the measure $\mathbb{P}_x^{(R)}$ assigns value 1 to the indicator function of $\pi_1^{-1}(x)\cap \mathrm{Sol}(r)$. But

$$\mathrm{Sol}(r) = \bigcup_{x\in B}\pi_1^{-1}(x)\cap \mathrm{Sol}(r).$$

So if $\mu(B)=1$, it follows from (2.51) that

$$\Sigma^{(\mu)}(\mathrm{Sol}(r)) = \int_{B^{\mathbb{N}}}\chi_{\mathrm{Sol}(r)}\,d\Sigma^{(\mu)} = 1;$$

and as a result that

$$\Sigma^{(\mu)}(B^{\mathbb{N}}\backslash \mathrm{Sol}) = 0$$

which is the desired conclusion. □

Corollary 3.12 *Let B, r, μ, R be as above and assume* (3.17). *Then Σ is supported on* $\mathrm{Sol}(r)$ *and $\widehat{r}$ is invertible on* $\mathrm{Sol}(r)$ *with $\widehat{r}^{-1}=\sigma$. The measure Σ is invariant (for $\widehat{r}$) if and only if*

$$\mu\circ R = \mu. \tag{3.24}$$

Proof It is enough to prove that

$$\int_{\mathrm{Sol}(r)} f\circ\sigma\,d\Sigma = \int_{\mathrm{Sol}(r)} f\,d\Sigma \tag{3.25}$$

holds for all $f\in L^1(\Sigma)$ if and only if (3.24) holds. But, by (3.18) we have

$$\int_{\mathrm{Sol}(r)} f \circ \sigma \, d\Sigma = \int_B \mathbb{E}_x(f \circ \sigma)\, d\mu(x) = \int_B R(\mathbb{E} \bullet (f))(x)\, d\mu(x),$$

and

$$\int_{\mathrm{Sol}(r)} f \, d\Sigma = \int_B \mathbb{E}_x(f)\, d\mu(x).$$

But the functions $x \mapsto \mathbb{E}_x(f)$ are dense in $L^1(B, \mu)$ as f varies in $L^1(\Sigma)$ (consider for example $f = g \circ \pi_1$ for $g \in C(B)$). Thus the equivalence of (3.24) and (3.25) is immediate from this. □

Corollary 3.13 *Let R be a positive operator in $C(B)$ satsifying the axioms above, $R1 = 1$, $R((\varphi \circ r)\psi) = \varphi R(\psi)$ for all $\varphi, \psi \in C(B)$. Let μ be a Borel measure on B, and set $\Sigma = \Sigma^{(\mu)}$*

$$\int_{\mathrm{Sol}(r)} f \, d\Sigma := \int_B \int_{\pi_1^{-1}(x)} f \, d\mathbb{P}_x^{(R)} \, d\mu(x);$$

then

$$\mathcal{U}^{(R)} f := f \circ \widehat{r} \tag{3.26}$$

defines a unitary operator on $L^2(\mathrm{Sol}(r), \Sigma)$ if and only if $\mu = \mu \circ R$.

Proof Since $\widehat{r}$ is invertible in $\mathrm{Sol}(r)$, we conclude that $\mathcal{U}^{(R)}$ maps onto $L^2(\mathrm{Sol}(r), \Sigma)$. Recall $\widehat{r}^{-1} = \sigma$

$$\sigma(x_1, x_2, x_3, \ldots) = (x_2, x_3, x_4, \ldots) \tag{3.27}$$

Since, by Proposition 3.11, the measure Σ is supported on $\mathrm{Sol}(r)$, the result follows from Corollary 3.12. □

Corollary 3.14 *Let B and $R : C(B) \to C(B)$ be as in* Corollary 2.12. *Let $\mu \in \mathcal{M}_1(B)$ and consider $\Sigma = \Sigma^{(\mu)}$. Let $r : B \to B$ be an endomorphism.*

(i) *For the operators $V_1 : L^2(B, \mu) \to L^2(\mathrm{Sol}(r), \Sigma)$ and $\mathcal{U}^{(R)} : f \mapsto f \circ \widehat{r}$ acting in $L^2(\mathrm{Sol}(r), \Sigma)$ we have the following covariance relation: V_1 is isometric and*

$$(V_1^* \mathcal{U}^{(R)} V_1)(\varphi) = \varphi \circ r, \quad (\varphi \in C(B)).$$

(ii) *Assume that $\mu = \mu \circ R$ and*

$$R((\varphi \circ r)\psi) = \varphi R(\psi), \quad (\varphi, \psi \in C(B)) \tag{3.28}$$

holds. For functions F on Ω, say $F \in L^\infty(\mathrm{Sol}(r))$, let M_F be the multiplication operator defined by F. Then $\mathcal{U}^{(R)}$ is unitary and the following covariance relation holds:

$$(\mathcal{U}^{(R)})^* M_F \mathcal{U}^{(R)} = M_{F\circ\sigma}.$$

Proof (i) We will make use of the formula (2.19) for V_1^*, see Corollary 2.12(ii). We now compute

$$(V_1^*\mathcal{U}^{(R)}V_1\varphi)(x) = \mathbb{E}_x\mathcal{U}^{(R)}V_1\varphi = \mathbb{E}_x(\varphi\circ\pi_1\circ\widehat{r}) = \mathbb{E}_x(\varphi\circ r\circ\pi_1) = (\varphi\circ r)(x),$$

which is the conclusion in (i). (ii) We proved in Corollary 3.13 that $\mathcal{U}^{(R)}$ is unitary. The covariance relation follows from a simple computation. □

For reference to earlier papers dealing with measures on infinite products, and shift-invariant systems; see e.g., [CGHU12, CH94].

3.2 Compact Groups

As a special case of our construction, we mention the compact groups; this will include the case of wavelet theory.

Proposition 3.15 *Assume B is a compact group with normalized Haar measure μ. Let $r : B \to B$ be a homomorphism $r(xy) = r(x)r(y)$ for all $x, y \in B$, and assume for $N \in \mathbb{N}$, $N > 1$*

$$\#r^{-1}(x) = N, \quad (x \in B).$$

Set

$$(R\varphi)(x) = \frac{1}{N}\sum_{r(y)=x}\varphi(y), \quad (\varphi \in C(B)); \tag{3.29}$$

then

(i) $\mathrm{Sol}(r)$ *is a compact subgroup of* $B^{\mathbb{N}}$.
(ii) *The induced measure* $\Sigma = \Sigma^{(\mu,R)}$ *is the Haar measure on the group* $\mathrm{Sol}(r)$.

Proof (i) If $\tilde{x} = (x_1, x_2, \dots)$, $\tilde{y} = (y_1, y_2, \dots) \in \mathrm{Sol}(r)$ then $r(x_{i+1}y_{i+1}) = r(x_{i+1})r(y_{i+1}) = x_i y_i$, so $\tilde{x}\tilde{y} \in \mathrm{Sol}(r)$.

(ii) From (3.29) we see that the measures $\mathbb{P}_x$ in the decomposition

$$d\Sigma = \int_B \mathbb{P}_x\, d\mu(x) \tag{3.30}$$

$\mathbb{P}_x \in \mathcal{M}(\pi_1^{-1}(x))$, $x \in B$, are random-walk measures with uniform distributions on the points in $r^{-1}(x)$, for all $x \in B$. If $\mathbb{E}$ denotes the Σ-expectation and $\mathbb{E}_x$ the $\mathbb{P}_x$-expectation, we have

$$\mathbb{E}_x(f) = \mathbb{E}(f \mid \pi_1 = x), \quad (x \in B).$$

For points $\tilde{y} = (y_1, y_2, \dots) \in \mathrm{Sol}(r)$, denote by $f(\cdot\tilde{y})$ the translated function on $\mathrm{Sol}(r)$. Then

$$\mathbb{E}(f(\cdot\tilde{y}) \mid \pi_1 = x) = \mathbb{E}(f \mid \pi_1 = xy_1) \tag{3.31}$$

where we use the terminology in Proposition 2.17.

Now, combining (3.30) and (3.31), we arrive at the formula:

$$\int_{\mathrm{Sol}(r)} f(\cdot\tilde{y})\, d\Sigma = \mathbb{E}(f(\cdot\tilde{y})) = \int_B \mathbb{E}(f(\cdot\tilde{y}) \mid \pi_1 = x)\, d\mu(x) = \int_B \mathbb{E}(f \mid \pi_1 = xy_1)\, d\mu(x)$$
$$= \int_B \mathbb{E}(f \mid \pi_1 = x)\, d\mu(x) = \int_{\mathrm{Sol}(r)} f\, d\Sigma.$$

□

Remark 3.16 In wavelet theory, one often takes $B = \mathbb{R}^n/\mathbb{Z}^n$, and a fixed $n \times n$ matrix A over $\mathbb{Z}$ such that the eigenvalues λ satisfy $|\lambda| > 1$. For $r : B \to B$, then take

$$r(x \bmod \mathbb{Z}^n) = Ax \bmod \mathbb{Z}^n, \quad (x \in \mathbb{R}^n)$$

and it is immediate that r satisfies the multiplicative property in Proposition 3.15.

There are important examples when $r : B \to B$ does not satisfy this property.

Example 3.17 (Non-group case: the Smale-Williams attractor) Take $B = \mathbb{T} \times \mathbb{D}$, where $\mathbb{T} = \mathbb{R}/\mathbb{Z}$ and $\mathbb{D} = \{z \in \mathbb{C} : |z| \leq 1\}$ the disk. For $(t, z) \in \mathbb{T} \times \mathbb{D}$, set

$$r(t, z) = (2t \bmod \mathbb{Z}, \frac{1}{4}z + \frac{1}{2}e^{2\pi i t}).$$

Then $\mathrm{Sol}(r)$ is the Smale-Williams attractor, see [KP07], a hyperbolic strange attractor.

4 Isometries

Below we study condition on functions $m : B \to \mathbb{C}$ which gurantees that $L^2(\mu) \ni f \mapsto m \cdot f \circ r \in L^2(\mu)$ defines an isometry in $L^2(B, \mu)$; and we will study the unitary dilations $L^2(\mathrm{Sol}(r), \Sigma) \ni \tilde{f} \mapsto \tilde{f} \circ \hat{r} \in L^2(\mathrm{Sol}(r), \Sigma)$.

Setting. B fixed compact Hausdorff space. We introduce

(i) $r : B \to B$ measurable, onto such that

$$1 \leq \#r^{-1}(x) < \infty, \quad (x \in B) \tag{4.1}$$

(ii) μ Borel measure on B, $\mu(B) = 1$.
(iii) $m : B \to \mathbb{C}$ a fixed function on B.

Question: Given two of them what are the conditions that the third should satisfy such that

$$L^2(\mu) \ni f \mapsto m \cdot f \circ r \in L^2(\mu) \tag{4.2}$$

is an isometry.

Definition 4.1 Transformations of measures. Given ν measure on B, $\nu \in \mathcal{M}_1(B)$ and $r : B \to B$, set $\nu \circ r^{-1} \in \mathcal{M}_1(B)$. For $A \in \mathcal{B}(B)$ a Borel set, $(\nu \circ r^{-1})(A) := \nu(r^{-1}(A))$ where $r^{-1}(A) := \{x \in B : r(x) \in A\}$.

Fact: $\nu \circ r^{-1}$ is determined uniquely by the condition

$$\int_B \varphi \circ r \, d\nu = \int \varphi \, d(\nu \circ r^{-1}), \quad (\varphi \in C(B)) \tag{4.3}$$

Lemma 4.2 *Fix r, μ, m; then* (4.2) *is satisfied iff*

$$(|m|^2 \, d\mu) \circ r^{-1} = \mu \tag{4.4}$$

Definition 4.3 Fix r, then we say that μ is *strongly invariant* iff

$$\int \varphi(x) \, d\mu(x) = \int \frac{1}{\#r^{-1}(x)} \sum_{r(y)=x} \varphi(y) \, d\mu(x) \tag{4.5}$$

Lemma 4.4 *Given r and assume μ is strongly invariant, then the isometry property* (4.2) *holds iff the corresponding positive operator*

$$(R\varphi)(x) := \frac{1}{\#r^{-1}(x)} \sum_{r(y)=x} |m(y)|^2 \varphi(y)$$

satisfies $R1 = 1$.

Proof Substitute (4.5) into (4.4). Note that then the equation

$$\int_B \varphi(r(x)) |m(x)|^2 \, d\mu(x) = \int_B \varphi(x) \frac{1}{\#r^{-1}(x)} \sum_{r(y)=x} |m(y)|^2 \, \mu(x), \quad (\varphi \in C(B))$$

holds; so $f \mapsto m \cdot f \circ r$ is isometric in $L^2(\mu)$ iff

$$\frac{1}{\# r^{-1}(x)} \sum_{r(y)=x} |m(y)|^2 = 1$$

μ-a.e. $x \in B$. □

The next corollary appears in [DJ07, Theorem 5.5].

Corollary 4.5 *Let B and r be as specified in* Sect. 3, *let μ be strongly invariant and let the function m be quadrature mirror filter, as in* Example 2.5. *Assume in addition that m is non-singular, i.e.*

$$\mu(\{x : m(x) = 0\}) = 0.$$

Then, with R as in Lemma 4.4, *we get a wavelet representation as in* Theorem 2.6 *to $L^2(\mathrm{Sol}(r), \Sigma^{(\mu)})$ with*

(i) $\mathcal{H} = L^2(\mathrm{Sol}(r), \Sigma^{(\mu)})$;
(ii) $\mathcal{U} f = (m \circ \pi_1)(f \circ \widehat{r})$, *for all* $f \in L^2(\mathrm{Sol}(r), \Sigma^{(\mu)})$;
(iii) $\pi(g) f = (g \circ \pi_1) f$, *for all* $g \in L^\infty(B)$, $f \in L^2(\mathrm{Sol}(r), \Sigma^{(\mu)})$;
(iv) $\varphi = 1$.

Proof The details are contained in [DJ07, Theorem 5.5] and require just some simple computations. We only have to check that our measure $\Sigma^{(\mu)}$ coincides with the one defined in [DJ07]. For this, we use [DJ07, Theorem 5.3] and we have to check that

$$\int \varphi \circ \pi_n \, d\Sigma^{(\mu)} = \int R^n(\varphi) \, d\mu, \quad (\varphi \in C(B)).$$

But this follows immediately from the definition of $\Sigma^{(\mu)}$ in (2.17). □

Proposition 4.6 *Let B, r, μ and m_0 as in* Example 2.5, *i.e., m_0 is a QMF and the measure μ on B is assumed strongly invariant with respect to r, and let R as in* Lemma 4.4. *Then the function $\rho = R^* 1$ in Corollary 2.20 is $\rho = |m_0|^2$.*

Proof The result follows if we verify the formula for R^*; we have

$$(R^* \psi)(x) = |m_0(x)|^2 \psi(r(x)). \tag{4.6}$$

The derivation of (4.6) may be obtained as a consequence of strong invariance as follows: for all $\varphi, \psi \in C(B)$, we have:

$$\int_B |m_0|^2 (\psi \circ r) \varphi \, d\mu = \int_B \psi(x) \frac{1}{N} \sum_{r(y)=x} |m_0(y)|^2 \varphi(y) \, d\mu(x) = \int_B \psi(x) (R\varphi)(x) \, d\mu(x);$$

and the assertion (4.6) follows. □

Acknowledgments One of the authors wishes to thank Professors Ka-Sing Lau, De-Jun Feng, and their colleagues, for organizing a wonderful conference in Hong-Kong, "The International Conference on Advances of Fractals and Related Topics", December 2012. Many discussions with participants at the conference inspired this paper. This work was partially supported by a grant from the Simons Foundation (#228539 to Dorin Dutkay).

References

[AJ12] Alpay, D., Jorgensen, P.E.T.: Stochastic processes induced by singular operators. Numer. Funct. Anal. Optim. **33**(7–9), 708–735 (2012)

[ABL11] Antonevich, A.B., Bakhtin, V.I., Lebedev, A.V.: Crossed product of a C^*-algebra by an endomorphism, coefficient algebras, and transfer operators. Mat. Sb. **202**(9), 3–34 (2011)

[ABL12] Antonevich, A.B., Bakhtin, V.I., Lebedev, A.V.: A road to the spectral radius of transfer operators. In: Dynamical Systems and Group Actions, volume 567 of Contemp. Math., pages 17–51. Amer. Math. Soc., Providence, RI, 2012

[Arv86] Arveson, William: Markov operators and OS-positive processes. J. Funct. Anal. **66**(2), 173–234 (1986)

[Bal00] Baladi, V.: Positive Transfer Operators and Decay of Correlations, Advanced Series in Nonlinear Dynamics, vol. 16. World Scientific Publishing Co., Inc., River Edge (2000)

[BCMN04] Beardon, A.F., Carne, T.K., Minda, D., Ng, T.W.: Random iteration of analytic maps. Ergodic Theor. Dyn. Syst. **24**(3), 659–675 (2004)

[BJ02] Bratteli, O., Jorgensen, P.: Wavelets Through a Looking Glass, Applied and Numerical Harmonic Analysis. Birkhäuser Boston Inc., Boston (2002). The world of the spectrum

[CGHU12] Chazottes, J.-R., Gambaudo, J.-M., Hochman, M., Ugalde, E.: On the finite-dimensional marginals of shift-invariant measures. Ergodic Theor. Dyn. Syst. **32**(5), 1485–1500 (2012)

[CH94] Courbage, M., Hamdan, D.: Chapman-Kolmogorov equation for non-Markovian shift-invariant measures. Ann. Prob. **22**(3), 1662–1677 (1994)

[Dau92] Daubechies, I.: Ten lectures on wavelets. In: CBMS-NSF Regional Conference Series in Applied Mathematics. Society for Industrial and Applied Mathematics (SIAM), vol. 61. Philadelphia, PA (1992)

[DMP08] D'Andrea, J., Merrill, K.D., Packer, J.: Fractal wavelets of Dutkay-Jorgensen type for the Sierpinski gasket space. In Frames and operator theory in analysis and signal processing, volume 451 of Contemp. Math., pages 69–88. Amer. Math. Soc., Providence, RI, (2008)

[DHSW11] Dutkay, D.E., Han, D., Sun, Q., Weber, E.: On the Beurling dimension of exponential frames. Adv. Math. **226**(1), 285–297 (2011)

[DJ05] Dutkay D.E., Jorgensen, P.E.T.: Hilbert spaces of martingales supporting certain substitution-dynamical systems. Conform. Geom. Dyn. **9**(electronic), 24–45 (2005)

[DJ06] Dutkay, D.E., Jorgensen, P.E.T.: Wavelets on fractals. Rev. Mat. Iberoam. **22**(1), 131–180 (2006)

[DJ07] Dutkay, D.E., Jorgensen, P.E.T.: Martingales, endomorphisms, and covariant systems of operators in Hilbert space. J. Oper. Theor. **58**(2), 269–310 (2007)

[DJ10] Dutkay, D.E., Jorgensen, P.E.T.: Spectral theory for discrete Laplacians. Complex Anal. Oper. Theory **4**(1), 1–38 (2010)

[DJ11a] Dutkay, D.E., Jorgensen, P.E.T.: Affine fractals as boundaries and their harmonic analysis. Proc. Amer. Math. Soc. **139**(9), 3291–3305 (2011)

[DJ11b] Dutkay, D.E., Jorgensen, P.E.T.: Spectral duality for unbounded operators. J. Oper. Theory **65**(2), 325–353 (2011)

[DJ12] Dutkay, D.E., Jorgensen, P.E.T.: Fourier duality for fractal measures with affine scales. Math. Comp. **81**(280), 2253–2273 (2012)

[DJP09] Dutkay, D.E., Jorgensen, P.E.T, Picioroaga, G.: Unitary representations of wavelet groups and encoding of iterated function systems in solenoids. Ergodic Theor. Dyn. Syst. **29**(6), 1815–1852 (2009)

[DJS12] Dutkay, D.E., Jorgensen, P.E.T., Silvestrov, S.: Decomposition of wavelet representations and Martin boundaries. J. Funct. Anal. **262**(3), 1043–1061 (2012)

[DL10] Dong, X.-H., Lau, K.-S.: Cantor boundary behavior of analytic functions. In Recent developments in fractals and related fields, Appl. Numer. Harmon. Anal., pages 283–294. Birkhäuser Boston Inc., Boston, MA, 2010

[DLS11] Dutkay, D.E., Larson, D.R., Silvestrov, S.: Irreducible wavelet representations and ergodic automorphisms on solenoids. Oper. Matrices **5**(2), 201–219 (2011)

[DS11] Dutkay, D.E., Silvestrov, S.: Reducibility of the wavelet representation associated to the Cantor set. Proc. Amer. Math. Soc. **139**(10), 3657–3664 (2011)

[Dut06] Dutkay, D.E.: Low-pass filters and representations of the Baumslag Solitar group. Trans. Amer. Math. Soc. **358**(12), 5271–5291 (electronic) (2006)

[FM12] Fraczek, M., Mayer, D.: Symmetries of the transfer operator for $\Gamma_0(N)$ and a character deformation of the Selberg zeta function for $\Gamma_0(4)$. Algebra Number Theory **6**(3), 587–610 (2012)

[Hen12] Hensley, D.: Continued fractions, Cantor sets, Hausdorff dimension, and transfer operators and their analytic extension. Discrete Contin. Dyn. Syst. **32**(7), 2417–2436 (2012)

[JP10] Jorgensen, P.E.T., Pearse, E.P.J.: A Hilbert space approach to effective resistance metric. Complex Anal. Oper. Theory **4**(4), 975–1013 (2010)

[JP11] Jorgensen, P.E.T., Pearse, E.P.J.: Ge'fand triples and boundaries of infinite networks. New York J. Math. **17**, 745–781 (2011)

[KP07] Kuznetsov, S.P., Pikovsky, A.: Autonomous coupled oscillators with hyperbolic strange attractors. Phys. D **232**(2), 87–102 (2007)

[Kuz10] Kuznetsov, S.P.: Example of blue sky catastrophe accompanied by a birth of Smale-Williams attractor. Regul. Chaotic Dyn. **15**(2–3), 348–353 (2010)

[Kwa12] Kwaśniewski, B.K.: On transfer operators for C^*-dynamical systems. Rocky Mountain J. Math. **42**(3), 919–936 (2012)

[LN12] Lau, K.-S., Ngai, S.-M.: Martin boundary and exit space on the Sierpinski gasket. Sci. China Math. **55**(3), 475–494 (2012)

[LR69] Lanford III, O.E., Ruelle, D.: Observables at infinity and states with short range correlations in statistical mechanics. Comm. Math. Phys. **13**, 194–215 (1969)

[LW09] Lau, K.-S., Wang, X.-Y.: Self-similar sets as hyperbolic boundaries. Indiana Univ. Math. J. **58**(4), 1777–1795 (2009)

[MP11] Marcolli, M., Paolucci, A.M.: Cuntz-Krieger algebras and wavelets on fractals. Complex Anal. Oper. Theory **5**(1), 41–81 (2011)

[MMS12] Mayer, D., Mühlenbruch, T., Strömberg, F.: The transfer operator for the Hecke triangle groups. Discrete Contin. Dyn. Syst. **32**(7), 2453–2484 (2012)

[Nau12] Naud, F.: The Ruelle spectrum of generic transfer operators. Discrete Contin. Dyn. Syst. **32**(7), 2521–2531 (2012)

[Rue96] Ruelle, D.: Sharp zeta functions for smooth interval maps. In: International Conference on Dynamical Systems (Montevideo, 1995), volume 362 of Pitman Res. Notes Math. Ser., pages 188–206. Longman, Harlow, 1996.

[Rue02] Ruelle, D.: Dynamical zeta functions and transfer operators. Notices Amer. Math. Soc. **49**(8), 887–895 (2002)

[Rue04] Ruelle, D.: Thermodynamic Formalism: The Mathematical Structures of Equilibrium Statistical Mechanics. Cambridge Mathematical Library, 2nd edn. Cambridge University Press, Cambridge (2004)

Generalized Energy Inequalities and Higher Multifractal Moments

Kenneth Falconer

Abstract We present a class of generalized energy inequalities which may be used to investigate higher multifractal moments, in particular L^q-dimensions of images of measures under Brownian-type processes, L^q-dimensions of almost self-affine measures, and moments of random cascade measures.

1 Introduction

Calculations in fractal geometry often fall into two parts: a geometric part and an analytic part. The geometric part may involve expressing geometric or metric aspects of a problem in mathematical terms leading to an analytic argument to estimate the integrals, sums, etc. so obtained. There are various analytic methods that are applicable to a range of problems across fractal geometry, for example, covering or potential theoretic methods for estimating dimensions. Here we look at an analytic technique which extends the potential theoretic method to higher moments and we indicate several applications.

2 L^q-Dimensions and Images of Measures

Coarse multifractal analysis reflects the asymptotic behavior of the moment sums of measures over small grid cubes. Let $\mathcal{M}_r$ be the set of mesh cubes of side r, that is cubes in $\mathbb{R}^n$ of the form $[j_1 r, (j_1+1)r) \times \cdots \times [j_n r, (j_n+1)r)$ where $j_1, \ldots, j_n \in \mathbb{Z}$. Let μ be a Borel measure of bounded support on $\mathbb{R}^n$. Define the *q-th power moment sum* of μ by

K. Falconer (✉)
Mathematical Institute, University of St Andrews, North Haugh, St Andrews, Fife KY 16 9SS, Scotland, UK
e-mail: kjf@st-andrews.ac.uk

D.-J. Feng and K.-S. Lau (eds.), *Geometry and Analysis of Fractals*, Springer Proceedings in Mathematics & Statistics 88,
DOI 10.1007/978-3-662-43920-3_4

$$M_r(q) = \sum_{C \in \mathcal{M}_r} \mu(C)^q. \tag{2.1}$$

The L^q*-dimension* or *generalized* q *dimension* of μ is given by

$$D_q(\mu) = \frac{1}{q-1} \lim_{r \searrow 0} \frac{\log M_r(q)}{\log r} \qquad (q > 0). \tag{2.2}$$

If this limit does not exist we may still take lower or upper limits to get the *lower* and *upper* L^q-dimensions:

$$\underline{D}_q(\mu) = \frac{1}{q-1} \underline{\lim}_{r \searrow 0} \frac{\log M_r(q)}{\log r} \quad \text{and} \quad \overline{D}_q(\mu) = \frac{1}{q-1} \overline{\lim}_{r \searrow 0} \frac{\log M_r(q)}{\log r}, \tag{2.3}$$

see, for example, [Fal97] The definitions (2.2) and (2.3) are unchanged if we replace the moment sum by a moment integral

$$M_r(q) = \int \mu(B(x,r))^{q-1} d\mu(x) \qquad (q > 0), \tag{2.4}$$

see [Lau95] for the case of $q > 1$ and [PS00] for $0 < q < 1$.

Often of interest are the dimensions of the image of a set or the generalized dimensions of the image of a measure under a parameterized family of mappings. Let X be a metric space, and let $x_\omega : X \to \mathbb{R}^n$ be a family of continuous mappings where $\omega \in \Omega$ for some parameter space Ω. Let μ be a Borel measure on X and let μ_ω be its image measure under x_ω, so

$$\mu_\omega(A) = \mu(x_\omega^{-1}(A)) \qquad (A \in \mathbb{R}^n)$$

or

$$\int f(x) d\mu_\omega(x) = \int f(x_\omega(t)) d\mu(t) \qquad (f : \mathbb{R}^n \to \mathbb{R}).$$

For a basic example, x_ω might be orthogonal projection from $\mathbb{R}^m$ onto a line L_ω (which we may identify with $\mathbb{R}$) in direction ω, with μ_ω the corresponding projection of the measure μ on $\mathbb{R}^m$ onto L_ω.

Now suppose $(\Omega, \mathsf{P}, \mathcal{F})$ is a probability space and write E for expectation. One way of obtaining lower estimates for L^q-dimensions of μ_ω valid for almost all ω is to bound the mean moment integrals. When $q \geq 2$ is an integer:

$$\begin{aligned}
&\mathsf{E} \int \mu_\omega(B(x,r))^{q-1} d\mu_\omega(x) \\
&= \mathsf{E} \int \mu_\omega\{y_1 : |x - y_1| \leq r\} \ldots \mu_\omega\{y_{q-1} : |x - y_{q-1}| \leq r\} d\mu_\omega(x)
\end{aligned}$$

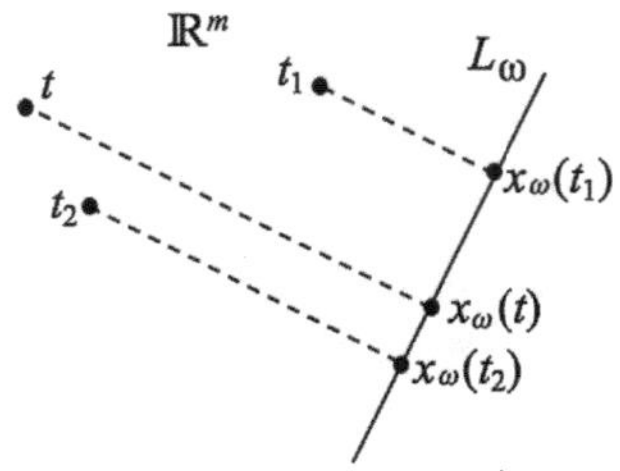

Fig. 1 Projection of three points onto a line parameterized by ω

$$= \mathsf{E}\int \mu\{t_1 : |x_\omega(t)-x_\omega(t_1)| \le r\} \dots \mu\{t_{q-1} : |x_\omega(t)-x_\omega(t_{q-1})| \le r\} d\mu(t)$$

$$= \mathsf{E}\int \cdots \int \chi_{\{|x_\omega(t)-x_\omega(t_j)|\le r \text{ for all } j\}}(t_1, \dots, t_{q-1}, t) d\mu(t_1) \dots d\mu(t_{q-1}) d\mu(t)$$

$$= \int \cdots \int \mathsf{P}\{|x_\omega(t) - x_\omega(t_j)| \le r \text{ for all } j\} d\mu(t_1) \dots d\mu(t_{q-1}) d\mu(t). \quad (2.5)$$

We may be able to use the geometry of the situation to estimate $\mathsf{P}\{|x_\omega(t) - x_\omega(t_j)| \le r \text{ for all } j\}$, which depends on the relative closeness of the $t_1, \dots, t_{q-1}, t$ in the metric space. For example, with $x_\omega : \mathbb{R}^m \to L_\omega$ as projection onto the line L_ω where $\omega \in \Omega$ is distributed according to the natural invariant measure on the space of directions Ω, the probability $\mathsf{P}\{|x_\omega(t) - x_\omega(t_j)| \le r$ for all $j\}$ is affected more, but not exclusively, by the t_j that are furthest from t, see Fig. 1. In particular, bounding (2.5) by const$\cdot r^{s(q-1)}$ may lead to a lower bound of s for the L^q-dimension of μ_ω for almost all ω.

In the case when $q = 2$ the integral (2.5) may be estimated by

$$\int\int \mathsf{P}\{|x_\omega(t) - x_\omega(t_1)| \le r\} d\mu(t_1) d\mu(t) \le \int\int \mathsf{E}\Big(\frac{r^s}{|x_\omega(t) - x_\omega(t_1)|^s}\Big) d\mu(t_1) d\mu(t)$$

for all $s > 0$. This expectation can often be estimated using a transversality argument which results in an energy-type integral. The classic case of this is in the projection theorems, see for example [Fal14, Mat99] for the projection case and [PYS00] for a more general setting.

3 The Main Inequality

In this section we consider an approach to estimating integrals such as (2.5) for $q > 1$ and present an inequality which may be applied in various settings. It is convenient to take X to be the symbolic space on a set of $m \ge 2$ symbols, $\Lambda \equiv \{1, \dots, m\}$. Thus Λ^k consists of the words of length k for $k \ge 0$ and we write $\Lambda^* \equiv \bigcup_{i=0}^k \Lambda^k$ which we identify with the vertices of the m-ary rooted tree in the usual way. The infinite sequences, identified with the boundary of the tree, are denoted by Λ^∞.

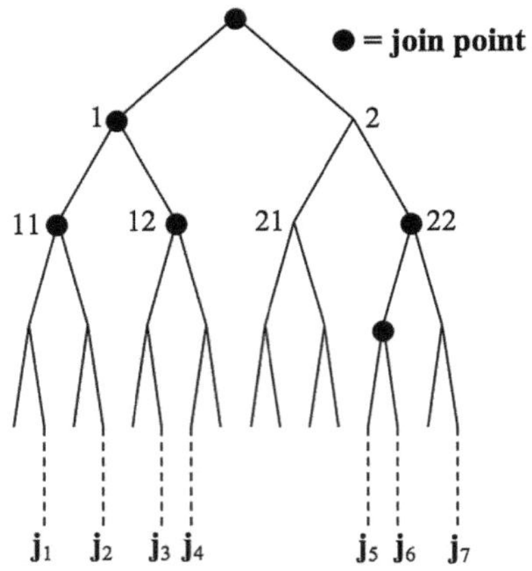

Fig. 2 A set of 7 points in Λ^∞ with their 6 join points in Λ^*

For $\mathbf{i} = i_1, \ldots, i_k \in \Lambda^*$ we write $|\,\mathbf{i}\,| = k$ for the length of the word $\mathbf{i}$. For $\mathbf{i} \in \Lambda^*$ and $\mathbf{j} \in \Lambda^* \cup \Lambda^\infty$ we write $\mathbf{j} \succ \mathbf{i}$ to mean that $\mathbf{i}$ is an initial segment of $\mathbf{j}$. The *cylinders* are the sets $C_\mathbf{i} = \{\mathbf{j} \in \Lambda^\infty : \mathbf{j} \succ \mathbf{i}\}$ for each $\mathbf{i} \in \Lambda^*$. The cylinders provide a basis for the natural topology on Λ^∞.

Write $\mathbf{j}_1 \wedge \mathbf{j}_2 \in \Lambda^*$ for the *join* of $\mathbf{j}_1, \mathbf{j}_2 \in \Lambda^\infty$, that is the longest $\mathbf{i} \in \Lambda^*$ such that $\mathbf{j}_1 \succ \mathbf{i}$ and $\mathbf{j}_2 \succ \mathbf{i}$. For an integer $q \geq 2$ we define the set of *join points* $\mathbf{i}_1, \ldots, \mathbf{i}_{q-1} \in \Lambda^*$ of $\mathbf{j}_1, \mathbf{j}_2, \ldots, \mathbf{j}_q \in \Lambda^\infty$ to be the set

$$J(\mathbf{j}_1, \mathbf{j}_2, \ldots, \mathbf{j}_q) = \{\mathbf{j}_i \wedge \mathbf{j}_j : 1 \leq i < j \leq m\},$$

see Fig. 2. This set will always consist of exactly $q-1$ points provided that they are counted according to multiplicity, that is if there are r distinct points $\mathbf{j}_{i_1}, \ldots, \mathbf{j}_{i_r}$ such that $\mathbf{i} = \mathbf{j}_{i_p} \wedge \mathbf{j}_{i_q}$ for all $1 \leq p < q \leq r$ then $\mathbf{i}$ is counted as a join point with multiplicity $r-1$. (If $m = 2$, corresponding to a binary tree, then all join points have multiplicity 1.)

In bounding expressions such as (2.5), where we now take $X = \Lambda^\infty$ so that the $t_i \in X$ are replaced by $\mathbf{j} \in \Lambda^\infty$, a generalised transversality argument may lead to an estimate of the form

$$\mathsf{P}\{|x_\omega(\mathbf{j}_q) - x_\omega(\mathbf{j}_j)| \leq r \text{ for all } j\} \leq F(\mathbf{j}_1, \mathbf{j}_2, \ldots, \mathbf{j}_q) \tag{3.1}$$

where F may be expressed as a product over the join points

$$F(\mathbf{j}_1, \mathbf{j}_2, \ldots, \mathbf{j}_q) = f(\mathbf{i}_1) f(\mathbf{i}_2) \ldots f(\mathbf{i}_{q-1}) \text{ where } \{\mathbf{i}_1, \ldots, \mathbf{i}_{q-1}\} = J(\mathbf{j}_1, \mathbf{j}_2, \ldots, \mathbf{j}_q),$$

for some $f : \Lambda^* \to \mathbb{R}^+$ defined on the vertices of the tree. Then (2.5) takes the form

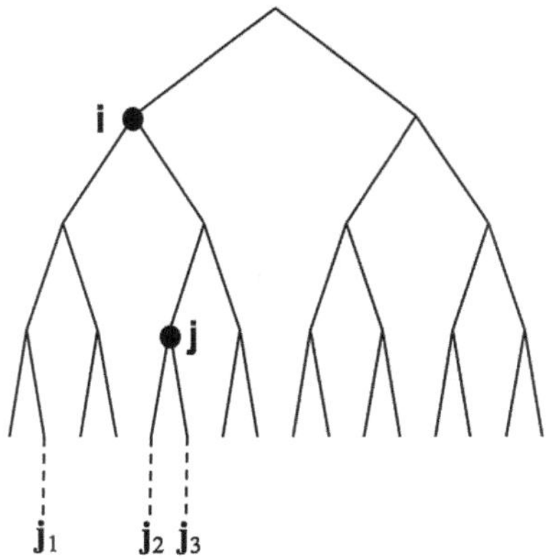

Fig. 3 The arrangement of three points in Λ^∞ the join points used in the proof of Theorem 3.1

$$\mathsf{E}\int \mu_\omega(B(x,r))^{q-1}d\mu_\omega(x) \le \int\cdots\int F(\mathbf{j}_1,\mathbf{j}_2,\ldots,\mathbf{j}_q)d\mu(\mathbf{j}_1)\ldots d\mu(\mathbf{j}_{q-1})d\mu(\mathbf{j}_q). \tag{3.2}$$

The following theorem estimates this integral in terms of f and the cylinder measures $\mu(C_\mathbf{i})$.

Theorem 3.1 *For each real number $q > 1$ there is a polynomial p such that*

$$\int\cdots\int F(\mathbf{j}_1,\mathbf{j}_2,\ldots,\mathbf{j}_q)d\mu(\mathbf{j}_1)\ldots d\mu(\mathbf{j}_q) \le \Big(\sum_{k=0}^{\infty} p(k)\Big[\sum_{|\mathbf{i}|=k} f(\mathbf{i})^{q-1}\mu(C_\mathbf{i})^q\Big]^{\frac{1}{q-1}}\Big)^{q-1}. \tag{3.3}$$

Proof We give the proof in the special case when $q = 3$, that is

$$\int\int\int F(\mathbf{j}_1,\mathbf{j}_2,\mathbf{j}_3)d\mu(\mathbf{j}_1)d\mu(\mathbf{j}_2)d\mu(\mathbf{j}_3) \le \Big(\sum_{k=0}^{\infty}\Big[\sum_{|\mathbf{i}|=k} f(\mathbf{i})^2\mu(C_\mathbf{i})^3\Big]^{1/2}\Big)^2. \tag{3.4}$$

Splitting this integral into a sum over possible pairs of join points, see Fig. 3,

$$\int\int\int F(\mathbf{j}_1,\mathbf{j}_2,\mathbf{j}_3)d\mu(\mathbf{j}_1)d\mu(\mathbf{j}_2)d\mu(\mathbf{j}_3) \le \sum_{\mathbf{i}\in\Lambda^*}\sum_{\mathbf{j}\in\Lambda^*,\,\mathbf{j}\succ\mathbf{i}} f(\mathbf{i})f(\mathbf{j})\mu(C_\mathbf{i})\mu(C_\mathbf{j})^2. \tag{3.5}$$

We first estimate the restriction of this double sum over vertices of the tree for given levels $|\mathbf{i}| = k$ and $|\mathbf{j}| = l$ where $0 \le k < l$; Cauchy's inequality is used at the places indicated.

$$\begin{aligned}
&\sum_{|\mathbf{i}|=k}\sum_{|\mathbf{j}|=l,\,\mathbf{j}\succ\mathbf{i}} f(\mathbf{i})f(\mathbf{j})\mu(C_{\mathbf{i}})\mu(C_{\mathbf{j}})^2\\
&\quad=\sum_{|\mathbf{i}|=k}\Big[f(\mathbf{i})\mu(C_{\mathbf{i}})\Big]\Big[\sum_{|\mathbf{j}|=l,\,\mathbf{j}\succ\mathbf{i}}\Big(f(\mathbf{j})\mu(C_{\mathbf{j}})^{3/2}\Big)\mu(C_{\mathbf{j}})^{1/2}\Big]\\
&\quad\leq\sum_{|\mathbf{i}|=k}\Big[f(\mathbf{i})\mu(C_{\mathbf{i}})\Big]\Big[\Big(\sum_{|\mathbf{j}|=l,\,\mathbf{j}\succ\mathbf{i}} f(\mathbf{j})^2\mu(C_{\mathbf{j}})^3\Big)^{1/2}\Big(\sum_{|\mathbf{j}|=l,\,\mathbf{j}\succ\mathbf{i}}\mu(C_{\mathbf{j}})\Big)^{1/2}\Big] \quad\text{(Cauchy)}\\
&\quad=\sum_{|\mathbf{i}|=k}\Big[f(\mathbf{i})\mu(C_{\mathbf{i}})\Big]\Big[\Big(\sum_{|\mathbf{j}|=l,\,\mathbf{j}\succ\mathbf{i}} f(\mathbf{j})^2\mu(C_{\mathbf{j}})^3\Big)^{1/2}\mu(C_{\mathbf{i}})^{1/2}\Big]\\
&\quad=\sum_{|\mathbf{i}|=k}\Big[f(\mathbf{i})\mu(C_{\mathbf{i}})^{3/2}\Big]\Big[\sum_{|\mathbf{j}|=l,\,\mathbf{j}\succ\mathbf{i}} f(\mathbf{j})^2\mu(C_{\mathbf{j}})^3\Big]^{1/2}\\
&\quad\leq\Big[\sum_{|\mathbf{i}|=k} f(\mathbf{i})^2\mu(C_{\mathbf{i}})^3\Big]^{1/2}\Big[\sum_{|\mathbf{i}|=k}\sum_{|\mathbf{j}|=l,\,\mathbf{j}\succ\mathbf{i}} f(\mathbf{j})^2\mu(C_{\mathbf{j}})^3\Big]^{1/2} \quad\text{(Cauchy)}\\
&\quad=\Big[\sum_{|\mathbf{i}|=k} f(\mathbf{i})^2\mu(C_{\mathbf{i}})^3\Big]^{1/2}\Big[\sum_{|\mathbf{j}|=l} f(\mathbf{j})^2\mu(C_{\mathbf{j}})^3\Big]^{1/2}
\end{aligned}$$

Summing over all levels $0 \leq k, l$ gives inequality (3.4). □

When q is a larger integer, (3.3) may be established using an induction on configurations of join points, requiring frequent uses of Hölder's inequality rather than Cauchy's inequality. A further extension of the calculation establishes that (3.3) remains valid for any real number $q > 1$, see [Fal10, FX14] for further details.

In applications $f(\mathbf{i}) \equiv f_s(\mathbf{i})$ typically depends on a parameter s such that

$$\sum_{|\mathbf{i}|=k} f_s(\mathbf{i})^{q-1}\mu(C_{\mathbf{i}})^q \asymp (\lambda_s)^k$$

for some $\lambda_s > 0$. Combining (3.3) with (3.2) gives

$$\mathsf{E}\int \mu_\omega(B(x,r))^{q-1}d\mu_\omega(x) \leq c\Big(\sum_{k=0}^{\infty} p(k)(\lambda_s)^{k/(q-1)}\Big)^{q-1},$$

so the value of s for which $\lambda_s = 1$ is critical for bounding the mean L^q dimensions of μ_ω.

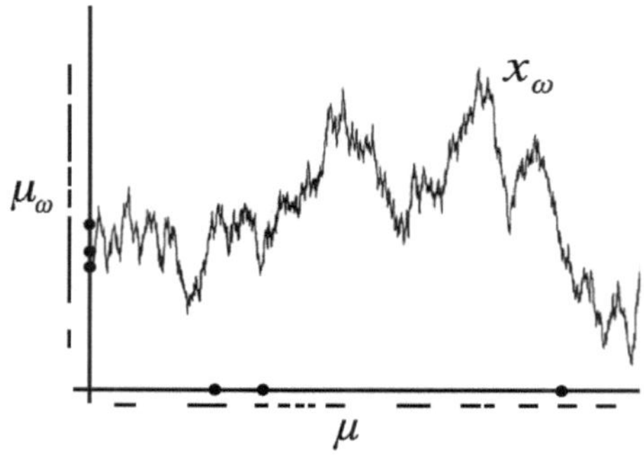

Fig. 4 A measure μ and its image μ_ω under a process x_ω

4 Images of Measures Under Gaussian Processes

For a first application of inequality (3.3), we examine images of measures under certain Gaussian processes. Let $\{x_\omega : [0,1] \to \mathbb{R},\ \ \omega \in \Omega\}$ be index-α fractional Brownian motion ($0 < \alpha \le 2$) defined on a suitable probability space Ω, see [Adl81, Ka85, MV68]. It was shown by Kahane [Ka85] that for a Borel set $E \subseteq \mathbb{R}$

$$\dim_{\rm H} X(E) = \min\left\{1, \frac{\dim_{\rm H} E}{\alpha}\right\} \quad \text{a.s.},$$

where $\dim_{\rm H}$ denotes Hausdorff dimension. It is natural to seek similar relationships between the generalized dimensions of measures and their images under such processes (Fig. 4). Full details of the following result are in [FX14].

Theorem 4.1 *Let $x_\omega : [0,1] \to \mathbb{R}$ be index-α fractional Brownian motion, let μ be a finite measure on $[0,1]$ and let μ_ω be the image of μ under x_ω. Let $q > 1$. Assuming that $D_q(\mu)$ exists then $D_q(\mu_\omega)$ exists almost surely and*

$$D_q(\mu_\omega) = \min\left\{1, \frac{D_q(\mu)}{\alpha}\right\} \quad \textit{a.s.}.$$

Sketch of proof Since index-α fractional Brownian motion almost surely satisfies an $(\alpha - \epsilon)$-Hölder condition for all $\epsilon > 0$, it follows easily from the definition of L^q-dimensions that $D_q(\mu_\omega) \le D_q(\mu)/\alpha$.

For the opposite inequality we use the local nondeterminism (LND) of fBm. Roughly this states that the variance of $x_\omega(t_1)$ conditional on $x_\omega(t_2), \ldots, x_\omega(t_q)$ is comparable with the variance of $x_\omega(t_1) - x_\omega(t_j)$ for the j for which $|t_1 - t_j|$ is least, see [Ber73, Xia06, Xia11]. It may be shown that the calculations are essentially unaffected if, for a suitably large m, we consider the numbers in $[0,1]$ to base m and identify the base m number $0.a_1a_2a_3\ldots$ with $(a_1+1, a_2+1, a_3+1, \ldots) \in \Lambda^\infty$, so that the hierarchy of m-ary subintervals of $[0,1]$ are the cylinders $C_{\mathbf{i}}$ in symbolic space. Using LND inductively we obtain, in symbolic space notation,

$$\mathsf{P}\{|x_\omega(\mathbf{j}_q) - x_\omega(\mathbf{j}_j)| \le r \text{ for all } j\} \le cF(\mathbf{j}_1, \mathbf{j}_2, \ldots, \mathbf{j}_q)$$

where F is a product over the join points $\mathbf{i}_1, \ldots, \mathbf{i}_{q-1} \in J(\mathbf{j}_1, \mathbf{j}_2, \ldots, \mathbf{j}_q)$ of the form

$$F(\mathbf{j}_1, \mathbf{j}_2, \ldots, \mathbf{j}_q) = cr^{s(q-1)} m^{|\mathbf{i}_1|\alpha s} m^{|\mathbf{i}_2|\alpha s} \ldots m^{|\mathbf{i}_{q-1}|\alpha s}$$

for any $s > 0$, where we have replaced Euclidean distance on [0, 1] by the m-ary ultrametric $d(\mathbf{j}_1, \mathbf{j}_2) = m^{-|\mathbf{j}_1 \wedge \mathbf{j}_2|}$. In this notation (2.5) becomes

$$\begin{aligned} &\mathsf{E} \int \mu_\omega(B(x,r))^{q-1} d\mu_\omega(x) \\ &\quad \le cr^{s(q-1)} \int \cdots \int_{\mathbf{i}_1, \ldots, \mathbf{i}_{q-1} \in J(\mathbf{j}_1, \ldots, \mathbf{j}_q)} m^{|\mathbf{i}_1|\alpha s} m^{|\mathbf{i}_2|\alpha s} \ldots m^{|\mathbf{i}_{q-1}|\alpha s} d\mu(\mathbf{j}_1) \ldots d\mu(\mathbf{j}_q). \end{aligned}$$

Inequality (3.3) with $f(\mathbf{i}) = f_s(\mathbf{i}) \equiv m^{|\mathbf{i}|\alpha s}$ now gives

$$\mathsf{E} \int \mu_\omega(B(x,r))^{q-1} d\mu_\omega(x) \le cr^{s(q-1)} \Big(\sum_{k=0}^{\infty} p(k) \Big[\sum_{|\mathbf{i}|=k} \lambda_{s,k} \Big]^{1/(q-1)} \Big)^{q-1} \tag{4.1}$$

where

$$\lambda_{s,k} \equiv \sum_{|\mathbf{i}|=k} f_s(\mathbf{i})^{q-1} \mu(C_\mathbf{i})^q = m^{k\alpha s(q-1)} \sum_{|\mathbf{i}|=k} \mu(C_\mathbf{i})^q. \tag{4.2}$$

The sum in (4.1) is finite if $\limsup_{k\to\infty} (\lambda_{s,k})^{1/k} < 1$, that is if $\alpha s < D_q(\mu)$ using (2.2) and noting that the sum in (4.2) is a sum over the m-ary mesh intervals of lengths m^{-k} that are identified with the cylinders $C_\mathbf{i}$ where $| i | = k$. It follows that if $s_1 < s < D_q(\mu)/\alpha$ then

$$\begin{aligned} &\mathsf{E} \sum_{k=1}^{\infty} 2^{-s_1(q-1)} \int \mu_\omega(B(x, 2^{-k}))^{q-1} d\mu_\omega(x) \\ &\qquad = \mathsf{E} \int \Big(\sum_{k=1}^{\infty} 2^{-s_1(q-1)} \mu_\omega(B(x, 2^{-k}))^{q-1} \Big) d\mu_\omega(x) < \infty, \end{aligned}$$

which implies that $\overline{D}_q(\mu_\omega) > s_1$ almost surely, since the generalized dimensions are determined by the sequence of $r = 2^{-k}$. □

This method yields similar conclusions for the L^q-dimensions of the images of measures under other classes of Gaussian process such as fractional Riesz-Bessel motion and infinity scale fractional Brownian motion, see [FX14].

5 Measures on Almost Self-affine Sets

Next we consider L^q-dimensions of measures on self-affine and almost self-affine sets. For $i = 1, \ldots, m$ let T_i be linear contractions on $\mathbb{R}^n$ and let ω_i be translation vectors. The iterated function system $\{T_j(x) + \omega_j\}_{j=1}^m$ has an non-empty compact attractor E satisfying $E = \cup_{j=1}^m (T_j(E) + \omega_j)$; such a set is termed a *self-affine set*. Writing $\boldsymbol{\omega} = (\omega_1, \ldots, \omega_m)$ for the set of translations, the attractor E may be characterised in terms of m-ary words: $E_{\boldsymbol{\omega}} = \bigcup_{\mathbf{j}\in\Lambda^\infty} x_{\boldsymbol{\omega}}(\mathbf{j})$, where $x_{\boldsymbol{\omega}} : \Lambda^\infty \to \mathbb{R}^n$ is given by the single point in the decreasing intersection

$$x_{\boldsymbol{\omega}}(\mathbf{j}) \equiv x_{\boldsymbol{\omega}}(j_1, j_2, \ldots) = \bigcap_{k=1}^{\infty} (T_{j_1} + \omega_{j_1})(T_{j_2} + \omega_{j_2}) \cdots (T_{j_k} + \omega_{j_k})(B), \quad (5.1)$$

where B is any ball such that $T_j(B) + \omega_i \subseteq B$ for all j.

Let $p_1, \ldots, p_m$ be probabilities, so that $0 < p_j < 1$ and $\sum_{j=1}^m p_j = 1$. Let μ be the Bernoulli probability measure on Λ^∞ defined on cylinders by

$$\mu(C_{\mathbf{j}}) = p_{j_1} p_{j_2} \ldots p_{j_k} \quad \text{where} \quad \mathbf{j} = (j_1, \ldots, j_k) \in \Lambda^*, \quad (5.2)$$

and extended to a Borel measure on Λ^∞. For each $\boldsymbol{\omega} \in \Omega$ let $\mu_{\boldsymbol{\omega}}$ be the image measure of μ under $x_{\boldsymbol{\omega}}$, which is supported by $E_{\boldsymbol{\omega}}$.

We wish to find the generalized dimensions $D_q(\mu_{\boldsymbol{\omega}})$. This is well-known in the case where the $T_j + \omega_j$ are similarities and $E_{\boldsymbol{\omega}}$ is a self-similar set. Provided the open set condition is satisfied (that is, there exists a non-empty open set U such that $\cup_{j=1}^m (T_j(U) + \omega_j) \subset U$ with this union disjoint), then the generalized dimension $D_q(\mu_{\boldsymbol{\omega}}) = d_0$ where d_0 satisfies the equation $\sum_{j=1}^m r_j^{(1-q)d_0} p_j^q = 1$, see [CM92, Fal14]. Closed formulae have also been obtained for the generalized dimensions or self-affine 'carpets' and 'sponges', where the T_j are all equal and the affine transformations $T_{j_i} + \omega_{j_i}$ map a given cube onto similarly-aligned rectangles or rectangular parallelepipeds [Kin95, Ols98].

In general it is difficult to obtain formulae for L^q-dimensions of measures on self-affine sets, or even for the Hausdorff dimension of the supporting self-affine sets, not least because they need not be continuous in $\boldsymbol{\omega}$. Nevertheless, using a potential-theoretic approach, one may obtain formulae that are valid for almost all $\boldsymbol{\omega} = (\omega_1, \ldots, \omega_m)$ in the sense of mn-dimensional Lebesgue measure in the case that $1 < q \leq 2$, see [Fal99]. However, in general there is 'not enough transversality' as $\boldsymbol{\omega}$ varies for the estimates to extend to $q > 2$.

One way of circumventing this difficulty is to introduce more randomness by allowing a random perturbation in the translation component at each stage of the construction. We let

$$\boldsymbol{\omega} = \{\omega_{j_1, j_2, \ldots, j_k} : (j_1, j_2, \ldots, j_k) \in \Lambda^*\} \in (\mathbb{R}^n)^{\Lambda^*} \quad (5.3)$$

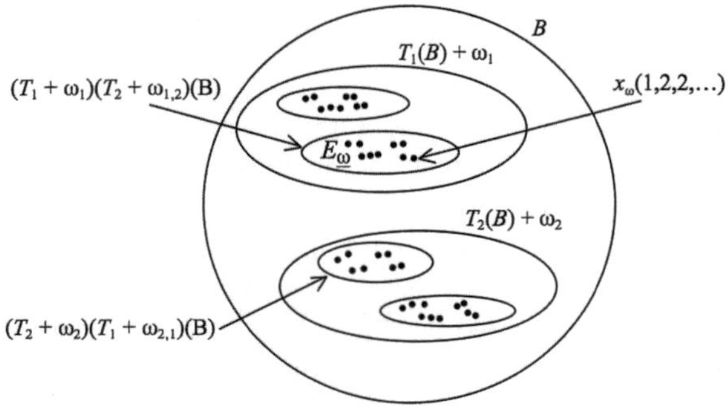

Fig. 5 Hierarchical construction of an almost self-affine set E_ω

be a family of translation vectors in $\mathbb{R}^n$ which we assume to be bounded. Analogously to (5.1) we let

$$\begin{aligned} x_{\omega}(\mathbf{j}) &= \bigcap_{k=1}^{\infty}(T_{j_1}+\omega_{j_1})(T_{j_2}+\omega_{j_1,j_2})(T_{j_3}+\omega_{j_1,j_2,j_3})\cdots(T_{j_k}+\omega_{j_1,j_2,\ldots j_k})(B) \\ &= \lim_{k\to\infty}(T_{j_1}+\omega_{j_1})(T_{j_2}+\omega_{j_1,j_2})(T_{j_3}+\omega_{j_1,j_2,j_3})\cdots(T_{j_k}+\omega_{j_1,j_2,\ldots j_k})(0) \\ &= \omega_{j_1}+T_{j_1}\omega_{j_1,j_2}+T_{j_1}T_{j_2}\omega_{j_1,j_2,j_3}+\cdots \end{aligned} \tag{5.4}$$

for each $\mathbf{j} \equiv (j_1, j_2, \ldots) \in \Lambda^\infty$, for some ball B large enough to ensure that $T_j(B) + \omega_{j_1,j_2,\ldots,j_k} \subseteq B$ for all $j_1, j_2, \ldots, j_k \in \Lambda^*$. We call

$$E_\omega = \bigcup_{\mathbf{j}\in\Lambda^\infty} x_\omega(\mathbf{j})$$

an *almost self-affine set*, see Fig. 5.

We may randomize the translation vectors in the self-affine construction. Assume now that $w_{j_1,j_2,\ldots,j_k}$ in (5.3) are independent identically distributed (i.i.d.) random vectors for $j_1, j_2, \ldots, j_k \in \Lambda^*$ with absolutely continuous density with respect to n-dimensional Lebesgue measure. We put the product probability measure on $(\mathbb{R}^n)^{\Lambda^*}$. We then term E_ω a *random almost self-affine set* [JPS07], see Fig. 6.

To analyse self-affine and almost self-affine sets we utilize the singular values of the mappings which control the geometry of the components in the construction. The *singular values* $\alpha_1 \geq \alpha_2 \geq \cdots \geq \alpha_n \geq 0$ of a linear mapping $T : \mathbb{R}^n \to \mathbb{R}^n$ are the positive square roots of the eigenvalues of TT^* or equivalently are the semi-axis lengths of the ellipsoid $T(B)$ where B is the unit ball. The *singular value function* of T is then defined by

$$\phi^s(T) = \alpha_1 \ldots \alpha_{p-1}\alpha_p^{s-p+1} \tag{5.5}$$

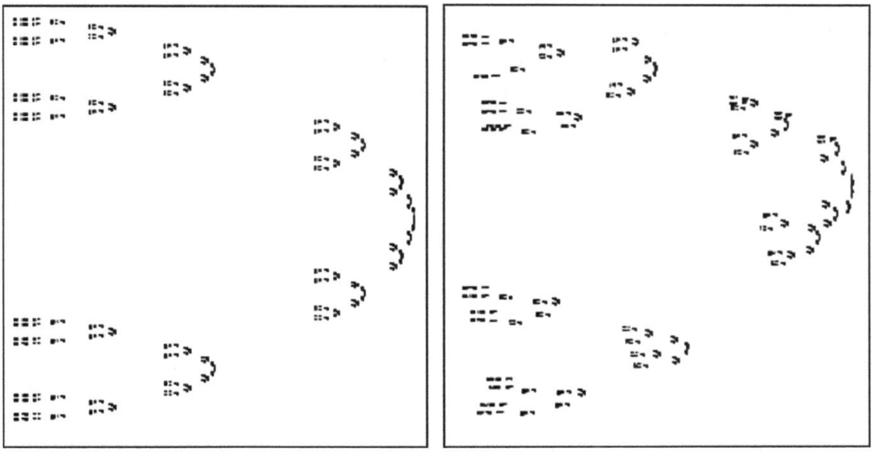

Fig. 6 A self-affine set and a random almost self-affine set with the same linear components in the defining transformations

where p is the integer such that $p - 1 \leq s \leq p$. (If T is a similarity then $\phi^s(T)$ is just the sth power of the scaling ratio of T.)

There are two important properties of ϕ^s. Firstly it is *submultiplicative*, that is

$$\phi^s(T_1 T_2) \leq \phi^s(T_1)\phi^s(T_2), \tag{5.6}$$

and secondly, if T is a contracting linear map, then $\phi^s(T)$ is continuous and strictly decreasing in s, see [Fal88]. It follows, writing

$$\Phi_k^s := \sum_{i_1,\ldots,i_k \in \Lambda^k} \phi^s(T_{i_1} \circ \cdots \circ T_{i_k}),$$

that Φ_k^s itself is also submultiplicative, that is $\Phi_{k+l}^s \leq \Phi_k^s \Phi_l^s$, so, by the standard property of submultiplicative sequences, the limit

$$\Phi^s := \lim_{k\to\infty} (\Phi_k^s)^{1/k}$$

exists and is decreasing in s.

The positive number d_0 that satisfies $\Phi^{d_0} = 1$ is called the *affinity dimension* $d_0 \equiv d_0(T_1, \ldots, T_m)$ of the self-affine set E_ω that is the attractor of the IFS of affine maps $\{T_i + \omega_i\}_{i=1}^m$. In other words d_0 is given by

$$\Phi^{d_0}(T_1, \ldots, T_m) \equiv \Phi^{d_0} = \lim_{k\to\infty} \Big(\sum_{i_1 \ldots i_k \in \Lambda^k} \phi^{d_0}(T_{i_1} \circ \cdots \circ T_{i_k}) \Big)^{1/k} = 1; \tag{5.7}$$

notice that the affinity dimension depends only on the linear parts of the IFS functions. Affinity dimensions provide 'generic' values for the Hausdorff and box-counting dimensions of self-affine sets. We write $\dim_{\mathrm{H}}$ and $\overline{\dim}_{\mathrm{B}}$ for Hausdorff and upper box-counting dimensions respectively.

Proposition 5.1 *Let E_ω be a self-affine or almost self-affine subset of $\mathbb{R}^n$. Then*

$$\dim_{\mathrm{H}} E_\omega \leq \overline{\dim}_{\mathrm{B}} E_\omega \leq d_0(T_1, \ldots, T_m) \tag{5.8}$$

where $(T_1, \ldots, T_m)$ are the linear parts of the affine contractions in the construction of E_ω. If E_ω is self-affine with $\|T_j\| < \frac{1}{2}$ for all j then there is equality in (5.8) *for almost all translation vectors $\omega \in (\mathbb{R}^n)^m$. If E_ω is a random almost self-affine set then there is equality in* (5.8) *for almost all $\omega \in (\mathbb{R}^n)^{\Lambda^*}$ with no restriction on $\|T_j\|$.*

Proof Inequality (5.8) is obtained by a covering method. Almost sure equality for self-affine sets and random almost self-affine sets may be derived from energy estimates for measures supported on the sets, see [Fal88] and [JPS07] for the two settings. □

To obtain generic formulae for L^q-dimensions, we adapt the definition of affinity dimension. With μ the Bernoulli measure on Λ^∞ defined by (5.2), let

$$\begin{aligned}\Phi_q^s &= \lim_{k\to\infty}\Big(\sum_{i_1,\ldots,i_k\in\Lambda^k} \phi^s(T_{i_1}\circ T_{i_2}\circ\cdots\circ T_{i_k})^{1-q}\mu(C_{i_1,i_2,\ldots,i_k})^q\Big)^{1/k} \quad (5.9)\\ &= \lim_{k\to\infty}\Big(\sum_{i_1,\ldots,i_k\in\Lambda^k} \phi^s(T_{i_1}\circ T_{i_2}\circ\cdots\circ T_{i_k})^{1-q}(p_{i_1}p_{i_2}\cdots p_{i_k})^q\Big)^{1/k}.\end{aligned}$$

Again the limits exist as a consequence of supermultiplicativity, and if $q > 1$ then Φ_q^s is strictly increasing and continuous in s. Thus we may define positive numbers d_q by the requirement that

$$\Phi_q^{d_q} = 1. \tag{5.10}$$

As before we write μ_ω for the image of the Bernoulli measure μ under x_ω. We refer to μ_ω as a *self-affine measure* when $\omega \in (\mathbb{R}^n)^m$ and the support E_ω is a self-affine set, and as a *(random) almost self-affine measure* when $\omega \in (\mathbb{R}^n)^{\Lambda^*}$ and the support is a (random) almost self-affine set.

Proposition 5.2 *Let $1 < q \leq 2$. Let μ be a Bernoulli measure on Λ^∞. For every self-affine or almost self-affine measure μ_ω on $\mathbb{R}^n$*

$$\overline{D}_q(\mu_\omega) \leq \min\{d_q, n\} \tag{5.11}$$

where d_q is given by (5.10)*. Moreover, $D_q(\mu_\omega)$ exists and*

$$D_q(\mu_\omega) = \min\{d_q, n\}$$

in the self-affine case provided $\|T_j\| < \frac{1}{2}$ *for all* j, *for almost all* $\boldsymbol{\omega} \in (\mathbb{R}^n)^m$, *and also in the random almost self-affine case for almost all* $\boldsymbol{\omega} \in (\mathbb{R}^n)^{\Lambda^*}$ *(with no restriction on the* $\|T_j\|$*).*

Note on proof This is proved in [Fal99] using a potential-theoretic method; the proof adapts easily to give equality in the random almost self-affine case. □

It is natural to ask whether the conclusion of Proposition 5.2 is valid for $q > 2$ when the basic potential-theoretic method is inadequate. However, for self-affine measures μ_ω there is, in general, not enough randomness or transversality to get an adequate estimate in (3.1) to lead to equality for almost all $\boldsymbol{\omega} \in (\mathbb{R}^n)^m$. Thus we consider the lower bound for random almost self-affine measures using the inequality of Sect. 3.

Theorem 5.1 *Let* $q > 1$. *Let* μ *be a Bernoulli measure on* Λ^∞. *For every self-affine or almost self-affine measure* μ_ω *on* $\mathbb{R}^n$

$$\overline{D}_q(\mu_\omega) \le \min\{d_q, n\} \tag{5.12}$$

where d_q *is given by* (5.10). *If* μ_ω *is a random almost self-affine measure then* $D_q(\mu_\omega)$ *exists and*

$$D_q(\mu_\omega) = \min\{d_q, n\}$$

for almost all $\boldsymbol{\omega} \in (\mathbb{R}^n)^{\Lambda^*}$.

Sketch of proof The upper bound (5.12) comes from splitting ellipses of the form that occur in the intersections in (5.4) into appropriate pieces and summing the powers of the measures, see [Fal99, Fal10].

For the case where $q \ge 2$ is an integer and μ_ω a random almost self-affine measure, let $\mathbf{j}_1, \dots, \mathbf{j}_q \in \Lambda^\infty$. Using the geometry and randomness or higher transversality available in the construction, we may obtain an estimate

$$\mathsf{P}\{|x_\omega(\mathbf{j}_q) - x_\omega(\mathbf{j}_j)| \le r \text{ for all } j\} \le c r^{s(q-1)} \phi^s(T_{\mathbf{i}_1})^{-1} \phi^s(T_{\mathbf{i}_2})^{-1} \dots \phi^s(T_{\mathbf{i}_{q-1}})^{-1} \tag{5.13}$$

where $\mathbf{i}_1, \dots, \mathbf{i}_{q-1}$ are the join points of $\mathbf{j}_1, \dots, \mathbf{j}_q$. Using (2.5) we get, for all $s > 0$,

$$\begin{aligned} \mathsf{E} \int \mu_\omega(B(x,r))^{q-1} d\mu_\omega(x) &\le c r^{s(q-1)} \int \dots \int \phi^s(T_{\mathbf{i}_1})^{-1} \phi^s(T_{\mathbf{i}_2})^{-1} \dots \phi^s(T_{\mathbf{i}_{q-1}})^{-1} d\mu(\mathbf{j}_1) \dots d\mu(\mathbf{j}_q) \\ &\le c r^{s(q-1)} \Big(\sum_{k=0}^{\infty} p(k) \Big[\sum_{|\mathbf{i}|=k} \phi^s(T_{\mathbf{i}})^{1-q} \mu(C_{\mathbf{i}})^q \Big]^{\frac{1}{q-1}} \Big)^{q-1}, \end{aligned}$$

for some polynomial p, taking $f(\mathbf{i}) = \phi^s(T_{\mathbf{i}})^{-1}$ in inequality (3.3). From the definition (5.9),(5.10) of Φ_q^s, this series converges if $0 < s < d_q$, in which case

$$\mathsf{E}\sum_{k=1}^{\infty} 2^{-s_1(q-1)} \int \mu_\omega(B(x,2^{-k}))^{q-1} d\mu_\omega(x)$$
$$= \mathsf{E} \int \Big(\sum_{k=1}^{\infty} 2^{-s_1(q-1)} \mu_\omega(B(x,2^{-k}))^{q-1} \Big) d\mu_\omega(x) < \infty,$$

for all $0 < s_1 < s$, giving $D_q(\mu_\omega) > s_1$ for almost all $\omega \in (\mathbb{R}^n)^{\Lambda^*}$ for all $s_1 < d_q$, as required.

For full details of this argument and the case of non-integer $q > 1$ see [Fal00]. □

6 Random Multiplicative Cascade Measures

Let $\Lambda = \{1, 2, \dots, m\}$, let $W_{\mathbf{i}}$ be independent positive random variables indexed by $\mathbf{i} = i_1, i_2, \dots, i_k \in \Lambda^*$ and let

$$X_{\mathbf{i}} = W_{i_1} W_{i_1,i_2} \dots W_{i_1,i_2,\dots,i_k}.$$

We may identify the cylinders in symbolic space with the hierarchy of m-ary subintervals of [0, 1] in the obvious way, see Fig. 7. We assume that $\mathsf{E}(W_{\mathbf{i}}) = 1$ for all $\mathbf{i} \in \Lambda^*$ in which case $(X_{\mathbf{j}|k}, \mathcal{F}_k)$ is a martingale for each $\mathbf{j} \in \Lambda^\infty$, where $\mathbf{j}|k$ denotes the curtailment of $\mathbf{j}$ after k terms and $\mathcal{F}_k$ is the σ-field generated by $\{W_{\mathbf{i}} : \mathbf{i} \in \cup_{l=1}^{k} \Lambda^l\}$.

These martingales, termed *random multiplicative cascade measures*, were introduced and studied in the 1970 s by Mandelbrot [Man74] and Kahane and Peyrière [Kah85, KP76] who obtained many properties in the 'self-similar' case, that is when the $W_{\mathbf{i}}$ are independent and identically distributed. Let μ be a Borel probability measure on Λ^∞. Of particular interest are the k-th level sums

$$\sum_{|\mathbf{i}|=k} X_{\mathbf{i}}\, \mu(C_{\mathbf{i}}) \equiv \int X_{\mathbf{j}|k} d\mu(\mathbf{j}),$$

which moments of the sums remain bounded as $k \to \infty$ and in what setting the integral converges. It follows from Minkowski's inequality that if $\mathsf{E}\big(\big(\sum_{|\mathbf{i}|=k} X_{\mathbf{i}}\, \mu(C_{\mathbf{i}})\big)^q\big)$ is bounded in k then so is $\sum_{|\mathbf{i}|=k} \mathsf{E}\big((X_{\mathbf{i}}\, \mu(C_{\mathbf{i}}))^q\big)$, but the opposite implications are more subtle. Using the inequality from Sect. 3 we get the following result.

Theorem 6.1 *Let $q > 1$ be a real number. If*

$$\limsup_{k\to\infty} \Big(\sum_{|\mathbf{i}|=k} \mathsf{E}(X_{\mathbf{i}}^q) \mu(C_{\mathbf{i}})^q \Big)^{1/k} < 1 \tag{6.1}$$

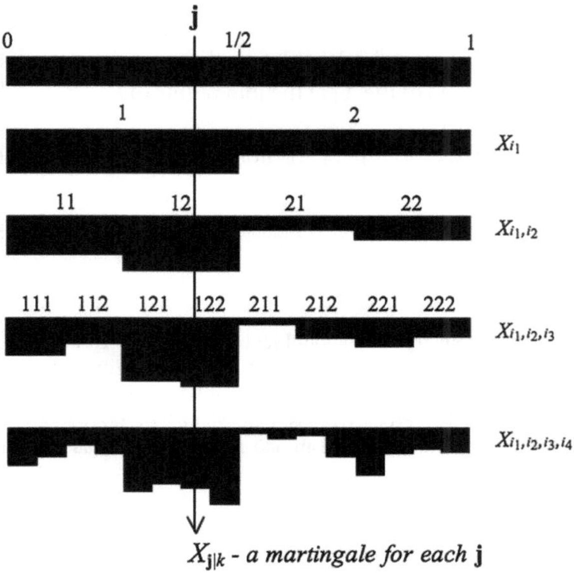

Fig. 7 A random multiplicative binary cascade measure represented on the interval [0, 1], with $X_{\mathbf{j}|k}$ a martingale for each $\mathbf{j} \in [0, 1]$

then

$$\limsup_{k\to\infty} \mathsf{E}\Big(\Big(\sum_{|\mathbf{i}|=k} X_\mathbf{i}\,\mu(C_\mathbf{i})\Big)^q\Big) < \infty \tag{6.2}$$

and $\int X_{\mathbf{j}|k} d\mu(\mathbf{j})$ *converges a.s. and in* L^q. *Note that we require the underlying* $W_\mathbf{i}$ *to be independent but not necessarily identically distributed.*

This result, along with many other properties of these martingales, was obtained by Kahane and Peyrière [Kah85, KP76] when the random cascade is 'self-similar', that is when the $W_\mathbf{i}$ are identically distributed, utilizing the self-similarity to show that the sums satisfy a random difference equation. There have been many subsequent extensions and variants, see [BM04, Liu00] which contain many further references. Barral [BM04] proved this result without the i.i.d. requirement on the $W_\mathbf{i}$ in the case $1 < q \le 2$, also with the martingales defined in a more general continuous setting.
Note on the proof of Theorem 6.1. When $q > 1$ is an integer we may expand

$$\mathsf{E}\Big(\Big(\sum_{|\mathbf{i}|=k} X_\mathbf{i}\,\mu(C_\mathbf{i})\Big)^q\Big) = \sum_{|\mathbf{i}_1|,|\mathbf{i}_2|,\ldots,|\mathbf{i}_q|=k} \mathsf{E}(X_{\mathbf{i}_1}X_{\mathbf{i}_2}\ldots X_{\mathbf{i}_q})\mu(C_{\mathbf{i}_1})\mu(C_{\mathbf{i}_2})\ldots\mu(C_{\mathbf{i}_q})$$
$$\le \Big(\sum_{k=0}^{\infty} p(k)\Big(\sum_{|\mathbf{i}|=k} \mathsf{E}(X_\mathbf{i}^q)\mu(C_\mathbf{i})^q\Big)^{\frac{1}{q-1}}\Big)^{q-1}.$$

for a polynomial p, where this inequality may be established using induction in a manner akin to that of Theorem 3.1 by relating the expectations of products of the $X_{\mathbf{i}_j}$ to expectations of powers of the $X_{\mathbf{i}}$ at the join points of $\mathbf{i}_1, \ldots, \mathbf{i}_q$. The conclusion (6.2) then follows from (6.1).

As with Theorem 3.1 the argument for non-integer q requires a more involved induction argument. □

References

[Adl81] Adler, R.J.: The Geometry of Random Fields. Wiley, New York (1981)

[BM04] Barral, J., Mandelbrot, B.B.: Random multiplicative multifractal measures I, II, III, in Fractal Geometry and Applications: A Jubilee of Benoît Mandelbrot (M. Lapidus and M. V. Frankenkuijsen, eds.), Proc. Sympos. Pure Math. 72 (2004) 3–90

[Ber73] Berman, S.M.: Local nondeterminism and local times of Gaussian processes. Indiana Univ. Math. J. **23**, 69–94 (1973)

[CM92] Cawley, R., Mauldin, R.D.: Multifractal decomposition of Moran fractals. Adv. Math. **92**, 196–236 (1992)

[Fal88] Falconer, K.J.: The Hausdorff dimension of self-affine fractals. Math. Proc. Cambridge Philos. Soc. **103**, 339–350 (1988)

[Fal97] Falconer, K.J.: Techniques in Fractal Geometry. Wiley, New York (1997)

[Fal99] Falconer, K.J.: Generalized dimensions of measures on self-affine sets. Nonlinearity **12**, 877–891 (1999)

[Fal00] Falconer, K.J.: Higher moments of random multiplicative cascade measures: to appear. J. Fractal geometry. arXiv:1311.0160

[Fal10] Falconer, K.J.: Generalised dimensions of measures on almost self-affine sets. Nonlinearity **23**, 1047–1069 (2010)

[Fal14] Falconer, K.J.: Fractal Geometry-Mathematical Foundations and Applications, 3rd Edn. Wiley, New York (2014)

[FX14] Falconer, K.J., Xiao, Y.: Generalized Dimensions of Images of Measures Under Gaussian Processes. Adv. Math. **252**, 492–517 (2014)

[JPS07] Jordan, T., Pollicott, M., Simon, K.: Hausdorff dimension for randomly perturbed self affine attractors. Commun. Math. Phys. **270**, 519–544 (2007)

[Kah85] Kahane, J.-P.: Sur le chaos multiplicatif. Ann. Sci. Math. Québec **9**, 105–150 (1985)

[Ka85] Kahane, J.-P.: Some Random Series of Functions, 2nd Edn. Cambridge University Press, Cambridge (1985)

[KP76] Kahane, J.-P., Peyrière, J.: Sur certaines martingales de Benoit Mandelbrot. Adv. Math. **22**, 131–145 (1976)

[Kin95] King, J.: The singularity spectrum for general Sierpinski carpets. Adv. Math. **116**, 1–8 (1995)

[Lau95] Lau, K.-S.: Self-similarity, L^p-spectrum and multifractal formalism. In: Bandt, C., Graf, S., Zähle, M. (eds.) Fractal Geometry and Stochastics, Progress in Probability 37, pp. 55–90. Birkhäuser, Basel (1995)

[Liu00] Liu, Quansheng: On generalized multiplicative cascades. Stoch. Process. Appl. **86**, 263–286 (2000)

[Man74] Mandelbrot, B.: Intermittent turbulence in self-similar cascades: divergence of higher moments and dimension of the carrier. J. Fluid. Mech. **62**, 331–358 (1974)

[MV68] Mandelbrot, B.B., Van Ness, J.W.: Fractional Brownian motions, fractional noises and applications. SIAM Rev. **10**, 422–437 (1968)

[Mat99] Mattila, P.: Geometry of Sets and Measures in Euclidean Spaces. Cambridge University Press, Cambridge (1999)

[Ols98] Olsen, L.: Self-affine multifractal Sierpinski sponges in $\mathbb{R}^d$. Pacific J. Math. **183**, 143–199 (1998)

[PYS00] Peres, Y., Schlag, B.: Smoothness of projections, Bernoulli convolutions, and the dimension of exceptions. Duke Math. J. **102**, 193–251 (2000)

[PS00] Peres, Y., Solomyak, B.: Existence of L^q dimensions and entropy dimension for self-conformal measures. Indiana Univ. Math. J. **49**, 1603–1621 (2000)

[Xia06] Xiao, Y.: Properties of local nondeterminism of Gaussian and stable random fields and their applications. Ann. Fac. Sci. Toulouse Math. **15**, 157–193 (2006)

[Xia11] Xiao, Y.: Properties of strong local nondeterminism and local times of stable random fields. In: Dalang, R.C., Dozzi, M., Russo, F. (eds.) Seminar on Stochastic Analysis, Random Fields and Applications VI, Progress in Probability, vol. 63, pp. 279–310. Birkhäuser, Basel (2011)

Some Aspects of Multifractal Analysis

Ai-Hua Fan

Abstract The aim of this survey is to present some aspects of multifractal analysis around the recently developed subject of multiple ergodic averages. Related topics include dimensions of measures, oriented walks, Riesz products etc. The exposition on the multifractal analysis of multiple ergodic averages is mainly based on [FLM12, KPS12, FSW00].

Keywords Dynamical systems · Ergodic averages · Multifractal analysis · Hausdorff dimension

2010 Mathematics Subject Classification Primary 37C45

1 Introduction

Multifractal problems can be put into the following frame. Let (X, d) be a metric space and $\mathcal{P}(x)$ a property (quantitative or qualitative) depending on a point x of the space X. For any prescribed property **P**, we look at the set of those points x which have the property **P**:

$$E(\mathbf{P}) = \{x \in X : \mathcal{P}(x) = \mathbf{P}\}.$$

The size of the sets $E(\mathbf{P})$ for different **P**'s is problematic in the multifractal analysis. According to popular folklore, the function $x \mapsto \mathcal{P}(x)$ is multifractal if $E(\mathbf{P})$ is not empty for uncountably many properties **P**. Usually the size of a set A in X is described by its Hausdorff dimension $\dim_H A$ or its packing dimension $\dim_P A$, or its topological entropy in a dynamical setting. See [Fal03, Mat95] for the dimension theory and [Bow73, Pes97] for the notion of topological entropy.

Seeds of multifractals were sown in Mandelbrot's works on multiplicative chaos in 1970's [Man74, Man74]. First rigorous results are due to Kahane and Peyrère [KP76]. The concept of multifractality came from geophysics and theoretical physics.

A.-H. Fan (✉)
LAMFA, UMR, 7352 CNRS, Université de Picardie, 33 rue Saint Leu, 80039 Amiens, France
e-mail: ai-hua.fan@u-picardie.fr

D.-J. Feng and K.-S. Lau (eds.), *Geometry and Analysis of Fractals*,
Springer Proceedings in Mathematics & Statistics 88,
DOI 10.1007/978-3-662-43920-3_5

At the beginning, Frisch and Parisi [FP85], Hentschel and Procaccia [HP83] had the rather vague idea of mixture of subsets of different dimensions each of which has a given Hölder singularity exponent. The multifractal formalism became clearer in the 1980–1990's in the works of Halsey et al. [HJKPS86], of Collet et al. [CLP87], and of Grown et al. [BMP92]. The multifractal formalism is tightly related to the thermodynamics and Ruelle [Rue78] was the first to use the thermodynamical formalism to compute the Hausdorff dimensions of some Julia sets.

The research on the subject has been very active and very fruitful since four decades. The first most studied multifractal quantity is the local dimension of a Borel measure μ on X. Recall that the local lower dimension of μ at $x \in X$ is defined by

$$\underline{D}_\mu(x) = \liminf_{r \to 0} \frac{\log \mu(B(x,r))}{\log r}$$

where $B(x,r)$ denotes the ball centered at x with radius r. The local upper dimension $\overline{D}_\mu(x)$ is similarly defined. There is a huge literature on this subject. Let us mention another example, Hölder exponent of a function (see [Jaf97]). Let $\alpha > 0$. Let $f : \mathbb{R}^d \to \mathbb{R}$ be a function and $x \in \mathbb{R}^d$ be a fixed point. We say f is α-Hölder at x and we write $f \in C^\alpha(x)$ if there exist two constants $\delta > 0$ and $C > 0$ and a polynomial $P(x)$ of degree strictly smaller than α such that

$$|y - x| < \delta \Rightarrow |f(y) - f(x) - P(y - x)| \le C|y - x|^\alpha.$$

The Hölder exponent of f at x is then defined by

$$h_f(x) = \sup\{\alpha > 0 : f \in C^\alpha(x)\}.$$

The multifractal analysis has now become a set of tools applicable in analysis, probability and stochastic processus, number theory, ergodic theory and dynamical systems etc if we don't account applications in physics and other sciences.

The main goal of this paper is to present some problems in the multifractal analysis of Birkhoff ergodic averages, especially of multiple Birkhoff ergodic averages, and some related topics like dimensions of measures and Riesz products as tools, and oriented walks as similar subject.

Let $T : X \to X$ be a map from X into X. We consider the dynamical system (X, T). The main concern about the system is the behavior of an orbit $\{T^n x\}$ of a given point $x \in X$. Some aspects of the behavior of the orbit may be described by the so-called Birkhoff averages

$$A_n f(x) = \frac{1}{n} \sum_{k=0}^{n-1} f(T^k x) \tag{1}$$

where $f : \mathbb{X}^d \to \mathbb{R}$ is a given function, called observable. We refer to [Wal82] for basic facts in the theory of dynamical system and ergodic theory.

The famous and fundamental Birkhoff Ergodic Theorem states that for any T-invariant ergodic probability measure μ on X and for any integrable function $f \in L^1(\mu)$, the limit

$$\lim_{n\to+\infty} A_n f(x) = \int_X f \, d\mu$$

holds for μ-almost all points $x \in X$. Even if μ is only T-invariant but not ergodic, the limit still exists for μ-almost all points $x \in X$. For many dynamical systems, there is a rich class of invariant measures and so that the limit of Birkhoff averages $A_n f(x)$ may vary for different points x. This variety reflects the chaotic feature of the dynamical system. A typical example is the doubling dynamics $x \mapsto 2x(\bmod 1)$ on the unit circle $\mathbb{T} := \mathbb{R}/\mathbb{Z}$.

The multifractal analysis of Birkhoff ergodic averages provides a way to study the chaotic feature of the dynamics. Let $\alpha \in \mathbb{R}$. We define the α-level set

$$E_f(\alpha) = \left\{ x \in X : \lim_{n\to+\infty} A_n f(x) = \alpha \right\}.$$

The purpose of the multifractal analysis is the determination of the size of sets $E_f(\alpha)$. If the Hausdorff dimension is used as measuring device, we are led to the Haudorff multifractal spectrum of f:

$$\mathbb{R} \ni \alpha \mapsto d_H(\alpha) := \dim_H E_f(\alpha).$$

The existing works show that in many cases it is possible to compute the spectrum $d_H(\cdot)$ and it is also possible to distinguish a nice invariant measure sitting on the set $F_f(\alpha)$ for each α. By "nice" we mean that the measure is supported by $E_f(\alpha)$ and its dimension is equal to that of $E_f(\alpha)$. The dimension of a measure is defined to be the dimension of the "smallest" Borel support of the measure. Therefore the nice measure is a maximal measure in the sense that it attains the maximum among all measures supported by $E_f(\alpha)$. This maximal measure may be invariant, ergodic and even mixing. Some other nice properties are also shared by the maximal measure. For this well studied classic ergodic averages, see for example [BSS02, FF99, FFW01, FLP08, FS03, FLW02, Oli99, TV03]. Let us also mention some useful tools for dimension estimation [BS07, BV06, Dur08, FST00].

The multiple ergodic theory started almost at the same time of the development of multifractal analysis. It started with Fürstenberg's proof of Szemerédi theorem on the existence of arbitrary long arithmetic sequence in a set of integers of positive density [Für77]. This theory involves several dynamics rather than one dynamics. Let T_1, T_2,..., T_d be d transformations on a space X. We assume that they are commuting each other and preserving a given probability measure μ. For d measurable functions $F = (f_1, \ldots, f_d)$ we define the multiple Birkhoff averages by

$$A_nF(x) := \frac{1}{n}\sum_{k=0}^{n-1} f_1(T_1^k x) f_2(T_2^k x) \cdots f_d(T_d^k x). \tag{2}$$

An inspiring example is the couple (τ_2, τ_3) on the circle $\mathbb{T}$ where

$$\tau_2 x = 2x \ (\mathrm{mod}\ 1), \quad \tau_3 x = 3x \ (\mathrm{mod}\ 1).$$

The mixture of the dynamics $T_1, T_2, \ldots, T_d$ is much more difficult to understand. After the first works of Fürstenberg-Weiss [FW96] and of Conze-Lesigne [LT84], Host and Kra [HK05] proved the L^2-convergence of A_nF when $T_j = T^j$ (powers of a fixed dynamics) and $f_j \in L^\infty(\mu)$. For the almost every convergence, results are sparse. Bourgain proved the almost everywhere convergence when $d = 2$. We should point out that even if the limit exists, it is not easy to recognize the limit. In particular, the limit may not be constant for some ergodic measures. For nilsystems, explicit formula for the limit was known to Lesigne [Les89] and Ziegler [Zie05]. Anyway, not like the "simple" ergodic theory, the multiple ergodic theory has not yet reached its maturity.

Although the multiple ergodic theory has been still developing, this situation doesn't prevent us from investigating the multifractal feature of multiple systems. Let us consider the following general set-up. For a given observable $\Phi : X^d \to \mathbb{R}$, we consider the Multiple Ergodic Averages

$$A_n\Phi(x) := \frac{1}{n}\sum_{k=0}^{n-1} \Phi(T_1^k x, T_2^k x, \ldots, T_d^k x). \tag{3}$$

The Birkhoff averages (2) correspond to the special case of tensor product $\Phi = f_1 \otimes f_2 \otimes \cdots \otimes f_d$.

It is natural to introduce the following generalization of multifractal spectrum. In this paper, we will only consider the case where $T_j = T^j$ for $1 \le j \le d$. So

$$A_n\Phi(x) = \frac{1}{n}\sum_{k=0}^{n-1} \Phi(T^k x, T^{2k} x, \ldots, T^{dk} x). \tag{4}$$

The multiple Hausdorff spectrum of the observable Φ is defined by

$$d_H(\alpha) = \dim_H E_\Phi(\alpha), \quad (\alpha \in \mathbb{R})$$

where

$$E_\Phi(\alpha) = \left\{ x \in X : \lim_{n\to+\infty} \frac{1}{n}\sum_{k=0}^{n-1} \Phi(T^k x, T^{2k} x, \ldots, T^{dk} x) = \alpha \right\}. \tag{5}$$

As we said, the classical theory ($d = 1$) is well developed. When $d > 1$, there are several results on the multifractal analysis of the limit of the averages $A_n\Phi$ in some special cases; but questions remain largely unanswered. In Sect. 2, we will present the first results obtained in [FLM12] in a very special case which give us a feeling of the problem and illustrate the difficulty of the problem. One result is the Hausdorff spectrum obtained by using Riesz products, a tool borrowed from Fourier analysis. Another result concerns the box dimension of a multiplicatively invariant set. These two kinds of result are respectively generalized in [FSW00] and [KPS12, PSSS00] in some general setting. In Sect. 5, we will present these results. As we mentioned, the local dimension of a measure was the first study object of multifractal analysis. In Sect. 3, we will give an account of dimensions of measures which are related to the local dimension and have their own interests. The idea of using Riesz products is inspired by a work on oriented walks [Fan00] to which Sect. 4 will be devoted. In the last section, we will collect some remarks and open problems.

2 First Multifractal Results on the Multiple Ergodic Averages

The question of computing the dimension of $E_\Phi(\alpha)$ was raised by Fan et al. in [FLM12], where the following special case was studied: $X =: \mathbb{M}_2 := \{-1, 1\}^{\mathbb{N}^*}$, T is the shift, $d \geq 1$ and Φ is the function

$$\Phi(x^{(1)}, x^{(2)}, \dots, x^{(d)}) = x_1^{(1)} x_1^{(2)} \dots x_1^{(d)}, \quad (x^{(1)}, x^{(2)}, \dots, x^{(d)}) \in \mathbb{M}_2^d.$$

Note that $x_1^{(j)}$ is the first coordinate of $x^{(j)}$. We consider $\mathbb{M}_2$ as the infinite product of the multiplicative group $\{-1, 1\}$. Then the function $x \mapsto x_j$ is a group character of $\mathbb{M}_2$, called Rademacher function and

$$\Phi(T^k x, T^{2d} x, \dots, T^{dk} x) = x_k x_{2k} \dots x_{dk}$$

is also a group character, called Walsh function. Recall that $\mathbb{M}_2 = \{-1, 1\}^{\mathbb{N}^*}$, considered as a symbolic space, is endowed with the metric

$$d(x, y) = 2^{-\min\{k : x_k \neq y_k\}}, \quad \text{for } x, y \in \mathbb{M}_2.$$

2.1 Multifractal Spectrum of a Sequence of Walsh Functions

Under the above assumption, we have the following result.

Theorem 1 [FLM12] *For every* $\alpha \in [-1, 1]$, *the set*

$$B_\alpha := \left\{ x \in \mathbb{M}_2 : \lim_{n \to +\infty} \frac{1}{n} \sum_{k=1}^{n} x_k x_{2k} \cdots x_{dk} = \alpha \right\}$$

has the Hausdorff dimension

$$\dim_H B_\alpha = 1 - \frac{1}{d} + \frac{1}{d \log 2} H\left(\frac{1+\alpha}{2}\right),$$

where $H(t) := -t \log t - (1-t)\log(1-t)$.

A key observation is that all these Walsh functions constitute a dissociated system in the sense of Hewitt-Zuckermann [HZ66]. This allows us to define probability measures called Riesz products on the group $\mathbb{M}_2$:

$$\mu_b := \prod_{k=1}^{\infty}(1 + b x_k x_{2k} \cdots x_{dk}) := w^* - \lim_{N\to\infty} \prod_{k=1}^{N}(1 + b x_k x_{2k} \cdots x_{dk})dx$$

for $b \in [-1, 1]$, where dx denote the Haar measure on $\mathbb{M}_2$. Fortunately, the Riesz product μ_α $(b = \alpha)$ is a maximizing measure on B_α. So, by computing the dimension of the measure μ_α, we get the stated formula. We will go back to Riesz products in Sect. 3 and to dimensions of measures in Sect. 4.

Note that the case $d = 1$ is nothing but the well-known Besicovich-Eggleston theorem dated back to 1940's, which would be considered as the first result of multifractal analysis.

2.2 Box Dimension of Some Multiplicatively Invariant Set

The very motivation of [FLM12] was the multiple ergodic averages in the following case where $X := \mathbb{D}_2 := \{0, 1\}^{\mathbb{N}^*}$, T is the shift on $\mathbb{D}_2$ and $\Phi(x, y) = x_1 y_1$ for $x = x_1 x_2 \ldots$ and $y = y_1 y_2 \ldots \in \mathbb{D}_2$. The space $\mathbb{D}_2$ may also be considered as the infinite product group of $\mathbb{Z}/2\mathbb{Z}$. But the function $x \mapsto x_1$ is no longer group character and the Fourier method fails. Then the authors of [FLM12] proposed to look at a subset of the 0-level set

$$E_\Phi(0) = \left\{ x \in \Sigma_2 : \lim_{n\to+\infty} \frac{1}{n} \sum_{k=1}^{n} x_k x_{2k} = 0 \right\}. \tag{6}$$

The proposed subset is

$$X_2 = \{x \in \Sigma_2 : \forall k \geq 1, \quad x_k x_{2k} = 0\}. \tag{7}$$

This set X_2 has a nicer structure than $E_\Phi(0)$. The condition $x_k x_{2k} = 0$ is imposed to all integers k without exception for all points x in X_2, while the same condition is imposed to "most" integers k for points x in $E_\Phi(0)$.

Theorem 2 [FLM12] *The box dimension of X_2 is equal to*

$$\dim_B X_2 = \frac{1}{2\log 2}\sum_{n=1}^{+\infty}\frac{\log F_n}{2^n},$$

where F_n is the Fibonacci sequence: $F_0 = 1$, $F_1 = 2$ and $F_{n+2} = F_{n+1} + F_n$ for $n \geq 0$.

The key idea to prove the formula is the following observation, which is also one of the key points for all the obtained results up to now in different cases. Look at the definition (7) of X_2. The value of the digit x_1 of an element $x = (x_k) \in X_2$ has an impact on the value of x_2, which in turn on the value of x_4,...and so forth on the values of x_{2^k} for all $k \geq 1$. But it has no influence on x_3, x_5,...Similarly, the value of x_i for an odd integer i only has influence on x_{i2^k}. This suggests us the following partition

$$\mathbb{N}^* = \bigsqcup_{i \text{ odd}} \Lambda_i, \qquad \text{with } \Lambda_i := \{i\,2^n : n \geq 0\}.$$

We could say that the defining conditions of X_2 restricted to different Λ_i are independent. We are then led to investigate, for each odd number i, the restriction of x to Λ_i which will be denoted by

$$x_{|\Lambda_i} = x_i x_{i2} x_{i2^2} \ldots x_{i2^n} \ldots .$$

If we rewrite $x_{|\Lambda_i} = z_1 z_2 \ldots$, which is considered as a point in $\mathbb{D}_2$, then (z_n) belongs to the subshift of finite type subjected to $z_k z_{k+1} = 0$.

It is clear that

$$\dim_B X_2 = \lim_{n\to\infty}\frac{\log_2 N_n}{n}$$

where N_n is the cardinality of the set

$$\{(x_1 x_2 \cdots x_n) : x_k x_{2k} = 0 \text{ for } k \geq 1 \text{ such that } 2k \leq n\}.$$

Let us decompose the set of the first n integers as follows $\{1, \ldots, n\} = C_0 \sqcup C_1 \sqcup \cdots \sqcup C_m$ with

$$\begin{aligned}
C_0 &:= \{1, 3, 5, \ldots, 2n_0 - 1\},\\
C_1 &:= \left\{1\cdot 2, 3\cdot 2, 5\cdot 2, \ldots, 2\cdot(2n_1 - 1)\right\},\\
&\ldots\\
C_k &:= \left\{1\cdot 2^k, 3\cdot 2^k, 5\cdot 2^k, \ldots, 2^k\cdot(2n_k - 1)\right\},\\
&\ldots\\
C_m &:= \left\{1\cdot 2^m\right\}.
\end{aligned}$$

These finite sequences have different length n_k $(0 \le k \le m)$. Actually n_k is the biggest integer such that $2^k(2n_k - 1) \le n$, i.e. $n_k = \left\lfloor \frac{n}{2^{k+1}} + \frac{1}{2} \right\rfloor$. The number m is the biggest integer such that $2^m \le n$, i.e. $m = \lfloor \log_2 n \rfloor$. It clear that $n_0 > n_1 > \cdots > n_{m-1} > n_m = 1$. The conditions $x_\ell x_{2\ell} = 0$ with ℓ in different columns in the table defining $C_0, \ldots, C_m$ are independent. This independence allows us to count the number of possible choices for $(x_1, \ldots, x_n)$ by the multiplication principle. First, we have $n_m (= 1)$ column which has $m + 1$ elements. We have F_{m+1} choices for those x_ℓ with ℓ in the first column because $x_\ell x_{2\ell}$ is conditioned to be different from the word 11. We repeat this argument for other columns. Each of the next $n_{m-1} - n_m$ columns has m elements, so we have $F_m^{n_{m-1}-n_m}$ choices for those x_ℓ with ℓ in these columns. By induction, we get

$$N_n = F_{m+1}^{n_m} F_m^{n_{m-1}-n_m} F_{m-1}^{n_{m-2}-n_{m-1}} \cdots F_1^{n_0-n_1}.$$

To finish the computation we need to note that $\frac{n_k}{n}$ tends to $2^{-(k+1)}$ as n tends to the infinity.

The set X_2 is not invariant under the shift. But as observed by Kenyon et al. [KPS12], it is multiplicatively invariant in the sense that $M_r X_2 \subset X_2$ for all integers $r \ge 1$ where

$$M_r(x_n) = (x_{rn}).$$

The Hausdorff dimension of X_2 was obtained in [KPS12] and the gauge function of X_2 was obtained in [PS13]. We will discuss the work done in [KPS12] in Sect. 5.

3 Dimensions of Measures

Multifractal properties were first investigated for measures. The multifractal analysis of a measure is the analysis of the local dimension on the whole space, while the dimensions of a measure concern with what happens on a Borel support of the measure. In [Fan89], lower and upper Hausdorff dimensions of a measure were introduced and systematically studied, inspired by Peyrière [Pey75] and Kahane [Kah87]. The lower and upper packing dimensions were later studied independently in [Tam95, Heu98]. Some aspects were also considered in [Haa92]. A fundamental theorem in the theory of fractals is Frostman theorem. Howroyd [How94] and Kaufman [Kau94] generalized it from Euclidean spaces to complete separable metric spaces. This fundamental theorem allows us to employ the potential theory.

3.1 Potential Theory

Let (X, d) be a complete separable metric space, called Polish space. Let $0 < \alpha < \infty$. For any locally finite Borel measure μ on X, we define its potential of order α by

$$U_\alpha^\mu(x) := \int_X \frac{\mathrm{d}\mu(y)}{(\mathrm{d}(x,y))^\alpha} \qquad (x \in X)$$

and its energy of order α by

$$I_\alpha^\mu := \int_X U_\alpha^\mu(x)\mathrm{d}\mu(x) = \int_X\int_X \frac{\mathrm{d}\mu(x)\mathrm{d}\mu(y)}{(\mathrm{d}(x,y))^\alpha}.$$

The capacity of order α of a compact set K in X is defined by

$$\mathrm{Cap}_\alpha K = \left(\inf_{\mu\in\mathcal{M}_1^+(K)} I_\alpha^\mu\right)^{-1}.$$

For an arbitrary set E of X, we define its capacity of order α by

$$\mathrm{Cap}_\alpha E = \sup\{\mathrm{Cap}_\alpha K : K \text{ compact contained in } E\}.$$

For a set E in X, we define its capacity dimension by

$$\dim_C E = \inf\{\alpha > 0 : \mathrm{Cap}_\alpha(E) = 0\} = \sup\{\alpha > 0 : \mathrm{Cap}_\alpha(E) > 0\}.$$

The following is the theorem of Frostman-Kaufman-Howroyd. Frostman initially dealt with the Euclidean space. Kaufman proved the result by generalizing a min-max theorem on quadratic function to the mutual potential energy functional, while Howroyd used the technique of weighted Hausdorff measures.

Theorem 3 [Kau94, How94] *Let (X,d) be a complete metric space. For any Borel $E \subset X$, we have* $\dim_C E = \dim_H E$.

3.2 *Hausdorff Dimensions of a Measures*

An important tool to study a measure is its dimensions, which attempt to estimate the size of the "supports" of the measure. The idea finds its origin in Peyrière's works on Riesz products [Pey75] and also in that of Kahane on Dvoretzky covering [Kah87]. The following definitions were introduced in [Fan89] (see also [Fan94a, Fal03]).

Let (X,d) be a complete separable metric space. Let μ be a Borel measure on X. The *lower Hausdorff dimension* and the *upper Hausdorff dimension* of a measure μ are respectively defined by

$$\dim_* \mu = \inf\{\dim_H A : \mu(A) > 0\}, \qquad \dim^* \mu = \inf\{\dim_H A : \mu(A^c) = 0\}.$$

It is evident that $\dim_* \mu \leq \dim^* \mu$. When the equality holds, μ is said to be *unidimensional* or *α-dimensional* where α is the common value of $\dim_* \mu$ and $\dim^* \mu$. The Hausdorff dimensions $\dim_* \mu$ and $\dim^* \mu$ are described by the lower local dimension function $\underline{D}_\mu(x)$ in the following way.

Theorem 4 [Fan89, Fan94a]

$$\dim_* \mu = \operatorname{ess\,inf}_\mu \underline{D}_\mu(x), \quad \dim^* \mu = \operatorname{ess\,sup}_\mu \underline{D}_\mu(x).$$

There are also a continuity-singularity criterion using Hausdorff measures and a energy-potential criterion. Sometimes these criteria are more practical.

Theorem 5 [Fan89, Fan94a]

$$\dim_* \mu = \sup\{\alpha > 0 : \mu \ll H^\alpha\} = \sup\{\alpha > 0 : \mu = \sum \mu_k, I_\alpha^{\mu_k} < \infty\}$$
$$\dim^* \mu = \inf\{\alpha > 0 : \ \mu \perp H^\alpha\} \ = \inf \{\alpha > 0 : U_\alpha^\mu(x) = \infty, \ \mu\text{−p.p.}\}.$$

Theorem 4 holds for lower and upper packing dimensions of a measures, similarly defined, if we replace $\underline{D}(\mu, x)$ by $\overline{D}(\mu, x)$ (see [Tam95, Heu98]). They are denoted by $\mathrm{Dim}_*\mu$ and $\mathrm{Dim}^*\mu$.

We say μ is exact if $\dim_* \mu = \dim^* \mu = \mathrm{Dim}_*\mu = \mathrm{Dim}^*\mu$.

3.3 Sums, Products, Convolutions, Projections of Measures

What we present in this section was in the first unpublished version of [Fan94a]. Some part was restated by in [Tam95] and some part was used in [Fan94b, Fan94c].

Sum The absolute continuity $\nu \ll \mu$ is a partial order on the space of positive Borel measures $M^+(X)$ on the space X. Using the continuity-singularity criterion, it is easy to see that

$$0 < \nu \ll \mu \Rightarrow \dim_* \mu \leq \dim_* \nu \leq \dim^* \nu \leq \dim^* \mu. \tag{8}$$

The relation $\nu \sim \mu$ (meaning $\nu \ll \mu \ll \nu$) is an equivalent relation. Since two equivalent measures have the same lower and upper Hausdorff dimensions, both $\dim_* \mu$ and $\dim^*$ are well defined for equivalent classes.

Given a family of positive measures $\{\mu_i\}_{i\in I}$ which is bounded under the order $\ll$, we denote its supremum by $\bigvee_{i\in I} \mu_i$. If the family is finite, we have $\bigvee_{i\in I} \mu_i \sim \sum_{i\in I} \mu_i$. Such equivalence also holds when the family is countable. In general, there is a countable sub-family μ_{i_k} such that $\bigvee_{i\in I} \mu_i \sim \sum_k \mu_{i_k}$. This is what we mean by sum of measures.

Theorem 6 *If $\{\mu_i\}_{i\in I}$ is a bounded family of measures in $M^+(X)$, we have*

$$\dim_* \bigvee_{i\in I} \mu_i = \inf_{i\in I} \dim_* \mu_i, \quad \dim^* \bigvee_{i\in I} \mu_i = \sup_{i\in I} \dim^* \mu_i.$$

Let us see how to prove the first formula for a family of two measures μ and ν. For any $\alpha < \dim_*(\mu+\nu)$, we have $U_\alpha^\mu(x) + U_\alpha^\mu(x) = U_\alpha^{\mu+\nu}(x) < \infty \quad (\mu+\nu)-a.e.$ This implies $\dim_*(\mu+\nu) \leq \min\{\dim_* \mu, \dim_* \nu\}$. The inverse inequality follows from the fact that if $\beta < \min\{\dim_* \mu, \dim_* \nu\}$, then $\mu \ll H^\beta$ and $\nu \ll H^\beta$ which implies $\mu + \nu \ll H^\beta$.

Let us consider now the infimum $\bigwedge_{i\in I}$ of a family of measure $\{\mu_i\}_{i\in I}$. Recall that by definition we have

$$\bigwedge_{i\in I} \mu_i \sim \bigvee_{\mu:\forall i\in I, \mu\ll\mu_i} \mu.$$

For a family of two measures μ and ν, we have $\mu \bigwedge \nu \sim \frac{d\mu}{d\nu} \bigvee \frac{d\nu}{d\mu}$.

Theorem 7 *If $\{\mu_i\}_{i\in I}$ is family of measures in $M^+(X)$ such that $\bigwedge_{i\in I} \mu_i \neq 0$, we have*

$$\sup_{i\in I} \dim_* \mu_i \leq \dim_* \bigwedge_{i\in I} \mu_i \leq \dim^* \bigwedge_{i\in I} \mu_i \leq \inf_{i\in I} \dim^* \mu_i.$$

Consequently, $\bigvee_{i\in I} \mu_i$ (resp. $\bigwedge_{i\in I} \mu_i$) is unidimensional if and only if all measures μ_i are unidimensional and have the same dimension.

Product Let (X, δ_X) and (Y, δ_Y) be two Polish spaces. Then $d := \delta_X \vee \delta_Y$ is a compatible metric on the product space. Let $\mu \in M^+(X)$ and $\nu \in M^+(Y)$. Let us consider the product measure $\mu \otimes \nu$.

Theorem 8 *For $\mu \in M^+(X)$ and $\nu \in M^+(Y)$, we have*

$$\dim_* \mu \otimes \nu \geq \dim_* \mu + \dim_* \nu, \qquad \dim^* \mu \otimes \nu \geq \dim^* \mu + \dim^* \nu.$$

We say a measure $\mu \in M^+(X)$ is regular if $\underline{D}(\mu, x) = \overline{D}(\mu, x)$ μ-a.e.

Theorem 9 *If $\mu \in M^+(X)$ or $\nu \in M^+(Y)$ is regular, we have*

$$\dim_* \mu \otimes \nu = \dim_* \mu + \dim_* \nu, \qquad \dim^* \mu \otimes \nu = \dim^* \mu + \dim^* \nu.$$

Convolution Assume that the Polish space X is a locally compact abelian group G. Assume further that G satisfies the following hypothesis

$$I_\alpha^{\mu*\nu} \leq C(\alpha, \nu) I_\alpha^\mu$$

for all measures $\mu, \nu \in M^+(X)$, where $C(\alpha, \nu)$ is a constant independent of μ. For example, if $G = \mathbb{R}^d$, we have

$$I_\alpha^\mu = \int |\xi|^{-(d-\alpha)} |\widehat{\mu}(\xi)|^2 d\xi$$

which implies that the hypothesis is satisfied by $\mathbb{R}^d$. The hypothesis is also satisfied by the group $\prod_{n=1}^{\infty} \mathbb{Z}/m_n\mathbb{Z}$ [Fan89].

Theorem 10 *For any measures* $\mu, \nu \in M^+(G)$, *we have*

$$\dim_* \mu * \nu \geq \max\{\dim_* \mu, \dim_* \nu\}, \qquad \dim^* \mu * \nu \geq \max\{\dim^* \mu, \dim^* \nu\}.$$

For a given measure $\mu \in M^+(G)$, we consider the following two subgroups

$$H_- := \{t \in G : \mu \ll \mu * \delta_t\}, \qquad H_+ := \{t \in G : \mu * \delta_t \ll \mu\}.$$

We could call $H = H_- \cap H_+$ the quasi-invariance group of μ.

Theorem 11 *Under the above assumption, if* $\nu(H_-) > 0$, *we have the equality* $\dim_* \mu * \nu = \dim_* \mu$ *and consequently* $\dim_* \nu \leq \dim_* \mu$ *and* $\dim_H H_- \leq \dim_* \mu$; *if* $\nu(H_+) > 0$, *we have the equality* $\dim^* \mu * \nu = \dim^* \mu$ *and consequently* $\dim^* \nu \leq \dim^* \mu$ *and* $\dim_H H_+ \leq \dim^* \mu$.

Projection Let $\mu \in M^+(\mathbb{R}^2)$ with $\mathbb{R}^2 = \mathbb{C}$. Let L_θ ($0 \leq \theta < 2\pi$) be the line passing the origin and having angle θ with the line of abscissa. The orthogonal projection P_θ on L_θ is defined by $P_\theta(x) = \langle x, e^{i\theta}\rangle$. Let $\mu_\theta := \mu \circ P_\theta^{-1}$ be the projection of μ on L_θ.

Theorem 12 *Let* $\mu \in M^+(\mathbb{R}^2)$. *For almost all* θ, *we have* $\dim_* \mu_\theta = \dim_* \mu \wedge 1$ *and* $\dim^* \mu_\theta = \dim^* \mu \wedge 1$.

3.4 Ergodicity and Dimension

Young [You82] considered diffeomorphisms of surfaces leaving invariant an ergodic Borel probability measure μ. She proved that μ is exact and found a formula relating $\dim \mu$ to the entropy and Lyapunov exponents of μ. One of the main problems in the interface of dimension theory and dynamical systems is the Eckmann-Ruelle conjecture on the dimension of hyperbolic ergodic measures: the local dimension of every hyperbolic measure invariant under a $C^{1+\alpha}$-diffeomorphism exists almost everywhere. This conjecture was proved by Barreira et al. [BPS99] based on the fundamental fact that such a measure possesses asymptotically "almost" local product structure. But, in general, the ergodicity of the measure doesn't imply that the measure is exact [Cut95]. In [Fan94b], D-ergodicity and unidimensionality were studied.

4 Oriented Walks and Riesz Products

4.1 Oriented Walks

Let $(\epsilon_n)_{n\geq 1} \subset [0, 2\pi)^{\mathbb{N}}$ be a sequence of angles. For $n \geq 1$, define

$$S_n(\epsilon) = \sum_{k=1}^{n} e^{i(\epsilon_1+\epsilon_2+\cdots+\epsilon_k)}.$$

We call $(S_n(\epsilon))_{n\geq 1}$ an oriented walk on the plan $\mathbb{C}$. In his book [Fel68] (vol. 1, pp. 240–241), Feller mentioned a model describing the length of long polymer molecules in chemistry. It is a random chain consisting of n links, each of unit length, and the angle between two consecutive links is $\pm\alpha$ where α is a positive constant. Then the distance L_n from the beginning to the end of the chain can be expressed by

$$L_n = |S_n(\epsilon)|$$

where (ϵ_n) is an i.i.d. sequence of random variables taking values in $\{-\alpha, \alpha\}$. If $\alpha = 0$, $L_n = n$ is deterministic. If $0 < \alpha < 2\pi$, the random variable L_n is not expressed as sums of independent variables. However Feller succeeded in computing the second order moment of L_n. It is actually proved in [Fel68] that $\|L_n\|_2$ is of order $\sqrt{n}$. More precisely, for $0 < \alpha < 2\pi$ we have

$$\mathbb{E}L_n(\alpha)^2 = n\frac{1+\cos\alpha}{1-\cos\alpha} - 2\cos\alpha\frac{1-\cos^n\alpha}{(1-\cos\alpha)^2}.$$

Observe that

$$\mathbb{E}L_n^2 = \frac{1-(-1)^n}{2} \text{ if } \alpha = \pi; \qquad \mathbb{E}L_n^2 \sim n\frac{1+\cos\alpha}{1-\cos\alpha} \text{ if } 0 < \alpha < 2\pi, \alpha \neq \pi.$$

What is the behavior of $S_n(\epsilon)$ as $n \to \infty$ for individuals ϵ? We could study the behavior from the multifractal point of view. Let us consider a more general setting. Fix $d \in \mathbb{N}^*$. Let $\tau \in GL(\mathbb{R}^d)$, $v \in \mathbb{R}^d$ and A a finite subset of $\mathbb{Z}$. For any $x = (x_n) \in \mathbb{D} := A^{\mathbb{N}}$, we define the oriented walk

$$S_0(x) = v, \qquad S_n(x) = \sum_{k=1}^{n} \tau^{x_1+x_2+\cdots+x_k} v.$$

For $\alpha \in \mathbb{R}^d$, we define the *α-level set*

$$E_\tau(\alpha) := \left\{x \in \mathbb{D} : \lim_{n\to\infty} \frac{1}{n} S_n(x) = \alpha\right\}.$$

Let $L_\tau := \{\alpha \in \mathbb{R}^d : E_\tau(\alpha) \neq \emptyset\}$.

The following two cases were first studied in [Fan00]. Case 1: $d = 2$, $\tau = -1$ and $A = \{0, 1\}$; Case 2: $d = 1$, $\tau = e^{i\pi/2}$ and $A = \{-1, 1\}$.

Theorem 13 [Fan00] *In the first case, we have* $L_\tau = [-1, 1]$ *and for* $\alpha \in L_\tau$ *we have*

$$\dim_H E_\tau(\alpha) = \dim_P E_\tau(\alpha) = H\left(\frac{1+\alpha}{2}\right).$$

In the seconde case, we have $L_\tau = \{z = a + ib : |a| \leq 1/2, |b| \leq 1/2\}$ *and for* $\alpha = a + bi \in L_\tau$ *we have*

$$\dim_H E_\tau(\alpha) = \dim_P E_\tau(\alpha) = \frac{1}{2\log 2}\left[H\left(\frac{1}{2}+a\right) + H\left(\frac{1}{2}+b\right)\right].$$

This theorem was proved by using Riesz products which will be described in the following section.

A new construction of measures allows us to deal with a class of oriented walks. We assume that $\tau \in GL(\mathbb{R}^d)$ is idempotent. That is to say $\tau^p = Id$ for some integer $p > 1$ (the case $p = 1$ is trivial). The least p is called the order of τ. The above two cases are special cases. In fact, $\theta = -1$ is idempotent with order $p = 2$ and $\theta = e^{i\pi/2}$ is idempotent with order $p = 4$. The following rotations in $\mathbb{R}^3$

$$\tau_1 = \begin{pmatrix} 0 & 0 & 1 \\ 1 & 0 & 0 \\ 0 & 1 & 0 \end{pmatrix}, \qquad \tau_2 = \begin{pmatrix} 0 & 0 & -1 \\ 1 & 0 & 0 \\ 0 & 1 & 0 \end{pmatrix}$$

are idempotent with order respectively equal to 3 and 6. Also remark that $\tau_1 \in SO_3(\mathbb{R})$ and $\tau_2 \in O_3(\mathbb{R}) \setminus SO_3(\mathbb{R})$.

Since $\tau^p = Id$, the sum $x_1 + x_2 + \cdots + x_k$ in the definition of $S_n(x)$ can be made modulo p. For $s \in \mathbb{R}^d$, we define a $p \times p$-matrix $M_s = (M_s(i, j))$: for $(i, j) \in \mathbb{Z}/p\mathbb{Z} \times \mathbb{Z}/p\mathbb{Z}$ define

$$M_s(i, j) = 1_A(j - i)\exp[\langle s, \tau^j \rangle v].$$

where $\langle \cdot, \cdot \rangle$ denotes the scalar product in $\mathbb{R}^d$. It is clear that M_s is irreducible iff M_0 is so. We consider A as a subset (modulo p) of $\mathbb{Z}/p\mathbb{Z}$. It is easy to see that M_0 is irreducible iff A generates the group $\mathbb{Z}/p\mathbb{Z}$.

Assume that A generates the group $\mathbb{Z}/p\mathbb{Z}$. Then M_s is irreducible and by the Perron-Frobenius theorem, the spectral radius $\lambda(s)$ of M_s is a simple eigenvalue and there is a unique corresponding probability eigenvector $t(s) = (t_s(0), t_s(1) \ldots, t_s(p-1)$. Let

$$P(s) = \log \lambda(s).$$

It is real analytic and strictly convex function on $\mathbb{R}^d$. We call it the pressure function associated to the oriented walk.

Theorem 14 [FW00] *Assume τ is idempotent with order p and A generates the group $\mathbb{Z}/p\mathbb{Z}$. Then $L_\tau = \overline{\{\nabla P(s) : s \in \mathbb{R}^d\}}$ and for $\alpha \in \Delta$, we have*

$$\dim_H E_\tau(\alpha) = \dim_P E_\tau(\alpha) = \frac{1}{\log p} \inf_{s\in\mathbb{R}^d} \{P(s) - \langle s, \alpha\rangle\} = \frac{P(s_\alpha) - \langle s_\alpha, \alpha\rangle}{\log p},$$

where s_α is the unique $s \in \mathbb{R}^d$ such that $\nabla P(s) = \alpha$.

4.2 Riesz Products

Theorem 1 was proved by using Riesz products. While Hausdorff introduced the Hausdorff dimension (1919), Riesz constructed a class of continuous but singular measures on the circle (1918), called Riesz products. Riesz products are used as tool in harmonic analysis and some of them are Gibbs measures in the sense of dynamical systems.

Let us recall the definition of Riesz product on a compact abelian group G, due to Hewitt-Zuckerman [HZ66] (1966). Let $\widehat{G}$ be the dual group of G. A sequence of characters $\Lambda = (\gamma_n)_{n\geq 1} \subset \widehat{G}$ is said to be *dissociated* if for any $n \geq 1$, the following characters are all distinct:

$$\gamma_1^{\epsilon_1}\gamma_2^{\epsilon_2}\cdots\gamma_n^{\epsilon_n}$$

where $\epsilon_j \in \{-1, 0, 1\}$ if γ_j is not of order 2, or $\epsilon_j \in \{0, 1\}$ otherwise. Given such a dissociated sequence $\Lambda = (\gamma_n)_{n\geq 1}$ and a sequence of complex numbers $a = (a_n)_{n\geq 1}$ such that $|a_n| \leq 1$, we can define a probability measure on G, called *Riesz product*,

$$\mu_a = \prod_{n=1}^{\infty} \big(1 + \mathrm{Re}\ a_n\gamma_n(t)\big) \tag{9}$$

as the weak* limit of $\prod_{n=1}^{N}\big(1 + \mathrm{Re}\ a_n\gamma_n(t)\big)\mathrm{d}t$ where $\mathrm{d}t$ denotes the Haar measure on G.

A very useful fact is that the Fourier coefficients of the Riesz product μ_a can be explicitly expressed in term of the coefficients a_n's:

$$\widehat{\mu}_a(\gamma) = \prod_{k=1}^{n} a_k^{(\epsilon_k)} \text{ if } \gamma = \gamma_1^{\epsilon_1}\gamma_2^{\epsilon_2}\cdots\gamma_n^{\epsilon_n}, \quad \widehat{\mu}_a(\gamma) = 0 \text{ otherwise.}$$

where $a_n^{(\epsilon)} = 1$, $a_n/2$ or $\bar{a}_n/2$ according to $\epsilon = 0$, 1 or -1. Consequently the sequence $\{\gamma_n - \bar{a}_n/2\}_{n\geq 1}$ is an orthogonal system in $L^2(\mu_a)$. Here are some properties of μ_a.

Theorem 15 [Zyg68] *The measure μ_a is either absolutely continuous or singular (with respect to the Haar measure) according to $\sum_{n=1}^{\infty} |a_n|^2 < \infty$ or $= \infty$.*

Theorem 16 [Fan93, Pey90] *Let $\{\alpha_n\}$ be a sequence of complex numbers. The orthogonal series $\sum_{n=1}^{\infty} \alpha_n(\gamma_n(t) - \bar{a}_n/2)$ converges μ_a-everywhere iff $\sum_{n=1}^{\infty} |\alpha_n|^2 < \infty$.*

The proof of Theorem 16 in [Fan93] involved the following Riesz product with $\omega = (\omega_n) \in G^{\mathbb{N}}$ as phase translation:

$$\mu_{a,\omega} = \prod_{n=1}^{\infty} (1 + \operatorname{Re} a_n \gamma_n(t + \omega_n)). \tag{10}$$

Actually $\mu_{a,\omega}$ was considered as a random measure and (ω_n) was considered as an i.i.d. random sequence with Haar measure as common probability law.

When two Riesz products μ_a and μ_b are singular or mutually absolutely continuous? It is a unsolved problem. Bernoulli infinite product measures can be viewed as Riesz products on the group $(\mathbb{Z}/m\mathbb{Z})^{\mathbb{N}}$. For these Bernoulli infinite measures, the Kakutani dichotomy theorem [Kak48] applies and there is a complete solution. But there is no complete solution for other groups.

The classical Riesz products are of the form

$$\mu_a = \prod_{k=1}^{\infty} (1 + \operatorname{Re} a_n e^{i\lambda_k x}) \tag{11}$$

where $\{\lambda_n\} \subset \mathbb{N}$ is a lacunary sequence in the sense that $\lambda_{n+1} \geq 3\lambda_n$. Peyrière has first studied the lower and upper dimensions of μ_a, without introducing the notion of dimension of measures. Let us mention an estimation for the energy integrals of μ_a [Fan89]:

$$\int\int \frac{d\mu_a(x) d\mu_a(y)}{|x-y|^\alpha} \approx \lambda_1^{\alpha-1}|a_1|^2 + \sum_{n=2}^{\infty} \lambda_n^{\alpha-1}|a_n|^2 \prod_{k=1}^{n-1}\left(1 + \frac{|a_n|^2}{2}\right).$$

4.3 Evolution Measures

The key for the proof of Theorem 14 is the construction of the following measures on $A^{\mathbb{N}}$, which describe the evolution of the oriented walk. It is similar to Markov measure but it is not. It plays the role of Gibbs measure but it is not Gibbs measure either.

Recall that $M_s t(s) = \lambda(s) t(s)$. In other words, for every $i \in \mathbb{Z}/p\mathbb{Z}$ we have

$$\lambda(s) t_i(s) = \sum_j 1_A(j-i) t_j(s) \exp[\langle s, \tau^j v\rangle].$$

Denote, for $a \in A$ and for $(x_1, \ldots, x_{k+1}) \in A^{k+1}$,

$$\pi(a) = \frac{t_a(s)}{\sum_{b\in A} t_b(s)};$$

$$Q_k(x_1, x_2, \ldots, x_{k+1}) = \frac{t_{x_1+x_2+\cdots+x_{k+1}}(s)\exp[\langle s, \tau^{x_1+x_2+\cdots+x_{k+1}}v\rangle]}{\lambda(s)t_{x_1+x_2+\cdots+x_k}(s)}.$$

Then we define a probability measure μ_s on $A^{\mathbb{N}}$ as follows. For any word $x_1x_2\cdots x_n \in A^n$, let

$$\mu_s([x_1x_2\cdots x_n]) = \pi(x_1)Q_1(x_1, x_2)Q_2(x_1, x_2, x_3)\cdots Q_{n-1}(x_1, x_2, \ldots, x_n).$$

For $x = (x_n) \in A^{\mathbb{N}}$, let

$$w_n(x) = x_1 + x_2 + \cdots + x_n \pmod{p}.$$

The mass $\mu_s([x_1x_2\cdots x_n])$ and the partial sum $S_n(x)$ are directly related as follows.

$$\log \mu_s([x_1x_2\cdots x_n]) = \langle s, S_n(x)\rangle - (n-1)\log\lambda(s) - \log\sum_{a\in A} t_a(s) + \log t_{w_n(x)}(s).$$

As the $t_i(s)$'s are bounded, we deduce the following relation between the measure μ_s and the oriented walk S_n.

Proposition 1 *For any $x \in \mathbb{D}$, we have*

$$\log \mu_s([x_1x_2\cdots x_n]) - \langle s, S_n(x)\rangle = -n\log\lambda(s) + O(1).$$

5 Multiple Birkhoff Averages

Let $\mathcal{A} = \{0, 1, 2, \ldots, m-1\}$ be a set of m symbols ($m \geq 2$). Denote $\Sigma_m = \mathcal{A}^{\mathbb{N}^*}$. Let $q \geq 2$ be a integer. Fan, Schmeling and Wu made a forward step in [FSW00] by obtaining a Hausdorff spectrum of multiple ergodic averages for a class of potentials. They consider an arbitrary function $\varphi : \mathcal{A}^d \to \mathbb{R}$ and study the sets

$$E(\alpha) = \left\{x \in \Sigma_m : \lim_{n\to\infty} A_n\varphi(x) = \alpha\right\}$$

for $\alpha \in \mathbb{R}$, where

$$A_n\varphi(x) = \frac{1}{n}\sum_{k=1}^{n} \varphi(x_k, x_{qk}, \ldots, x_{q^{d-1}k}). \qquad (12)$$

Let

$$\alpha_{\min} = \min_{a_1,\dots,a_d \in \mathcal{A}} \varphi(a_1,\dots,a_d), \quad \alpha_{\max} = \max_{a_1,\dots,a_d \in \mathcal{A}} \varphi(a_1,\dots,a_d).$$

It is assumed that $\alpha_{\min} < \alpha_{\max}$ (otherwise φ is constant and the problem is trivial). A key ingredient of the proof is a class of measures constructed by Kenyon, Peres and Solomyak [KPS12] that we call telescopic product measures. In [FSW00], a nonlinear thermodynamic formalism was developed.

5.1 Thermodynamic Formalism

The Hausdorff dimension of $E(\alpha)$ is determined through the following thermodynamic formalism. Let $\mathcal{F}(\mathcal{A}^{d-1}, \mathbb{R}^+)$ be the cone of functions defined on $\mathcal{A}^{d-1}$ taking non-negative real values. For any $s \in \mathbb{R}$, consider the transfer operator $\mathcal{L}_s$ defined on $\mathcal{F}(\mathcal{A}^{d-1}, \mathbb{R}^+)$ by

$$\mathcal{L}_s \psi(a) = \sum_{j \in \mathcal{A}} e^{s\varphi(a,j)} \psi(Ta, j) \tag{13}$$

where $T : \mathcal{A}^{d-1} \to \mathcal{A}^{d-2}$ is defined by $T(a_1,\dots,a_{d-1}) = (a_2,\dots,a_{d-1})$. Then define the non-linear operator $\mathcal{N}_s$ on $\mathcal{F}(\mathcal{A}^{d-1}, \mathbb{R}^+)$ by $\mathcal{N}_s\psi(a) = (\mathcal{L}_s\psi(a))^{1/q}$. It is proved in [FSW00] that the equation

$$\mathcal{N}_s \psi_s = \psi_s \tag{14}$$

admits a unique strictly positive solution $\psi_s = \psi_s^{(d-1)} : \mathcal{A}^{d-1} \to \mathbb{R}_+^*$. Extend the function ψ_s onto $\mathcal{A}^k$ for all $1 \le k \le d-2$ by induction:

$$\psi_s^{(k)}(a) = \left(\sum_{j \in \mathcal{A}} \psi_s^{(k+1)}(a, j) \right)^{\frac{1}{q}}, \quad (a \in \mathcal{A}^k). \tag{15}$$

For simplicity, we write $\psi_s(a) = \psi_s^{(k)}(a)$ for $a \in \mathcal{A}^k$ with $1 \le k \le d-1$. Then the pressure function is defined by

$$P_\varphi(s) = (q-1)q^{d-2} \log \sum_{j \in \mathcal{A}} \psi_s(j). \tag{16}$$

It is proved [FSW00] that $P_\varphi(s)$ is an analytic and convex function of $s \in \mathbb{R}$ and even strictly convex when $\alpha_{\min} < \alpha_{\max}$. The Legendre transform of P_φ is defined as

$$P_\varphi^*(\alpha) = \inf_{s \in \mathbb{R}} (P_\varphi(s) - s\alpha).$$

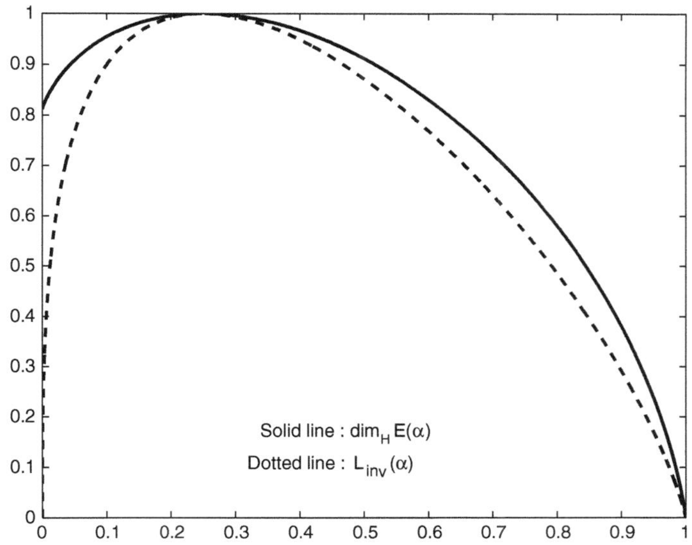

Fig. 1 Spectra $\alpha \mapsto \dim_H E(\alpha)$ and $\alpha \mapsto F_{\rm inv}(\alpha)$ for φ_1

We denote by L_φ the set of levels $\alpha \in \mathbb{R}$ such that $E(\alpha) \neq \emptyset$.

Theorem 17 [FSW00] *We have* $L_\varphi = [P'_\varphi(-\infty), P'_\varphi(+\infty)]$. *If* $\alpha = P'_\varphi(s_\alpha)$ *for some* $s_\alpha \in \mathbb{R} \cup \{-\infty, +\infty\}$, *then* $E(\alpha) \neq \emptyset$ *and the Hausdorff dimension of* $E(\alpha)$ *is equal to*

$$\dim_H E(\alpha) = \frac{P^*_\varphi(\alpha)}{q^{d-1} \log m}.$$

Similar results hold for vector valued functions φ [FSW00]. Peres and Solomyak [PS13] have obtained a result for the special case $\varphi(x, y) = x_1 y_1$ on Σ_2. Kifer [Kif12] has obtained a result on the multiple recurrence sets for some frequency of product form.

Let us consider two examples. Let $q = 2$ and $\ell = 2$ and let $\varphi_1(x, y) = x_1 y_1$ (See Fig. 1) and $\varphi_2(x, y) = (2x_1 - 1)(2y_1 - 1)$ (see Fig. 2) be two potentials on Σ_2. The invariant spectra (see Sect. 6.2) are also shown in the figures.

5.2 Telescopic Product Measures

One of the key points in the proof of the Hausdorff spectrum (Theorem 17) is the observation that the coordinates $x_1, \ldots, x_n, \ldots$ of x appearing in the definition of $A_n\varphi(x)$ share the following independence. Consider the partition of $\mathbb{N}^*$:

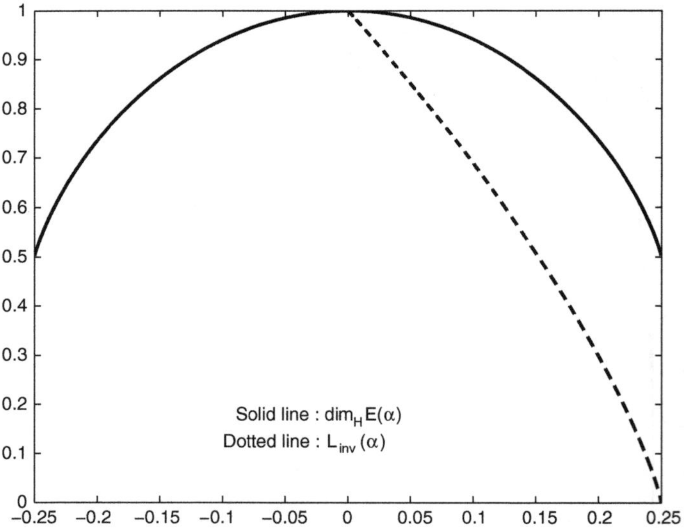

Fig. 2 Spectra $\alpha \mapsto \dim_H E(\alpha)$ and $\alpha \mapsto F_{\text{inv}}(\alpha)$ for φ_2

$$\mathbb{N}^* = \bigsqcup_{i \geq 1, q \nmid i} \Lambda_i \quad \text{with } \Lambda_i = \{iq^j\}_{j \geq 0}.$$

Observe that if $k = iq^j$ with $q \nmid i$, then $\varphi(x_k, x_{kq}, \ldots, x_{kq^{d-1}})$ depends only on $x_{|\Lambda_i}$, the restriction of x on Λ_i. So the summands in the definition of $A_n\varphi(x)$ can be put into different groups, each of which depends on one restriction $x_{|\Lambda_i}$. For this reason, we decompose Σ_m as follows:

$$\Sigma_m = \prod_{i \geq 1, q \nmid i} \mathcal{A}^{\Lambda_i}.$$

Telescopic product measures are now constructed as follows. Let μ be a probability measure on Σ_m. Notice that $\mathcal{A}^{\Lambda_i}$ is nothing but a copy of Σ_m. We consider μ as a measure on $\mathcal{A}^{\Lambda_i}$ for every i with $q \nmid i$. Then we define the infinite product measure $\mathbb{P}_\mu$ on $\prod_{i \geq 1, q \nmid i} \mathcal{A}^{\Lambda_i}$ of the copies of μ. More precisely, for any word u of length n we define

$$\mathbb{P}_\mu([u]) = \prod_{i \leq n, q \nmid i} \mu([u_{|\Lambda_i}]),$$

where $[u]$ denotes the cylinder of all sequences starting with u. The probability measure $\mathbb{P}_\mu$ is called *telescopic product measure*. Kenyon, Peres and Solomyak [KPS12] have first constructed these measures.

The Hausdorff dimension of every telescopic product measure is computable.

Theorem 18 [KPS12, FSW00] *For any given measure μ, the telescopic product measure $\mathbb{P}_\mu$ is exact and its dimension is equal to*

$$\dim_H \mathbb{P}_\mu = \dim_P \mathbb{P}_\mu = \frac{(q-1)^2}{\log m} \sum_{k=1}^{\infty} \frac{H_k(\mu)}{q^{k+1}}$$

where $H_k(\mu) = -\sum_{a_1,\ldots,a_k \in S} \mu([a_1 \cdots a_k]) \log \mu([a_1 \cdots a_k])$.

5.3 Dimension Formula of Ruelle-Type

The function ψ_s defined by (14) and (15) determine a special telescopic product measure which plays the role of Gibbs measure in the proof of the Hausdorff spectrum.

First we define a $(d-1)$-step Markov measure on Σ_m, which will be denoted by μ_s, with the initial law

$$\pi_s([a_1, \ldots, a_{d-1}]) = \prod_{j=1}^{d-1} \frac{\psi_s(a_1, \ldots, a_j)}{\psi_s^q(a_1, \ldots, a_{j-1})} \tag{17}$$

and the transition probability

$$Q_s\left([a_1, \ldots, a_{d-1}], [a_2, \ldots, a_d]\right) = e^{s\varphi(a_1,\ldots,a_d)} \frac{\psi_s(a_2, \ldots, a_d)}{\psi_s^q(a_1, \ldots, a_{d-1})}. \tag{18}$$

The corresponding telescopic product measure $\mathbb{P}_{\mu_s}$ is proved to be a dimension maximizing measure of $E(\alpha)$ if s is chosen to be the solution of $P'_\varphi(s) = \alpha$. The dimension of $\mathbb{P}_{\mu_s}$ is simply expressed by the pressure function. In other words, we have the following formula of Ruelle-type.

Theorem 19 [FSW00] *For any $s \in \mathbb{R}$, we have*

$$\dim_H \mathbb{P}_{\mu_s} = \frac{1}{q^{d-1}} [P_\varphi(s) - s P'_\varphi(s)].$$

5.4 Multiplicatively Invariant Sets

Kenyon et al. [KPS12] were able to compute both the Hausdorff dimension and the box dimension of X_2, already considered in Sect. 2.2, and of a class of generalizations of X_2. Peres et al. [PSSS00] generalized the results to a more general class of sets.

Recall that $\Sigma_m = \{0, 1, \ldots, m-1\}^{\mathbb{N}^*}$, q is an integer greater than 2 and $\Lambda_i := \{i\, q^n : n \geq 0\}$. For any subset $\Omega \subset \Sigma_m$, define

$$X_\Omega := \{x = (x_k)_{k\geq 1} \in \Sigma_m : \forall i \neq 0 \mod q,\ x_{|\Lambda_i} \in \Omega\}. \tag{19}$$

We get $X_\Omega = X_2$ when $q = 2$ and Ω is the Fibonacci set $\{y \in \Sigma_m : \forall k \geq 1,\ y_k y_{k+1} = 0\}$.

The set X_Ω is not shift-invariant but is multiplicatively invariant in the sense that $M_r X_\Omega \subset X_\Omega$ for every integer $r \in \mathbb{N}^*$ where M_r maps (x_n) to (x_{rn}).

The generating set Ω has a tree of prefixes, which is a directed graph Γ. The set $V(\Gamma)$ of vertices consists of all possible prefixes of finite length in Ω, i.e.

$$V(\Gamma) := \bigcup_{k\geq 0} \text{Pref}_k(\Omega),$$

where $\text{Pref}_0(\Omega) = \{\emptyset\}$, $\text{Pref}_k(\Omega) := \{u \in \{0, 1, \ldots, m-1\}^k : \Omega \cap [u] \neq \emptyset\}$. There is a directed edge from a vertex u to another v if and only if $v = ui$ for some $i \in \{0, 1, \ldots, m-1\}$.

Theorem 20 [KPS12] *There exists a unique vector $\bar{t} = (t_v)_{v\in\Gamma} \in [1, m^{\frac{1}{q-1}}]^{V(\Gamma)}$ defined on the tree such that*

$$\forall v \in V(\Gamma), \quad (t_v)^q = \sum_{i\in\{0,1,\ldots,m-1\}:\ vi\in\Omega} t_{vi}. \tag{20}$$

The Hausdorff dimension and the box dimension of X_Ω are respectively equal to

$$\dim_H(X_\Omega) = (q-1)\log_m t_\emptyset \tag{21}$$

$$\dim_B(X_\Omega) = (q-1)^2 \sum_{k=1}^{+\infty} \frac{\log_m |\text{Pref}_k(\Omega)|}{q^{k+1}}. \tag{22}$$

The two dimensions coincide if and only if the tree Γ is spherically symmetric, i.e. all prefixes of length k in Ω have the same number of continuations of length $k+1$ in Ω.

The vector $\bar{t}$ defines a measure μ on $\Omega \subset \{0, 1, \ldots, m-1\}^{\mathbb{N}^*}$. Then a telescopic product measure can be built on X_Ω. It is proved in [KPS12] that there is a maximizing measure on X_Ω of this form.

A typical example of the class of sets studied by Peres et al. [PSSS00] is

$$X_{2,3} = \{x \in \Sigma_m : x_k x_{2k} x_{3k} = 0\}.$$

The construction of the sets is as follows. Let $\kappa \geq 1$ be an integer and let $p_1, \ldots, p_\kappa$ be κ primes, which generates a semigroup S of $\mathbb{N}^*$:

$$S = \langle p_1, p_2, \ldots, p_\kappa\rangle = \{p_1^{\alpha_1} p_2^{\alpha_2} \cdots p_\kappa^{\alpha_\kappa} : \alpha_1, \ldots, \alpha_\kappa \in \mathbb{N}\}.$$

The elements of S are arranged in increasing order and denoted by ℓ_k the k-th element of $S = \{\ell_k\}_{k=1}^{\infty} : 1 = \ell_1 < \ell_2 < \cdots$. Define

$$\gamma(S) := \sum_{k=1}^{\infty} \frac{1}{\ell_k}. \tag{23}$$

Write $(i, S) = 1$ when $(i, p_j) = 1$ for all $j \le \kappa$. We have the following partition of $\mathbb{N}^*$:

$$\mathbb{N}^* = \bigsqcup_{(i,S)=1} iS. \tag{24}$$

For each element $x = (x_k)_{k=1}^{\infty}$, $x|_{iS}$ denotes the restriction $x|_{iS} := (x_{i\ell_k})_{k=1}^{\infty}$, which is also viewed as an element of Σ_m. Given a closed subset $\Omega \subset \Sigma_m$, we define a new subset of Σ_m:

$$X_{\Omega}^{(S)} := \left\{ x = (x_k)_{k=1}^{\infty} \in \Sigma_m : \ x|_{iS} \in \Omega \ \text{ for all } i, \ (i, S) = 1 \right\}. \tag{25}$$

Theorem 21 [PSSS00] *There exists a vector* $\overline{t} = (t(u))_{u \in \mathrm{Pref}(\Omega)} \in [1, +\infty)^{\mathrm{Pref}(\Omega)}$ *defined on the tree of prefixes of* Ω *such that*

$$t(\varnothing) \in [1, m], \quad t(u) \in [1, m^{\ell_k(\ell_{k+1}^{-1} + \ell_{k+2}^{-1} + \cdots)}], \ |u| = k, \ k \ge 1,$$

which is the solution of the system

$$\begin{aligned} t(\varnothing)^{\gamma(S)} &= \sum_{j=0}^{m-1} t(j), \\ t(u)^{\ell_{k+1}/\ell_k} &= \sum_{j:\ uj \in \mathrm{Pref}_{k+1}(\Omega)} t(uj), \quad \forall\, u \in \mathrm{Pref}_k(\Omega), \ \forall k \ge 1. \end{aligned}$$

The Hausdorff dimension and the box dimension of $X_{\Omega}^{(S)}$ *are respectively equal to*

$$\begin{aligned} \dim_H \left(X_{\Omega}^{(S)} \right) &= \log_m t(\varnothing) \\ \dim_B \left(X_{\Omega}^{(S)} \right) &= \gamma(S)^{-1} \sum_{k=1}^{\infty} \Big(\frac{1}{\ell_k} - \frac{1}{\ell_{k+1}} \Big) \log_m |\mathrm{Pref}_k(\Omega)|. \end{aligned}$$

We have $\dim_H \left(X_{\Omega}^{(S)} \right) = \dim_B \left(X_{\Omega}^{(S)} \right)$ *if and only if the tree of prefixes of* Ω *is spherically symmetric.*

Ban et al. [BHL00] studied the Minkowski dimension of $X_{2,3}$ and of some other multiplicative sets as pattern generating problem.

6 Remarks and Problems

6.1 Vector Valued Potential

The non-linear thermodynamic formalism can be generalized to vectorial potentials. Let φ, γ be two functions defined on $\mathcal{A}^\ell$ taking real values. Instead of considering the transfer operator $\mathcal{L}_s$ as defined in (13), we consider the following one:

$$\mathcal{L}_s\psi(a) = \sum_{j\in S} e^{s\varphi(a,j)+\gamma(a,j)}\psi(Ta, j), \ a \in S^{\ell-1}, \ s \in \mathbb{R}.$$

There exists a unique solution to the equation

$$(\mathcal{L}_s\psi)^{\frac{1}{q}} = \psi.$$

Then we define the pressure function $P_{\varphi,\gamma}(s)$ as indicated in $P_{\varphi,\gamma}(s)$. The function $s \mapsto P_{\varphi,\gamma}(s)$ is convex and analytic. Now, let $\underline{\varphi} = (\varphi_1, \dots, \varphi_d)$ be a function defined on S^ℓ taking values in $\mathbb{R}^d$. For $\underline{s} = (s_1, \dots, s_d) \in \mathbb{R}^d$, we consider the following transfer operator.

$$\mathcal{L}_{\underline{s}}\psi(a) = \sum_{j\in S} e^{\langle \underline{s}, \underline{\varphi}\rangle}\psi(Ta, j), \ a \in S^{\ell-1},$$

where $\langle \cdot, \cdot \rangle$ denotes the scalar product in $\mathbb{R}^d$. We denote the associated pressure function by $P(\underline{\varphi})(\underline{s})$. Then for any vectors $u, v \in \mathbb{R}^d$ the function

$$\mathbb{R} \ni s \longmapsto P(\underline{\varphi})(us + v)$$

is analytic and convex. We deduce from this that the function $\underline{s} \mapsto P(\underline{\varphi})(\underline{s})$ is infinitely differentiable and convex on $\mathbb{R}^d$.

Similarly, we define the level sets $E(\underline{\alpha})$ ($\underline{\alpha} \in \mathbb{R}^d$) of $\underline{\varphi}$. A vector version of Theorem 17 is stated by just replacing the derivative of the pressure function by the gradient.

6.2 Invariant Spectrum and Mixing Spectrum

The set $E_\Phi(\alpha)$ defined by (4) is not invariant. The size of the invariant part of $E_\Phi(\alpha)$ could be considered to be

$$d_{\mathrm{inv}}(\alpha) = \sup\{\dim^* \mu : \mu \text{ invariant}, \ \mu(E_\Phi(\alpha)) = 1\}.$$

The function $\alpha \mapsto d_{\text{inv}}(\alpha)$ is called the invariant spectrum of Φ. Similarly we define the mixing spectrum of Φ by

$$d_{\text{mix}}(\alpha) = \sup\{\dim^* \mu : \mu \text{ mixing}, \ \mu(E_\Phi(\alpha)) = 1\}.$$

Examples in [FSW00] show that it is possible to have

$$d_{\text{mix}}(\alpha) < d_{\text{inv}}(\alpha) < d_H(\alpha).$$

6.3 Semigroups

The semigroup $\{q^n\}_{n\geq 0}$ of $\mathbb{N}^*$ appeared in [FSW00, KPS12]. Other semigroup structures appeared in [PSSS00]. Combining the ideas in [PSSS00, FSW00], averages like

$$\lim_{n\to\infty} \frac{1}{n} \sum_{k=1}^{n} \varphi(x_k, x_{2k}, x_{3k})$$

can be treated [Wu13]. The Riesz product method used in [FLM12] is well adapted to the study of the special limit on Σ_2:

$$\lim_{n\to\infty} \frac{1}{n} \sum_{k=1}^{n} (2x_k - 1)(2x_{2k} - 1) \cdots (2x_{\ell k} - 1)$$

where $\ell \geq 2$ is any integer.

6.4 Subshifts of Finite Type

What we have presented is strictly restricted to the full shift dynamics. It is a challenging problem to study the dynamics of subshift of finite type and the dynamics with Markov property. New ideas are needed to deal with these dynamics. It is also a challenging problem to deal with potential depending more than one coordinates.

The doubling dynamics $Tx = 2x \bmod 1$ on the interval $[0, 1)$ is essentially a shift dynamics. Cookie cutters are the first interval maps coming into the mind after the doubling map. If the cookie cutter maps are not linear, it is a difficult problem. A cookie-cutter can be coded, but the non-linearity means that the derivative is a potential depending more than one codes.

Based on the computation made in [PS13], Liao and Rams [LR00] considered a special piecewise linear map of two branches defined on two intervals I_0 and I_1 and studied the following limit

$$\lim_{n\to\infty} \frac{1}{n}\sum_{k=1}^{n} 1_{I_1}(T^k x) 1_{I_1}(T^{2k} x).$$

The techniques presented in [FSW00] can be used to treat the problem for general piecewise *linear* cookie cutter dynamics [FLW00, Wu13].

6.5 Discontinuity of Spectrum for V-Statistics

The limit of V-statistics

$$\lim_{n\to\infty} n^{-r} \sum_{1\le i_1,\dots,i_r\le n} \Phi(T^{i_1}x,\dots,T^{i_r}x).$$

was studied in [FSW13] where it is proved that the multifractal spectrum of topological entropy of the above limit is expressed by an variational principle when the system satisfies the specification property. Unlike the classical case ($r = 1$) where the spectrum is an analytic function when Φ is Hölder continuous, the spectrum of the limit of higher order V-statistics ($r \ge 2$) may be discontinuous even for very regular kernel Φ. It is an interesting problem to determine the number of discontinuities. Rauch [Rau13] has recently established a variational principe relative to V-statistics.

6.6 Mutual Absolute Continuity of Two Riesz Products

Let us state two conjectures. See [BFP10] for the discussion on these conjectures.

Conjecture 1. Let μ_a *and* μ_b *be two Riesz products and let* $\omega := (\omega_n) \in G^{\mathbb{N}}$. *Then* $\mu_a \ll \mu_b \Rightarrow \mu_{a,\omega} \ll \mu_{b,\omega}$, *and* $\mu_a \perp \mu_b \Rightarrow \mu_{a,\omega} \perp \mu_{b,\omega}$.

For a function f defined on G, we use $\mathbb{E}f$ to denote the integral of f with respect to the Haar measure. The truthfulness is that the preceding conjecture implies the following one.

Conjecture 2. Let μ_a *and* μ_b *be two Riesz products. Then*

$$\prod_{n=1}^{\infty} \mathbb{E}\sqrt{(1+\mathrm{Re}\, a_n\gamma_n)(1+\mathrm{Re}\, b_n\gamma_n)} > 0 \Longrightarrow \mu_a \ll \mu_b;$$

$$\prod_{n=1}^{\infty} \mathbb{E}\sqrt{(1+\mathrm{Re}\, a_n\gamma_n)(1+\mathrm{Re}\, b_n\gamma_n)} = 0 \Longrightarrow \mu_a \perp \mu_b.$$

6.7 Doubling and Tripling

For any integer $m \geq 2$, we define the dynamics $\tau_m x = mx \pmod 1$ on $[0, 1)$. A typical couple of commuting transformations is the couple (τ_2, τ_3). Let us take, for example, $\Phi(x, y) = e^{2\pi i(ax+by)}$ with a, b being two fixed integers. We are then led to the multiple ergodic averages, a special case of (3),

$$A_n^{(2,3)}(x) := \frac{1}{n}\sum_{k=1}^{n} e^{2\pi i(a2^k+b3^k)x}. \tag{26}$$

This is an object not yet well studied in the literature (but if $a = 0$, we get a classical Birkhoff average). We propose to develop a thermodynamic formalism by studying Gibbs type measures which are weak limits $\mu_{s,t}$ $(s, t \in \mathbb{R})$ of

$$Z_n(s,t)^{-1}Q_n(x)dx$$

where

$$Q_n(x) := \prod_{k=1}^{n} e^{s\cos(2\pi(a2^k+b3^k)x)+t\sin(2\pi(a2^k+b3^k)x)}.$$

The pressure function defined by

$$P(s,t) := \lim_{n\to\infty}\frac{\log Z_n(s,t)}{n}$$

would be differentiable. But first we have to prove the existence of the limit defining $P(s,t)$.

More generally, let (c_n) be a sequence of complex numbers and (λ_n) a lacunary sequence of positive integers (by lacunary we mean $\inf_n \frac{\lambda_{n+1}}{\lambda_n} > 1$). We can consider the following weighted lacunary trigonometric averages

$$\frac{1}{n}\sum_{k=1}^{n} c_k e^{2\pi i\lambda_k x}.$$

Under the divisibility condition $\lambda_n | \lambda_{n+1}$, such averages and more general averages were studied in [Fan97]. For example, if $c_k = e^{2\pi i\omega_k}$ with (ω_k) being an i.i.d. sequence of Lebesgue distributed random variables, from the results obtained in [Fan97] we deduce that almost surely the pressure is well defined and equal to the following deterministic function

$$P(s,t) = \log\int_0^{2\pi} e^{\sqrt{t^2+s^2}\cos x}\frac{dx}{2\pi}.$$

Recall that

$$J_0(r) = \frac{1}{2\pi}\int_0^{2\pi} e^{r\cos x}dx = \sum_{n=0}^{\infty}\frac{r^{2n}}{(n!)^2 2^{2n}}.$$

is the Bessel function.

But, the condition $\lambda_n|\lambda_{n+1}$ for $\lambda_n = 2^n + 3^n$ is not satisfied. Neither the condition is satisfied for $\lambda_n = 2^n + 4^n$. No rigorous results are known for the multifractal analysis of the averages defined by (26).

As conjectured by Fürstenberg, the Lebesgue measure is the unique continuous probability measure which is both τ_2-invariant and τ_3-invariant. However, common τ_2- and τ_3-periodic points (different from the trivial one 0) do exists. Given two integers $n \geq 1$ and $m \geq 1$. We can prove that there is a point $x(\neq 0)$ which is n-periodic with respect to τ_2 and m-periodic with respect to τ_3 if and only if

$$(2^n - 1, 3^m - 1) > 1. \tag{27}$$

Let $d = (2^n - 1, 3^m - 1)$. When the above condition on GCD is satisfied, there are $d - 1$ such common periodic points. These common periodic points $x(\neq 0)$ are of the form $x = \frac{k}{2^n-1} = \frac{j}{3^m-1}$ for some $1 \leq k < 2^n - 2$ and $1 \leq j < 3^m - 2$. Actually choices for k are

$$1\cdot\frac{2^n-1}{d},\quad 2\cdot\frac{2^n-1}{d},\ldots,\ (d-1)\cdot\frac{2^n-1}{d}.$$

Choices for j are $1\cdot\frac{3^m-1}{d}$, $2\cdot\frac{3^m-1}{d}, \ldots, (d-1)\cdot\frac{3^m-1}{d}$. Thus the $d-1$ common periodic points are $\frac{1}{d}, \frac{2}{d}, \ldots, \frac{d-1}{d}$. For such a point x, the following limit exists

$$\lim_{N\to\infty} A_N^{(2,3)}(x) = A_{nm}^{(2,3)}(x).$$

Note that there is an infinite number of such couples n and m such that (27) holds. There would be some relation between these common periodic points and the multifractal behavior of $A_N^{(2,3)}(x)$.

References

[BHL00] Ban, J.C., Hu, W.G., Lin, S.S.: Pattern Generating Problems Arising in Multiplicative Interger Systems (preprint)

[BS07] Barral, J., Seuret, S.: Heterogeneous ubiquitous systems in $\mathbb{R}^d$ and Hausdorff dimension. Bull. Braz. Math. Soc. **38**(3), 467–515 (2007)

[BFP10] Barral, J., Fan, A.H., Peyrière, J.: Mesures engendrées par multiplication, Quelques Interactions Entre Analyse, Probabilités et Fractales, pp. 57–189. Panorama et Synthèse/SMF, No. 32 (2010)

[BPS99] Barreira, L., Pesin, Y., Schmeling, J.: Dimension and product structure of hyperbolic measures. Ann. Math. **149**(3), 755–783 (1999) (2nd ser)

[BSS02] Barreira, L., Saussol, B., Schmeling, J.: Higher dimensional multifractal analysis. J. Math. Pure. Appl. **81**, 67–91 (2002)

[BV06] Beresnevich, V., Velani, S.: A mass transference principle and the Duffin-Schaeffer conjecture for Hausdorff measures. Ann. Math. **164**, 971–992 (2006)

[Bou90] Bourgain, J.: Double recurrence and almost sure convergence. J. Reine. Angew. Math. **404**, 140–161 (1990)

[Bow73] Bowen, R.: Topological entropy for noncompact sets. Trans. Amer. Math. Soc. **184**, 125–136 (1973)

[BMP92] Brown, G., Michon, G., Peyrière, J.: On the multifractal analysis of measures. J. Statist. Phys. **66**, 775–790 (1992)

[CLP87] Collet, P., Lebowitz, J.L., Porzio, A.: The dimension spectrum of some dynamical systems. J. Stat. Phys. **47**, 609–644 (1987)

[LT84] Conze, J.P., Lesigne, E.: Théorèmes ergodiques pour des mesures diagonales. Bull. Soc. Math. France **112**(2), 143–175 (1984)

[Cut95] Cutler, C.: Connecting ergodicity and dimension in dynamical systems. Ergodic. Theor. Dynam. Syst. **118**, 393–410 (1995)

[Dur08] Durand, A.: Ubiquitous systems and metric number theory. Adv. Math. **218**, 368–394 (2008)

[Fal03] Falconer, K.J.: Fractal Geometry: Mathematical Foundations and Applications, 2nd ed. Wiley, Chichester (2003)

[Fan89] Fan, A.H.: Recouvrement aléatoire et Décompositions de Mesures. Publication d'Orsay, Orsay (1989)

[Fan93] Fan, A.H.: Quelques propriétés de produits de Riesz. Bull. Sci. Math. **117**(3), 421–439 (1993)

[Fan94a] Fan, A.H.: Sur les dimensions de mesures. Studia Math. **111**(1), 1–17 (1994)

[Fan94b] Fan, A.H.: On ergodicity and unidimensionality. Kyushu J. Math. **48**(2), 249–255 (1994)

[Fan94c] Fan, A.H.: Une formule approximative de dimension pour certains produits de Riesz. Monatsh. für Math. **118**, 83–89 (1994)

[Fan97] Fan, A.H.: Multifractal analysis of infinite products. J. Stat. Phys. **86**(5/6), 1313–1336 (1997)

[Fan00] Fan, A.H.: Individual behaviors of oriented walks. Stoc. Proc. Appl. Stoc. Proc. Appl. **90**, 263–275 (2000)

[FF99] Fan, A.H., Feng, D.J.: On the distribution of long-term time averages on symbolic space. J. Statist. Phys. **99**, 813–856 (1999)

[FFW01] Fan, A.H., Feng, D.J., Wu, J.: Recurrence, dimension and entropy. J. Lond. Math. Soc. **64**, 229–244 (2001)

[FLR02] Fan, A.H., Lau, K.S., Rao, H.: Relationship between different dimension of a measure. Monatsch. Math. **135**, 191–201 (2002)

[FLP08] Fan, A.H., Liao, L.M., Peyrière, J.: Generic points in systems of specification and Banach valued Birkhoff ergodic average. Discr. Cont. Dynam. Syst. **21**(4), 1103–1128 (2008)

[FLM12] Fan, A.-H., Liao, L., Ma, J.-H.: Level sets of multiple ergodic averages. Monatsh. Math. **168**(1), 17–26 (2012)

[FLW00] Fan, A.H., Liao, L.M., Wu, M.: Multifractal Analysis of Some Multiple Ergodic Averages in Linear Cookie-Cutter Dynamical Systems (preprint)

[FS03] Fan, A.H., Schmeling, J.: On fast Birkhoff averaging. Math. Proc. Camb. Phil. Soc. **135**, 443–467 (2003)

[FSW00] Fan, A.-H., Schmeling, J., Wu, M.: Multifractal Analysis of Some Multiple Ergodic Average (preprint)

[FSW13] Fan, A.-H., Schmeling, J., Wu, M.: Multifractal analysis of V-statistics, in further developments in fractals and related fields. Barral, J., Seuret, S (Eds.) Trends in Mathematics, pp. 135–151 (2013)

[FST00] Fan, A.H., Schmeling, J., Troubetzkoy, S.: A multifractal mass transference principle for Gibbs measures with applications to dynamical Diophantine approximation. Proc. London Math. Soc. **107**(3), 1173–1219 (2013)

[FW00] Fan, A.H., Wu, M.: Oriented Walks on Euclidean Spaces (preprint)

[Fel68] Feller, W.: An Introduction to Probability Theory and Its Applications. 3rd Edn, vol. 1. Wiley, New York (1968)

[FLW02] Feng, D.J., Lau, K.-S., Wu, J.: Ergodic limits on the conformal repellers. Adv. Math. **169**, 58–91 (2002)

[FP85] Frisch, U., Parisi, G.: Fully developped turbulence and intermittency in turbulence, and predictability in geophysical fluid dynamics and climate dymnamics. In: Ghil, M. (ed.) International School of Physics "Enrico Fermi": Course 88, p. 84. North Holland (1985)

[Für77] Fürstenberg, H.: Ergodic behavior of diagonal measures and a theorem of Szemerédi on arithmetic progressions. J. Anal. Math. **31**, 204–256 (1977)

[FW96] Fürstenberg, H., Weiss, B.: A mean ergodic theorem for $(1/N)\sum_{n=1}^{N} f(T^n x)g(T^{n^2} x)$. Convergence in ergodic theory and probability, Ohio State University, Mathematical Sciences Reseach Institute Publication, vol 5, pp. 193–227. de Gruyter (1996)

[Haa92] Haase, H.: A survey on the dimension of measures, topology, measures and fractals. Warnemünde Math. Res. **66**(1992), 66–75 (1991)

[HJKPS86] Halsey, T.C., Jensen, M.H., Kadanoff, L.P., Procaccia, I., Shraiman, B.I.: Fractal measures and their singularities: the characterisation of strange sets. Phys. Rev. A **33**, 1141 (1986)

[HP83] Hentschel, H.G., Procaccia, I.: The infinite number of generalized dimensions of fractals and strange attractors. Physica **8D**, 435 (1983)

[Heu98] Heurteaux, Y.: Estimations de la dimension inférieure et de la dimension supérieure des mesures. Ann. Inst. H. Poincaré Probab. Statist. **34**, 309–338 (1998)

[HZ66] Hewitt, E., Zuckerman, H.S.: Singular measures with absolutely continuous convolution squares. Proc. Camb. Phil. Soc. **62**, 399–420 (1966)

[HK05] Host, B., Kra, B.: Nonconventional ergodci averages and nilmanifolds. Ann. Math. **161**, 397–488 (2005)

[How94] Howroyd, J.D.: On the theory of Hausdorff measures in metric spaces. Thesis, University of London (1994)

[Jaf97] Jaffard, S.: Multifractal formalism for functions I. Results valid for all functions II. Self-similar functions. SIAM J. Math. Anal. **28**, 944–970, 971–998 (1997)

[Kah87] Kahane, J.-P.: Intervalles aléatoires et décomposition des mesures. C. R. Acad. Sci. Paris Sér. I **304**, 551–554 (1987)

[KP76] Kahane, J.-P., Peyrière, J.: Sur certaines martingales de Mandelbrot. Adv. Math. **22**, 131–145 (1976)

[Kak48] Kakutani, S.: On equivalence of infinite product measures. Ann. Math. **49**, 214–224 (1948)

[Kau94] Kaufman, R.: A "min-max" theorem on potentials. C. R. Acad. Sci. Paris Sér. I Math. **319**(8), 799–800 (1994)

[KPS12] Kenyon, R., Peres, Y., Solomyak, B.: Hausdorff dimension for fractals invariant under the multiplicative integers. Erg. Th. Dyn. Syst. **32**(5), 1567–1584 (2012)

[Kif12] Kifer, Y.: A nonconventional strong law of large numbers and fractal dimensions of some multiple recurrence sets. Stochast. Dyn. **12**(3), 21 (2012)

[Les89] Lesigne, E.: Theorems ergodiques pour une translation sur une nilvariete. Erg. Thm. Dyn. Syst. **9**(1), 115–126 (1989)

[LR00] Liao, L.M., Rams, M.: Multifractal Analysis of Some Multiple Ergodic Averages for the Systems with Non-constant Lyapunov Exponents Real Analysis Exchange, to appear

[Man72] Mandelbrot, B.: Possible refinement of the log-normal hypothesis concerning the distribution of energy dissipation in intermittent trubulence in statistical models and trubulence. In: Abstracts of the Symposium at University of California, San Diego 1972. Lecture Notes in Physics, pp. 333–351 (1971)

[Man74] Mandelbrot, B.: Multiplications aléatoires itérées et distributions invariantes par moyennes pondérée aléatoire. C. R. Acad. Sci. Paris **278**, 289–292 (1974)

[Man74] Mandelbrot, B.: Multiplications aléatoires itérées et distributions invariantes par moyennes pondérée aléatoire: quelques extensions. C. R. Acad. Sci. Paris **278**, 355–358 (1974)

[Mat95] Mattila, P.: Geometry of sets and measures in Euclidean spaces. Fractals and rectifiability. Cambridge Studies in Advanced Mathematics, vol. 44. Cambridge University Press, Cambridge (1995)

[Oli99] Olivier, E.: Multifractal analysis in symbolic dynamics and distribution of pointwise dimension for g-measures. Nonlinearity **12**, 1571–1585 (1999)

[Ols95] Olsen, L.: A multifractal formalism. Adv. Math. **116**, 92–195 (1995)

[PSSS00] Peres, Y., Schmeling, J., Seuret, S., Solomyak, B.: Hausdorff Dimension for Fractals Generated by Multiplication by 2 and 3 Israel Journal Math., to appear

[PS11-12] Peres, Y., Solomyak, B.: Dimension spectrum for a nonconventional erdofic average. Real Anal. Exch. **37**(2), 375–388 (2011/2012)

[PS13] Peres, Y., Solomyak, B.: The multiplicative golden mean shift has infinite Hausdorff measure: further developments in fractals and related fields. Barral, J., Seuret, S. (eds.) Trends in Mathematics, pp. 193–212 (2013)

[Pes97] Pesin, Y.: Dimension Theory in Dynamical Systems: Contemporary Views and Applications. The University of Chicago Press, Chicago (1997). Chicago lectures in Mathematics

[Pey75] Peyrière, J.: Étude de quelques propriétés des produits de Riesz. Ann. Inst. Fourier **25**, 127–169 (1975)

[Pey90] Peyrière, J.: Almost everywhere convergence of lacunary trigonometric series with respect to Riesz products. Aust. J. Math. Ser. A **48**, 376–383 (1990)

[Ran89] Rand, D.A.: The singularity spectrum $f(\alpha)$ for cookie-cutters. Ergod. Th. Dynam. Syst. **9**, 527–541 (1989)

[Rau13] Rauch, M.: A multivariate variational principle. Diploma Thesis, University of Jena (2013)

[Rue78] Ruelle, D.: Thermodynamic Formalism, Encyclopedia of Mathematics and its Applications, vol. 5. Addison-Weysley, Reading (1978)

[TV03] Takens, F., Verbitskiy, E.: On the variational principle for the topological entropy of certain non-compact sets. Ergod. Th. Dyn. Syst. **23**(1), 317–348 (2003)

[Tam95] Tamashira, M.: Dimensions in a separable metric space. Kyushu J. Math. **49**(1), 143–162 (1995)

[Wal82] Walters, P.: An Introduction to Ergodic Theory, GTM 79. Springer, New York (1982)

[Wu13] Wu, M.: Multifractal Analysis of Some Nonconventional Ergodic Averages. Ph. D. Thesis, University of Picardie, Jules Verne (2013)

[You82] Young, L.S.: Dimension, entropy and Lyapounov exponents. Erg. Th. Dyn. Syst. **2**, 109–124 (1982)

[Zie05] Ziegler, T.: A non-conventional ergodic theorem for a nilsystem. Erg. Th. Dyn. Syst. **25**(4), 1357–1370 (2005)

[Zyg68] Zygmund, A.: Trigonometric Series. Cambridge University Press, Cambridge (1968)

Heat Kernels on Metric Measure Spaces

Alexander Grigor'yan, Jiaxin Hu and Ka-Sing Lau

Contents

Supported by SFB 701 of the German Research Council and the Grants from the Department of Mathematics and IMS of CUHK.
Corresponding author. Supported by NSFC (Grant No. 11271122), SFB 701 and the HKRGC Grant of CUHK.
Supported by the HKRGC Grant and the Focus Investment Scheme of CUHK

A. Grigor'yan
Department of Mathematics, University of Bielefeld, 33501 Bielefeld, Germany

J. Hu (✉)
Department of Mathematical Sciences, Tsinghua University, Beijing 100084, China
e-mail: hujiaxin@mail.tsinghua.edu.cn

K.-S. Lau
Department of Mathematics, The Chinese University of Hong Kong, Shatin, N.T., Hong Kong

D.-J. Feng and K.-S. Lau (eds.), *Geometry and Analysis of Fractals*, Springer Proceedings in Mathematics & Statistics 88,
DOI 10.1007/978-3-662-43920-3_6

1 What is the Heat Kernel

In this section we shall discuss the notion of the heat kernel on a metric measure space (M, d, μ). Loosely speaking, a heat kernel $p_t(x, y)$ is a family of measurable functions in $x, y \in M$ for each $t > 0$ that is *symmetric*, *Markovian* and satisfies the *semigroup* property and the *approximation of identity* property. It turns out that the heat kernel coincides with the *integral kernel* of the *heat semigroup* associated with the *Dirichlet form* in $L^2(M, \mu)$.

Let us start with some basic examples of the heat kernels.

1.1 Examples of Heat Kernels

1.1.1 Heat Kernel in Euclidean Spaces

The classical Gauss-Weierstrass heat kernel is the following function

$$p_t(x, y) = \frac{1}{(4\pi t)^{n/2}} \exp\left(-\frac{|x-y|^2}{4t}\right), \tag{1.1}$$

where $x, y \in \mathbb{R}^n$ and $t > 0$. This function is a fundamental solution of the heat equation

$$\frac{\partial u}{\partial t} = \Delta u,$$

where $\Delta = \sum_{i=1}^n \frac{\partial^2}{\partial x_i^2}$ is the Laplace operator. Moreover, if f is a continuous bounded function on $\mathbb{R}^n$, then the function

$$u(t, x) = \int_{\mathbb{R}^n} p_t(x, y) f(y)\, dy$$

solves the Cauchy problem

$$\begin{cases} \frac{\partial u}{\partial t} = \Delta u, \\ u(0, x) = f(x). \end{cases}$$

This can be also written in the form

$$u(t, \cdot) = \exp(-t\mathcal{L}) f,$$

where $\mathcal{L}$ here is a self-adjoint extension of $-\Delta$ in $L^2(\mathbb{R}^n)$ and $\exp(-t\mathcal{L})$ is understood in the sense of the functional calculus of self-adjoint operators. That means that $p_t(x, y)$ is the integral kernel of the operator $\exp(-t\mathcal{L})$.

The function $p_t(x, y)$ has also a probabilistic meaning: it is the transition density of Brownian motion $\{X_t\}_{t \geq 0}$ in $\mathbb{R}^n$ (Fig. 1). The graph of $p_t(x, 0)$ as a function of x is shown here:

The term $\frac{|x-y|^2}{t}$ determines the *space/time scaling*: if $|x-y|^2 \leq Ct$, then $p_t(x, y)$ is comparable with $p_t(x, x)$, that is, the probability density in the $C\sqrt{t}$-neighborhood of x is nearly constant.

1.1.2 Heat Kernels on Riemannian Manifolds

Let (M, g) be a connected Riemannian manifold, and Δ be the Laplace-Beltrami operator on M. Then the heat kernel $p_t(x, y)$ can be defined as the integral kernel of the heat semigroup $\{\exp(-t\mathcal{L})\}_{t \geq 0}$, where $\mathcal{L}$ is the Dirichlet Laplace operator, that is, the minimal self-adjoint extension of $-\Delta$ in $L^2(M, \mu)$, and μ is the Riemannian volume. Alternatively, $p_t(x, y)$ is the minimal positive fundamental solution of the heat equation

$$\frac{\partial u}{\partial t} = \Delta u.$$

The function $p_t(x, y)$ can be used to define Brownian motion $\{X_t\}_{t \geq 0}$ on M. Namely, $\{X_t\}_{t \geq 0}$ is a *diffusion process* (that is, a Markov process with continuous trajectories), such that

$$\mathbb{P}_x(X_t \in A) = \int_A p_t(x, y)\, d\mu(y)$$

for any Borel set $A \subset M$ (Fig. 2).

Let $d(x, y)$ be the geodesic distance on M. It turns out that the *Gaussian* type space/time scaling $\frac{d^2(x,y)}{t}$ appears in heat kernel estimates on general Riemannian manifolds:

1. (*Varadhan*) For an arbitrary Riemannian manifold,

$$\log p_t(x, y) \sim -\frac{d^2(x, y)}{4t} \quad \text{as } t \to 0.$$

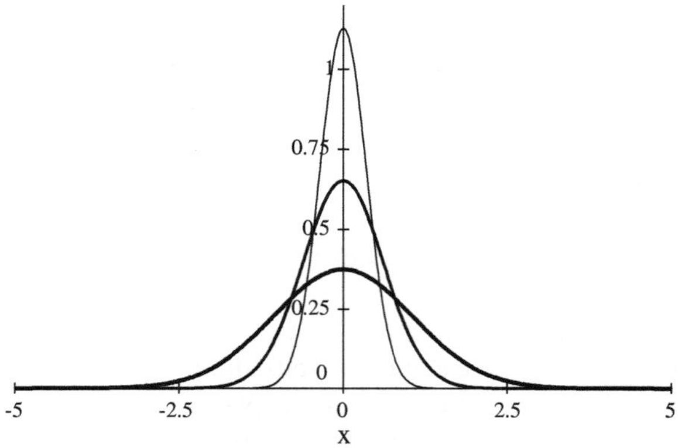

Fig. 1 The Gauss-Weierstrass heat kernel at different values of t

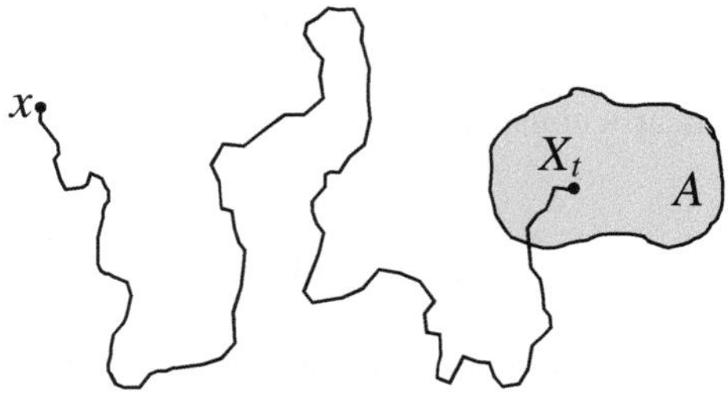

Fig. 2 The Brownian motion X_t hits a set A

2. (*Davies*) For an arbitrary manifold M, for any two measurable sets $A, B \subset M$

$$\int_A \int_B p_t(x, y)\, d\mu(x)\, d\mu(y) \le \sqrt{\mu(A)\,\mu(B)} \exp\left(-\frac{d^2(A, B)}{4t}\right).$$

Technically, all these results depend upon the property of the geodesic distance: $|\nabla d| \le 1$.

It is natural to ask the following question:

Are there settings where the space/time scaling is different from Gaussian?

1.1.3 Heat Kernels of Fractional Powers of Laplacian

Easy examples can be constructed using another operator instead of the Laplacian. As above, let $\mathcal{L}$ be the Dirichlet Laplace operator on a Riemannian manifold M, and consider the evolution equation

$$\frac{\partial u}{\partial t} + \mathcal{L}^{\beta/2} u = 0,$$

where $\beta \in (0,2)$. The operator $\mathcal{L}^{\beta/2}$ is understood in the sense of the functional calculus in $L^2(M,\mu)$. Let $p_t(x,y)$ be now the heat kernel of $\mathcal{L}^{\beta/2}$, that is, the integral kernel of $\exp\left(-t\mathcal{L}^{\beta/2}\right)$.

The condition $\beta < 2$ leads to the fact that the semigroup $\exp\left(-t\mathcal{L}^{\beta/2}\right)$ is *Markovian*, which, in particular, means that $p_t(x,y) > 0$ (if $\beta > 2$ then $p_t(x,y)$ may be signed). Using the techniques of subordinators, one obtains the following estimate for the heat kernel of $\mathcal{L}^{\beta/2}$ in $\mathbb{R}^n$:

$$p_t(x,y) \asymp \frac{C}{t^{n/\beta}}\left(1+\frac{|x-y|}{t^{1/\beta}}\right)^{-(n+\beta)} \asymp \frac{C}{t^{n/\beta}}\left(1+\frac{|x-y|^\beta}{t}\right)^{-\frac{n+\beta}{\beta}}. \quad (1.2)$$

(the symbol $\asymp$ means that both $\leq$ and $\geq$ are valid but with different values of the constant C).

The heat kernel of $\sqrt{\mathcal{L}} = (-\Delta)^{1/2}$ in $\mathbb{R}^n$ (that is, the case $\beta = 1$) is known explicitly:

$$p_t(x,y) = \frac{c_n}{t^n}\left(1+\frac{|x-y|^2}{t^2}\right)^{-\frac{n+1}{2}} = \frac{c_n t}{\left(t^2+|x-y|^2\right)^{\frac{n+1}{2}}},$$

where $c_n = \Gamma\left(\frac{n+1}{2}\right)/\pi^{(n+1)/2}$. This function coincides with the Poisson kernel in the half-space $\mathbb{R}^{n+1}_+$ and with the density of the Cauchy distribution in $\mathbb{R}^n$ with the parameter t.

As we have seen, the space/time scaling is given by the term $\frac{d^\beta(x,y)}{t}$ where $\beta < 2$. The heat kernel of the operator $\mathcal{L}^{\beta/2}$ is the transition density of a *symmetric stable process of index* β that belongs to the family of Lévy processes. The trajectories of this process are discontinuous, thus allowing jumps. The heat kernel $p_t(x,y)$ of such process is nearly constant in some $Ct^{1/\beta}$-neighborhood of y. If t is large, then

$$t^{1/\beta} \gg t^{1/2},$$

that is, this neighborhood is much larger than that for the diffusion process, which is not surprising because of the presence of jumps. The space/time scaling with $\beta < 2$ is called *super-Gaussian.*

1.1.4 Heat Kernels on Fractal Spaces

A rich family of heat kernels for diffusion processes has come from Analysis on *fractals*. Loosely speaking, fractals are subsets of $\mathbb{R}^n$ with certain self-similarity

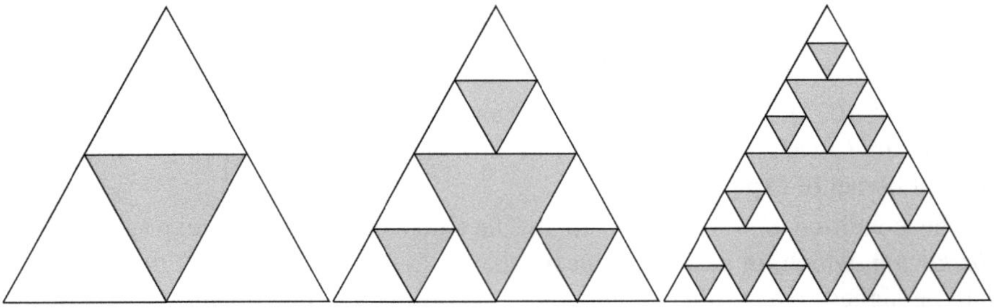

Fig. 3 Construction of the Sierpinski gasket

Fig. 4 The unbounded SG is obtained from SG by merging the latter (at the *left lower corner* of the diagram) with two shifted copies and then by repeating this procedure at larger scales

properties. One of the best understood fractals is *the Sierpinski gasket* (SG). The construction of the Sierpinski gasket is similar to the Cantor set: one starts with a triangle as a closed subset of $\mathbb{R}^2$, then eliminates the open middle triangle (shaded on the diagram), then repeats this procedure for the remaining triangles, and so on (Fig. 3).

Hence, SG is a compact connected subset of $\mathbb{R}^2$. The *unbounded* SG is obtained from SG by merging the latter (at the left lower corner of the next diagram) with two shifted copies and then by repeating this procedure at larger scales (Fig. 4).

Barlow and Perkins [BP88], Goldstein [Gol87] and Kusuoka [Kus87] have independently constructed by different methods a natural diffusion process on SG that has the same self-similarity as SG. Barlow and Perkins considered random walks on the graph approximations of SG and showed that, with an appropriate scaling, the random walks converge to a diffusion process. Moreover, they proved that this process has a transition density $p_t(x, y)$ with respect to a proper Hausdorff measure μ of SG, and that p_t satisfies the following elegant estimate:

$$p_t(x, y) \asymp \frac{C}{t^{\alpha/\beta}} \exp\left(-c\left(\frac{d^\beta(x, y)}{t}\right)^{\frac{1}{\beta-1}}\right), \tag{1.3}$$

where $d(x, y) = |x - y|$ and

$$\alpha = \dim_H SG = \frac{\log 3}{\log 2}, \quad \beta = \frac{\log 5}{\log 2} > 2.$$

Similar estimates were proved by Barlow and Bass for other families of fractals, including Sierpinski carpets, and the parameters α and β in (1.3) are determined by the intrinsic properties of the fractal. In all cases, α is the Hausdorff dimension (which is also called the *fractal dimension*). The parameter β, that is called the *walk dimension*, is larger than 2 in all interesting examples.

The heat kernel $p_t(x, y)$, satisfying (1.3) is nearly constant in some $Ct^{1/\beta}$-neighborhood of y. If t is large, then

$$t^{1/\beta} \ll t^{1/2},$$

that is, this neighborhood is much smaller than that for the diffusion process, which is due to the presence of numerous holes-obstacles that the Brownian particle must bypass. The space/time scaling with $\beta > 2$ is called *sub-Gaussian*.

1.1.5 Summary of Examples

Observe now that in all the above examples, the heat kernel estimates can be unified as follows:

$$p_t(x, y) \asymp \frac{C}{t^{\alpha/\beta}} \Phi\left(c\frac{d(x, y)}{t^{1/\beta}}\right), \tag{1.4}$$

where α, β are positive parameters and $\Phi(s)$ is a positive decreasing function on $[0, +\infty)$. For example, the Gauss-Weierstrass function (1.1) satisfies (1.4) with $\alpha = n, \beta = 2$ and

$$\Phi(s) = \exp\left(-s^2\right)$$

(Gaussian estimate).

The heat kernel (1.2) of the symmetric stable process in $\mathbb{R}^n$ satisfies (1.4) with $\alpha = n, 0 < \beta < 2$, and

$$\Phi(s) = (1 + s)^{-(\alpha+\beta)}$$

(super-Gaussian estimate).

The heat kernel (1.3) of diffusions on fractals satisfies (1.4) with $\beta > 2$ and

$$\Phi(s) = \exp\left(-s^{\frac{\beta}{\beta-1}}\right)$$

(sub-Gaussian estimate).

There are at least two questions related to the estimates of the type (1.4):

1. What values of the parameters α, β and what functions Φ can actually occur in the estimate (1.4)?
2. How to obtain estimates of the type (1.4)?

To give these questions a precise meaning, we must define what is a heat kernel.

1.2 Abstract Heat Kernels

Let (M, d) be a locally compact, separable metric space and let μ be a Radon measure on M with full support. The triple (M, d, μ) will be called a *metric measure space.*

Definition 1.1 (*heat kernel*) A family $\{p_t\}_{t>0}$ of measurable functions $p_t(x, y)$ on $M \times M$ is called a *heat kernel* if the following conditions are satisfied, for μ-almost all $x, y \in M$ and all $s, t > 0$:

(i) Positivity: $p_t(x, y) \geq 0$.
(ii) The total mass inequality:

$$\int_M p_t(x, y)d\mu(y) \leq 1.$$

(iii) Symmetry: $p_t(x, y) = p_t(y, x)$.
(iv) The semigroup property:

$$p_{s+t}(x, y) = \int_M p_s(x, z)p_t(z, y)d\mu(z).$$

(v) Approximation of identity: for any $f \in L^2 := L^2(M, \mu)$,

$$\int_M p_t(x, y)f(y)d\mu(y) \xrightarrow{L^2} f(x) \text{ as } t \to 0+.$$

If in addition we have, for all $t > 0$ and almost all $x \in M$,

$$\int_M p_t(x, y)d\mu(y) = 1,$$

then the heat kernel p_t is called *stochastically complete* (or *conservative*).

1.3 Heat Semigroups

Any heat kernel gives rise to the family of operators $\{P_t\}_{t\geq 0}$ where $P_0 = \text{id}$ and P_t for $t > 0$ is defined by

$$P_t f(x) = \int_M p_t(x, y) f(y) d\mu(y),$$

where f is a measurable function on M. It follows from (i) –(ii) that the operator P_t is *Markovian*, that is, $f \geq 0$ implies $P_t f \geq 0$ and $f \leq 1$ implies $P_t f \leq 1$. It follows that P_t is a bounded operator in L^2 and, moreover, is a contraction, that is, $\|P_t f\|_2 \leq \|f\|_2$.

The symmetry property (iii) implies that the operator P_t is *symmetric* and, hence, self-adjoint. The semigroup property (iv) implies that $P_t P_s = P_{t+s}$, that is, the family $\{P_t\}_{t\geq 0}$ is a *semigroup* of operators. It follows from (v) that

$$s\text{-}\lim_{t\to 0} P_t = \text{id} = P_0$$

where s-lim stands for the *strong* limit. Hence, $\{P_t\}_{t\geq 0}$ is a strongly continuous, symmetric, Markovian semigroup in L^2. In short, we call that $\{P_t\}$ is a *heat semigroup*.

Conversely, if $\{P_t\}$ is a heat semigroup and if it has an integral kernel $p_t(x, y)$, then the latter is a heat kernel in the sense of the above Definition.

Given a heat semigroup P_t in L^2, define the *infinitesimal generator* $\mathcal{L}$ of the semigroup by

$$\mathcal{L}f := \lim_{t\to 0} \frac{f - P_t f}{t},$$

where the limit is understood in the L^2-norm. The *domain* $\text{dom}(\mathcal{L})$ of the generator $\mathcal{L}$ is the space of functions $f \in L^2$ for which the above limit exists. By the Hille–Yosida theorem, $\text{dom}(\mathcal{L})$ is dense in L^2. Furthermore, $\mathcal{L}$ is a self-adjoint, positive definite operator, which immediately follows from the fact that the semigroup $\{P_t\}$ is self-adjoint and contractive. Moreover, P_t can be recovered from $\mathcal{L}$ as follows

$$P_t = \exp(-t\mathcal{L}),$$

where the right hand side is understood in the sense of spectral theory.

Heat kernels and heat semigroups arise naturally from Markov processes. Let $(\{X_t\}_{t\geq 0}, \{\mathbb{P}_x\}_{x\in M})$ be a Markov process on M, that is reversible with respect to measure μ. Assume that it has the transition density $p_t(x, y)$, that is, a function such that, for all $x \in M, t > 0$, and all Borel sets $A \subset M$,

$$\mathbb{P}_x (X_t \in A) = \int_M p_t (x, y) \, d\mu (y) \, .$$

Then $p_t (x, y)$ is a heat kernel in the sense of the above Definition.

1.4 Dirichlet Forms

Given a heat semigroup $\{P_t\}$ on a metric measure space (M, d, μ), for any $t > 0$, we define a bilinear form $\mathcal{E}_t$ on L^2 by

$$\mathcal{E}_t (u, v) := \left(\frac{u - P_t u}{t}, v \right) = \frac{1}{t} ((u, v) - (P_t u, v)),$$

where $(\cdot, \cdot)$ is the inner product in L^2. Since P_t is symmetric, the form $\mathcal{E}_t$ is also symmetric. Since P_t is a contraction, it follows that

$$\mathcal{E}_t (u) := \mathcal{E}_t (u, u) = \frac{1}{t} ((u, u) - (P_t u, u)) \geq 0,$$

that is, $\mathcal{E}_t$ is a positive definite form.

In terms of the spectral resolution $\{E_\lambda\}$ of the generator $\mathcal{L}$, $\mathcal{E}_t$ can be expressed as follows

$$\begin{aligned} \mathcal{E}_t (u) &= \frac{1}{t} ((u, u) - (P_t u, u)) = \frac{1}{t} \left(\int_0^\infty d\|E_\lambda u\|_2^2 - \int_0^\infty e^{-t\lambda} d\|E_\lambda u\|_2^2 \right) \\ &= \int_0^\infty \frac{1 - e^{-t\lambda}}{t} d\|E_\lambda u\|_2^2, \end{aligned}$$

which implies that $\mathcal{E}_t (u)$ is decreasing in t, since the function $t \mapsto \frac{1-e^{-t\lambda}}{t}$ is decreasing. Define for any $u \in L^2$

$$\mathcal{E} (u) = \lim_{t \downarrow 0} \mathcal{E}_t (u)$$

where the limit (finite or infinite) exists by the monotonicity, so that $\mathcal{E} (u) \geq \mathcal{E}_t (u)$. Since $\frac{1-e^{-t\lambda}}{t} \to \lambda$ as $t \to 0$, we have

$$\mathcal{E}(u) = \int_0^\infty \lambda d\|E_\lambda u\|_2^2.$$

Set

$$\mathcal{F} := \{u \in L^2 : \mathcal{E}(u) < \infty\} = \operatorname{dom}\left(\mathcal{L}^{1/2}\right) \supset \operatorname{dom}(\mathcal{L})$$

and define a bilinear form $\mathcal{E}(u, v)$ on $\mathcal{F}$ by the polarization identity

$$\mathcal{E}(u, v) := \frac{1}{4}\left(\mathcal{E}(u + v) - \mathcal{E}(u - v)\right),$$

which is equivalent to

$$\mathcal{E}(u, v) = \lim_{t \to 0} \mathcal{E}_t(u, v).$$

Note that $\mathcal{F}$ contains $\operatorname{dom}(\mathcal{L})$. Indeed, if $u \in \operatorname{dom}(\mathcal{L})$, then we have for all $v \in L^2$

$$\lim_{t \to 0} \mathcal{E}_t(u, v) = \left(\lim_{t \to 0} \frac{u - P_t u}{t}, v\right) = (\mathcal{L}u, v).$$

Setting $v = u$ we obtain $u \in \mathcal{F}$. Then choosing any $v \in \mathcal{F}$ we obtain the identity

$$\mathcal{E}(u, v) = (\mathcal{L}u, v) \text{ for all } u \in \operatorname{dom}(\mathcal{L}) \text{ and } v \in \mathcal{F}.$$

The space $\mathcal{F}$ is naturally endowed with the inner product

$$[u, v] := (u, v) + \mathcal{E}(u, v).$$

It is possible to show that the form $\mathcal{E}$ is *closed*, that is, the space $\mathcal{F}$ is *Hilbert*. Furthermore, $\operatorname{dom}(\mathcal{L})$ is dense in $\mathcal{F}$.

The fact that P_t is Markovian implies that the form $\mathcal{E}$ is also *Markovian*, that is

$$u \in \mathcal{F} \Rightarrow \widetilde{u} := \min(u_+, 1) \in \mathcal{F} \text{ and } \mathcal{E}(\widetilde{u}) \leq \mathcal{E}(u).$$

Indeed, let us first show that for any $u \in L^2$

$$\mathcal{E}_t(u_+) \leq \mathcal{E}_t(u).$$

We have

$$\mathcal{E}_t(u) = \mathcal{E}_t(u_+ - u_-) = \mathcal{E}_t(u_+) + \mathcal{E}_t(u_-) - 2\mathcal{E}_t(u_+, u_-) \geq \mathcal{E}_t(u_+)$$

because $\mathcal{E}_t(u_-) \geq 0$ and

$$\mathcal{E}_t(u_+, u_-) = \frac{1}{t}(u_+, u_-) - \frac{1}{t}(P_t u_+, u_-) \le 0.$$

Assuming $u \in \mathcal{F}$ and letting $t \to 0$, we obtain

$$\mathcal{E}(u_+) = \lim_{t\to 0} \mathcal{E}_t(u_+) \le \lim_{t\to 0} \mathcal{E}_t(u) = \mathcal{E}(u) < \infty$$

whence $\mathcal{E}(u_+) \le \mathcal{E}(u)$ and, hence, $u_+ \in \mathcal{F}$.

Similarly one proves that $\widetilde{u} = \min(u_+, 1)$ belongs to $\mathcal{F}$ and $\mathcal{E}(\widetilde{u}) \le \mathcal{E}(u_+)$.

Conclusion *Hence,* $(\mathcal{E}, \mathcal{F})$ *is a* Dirichlet form, *that is, a bilinear, symmetric, positive definite, closed, densely defined form in* L^2 *with Markovian property.*

If the heat semigroup is defined by means of a heat kernel p_t, then $\mathcal{E}_t$ can be equivalently defined by

$$\mathcal{E}_t(u) = \frac{1}{2t}\int_M\int_M (u(x) - u(y))^2 p_t(x,y)d\mu(y)d\mu(x) + \frac{1}{t}\int_M (1 - P_t 1(x))\, u^2(x)d\mu(x). \tag{1.5}$$

Indeed, we have

$$\begin{aligned} u(x) - P_t u(x) &= u(x) P_t 1(x) - P_t u(x) + (1 - P_t 1(x))\, u(x) \\ &= \int_M (u(x) - u(y))\, p_t(x,y)d\mu(y) + (1 - P_t 1(x))\, u(x), \end{aligned}$$

whence

$$\mathcal{E}_t(u) = \frac{1}{t}\int_M\int_M (u(x) - u(y))\, u(x) p_t(x,y)d\mu(y)d\mu(x) + \frac{1}{t}\int_M (1 - P_t 1(x))\, u^2(x)d\mu(x).$$

Interchanging the variables x and y in the first integral and using the symmetry of the heat kernel, we obtain also

$$\mathcal{E}_t(u) = \frac{1}{t}\int_M\int_M (u(y) - u(x))\, u(y) p_t(x,y)d\mu(y)d\mu(x) + \frac{1}{t}\int_M (1 - P_t 1(x))\, u^2(x)d\mu(x),$$

and (1.5) follows by adding up the two previous lines.

Since $P_t 1 \leq 1$, the second term in the right hand side of (1.5) is non-negative. If the heat kernel is stochastically complete, that is, $P_t 1 = 1$, then that term vanishes and we obtain

$$\mathcal{E}_t(u) = \frac{1}{2t} \int_M \int_M (u(x) - u(y))^2 \, p_t(x, y) d\mu(y) d\mu(x). \tag{1.6}$$

Definition 1.2 The form $(\mathcal{E}, \mathcal{F})$ is called *local* if $\mathcal{E}(u, v) = 0$ whenever the functions $u, v \in \mathcal{F}$ have compact disjoint supports. The form $(\mathcal{E}, \mathcal{F})$ is called *strongly local* if $\mathcal{E}(u, v) = 0$ whenever the functions $u, v \in \mathcal{F}$ have compact supports and $u \equiv \text{const}$ in an open neighborhood of $\operatorname{supp} v$.

For example, if $p_t(x, y)$ is the heat kernel of the Laplace-Beltrami operator on a complete Riemannian manifold, then the associated Dirichlet form is given by

$$\mathcal{E}(u, v) = \int_M \langle \nabla u, \nabla v \rangle d\mu, \tag{1.7}$$

and $\mathcal{F}$ is the Sobolev space $W_2^1(M)$. Note that this Dirichlet form is strongly local because $u = \text{const}$ on $\operatorname{supp} v$ implies $\nabla u = 0$ on $\operatorname{supp} v$ and, hence, $\mathcal{E}(u, v) = 0$.

If $p_t(x, y)$ is the heat kernel of the symmetric stable process of index β in $\mathbb{R}^n$, that is, $\mathcal{L} = (-\Delta)^{\beta/2}$, then

$$\mathcal{E}(u, v) = c_{n,\beta} \int_{\mathbb{R}^n} \int_{\mathbb{R}^n} \frac{(u(x) - u(y))(v(x) - v(y))}{|x - y|^{n+\beta}} dx dy,$$

and $\mathcal{F}$ is the Besov space $B_{2,2}^{\beta/2}(\mathbb{R}^n) = \{u \in L^2 : \mathcal{E}(u, u) < \infty\}$. This form is clearly non-local.

Denote by $C_0(M)$ the space of continuous functions on M with compact supports, endowed with sup-norm.

Definition 1.3 The form $(\mathcal{E}, \mathcal{F})$ is called *regular* if $\mathcal{F} \cap C_0(M)$ is dense both in $\mathcal{F}$ (with $[\cdot, \cdot]$-norm) and in $C_0(M)$ (with sup-norm).

All the Dirichlet forms in the above examples are regular.

Assume that we are given a Dirichlet form $(\mathcal{E}, \mathcal{F})$ in $L^2(M, \mu)$. Then one can define the generator $\mathcal{L}$ of $(\mathcal{E}, \mathcal{F})$ by the identity

$$(\mathcal{L}u, v) = \mathcal{E}(u, v) \quad \text{for all } u \in \operatorname{dom}(\mathcal{L}), \ v \in \mathcal{F}, \tag{1.8}$$

where $\operatorname{dom}(\mathcal{L}) \subset \mathcal{F}$ must satisfy one of the following two equivalent requirements:

1. $\operatorname{dom}(\mathcal{L})$ is a maximal possible subspace of $\mathcal{F}$ such that (1.8) holds
2. $\mathcal{L}$ is a densely defined self-adjoint operator.

Clearly, $\mathcal{L}$ is positive definite so that spec$\mathcal{L} \subset [0, +\infty)$. Hence, the family of operators $P_t = e^{-t\mathcal{L}}$, $t \geq 0$, forms a strongly continuous, symmetric, contraction semigroup in L^2. Moreover, using the Markovian property of the Dirichlet form $(\mathcal{E}, \mathcal{F})$, it is possible to prove that $\{P_t\}$ is Markovian, that is, $\{P_t\}$ is a heat semigroup. The question whether and when P_t has the heat kernel requires a further investigation.

1.5 More Examples of Heat Kernels

Let us give some examples of stochastically complete heat kernels that do not satisfy (1.4).

Example 1.4 (*A frozen heat kernel*) Let M be a countable set and let $\{x_k\}_{k=1}^{\infty}$ be the sequence of all distinct points from M. Let $\{\mu_k\}_{k=1}^{\infty}$ be a sequence of positive reals and define measure μ on M by $\mu(\{x_k\}) = \mu_k$. Define a function $p_t(x, y)$ on $M \times M$ by

$$p_t(x, y) = \begin{cases} \frac{1}{\mu_k}, & x = y = x_k \\ 0, & \text{otherwise.} \end{cases}$$

It is easy to check that $p_t(x, y)$ is a heat kernel. For example, let us check the approximation of identity: for any function $f \in L^2(M, \mu)$, we have

$$P_t f(x) = \int_M p_t(x, y) f(y)\, d\mu(y) = p_t(x, x) f(x) \mu(\{x\}) = f(x).$$

This identity also implies the stochastic completeness. The Dirichlet form is

$$\mathcal{E}(f) = \lim_{t \to 0} \left(\frac{f - P_t f}{t}, f \right) = 0.$$

The Markov process associated with the frozen heat kernel is very simple: $X_t = X_0$ for all $t \geq 0$ so that it is a frozen diffusion.

Example 1.5 (*The heat kernel in* $\mathbb{H}^3$) The heat kernel of the Laplace-Beltrami operator on the 3-dimensional hyperbolic space $\mathbb{H}^3$ is given by the formula

$$p_t(x, y) = \frac{1}{(4\pi t)^{3/2}} \frac{r}{\sinh r} \exp\left(-\frac{r^2}{4t} - t \right),$$

where $r = d(x, y)$ is the geodesic distance between x, y. The Dirichlet form is given by (1.7).

Example 1.6 (*The Mehler heat kernel*) Let $M = \mathbb{R}$, measure μ be defined by

$$d\mu = e^{x^2} dx,$$

and let $\mathcal{L}$ be given by

$$\mathcal{L} = -e^{-x^2} \frac{d}{dx} \left(e^{x^2} \frac{d}{dx} \right) = -\frac{d^2}{dx^2} - 2x \frac{d}{dx}.$$

Then the heat kernel of $\mathcal{L}$ is given by the formula

$$p_t(x, y) = \frac{1}{(2\pi \sinh 2t)^{1/2}} \exp\left(\frac{2xye^{-2t} - x^2 - y^2}{1 - e^{-4t}} - t \right).$$

The associated Dirichlet form is also given by (1.7).

Similarly, for the measure

$$d\mu = e^{-x^2} dx$$

and for the operator

$$\mathcal{L} = -e^{x^2} \frac{d}{dx} \left(e^{-x^2} \frac{d}{dx} \right) = -\frac{d^2}{dx^2} + 2x \frac{d}{dx},$$

we have

$$p_t(x, y) = \frac{1}{(2\pi \sinh 2t)^{1/2}} \exp\left(\frac{2xye^{-2t} - \left(x^2 + y^2\right) e^{-4t}}{1 - e^{-4t}} + t \right).$$

2 Necessary Conditions for Heat Kernel Bounds

In this Chapter we assume that $p_t(x, y)$ is a heat kernel on a metric measure space (M, d, μ) that satisfies certain upper and/or lower estimates, and state the consequences of these estimates. The reader may consult [GK08, GHL03] or [GHL09] for the proofs.

Fix two positive parameters α and β, and let $\Phi : [0, +\infty) \to [0, +\infty)$ be a monotone decreasing function. We will consider the bounds of the heat kernel via the function $\frac{1}{t^{\alpha/\beta}} \Phi \left(\frac{d(x,y)}{t^{1/\beta}} \right)$.

2.1 *Identifying Φ in the Non-local Case*

Theorem 2.1 (Grigor'yan and Kumagai [GK08]) *Let $p_t(x, y)$ be a heat kernel on (M, d, μ).*

(a) If the heat kernel satisfies the estimate

$$p_t(x,y) \leq \frac{1}{t^{\alpha/\beta}} \Phi\left(\frac{d(x,y)}{t^{1/\beta}}\right),$$

for all $t > 0$ and almost all $x, y \in M$, then either the associated Dirichlet form $(\mathcal{E}, \mathcal{F})$ is local or

$$\Phi(s) \geq c(1+s)^{-(\alpha+\beta)}$$

for all $s > 0$ and some $c > 0$.

(b) *If the heat kernel satisfies the estimate*

$$p_t(x,y) \geq \frac{1}{t^{\alpha/\beta}} \Phi\left(\frac{d(x,y)}{t^{1/\beta}}\right),$$

then we have

$$\Phi(s) \leq C(1+s)^{-(\alpha+\beta)}$$

for all $s > 0$ and some $C > 0$.

(c) *Consequently, if the heat kernel satisfies the estimate*

$$p_t(x,y) \asymp \frac{C}{t^{\alpha/\beta}} \Phi\left(c\frac{d(x,y)}{t^{1/\beta}}\right),$$

then either the Dirichlet form $\mathcal{E}$ is local or

$$\Phi(s) \asymp (1+s)^{-(\alpha+\beta)}.$$

2.2 Volume of Balls

Denote by $B(x,r)$ a metric ball in (M,d), that is

$$B(x,r) := \{y \in M : d(x,y) < r\}.$$

Theorem 2.2 (Grigor'yan et al. [GHL03]) *Let p_t be a heat kernel on (M,d,μ). Assume that it is stochastically complete and that it satisfies the two-sided estimate*

$$p_t(x,y) \asymp \frac{C}{t^{\alpha/\beta}} \Phi\left(c\frac{d(x,y)}{t^{1/\beta}}\right). \tag{2.1}$$

Then, for all $x \in M$ and $r > 0$,

$$\mu(B(x,r)) \asymp r^{\alpha},$$

that is, μ is α-regular.

Consequently, $\dim_H(M,d)=\alpha$ and $\mu \asymp H^\alpha$ on all Borel subsets of M, where H^α is the α-dimensional Hausdorff measure in M.

In particular, the parameter α is the invariant of the metric space (M,d), and measure μ is determined (up to a factor $\asymp 1$) by the metric space (M,d).

2.3 Besov Spaces

Fix $\alpha > 0$, $\sigma > 0$. We introduce the following seminorms on $L^2 = L^2(M,\mu)$:

$$N_{2,\infty}^{\alpha,\sigma}(u) = \sup_{0<r\leq 1} \frac{1}{r^{\alpha+2\sigma}} \iint\limits_{\{x,y\in M:d(x,y)<r\}} |u(x)-u(y)|^2\, d\mu(x)d\mu(y), \tag{2.2}$$

and

$$N_{2,2}^{\alpha,\sigma}(u) = \int_0^\infty \frac{dr}{r}\frac{1}{r^{\alpha+2\sigma}} \iint\limits_{\{x,y\in M:d(x,y)<r\}} |u(x)-u(y)|^2\, d\mu(x)d\mu(y). \tag{2.3}$$

Define the space $\Lambda_{2,\infty}^{\alpha,\sigma}$ by

$$\Lambda_{2,\infty}^{\alpha,\sigma} = \left\{u \in L^2 : N_{2,\infty}^{\alpha,\sigma}(u) < \infty\right\},$$

and the norm by

$$\|u\|^2_{\Lambda_{2,\infty}^{\alpha,\sigma}} = \|u\|_2^2 + N_{2,\infty}^{\alpha,\sigma}(u).$$

Similarly, one defines the space $\Lambda_{2,2}^{\alpha,\sigma}$. More generally one can define $\Lambda_{p,q}^{\alpha,\sigma}$ for $p \in [1,+\infty)$ and $q \in [1,+\infty]$.

In the case of $\mathbb{R}^n$, we have the following relations

$$\begin{aligned}
\Lambda_{p,q}^{n,\sigma}\left(\mathbb{R}^n\right) &= B_{p,q}^{\sigma}\left(\mathbb{R}^n\right),\ 0<\sigma<1,\\
\Lambda_{2,\infty}^{n,1}\left(\mathbb{R}^n\right) &= W_p^1\left(\mathbb{R}^n\right),\\
\Lambda_{2,2}^{n,1}\left(\mathbb{R}^n\right) &= \{0\},\\
\Lambda_{p,q}^{n,\sigma}\left(\mathbb{R}^n\right) &= \{0\},\ \sigma>1.
\end{aligned}$$

where $B_{p,q}^{\sigma}$ is the Besov space and W_p^1 is the Sobolev space. The spaces $\Lambda_{p,q}^{\alpha,\sigma}$ will also be called Besov spaces.

Theorem 2.3 (Jonsson [Jon96], Pietruska-Pałuba [Pie00], Grigor'yan et al. [GHL03]) *Let p_t be a heat kernel on (M,d,μ). Assume that it is stochastically* *complete and that it satisfies the following estimate: for all $t > 0$ and almost all $x, y \in M$,*

$$\frac{1}{t^{\alpha/\beta}}\Phi_1\left(\frac{d(x,y)}{t^{1/\beta}}\right) \le p_t(x,y) \le \frac{1}{t^{\alpha/\beta}}\Phi_2\left(\frac{d(x,y)}{t^{1/\beta}}\right), \tag{2.4}$$

where α, β be positive constants, and Φ_1, Φ_2 are monotone decreasing functions from $[0,+\infty)$ to $[0,+\infty)$ such that $\Phi_1(s) > 0$ for some $s > 0$ and

$$\int_0^\infty s^{\alpha+\beta}\Phi_2(s)\frac{ds}{s} < \infty. \tag{2.5}$$

Then, for any $u \in L^2$,

$$\mathcal{E}(u) \asymp N_{2,\infty}^{\alpha,\beta/2}(u),$$

and, consequently, $\mathcal{F} = \Lambda_{2,\infty}^{\alpha,\beta/2}$.

By Theorem 2.1, the upper bound in (2.4) implies that either $(\mathcal{E},\mathcal{F})$ is local or

$$\Phi_2(s) \ge c(1+s)^{-(\alpha+\beta)}.$$

Since the latter contradicts condition (2.5), the form $(\mathcal{E},\mathcal{F})$ must be local. For non-local forms the statement is not true. For example, for the operator $(-\Delta)^{\beta/2}$ in $\mathbb{R}^n$, we have $\mathcal{F} = B_{2,2}^{\beta/2} = \Lambda_{2,2}^{n,\beta/2}$ that is strictly smaller than $B_{2,\infty}^{\beta/2} = \Lambda_{2,\infty}^{n,\beta/2}$. This case will be covered by the following theorem.

Theorem 2.4 (Stós [Sto00]) *Let p_t be a stochastically complete heat kernel on (M,d,μ) satisfying estimate* (2.4) *with functions*

$$\Phi_1(s) \asymp \Phi_2(s) \asymp (1+s)^{-(\alpha+\beta)}.$$

Then, for any $u \in L^2$,

$$\mathcal{E}(u) \asymp N_{2,2}^{\alpha,\beta/2}(u).$$

Consequently, we have $\mathcal{F} = \Lambda_{2,2}^{\alpha,\beta/2}$.

2.4 Subordinated Semigroups

Let $\mathcal{L}$ be the generator of a heat semigroup $\{P_t\}$. Then, for any $\delta \in (0,1)$, the operator $\mathcal{L}^\delta$ is also a generator of a heat semigroup, that is, the semigroup $\left\{e^{-t\mathcal{L}^\sigma}\right\}$

is a heat semigroup. Furthermore, there is the following relation between the two semigroups

$$e^{-t\mathcal{L}^{\delta}} = \int_0^{\infty} e^{-s\mathcal{L}} \eta_t(s)\, ds,$$

where $\eta_t(s)$ is a *subordinator* whose Laplace transform is given by

$$e^{-t\lambda^{\delta}} = \int_0^{\infty} e^{-s\lambda} \eta_t(s)\, ds, \ \lambda > 0.$$

Using the known estimates for $\eta_t(s)$, one can obtain the following result.

Theorem 2.5 *Let a heat kernel p_t satisfy the estimate* (2.4) *where $\Phi_1(s) > 0$ for some $s > 0$ and*

$$\int_0^{\infty} s^{\alpha+\beta'} \Phi_2(s) \frac{ds}{s} < \infty,$$

where $\beta' = \delta\beta$, $0 < \delta < 1$. Then the heat kernel $q_t(x, y)$ of operator $\mathcal{L}^{\delta}$ satisfies the estimate

$$q_t(x, y) \asymp \frac{1}{t^{\alpha/\beta'}} \left(1 + \frac{d(x, y)}{t^{1/\beta'}}\right)^{-(\alpha+\beta')} \asymp \min\left(t^{-\alpha/\beta'}, \frac{t}{d(x, y)^{\alpha+\beta'}}\right),$$

for all $t > 0$ and almost all $x, y \in M$.

2.5 The Walk Dimension

It follows from definition that the Besov seminorm

$$N_{2,\infty}^{\alpha,\sigma}(u) = \sup_{0<r\leq 1} \frac{1}{r^{\alpha+2\sigma}} \iint_{\{x,y\in M: d(x,y)<r\}} |u(x) - u(y)|^2 \, d\mu(x) d\mu(y)$$

increases when σ increases, which implies that the space

$$\Lambda_{2,\infty}^{\alpha,\sigma} := \left\{u \in L^2 : N_{2,\infty}^{\alpha,\sigma}(u) < \infty\right\}$$

shrinks. For a certain value of σ, this space may become trivial. For example, as was already mentioned, $\Lambda_{2,\infty}^{n,\sigma}(\mathbb{R}^n) = \{0\}$ for $\sigma > 1$, while $\Lambda_{2,\infty}^{n,\sigma}(\mathbb{R}^n)$ is non-trivial for $\sigma \leq 1$.

Definition 2.6 Fix $\alpha > 0$ and set

$$\beta^* := \sup \left\{ \beta > 0 : \Lambda_{2,\infty}^{\alpha,\beta/2} \text{ is dense in } L^2(M, \mu) \right\}. \tag{2.6}$$

The number $\beta^* \in [0, +\infty]$ is called the *critical exponent* of the family $\left\{ \Lambda_{2,\infty}^{\alpha,\beta/2} \right\}_{\beta>0}$ of Besov spaces.

Note that the value of β^* is an intrinsic property of the space (M, d, μ), which is defined independently of any heat kernel. For example, for $\mathbb{R}^n$ with $\alpha = n$ we have $\beta^* = 2$.

Theorem 2.7 (Jonsson [Jon96], Pietruska-Pałuba [Pie00], Grigor'yan et al. [GHL03]) *Let p_t be a heat kernel on a metric measure space (M, d, μ). If the heat kernel is stochastically complete and satisfies* (2.4), *where $\Phi_1(s) > 0$ for some $s > 0$ and*

$$\int_0^\infty s^{\alpha+\beta+\varepsilon} \Phi_2(s) \frac{ds}{s} < \infty \tag{2.7}$$

for some $\varepsilon > 0$, then $\beta = \beta^*$.

By Theorem 2.1, condition (2.7) implies that the Dirichlet form $(\mathcal{E}, \mathcal{F})$ is local. For non-local forms the statement is not true: for example, in $\mathbb{R}^n$ for symmetric stable processes we have $\beta < 2 = \beta^*$.

Theorem 2.8 *Under the hypotheses of Theorem 2.7, the values of the parameters α and β are the invariants of the metric space (M, d) alone. Moreover, we have*

$$\mu \asymp H^\alpha \text{ and } \mathcal{E} \asymp N_{2,\infty}^{\alpha,\beta/2}.$$

Consequently, both the measure μ and the energy form $\mathcal{E}$ are determined (up to a factor $\asymp 1$) by the metric space (M, d) alone.

Example 2.9 Consider in $\mathbb{R}^n$ the Gauss-Weierstrass heat kernel

$$p_t(x, y) = \frac{1}{(4\pi t)^{n/2}} \exp\left(-\frac{|x-y|^2}{4t} \right)$$

and its generator $\mathcal{L} = -\Delta$ in $L^2(\mathbb{R}^n)$ with the Lebesgue measure. Then $\alpha = n$, $\beta = 2$, and

$$\mathcal{E}(u) = \int_{\mathbb{R}^n} |\nabla u|^2 \, dx.$$

Consider now another elliptic operator in $\mathbb{R}^n$:

$$\mathcal{L} = -\frac{1}{m(x)} \sum_{i,j=1}^{n} \frac{\partial}{\partial x_i} \left(a_{ij}(x) \frac{\partial}{\partial x_j} \right),$$

where $m(x)$ and $a_{ij}(x)$ are continuous functions, $m(x) > 0$ and the matrix $\left(a_{ij}(x)\right)$ is positive definite. The operator $\mathcal{L}$ is symmetric with respect to measure

$$d\mu = m(x)\, dx,$$

and its Dirichlet form is

$$\mathcal{E}(u) = \int_{\mathbb{R}^n} a_{ij}(x) \frac{\partial u}{\partial x_i} \frac{\partial u}{\partial x_j} dx.$$

Let $d(x, y) = |x - y|$ and assume that the heat kernel $p_t(x, y)$ of $\mathcal{L}$ satisfies the conditions of Theorem 2.7. Then we conclude by Corollary 2.8 that α and β must be the same as in the Gauss-Weierstrass heat kernel, that is, $\alpha = n$ and $\beta = 2$; moreover, measure μ must be comparable to the Lebesgue measure, which implies that $m \asymp 1$, and the energy form must admit the estimate

$$\mathcal{E}(u) \asymp \int_{\mathbb{R}^n} |\nabla u|^2 \, dx,$$

which implies that the matrix $\left(a_{ij}(x)\right)$ is uniformly elliptic. Hence, the operator $\mathcal{L}$ is uniformly elliptic.

By Aronson's theorem [Aro67, Aro68] the heat kernel for uniformly elliptic operators satisfies the estimate

$$p_t(x, y) \asymp \frac{C}{t^{n/2}} \exp\left(-c \frac{|x - y|^2}{t} \right).$$

What we have proved here implies the converse to Aronson's theorem: if the Aronson estimate holds for the operator $\mathcal{L}$, then $\mathcal{L}$ is uniformly elliptic.

The next theorem handles the non-local case.

Theorem 2.10 *Let p_t be a heat kernel on a metric measure space (M, d, μ). If the heat kernel satisfies the lower bound*

$$p_t(x, y) \geq \frac{1}{t^{\alpha/\beta}} \Phi_1 \left(\frac{d(x, y)}{t^{1/\beta}} \right),$$

where $\Phi_1(s) > 0$ *for some* $s > 0$, *then* $\beta \leq \beta^*$.

Proof In the proof of Theorem 2.3 one shows that the lower bound of the heat kernel implies $\mathcal{F} \subset \Lambda_{2,\infty}^{\alpha,\beta/2}$ (and the opposite inclusion follows from the upper bound and the stochastic completeness). Since $\mathcal{F}$ is dense in L^2, it follows that $\beta \leq \beta^*$. ∎

As a conclusion of this part, we briefly explain the walk dimension from three different points of view. As we have seen, there is a parameter appears in three different places:

- A parameter β in heat kernel bounds (2.4).
- A parameter θ in Markov processes: for a process X_t, one may have (cf. [Bar98, formula (1.1)])

$$\mathbb{E}_x \left(|X_t - x|^2 \right) \asymp t^{2/\theta}.$$

  Then θ is a parameter that measures how fast the process X_t goes away from the starting point x in time t. Alternatively, one may have that, for any ball $B(x, r) \subset M$,

$$\mathbb{E}_x \left(\tau_{B(x,r)} \right) \asymp r^\theta,$$

  where $\tau_{B(x,r)}$ is the *first exit time* of X_t from the ball

$$\tau_B = \inf \{ t > 0 : X_t \notin B(x, r) \}.$$

- A parameter σ in function spaces $N_{2,\infty}^{\alpha,\sigma}$ or $N_{2,2}^{\alpha,\sigma}$. By (2.2) or by (2.3), it is not hard to see that σ measures how much smooth of the functions in the space $N_{2,\infty}^{\alpha,\sigma}$ or $N_{2,2}^{\alpha,\sigma}$.

In general the three parameters $\beta, \theta, 2\sigma$ may be different. However, it turns out that, under some certain conditions, all these parameters are the same:

$$\beta = \theta = 2\sigma. \tag{2.8}$$

For examples, by Theorems 2.3 and 2.4, we see that $\sigma = \frac{\beta}{2}$, whilst by Theorems 3.8 and 4.3 below, we will see that $\beta = \theta$.

2.6 Inequalities for the Walk Dimension

Definition 2.11 We say that a metric space (M, d) satisfies the *chain condition* if there exists a (large) constant C such that for any two points $x, y \in M$ and for any positive integer n there exists a sequence $\{x_i\}_{i=0}^n$ of points in M such that $x_0 = x$, $x_n = y$, and

$$d(x_i, x_{i+1}) \le C\frac{d(x, y)}{n}, \quad \text{for all } i = 0, 1, \dots, n-1. \tag{2.9}$$

The sequence $\{x_i\}_{i=0}^n$ is referred to as a *chain* connecting x and y.

Theorem 2.12 (Grigor'yan et al. [GHL03]) *Let (M, d, μ) be a metric measure space. Assume that*

$$\mu(B(x, r)) \asymp r^\alpha \tag{2.10}$$

for all $x \in M$ and $0 < r \le 1$. Then

$$\beta^* \ge 2.$$

If in addition (M, d) satisfies the chain condition, then

$$\beta^* \le \alpha + 1.$$

Observe that the chain condition is essential for the inequality $\beta^* \le \alpha + 1$ to be true. Indeed, assume for a moment that the claim of Theorem 2.12 holds without the chain condition, and consider a new metric d' on M given by $d' = d^{1/\gamma}$ where $\gamma > 1$. Let us mark by a dash all notions related to the space (M, d', μ) as opposed to those of (M, d, μ). It is easy to see that $\alpha' = \alpha\gamma$ and $\beta^{*\prime} = \beta^*\gamma$. Hence, if Theorem 2.12 could apply to the space (M, d', μ) it would yield $\beta^*\gamma \le \alpha\gamma + 1$ which implies $\beta^* \le \alpha$ because γ may be taken arbitrarily large. However, there are spaces with $\beta^* > \alpha$, for example on SG.

Clearly, the metric d' does not satisfy the chain condition; indeed the inequality (2.9) implies

$$d'(x_i, x_{i+1}) \le C\frac{d'(x, y)}{n^{1/\gamma}},$$

which is not good enough. Note that if in the inequality (2.9) we replace n by $n^{1/\gamma}$, then the proof of Theorem 2.12 will give that $\beta^* \le \alpha + \gamma$ instead of $\beta^* \le \alpha + 1$.

Theorem 2.13 (Grigor'yan et al. [GHL03]) *Let p_t be a stochastically complete heat kernel on (M, d, μ) such that*

$$p_t(x,y) \asymp \frac{C}{t^{\alpha/\beta}} \Phi\left(c\frac{d(x,y)}{t^{1/\beta}}\right).$$

(a) If for some $\varepsilon > 0$

$$\int_0^\infty s^{\alpha+\beta+\varepsilon} \Phi(s) \frac{ds}{s} < \infty, \tag{2.11}$$

then $\beta \geq 2$.

(b) If (M,d) *satisfies the chain condition, then* $\beta \leq \alpha + 1$.

Proof By Theorem 2.2 μ is α-regular so that Theorem 2.12 applies.

(a) By Theorem 2.12, $\beta^* \geq 2$, and by Theorem 2.12, $\beta = \beta^*$, whence $\beta \geq 2$.

(b) By Theorem 2.12, $\beta^* \leq \alpha + 1$, and by Theorem 2.10, $\beta \leq \beta^*$, whence $\beta \leq \alpha + 1$. ■

Note that the condition (2.11) can occur only for a local Dirichlet form $\mathcal{E}$. If both (2.11) and the chain condition are satisfied, then we obtain

$$2 \leq \beta \leq \alpha + 1. \tag{2.12}$$

This inequality was stated by Barlow [Bar98] without proof.

The set of couples (α, β) satisfying (2.12) is shown on the diagram (Fig. 5):

Barlow [Bar04] proved that any couple of α, β satisfying (2.12) can be realized for the heat kernel estimate

$$p_t(x,y) \asymp \frac{C}{t^{\alpha/\beta}} \exp\left(-c\left(\frac{d^\beta(x,y)}{t}\right)^{\frac{1}{\beta-1}}\right) \tag{2.13}$$

For a non-local form, we can only claim that

$$0 < \beta \leq \alpha + 1$$

(under the chain condition). In fact, any couple α, β in the range

$$0 < \beta < \alpha + 1$$

can be realized for the estimate

$$p_t(x,y) \asymp \frac{1}{t^{\alpha/\beta'}} \left(1 + \frac{d(x,y)}{t^{1/\beta'}}\right)^{-(\alpha+\beta')}.$$

Indeed, if $\mathcal{L}$ is the generator of a diffusion with parameters α and β satisfying (2.13), then the operator $\mathcal{L}^\delta$, $\delta \in (0, 1)$, generates a jump process with the walk dimension

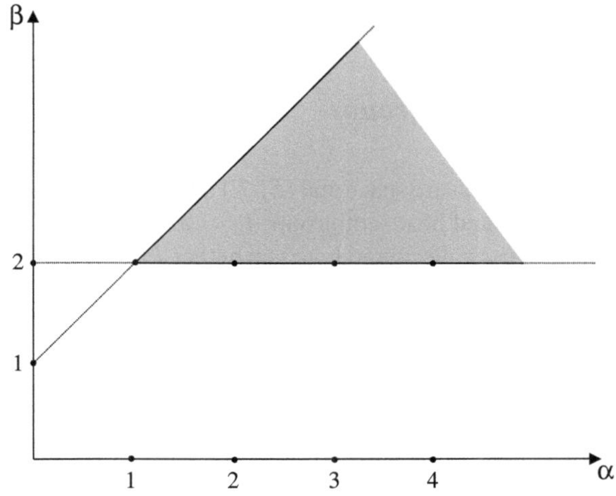

Fig. 5 The set $2 \le \beta \le \alpha + 1$

$\beta' = \delta\beta$ and the same α (cf. Theorem 2.5). Clearly, β' can take any value from $(0, \alpha + 1)$.

It is not known whether the walk dimension for a non-local form can be equal to $\alpha + 1$.

2.7 Identifying Φ in the Local Case

Theorem 2.14 (Grigor'yan and Kumagai [GK08]) *Assume that the metric space (M, d) satisfies the chain condition and all metric balls are precompact. Let p_t be a stochastically complete heat kernel in (M, d, μ). Assume that the associated Dirichlet form $(\mathcal{E}, \mathcal{F})$ is regular, and the following estimate holds with some $\alpha, \beta > 0$ and $\Phi : [0, +\infty) \to [0, +\infty)$:*

$$p_t(x, y) \asymp \frac{C}{t^{\alpha/\beta}} \Phi\left(c \frac{d(x, y)}{t^{1/\beta}}\right).$$

Then the following dichotomy holds:

- *either the Dirichlet form $\mathcal{E}$ is local, $2 \le \beta \le \alpha + 1$, and $\Phi(s) \asymp C \exp(-c s^{\frac{\beta}{\beta-1}})$.*
- *or the Dirichlet form $\mathcal{E}$ is non-local, $\beta \le \alpha + 1$, and $\Phi(s) \asymp (1 + s)^{-(\alpha+\beta)}$.*

3 Sub-Gaussian Upper Bounds

3.1 Ultracontractive Semigroups

Let (M, d, μ) be a metric measure space and $(\mathcal{E}, \mathcal{F})$ be a Dirichlet form in $L^2(M, \mu)$, and let $\{P_t\}$ be the associated heat semigroup, $P_t = e^{-t\mathcal{L}}$ where $\mathcal{L}$ is the generator of $(\mathcal{E}, \mathcal{F})$. The question to be discussed here is whether P_t possesses the heat kernel, that is, a function $p_t(x, y)$ that is non-negative, jointly measurable in (x, y), and satisfies the identity

$$P_t f(x) = \int_M p_t(x, y) f(y) \, d\mu(y)$$

for all $f \in L^2$, $t > 0$, and almost all $x \in M$. Usually the conditions that ensure the existence of the heat kernel give at the same token some upper bounds.

Given two parameters $p, q \in [0, +\infty]$, define the $L^p \to L^q$ norm of P_t by

$$\|P_t\|_{L^p \to L^q} = \sup_{f \in L^p \cap L^2 \setminus \{0\}} \frac{\|P_t f\|_q}{\|f\|_p}.$$

In fact, the Markovian property allows to extend P_t to an operator in L^p so that the range $L^p \cap L^2$ of f can be replaced by L^p. Also, it follows from the Markovian property that $\|P_t\|_{L^p \to L^p} \le 1$ for any p.

Definition 3.1 The semigroup $\{P_t\}$ is said to be $L^p \to L^q$ *ultracontractive* if there exists a positive decreasing function γ on $(0, +\infty)$, called the *rate* function, such that, for each $t > 0$

$$\|P_t\|_{L^p \to L^q} \le \gamma(t).$$

By the symmetry of P_t, if P_t is $L^p \to L^q$ ultracontractive, then P_t is also $L^{q^*} \to L^{p^*}$ ultracontractive with the same rate function, where p^* and q^* are the Hölder conjugates to p and q, respectively. In particular, P_t is $L^1 \to L^2$ ultracontractive if and only if it is $L^2 \to L^\infty$ ultracontractive.

Theorem 3.2 *(a) The heat semigroup $\{P_t\}$ is $L^1 \to L^2$ ultracontractive with a rate function γ, if and only if $\{P_t\}$ has the heat kernel p_t satisfying the estimate*

$$\operatorname*{esup}_{x,y \in M} p_t(x, y) \le \gamma(t/2)^2 \quad \text{for all } t > 0.$$

(b) The heat semigroup $\{P_t\}$ is $L^1 \to L^\infty$ ultracontractive with a rate function γ, if and only if $\{P_t\}$ has the heat kernel p_t satisfying the estimate

$$\operatorname*{esup}_{x,y \in M} p_t(x, y) \le \gamma(t) \quad \text{for all } t > 0.$$

This result is "well-known" and can be found in many sources. However, there are hardly complete proofs of the measurability of the function $p_t(x, y)$ in (x, y), which is necessary for many applications, for example, to use Fubini. Normally the existence of the heat kernel is proved in some specific setting where $p_t(x, y)$ is continuous in (x, y), or one just proves the existence of a family of functions $p_{t,x} \in L^2$ so that

$$P_t f(x) = (p_{t,x}, f) = \int_M p_{t,x}(y) f(y) \, d\mu(y)$$

for all $t > 0$ and almost all x. However, if one defines $p_t(x, y) = p_{t,x}(y)$, then this function does not have to be jointly measurable. The proof of the existence of a jointly measurable version can be found in [GH10]. Most of the material of this section can also be found there.

3.2 Restriction of the Dirichlet Form

Let Ω be an open subset of M. Define the function space $\mathcal{F}(\Omega)$ by

$$\mathcal{F}(\Omega) = \overline{\{f \in \mathcal{F} : \operatorname{supp} f \subset \Omega\}}^{\mathcal{F}}.$$

Clearly, $\mathcal{F}(\Omega)$ is a closed subspace of $\mathcal{F}$ and a subspace of $L^2(\Omega)$.

Theorem 3.3 *If $(\mathcal{E}, \mathcal{F})$ is a regular Dirichlet form in $L^2(M)$, then $(\mathcal{E}, \mathcal{F}(\Omega))$ is a regular Dirichlet form in $L^2(\Omega)$. If $(\mathcal{E}, \mathcal{F})$ is (strongly) local then so is $(\mathcal{E}, \mathcal{F}(\Omega))$.*

The regularity is used, in particular, to ensure that $\mathcal{F}(\Omega)$ is dense in $L^2(\Omega)$. From now on let us assume that $(\mathcal{E}, \mathcal{F})$ is a regular Dirichlet form. Other consequences of this assumptions are as follows (cf. [FOT11]):

1. The existence of cutoff functions: for any compact set K and any open set $U \supset K$, there is a function $\varphi \in \mathcal{F} \cap C_0(U)$ such that $0 \leq \varphi \leq 1$ and $\varphi \equiv 1$ in an open neighborhood of K.
2. The existence of a Hunt process $\left(\{X_t\}_{t \geq 0}, \{\mathbb{P}_x\}_{x \in M}\right)$ associated with $(\mathcal{E}, \mathcal{F})$.

Hence, for any open subset $\Omega \subset M$, we have the Dirichlet form $(\mathcal{E}, \mathcal{F}(\Omega))$ that is called a *restriction* of $(\mathcal{E}, \mathcal{F})$ to Ω.

Example 3.4 Consider in $\mathbb{R}^n$ the canonical Dirichlet form

$$\mathcal{E}(u) = \int_{\mathbb{R}^n} |\nabla u|^2 \, dx$$

in $\mathcal{F} = W_2^1(\mathbb{R}^n)$. Then $\mathcal{F}(\Omega) = \overline{C_0^1(\Omega)}^{W_2^1} =: H_0^1(\Omega)$.

Using the restricted form $(\mathcal{E}, \mathcal{F}(\Omega))$ corresponds to imposing the Dirichlet boundary conditions on $\partial\Omega$ (or on Ω^c), so that the form $(\mathcal{E}, \mathcal{F}(\Omega))$ could be called the Dirichlet form with the Dirichlet boundary condition.

Denote by $\mathcal{L}_\Omega$ the generator of $(\mathcal{E}, \mathcal{F}(\Omega))$ and set

$$\lambda_{\min}(\Omega) := \inf \operatorname{spec} \mathcal{L}_\Omega = \inf_{u \in \mathcal{F}(\Omega)\setminus\{0\}} \frac{\mathcal{E}(u)}{\|u\|_2^2}. \tag{3.1}$$

Clearly, $\lambda_{\min}(\Omega) \geq 0$ and $\lambda_{\min}(\Omega)$ is decreasing when Ω expands.

Example 3.5 If $(\mathcal{E}, \mathcal{F})$ is the canonical Dirichlet form in $\mathbb{R}^n$ and Ω is the bounded domain in $\mathbb{R}^n$, then the operator $\mathcal{L}_\Omega$ has the discrete spectrum $\lambda_1(\Omega) \leq \lambda_2(\Omega) \leq \lambda_3(\Omega) \leq \ldots$ that coincides with the eigenvalues of the Dirichlet problem

$$\begin{cases} \Delta u + \lambda u = 0, \\ u|_{\partial\Omega} = 0, \end{cases}$$

so that $\lambda_1(\Omega) = \lambda_{\min}(\Omega)$.

3.3 Faber-Krahn and Nash Inequalities

Continuing the above example, we have by a theorem of Faber-Krahn

$$\lambda_1(\Omega) \geq \lambda_1(\Omega^*),$$

where Ω^* is the ball of the same volume as Ω. If r is the radius of Ω^*, then we have

$$\lambda_1(\Omega^*) = \frac{c'}{r^2} = \frac{c}{|\Omega^*|^{2/n}} = \frac{c}{|\Omega|^{2/n}},$$

whence

$$\lambda_1(\Omega) \geq c_n |\Omega|^{-2/n}.$$

It turns out that this inequality, that we call *the Faber-Krahn inequality*, is intimately related to the existence of the heat kernel and its upper bound.

Theorem 3.6 *Let $(\mathcal{E}, \mathcal{F})$ be a regular Dirichlet form in $L^2(M, \mu)$. Fix some constant $\nu > 0$. Then the following conditions are equivalent:*

(i) (The Faber-Krahn inequality) *There is a constant $a > 0$ such that, for all non-empty open sets $\Omega \subset M$,*

$$\lambda_{\min}(\Omega) \geq a\mu(\Omega)^{-\nu}. \tag{3.2}$$

(ii) (The Nash inequality) *There exists a constant $b > 0$ such that*

$$\mathcal{E}(u) \geq b\|u\|_2^{2+2\nu}\|u\|_1^{-2\nu}, \tag{3.3}$$

for any function $u \in \mathcal{F} \setminus \{0\}$.

(iii) (On-diagonal estimate of the heat kernel) *The heat kernel exists and satisfies the upper bound*

$$\operatorname*{esup}_{x,y \in M} p_t(x,y) \leq ct^{-1/\nu} \tag{3.4}$$

for some constant c and for all $t > 0$.

The relation between the parameters a, b, c is as follows:

$$a \asymp b \asymp c^{-\nu}$$

where the ratio of any two of these parameters is bounded by constants depending only on ν.

In $\mathbb{R}^n$, we see that $\nu = 2/n$.

The implication (ii) $\Rightarrow$ (iii) was proved by Nash [Nas58], and (iii) $\Rightarrow$ (ii) by Carlen-Kusuoka-Stroock [CKS87], and (i) $\Leftrightarrow$ (iii) by

Grigor'yan [Gri94] and Carron [Car96].

Proof of (i) $\Rightarrow$ (ii) $\Rightarrow$ (iii) Observe first that (ii) $\Rightarrow$ (i) is trivial: choosing in (3.3) a function $u \in \mathcal{F}(\Omega) \setminus \{0\}$ and applying the Cauchy-Schwarz inequality

$$\|u\|_1 \leq \mu(\Omega)^{1/2}\|u\|_2,$$

we obtain

$$\mathcal{E}(u) \geq b\mu(\Omega)^{-\nu}\|u\|_2^2,$$

whence (3.2) follow by the variational principle (3.1).

The opposite inequality (i) $\Rightarrow$ (ii) is a bit more involved, and we prove it for functions $0 \leq u \in \mathcal{F} \cap C_0(M)$ (a general $u \in \mathcal{F}$ requires some approximation argument). By the Markovian property, we have $(u-t)_+ \in \mathcal{F} \cap C_0(M)$ for any $t > 0$ and

$$\mathcal{E}(u) \geq \mathcal{E}\left((u-t)_+\right). \tag{3.5}$$

For any $s > 0$, consider the set

$$U_s := \{x \in M : u(x) > s\},$$

which is clearly open and precompact. If $t > s$, then $(u-t)_+$ is supported in U_s, and whence, $(u-t)_+ \in \mathcal{F}(U_s)$. It follows from (3.1)

$$\mathcal{E}\left((u-t)_+\right) \geq \lambda_{\min}(U_s) \int\limits_{U_s} (u-t)_+^2 \, d\mu. \tag{3.6}$$

For simplicity, set $A = \|u\|_1$ and $B = \|u\|_2^2$. Since $u \geq 0$, we have

$$(u-t)_+^2 \geq u^2 - 2tu,$$

which implies that

$$\int\limits_{U_s} (u-t)_+^2 d\mu = \int\limits_{M} (u-t)_+^2 d\mu \geq B - 2tA. \tag{3.7}$$

On the other hand, we have

$$\mu(U_s) \leq \frac{1}{s} \int\limits_{U_s} u \, d\mu \leq \frac{A}{s},$$

which together with the Faber-Krahn inequality implies

$$\lambda_{\min}(U_s) \geq a\mu(U_s)^{-\nu} \geq a\left(\frac{s}{A}\right)^{\nu}. \tag{3.8}$$

Combining (3.5)–(3.8), we obtain

$$\mathcal{E}(u) \geq \lambda_{\min}(U_s) \int\limits_{U_s} (u-t)_+^2 \, d\mu \geq a\left(\frac{s}{A}\right)^{\nu} (B - 2tA).$$

Letting $t \to s+$ and then choosing $s = \frac{B}{4A}$, we obtain

$$\mathcal{E}(u) \geq a\left(\frac{s}{A}\right)^{\nu} (B - 2sA) = a\left(\frac{B}{4A^2}\right)^{\nu} \frac{B}{2} = \frac{a}{4^{\nu}2} B^{\nu+1} A^{-2\nu},$$

which is exactly (3.3).

To prove (ii)$\Rightarrow$(iii), choose $f \in L^2 \cap L^1$, and consider $u = P_t f$. Since $u = e^{-t\mathcal{L}} f$ and $\frac{d}{dt} u = -\mathcal{L} u$, we have

$$\begin{aligned}\frac{d}{dt}\|u\|_2^2 = \frac{d}{dt}(u,u) &= -2(\mathcal{L}u,u) = -2\mathcal{E}(u,u) \\ &\leq -2b\|u\|_2^{2+2\nu}\|u\|_1^{-2\nu} \leq -2b\|u\|_2^{2+2\nu}\|f\|_1^{-2\nu},\end{aligned}$$

since $\|u\|_1 \leq \|f\|_1$. Solving this differential inequality, we obtain

$$\|P_t f\|_2^2 \le ct^{-1/v} \|f\|_1^2,$$

that is, the semigroup P_t is $L^1 \to L^2$ ultracontractive with the rate function $\gamma(t) = \sqrt{ct^{-1/v}}$. By Theorem 3.2 we conclude that the heat kernel exists and satisfies (3.4). ■

Let M be a Riemannian manifold with the geodesic distance d and the Riemannian volume μ. Let $(\mathcal{E}, \mathcal{F})$ be the canonical Dirichlet form on M. The heat kernel on manifolds always exists and is a smooth function. In this case the estimate (3.4) is equivalent to the on-diagonal upper bound

$$\sup_{x \in M} p_t(x, x) \le ct^{-1/\nu}.$$

It is known (but non-trivial) that the on-diagonal estimate implies the Gaussian upper bound

$$p_t(x, y) \le Ct^{-1/\nu} \exp\left(-\frac{d^2(x, y)}{(4+\varepsilon)t}\right),$$

for all $t > 0$ and $x, y \in M$, which is due to the specific property of the geodesic distance function that $|\nabla d| \le 1$.

In the context of abstract metric measure space, the distance function does not have to satisfy this property, and typically it does not (say, on fractals). Consequently, one needs some additional conditions that would relate the distance function to the Dirichlet form and imply the off-diagonal bounds.

3.4 Off-diagonal Upper Bounds

From now on, let $(\mathcal{E}, \mathcal{F})$ be a regular *local* Dirichlet form, so that the associated Hunt process $\left(\{X_t\}_{t \ge 0}, \{\mathbb{P}_x\}_{x \in M}\right)$ is a diffusion. Recall that it is related to the heat semigroup $\{P_t\}$ of $(\mathcal{E}, \mathcal{F})$ by means of the identity

$$\mathbb{E}_x(f(X_t)) = P_t f(x)$$

for all $f \in \mathcal{B}_b(M)$, $t > 0$ and almost all $x \in M$ (Fig. 6).

Fix two parameters $\alpha > 0$ and $\beta > 1$ and introduce some conditions.

(V_α) (Volume regularity) For all $x \in M$ and $r > 0$,

$$\mu(B(x, r)) \asymp r^\alpha.$$

(FK) (The Faber-Krahn inequality) For any open set $\Omega \subset M$,

$$\lambda_{\min}(\Omega) \ge c\mu(\Omega)^{-\beta/\alpha}.$$

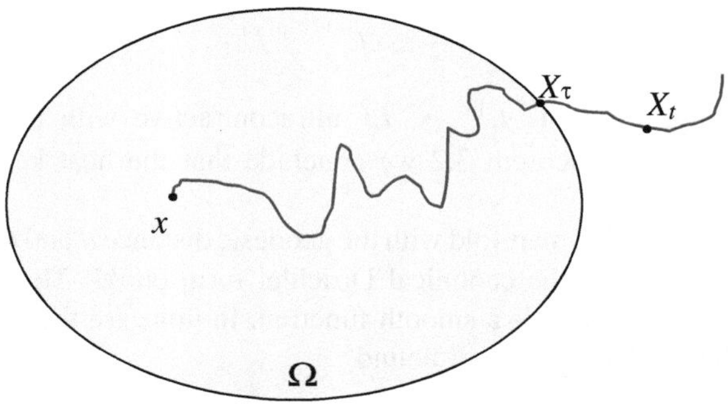

Fig. 6 First exit time τ

For any open set $\Omega \subset M$, define the *first exist time* from Ω by

$$\tau_\Omega = \inf\{t > 0 : X_t \notin \Omega\}.$$

A set $N \subset M$ is called *properly exceptional*, if it is a Borel set of measure 0 that is almost never hit by the process X_t starting outside N. In the next conditions N denotes some properly exceptional set.

(E_β) (*An estimate for the mean exit time from balls*) For all $x \in M \setminus N$ and $r > 0$

$$\mathbb{E}_x\left[\tau_{B(x,r)}\right] \asymp r^\beta$$

(the parameter β is called the walk dimension of the process).

(P_β) (*The exit probability estimate*) There exist constants $\varepsilon \in (0,1)$, $\delta > 0$ such that, for all $x \in M \setminus N$ and $r > 0$,

$$\mathbb{P}_x\left(\tau_{B(x,r)} \leq \delta r^\beta\right) \leq \varepsilon.$$

$(E\Omega)$ (*An isoperimetric estimate for the mean exit time*) For any open subset $\Omega \subset M$,

$$\sup_{x\in\Omega\setminus N} \mathbb{E}_x(\tau_\Omega) \leq C\mu(\Omega)^{\beta/\alpha}.$$

If both (V_α) and (E_β) are satisfied, then we obtain for any ball $B \subset M$

$$\sup_{x\in B\setminus N} \mathbb{E}_x(\tau_B) \asymp r^\beta \asymp \mu(B)^{\beta/\alpha}.$$

It follows that the balls are in some sense optimal sets for the condition $(E\Omega)$.

Example 3.7 If X_t is Brownian motion in $\mathbb{R}^n$, then it is known that

$$\mathbb{E}_x \tau_{B(x,r)} = c_n r^2,$$

so that (E_β) holds with $\beta = 2$. This can also be rewritten in the form

$$\mathbb{E}_x \tau_B = c_n |B|^{2/n},$$

where $B = B(x, r)$.

It is also known that for any open set $\Omega \subset \mathbb{R}^n$ with finite volume and for any $x \in \Omega$,

$$\mathbb{E}_x (\tau_\Omega) \le \mathbb{E}_x \left(\tau_{B(x,r)}\right),$$

provided that ball $B(x, r)$ has the same volume as Ω; that is, for a fixed value of $|\Omega|$, the mean exist time is maximal when Ω is a ball and x is the center. It follows that

$$\mathbb{E}_x (\tau_\Omega) \le c_n |\Omega|^{2/n}$$

so that $(E\Omega)$ is satisfied with $\beta = 2$ and $\alpha = n$.

Finally, introduce notation for the following estimates of the heat kernel:

(UE_{loc}) (*Sub-Gaussian upper estimate*) The heat kernel exists and satisfies the estimate

$$p_t(x, y) \le \frac{C}{t^{\alpha/\beta}} \exp\left(-c\left(\frac{d^\beta(x, y)}{t}\right)^{\frac{1}{\beta-1}}\right)$$

for all $t > 0$ and almost all $x, y \in M$.

(ΦUE) (Φ-*upper estimate*) The heat kernel exists and satisfies the estimate

$$p_t(x, y) \le \frac{1}{t^{\alpha/\beta}} \Phi\left(\frac{d(x, y)}{t^{1/\beta}}\right)$$

for all $t > 0$ and almost all $x, y \in M$, where Φ is a decreasing positive function on $[0, +\infty)$ such that

$$\int_0^\infty s^\alpha \Phi(s) \frac{ds}{s} < \infty.$$

(*DUE*) (*On-diagonal upper estimate*) The heat kernel exists and satisfies the estimate

$$p_t(x, y) \le \frac{C}{t^{\alpha/\beta}}$$

for all $t > 0$ and almost all $x, y \in M$.

$(T_{\exp})$ *(The exponential tail estimate)* The heat kernel p_t exists and satisfies the estimate

$$\int\limits_{B(x,r)^c} p_t(x,y)\,d\mu(y) \le C \exp\left(-c\left(\frac{r}{t^{1/\beta}}\right)^{\frac{\beta}{\beta-1}}\right), \tag{3.9}$$

for some constants $C, c > 0$, all $t > 0, r > 0$ and μ-almost all $x \in M$.
Note that it is easy to show that (3.9) is equivalent to the following inequality: for any ball $B = B(x_0, r)$ and $t > 0$,

$$P_t \mathbf{1}_{B^c}(x) \le C \exp\left(-c\left(\frac{r}{t^{1/\beta}}\right)^{\frac{\beta}{\beta-1}}\right) \text{ for } \mu\text{-almost all } x \in \frac{1}{4}B$$

(see [GH08, Remark 3.3]).
(T_β) *(The tail estimate)* There exist $0 < \varepsilon < \frac{1}{2}$ and $C > 0$ such that, for all $t > 0$ and all balls $B = B(x_0, r)$ with $r \ge Ct^{1/\beta}$,

$$P_t \mathbf{1}_{B^c}(x) \le \varepsilon \quad \text{for } \mu \text{-almost all } x \in \frac{1}{4}B.$$

(S_β) *(The survival estimate)* There exist $0 < \varepsilon < 1$ and $C > 0$ such that, for all $t > 0$ and all balls $B = B(x_0, r)$ with $r \ge Ct^{1/\beta}$,

$$1 - P_t^B \mathbf{1}_B(x) \le \varepsilon \quad \text{for } \mu \text{-almost all } x \in \frac{1}{4}B.$$

Clearly, we have

$$(UE_{loc}) \Rightarrow (\Phi UE) \Rightarrow (DUE).$$

Theorem 3.8 (Grigor'yan and Hu [GH10]) *Let (M, d, μ) be a metric measure space and let (V_α) hold. Let $(\mathcal{E}, \mathcal{F})$ be a regular, local, conservative Dirichlet form in $L^2(M, \mu)$. Then, the following equivalences are true:*

$$\begin{aligned}
(UE_{loc}) &\Leftrightarrow (FK) + (E_\beta) \Leftrightarrow (E\Omega) + (E_\beta) \\
&\Leftrightarrow (FK) + (P_\beta) \Leftrightarrow (E\Omega) + (P_\beta) \\
&\Leftrightarrow (DUE) + (E_\beta) \Leftrightarrow (DUE) + (P_\beta), \\
&\Leftrightarrow (\Phi UE) \\
&\Leftrightarrow (FK) + (S_\beta) \Leftrightarrow (FK) + (T_\beta) \\
&\Leftrightarrow (DUE) + (S_\beta) \Leftrightarrow (DUE) + (T_\beta) \\
&\Leftrightarrow (DUE) + (T_{\exp}).
\end{aligned}$$

Let us emphasize the equivalence

$$(UE_{loc}) \Leftrightarrow (E\Omega) + \left(E_\beta\right)$$

where the right hand side means the following: the mean exit time from all sets Ω satisfies the isoperimetric inequality, and this inequality is optimal for balls (up to a constant multiple). Note that the latter condition relates the properties of the diffusion (and, hence, of the Dirichlet form) to the distance function.

Conjecture 3.9 *Under the hypotheses of Theorem 3.8,*

$$(UE_{loc}) \Leftrightarrow (FK) + \left\{\lambda_{\min}(B_r) \asymp r^{-\beta}\right\}$$

Indeed, the Faber-Krahn inequality *(FK)* can be regarded as an isoperimetric inequality for $\lambda_{\min}(\Omega)$, and the condition

$$\lambda_{\min}(B_r) \asymp r^{-\beta}$$

means that *(FK)* is optimal for balls (up to a constant multiple).

Theorem 3.8 is an oversimplified version of a result of [GH10], where instead of (V_α) one uses the volume doubling condition, and other hypotheses must be appropriately changed.

The following lemma is used in the proof of Theorem 3.8.

Lemma 3.10 *For any open set $\Omega \subset M$*

$$\lambda_{\min}(\Omega) \geq \frac{1}{\operatorname{esup}_{x\in\Omega}\mathbb{E}_x(\tau_\Omega)}.$$

Proof Let G_Ω be the Green operator in Ω, that is,

$$G_\Omega = \mathcal{L}_\Omega^{-1} = \int_0^\infty e^{-t\mathcal{L}_\Omega} dt.$$

We claim that

$$\mathbb{E}_x(\tau_\Omega) = G_\Omega 1(x)$$

for almost all $x \in \Omega$. We have

$$\begin{aligned} G_\Omega 1(x) &= \int_0^\infty e^{-t\mathcal{L}_\Omega} 1_\Omega(x)\, dt = \int_0^\infty \mathbb{E}_x\left(1_\Omega\left(X_t^\Omega\right)\right) \\ &= \int_0^\infty \mathbb{E}_x\left(\mathbf{1}_{\{t<\tau_\Omega\}}\right) dt = \mathbb{E}_x \int_0^\infty \left(\mathbf{1}_{\{t<\tau_\Omega\}}\right) dt = \mathbb{E}_x(\tau_\Omega). \end{aligned}$$

Setting

$$m = \operatorname*{esup}_{x\in\Omega} \mathbb{E}_x\left(\tau_\Omega\right),$$

we obtain that $G_\Omega 1 \leq m$, so that $m^{-1}G_\Omega$ is a Markovian operator. Therefore, $\left\|m^{-1}G_\Omega\right\|_{L^2\to L^2} \leq 1$ whence spec$G_\Omega \in [0,m]$. It follows that spec $\mathcal{L}_\Omega \subset [m^{-1},\infty)$ and $\lambda_{\min}(\Omega) \geq m^{-1}$. ■

A new analytical approach is developed in [GH10] to prove Theorem 3.8, which is different from the Davies-Gaffney approach [Dav92] . The difficult part in proving Theorem 3.8 is to deduce (UE_{loc}) from various conditions.

Sketch of proof for Theorem 3.8 We sketch the main steps.

- By a direct integration, we have

$$(\Phi UE) \Rightarrow \left(T_\beta\right).$$

Indeed, for any $x \in \frac{1}{4}B$, we see that $B(x,\frac{1}{2}r) \subset B$. Thus, setting $r_k = 2^k(r/2)$ and using condition (ΦUE) and the monotonicity of Φ, we obtain that

$$\begin{aligned}
\int\limits_{M\setminus B} p_t(x,y)d\mu(y) &\leq \int\limits_{M\setminus B(x,r/2)} p_t(x,y)d\mu(y) \qquad (3.10)\\
&= \sum_{k=0}^{\infty}\int\limits_{B(x,r_{k+1})\setminus B(x,r_k)} p_t(x,y)\,d\mu(y)\\
&\leq \sum_{k=0}^{\infty}\int\limits_{B(x,r_{k+1})\setminus B(x,r_k)} Ct^{-\alpha/\beta}\Phi\left(\frac{r_k}{t^{1/\beta}}\right)d\mu(y)\\
&\leq \sum_{k=0}^{\infty} Cr_{k+1}^{\alpha}t^{-\alpha/\beta}\Phi\left(\frac{r_k}{t^{1/\beta}}\right)\\
&= C'\sum_{k=0}^{\infty}\left(\frac{2^{k-1}r}{t^{1/\beta}}\right)^{\alpha}\Phi\left(\frac{2^{k-1}r}{t^{1/\beta}}\right)\\
&\leq C'\int\limits_{\frac{1}{4}r/t^{1/\beta}}^{\infty} s^{\alpha}\Phi(s)\frac{ds}{s}.
\end{aligned}$$

The integral (3.10) converges, and its value can be made arbitrarily small provided that r^β/t is large enough. Hence, condition $\left(T_\beta\right)$ follows.

- The following implications hold:

$$(E\Omega) \overset{\text{L. 3.10}}{\Rightarrow} (FK) \overset{\text{T. 3.6}}{\Rightarrow} (DUE).$$

In particular, we see that the heat kernel exists under any of the hypotheses of Theorem 3.8.

- We can also show that
$$\left(E_\beta\right) \Rightarrow \left(P_\beta\right) \Longrightarrow \left(T_\beta\right)$$
(the implication $\left(E_\beta\right) \Rightarrow \left(P_\beta\right)$ was pointed out in [Bar98]).
- By a bootstrapping technique, we obtain (hard!) the implication
$$\left(T_\beta\right) \Longrightarrow \left(T_{\exp}\right)$$
(see also [GH08]). Hence, any set of the hypothesis of Theorem 3.8 imply both (DUE) and $\left(T_{\exp}\right)$.
- Finally, it is easy to check the implication
$$(DUE) + \left(T_{\exp}\right) \Rightarrow (UE_{loc})\,. \tag{3.11}$$

Indeed, using the semigroup identity, we have that, for all $t > 0$, almost all $x, y \in M$, and $r := \frac{1}{2}d\,(x, y)$,

$$\begin{aligned} p_t\,(x, y) &= \int_M p_{\frac{t}{2}}\,(x, z)\, p_{\frac{t}{2}}\,(z, y)\, d\mu(z) \\ &\le \left(\int_{B(x,r)^c} + \int_{B(y,r)^c} \right) p_{\frac{t}{2}}\,(x, z)\, p_{\frac{t}{2}}\,(z, y)\, d\mu(z) \\ &\le \operatorname*{esup}_{z\in M} p_{\frac{t}{2}}\,(z, y) \int_{B(x,r)^c} p_{\frac{t}{2}}\,(x, z)\, d\mu(z) \\ &\quad + \operatorname*{esup}_{z\in M} p_{\frac{t}{2}}\,(x, z) \int_{B(y,r)^c} p_{\frac{t}{2}}\,(y, z)\, d\mu(z). \end{aligned} \tag{3.12}$$

On the other hand, by condition (DUE),

$$\operatorname{esup} p_t \le Ct^{-\alpha/\beta},$$

whilst by condition $\left(T_\beta\right)$,

$$\int_{B(x,r)^c} p_{\frac{t}{2}}\,(x, z)\, d\mu(z) \le C \exp\left(-c \left(\frac{d^\beta\,(x, y)}{t} \right)^{\frac{1}{\beta-1}} \right).$$

Therefore, it follows from (3.12) that, for almost all $x, y \in M$,

$$p_t(x,y) \le \frac{C}{t^{\alpha/\beta}} \exp\left(-c\left(\frac{d^\beta(x,y)}{t}\right)^{\frac{1}{\beta-1}}\right),$$

proving the implication (3.11). ■

Recently, Andres and Barlow [AB] gave a new equivalence condition for (UE_{loc}). Consider the following functional inequality.

(CSA_β) (*The cutoff Sobolev annulus inequality*) There exists a constant $C > 0$ such that, for all two concentric balls $B(x,R)$, $B(x,R+r)$, there exists a cutoff function φ satisfying

$$\int_U f^2 d\mu_{\langle\varphi\rangle} \le \frac{1}{8}\int_U \varphi^2 d\mu_{\langle f\rangle} + Cr^{-\beta}\int_U f^2 d\mu$$

for any $f \in \mathcal{F}$, where $U = B(x,R+r) \setminus B(x,R)$ is the annulus and $\mu_{\langle\varphi\rangle}$ is the energy measure associated with φ:

$$\int_M u d\mu_{\langle\varphi\rangle} = 2\mathcal{E}(u\varphi,\varphi) - \mathcal{E}(\varphi^2,u) \text{ for any } u \in \mathcal{F} \cap C_0(M).$$

We remark here that constant C is universal that is independent of two concentric balls $B(x,R)$, $B(x,R+r)$ and function f, whilst the cutoff function φ may depend on the balls but is independent of function f. The coefficient $\frac{1}{8}$ is not essential and is chosen for technical reasons.

Theorem 3.11 (Andres, Barlow [AB]) *Let (M,d,μ) be an unbounded metric measure space and let (V_α) hold. Let $(\mathcal{E},\mathcal{F})$ be a regular, local Dirichlet form in $L^2(M,\mu)$. Then, the following equivalence is true:*

$$(UE_{loc}) \Leftrightarrow (FK) + (CSA_\beta).$$

We mention that here the Dirichlet form is not required to be conservative as in Theorem 3.8.

The key point in proving Theorem 3.11 is to derive the "Davies-Gaffney" bound [Dav92], and then use the technique developed in [Gri92, CG98] to show a mean value inequality for weak solutions of the heat equation. It is quite surprising that the Davies-Gaffney method still works when the walk dimension β may be greater than 2.

4 Two-Sided Sub-Gaussian Bounds

4.1 Using Elliptic Harnack Inequality

Now we would like to extend the results of Theorems 3.8, 3.11, and obtain also the lower estimates and the Hölder continuity of the heat kernel. As before, (M, d, μ) is a metric measure space, and assume in addition that all metric balls are precompact. Let $(\mathcal{E}, \mathcal{F})$ is a local regular conservative Dirichlet form in $L^2(M, \mu)$.

Definition 4.1 We say that a function $u \in \mathcal{F}$ is *harmonic* in an open set $\Omega \subset M$ if

$$\mathcal{E}(u, v) = 0 \text{ for all } v \in \mathcal{F}(\Omega).$$

For example, if $M = \mathbb{R}^n$ and $(\mathcal{E}, \mathcal{F})$ is the canonical Dirichlet form in $\mathbb{R}^n$, then a function $u \in W_2^1(\mathbb{R}^n)$ is harmonic in an open set $\Omega \subset \mathbb{R}^n$ if

$$\int_{\mathbb{R}^n} \langle \nabla u, \nabla v \rangle dx = 0$$

for all $v \in H_0^1(\Omega)$ or for $v \in C_0^\infty(\Omega)$. This of course implies that $\Delta u = 0$ in a weak sense in Ω and, hence, u is harmonic in Ω in the classical sense. However, unlike the classical definition, we a priori require $u \in W_2^1(\mathbb{R}^n)$.

Definition 4.2 (*Elliptic Harnack inequality* (H)) We say that M satisfies the *elliptic Harnack inequality* (H) if there exist constants $C > 1$ and $\delta \in (0, 1)$ such that for any ball $B(x, r)$ and for any function $u \in \mathcal{F}$ that is non-negative and harmonic in $B(x, r)$,

$$\operatorname*{esup}_{B(x,\delta r)} u \leq C \operatorname*{einf}_{B(x,\delta r)} u.$$

We remark that constants C and δ are independent of ball $B(x, r)$ and function u. We introduce the near-diagonal lower estimate of heat kernel.

(NLE) (*Near-diagonal lower estimate*) The heat kernel $p_t(x, y)$ exists, and satisfies

$$p_t(x, y) \geq \frac{c}{t^{\alpha/\beta}}$$

for all $t > 0$ and $\mu \times \mu$-almost all $x, y \in M$ such that $d(x, y) \leq \delta t^{1/\beta}$, where $\delta > 0$ is a sufficiently small constant.

Denote by (UE_{strong}) a modification of condition (UE_{loc}) that is obtained by adding the Hölder continuity of $p_t(x, y)$ and by restricting inequality in (UE_{loc}) to all $x, y \in M$. In a similar way, we can define *condition* (NLE_{strong}).

Theorem 4.3 (Grigor'yan, Telcs [GT12, Theorem 7.4]) *Let (M, d, μ) be a metric measure space and let (V_α) hold. Let $(\mathcal{E}, \mathcal{F})$ be a regular, strongly local Dirichlet form in $L^2(M, \mu)$. Then, the following equivalences are true:*

$$\begin{aligned}(H) + \left(E_\beta\right) &\Leftrightarrow (UE_{loc}) + (NLE)\\ &\Leftrightarrow \left(UE_{strong}\right) + \left(NLE_{strong}\right).\end{aligned}$$

This theorem is proved in [GT12] for a more general setting of volume doubling instead of (V_α).

Observe that the following implications hold [GT12, Lemma 7.3]:

$$\begin{aligned}(H) &\Rightarrow (M, d) \text{ is connected,}\\ \left(E_\beta\right) &\Rightarrow (\mathcal{E}, \mathcal{F}) \text{ is conservative,}\\ \left(E_\beta\right) &\Rightarrow diam\,(M) = \infty.\end{aligned}$$

Proof Sketch of proof for Theorem 4.3 First one shows that

$$(V_\alpha) + \left(E_\beta\right) + (H) \Rightarrow (FK),$$

which is quite involved and uses, in particular, Lemma 3.10. Once having $(V_\alpha) + \left(E_\beta\right) + (FK)$, we obtain (UE_{loc}) by Theorem 3.8.

Using the elliptic Harnack inequality, one obtains in a standard way the oscillating inequality for harmonic functions and then for functions of the form $u = G_\Omega f$ (that solves the equation $\mathcal{L}_\Omega u = f$) in terms of $\|f\|_\infty$.

If now $u = P_t^\Omega f$ then u satisfies the equation

$$\frac{d}{dt}u = -\mathcal{L}_\Omega u,$$

and whence

$$u = -G_\Omega\left(\frac{d}{dt}u\right).$$

Knowing an upper bound for u, which follows from the upper bound of the heat kernel, one obtains also an upper bound for $\frac{d}{dt}u$ in terms of u. Applying the oscillation inequality one obtains the Hölder continuity of u and, hence, of the heat kernel.

Let us prove the on-diagonal lower bound

$$p_t(x, x) \geq ct^{-\alpha/\beta}.$$

Note that (UE_{loc}) and (V_α) imply that

$$\int_{B(x,r)} p_t(x,y)\,d\mu(y) \geq \frac{1}{2}$$

provided $r \geq Kt^{1/\beta}$ (cf. [GHL03, formula (3.8)]). Choosing $r = Kt^{1/\beta}$, we obtain

$$\begin{aligned} p_{2t}(x,x) &= \int_M p_t^2(x,y)\,d\mu(y) \\ &\geq \frac{1}{\mu(B(x,r))}\left(\int_{B(x,r)} p_t(x,y)\,d\mu(y)\right)^2 \\ &\geq \frac{c}{r^\alpha} = \frac{c'}{t^{\alpha/\beta}}. \end{aligned}$$

Then (NLE) follows from the upper estimate for

$$|p_t(x,x) - p_t(x,y)|$$

when y close to x, which follows from the oscillation inequality. ■

We next characterize $(UE_{loc}) + (NLE)$ by using the estimates of the capacity and of the Green function.

Definition 4.4 *(capacity)* Let Ω be an open set in M and $A \Subset \Omega$ be a Borel set. Define the *capacity* $\text{cap}(A,\Omega)$ by

$$\text{cap}(A,\Omega) := \inf\{\mathcal{E}(\varphi) : \varphi \text{ is a cutoff function of } (A,\Omega)\}. \tag{4.1}$$

It follows from the definition that the capacity $\text{cap}(A,\Omega)$ is increasing in A, and decreasing in Ω, namely, if $A_1 \subset A_2, \Omega_1 \supset \Omega_2$, then $\text{cap}(A_1,\Omega_1) \leq \text{cap}(A_2,\Omega_2)$. Using the latter property, let us extend the definition of capacity when $A \subset \Omega$ as follows:

$$\text{cap}(A,\Omega) = \lim_{n\to\infty} \text{cap}(A \cap \Omega_n, \Omega) \tag{4.2}$$

where $\{\Omega_n\}$ is any increasing sequence of precompact open subsets of Ω exhausting Ω (in particular, $A \cap \Omega_n \Subset \Omega$).

Note that by the monotonicity property of the capacity, the limit in the right hand side of (4.2) exists (finite or infinite) and is independent of the choice of the exhausting sequence $\{\Omega_n\}$.

Next, define the *resistance* $\text{res}(A,\Omega)$ by

$$\text{res}(A,\Omega) = \frac{1}{\text{cap}(A,\Omega)}. \tag{4.3}$$

We introduce the notions of the Green operator and the Green function.

Definition 4.5 For an open $\Omega \subset M$, a linear operator $G^{\Omega} : L^2(\Omega) \to \mathcal{F}(\Omega)$ is called a *Green operator* if, for any $\varphi \in \mathcal{F}(\Omega)$ and any $f \in L^2(\Omega)$,

$$\mathcal{E}(G^{\Omega} f, \varphi) = (f, \varphi). \tag{4.4}$$

If G^{Ω} admits an integral kernel g^{Ω}, that is,

$$G^{\Omega} f(x) = \int_{\Omega} g^{\Omega}(x, y) f(y) d\mu(y) \text{ for any } f \in L^2(\Omega), \tag{4.5}$$

then g^{Ω} is called a *Green function.*

It is known (cf. [GH00, Lemma 5.1]) that if $(\mathcal{E}, \mathcal{F})$ is regular and if $\Omega \subset M$ is open such that $\lambda_{\min}(\Omega) > 0$, then the Green operator G^{Ω} exists, and in fact, $G^{\Omega} = (-\mathcal{L}^{\Omega})^{-1}$, the inverse of $-\mathcal{L}^{\Omega}$, where $\mathcal{L}^{\Omega}$ is the generator of $(\mathcal{E}, \mathcal{F}(\Omega))$. However, the issue of the Green function g^{Ω} is much more involved, and is one of the key topics in [GH00].

For an open set $\Omega \subset M$, function E^{Ω} is defined by

$$E^{\Omega}(x) := G^{\Omega}\mathbf{1}(x) \; (x \in M), \tag{4.6}$$

namely, the function E^{Ω} is a unique weak solution of the following Poisson-type equation

$$-\mathcal{L}^{\Omega} E^{\Omega} = 1, \tag{4.7}$$

provided that $\lambda_{\min}(\Omega) > 0$.

It is known that

$$E^{\Omega}(x) = \mathbb{E}_x(\tau_{\Omega}) \text{ for } \mu\text{-a.a. } x \in M. \tag{4.8}$$

Clearly, if the Green function g^{Ω} exists, then

$$E^{\Omega}(x) = G^{\Omega}\mathbf{1}(x) = \int_{\Omega} g^{\Omega}(x, y) \, d\mu(y) \tag{4.9}$$

for μ-almost all $x \in M$.

We introduce the following hypothesis.

(R_β) (*Resistance condition* (R_β)) We say that the *resistance condition* (R_β) is satisfied if, there exist constants $K, C > 1$ such that, for any ball B of radius $r > 0$,

$$C^{-1}\frac{r^{\beta}}{\mu(B)} \le \operatorname{res}(B, KB) \le C\frac{r^{\beta}}{\mu(B)}, \tag{4.10}$$

where constants K and C are independent of the ball B. Equivalently, (4.10) can be written in the form

$$\operatorname{res}(B, KB) \asymp \frac{r^\beta}{\mu(B)}.$$

$\left(E'_\beta\right)$ (*Condition* $\left(E'_\beta\right)$) We say that *condition* $\left(E'_\beta\right)$ holds if, there exist two constants $C > 1$ and $\delta_1 \in (0, 1)$ such that, for any ball B of radius $r > 0$,

$$\begin{aligned} \operatorname*{esup}_B E^B &\leq Cr^\beta, \\ \operatorname*{einf}_{\delta_1 B} E^B &\geq C^{-1} r^\beta. \end{aligned}$$

(G_β) (*Condition* (G_β)) We say that *condition* (G_β) holds if, there exist constants $K > 1$ and $\dot{C} > 0$ such that, for any ball $B := B(x_0, R)$, the Green kernel g^B exists and is jointly continuous off the diagonal, and satisfies

$$g^B(x_0, y) \leq C \int_{K^{-1} d(x_0, y)}^{R} \frac{s^\beta ds}{s V(x, s)} \text{ for all } y \in B \setminus \{x_0\},$$

$$g^B(x_0, y) \geq C^{-1} \int_{K^{-1} d(x_0, y)}^{R} \frac{s^\beta ds}{s V(x, s)} \text{ for all } y \in K^{-1} B \setminus \{x_0\},$$

where $V(x, r) = \mu(B(x, r))$ as before.

Theorem 4.6 (Grigor'yan and Hu) [GH00, Theorem 3.14]) *Let* (M, d, μ) *be a metric measure space and let* (V_α) *hold. Let* $(\mathcal{E}, \mathcal{F})$ *be a regular, strongly local Dirichlet form in* $L^2(M, \mu)$. *Then, the following equivalences are true:*

$$\begin{aligned} (H) + \left(E'_\beta\right) &\Leftrightarrow (G_\beta) \Leftrightarrow (H) + (R_\beta) \\ &\Leftrightarrow (UE_{loc}) + (NLE) \\ &\Leftrightarrow \left(UE_{strong}\right) + \left(NLE_{strong}\right). \end{aligned}$$

We mention that condition (V_α) can be replaced by conditions (*VD*) and (*RVD*), the latter refers to the reverse doubling condition (cf. [GH00]).

Sketch of proof for Theorem 4.6 The proofs of Theorem 4.6 consists of two parts.

- *Part One*. Firstly, the following implications hold:

$$\boxed{\left(UE_{strong}\right)+\left(NLE_{strong}\right)}$$

$$\Updownarrow$$

$$\boxed{(UE_{loc})+(NLE)}$$

$$\Updownarrow \qquad\qquad \Uparrow$$

$$\boxed{(H)+\left(E_\beta\right)} \Rightarrow \boxed{(H)+\left(E'_\beta\right)}$$

In fact, by Theorem 4.3, we only need to show that

$$\left(E_\beta\right) \Rightarrow \left(E'_\beta\right), \tag{4.11}$$

$$(H)+\left(E'_\beta\right) \Longrightarrow (UE_{loc})+(NLE). \tag{4.12}$$

The implication (4.11) can be proved directly by using the probability argument, see [GH00, Theorem 3.14]. And the implication (4.12) can be done by showing the following

$$(H)+\left(E'_\beta\right) \Rightarrow (FK) \;\; \text{([GT12, formula (3.17) and T.3.11])}$$
$$\left(E'_\beta\right) \Rightarrow \left(S_\beta\right) \;\; \text{(by [GHL00, formula (6.34)])}$$
$$(FK)+\left(S_\beta\right) \Rightarrow (UE_{loc}) \;\; \text{(by Theorem 3.8)}$$
$$(H)+\left(E'_\beta\right) \Rightarrow (NLE) \;\; \text{(by [GT12, Section 5.4]).}$$

- *Part Two*. Secondly, we need to show that

$$(H)+\left(E'_\beta\right) \Leftrightarrow \left(G_\beta\right) \Leftrightarrow (H)+\left(R_\beta\right).$$

This is the hard part. The cycle implications are obtained in [GH00, Section 8] as follows:

$$(H)+\left(R_\beta\right) \Longrightarrow \left(G_\beta\right) \Longrightarrow (H)+\left(E'_\beta\right) \Longrightarrow (H)+\left(R_\beta\right).$$

One of the most challenging results (cf. [GH00, Lemma 5.7]) is to obtain an *annulus Harnack inequality* for the Green function, without assuming any specific properties of the metric d, unlike previously known similar results in [Bar05], [GT02] where the geodesic property of the distance function was used. ■

4.2 Matching Upper and Lower Bounds

The purpose of this subsection is to improve both (UE_{loc}) and (NLE) in order to obtain matching upper and lower bounds for the heat kernel. The reason why (UE_{loc}) and (NLE) do not match, in particular, why (NLE) contains no information about lower bound of $p_t(x, y)$ for distant x, y is the lack of *chaining properties* of the distance function, that is an ability to connect any two points $x, y \in M$ by a chain of balls of controllable radii so that the number of balls in this chain is also under control.

For example, the chain condition considered above is one of such properties. If (M, d) satisfies the chain condition, then as we have already mentioned, (NLE) implies the full sun-Gaussian lower estimate by the chain argument and the semigroup property (see for example [GHL03, Corollary 3.5]).

Here we consider a setting with weaker chaining properties. For any $\varepsilon > 0$, we introduce a modified distance $d_\varepsilon(x, y)$ by

$$d_\varepsilon(x, y) = \inf_{\{x_i\} \text{ is } \varepsilon\text{-chain}} \sum_{i=1}^{N} d(x_i, x_{i-1}), \tag{4.13}$$

where an *ε-chain* is a sequence $\{x_i\}_{i=0}^N$ of points in M such that

$$x_0 = x, \quad x_N = y, \quad \text{and } d(x_i, x_{i-1}) < \varepsilon \text{ for all } i = 1, 2, ..., N.$$

Clearly, $d_\varepsilon(x, y)$ is decreases as ε increases and $d_\varepsilon(x, y) = d(x, y)$ if $\varepsilon > d(x, y)$. As $\varepsilon \downarrow 0$, $d_\varepsilon(x, y)$ increases and can go to ∞ or even become equal to ∞. It is easy to see that $d_\varepsilon(x, y)$ satisfies all properties of a distance function except for finiteness, so that it is a distance function with possible value $+\infty$.

It is easy to show that

$$d_\varepsilon(x, y) \asymp \varepsilon N_\varepsilon(x, y),$$

where $N_\varepsilon(x, y)$ is the smallest number of balls in a chain of balls of radius ε connecting x and y (Fig. 7):

N_ε can be regarded as the graph distance on a graph approximation of M by an ε-net.

If d is geodesic, then the points $\{x_i\}$ of an ε-chain can be chosen on the shortest geodesic, whence $d_\varepsilon(x, y) = d(x, y)$ for any $\varepsilon > 0$. If the distance function d satisfies the chain condition, then one can choose in (4.13) an ε-chain so that $d(x_i, x_{i+1}) \le C\frac{d(x,y)}{N}$, whence $d_\varepsilon(x, y) \le Cd(x, y)$. In general, $d_\varepsilon(x, y)$ may go to ∞ as $\varepsilon \to 0$, and the rate of growth of $d_\varepsilon(x, y)$ as $\varepsilon \to 0$ can be regarded as a quantitative description of the chaining properties of d.

We need the following hypothesis

C_β (*Chaining property*) For all $x, y \in M$,

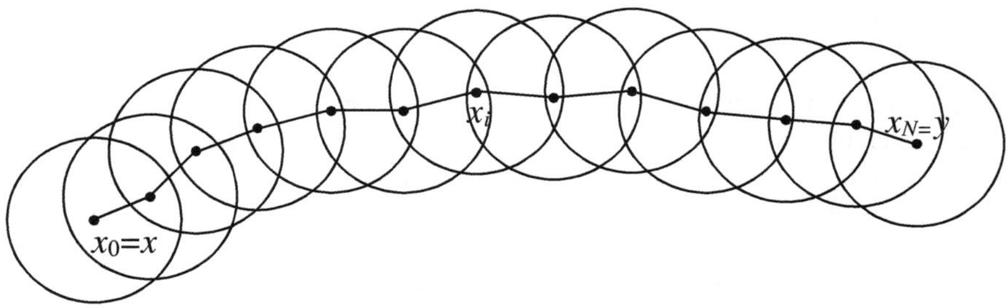

Fig. 7 Chain of balls connecting x and y

$$\varepsilon^{\beta-1} d_{\varepsilon}(x, y) \to 0 \text{ as } \varepsilon \to 0,$$

or equivalently,

$$\varepsilon^{\beta} N_{\varepsilon}(x, y) \to 0 \text{ as } \varepsilon \to 0.$$

For $x \neq y$ we have $\varepsilon^{\beta-1} d_{\varepsilon}(x, y) \to \infty$ as $\varepsilon \to \infty$, which implies under (C_β) that for any $t > 0$, there is $\varepsilon = \varepsilon(t, x, y)$ satisfying the identity

$$\varepsilon^{\beta-1} d_{\varepsilon}(x, y) = t \tag{4.14}$$

(always take the maximal possible value of ε). If $x = y$, then set $\varepsilon(t, x, x) = \infty$.

Theorem 4.7 (Grigor'yan, Telcs [GT12, Section 6]) *Assume that all the hypothesis of Theorem 4.6 hold. If $(E_\beta) + (H)$ and (C_β) are satisfied, then*

$$p_t(x, y) \asymp \frac{C}{t^{\alpha/\beta}} \exp\left(-c\left(\frac{d_{\varepsilon}^{\beta}(x, y)}{t}\right)^{\frac{1}{\beta-1}}\right) \tag{4.15}$$

$$\asymp \frac{C}{t^{\alpha/\beta}} \exp\left(-cN_{\varepsilon}(x, y)\right), \tag{4.16}$$

where $\varepsilon = \varepsilon(t, x, y)$.

Since $d_{\varepsilon}(x, y) \geq d(x, y)$, the upper bound in (4.15) is an improvement of (UE_{loc}); similarly the lower bound in (4.15) is an improvement of (NLE). The proof of the upper bound in (4.15) follows the same line as the proof of (UE_{loc}) with careful tracing all places where the distance $d(x, y)$ is used and making sure that it can be replaced by $d_{\varepsilon}(x, y)$. The proof of the lower bound in (4.16) uses (NLE) and the semigroup identity along the chain with N_{ε} balls connecting x and y . Finally, observe that (4.15) and (4.16) are equivalent, that is

$$N_\varepsilon \asymp \left(\frac{d_\varepsilon^\beta (x, y)}{t}\right)^{\frac{1}{\beta-1}},$$

which follows by substituting here $N_\varepsilon \asymp d_\varepsilon/\varepsilon$ and $t = \varepsilon^{\beta-1} d_\varepsilon (x, y)$.

By Theorem 4.6, the same conclusion in Theorem 4.7 is true if $(E_\beta) + (H)$ is instead replaced by the one of conditions $(H) + \left(E'_\beta\right)$, (G_β) and $(H) + (R_\beta)$.

Example 4.8 A good example to illustrate Theorem 4.7 is the class of post critically finite (p.c.f.) fractals. For connected p.c.f. fractals with regular harmonic structure, the heat kernel estimate (4.16) was proved by Hambly and Kumagai [HK99], see also [KS05, Theorem 5.2]. In this setting $d (x, y)$ is the resistance metric of the fractal M and μ is the Hausdorff measure of M of dimension $\alpha := \dim_H M$. Hambly and Kumagai proved that (V_α) and (E_β) are satisfied with $\beta = \alpha + 1$. The condition (C_β) follows from their estimate

$$N_\varepsilon (x, y) \le C \left(\frac{d (x, y)}{\varepsilon}\right)^{\beta/2},$$

because

$$\varepsilon^\beta N_\varepsilon (x, y) \le C d (x, y)^{\beta/2} \varepsilon^{\beta/2} \to 0 \text{ as } \varepsilon \to 0.$$

The Harnack inequality (H) on p.c.f. fractals was proved by Kigami [Kig01, Proposition 3.2.7, p.78]. Hence, Theorem 4.7 applies and gives the estimates (4.15) and (4.16).

The estimate (4.16) means that the diffusion process goes from x to y in time t in the following way. The process firstly "computes" the value $\varepsilon (t, x, y)$, secondly "detects" a shortest chain of ε-balls connecting x and y, and then goes along that chain (Fig. 8).

This phenomenon was first observed by Hambly and Kumagai on p.c.f. fractals, but it seems to be generic. Hence, to obtain matching upper and lower bounds, one needs in addition to the usual hypotheses also the following information, encoded in the function $N_\varepsilon (x, y)$: the graph distance between x and y on any ε-net approximation of M.

Example 4.9 (*Computation of* ε) Assume that the following bound is known for all $x, y \in M$ and $\varepsilon > 0$

$$N_\varepsilon (x, y) \le C \left(\frac{d (x, y)}{\varepsilon}\right)^\gamma,$$

where $0 < \gamma < \beta$, so that (C_β) is satisfied (since $N_\varepsilon \ge d (x, y) / \varepsilon$, one must have $\gamma \ge 1$). Since by (4.14) we have $\varepsilon^\beta N_\varepsilon \asymp t$, it follows that

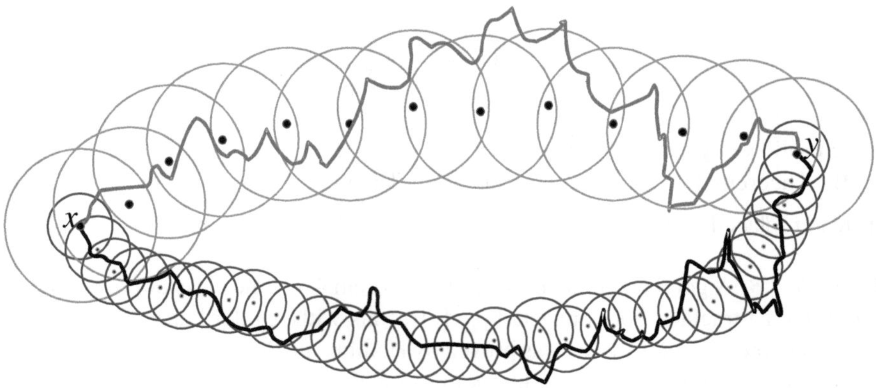

Fig. 8 Two shortest chains of ε-ball for two distinct values of ε provide different routes for the diffusion from x to y for two distinct values of t

$$\varepsilon^{\beta}\left(\frac{d(x,y)}{\varepsilon}\right)^{\gamma} \geq ct,$$

whence

$$\varepsilon \geq c\left(\frac{t}{d(x,y)^{\gamma}}\right)^{\frac{1}{\beta-\gamma}}.$$

Consequently, we obtain

$$N_{\varepsilon}(x,y) \leq Cd(x,y)^{\gamma}\varepsilon^{-\gamma} \leq Cd(x,y)^{\gamma}\left(\frac{d(x,y)^{\gamma}}{t}\right)^{\frac{\gamma}{\beta-\gamma}} = C\left(\frac{d(x,y)^{\beta}}{t}\right)^{\frac{\gamma}{\beta-\gamma}},$$

and so

$$p_t(x,y) \geq \frac{c}{t^{\alpha/\beta}}\exp\left(-\left(\frac{d(x,y)^{\beta}}{ct}\right)^{\frac{\gamma}{\beta-\gamma}}\right).$$

Similarly, the lower estimate of N_{ε}

$$N_{\varepsilon}(x,y) \geq c\left(\frac{d(x,y)}{\varepsilon}\right)^{\gamma}$$

implies an upper bound for the heat kernel

$$p_t(x,y) \leq \frac{C}{t^{\alpha/\beta}}\exp\left(-\left(\frac{d(x,y)^{\beta}}{Ct}\right)^{\frac{\gamma}{\beta-\gamma}}\right).$$

Remark 4.10 Assume that (V_α) holds and all balls in M of radius $\geq r_0$ are connected, for some $r_0 > 0$. We claim that (C_β) holds with any $\beta > \alpha$. The α-regularity of measure μ implies, by the classical ball covering argument, that any ball B_r of radius r can be covered by at most $C\left(\frac{r}{\varepsilon}\right)^\alpha$ balls of radii $\varepsilon \in (0, r)$. Consequently, if B_r is connected then any two points $x, y \in B_r$ can be connected by a chain of ε-balls containing at most $C\left(\frac{r}{\varepsilon}\right)^\alpha$ balls, so that

$$N_\varepsilon(x, y) \leq C\left(\frac{r}{\varepsilon}\right)^\alpha.$$

Since any two points $x, y \in M$ are contained in a connected ball B_r (say, with $r = r_0 + d(x, y)$), we obtain

$$\varepsilon^\beta N_\varepsilon(x, y) \leq C\varepsilon^{\beta-\alpha} r^\alpha \to 0$$

as $\varepsilon \to 0$, which was claimed.

4.3 Further Results

We discuss here some consequences and extensions of the above results. For this, we introduce two-sided estimates of the heat kernel.

(ULE_{loc}) (*Upper and lower estimates*) The heat kernel $p_t(x, y)$ exists and satisfies

$$p_t(x, y) \asymp \frac{C}{t^{\alpha/\beta}} \exp\left(-c\left(\frac{d^\beta(x, y)}{t}\right)^{\frac{1}{\beta-1}}\right). \tag{4.17}$$

Theorem 4.11 *Let (M, d, μ) be a metric measure space, and let $(\mathcal{E}, \mathcal{F})$ be a regular, conservative Dirichlet form in $L^2(M, \mu)$. If (M, d) satisfies the chain condition, then the following equivalences take place:*

$$(V_\alpha) + \begin{cases} (E_\beta) + (H) \\ (E'_\beta) + (H) \\ (R_\beta) + (H) \\ (G_\beta) \end{cases} + (\text{locality}) \Longleftrightarrow (ULE_{loc}),$$

where condition (locality) means that $(\mathcal{E}, \mathcal{F})$ is local.

Remark 4.12 Observe that if $(\mathcal{E}, \mathcal{F})$ is regular, conservative and local, then $(\mathcal{E}, \mathcal{F})$ is strongly local; this is easily seen by using the Beuling-Deny decomposition [FOT11, Theorem 3.2.1, p. 120] and by noting that both killing and jump measures disappear.

Remark 4.13 Observe also that $(V_\alpha) + (NLE)+$ (chain condition) implies that the off-diagonal lower estimate

$$p_t(x, y) \geq \frac{C'}{t^{\alpha/\beta}} \exp\left(-c'\left(\frac{d^\beta(x, y)}{t}\right)^{\frac{1}{\beta-1}}\right) \tag{4.18}$$

for μ-almost all $x, y \in M$ and all $t > 0$, see for example [GHL08, Proposition 3.1] or [Bar98], [GHL03, Corollary 3.5].

Sketch of proof for Theorem 4.11 (1) "$\Rightarrow$".
Let us show the implication

$$(V_\alpha) + (E_\beta) + (H) + (\text{ locality}) \Rightarrow (ULE_{loc}). \tag{4.19}$$

Indeed, by Remark 4.12, we have that $(\mathcal{E}, \mathcal{F})$ is strongly local. Now, using Theorem 4.3, we obtain $(UE_{loc}) + (NLE)$. Using Remark 4.13, we see that (4.18) holds, showing that (ULE_{loc}) is true.

Similarly, using Theorem 4.6, we obtain the other three implications "$\Rightarrow$".
(2) "$\Leftarrow$".
Let us show the opposite implication

$$(ULE_{loc}) \Rightarrow (V_\alpha) + (E_\beta) + (H) + (\text{locality}). \tag{4.20}$$

Indeed, note that

$$\begin{aligned} (ULE_{loc}) &\Rightarrow (V_\alpha) \text{ (by Theorem 2.2)} \\ (UE_{loc}) &\Rightarrow \text{(locality) (by Theorem 2.14)} \\ (UE_{loc}) + (NLE) &\Rightarrow (E_\beta) + (H) \text{ (by Theorem 4.3)} \end{aligned}$$

showing that the implication (4.20) holds.
Similarly, all the other three implications "$\Longleftarrow$" also hold. ■

Remark 4.14 The implication (4.19) can also be proved by using Theorem 4.7 and the fact that $d_\varepsilon \asymp d$.

Conjecture 4.15 *The condition* (E_β) *above may be replaced by*

$$\lambda_{\min}(B(x, r)) \asymp r^{-\beta}. \tag{λ_β}$$

In fact, (E_β) in all statements can be replaced by the resistance condition:

$$\text{res}(B_r, B_{2r}) \asymp r^{\beta-\alpha} \tag{R_β}$$

where $B_r = B(x, r)$. In the strongly recurrent case $\alpha < \beta$, it alone implies the elliptic Harnack inequality (H) so that two sided heat kernel estimates are equivalent to $(V_\alpha) + (R_\beta)$ as was proved by Barlow, Coulhon, Kumagai [BCK05] (in a setting of graphs) and was discussed in M. Barlow's lectures.

An interesting (and obviously hard) question is the characterization of the elliptic Harnack inequality (H) in more geometric terms—so far nothing is known, not even a conjecture.

One can consider also a *parabolic* Harnack inequality *(PHI)*, which uses caloric functions instead of harmonic functions. Then in a general setting and assuming the volume doubling condition *(VD)* (instead of (V_α)), the following holds (cf. [BGK12]):

$$(PHI) \Leftrightarrow (UE_{loc}) + (NLE)\,.$$

On the other hand, *(PHI)* is equivalent to

Poincaré inequality + cutoff Sobolev inequality,

see [BBK06].

Conjecture 4.16 *The cutoff Sobolev inequality here can be replaced by* (λ_β) *and/or* (R_β).

5 Upper Bounds for Jump Processes

We have investigated above the heat kernel for the *local* Dirichlet form. In this section we shall study the *non-local* Dirichlet form and present the equivalence conditions for upper bounds of the associated heat kernel. As an interesting example, we discuss the heat kernel estimates for effective metric spaces.

A non-local Dirichlet form will give arise to a jump process, that is, the trajectories of this process are discontinuous, as we have already seen for a symmetric stable process of index β (Lévy process). And the heat kernel decays at a polynomial rate (cf. 1.2), instead of an exponential rate as for a local Dirichlet form.

Jump process have found various applications in science. For instance, a Lévy flight is a jump process and can be used to describe animal foraging patterns, the distribution of human travel and some aspects of earthquake behavior (cf. [BBW08]).

5.1 Upper Bounds for Non-local Dirichlet Forms

The techniques for obtaining heat kernel bounds for non-local Dirichlet forms has been developed by a number of authors, see for example [BBCK09, BGK09, BL02, CK03, CK08] and the references therein. The basic approach to obtaining heat kernel upper estimates used in these papers consists of the two steps. The first step is to obtain the heat kernel upper bounds for a *truncated* Dirichlet form, that is, in the case when the jump density $J(x, y)$ has a bounded range. In this case one uses the Davies method as it was presented in the seminal work [CKS87] and where the cut-off functions of form $(\lambda - d(x_0, x))_+$ were used (where λ is a positive constant). This method can be used as long as the cut-off functions belong to the domain of the Dirichlet form, which is the case only when $\beta < 2$ (hence, if $\beta \geq 2$ then this method does not work).

The second step is to obtain heat kernel estimates for the original Dirichlet form by comparing the heat semigroup of the truncated Dirichlet form with the original heat semigroup. We remark that while the first step was done by purely analytic means, the second step in the above-mentioned papers used a probabilistic argument.

Here we describe an alternative new approach of [GHL00] for obtaining upper bounds.

Recall that by a theorem of Beurling and Deny, any regular conservative Dirichlet form admits a decomposition

$$\mathcal{E}(u, v) = \mathcal{E}^{(L)}(u, v) + \mathcal{E}^{(J)}(u, v), \tag{5.1}$$

where $\mathcal{E}^{(L)}$ is a *local part* and

$$\mathcal{E}^{(J)}(u, v) = \iint\limits_{M \times M \setminus \text{diag}} (u(x) - u(y))(v(x) - v(y))\, dj(x, y) \tag{5.2}$$

is a *jump part* with a jump measure j defined on $M \times M \setminus \text{diag}$. In our setting the jump measure j will have a density with respect to $\mu \times \mu$, which will be denoted by $J(x, y)$, and so the jump part $\mathcal{E}^{(J)}$ becomes

$$\mathcal{E}^{(J)}(u, v) = \iint\limits_{M \times M} (u(x) - u(y))(v(x) - v(y))\, J(x, y) d\mu(y) d\mu(x). \tag{5.3}$$

We introduce the following hypothesis.

($\mathbf{V}_{\leq}$) (*Upper α- regularity*) For all $x \in M$ and all $r > 0$,

$$V(x, r) \leq C r^{\alpha}.$$

(**UE**) (*Upper estimate of non-local type*) The heat kernel p_t exists and satisfies the off-diagonal upper estimate

$$p_t(x, y) \leq \frac{C}{t^{\alpha/\beta}} \left(1 + \frac{d(x, y)}{t^{1/\beta}}\right)^{-(\alpha+\beta)}$$

for all $t > 0$ and μ-almost all $x, y \in M$.

($\mathbf{J}_{\leq}$) (*Upper bound of jump density*) The jump density exists and admits the estimate

$$J(x,y) \le Cd(x,y)^{-(\alpha+\beta)},$$

for μ-almost all $x, y \in M$.

($\mathbf{T}_{strong}$) (*Strong tail estimate*) There exist constants $c > 0$ and $\beta > 0$ such that, for all balls $B = B(x_0, r)$ and for all $t > 0$,

$$P_t \mathbf{1}_{B^c}(x) \le \frac{ct}{r^\beta} \text{ for } \mu \text{ -almost all } x \in \frac{1}{4}B.$$

Clearly, we have that $\left(T_{\exp}\right) \Rightarrow \left(T_{strong}\right) \Rightarrow \left(T_\beta\right)$.

We now state the main technical result of [GHL00].

Theorem 5.1 (Grigor'yan et al. [GHL00]) *Let (M, d, μ) be a metric measure space with precompact balls, and let $(\mathcal{E}, \mathcal{F})$ be a regular conservative Dirichlet form in $L^2(M, \mu)$ with jump density J. Then the following implication holds:*

$$\left(V_\le\right) + (DUE) + (J_\le) + \left(S_\beta\right) \Rightarrow (UE). \tag{5.4}$$

We remark that by [GHL03, Theorem 3.2], if $(\mathcal{E}, \mathcal{F})$ is conservative then

$$\left(V_\le\right) + (UE) \Rightarrow (V_\alpha).$$

Hence, the hypotheses of Theorem 5.1 imply that μ is α-regular.

Sketch of proof for Theorem 5.1 We sketch the ideas of the proof.

- *Step* 1. We decompose $\mathcal{E}(u)$ into two parts:

$$\mathcal{E}(u) = \mathcal{E}^{(\rho)}(u) + \int_M \int_{M \setminus B(x,\rho)} (u(x) - u(y))^2 J(x,y) d\mu(y) d\mu(x),$$

where $\rho \in (0, \infty)$ is any fixed number. Then the form $\left(\mathcal{E}^{(\rho)}, \mathcal{F}\right)$ can be extended to a regular Dirichlet form $\left(\mathcal{E}^{(\rho)}, \mathcal{F}^{(\rho)}\right)$. Indeed, since using condition $(J_\le)$,

$$\operatorname*{esup}_{x \in M} \int_{B(x,\rho)^c} J(x,y) d\mu(y) < \infty,$$

the form $\left(\mathcal{E}^{(\rho)}, \mathcal{F}\right)$ is closable, and its closure $\left(\mathcal{E}^{(\rho)}, \mathcal{F}^{(\rho)}\right)$ in L^2 is a regular Dirichlet form in L^2. Note that $\left(\mathcal{E}^{(\rho)}, \mathcal{F}^{(\rho)}\right)$ is ρ-*local* (non-local): $\mathcal{E}^{(\rho)}(f, g) = 0$ for any two functions $f, g \in \mathcal{F}^{(\rho)}$ with compact supports such that

$$\operatorname{dist}(\operatorname{supp} f, \operatorname{supp} g) > \rho.$$

- *Step* 2. We need to obtain upper estimates of the heat kernel $q_t(x,y)$ of the *truncated* Dirichlet form $\left(\mathcal{E}^{(\rho)}, \mathcal{F}^{(\rho)}\right)$. Indeed, conditions (DUE), $\left(J_\le\right)$, (S_β) and $\left(V_\le\right)$

imply the following estimate of $q_t(x, y)$:

$$q_t(x, y) \le \frac{C}{t^{\alpha/\beta}} \exp\left(4\rho^{-\beta}t\right) \exp\left(-c\left(\frac{d(x, y)}{\rho} \wedge \frac{\rho}{t^{1/\beta}}\right)\right) \tag{5.5}$$

for all $t > 0$ and μ-almost all $x, y \in M$, where constants $C, c > 0$ depend on the constants in the hypotheses but are independent of ρ. This can be done with a certain amount of effort, by using the bootstrapping technique where the comparison inequality [GHL10, Corollary 4.8, Remark 4.10] for heat semigroups play an important rôle.

- *Step* 3. Next we apply the following useful inequality between two heat kernels:

$$p_t(x, y) \le q_t(x, y) + 2t \operatorname*{esup}_{x\in M, y\in B(x,\rho)^c} J(x, y) \tag{5.6}$$

for all $t > 0$ and almost all $x, y \in M$; this inequality follows from the parabolic maximum principle alone. Therefore, by choosing an appropriate ρ, it follows from (5.5), (5.6) that, for any real $n \ge 0$,

$$p_t(x, y) \le \frac{c(n)}{t^{\alpha/\beta}} \left(1 + \frac{d(x, y)}{t^{1/\beta}}\right)^{-\frac{(\alpha+\beta)n}{n+\alpha+\beta}} \tag{5.7}$$

for almost all $x, y \in M$ and all $t > 0$.
Note that (5.7) is nearly close to our desired estimate (UE). However, one can not just obtain (UE) by directly taking the limit as $n \to \infty$, since we do not know whether the coefficient $c(n)$ is bounded uniformly in n. We need the second iteration.

- *Step* 4. Finally, we will obtain (UE) by a self-improvement of (5.7). Indeed, one can use (5.7) to obtain

$$\int_{B(x,r)^c} p_t(x, y)d\mu(y) \le C(n)\left(rt^{-1/\beta}\right)^{-\theta},$$

where $\theta = \frac{n\beta-\alpha(\alpha+\beta)}{n+\alpha+\beta} \in (0, \beta)$ (note that this estimate is sharper than condition (S_β)), and then repeating the above procedure, we arrive at (UE). ∎

Now we can state some equivalences for (UE).

Theorem 5.2 (Grigor'yan et al. [GHL00]) *Let (M, d, μ) be a metric measure space with precompact balls, and let $(\mathcal{E}, \mathcal{F})$ be a regular conservative Dirichlet form in $L^2(M, \mu)$ with jump density J. If $(V_\le)$ holds, then the following equivalences are true:*

$$\begin{aligned}(UE) &\Leftrightarrow (UE\Phi) + (J_{\leq}) \\ &\Leftrightarrow (DUE) + (J_{\leq}) + \left(T_{\beta}\right) \\ &\Leftrightarrow (DUE) + (J_{\leq}) + \left(S_{\beta}\right) \\ &\Leftrightarrow (DUE) + (J_{\leq}) + \left(T_{strong}\right).\end{aligned} \tag{5.8}$$

Proof Observe that the implication $(UE) \Rightarrow (J_{\leq})$ holds by [BGK09, p. 150], and $(UE) \Rightarrow (UE\Phi)$ is trivial by taking $\Phi(s) = (1+s)^{-(\alpha+\beta)}$. The implication $(UE\Phi) \Rightarrow (DUE)$ is obvious. The implication $(UE\Phi) \Rightarrow \left(T_{\beta}\right)$ was proved in (3.10) (see also [GHL03, formula (3.6), p. 2072]). Since $(\mathcal{E}, \mathcal{F})$ is conservative, the equivalence $\left(T_{\beta}\right) \Leftrightarrow \left(S_{\beta}\right)$ holds by [GH08, Theorem 3.1, p. 96]. By Theorem 5.1 we have

$$(DUE) + (J_{\leq}) + \left(S_{\beta}\right) \Rightarrow (UE),$$

which closes the cycle of implications, thus proving the first three equivalences.

Finally, the implication $(UE) \Rightarrow \left(T_{strong}\right)$ is true by using (3.10), and hence

$$\begin{aligned}(UE) &\Rightarrow (DUE) + (J_{\leq}) + \left(T_{strong}\right) \\ &\Rightarrow (DUE) + (J_{\leq}) + \left(T_{\beta}\right) \Rightarrow (UE),\end{aligned}$$

which finishes the proof. ∎

Remark 5.3 The upper estimate (UE) is best possible for non-local forms in the following sense: if the heat kernel p_t satisfies the estimate

$$p_t(x,y) \leq \frac{1}{t^{\alpha/\beta}} \Phi\left(\frac{d(x,y)}{t^{1/\beta}}\right)$$

for all $t > 0$ and μ-almost all $x, y \in M$, where Φ is a continuous decreasing function on $[0, +\infty)$, then necessarily

$$\Phi(s) \geq c(1+s)^{-(\alpha+\beta)}$$

for some $c > 0$ (see Theorem 2.14).

Remark 5.4 Under the standing assumptions of Theorem 5.2, the following equivalence is true

$$(UE_{loc}) \Leftrightarrow (DUE) + (\text{"locality"}) + \left(S_{\beta}\right).$$

Indeed, since (UE_{loc}) is stronger than (UE), it implies (DUE) and $\left(S_{\beta}\right)$ by Theorem 5.2. Next, $(UE_{loc}) \Rightarrow$("locality") by Theorem 2.14 above. The opposite implication

$$(DUE) + (\text{"locality"}) + \left(S_{\beta}\right) \Rightarrow (UE_{loc})$$

was stated in Theorem 3.8.

In order to state some consequence of Theorem 5.2, we need the following Proposition.

Define first the following condition:

($J_{\geq}$) (*Lower bound of jump density*) There exist constants $C, \alpha, \beta > 0$ such that, for μ -almost all $x \neq y$,

$$J(x,y) \geq C^{-1} d(x,y)^{-(\alpha+\beta)}.$$

Proposition 5.5 *Let (M, d, μ) be a metric measure space, and let $(\mathcal{E}, \mathcal{F})$ be a regular Dirichlet form in $L^2(M, \mu)$ with jump density J. Then*

$$(V_\alpha) + (J_{\geq}) \Rightarrow (DUE). \tag{5.9}$$

Proof As was proved in [HuK06, Theorem 3.1], under (V_α) the following inequality holds for all non-zero functions $u \in L^1 \cap L^2$:

$$\int_M \int_M \frac{(u(x) - u(y))^2}{d(x,y)^{\alpha+\beta}} d\mu(x) d\mu(y) \geq c||u||_2^{2(1+\beta/\alpha)} \|u\|_1^{-2\beta/\alpha},$$

where c is a positive constant. Using (5.1), (5.3) and $(J_{\geq})$ we obtain

$$\begin{aligned} \mathcal{E}(u) &= \mathcal{E}^{(L)}(u) + \mathcal{E}^{(J)}(u) \\ &\geq C \int_M \int_M \frac{(u(x) - u(y))^2}{d(x,y)^{\alpha+\beta}} d\mu(x) d\mu(y) \\ &\geq c||u||_2^{2(1+\beta/\alpha)} \|u\|_1^{-2\beta/\alpha} \end{aligned}$$

for all $u \in \mathcal{F} \cap L^1$. Hence, (*DUE*) follows by Theorem 3.6. ■

We obtain the following consequence of Theorem 5.2.

Theorem 5.6 (Grigor'yan et al. [GHL00]) *Let (M, d, μ) be a metric measure space with precompact balls, and let $(\mathcal{E}, \mathcal{F})$ be a regular conservative Dirichlet form in $L^2(M, \mu)$ with jump density J. If (V_α) holds and $J(x,y) \asymp d(x,y)^{-(\alpha+\beta)}$, then*

$$(UE) \Leftrightarrow (S_\beta). \tag{5.10}$$

Proof Let us show that $(S_\beta) \Rightarrow (UE)$. Indeed, (*DUE*) holds by Proposition 5.5. Hence, (UE) is satisfied by Theorem 5.2. The opposite implication $(UE) \Rightarrow (S_\beta)$ holds also by Theorem 5.2. ■

Therefore, if (V_α) holds and $J(x,y) \asymp d(x,y)^{-(\alpha+\beta)}$, then in order to obtain off-diagonal upper bounds of heat kernels, one needs only to verify the survival condition (S_β). In the sequel, we will show that the survival condition (S_β) holds for a class of measure spaces with effective resistance metrics.

5.2 Upper Bounds Using Effective Resistance

We will show how Theorem 5.2 can be applied for a certain class of metric measure spaces with effective resistance.

Let $(\mathcal{E}, \mathcal{F})$ be a regular Dirichlet form in $L^2(M, \mu)$ as before. Recall that the *effective resistance* $R(A, B)$ between two disjoint non-empty closed subsets A and B of M is defined by

$$R(A, B)^{-1} = \inf \{\mathcal{E}(u) : u \in \mathcal{F} \cap C_0, \ u|_A = 1 \text{ and } u|_B = 0\}. \tag{5.1}$$

It follows from (5.1) that, for any fixed A, $R(A, B)$ is a non-increasing function of B. Denote by

$$R(x, B) := R(\{x\}, B) \text{ and } R(x, y) := R(\{x\}, \{y\}).$$

In general, it may happen that $R(x, y) = \infty$ for some points $x, y \in M$. Below we will exclude this case.

Fix a parameter $\gamma > 0$, and introduce conditions (R_1) and (R_2).

$(\mathbf{R}_1)$: For all $u \in \mathcal{F} \cap C_0(M)$ and all $x, y \in M$, the following inequality holds:

$$|u(x) - u(y)|^2 \leq C d(x, y)^{\gamma} \mathcal{E}(u).$$

$(\mathbf{R}_2)$: For all $x \in M$ and $r > 0$,

$$R(x, B(x, r)^c) \geq C^{-1} r^{\gamma}.$$

Theorem 5.7 (Grigor'yan et al. [GHL00]) *Let $(\mathcal{E}, \mathcal{F})$ be a regular Dirichlet form in $L^2(M, \mu)$. Then*

$$(V_\alpha) + (R_1) + (R_2) \Rightarrow \left(S_\beta\right) + (DUE),$$

where $\beta = \alpha + \gamma$. Consequently, under the standing conditions $(V_\alpha) + (R_1) + (R_2)$, we have that

$$(UE) \Leftrightarrow (J_{\leq}). \tag{5.2}$$

Sketch of proof for Theorem 5.7 The proof consists of the following five steps.

- *Step* 1. For any ball $B := B(x_0, r)$, using conditions (R_1) and (R_2), we can obtain the two-sided estimate of the Green functions $g_B(x, y)$:

$$\sup_{x,y \in B} g_B(x, y) \leq C r^{\gamma}, \tag{5.3}$$

$$\inf_{y \in B(x_0, \eta r)} g_B(x_0, y) \geq C^{-1} r^{\gamma}, \tag{5.4}$$

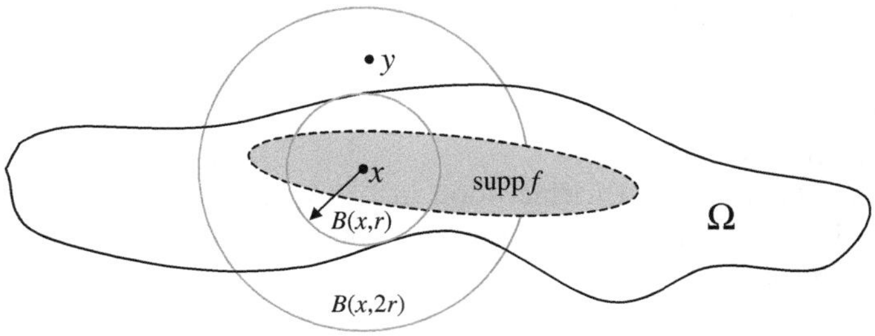

Fig. 9 Points x, y

where $C > 0$ and $\eta \in (0, 1)$.

- *Step* 2. Therefore, under condition (V_α), it follows from (5.3), (5.4) that condition $\left(E'_\beta\right)$ holds:

$$\operatorname*{esup}_{B} E^B \leq C r^{\alpha+\gamma}, \tag{5.5}$$

$$\operatorname*{einf}_{\delta_1 B} E^B \geq C^{-1} r^{\alpha+\gamma}, \tag{5.6}$$

where E^B is the weak solution of the Poisson-type equation (4.7) as before, and $C > 0$ and $\delta_1 \in (0, 1/2)$.

- *Step* 3. To show condition $\left(S_\beta\right)$, observe that, for all $t > 0$ and μ-almost all $x \in B$,

$$P_t^B 1_B(x) \geq \frac{E^B(x) - t}{\left\|E^B\right\|_\infty}, \tag{5.7}$$

which follows by using the parabolic maximum principle, nothing else. Hence, using (5.5), (5.6),

$$\begin{aligned} P_t^B 1_B(x) &\geq \frac{E^B(x) - t}{\left\|E^B\right\|_\infty} \\ &\geq c - c_1 t r^{-\beta} \\ &\geq \frac{c}{2}, \end{aligned}$$

for all $t > 0$ and μ-almost all $x \in B(x_0, \delta_1 r)$, provided that $tr^{-\beta}$ is small enough, thus proving $\left(S_\beta\right)$.

- *Step* 4. We show that $(R_1) \Rightarrow (DUE)$. Consider a function $f \in \mathcal{F} \cap C_0(\Omega)$ normalized so that $\sup|f| = 1$, and let $x \in \Omega$ be a point such that $|f(x)| = 1$. Let r be the largest radius such that $B(x, r) \subset \Omega$. Then the ball $B(x, 2r)$ is not covered by Ω so that there exists a point $y \in B(x, 2r) \setminus \Omega$ (note that M is unbounded by condition (V_α)). In particular, $y \notin \operatorname{supp} f$ (see Fig. 9). Noting that

$\mathcal{E}^{(J)}(f) \le \mathcal{E}(f)$ and by the α-regularity of μ

$$r \le C\left[\mu\left(B(x,r)\right)\right]^{1/\alpha} \le C\left[\mu\left(\Omega\right)\right]^{1/\alpha},$$

we obtain from (R_1) that

$$\begin{aligned} 1 &= |f(y) - f(x)|^2 \\ &\le Cd(y,x)^{\beta-\alpha}\mathcal{E}^{(J)}(f) \\ &\le C\,(2r)^{\beta-\alpha}\,\mathcal{E}(f) \le C2^{\beta-\alpha}\left[\mu(\Omega)\right]^{\beta/\alpha-1}\mathcal{E}(f). \end{aligned}$$

Since $\|f\|_2^2 \le \mu(\Omega)$, it follows that

$$\frac{\mathcal{E}(f)}{\|f\|_2^2} \ge c\left[\mu(\Omega)\right]^{-\beta/\alpha},$$

for some $c > 0$, thus proving the Faber-Krahn inequality. Hence, condition (DUE) follows by using Theorem 3.6.

- *Step* 5. Finally, with a certain amount of effort [GHL00, Proposition 6.5, Lemma 6.4], one can show that

$$(R_1) + (R_2) \Rightarrow \text{conservativeness of } (\mathcal{E}, \mathcal{F}).$$

Therefore, the equivalence (5.2) follows directly by using Theorem 5.2.

References

[AB] Andres, S., Barlow, M.: Energy inequalities for cutoff fucntions and some applications. J. Reine Angew. Math. (preprint)

[Aro67] Aronson, D.: Bounds for the fundamental solution of a parabolic equation. Bull. Amer. Math. Soc. **73**, 890–896 (1967)

[Aro68] Aronson, D.: Non-negative solutions of linear parabolic equations. Ann. Scuola Norm. Sup. Pisa. Cl. Sci. **22**, 607–694 (1968) (Addendum **25**:221–228, 1971)

[Bar98] Barlow, M.: Diffusions on fractals. Lecture Notes of Mathematics, vol. 1690, pp. 1–121. Springer, Berlin (1998)

[Bar04] Barlow, M.: Which values of the volume growth and escape time exponents are possible for graphs? Rev. Mat. Iberoamericana **20**, 1–31 (2004)

[Bar05] Barlow, M.: Some remarks on the elliptic harnack inequality. Bull. London Math. Soc. **37**, 200–208 (2005)

[BBCK09] Barlow, M., Bass, R., Chen, Z.-Q., Kassmann, M.: Non-local Dirichlet forms and symmetric jump processes. Trans. Amer. Math. Soc. **361**, 1963–1999 (2009)

[BBK06] Barlow, M., Bass, R., Kumagai, T.: Stability of parabolic Harnack inequalities on metric measure spaces. J. Math. Soc. Japan **58**, 485–519 (2006)

[BCK05] Barlow, M., Coulhon, T., Kumagai, T.: Characterization of sub-Gaussian heat kernel estimates on strongly recurrent graphs. Comm. Pure Appl. Math. **58**, 1642–1677 (2005)

[BGK09] Barlow, M., Grigor'yan, A., Kumagai, T.: Heat kernel upper bounds for jump processes and the first exit time. J. Reine Angew. Math. **626**, 135–157 (2009)

[BGK12] Barlow, M., Grigor'yan, A., Kumagai, T.: On the equivalence of parabolic harnack inequalities and heat kernel estimates. J. Math. Soc. Japan **64**, 1091–1146 (2012)

[BP88] Barlow, M., Perkins, E.: Brownian motion on the Sierpínski gasket. Probab. Theory Relat. Fields **79**, 543–623 (1988)

[BBW08] Barthelemy, P., Bertolotti, J., Wiersma, D.: A L*é*vy flight for light. Nature **453**, 495–498 (2008)

[BL02] Bass, R., Levin, D.: Transition probabilities for symmetric jump processes. Trans. Amer. Math. Soc. **354**, 2933–2953 (2002)

[CKS87] Carlen, E., Kusuoka, S., Stroock, D.: Upper bounds for symmetric Markov transition functions. Ann. Inst. H. Poincaré-Probab. Statist. **23**, 245–287 (1987)

[Car96] Carron, G.: Inégalités isopérimétriques de Faber-Krahn et conséquences. In: Actes de la table ronde de géométrie différentielle (Luminy, 1992) Collection SMF Séminaires et Congrès, **1** (1996), pp. 205–232

[CK03] Chen, Z., Kumagai, T.: Heat kernel estimates for stable-like processes on d-sets. Stoch. Process. Appl. **108**, 27–62 (2003)

[CK08] Chen, Z., Kumagai, T.: Heat kernel estimates for jump processes of mixed types on metric measure spaces. Probab. Theory Relat. Fields **140**, 277–317 (2008)

[CG98] Coulhon, T., Grigor'yan, A.: Random walks on graphs with regular volume growth. Geom. Funct. Anal. **8**, 656–701 (1998)

[Dav92] Davies, E.: Heat kernel bounds, conservation of probability and the Feller property. J. d'Analyse Math. **58**, 99–119 (1992)

[FOT11] Fukushima, M., Oshima, Y. Takeda, M.: Dirichlet forms and symmetric Markov processes. In: de Gruyter, W. (ed.) Studies in Mathematics, vol. 19, 2nd edn. Springer, Berlin (2011)

[Gol87] Goldstein, S.: Random walks and diffusion on fractals. In: Kesten, H. (ed.) Percolation Theory and Ergodic Theory of Infinite Particle Systems, vol. 8, pp. 121–129, IMA Math. Appl. Springer, New York (1987)

[Gri92] Grigor'yan, A.: The heat equation on noncompact riemannian manifolds. Mat. Sb. **182**(1), 55–87 (1991) (translation in Math. USSR-Sb. **72**(1) (1992), 47–77)

[Gri94] Grigor'yan, A.: Heat kernel upper bounds on a complete non-compact manifold. Rev. Mat. Iberoamericana **10**, 395–452 (1994)

[GH08] Grigor'yan, A., Hu, J.: Off-diagonal upper estimates for the heat kernel of the dirichlet forms on metric spaces. Invent. Math. **174**, 81–126 (2008)

[GH10] Grigor'yan, A., Hu, J.: Upper bounds of heat kernels on doubling spaces. Moscow Math. J. **14**, 505–563 (2014)

[GH00] Grigor'yan, A., Hu, J.: Heat kernels and green functions on metric measure spaces. Canad. J. Math. **66**, 641–699 (2014)

[GHL03] Grigor'yan, A., Hu, J., Lau, K.-S.: Heat kernels on metric-measure spaces and an application to semilinear elliptic equations. Trans. Amer. Math. Soc. **355**, 2065–2095 (2003)

[GHL08] Grigor'yan, A., Hu, J., Lau, K.-S.: Obtaining upper bounds of heat kernels from lower bounds. Comm. Pure Appl. Math. **61**, 639–660 (2008)

[GHL09] Grigor'yan, A., Hu, J., Lau, K.-S.: Heat kernels on metric spaces with doubling measure. In: Bandt, C., Mörters, P., Zähle, M. (eds.) Fractal Geometry and Stochastics IV, Progress in Probability, vol. 61, pp. 3–44. Birkh*ä*user, Basel (2009)

[GHL10] Grigor'yan, A., Hu, J., Lau, K.-S.: Comparison inequalities for heat semigroups and heat kernels on metric measure spaces. J. Funct. Anal. **259**, 2613–2641 (2010)

[GHL00] Grigor'yan, A., Hu, J., Lau, K.-S.: Estimates of heat kernels for non-local regular dirichlet forms. Trans. AMS (to appear)

[GK08] Grigor'yan, A., Kumagai, T.: On the dichotomy in the heat kernel two sided estimates. In: P. E. et al. (ed.) Analysis on Graphs and its Applications, Procedings of the Symposia in Pure Mathematics, American Mathematical Society, vol. 77, pp. 199–210 (2008)

[GT02] Grigor'yan, A., Telcs, A.: Harnack inequalities and sub-Gaussian estimates for random walks. Math. Ann. **324**, 521–556 (2002)

[GT12] Grigor'yan, A., Telcs, A.: Two-sided estimates of heat kernels on metric measure spaces. Ann. Probab. **40**, 1212–1284 (2012)

[HK99] Hambly, B., Kumagai, T.: Transition density estimates for diffusion processes on post critically finite self-similar fractals. Proc. London Math. Soc. **79**, 431–458 (1999)

[HuK06] Hu, J., Kumagai, T.: Nash-type inequalities and heat kernels for non-local Dirichlet forms. Kyushu J. Math. **60**, 245–265 (2006)

[Jon96] Jonsson, A.: Brownian motion on fractals and function spaces. Math. Zeit. **222**, 495–504 (1996)

[Kig01] Kigami, J.: Analysis on fractals. Cambridge University Press, Cambridge (2001)

[KS05] Kumagai, T., Sturm, K.: Construction of diffusion processes on fractals, d-sets, and general metric measure spaces. J. Math. Kyoto Univ. **45**, 307–327 (2005)

[Kus87] Kusuoka, S.: A diffusion process on a fractal. In: Symposium on Probabilistic Methods in Mathematical Physics, pp. 251–274. Katata, Academic Press, Amsterdam, Taniguchi (1987)

[Nas58] Nash, J.: Continuity of solutions of parabolic and elliptic equations. Amer. J. Math. **80**, 931–954 (1958)

[Pie00] Pietruska-PaŁuba, K.: On function spaces related to fractional diffusion on d-sets. Stoch. Stoch. Rep. **70**, 153–164 (2000)

[Sto00] Stós, A.: Symmetric α-stable processes on d-sets. Bull. Pol. Acad. Sci. Math. **48**, 237–245 (2000)

Stochastic Completeness of Jump Processes on Metric Measure Spaces

Alexander Grigor'yan and Xueping Huang

Abstract We give criteria for stochastic completeness of jump processes on metric measure spaces and on graphs in terms of volume growth.

1 Stochastic Completeness of a Diffusion

Let $\{X_t\}_{t\geq 0}$ be a reversible Markov process on a state space M. This process is called *stochastically complete* if its lifetime is almost surely ∞, that is

$$\mathbb{P}_x (X_t \in M) = 1.$$

If the process has no interior killing (which will be assumed) then the only way the stochastic incompleteness can occur is if the process leaves the state space in finite time. For example, diffusion in a bounded domain with the Dirichlet boundary condition is stochastically incomplete (Fig. 1).

A by far less trivial example was discovered by Azencott [Aze74] in 1974: he showed that Brownian motion on a *geodesically complete* non-compact manifold can be stochastically incomplete. In his example the manifold has negative sectional curvature that grows to $-\infty$ very fast with the distance to an origin. The stochastic incompleteness occurs because negative curvature plays the role of a drift towards infinity, and a very high negative curvature produces an extremely fast drift that sweeps the Brownian particle away to infinity in a finite time.

A. Grigor'yan (✉)
Department of Mathematics, University of Bielefeld, Bielefeld, Germany
e-mail: alexander.grigorian@googlemail.com

X. Huang
Department of Mathematics and Computer Science, Friedrich Schiller University of Jena, Jena, Germany

D.-J. Feng and K.-S. Lau (eds.), *Geometry and Analysis of Fractals*,
Springer Proceedings in Mathematics & Statistics 88,
DOI 10.1007/978-3-662-43920-3_7

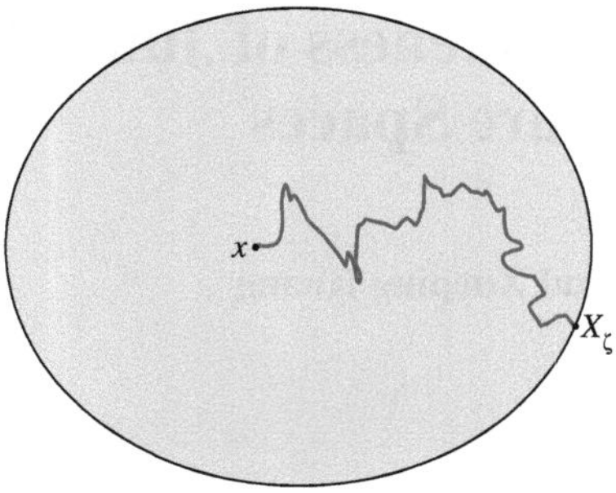

Fig. 1 Diffusion in a bounded domain

Various sufficient conditions in terms of curvature bounds were obtained by Yau [Yau78], Hsu [Hsu89], etc. It is somewhat surprising that one can obtain a sufficient condition for stochastic completeness in terms of the volume growth. Let $V(x,r)$ be the volume of the geodesic ball of radius r centered at some fixed x. Then

$$V(x,r) \leq \exp\left(Cr^2\right) \Rightarrow \text{stochastic completeness.}$$

Moreover,

$$\int^{\infty} \frac{rdr}{\ln V(x,r)} = \infty \Rightarrow \text{stochastic completeness.} \tag{1}$$

Let us sketch the construction of Brownian motion on a Riemannian manifold M and approach to the proof of the volume test for stochastic completeness (cf. [Gri09] for more details). Let M be a Riemannian manifold, μ be the Riemannian measure on M and Δ be the Laplace-Beltrami operator on M. By the Green formula, Δ is a symmetric operator on $C_0^\infty(M)$ with respect to μ, which allows to extend Δ to a a self-adjoint operator in $L^2(M,\mu)$. Assuming that M is geodesically complete, it is possible to prove that this extension is unique. Hence, Δ can be regarded as a (non-positive definite) self-adjoint operator in L^2.

By functional calculus, the operator $P_t := e^{t\Delta}$ is a bounded self-adjoint operator for any $t \geq 0$. The family $\{P_t\}_{t\geq 0}$ is called the *heat semigroup* of Δ. It can be used to solve the Cauchy problem in $\mathbb{R}_+ \times M$:

$$\begin{cases} \frac{\partial u}{\partial t} = \Delta u, \\ u|_{t=0} = f, \end{cases}$$

since $u(t,\cdot) = P_t f$ is solution for any $f \in L^2$.

Local regularity theory implies that P_t is an integral operator, whose kernel $p_t(x,y)$ is a positive smooth function of (t,x,y). In fact, $p_t(x,y)$ is the minimal positive fundamental solution to the heat equation.

The heat kernel can be used to construct a diffusion process $\{X_t\}$ on M with transition density $p_t(x, y)$. For example, in $\mathbb{R}^n$ one has

$$p_t(x, y) = \frac{1}{(4\pi t)^{n/2}} \exp\left(-\frac{|x-y|^2}{4t}\right),$$

and the process $\{X_t\}$ with this transition density is Brownian motion.

In terms of the heat kernel the stochastic completeness of diffusion $\{X_t\}$ is equivalent to the following identity:

$$\int_M p_t(x, y)\, d\mu(y) = 1,$$

for all $t > 0$ and $x \in M$.

Another useful criterion for stochastic completeness is as follows: M is stochastically complete if the homogeneous Cauchy problem

$$\begin{cases} \frac{\partial u}{\partial t} = \Delta u \\ u|_{t=0} = 0 \end{cases} \tag{2}$$

has a unique solution $u \equiv 0$ in the class of bounded functions (Khas'minskii [Kha60]).

By classical results, in $\mathbb{R}^n$ the uniqueness for (2) holds even in the class

$$|u(t, x)| \leq \exp\left(C |x|^2\right)$$

(Tikhonov class [Tic35]), but not in

$$|u(t, x)| \leq \exp\left(C |x|^{2+\varepsilon}\right).$$

More generally, uniqueness holds in the class

$$|u(t, x)| \leq \exp(f |x|)$$

provided the positive increasing function f satisfies

$$\int^{\infty} \frac{r dr}{f(r)} = \infty$$

(Täcklind class [Täc36]).

The following result can be regarded as an analogue of the latter uniqueness class.

Theorem 1 (AG [Gri87]) *Let M be a complete connected Riemannian manifold, and let $u(x,t)$ be a solution to the Cauchy problem (2). Assume that, for some $x \in M$ and for some $T > 0$ and all $r > 0$,*

$$\int_0^T \int_{B(x,r)} u^2(y,t)\, d\mu(y) dt \leq \exp(f(r)), \tag{3}$$

where $f(r)$ is a positive increasing function on $(0, +\infty)$ such that

$$\int^{\infty} \frac{r dr}{f(r)} = \infty.$$

Then $u \equiv 0$ in $(0,T) \times M$.

If u is a bounded solution, then replacing in (3) u by const we obtain that if

$$V(x,r) \leq \exp(f(r))$$

then $u \equiv 0$, that is, M is stochastically complete. Setting

$$f(r) = \ln V(x,r)$$

we obtain the volume test for stochastic completeness:

$$\int^{\infty} \frac{r dr}{\ln V(x,r)} = \infty.$$

The latter condition cannot be further improved: if $W(r)$ is an increasing function such that

$$\int^{\infty} \frac{r dr}{\ln W(r)} < \infty$$

then there exists a geodesically complete but stochastically incomplete manifold with $V(x,r) \leq W(r)$.

One may wonder why the geodesic balls can be used to determine the stochastic completeness, as the latter condition does not depend on the distance function at all. The reason is that the geodesic distance d is by definition related to the gradient ∇ (and, hence, to the Laplacian) by $|\nabla d| \leq 1$. An analogue of this condition will appear later also in jump processes.

2 Jump Processes

Let (M, d) be a metric space such that all closed metric balls

$$B(x, r) = \{y \in M : d(x, y) \leq r\}$$

are compact. In particular, (M, d) is locally compact and separable. Let μ be a Radon measure on M with a full support.

Recall that a *Dirichlet form* $(\mathcal{E}, \mathcal{F})$ in $L^2(M, \mu)$ is a symmetric, non-negative definite, bilinear form $\mathcal{E} : \mathcal{F} \times \mathcal{F} \to \mathbb{R}$ defined on a dense subspace $\mathcal{F}$ of $L^2(M, \mu)$, that satisfies in addition the following properties:

- Closedness: $\mathcal{F}$ is a Hilbert space with respect to the following inner product:

$$\mathcal{E}_1(f, g) := \mathcal{E}(f, g) + (f, g).$$

- The Markov property: if $f \in \mathcal{F}$ then also $\widetilde{f} := (f \wedge 1)_+$ belongs to $\mathcal{F}$ and $\mathcal{E}(\widetilde{f}) \leq \mathcal{E}(f)$, where $\mathcal{E}(f) := \mathcal{E}(f, f)$.

For example, the classical Dirichlet form in $\mathbb{R}^n$ is

$$\mathcal{E}(f, g) = \int_{\mathbb{R}^n} \nabla f \cdot \nabla g \, dx$$

in $\mathcal{F} = W^{1,2}(\mathbb{R}^n)$.

A general Dirichlet form $(\mathcal{E}, \mathcal{F})$ has the *generator* $\mathcal{L}$ that is a non-positive definite, self-adjoint operator on $L^2(M, \mu)$ with domain $\mathcal{D} \subset \mathcal{F}$ such that

$$\mathcal{E}(f, g) = (-\mathcal{L}f, g)$$

for all $f \in \mathcal{D}$ and $g \in \mathcal{F}$. The generator $\mathcal{L}$ determines the *heat semigroup* $\{P_t\}_{t \geq 0}$ by $P_t = e^{t\mathcal{L}}$ in the sense of functional calculus of self-adjoint operators. It is known that $\{P_t\}_{t \geq 0}$ is a strongly continuous, contractive, symmetric semigroup in L^2, and is *Markovian*, that is, $0 \leq P_t f \leq 1$ for any $t > 0$ if $0 \leq f \leq 1$.

The Markovian property of the heat semigroup implies that the operator P_t preserves the inequalities between functions, which allows to use monotone limits to extend P_t from L^2 to L^∞. In particular, $P_t 1$ is defined.

Definition The form $(\mathcal{E}, \mathcal{F})$ is called *conservative* or *stochastically complete* if $P_t 1 = 1$ for every $t > 0$.

Assume in addition that $(\mathcal{E}, \mathcal{F})$ is *regular*, that is, the set $\mathcal{F} \cap C_0(M)$ is dense both in $\mathcal{F}$ with respect to the norm $\mathcal{E}_1$ and in $C_0(M)$ "(compactly supported continuous functions)" with respect to the sup-norm. By a theory of Fukushima et al. [FOT11], for any regular Dirichlet form there exists a Hunt process $\{X_t\}_{t\geq 0}$ such that, for all bounded Borel functions f on M,

$$\mathbb{E}_x f(X_t) = P_t f(x) \tag{4}$$

for all $t > 0$ and almost all $x \in M$, where $\mathbb{E}_x$ is expectation associated with the law of $\{X_t\}$ started at x.

Using the identity (4), one can show that the lifetime of X_t is almost surely ∞ if and only if $P_t 1 = 1$ for all $t > 0$, which motivates the term "stochastic completeness" in the above definition.

One distinguishes local and non-local Dirichlet forms. The Dirichlet form $(\mathcal{E}, \mathcal{F})$ is called *local* if $\mathcal{E}(f, g) = 0$ for all functions $f, g \in \mathcal{F}$ with disjoint compact support. It is called *strongly local* if the same is true under a milder assumption that $f = \text{const}$ on a neighborhood of $\operatorname{supp} g$.

For example, the following Dirichlet form on a Riemannian manifold

$$\mathcal{E}(f, g) = \int_M \nabla f \cdot \nabla g d\mu$$

is strongly local. The generator of this form the self-adjoint Laplace-Beltrami operator Δ, and the Hunt process is Brownian motion on M.

A well-studied non-local Dirichlet form in $\mathbb{R}^n$ is given by

$$\mathcal{E}(f, g) = \int_{\mathbb{R}^n \times \mathbb{R}^n} \frac{(f(x) - f(y))(g(x) - g(y))}{|x - y|^{n+\alpha}} dxdy \tag{5}$$

where $0 < \alpha < 2$. The domain of this form is the Besov space $B_{2,2}^{\alpha/2}$, the generator is (up to a constant multiple) the operator $-(-\Delta)^{\alpha/2}$, where Δ is the Laplace operator in $\mathbb{R}^n$, and the Hunt process is the symmetric stable process of index α.

By a theorem of Beurling and Deny (cf. [FOT11]), any regular Dirichlet form can be represented in the form

$$\mathcal{E} = \mathcal{E}^{(c)} + \mathcal{E}^{(j)} + \mathcal{E}^{(k)},$$

where $\mathcal{E}^{(c)}$ is a strongly local part that has the form (assuming absolute continuity of energy measure for simplicity)

$$\mathcal{E}^{(c)}(f, g) = \int_M \Gamma(f, g) d\mu,$$

where $\Gamma(f, g)$ is the so called *energy density* (generalizing $\nabla f \cdot \nabla g$ on manifolds); $\mathcal{E}^{(j)}$ is a jump part that has the form

$$\mathcal{E}^{(j)}(f, g) = \frac{1}{2} \iint_{X \times X} (f(x) - f(y))(g(x) - g(y))\, dJ(x, y)$$

with some measure J on $X \times X$ that is called a *jump measure*; and $\mathcal{E}^{(k)}$ is a killing part that has the form

$$\mathcal{E}^{(k)}(f, g) = \int_X fg\,dk$$

where k is a measure on X that is called a *killing measure*.

In terms of the associated process this means that X_t is in some sense a mixture of diffusion and jump processes with a killing condition.

The ln-volume test of stochastic completeness of manifolds can be extended to strongly local Dirichlet forms as follows. Set as before $V(x, r) = \mu(B(x, r))$.

Theorem 2 (Sturm [Stu94]) *Let $(\mathcal{E}, \mathcal{F})$ be a regular strongly local Dirichlet form. Assume that the distance function $\rho(x) = d(x, x_0)$ on M satisfies the condition*

$$\Gamma(\rho, \rho) \leq C,$$

for some constant C. If, for some $x \in M$,

$$\int^{\infty} \frac{rdr}{\ln V(x, r)} = \infty$$

then the Dirichlet form $(\mathcal{E}, \mathcal{F})$ is stochastically complete.

The method of proof is basically the same as for manifolds because for strongly local forms the same chain rule and product rules are available. The condition $\Gamma(\rho, \rho) \leq C$ is analogous to $|\nabla \rho| \leq 1$ that is automatically satisfied for the geodesic distance on any manifold.

Now let us turn to jump processes. For simplicity let us assume that the jump measure J has a density $j(x, y)$. Namely, let $j(x, y)$ be a non-negative Borel function on $M \times M$ that satisfies the following two conditions:

(a) $j(x, y)$ is symmetric: $j(x, y) = j(y, x)$;
(b) there is a positive constant C such that

$$\int_M (1 \wedge d(x, y)^2) j(x, y) d\mu(y) \leq C \text{ for all } x \in M.$$

Definition We say that a distance function d is *adapted* to a kernel $j(x, y)$ (or j is adapted to d) if (b) is satisfied.

The condition (b) relates the distance function to the Dirichlet form and plays the same role as $\Gamma(\rho, \rho) \leq C$ does for diffusion.

Consider the following bilinear functional

$$\mathcal{E}(f,g)=\frac{1}{2}\int\int_{X\times X}(f(x)-f(y))(g(x)-g(y))j(x,y)d\mu(x)d\mu(y)$$

that is defined on Borel functions f and g whenever the integral makes sense. Define the maximal domain of $\mathcal{E}$ by

$$\mathcal{F}_{\max}=\left\{f\in L^2:\mathcal{E}(f,f)<\infty\right\},$$

where $L^2=L^2(M,\mu)$. By the polarization identity, $\mathcal{E}(f,g)$ is finite for all $f,g\in\mathcal{F}_{\max}$. Moreover, $\mathcal{F}_{\max}$ is a Hilbert space with the norm $\mathcal{E}_1$.

Denote by $\mathrm{Lip}_0(M)$ the class of Lipschitz functions on M with compact support. It follows from (b) that

$$\mathrm{Lip}_0(M)\subset\mathcal{F}_{\max}.$$

Define the space $\mathcal{F}$ as the closure of $\mathrm{Lip}_0(M)$ in $(\mathcal{F}_{\max},\|\cdot\|_{\mathcal{E}_1})$. Under the above hypothesis, $(\mathcal{E},\mathcal{F})$ is a regular Dirichlet form in $L^2(M,\mu)$. The associated Hunt process $\{X_t\}$ is a pure jump process with the jump density $j(x,y)$.

Many examples of jump processes in $\mathbb{R}$ are provided by Lévy-Khintchine theorem where the Lévy measure $W(dy)$ corresponds to $j(x,y)d\mu(y)$. The condition (b) appears also in Lévy-Khintchine theorem in the form

$$\int_{\mathbb{R}\setminus\{0\}}\left(1\wedge|y|^2\right)W(dy)<\infty.$$

Hence, the Euclidean distance in $\mathbb{R}$ is adapted to any Lévy process.

An explicit example of a jump density in $\mathbb{R}^n$ is

$$j(x,y)=\frac{\mathrm{const}}{|x-y|^{n+\alpha}},$$

where $\alpha\in(0,2)$, which defines the Dirichlet form (5).

The next theorem is the main result.

Theorem 3 *Assume that j satisfies (a) and (b) and let $(\mathcal{E},\mathcal{F})$ be the jump form defined as above. If, for some $x\in M$, $c>0$ and for all large enough r,*

$$V(x,r)\leq\exp(cr\ln r),\tag{6}$$

then the Dirichlet form $(\mathcal{E},\mathcal{F})$ is stochastically complete.

This theorem was proved by Grigor'yan et al. [GHM12] for $c < \frac{1}{2}$, improving the work of Masamune and Uemura [MU11] for the sub-exponential volume growth case. Then it was observed ([MUW12]) that a minor modification of the proof of [GHM12] works for all c.

For the proof of Theorem 3 we split the jump kernel $j(x, y)$ into the sum of two parts:

$$j'(x, y) = j(x, y)\mathbf{1}_{\{d(x,y)\leq\varepsilon\}} \text{ and } j''(x, y) = j(x, y)\mathbf{1}_{\{d(x,y)>\varepsilon\}} \tag{7}$$

and show first the stochastic completeness of the Dirichlet form $(\mathcal{E}', \mathcal{F})$ associated with j'. For that we adapt the methods used for stochastic completeness for the local form.

The bounded range of j' allows to treat the Dirichlet form $(\mathcal{E}', \mathcal{F})$ as "almost" local: if f, g are two functions from $\mathcal{F}$ such that $d\left(\operatorname{supp} f, \operatorname{supp} g\right) > \varepsilon$ then $\mathcal{E}\left(f, g\right) = 0$. The condition (b) plays in the proof the same role as the condition $|\nabla d| \leq 1$ in the local case. However, the lack of locality brings up in the estimates various additional terms that have to be compensated by a stronger hypothesis of the volume growth (6).

The tail j'' can be regarded as a small perturbation of j' in the following sense: $(\mathcal{E}, \mathcal{F})$ is stochastically complete if and only if $(\mathcal{E}', \mathcal{F})$ is so. The proof is based on the fact that the integral operator with the kernel j'' is a bounded operator in $L^2\left(M, \mu\right)$, because by (b)

$$\int_M j''\left(x, y\right) d\mu\left(y\right) \leq C.$$

It is not yet clear if the volume growth condition (6) in Theorem 3 is sharp.

In contrast to the manifold case, we can not expect a corresponding uniqueness class result. Let us briefly mention a result about uniqueness class for the heat equation associated with the jump Dirichlet form on graphs satisfying (a) and (b).

Namely, Huang [Hua12] proved in 2011 that, for any $b < \frac{1}{2}$ the following inequality determines a uniqueness class

$$\int_0^T \int_{B(x,r)} u^2\left(t, x\right) d\mu\left(x\right) dt \leq \exp\left(br \ln r\right). \tag{8}$$

What is more surprising, that for $b > 2\sqrt{2}$ this statement fails even on the graph $\mathbb{Z}$.

The optimal value of b in (8) is unknown. If $b < \frac{1}{2}$ then Huang's result can be used to obtain Theorem 3 on graphs provided the constant c in (6) is smaller than $\frac{1}{2}$. However, in general the stochastic completeness test (6) does not follow from the uniqueness class (8), as can be seen from the range of constants. Indeed, even better results for stochastic completeness are known in the graph case, which we will discuss in the next section.

3 Random Walks on Graphs

Let us now turn to random walks on graphs. Let (X, E) be a locally finite, infinite, connected graph, where X is the set of vertices and E is the set of edges. We assume that the graph is undirected, simple, without loops. Let μ be the counting measure on X. Define the jump kernel by $j(x, y) = 1_{\{x\sim y\}}$, where $x \sim y$ means that x, y are neighbors, that is, $(x, y) \in E$. The corresponding Dirichlet form is

$$\mathcal{E}(f) = \frac{1}{2} \sum_{x,y:\, x\sim y} (f(x) - f(y))^2,$$

and its generator is

$$\Delta f(x) = \sum_{y,y\sim x} (f(y) - f(x)).$$

The operator Δ is called *unnormalized* (or *physical)* Laplace operator on (X, E). This is to distinguish from the *normalized* or *combinatorial* Laplace operator

$$\hat{\Delta} f(x) = \frac{1}{\deg(x)} \sum_{y,y\sim x} (f(y) - f(x)),$$

where $\deg(x)$ is the number of neighbors of x. The normalized Laplacian $\hat{\Delta}$ is the generator of the same Dirichlet form but with respect to the degree measure $\deg(x)$.

Both Δ and $\hat{\Delta}$ generate the heat semigroups $e^{t\Delta}$ and $e^{t\hat{\Delta}}$ and, hence, associated continuous time random walks on X. It is easy to prove that $\hat{\Delta}$ is a bounded operator in $L^2(X, \deg)$, which then implies that the associated random walk is always stochastically complete. On the contrary, the random walk associated with the unnormalized Laplace operator can be stochastically incomplete.

We say that the graph (X, E) is stochastically complete if the heat semigroup $e^{t\Delta}$ is stochastically complete.

Denote by $\rho(x, y)$ the graph distance on X, that is the minimal number of edges in an edge chain connecting x and y. Let $B_\rho(x, r)$ be closed metric balls with respect to this distance ρ and set $V_\rho(x, r) = |B_\rho(x, r)|$ where $|\cdot| := \mu(\cdot)$ denotes the number of vertices in a given set.

Theorem 4 *If there is a point $x_0 \in X$ and a constant $c > 0$ such that*

$$V_\rho(x_0, r) \le cr^3 \ln r \tag{9}$$

for all large enough r, then the graph (X, E) is stochastically complete.

Note that the function $r^3 \ln r$ is sharp here in the sense that it cannot be replaced by $r^3 \ln^{1+\varepsilon} r$. For any non-negative integer r, set

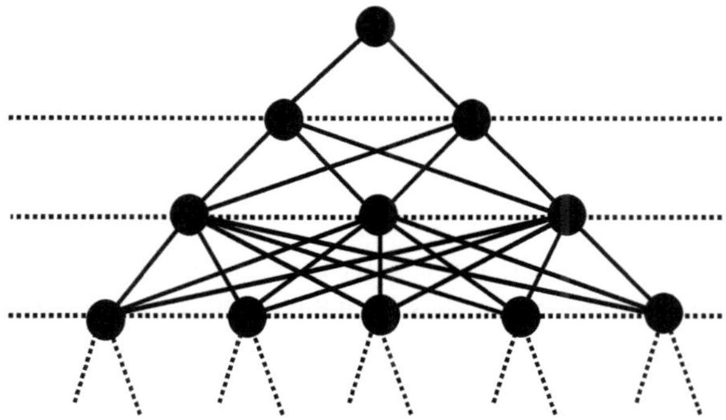

Fig. 2 Anti-tree of Wojciechowski

$$S_r = \{x \in X : \rho(x_0, x) = r\}.$$

Wojciechowski [Woj11] considered the graph where every vertex on S_r is connected to all vertices on S_{r-1} and S_r (see Fig. 2).

He proved that for such graphs the stochastic incompleteness is equivalent to the following condition:

$$\sum_{r=1}^{\infty} \frac{V_\rho(x_0, r)}{|S_{r+1}|\,|S_r|} < \infty. \tag{10}$$

Taking $|S_r| \simeq r^2 \ln^{1+\varepsilon} r$ we obtain $V_\rho(x_0, r) \simeq r^3 \ln^{1+\varepsilon}$ so that the condition (10) is satisfied and, hence, the graph is stochastically incomplete.

The proof of Theorem 4 is based on the following ideas. Observe first that the graph distance ρ is in general not adapted. Indeed, the integral in (b) is equal to

$$\sum_y \left(1 \wedge \rho^2(x, y)\right) j(x, y) = \sum_y j(x, y) = \deg(x)$$

so that (b) holds if and only if the graph has uniformly bounded degree, which is not interesting as all graphs with bounded degree are automatically stochastically complete.

Let us construct an adapted distance as follows. For any edge $x \sim y$ define first its length $\sigma(x, y)$ by

$$\sigma(x, y) = \frac{1}{\sqrt{\deg(x)}} \wedge \frac{1}{\sqrt{\deg(y)}}.$$

Then, for all $x, y \in X$ define $d(x, y)$ as the smallest total length of all edges in an edge chain connecting x and y. It is easy to verify that d satisfies (b):

$$\sum_y \left(1 \wedge d^2(x,y)\right) j(x,y) \le \sum_y \left(\frac{1}{\deg(x)} \wedge \frac{1}{\deg(y)}\right) j(x,y)$$
$$\le \sum_{y \sim x} \frac{1}{\deg(x)} = 1.$$

Then we will show that (9) for ρ-balls implies that the d-balls have at most quadratic exponential volume growth, so that the stochastic completeness will follow by the following result of Folz (stated in the current specific setting).

Theorem 5 (Folz [Fol00]) *Let (X, E) be a graph as above, with an adapted distance d. If the volume growth $V_d(x_0, r) = \mu(B_d(x_0, r))$ with respect to d satisfies:*

$$\int^{\infty} \frac{r dr}{\ln V_d(x_0, r)} = \infty, \tag{11}$$

for some reference point $x_0 \in X$, then the graph (X, E) is stochastically complete.

Roughly speaking, for a graph (X, E) with an adapted distance d, Folz constructed a corresponding metric graph Y, which is enriched from X by attaching intervals to the edges. The length and measure of intervals, which are used to define a strongly local Dirichlet form on Y, are determined by the adapted distance. Folz proved two significant relations between the metric graph Y with the original graph X. First, the volume growth of Y is controlled by that of X. More importantly, X is stochastically complete if so is the diffusion on Y. Theorem 5 is then obtained as a consequence of Theorem 2. The second relation is the key to overcome the difficulty coming from lack of chain rule. It was first proven by Folz using probabilistic arguments. Two analytic proofs of this comparison result are obtained by Huang [Hua13]. We briefly describe one of them as it is rather concise.

By a well-known result in [FOT11], the stochastic completeness of a Dirichlet form $(\mathcal{E}, \mathcal{F})$ on a measure space (M, μ) is equivalent to the existence of a sequence of functions $\{v_n\} \subset \mathcal{F}$ such that

$$0 \le v_n \le 1, \lim_{n\to\infty} v_n = 1 \quad \mu\text{-a.e.}$$

and such that

$$\lim_{n\to\infty} \mathcal{E}(v_n, w) = 0$$

holds for any $w \in \mathcal{F} \cap L^1(M, \mu)$. Thus comparison of stochastic completeness boils down to comparing the existence of certain functions. There are natural ways to transfer back and forth between a function space on a graph and that on the corresponding metric graph. Assume for a graph X that the corresponding metric graph Y is stochastically complete, with a sequence $\{v_n\}$ as above. The sequence $\{\tilde{v}_n\}$ on X, as restrictions of $\{v_n\}$, is naturally expected to satisfy the conditions above. The condition

$$\lim_{n\to\infty} \mathcal{E}(\tilde{v}_n, \tilde{w}) = 0$$

for $\tilde{w}$ on X, can be checked by extending $\tilde{w}$ to w on Y through linear interpolation. The rest are simple calculations to make sure that $\tilde{v}_n$ and w are in the correct function spaces.

Now we deduce Theorem 4 from Theorem 5. Without loss of generality, we assume that

$$V_\rho(x_0, r) \le c(r+1)^3 \ln(r+3), \tag{12}$$

for all $r \ge 0$. Observe that

$$V_\rho\,(x_0, n) = \sum_{r=0}^{n} \mu\left(S_\rho\,(r)\right).$$

Put $\varepsilon = \frac{1}{5}$ and $\alpha = 200c$ where c is the constant in (12). It follows from (12) that, for any $n \ge 1$,

$$\begin{aligned}&\left|\{r \in [n-1, 2n+1] : \mu(S_r) > \alpha(n+1)^2 \ln(n+3)\}\right|\\ &\quad \le \frac{c(2n+2)^3 \ln(2n+4)}{\alpha(n+1)^2 \ln(n+3)} \le \varepsilon n.\end{aligned}$$

Therefore,

$$\left|\{r \in [n+1, 2n] : \max_{i=-2,-1,0,1} \mu(S_{r+i}) > \alpha(n+1)^2 \ln(n+3)\}\right| \le 4\varepsilon n$$

and, hence,

$$\left|\{r \in [n+1, 2n] : \max_{i=-2,-1,0,1} \mu(S_{r+i}) \le \alpha(n+1)^2 \ln(n+3)\}\right| \ge (1-4\varepsilon)n. \tag{13}$$

For any point $x \in S_r$ we have

$$\deg x \le \mu\,(S_{r-1}) + \mu\,(S_r) + \mu\,(S_{r+1}) \tag{14}$$

(see Fig. 3).

It follows from (13) and (14) that

$$\left|\{r \in [n+1, 2n] : \max_{x\in S_{r-1}\cup S_r} \deg x \le 3\alpha(n+1)^2 \ln(n+3)\}\right| \ge (1-4\varepsilon)n. \tag{15}$$

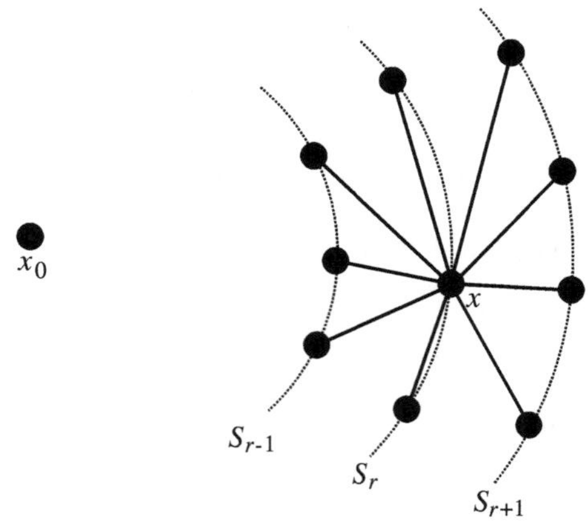

Fig. 3 Neighbors of a vertex x on S_r

It follows that, for r as in (15), any pair of $x \sim y$ with $x \in S_{r-1}$, $y \in S_r$ necessarily satisfies

$$\sigma(x, y) \geq \frac{1}{\sqrt{3\alpha}(n+1)\sqrt{\ln(n+3)}}. \tag{16}$$

Fix a positive integer n and two vertices $x \in S_n$ and $y \in S_{2n}$. Consider a chain of vertices connecting x and $y \in S_{2n}$, and let us estimate from below the length L of this chain. For any $r \in [n+1, 2n]$ there is an edge $x_r \sim y_r$ from this chain such $x_r \in S_{r-1}$ and $y_r \in S_r$. Clearly, we have

$$L \geq \sum_{r=n+1}^{2n} \sigma(x_r, y_r).$$

Restricting the summation to those r that satisfy (15) and noticing that for any such r,

$$\sigma(x_r, y_r) \geq \frac{1}{\sqrt{3\alpha}(n+1)\sqrt{\ln(n+3)}},$$

we obtain

$$L \geq \frac{1}{\sqrt{3\alpha}(n+1)\sqrt{\ln(n+3)}}(1-4\varepsilon)\,n \geq \frac{\delta}{\sqrt{\ln(n+3)}} \geq \frac{\delta}{\sqrt{2+\ln n}}, \tag{17}$$

where $\delta = \frac{1-4\varepsilon.}{2\sqrt{3\alpha}}$.

Now we can estimate $d(x_0, x)$ for any vertex $x \notin B_\rho(x_0, R)$, where $R > 4$. Choose a positive integer k so that

$$2^k \le R < 2^{k+1}.$$

Any chain connecting x_0 and x contains a subsequence $\{x_i\}_{i=1}^k$ of vertices such that $x_i \in S_{2^i}$. By (17) the length of the chain between x_i and x_{i+1} is bounded below by $\frac{\delta}{\sqrt{i+2}}$, for any $i = 1, ..., k-1$. It follows that the length of the whole chain is bounded below by

$$\delta \sum_{i=1}^{k-1} \frac{1}{\sqrt{i+2}},$$

whence

$$d(x_0, x) \ge \delta' \sqrt{k+1} \ge \delta' \sqrt{\ln R},$$

for some constant $\delta' > 0$. It follows that

$$B_d(x_0, \delta' \sqrt{\ln R}) \subset B_\rho(x_0, R).$$

Given a large enough r, define R from the identity $r = \delta' \sqrt{\ln R}$, that is, $R = \exp(r^2/\delta'^2)$. Then we obtain

$$\mu\left(B_d(x_0, r)\right) \le \mu\left(B_\rho(x_0, R)\right) \le c(R+1)^3 \ln(R+3) \le C \exp\left(br^2\right),$$

for some constants C and b, which finishes the proof.

References

[Aze74] Azencott, R.: Behavior of diffusion semi-groups at infinity. Bull. Soc. Math. (France) **102**, 193–240 (1974)

[Fol00] Folz, M.: Volume growth and stochastic completeness of graphs. Trans. Amer. Math. Soc. **366**(4), 2089–2119. ISSN: 0002-9947 (2014)

[FOT11] Fukushima, M., Oshima, Y., Takeda, M.: Dirichlet forms and symmetric Markov processes. In: de Gruyter (ed.) Studies in Mathematics, vol. 19. Walter de Gruyter and Co., Berlin (2011) [MR 2778606 (2011k:60249)]

[Gri87] Grigor'yan, A.: On stochastically complete manifolds. DAN SSSR **290** 534–537 (1986) (in Russian) [English translation: Soviet Math. Dokl. **34**(2) 310–313 (1987)]

[Gri09] Grigor'yan, A.: Heat kernel and analysis on manifolds. In: AMS-IP Studies in Advanced Mathematics, vol. 47. American Mathematical Society, Providence (2009)

[GHM12] Grigor'yan, A., Huang, X., Masamune, J.: On stochastic completeness of jump processes. Math. Z. **271**(3–4), 1211–1239 (2012)

[Hsu89] Hsu, E.P.: Heat semigroup on a complete Riemannian manifold. Ann. Probab. **17**, 1248–1254 (1989)

[Hua12] Huang, X.: On uniqueness class for a heat equation on graphs. J. Math. Anal. Appl. **393**(2), 377–388 (2012)

[Hua13] Huang, X.: A note on the volume growth criterion for stochastic completeness of weighted graphs. Potential Anal. **40**(2), 117–142 (2014)

[Kha60] Khas'minskii, R.Z.: Ergodic properties of recurrent diffusionprocesses and stabilization of solutions to the Cauchy problem forparabolic equations. Theor. Prob. Appl. **5**, 179–195 (1960)

[MU11] Masamune, J., Uemura, T.: Conservation property of symmetric jump processes. Ann. Inst. Henri. Poincaré Probab. Statist. **47**(3), 650–662 (2011)

[MUW12] Masamune, J., Uemura, T., Wang, J.: On the conservativeness and the recurrence of symmetric jump-diffusions, J. Funct. Anal. **263**(12) 3984–4008 (2012)

[Stu94] Sturm, K.T.: Analysis on local Dirichlet spaces. I. Recurrence, conservativeness and L^p-Liouville properties. J. Reine Angew. Math. **456**, 173–196 (1994)

[Täc36] Täcklind, S.: Sur les classes quasianalytiques des solutions deséquations aux dérivées partielles du type parabolique. Nova Acta Soc. Sci. Upsal., IV. Ser. **10**(3) 1–57 (1936)

[Tic35] Tichonov, A.N.: Uniqueness theorems for the equation of heat conduction, (in Russian) Matem. Sbornik **42**, 199–215 (1935)

[Woj11] Wojciechowski, R.K.: Stochastically incomplete manifolds and graphs. Prog. Probab. **64**, 163–179 (2011)

[Yau78] Yau, S.T.: On the heat kernel of a complete Riemannian manifold. J. Math. Pures Appl. **57**, 191–201 (1978)

Self Similar Sets, Entropy and Additive Combinatorics

Michael Hochman

Abstract This article is an exposition of the main result of [Hoc12], that self-similar sets whose dimension is smaller than the trivial upper bound have "almost overlaps" between cylinders. We give a heuristic derivation of the theorem using elementary arguments about covering numbers. We also give a short introduction to additive combinatorics, focusing on inverse theorems, which play a pivotal role in the proof. Our elementary approach avoids many of the technicalities in [Hoc12], but also falls short of a complete proof; in the last section we discuss how the heuristic argument is turned into a rigorous one.

1 Introduction

1.1 Self-similar Sets

Self-similar sets in the line are compact sets that are composed of finitely many scaled copies of themselves. These are the simplest fractal sets, the prototypical example being the famous middle-$\frac{1}{3}$ Cantor set $X \subseteq [0,1]$, which satisfies the "geometric recursion"[1] relation $X = \frac{1}{3}X \cup (\frac{1}{3}X + \frac{2}{3})$, using the obvious notation for scaling and translation of a set. In general, a self-similar set is defined by a finite family $\Phi = \{f_i\}_{i\in\Lambda}$ of maps of the form $f_i(x) = r_i x + a_i$, where $0 < |r_i| < 1$ and $a_i \in \mathbb{R}$. The family Φ is called an *iterated function system* (or *IFS*),[2,3] and the self-similar set they define is unique compact set $X \neq \emptyset$ satisfying

[1] The mddle-1/3 Cantor set can also be described in other ways, e.g. by a recursive construction, or symbolically as the points in [0, 1] that can be written in base 3 without the digit 1. General self-similar sets also have representations of this kind, but in this paper we shall not use them.
[2] Iterated function systems consisting of non-affine maps and on other metric spaces than $\mathbb{R}$ are also of interest, but we do not discuss them here.
[3] Supported by ERC grant 306494.

M. Hochman (✉)
Department of Mathematics,The Hebrew University of Jerusalem, Jerusalem, Israel
e-mail: mhochman@math.huji.ac.il

D.-J. Feng and K.-S. Lau (eds.), *Geometry and Analysis of Fractals*,
Springer Proceedings in Mathematics & Statistics 88,
DOI 10.1007/978-3-662-43920-3_8

$$X = \bigcup_{i \in \Lambda} f_i X. \tag{1}$$

(existence and uniqueness are due to Hutchinson [Hut81]).

Throughout this paper we make a few simplifying assumptions. To avoid trivialities, we always assume that Φ contains at least two distinct maps, otherwise X is just the common fixed point of the maps. We assume that Φ has *uniform contraction*, i.e. all the contraction ratios r_i are equal to the same value r. Finally, we assume that $r > 0$, so the maps preserve orientation. These assumptions are not necessary but they simplify the statements and arguments considerably.

1.2 Dimension of Self-similar Sets

Despite the apparent simplicity of the definition, and of some of the better known examples, there are still large gaps in our understanding of the geometry of self-similar sets. In general, we do not even know how to compute their dimension. Usually one should be careful to specify the notion of dimension that one means, but it is a classical fact that, for self-similar sets, all the major notions of dimension coincide, and in particular the Hausdorff and box (Minkowski) dimensions agree (e.g. [Fal89, Theorem 4 and Example 2]). Thus we are free to choose either one of these, and we shall choose the latter, whose definition we now recall. For a subset $Y \subseteq \mathbb{R}$ denote its covering number at scale ε by

$$N_\varepsilon(Y) = \min\{k \ : \ Y \text{ can be covered by } k \text{ sets of diameter } \leq \varepsilon\}$$

The *box dimension* of Y, if it exists, is the exponential growth rate of $N_\varepsilon(Y)$:

$$\dim_{\mathrm{B}} Y = \lim_{\varepsilon \to 0} \frac{\log N_\varepsilon(Y)}{\log(1/\varepsilon)}$$

Thus $\dim_{\mathrm{B}} Y = \alpha$ means that $N_\varepsilon(Y) = \varepsilon^{-\alpha + o(1)}$ as $\varepsilon \to 0$. It is again well known that the limit exists when Y is self-similar, we shall see a short proof in Sect. 3.1.

It is easy to give upper bounds for the dimension of a self-similar set. Taking X as in (1) and iterating the relation we get

$$X = \bigcup_{i \in \Lambda} f_i(\bigcup_{j \in \Lambda} f_j X) = \bigcup_{i,j \in \Lambda} f_i \circ f_j(X)$$

Writing $f_{i_1 \ldots i_n} = f_{i_1} \circ \ldots \circ f_{i_n}$ for $\underline{i} = i_1 \ldots i_n \in \Lambda^n$ and iterating n times, we have

$$X = \bigcup_{\underline{i} \in \Lambda^n} f_{\underline{i}} X \tag{2}$$

This union consists of $|\Lambda|^n$ sets of diameter $r^n|X|$, so by definition,

$$N_{r^n \mathrm{diam}(X)}(X) \leq |\Lambda|^n$$

Hence

$$\dim_B(X) = \lim_{n\to\infty} \frac{\log N_{r^n \mathrm{diam}(X)}(X)}{\log(1/r^n \mathrm{diam}(X))} \leq \frac{\log|\Lambda|}{\log(1/r)} \tag{3}$$

The right hand side of (4) is called the *similarity dimension* of X and is denoted $\mathrm{sdim} X$.[4]

Is this upper bound an equality? Note that the bound is purely combinatorial and does not take into account the parameters a_i at all. Equality is known to hold under some assumptions on the separation of the "pieces" $f_i X$, $i \in \Lambda$, for instance assuming *strong separation* (that the union (1) is disjoint), or the open set condition (that there exists open set $\emptyset \neq U \subseteq \mathbb{R}$ such that $f_i U \subseteq U$ and $f_i U \cap f_j U = \emptyset$ for $i \neq j$).

Without separation conditions, however, the inequality (3) can be strict. There are two trivial ways this can occur. First, there could be too many maps: if $|\Lambda| > 1/r$ then the right hand side of (3) is greater then 1, whereas $\dim_B X \leq 1$ due to the trivial bound $N(X,\varepsilon) \leq \lceil \mathrm{diam}(X)/\varepsilon \rceil$. Thus we should adjust (3) to read

$$\dim_B(X) \leq \min\{1, \frac{\log|\Lambda|}{\log(1/r)}\} \tag{4}$$

Second, the combinatorial bound may be over-counting if some of the sets in the union (2) coincide, that is, for some n we have $f_{\underline{i}} = f_{\underline{j}}$ for some distinct $\underline{i}, \underline{j} \in \Lambda^n$. This situation is known as *exact overlaps*. If such $\underline{i}, \underline{j}$ exist then we can rewrite (2) as $X = \bigcup_{\underline{u}\in\Lambda^n\setminus\{\underline{i}\}} f_{\underline{u}} X$, which presents X as the attractor of the IFS $\Phi' = \{f_{\underline{u}}\}_{\underline{u}\in\Lambda'}$ for $\Lambda' = \Lambda^n \setminus \{\underline{i}\}$. This IFS consists of $|\Lambda'| = |\Lambda|^n - 1$ maps that contract by r^n, so, applying the trivial bound (3) to this IFS, we have $\dim_B X \leq \log(|\Lambda|^n - 1)/\log(1/r^n)$, which is better than the previous bound of $\log|\Lambda|/\log(1/r)$. To take an extreme example, if all the maps f_i coincide then the attractor X is just the unique fixed point of the map, and its dimension is 0.

Are there other situations where a strict inequality occurs in (4)? A-priori, one does not need *exact* coincidences between sets in (2) to make the combinatorial bound very inefficient. It could happen, for example, that many of the sets $f_{\underline{i}} X$, $\underline{i} \in \Lambda^n$, align almost exactly, in which case one may need significantly fewer than $|\Lambda|^n$ ε-intervals to cover them. Nevertheless, although such a situation can easily be arranged for a fixed n, to get a drop in dimension one would need this to happen at all sufficiently small scales. No such examples are known, and the main subject of this paper is the conjecture that this cannot happen:

---

[4] It would be better to write $\mathrm{sdim}\Phi$, since this quantity depends on the presentation of X and not on X itself, but generally there is only one IFS given and no confusion should arise.

Conjecture 1.1 A strict inequality in (4) can occur only in the presence of exact overlaps.

This conjecture appears in [PS00, Question 2.6], though special cases of it have received attention for decades, in particular Furstenberg's projection problem for the 1-dimensional Sierpinski gasket (see e.g. [Ken97]), the 0,1,3-problem (see e.g. [PS95]) and, for self-similar measures instead of sets, the Bernoulli convolutions problem, (e.g. [PSS00]).

One may also draw an analogy between this conjecture and rigidity statements in ergodic theory. Rigidity is the phenomenon that, for certain group actions of algebraic origin, the orbit of the point is as large as it can be (dense or possibly even equidistributed for the volume measure) unless there is an algebraic obstruction to this happening. To see the connection with the conjecture above, note that X is just the orbit closure of (any) $x \in \mathbb{R}$ under the semigroup $\{f_{\underline{i}} : \underline{i} \in \bigcup_{n=1}^{\infty} \Lambda^N\}$ of affine maps, and that exact overlaps occur if and only if this semigroup is not generated freely by $\{f_i\}_{i\in\Lambda}$. Thus the conjecture predicts that the orbit closure of any point is as large as it can be unless there are algebraic obstructions.

1.3 Progress Towards the Conjecture

Our main subject here is a weakened form of the conjecture which proves the full conjecture in some important examples and special classes of IFSs. In order to state it we must first quantify the degree to which the sets $f_{\underline{i}}(X)$ are separated from each other. Since all of the maps in Φ contract by the same ratio, any two of the sets $f_{\underline{i}}(X)$ and $f_{\underline{j}}(X)$ for $\underline{i}.\underline{j} \in \Lambda^n$ are translates of each other. We define the distance between them as the magnitude of this translation, which is given by $f_{\underline{i}}(x) - f_{\underline{j}}(x)$ for any $x \in \mathbb{R}$; we shall choose $x = 0$ for concreteness. Thus a measure of the degree of concentration of cylinders $f_{\underline{i}}(X)$, $\underline{i} \in \Lambda^n$, is provided by

$$\Delta_n = \min\{|f_{\underline{i}}(0) - f_{\underline{j}}(0)| : \underline{i}, \underline{j} \in \Lambda^n , \underline{i} \neq \underline{j}\}$$

Evidently, exact overlaps occur if and only if there exists an n such that $\Delta_n = 0$. Fixing $x \in X$, the points $f_{\underline{i}}(x)$, $\underline{i} \in \Lambda^n$, all lie in X, and so there must be a distinct pair $\underline{i}, \underline{j} \in \Lambda^n$ with $|f_{\underline{i}}(x) - f_{\underline{j}}(x)| \leq \text{diam}(X)/|\Lambda|^n$; hence $\Delta_n \to 0$ at least exponentially. In general there may be an exponential lower bound on Δ_n as well, i.e. $\Delta_n \geq cr^n$ for some $c, r > 0$. This is always the case when the IFS satisfies strong separation or the open set condition, but there are examples where it holds even when these conditions fail (see Garsia [Gar62]). Therefore the following theorem from [Hoc12] gives nontrivial information and should be understood as a weak form of Conjecture 1.1.

Theorem 1.2 *If $X \subseteq \mathbb{R}$ is a self-similar set and* $\dim X < \min\{1, sdimX\}$, *then* $\Delta_n \to 0$ *super-exponentially, that is,* $-\frac{1}{n}\log \Delta_n \to \infty$.

In practice, one applies the theorem after establishing an exponential lower bound on Δ_n to deduce that $\dim X = \min\{1, \text{sdim}X\}$. For example,

Proposition 1.3 *Let $\mathcal{R}$ denote the set of rational IFSs, i.e. such that $r, a_i \in \mathbb{Q}$. Then Conjecture 1.1 holds in $\mathcal{R}$.*

Proof First, a useful identity: For $\underline{i} \in \Lambda^n$, a direct calculation shows that

$$f_{\underline{i}}(x) = r^n x + \sum_{k=1}^{n} a_{i_k} r^{k-1} \tag{5}$$

$$= r^n x + f_{\underline{i}}(0) \tag{6}$$

Now let that $f_i(x) = rx + a_i$ where $r = p/q$ and $a_i = p_i/q_i$ for p, p_i, q, q_i integers and write $Q = \prod_{i\in\Lambda} q_i$. Then $f_{\underline{i}}(0) = \sum_{k=1}^{n} a_{i_k} r^{n-k}$ is a rational number with denominator Qq^n. Suppose that no overlaps occur, so that $\Delta_n > 0$ for all n. Given n, by definition there exist distinct $\underline{i}, \underline{j} \in \Lambda^n$ such that $\Delta_n = f_{\underline{i}}(0) - f_{\underline{j}}(0)$. Therefore Δ_n is a non-zero rational number with denominator Qq^n so we must have $\Delta_n \geq 1/Qq^n$. By Theorem 1.2 we conclude that $\dim X = \min\{1, \text{sdim}X\}$. □

The same argument works in the class of IFSs with algebraic coefficients, using a similar lower bound on polynomial expressions in a given set of algebraic numbers. See [Hoc12, Theorem 1.5]. A simple (but non-trivial) calculation, due to B. Solomyak and P. Shmerkin, also allows one to deal with the case that one of the translation parameters a_i is irrational, resolving Furstenberg's question about linear projections of the one-dimensional Sierpinski gasket [Hoc12, Theorem 1.6]. Theorem 1.2 leads to strong results about parametric families of self-similar sets [Hoc12, Theorem 1.8], and there is a version for measures which has also led to substantial progress on the Bernoulli convolutions problem, see [Hoc12, Theorem 1.9] and the recent advance by Shmerkin [Shm13]. Another interesting application is given in [Orp13].

The rest of this paper is an exposition of the proof of the theorem. Our goal is to present the ideas as transparently as possible, and to this end we frame the argument in terms of covering numbers (rather than entropy as in [Hoc12]). This leads to simpler statements and to an argument that is conceptually correct but, unfortunately, incomplete; some crucial steps of this simplified argument are flawed. In spite of this deficiency we believe that such an exposition will be useful as a guide to the more technical proof in [Hoc12]. To avoid any possible misunderstandings, we have indicated the false statements in quotation marks ("Lemma", "Proof", etc.).

As we shall see, the main idea is to reduce (the negation of) the theorem to a statement about sums of self-similar sets with other sets. Problems about sums of sets fall under the general title of additive combinatorics, and in the next section we give a brief introduction to the parts of this theory that are relevant to us. In Sect. 3 we explain the reduction to a statement about sumsets, and show how an appropriate inverse theorem essentially settles the matter. Finally, in Sect. 4, we discuss how the heuristic argument can be made rigorous.

2 A Birds-Eye View of Additive Combinatorics

2.1 *Sumsets and Inverse Theorems*

The sum (or sumset) of non-empty sets $A, B \subseteq \mathbb{R}^d$ is

$$A + B = \{a + b \,:\, a \in A\,,\, b \in B\}$$

Additive combinatorics, or at least an important chapter of it, is devoted to the study of sumsets and the relation between the structure of A, B and $A + B$. We focus here on so-called inverse problems, that is the problem of describing the structure of sets A, B such that $A + B$ is "small" relative to the sizes of the original sets. The general flavor of results of this kind is that, if the sumset is small, there must be an algebraic or geometric reason for it. It will become evident in later sections that this question comes up naturally in the study of self-similar sets.

To better interpret what "small" means, first consider the trivial bounds. Assume that A, B are finite and non-empty. Then $|A + B| \geq \max\{|A|, |B|\}$, with equality if and only if at least one of the sets is a singleton. In the other direction, $|A + B| \leq |A||B|$, and equality can occur (consider $A = \{0, 10, 20, 30, \ldots, 10n\}$ and $B = \{0, 1, \ldots, 9\}$). For "generic" pairs of sets the upper bound is close to the truth. For example, when $A, B \subseteq \{1, \ldots, n\}$ are chosen randomly by including each $1 \leq i \leq n$ in A with probability p and similarly for B, with all choices independent, there is high probability that $|A + B| \geq c|A||B|$. The question becomes, what can be said between these two extremes.

2.2 *Minimal Growth*

One of the earliest inverse theorems is the Brunn-Minkowski inequality of the late 19th century. The setting is $\mathbb{R}^d$ with the volume measure, and it states that if $A, B \subseteq \mathbb{R}^d$ are convex sets then, given the volumes of A, B, the volume of $A+B$ is minimized when A, B are balls with respect to some common norm. Since the volume of a ball scales like the d-th power of the radius, this means that $\mathrm{vol}(A + B) \geq (\mathrm{vol}(A)^{1/d} + \mathrm{vol}(B)^{1/d})^d$, and equality occurs if and only if, up to a nullset, A, B are dilates of the same convex set. The inequality was later extended to arbitrary Borel sets (note that $A+B$ may not be a Borel set but it is an analytic set and hence Lebesgue measurable). For a survey of this topic see Gardner [Gar02].

Similar tight statements hold in the discrete setting. The analog of a convex body is an *arithmetic progression (AP)*, namely a set of the form $P = \{a, a + d, a + 2d, \ldots, a + (1 - k)d\}$, where d is called the gap (we assume $d \neq 0$) and k is called the length of P. Then for finite sets $A, B \subseteq \mathbb{Z}^d$ with $|A|, |B| \geq 2$ we always have $|A + B| \geq |A| + |B| - 1$, with equality if and only if A, B are APs of the same gap [TV06, Proposition 5.8].

2.3 Linear Growth: Small Doubling and Freiman's Theorem

Now suppose that $A = B \subseteq \mathbb{Z}^d$ but weaken the hypothesis, assuming only that

$$|A + A| \leq C|A| \tag{7}$$

where we think of A as large and C as constant. Such sets are said to have *small doubling*.

The simplest example of small doubling in $\mathbb{Z}^d$ is when $A = \{1, \ldots, n\}^d$, in which case $|A + A| \leq 2^d|A|$. This example can be pushed down to any lower dimension as follows. For $i = 1, \ldots, k$, take intervals of integers $I_i = \{1, 2, \ldots, n_i\}$, and let $T : \mathbb{Z}^k \to \mathbb{Z}^d$ be an affine map given by integer parameters. Suppose that T is injective on $I = I_1 \times \ldots \times I_k$. Then $A = T(I) \subseteq \mathbb{Z}^d$ has the property that

$$|A + A| = |T(I) + T(I)| = |T(I + I)| \leq |I + I| \leq 2^k|I| = 2^k|A|$$

A set A as above is called a *(proper) generalized arithmetic progression (GAP) of rank k*.

GAPs are still extremely algebraic objects but one can get away from this a little using another cheap trick: Begin with a set A satisfying $|A+A| \leq C|A|$ (e.g. a GAP) and choose any $A' \subseteq A$ with cardinality $|A'| \geq D^{-1}|A|$ for some $D > 1$. Then

$$|A' + A'| \leq |A + A| \leq C|A| \leq CD|A'|$$

One of the central results of additive combinatorics is Freiman's theorem, which says that, remarkably, these are the only ways to get small doubling.

Theorem 2.1 (Freiman) *If $A \subseteq \mathbb{Z}^d$ and $|A + A| \leq C|A|$, then $A \subseteq P$ for a GAP P of rank C' and satisfying $|P| \leq C''|A|$. The constants satisfy $C' = O(C(1 + \log C))$ and $C'' = C^{O(1)}$.*

For more information see [TV06, Theorem 5.32 and Theorem 5.33].

Combined with some standard arguments (e.g. the Plünnecke-Rusza inequality), the symmetric version leads to an asymmetric versions: assuming $A, B \subseteq \mathbb{Z}^d$ and $C^{-1} \leq |A|/|B| \leq C$, if $|A + B| \leq C|A|$ then A, B are contained in a GAP P of rank and $\leq C'$ and size $|P| \leq C'|A|$, with similar bounds on the constants.

2.4 Power Growth, the "Fractal" Regime

Now relax the growth condition even more and consider finite sets $A \subseteq \mathbb{Z}$ (or $A \subseteq \mathbb{R}$) such that

$$|A + A| \leq |A|^{1+\delta} \tag{8}$$

This is the discrete analog of the condition

$$\dim_B(X + X) \leq (1 + \delta)\dim_B X \tag{9}$$

for $X \subseteq \mathbb{R}$. Indeed, given $X \subseteq \mathbb{R}$ and $n \in \mathbb{N}$ let X_n denote the set obtained by replacing each $x \in X$ with the closest point $k/2^n$, $k \in \mathbb{Z}$. Then $|X_n| \sim 2^{n(\dim_B X + o(1))}$ and $|X_n + X_n| \sim 2^{n(\dim_B(X+X)+o(1))}$ for large n, so (9) is equivalent to $|X_n + X_n| \lesssim |X_n|^{1+o(1)}$. Thus, the difference between (7) and (8) is roughly the difference between using Lebesgue measure or dimension to quantify the size of a set $X \subseteq \mathbb{R}$.

Here is a typical example of a set satisfying (8). Write $P_n = \{0, \dots, n-1\}$ and let

$$\begin{aligned} A_n &= \sum_{i=1}^{n} \frac{1}{2^{i^2}} P_{2^i} \\ &= \{\sum_{i=1}^{n} a_i 2^{-i^2} \ : \ 1 \leq a_i \leq 2^i\} \end{aligned}$$

(again, one can think of this either as a subset of $\mathbb{R}$, or of $\frac{1}{4^{n^2}}\mathbb{Z}$). It is easy to verify that the distance between distinct points $x, x' \in A_n$ is at least $1/4^{n^2}$, and that such x has a unique representation as a sum $\sum_{i=1}^{n} a_i 4^{-i^2} \ : \ 1 \leq a_i \leq 2^i$. Indeed, each term in the sum $\sum_{i=1}^{n} a_i 2^{-i^2}$ determines a distinct block of binary digits. Thus A_n is a GAP, being the image of $P_2 \times P_4 \times \ldots \times P_{2^n}$ by the map $(x_1, \dots, x_n) \mapsto \sum \frac{1}{2^{i^2}} x_i$. The rank is n, and so, as we saw in the previous section,

$$|A_n + A_n| \leq 2^n |A_n|$$

Since

$$|A_n| = \prod_{i=1}^{n} |P_n| = 2^{\sum_{i=1}^{n} i} = 2^{n(n+1)/2}$$

we have

$$|A_n + A_n| = |A_n|^{1+o(1)} \qquad \text{as } n \to \infty$$

The reader may recognize the example above as the discrete analog of a Cantor set construction, where at stage n we have a collection of intervals $2^{n(n+1)/2}$ of length 2^{-n^2}, and from each of these intervals we keep 2^{n+1} sub-intervals of length $2^{-(n+1)^2}$, separated by gaps of length $2^{-n^2-(n+1)}$. For the resulting Cantor set X it is a standard exercise to see that $\dim X = \dim_B X = 1/2$, and the calculation above shows that $\dim X + X = 1/2$ as well. Such constructions appear in the work of Erdős-Volkmann [EV66], and also in the papers of Schmeling-Shmerkin [SS10] and

Körner [Kör08], who showed that for any sequence $\alpha_1 \leq \alpha_2 \leq \ldots$ there is a set X with $\dim \sum_{i=1}^{n} X = \alpha_n$.

Do all examples of (9) look essentially like this one? In principle one can apply Freiman's theorem, since the hypothesis (9) can be written as $|A+A| \leq C|A|$ for $C = |A|^{\delta}$. What one gets, however, is that A is a $|A|^{O(\delta)}$-fraction of a GAP or rank $|A|^{O(\delta)}$, and this gives rather coarse information about A (note that, trivially, every set is a GAP of rank $|A|$).

Instead, it is possible to apply a multi-scale analysis, showing that at some scales the set looks quite "dense" and at others quite "sparse". The best way to explain this is in the language of trees, which we introduce next.

2.5 Trees and Tree-Measures

Denote the length of a finite sequence $\sigma = \sigma_1 \ldots \sigma_n$ by $|\sigma| = n$ and write $\emptyset$ for the empty word, which by definition has $|\emptyset| = 0$. Denote the concatenation of words σ and τ by $\sigma\tau$, in which case we say that σ is a prefix of τ, and that $\sigma\tau$ extends σ.

The *full binary tree* of height h is the set $\{0,1\}^{\leq h} = \bigcup_{k=0}^{h}\{0,1\}^k$ of 0, 1-valued sequences of length $\leq h$, where our convention is that $\{0,1\}^0 = \{\emptyset\}$, so the empty word is included. We define a *tree of height h* is a subset $T \subseteq \bigcup_{i=0}^{h}\{0,1\}^i$ satisfying

(T1) $\emptyset \in T$.
(T2) If $\sigma \in T$ and η is an initial segment of σ then $\eta \in T$.
(T3) If $\sigma \in T$ then there is an $\eta \in T$ which extends σ and $|\eta| = h$.

One may think of T as a set of vertices and introduce edges between every pair $\sigma_1 \ldots \sigma_i, \sigma_1 \ldots \sigma_i\sigma_{i+1} \in T$. Then T is a tree if $\emptyset \in T$ and in the associated graph there is a path from $\emptyset$ to every node, and all maximal paths are of length h.

The *level (or depth)* of $\sigma \in T$ is its length (the graph-distance from $\emptyset$ to σ). The *leaves* of a tree T of height h are the elements of the lowest (deepest) level, namely h:

$$\partial T = T \cap \{0,1\}^h$$

The *descendants* of $\sigma \in T$ are the nodes $\eta \in T$ that extend σ. The nodes m generations below σ in T are the nodes of the form $\eta = \sigma\sigma' \in T$ for $\sigma' \in \{0,1\}^m$.

We also shall need to work with measures "on trees", or, rather, measures on their leaves. For notational purposes it is better to introduce the notion of a *tree-measure*[5] on the full tree $\{0,1\}^{\leq h}$, namely, a function $\mu : \{0,1\}^{\leq h} \to [0,1]$ satisfying

(M1) $\mu(\emptyset) = 1$.
(M2) $\mu(\sigma) = \sum_{i \in \{0,1\}} \mu(\sigma i)$

It is easily to derive from (M1) and (M2) that $\sum_{\sigma \in \{0,1\}^k} \mu(\sigma) = 1$ for every $1 \leq k \leq h$, so a tree-measure induces genuine probability measures on every level

[5] This notion is identical to a flow on the tree in the sense of network theory.

of the full tree, and in particular on ∂T. Conversely, if we have a genuine probability measure μ on the set of leaves $\{0,1\}^h$ of the full tree of height h then it induces a tree-measure by $\mu(\sigma) = \sum_{\eta:\sigma\eta\in\partial T} \mu(\{\sigma\eta\})$. Given a tree-measure, the set $T = \{\sigma : \mu(\sigma) > 0\}$ is a tree which might be called the support of μ.

Every tree-measure μ on $\{0,1\}^{\leq h}$ defines a distribution on the nodes of the tree as follows: first choose a level $0 \leq i \leq h$ uniformly, and then choose a node $\sigma \in \{0,1\}^i$ in level i with the probability given by μ, i.e. $\mu(\sigma)$ (we have already noted that at each level the masses sum to 1). Thus the probability of $A \subseteq T$ is

$$\mathbb{P}_\mu(A) = \frac{1}{h+1}\sum_{\sigma\in A}\mu(\sigma)$$

and the expectation of $f : \{0,1\}^{\leq h} \to \mathbb{R}$ is

$$\mathbb{E}_\mu(f) = \frac{1}{h+1}\sum_{k=0}^{n}\sum_{\sigma\in\{0,1\}^k}\mu(\sigma)f(\sigma)$$

Sometimes we write $\mathbb{P}_{\sigma\sim\mu}$ or $\mathbb{E}_{\sigma\sim\mu}$ to define σ as a random node, as in the expression

$$\mathbb{P}_{\sigma\sim\mu}(\sigma \in T \text{ and } \sigma \text{ has two children in } T) = \frac{1}{h+1}\sum_{\sigma\in T}\mu(\sigma)1_{\{\sigma 0,\sigma 1\in T\}}$$

Given the tree T of height h, it is natural to consider the uniform probability measure $\mu_{\partial T}$ on ∂T and, as described above, extend it to a tree-measure, which we denote μ_T. In this case we abbreviate the probability and expectation operators above by $\mathbb{P}_T$ and $\mathbb{E}_T$, etc. It is important to note that *choosing a node according to μ_T is not the same as choosing a node uniformly from T*. The latter procedure is usually heavily biased towards sampling from the leaves, since these generally constitute a large fraction of the nodes (in the full binary tree, sampling this way gives a leaf with probability $> 1/2$). In contrast, μ_T samples uniformly from the levels, and within each level we sample according to the relative number of leaves descended from each node.

Trees and tree-measures are naturally related to sets and measures on $[0,1)$ using binary coding. Given a set $X \subseteq [0,1)$ and $h \in \mathbb{N}$, we lift X to a tree T of height h by taking all the initial sequences of length $\leq h$ of binary expansions of points in X, with the convention that the expansion terminates in 1s if there is an ambiguity. We remark that for $k \leq h$,

$$N_{1/2^k}(X) \leq \left|T \cap \{0,1\}^k\right| \leq 2N_{1/2^k}(X)$$

Similarly, a probability measure μ on $[0,1)$ can be lifted to a tree-measure $\widetilde{\mu}$ on $\{0,1\}^{\leq h}$ by defining $\widetilde{\mu}(\sigma)$ equal to the mass of the interval of numbers whose binary expansion begins with σ.

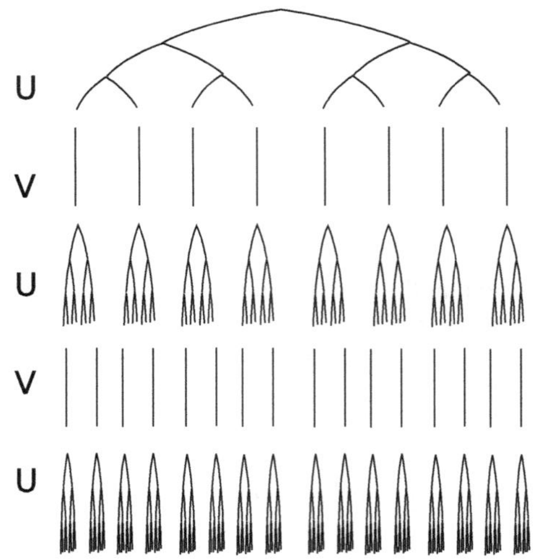

Fig. 1 A tree with alternating levels having full branching at some levels, full concentration at others, and a few levels omitted. Schematically this is what the tree associated to A_n from Sect. 2.4 looks like, as well as the conclusion of Theorem 2.2 (with W indicated by the small space between levels)

2.6 Inverse Theorems in the Power-Growth Regime

We need some terminology for describing the local structure of trees. We say that T has *full branching for m generations at σ* if σ has all 2^m possible descendants m generations below it, that is, $\sigma\eta \in T$ for all $\eta \in \{0, 1\}^m$. At the other extreme, we say that T is *fully concentrated for m generations at σ* if σ has a single descendant m generations down, that is, there is a unique $\eta \in \{0, 1\}^m$ with $\sigma\eta \in T$.

Let us return to the example A_n from Sect. 2.4 and examine the associated tree T_n of height n^2. For every $i < n$, every node at level i^2 has full branching for i generations; and every node at level i^2+i is fully concentrated for $i+1$ generations. Consequently, for every $j \in [i^2, i^2+i)$ every node of level j has full branching for one generation; for $j \in [i^2+i, (i+1)^2)$, every node at level j is fully concentrated for one generation. We also have the following statement: For every m we can partition the levels $0, 1, \ldots, n^2$ into three sets U, V, W, such that (a) For every $i \in U$, every level-i node has full branching for m generations; (b) For every $j \in V$, every level-j node is fully concentrated for m generations; and (c) W is a negligible fraction of the levels, specifically $|W|/n^2 = o(1)$ as $n \to \infty$ (with m fixed). Of course, $U = \bigcup_{i>m}[i^2, i^2+i-m)$, $V = \bigcup_{i>m}[i^2+i, (i+1)^2-m)$, and W is the set of remaining levels. This is pictured schematically in Fig. 1.

Does this picture hold in general when $|A+A| \leq |A|^{1+\delta}$? Certainly not exactly, since we can always pass to a subset $A' \subseteq A$ of size $|A'| \geq |A|^{1-\delta}$ and get a set with similar doubling behavior (for a constant loss in δ), but much less structure. One can also perturb it in other ways. However, in a looser sense, the picture above is quite general. One approach is to pass to a subtree of reasonably large relative size. Such an approach was taken by Bourgain in [Bou03, Bou10]. The approach taken in [Hoc12] is more statistical, and in a sense it gives a description of the entire tree, but requires us to weaken the notion of concentration. Given $\delta > 0$, we say that T is

δ-concentrated for m generations at $\sigma \in T$ if there exists $\eta \in \{0,1\}^m$ such that

$$\mu_T(\sigma\eta) \geq (1-\delta)\mu_T(\sigma)$$

where μ_T is the tree-measure associated to T. In other words, T is δ-concentrated at σ if it is possible to remove an δ-fraction of the leaves descended from σ in such a way that the resulting tree becomes fully concentrated for m generations at σ. Note that this definition is not purely local, since it depends not only on the depth-m subtree of T rooted at σ, but on the entire subtree rooted at σ, since the weights on $S = \{\sigma\eta : \eta \in \{0,1\}^m\}$ are determined by the number of leaves of T, not by S itself.

Theorem 2.2 *For every $\varepsilon > 0$ and $m > 1$, there is a $\delta > 0$ such that for all sufficiently small $\rho > 0$ the following holds. Let $X \subseteq [0,1]$ be a finite set such that*

$$N_\rho(X+X) \leq N_\rho(X)^{1+\delta}$$

and let T be the associated tree of height $h = \lceil \log(1/\rho) \rceil$. Then the levels $0, 1, \ldots, h$ can be partitioned into sets U, V, W such that

1. *For every $i \in U$,*

$$\mathbb{P}_{\sigma \sim T}(T \text{ has full branching at } \sigma \text{ for } m \text{ generations} \mid \sigma \text{ is in level } i) > 1-\varepsilon.$$

2. *For every $j \in V$,*

$$\mathbb{P}_{\sigma \sim T}(T \text{ is } \varepsilon\text{-concentrated at } \sigma \text{ for } m \text{ generations} \mid \sigma \text{ is in level } j) > 1-\varepsilon.$$

3. *$|W| < \varepsilon h$.*

Note that if X is ρ-separated, the hypothesis is essentially the same as $|X+X| \leq |X|^{1+\varepsilon}$.

Our analysis of self-similar sets requires the following asymmetric variant, which is easily seen to imply the symmetric one above. To motivate it, note that $|A+B| \leq C|A|$ can occur for two trivial reasons: One is that $A = \{1,\ldots,n\}$ for some n and $B \subseteq \{1,\ldots,n\}$ is arbitrary. The second is that $B = \{b\}$, a singleton, and A is arbitrary. The following theorem says that when $|A+B| \leq |A|^{1+\delta}$ then there are essentially two kinds of scales: those where, locally, the sets A, B look like in the first trivial case, and those where, locally, A, B look like the second trivial case. See Fig. 2.

Theorem 2.3 *For every $\varepsilon > 0$ and $m > 1$, there is a $\delta > 0$ such that for all sufficiently small $\rho > 0$ the following holds. Let $X, Y \subseteq [0,1]$ be finite sets such that*

$$N_\rho(X+Y) \leq N_\rho(X)^{1+\delta}$$

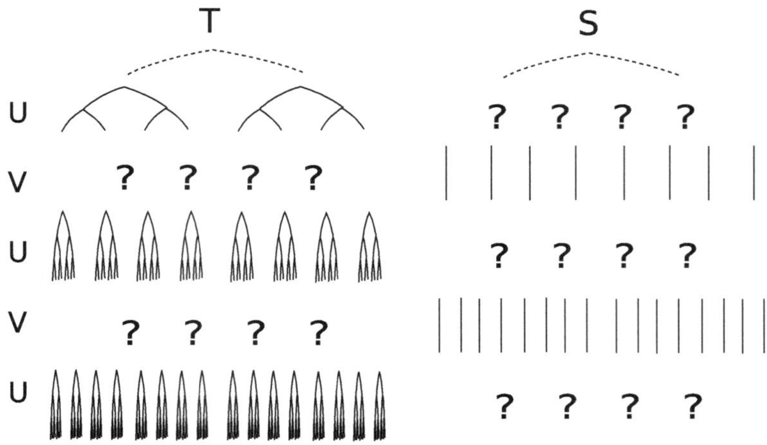

Fig. 2 Schematic representation of the conclusion of Theorem 2.3 (with W indicated by the small space between levels)

Let T, S be the associated trees of height $h = \lceil \log(1/\rho) \rceil$, respectively. Then the levels $0, 1, \ldots, h$ can be partitioned into sets U, V, W such that

1. *For every $i \in U$,*

$$\mathbb{P}_{\sigma \sim T}(T \text{ has full branching at } \sigma \text{ for } m \text{ generations} \mid \sigma \text{ is in level } i) > 1 - \varepsilon$$

   (but we know nothing about S at level i).
2. *For every $j \in V$,*

$$\mathbb{P}_{\sigma \sim S}(S \text{ is } \varepsilon\text{-concentrated at } \sigma \text{ for } m \text{ generations} \mid \sigma \text{ is in level } j) > 1 - \varepsilon$$

   (but we know nothing about T at level j).
3. *$|W| < \varepsilon h$.*

The theorems above follow from [Hoc12, Theorems 2.7 and 2.9], using the fact that high enough entropy at a given scale implies full branching, and small enough entropy at a given scale implies δ-concentration.

3 A Conceptual Proof of Theorem 1.2

In this section we give a heuristic proof of Theorem 1.2. We begin with some general observations about self-similar sets. Then we explain how the theorem is reduced to a statement about sumset growth. Finally, we demonstrate how the inverse theorems of the previous section are applied.

From now on let $\Phi = \{f_i\}_{i\in\Lambda}$ be an IFS with attractor X, as in the introduction. We assume that $0 \in X \subseteq [0, 1)$; this can always be achieved by a change of coordinates, which does not affect the statement of Theorem 1.2.

3.1 Sumset Structure of Self-similar Sets

Our analysis will focus on finite approximations of X. Define the n-th approximations by

$$X_n = \{f_{\underline{i}}(0) \,:\, \underline{i} \in \Lambda^n\}$$

Clearly $X_n \subseteq X$. Also note that $|X_n| \leq |\Lambda|^n$, with a strict inequality for some n if and only if exact overlaps occur. Self similarity enters our argument via the following lemma.

Lemma 3.1 *For any* $m, n \in \mathbb{N}$,

$$X = X_m + r^m X \tag{10}$$

$$X_{m+n} = X_m + r^m X_n \tag{11}$$

Proof By (2) and (6),

$$\begin{aligned} X &= \bigcup_{\underline{i}\in\Lambda^n} f_{\underline{i}}(X) \\ &= \bigcup_{\underline{i}\in\Lambda^n} \{f_{\underline{i}}(0) + r^m x \,:\, x \in X\} \\ &= X_m + r^m X \end{aligned}$$

which is the first identity. To prove the second, for $\underline{i} \in \Lambda^m$ and $\underline{j} \in \Lambda^n$ denote by $\underline{i}\,\underline{j}$ their concatenation. By (5),

$$\begin{aligned} f_{\underline{i}\,\underline{j}}(0) &= \sum_{k=1}^{m} a_{i_k} r^{k-1} + r^m \sum_{k=1}^{n} a_{j_k} r^{k-1} \\ &= f_{\underline{i}}(0) + r^m f_{\underline{j}}(0) \end{aligned}$$

hence

$$\begin{aligned} X_{m+n} &= \{f_{\underline{i}\,\underline{j}}(0) \,:\, \underline{ij} \in \Lambda^{m+n}\} \\ &= \{f_{\underline{i}}(0) + r^m f_{\underline{j}}(0) \,:\, \underline{i} \in \Lambda^m, \underline{j} \in \Lambda^n\} \\ &= X_m + r^m X_n \end{aligned}$$

□

Let us demonstrate the usefulness of this lemma by showing that $\dim_B(X)$ exists. First, since $r^m X$ is of diameter $\leq r^m$, it is easy to deduce from (10) that $N_{r^n}(X_n)$, $N_{r^n}(X)$ differ by at most a factor of 2. Thus the existence of $\dim_B X$ is equivalent to existence of the limit $\frac{1}{m} \log N_{r^m}(X_m)$ as $n \to \infty$. Next, we have a combinatorial lemma.

Lemma 3.2 *Let $A, B \subseteq \mathbb{R}$ with $B \subseteq [0, \varepsilon)$. Then for any $\gamma < \varepsilon$,*

$$N_\gamma(A + B) \geq \frac{1}{3} \cdot N_\varepsilon(A) \cdot N_\gamma(B)$$

Proof Let $\mathcal{I} = \{I_i\}_{i=1}^{N_\varepsilon(A)}$ be an optimal cover of A by disjoint intervals of length ε. Let $\mathcal{J} = \{J_j\}_{j=1}^{N_\gamma(A+B)}$ be an optimal cover of $A + B$ by intervals of length γ. For each $I_i \in \mathcal{I}$ fix a point $a_i \in A \cap I_i$ and note that $a_i + B \subseteq A + B$ is covered by $\mathcal{J}$, so $a_i + B$ intersects at least $N_\gamma(B)$ intervals in $\mathcal{J}$. If each interval J_j intersects a unique translate $a_i + B$, we would conclude that $N_\gamma(A+B) \geq N_\varepsilon(A)N_\gamma(B)$. While a_i may not be unique, we can argue as follows: Since $B \subseteq [0, \varepsilon)$, if $J_j = [u, u + \varepsilon]$ and intersects $a + B$ for some $a \in A$, then $a \in [u - \varepsilon, u + 2\varepsilon)$. Since the intervals I_i are disjoint and of length ε, there are most 3 intervals $I_i \in \mathcal{I}$ that a could belong to. The claim follows. □

Since $X \subseteq [0, 1)$ we have $r^m X_n \subseteq [0, r^m)$, so by the lemma,

$$\begin{aligned} N_{r^{m+n}}(X_{m+n}) &= N_{r^{m+n}}(X_m + r^m X_n) \\ &\geq \frac{1}{3} \cdot N_{r^m}(X_m) \cdot N_{r^{m+n}}(r^m X_n) \\ &= \frac{1}{3} \cdot N_{r^m}(X_m) \cdot N_{r^n}(X_n) \end{aligned}$$

where in the last equality we used the identity $N_{t\varepsilon}(tZ) = N_\varepsilon(Z)$. Taking logarithms, this shows that the sequence $s_n = \log N_{r^m}(X_m)$ is approximately super-additive in the sense that $s_{m+n} \geq s_m + s_n - C$ for a constant C. The existence of the limit of $\frac{1}{n}s_n$ as $n \to \infty$ is then well known (perhaps it is better known when $C = 0$ and s_n is (really) super-additive. The proof for $C = 0$ works also in the $C > 0$ case; alternatively, note that $s'_n = s_n - \log n$ becomes super-additive after excluding finitely many terms, so $\lim \frac{1}{n}s'_n$ exists, and $\frac{1}{n}s'_n - \frac{1}{n}s_n \to 0$).

3.2 From Theorem 1.2 to Additive Combinatorics

Let us return to our main objective, Theorem 1.2. Continuing with the previous notation, write

$$\begin{aligned} \alpha &= \dim_B X \\ \beta &= \min\{1, \mathrm{sdim} X\} \end{aligned}$$

and suppose, by way of contradiction, that $\alpha < \beta$ and that for some $k \in \mathbb{N}$ we have $\Delta_n \geq 2^{-kn}$ for all n (in particular, there are no exact overlaps). We make a number of observations. The first is rather trivial: that "too small" dimension means that there are intervals of length r^m containing exponentially many points from X_m. Precisely,

Proposition 3.3 *Let $\sigma = \frac{1}{2}(\beta - \alpha) > 0$. Then for every large enough m, there is an interval I_m of length r^m such that $|X_m \cap I_m| > r^{-\sigma m}$.*

Proof As we have already noted, $\frac{1}{m\log(1/r)} \log N_{r^m}(X_m) \to \alpha$ as $m \to \infty$. Thus for large enough m,

$$N_{r^m}(X_m) < r^{-(\beta-\sigma)m}$$

On the other hand, since there are no exact overlaps,

$$|X_m| = |\Lambda|^m = r^{-m\text{sdim}(X)} \geq r^{-m\beta}$$

Thus in an optimal cover of X_m by r^m-intervals, at least one must contain $|X_m|/N_{r^m}(X_m) \geq (1/r)^{\sigma m}$ points. □

We now wish to extract more information from the sumset identity $X_{m+n} = X_m + r^m X_n$. In itself it provides limited information about the covering number $N_{m+n}(X_n)$, since the summands live at different scales. This is what was used earlier in proving super-additivity of $s_n = \log N_{r^m}(X_m)$. The next step is to localize the sumset relation.

Proposition 3.4 *For all $\delta > 0$, for all large m there exists an interval J_m of length r^m such that $X_m \cap J_m \neq \emptyset$ and, writing $n = km$,*

$$N_{r^{m+n}}((X_m \cap J_m) + r^m X_n) < r^{-(1+\delta)\alpha n} \tag{12}$$

Proof Fix m, set $n = km$, and let $\mathcal{J}$ denote the partition of $\mathbb{R}$ into intervals $[ur^m, (u+1)r^m)$, $u \in \mathbb{Z}$, whose lengths are r^m. Since $X_m = \bigcup_{J\in\mathcal{J}}(X_m \cap J)$, we can re-write (11) as

$$\begin{aligned} X_{m+n} &= X_m + r^m X_n \\ &= \bigcup_{J\in\mathcal{J}} ((X_m \cap J) + r^m X_n) \end{aligned} \tag{13}$$

Since $X_m \cap J \subseteq [ur^m, (u+1)r^m)$ for some u and $r^m X_n \subseteq [0, r^m)$, we have $(X_m \cap J) + r^m X_n \subseteq [ur^m, (u+2)r^m)$ and in particular each set in the union (13) is of diameter $\leq 2r^m$. On the other hand, no interval of length r^{m+n} intersects more than three of the sets $[ur^m, (u+2)r^m)$. Therefore, arguing as in the proof of Lemma (3.2),

$$N_{r^{m+n}}(X_{m+n}) \geq \frac{1}{3} \cdot N_{r^m}(X_m) \cdot \min_{J\in\mathcal{J}\,:\,X_m\cap J\neq\emptyset} N_{r^{m+n}}((X_{m+n} \cap J) + r^m X_n)$$

so

$$\begin{aligned}\min_{J\in\mathcal{J}\,:\,X_m\cap J\neq\emptyset} N_{r^{m+n}}((X_{m+n}\cap J)+r^mX_n) &\leq 3\cdot\frac{N_{r^{m+n}}(X_{m+n})}{N_{r^m}(X_m)}\\ &\leq 3\cdot\frac{r^{-(\alpha+o(1))(m+n)}}{r^{-(\alpha+o(1))m}}\\ &= r^{-(\alpha+o(1))n} \qquad \text{as } m\to\infty\end{aligned}$$

The proposition follows. □

Now suppose that it so happens that, for large m, the propositions above produce the *same* interval: $I_m = J_m$. Then we would have the following:

Proposition 3.5 *Suppose that* $\dim X < \min\{1, sdimX\}$ *and* $\Delta_n \geq 2^{-kn}$ *for all* n. *Then there is a constant* $\tau > 0$ *such that, for every* $\delta > 0$ *and all suitably large* n, *there is a subset* $Y_n \subseteq [0, 1]$ *with*

$$N_{r^n}(Y_n) \geq 2^{\tau n} \tag{14}$$

$$N_{r^n}(X_n + Y_n) \leq N_{r^n}(X_n)^{1+\delta} \tag{15}$$

Proof Let σ be as in Proposition 3.3 and take $\tau = \sigma/(k\log(1/r))$. As before write $n = (k+1)m$, and assume that the intervals I_m, J_m provided by the two previous propositions coincide for arbitrarily large m: $I_m = J_m = [a_m, b_m)$. Let

$$Y_m = r^{-m}(X_m \cap I_m - a_m)$$

and note that $Y_m \subseteq [0, 1)$. Now, by choice of I_m we know that $|X_m\cap I_m| \geq r^{-\sigma m}$, and since $\Delta_m \geq 2^{-km} = 2^{-n}$, we know that every two points in $X_m \cap I_m$ are separated by at least 2^{-n}. Therefore,

$$\begin{aligned}N_{r^n}(Y_m) &= N_{r^{m+n}}(X_m\cap I_m)\\ &\geq r^{-\sigma m}\end{aligned}$$

Using the identity $N_{t\varepsilon}(tZ) = N_\varepsilon(Z)$ with $t = r^m$ and $Z = Y_m$, we conclude that

$$N_{r^{-n}}(Y_m) \geq r^{-\sigma m} = r^{\tau n}$$

Similarly, since $X_n+Y_m = r^{-m}((X_m\cap I_m)+r^mX_n)$, from the definition of J_m and the identity $N_{t\varepsilon}(tZ) = N_\varepsilon(Z)$ again, we find that for large enough n (equivalently, m),

$$\begin{aligned}N_{r^n}(X_n+Y_n) &= N_{r^{m+n}}((X_m\cap I_m)+r^mX_n)\\ &\leq r^{-(1+\delta)\alpha n}\\ &\leq N_{r^n}(X_n)^{(1+2\delta)}\end{aligned}$$

where in the last inequality we again used the fact that $N_{r^m}(X_m) \sim r^{-n\alpha}$. □

The task of showing that the conclusion of the "Proposition" is impossible falls within the scope of additive combinatorics. Heuristically, it cannot happen because, being a fractal, X_n has very little "additive structure". This intuition is correct, as we discuss in the next section.

But can one really ensure that I_m,J_m coincide? A natural attempt would be to show that, for a fixed optimal r^m-cover of X_m, "most" intervals of length r^m can play each of the roles, and hence a positive fraction can play both. In fact, for every $\eta > 0$, for large m at least a $(1-\eta)$-proportion of these intervals will be a good choice for J_m. Unfortunately, although the number of candidates for I_m can be shown to be exponential in m, it could still be exponentially small compared to $N_{r^m}(X_m)$, and so we cannot conclude that the two families of "good" intervals have members in common. It is possible that more sophisticated counting can make this work, but the approach that is currently simplest is to replace covering numbers by the entropy, at an appropriate scale, of the uniform measure on X_m. We return to this in Sect. 4.

3.3 Getting a Contradiction

Our goal now is to demonstrate that the conclusion of "Proposition" 3.5 is impossible. The argument we give again falls short of this goal, but it gives the essential ideas of the proof. Thus, we ask the reader to suspend his disbelief a little longer.

Let $\tau > 0$ be as given in "Proposition" 3.5. Choose a very small parameter $\varepsilon > 0$ which we shall later assume is small compared to τ. Choose m large enough that

$$N_{r^m}(X_m) \geq r^{-m(1-\varepsilon)\alpha}$$

Apply the inverse Theorem 2.3 with parameters ε, m and obtain the promised $\delta > 0$. From "Proposition" 3.5 obtain the corresponding $Y_n \subseteq [0, 1)$ satisfying (14) and (15).

Write T^n for the tree of height $h_n = [1/r^n]$ associated to X_n and S^n for the tree of the same height associated to Y_n. From our choice of δ and (15), by the inverse theorem there is a partition $U_n \cup V_n \cup W_n$ of $\{1, \dots, h_n\}$ such that

(I) At scales $i \in U_n$, a $1-\varepsilon$ fraction of nodes of T^n at level i have full branching for m-generations.
(II) At scales $j \in V_n$, a $1-\varepsilon$ fraction of nodes of S^n at level j are ε-concentrated for m generations.
(III) $|W_n| \leq \varepsilon h_n$.

Our first task is to show that V_n is not too large. It is quite clear (or at least believable) that if a tree has few nodes with more than one child, then it can have only an exponentially small number of leaves. The same is true if we only assume, for a small $\lambda > 0$, that most nodes are λ-concentrated. More precisely,

Lemma 3.6 *Let S be a tree of height h, let $\lambda > 0$ and $\ell \geq 1$. Suppose that*

$$\mathbb{P}_S(\sigma \in S : S \text{ is } \lambda\text{-concentrated at } \sigma \text{ for } \ell \text{ generations}\} > 1 - \lambda$$

Then $|\partial S| \leq 2^{\lambda' \cdot h}$ where $\lambda' \to 0$ as $\lambda \to 0$ and $h/\ell \to \infty$.

We leave the proof to the motivated reader. We note that this lemma is superseded by Proposition 4.5, which gives a stronger statement and has a simpler proof.

We apply the lemma to $S = S^n$ with $\ell = m$. Choose λ small enough that $\lambda' < \tau$ for large n (hence large h_n). Thus λ depends only on τ and we may assume that at the start we chose $\varepsilon < \frac{1}{2}\lambda$. Suppose that we had $|V_n| > (1 - \lambda/2)h_n$. Since in each level $j \in V_n$ a $(1-\varepsilon)$-fraction of the nodes (with respect to the tree measure μ_{S^n}) is ε-concentrated, at least the same fraction is λ-concentrated, and we would conclude

$$\begin{aligned}\mathbb{P}_{S^n}(\sigma \in S^n : S^n \text{ is } \lambda\text{-concentrated at } \sigma \text{ for } m \text{ generations}) &\geq \frac{1}{h_n}|V_n| \cdot (1-\varepsilon)\\ &> (1 - \frac{\lambda}{2})(1-\varepsilon)\\ &> 1 - \lambda\end{aligned}$$

From the lemma we would have $N_{r^n}(Y_n) \leq |\partial S^n| < 2^{\lambda' h_n} < 2^{\tau h_n}$, contradicting (14). Thus, we conclude that

$$|V_n| < (1 - \frac{\lambda}{2})h_n$$

Consequently, assuming as we may that $\varepsilon < \lambda/6$,

$$|U_n| = h_n - |V_n| - |W_n| \geq (\frac{\lambda}{2} - \varepsilon)h_n > \frac{\lambda}{3}h_n \tag{16}$$

So far we have seen that U_n consists of a positive fraction of the levels of T^n, and hence a positive fraction of nodes in T^n have full branching for m generations. Our next task will be to show that most of the remaining nodes have roughly $r^{-\alpha m}$ descendants m generations down. This is where we use self-similarity again in an essential way.

Proposition 3.7 *If m is large enough, then for all large enough n,*

$$\mathbb{P}_{\sigma \sim \mu_{T^n}}\begin{pmatrix}\sigma \text{ has } \geq 2^{(1-\varepsilon)\alpha m} \text{ descendants}\\ m \text{ generations down in } T\end{pmatrix} > 1 - \varepsilon$$

Proof (sketch.) A node $\sigma \in T^n$ of level ℓ corresponds to an interval $I = [\frac{u}{2^\ell}, \frac{u+1}{2^\ell})$. We call such intervals level-ℓ intervals, and recall that the probability induced from μ_{T^n} on level-ℓ intervals is just proportional to $|I \cap X_n|$. The claim is then that if we choose $0 \leq \ell \leq h_n$ uniformly and then choose a level-ℓ interval I at random, then with probability at least $1-\varepsilon$, we will have $N_{2^{-\ell-m}}(I \cap X_n) \geq 2^{-(1-\varepsilon)\alpha m}$. In order to prove this, it is enough to show that for all levels $0 \leq \ell \leq (1-\frac{\varepsilon}{2})h_n$, if we choose a level-$\ell$ interval I at random, then with probability at least $1-\frac{\varepsilon}{2}$ we have $N_{2^{-\ell-m}}(I \cap X_n) \geq 2^{-(1-\varepsilon)\alpha m}$.

Fix a parameter m_0 depending on ε and assume m, n large with respect to it. Observe that X_n decomposes into a union of copies of $X_{n'}$ scaled by approximately $2^{-\ell-m_0}$. More precisely, choosing $u \in \mathbb{N}$ such that $r^u \approx 2^{-\ell-m_0}$, by (11) we have

$$X = X_u + r^u X_{n-u} = \bigcup_{x \in X_u} (x + r^u X_{n-u})$$

The idea is now the following. The translates $x + r^u X_{n-u}$ in the union are of diameter $r^u \approx 2^{-\ell}/2^{m_0}$, which is much smaller than $2^{-\ell}$, and hence with probability at least $1-\frac{\varepsilon}{2}$ a level-ℓ interval I will contain an entire translate $x + r^u X_{n-u}$ from the union above. The details of the proof are somewhat tedious and we omit them. The point is that, if $x + r^u X_{n-u} \subseteq I$, and assuming that m is large enough relative to ε, m_0, we have

$$\begin{aligned} N_{2^{-\ell-m}}(I \cap X_n) &\geq N_{2^{-\ell-m}}(x + r^u X_{n-u}) \\ &= N_{2^{-\ell-m} r^{-u}}(X_{n-u}) \\ &\approx N_{2^{-(m-m_0)}}(X_{n-u}) \\ &> 2^{(1-\varepsilon)\alpha m} \end{aligned}$$

which is what we wanted to prove. □

Now that we know that most nodes in T^n have many descendants, and a positive fraction have the maximal possible number of descendants, m generations down, the last ingredient we need is a way to use this information to get a lower bound on the number of leaves in T^n. Heuristically, this is the analog of the upper bound we had in Lemma 3.6.

Proposition 3.8 *Let T be a tree of height h, let $m \geq 0$, and suppose that the nodes of T can be partitioned into disjoint sets $A_1, \ldots, A_\ell$ such that each node $\sigma \in A_i$ has $2^{c_i m}$ descendants m generations down. Write $p_i = \mathbb{P}_{\mu_T}(A_i)$. Then*

$$|\partial T| \geq \prod_{i=1}^{\ell} 2^{c_i \cdot p_i h}$$

This "Proposition" is, unfortunately, incorrect, and the reader may find it instructive to look for a counterexample. The statement could be fixed if we made stronger assumptions than just bounding the branching in each of the sets A_i, but the resulting argument would almost certainly be more complicated than the proof in [Hoc12], and we do not pursue it. The correct statement is given in Proposition 4.5 below.

We can now put the pieces together. By the defining property (I) of U_n and equation (16), the set $A_1^n \subseteq T^n$ of nodes with full branching for m-generations satisfies

$$\begin{aligned}\mathbb{P}_{T^n}(A_1^n) &\geq \frac{1}{h_n}|U_n| \cdot (1-\varepsilon)\\ &\geq \frac{\lambda}{3}(1-\varepsilon)\\ &\geq \frac{\lambda}{4}\end{aligned}$$

assuming again ε small compared to λ (equivalently τ). Let A_2^n denote the set of nodes of $T^n \setminus A_1^n$ which do *not* have at least $2^{m(1-2\varepsilon)\dim X}$ descendants m generations down; by Proposition 3.7,

$$\mathbb{P}_{T^n}(A_2^n) < \varepsilon$$

Therefore if we define $A_3^n = T^n \setminus \{A_1^n \cup A_2^n \backslash\}$ then all nodes in A_3^n have at least $2^{m(1-2\varepsilon)\dim X}$ descendants m generations down and

$$\mathbb{P}_{T^n}(A_3^n) = 1 - \mathbb{P}_{T^n}(A_1^n) - \mathbb{P}_{T^n}(A_2^n)$$

In the terminology of the "Proposition", we have $p_1 \geq \lambda/4$ and $p_2 < \varepsilon$, hence $p_3 \geq 1 - p_1 - \varepsilon$. Also $c_1 = 1$, $c_3 = (1-2\varepsilon)\dim X$ and by default $c_2 \geq 0$. From the "Proposition" we find that

$$\begin{aligned}|\partial T^n| &\geq 2^{p_1 h_n} \cdot 2^{p_2 \cdot c_2 h_n} \cdot 2^{p_3 (1-2\varepsilon)\dim X \cdot h_n}\\ &\geq 2^{p_1 h_n + (1-2\varepsilon)\dim X \cdot (1-p_1-\varepsilon) h_n}\\ &\geq 2^{(\dim X + \varepsilon) h_n}\end{aligned}$$

where in the last inequality we assumed that ε is small compared to p_1 and $\dim X$. Since $N_{r^n}(X) = |\partial T^n|^{1+o(1)}$ as $n \to \infty$, this contradicts the definition of $\dim X$.

3.4 Sums with Self-similar Sets

What we "proved" above is the following statement which is of independent interest, and is, moreover, true (a proof follows easily from the methods of [Hoc12]).

Theorem 3.9 *For every any self-similar set X with $\dim X < 1$ and every $\tau > 0$ there is a $\delta > 0$ such that for all small enough $\rho > 0$ and any set $Y \subseteq \mathbb{R}$,*

$$N_\rho(Y) > (1/\rho)^\tau \quad \Longrightarrow \quad N_\rho(X+Y) > N_\rho(X)^{1+\delta}$$

There is also a fractal version for Hausdorff dimension:

Theorem 3.10 *For every any self-similar set X with* $\dim X < 1$ *and every $\tau > 0$ there is a $\delta > 0$ such that for any set $Y \subseteq \mathbb{R}$,*

$$\dim Y > \tau \qquad \Longrightarrow \qquad \dim(X + Y) > \dim X + \delta$$

For box dimension (lower or upper) the analogous statement follows directly from the previous theorem. The version for Hausdorff dimension requires slightly more effort and will appear in [Hoc13] along with the analog for measures.

4 Entropy

In this final section we discuss how to turn the outline above into a valid proof. The main change is to replace sets by measures and covering numbers by entropy. Each of the three parts of the argument (inverse theorem, reduction to a statement about sumsets, and the analysis of the sums) has an entropy analog which we indicate below, along with a reference to the relevant part of [Hoc12].

The reader should note that the outline given below is designed to match as closely as possible the argument from the previous section, rather than the proof from [Hoc12]. Although the ideas and many of the details are the same, the original proof is direct, whereas the one here is by contradiction. For this reason not all of the statements below have exact analogs in [Hoc12].

4.1 Entropy

We assume that the reader is familiar with the basic properties of Shannon entropy, see for example [CT06]. Let $\mathcal{I}_\varepsilon = \{[k\varepsilon, (k+1)\varepsilon)\}_{k\in\mathbb{Z}}$, which is a partition of $\mathbb{R}$ into intervals of length ε. The entropy $H(\mu, \mathcal{I}_\varepsilon)$ of μ at scale ε is the natural measure-analog of the covering number $N_\varepsilon(X)$, albeit in a logarithmic scale. For a measure ν supported on a set X, the two quantities are related by the basic inequality

$$0 \le H(\nu, \mathcal{I}_\varepsilon) \le \log \#\{I \in \mathcal{I}_\varepsilon \,:\, X \cap I \ne \emptyset\} \le \log N_\varepsilon(X) + O(1)$$

(the $O(1)$ error is because we are choosing a sub-cover of X from a fixed cover of $\mathbb{R}$ rather than allowing arbitrary ε-intervals). We introduce the normalized ε-scale entropy:

$$H_\varepsilon(\nu) = \frac{1}{\log(1/\varepsilon)} H(\nu, \mathcal{I}_\varepsilon)$$

Thus, for ν supported on a set X with well-defined box dimension, the previous inequality implies

$$\limsup_{\varepsilon\to 0} H_{\varepsilon}(\nu) \leq \dim_{\mathrm{B}} X \tag{17}$$

4.2 Inverse Theorems for Entropy

The measure-analog of the sumset operation is convolution, which for discrete probability measures $\mu = \sum p_i \delta_{x_i}$ and $\nu = \sum q_j \delta_{y_j}$ is[6]

$$\mu * \nu = \sum_{i,j} p_i q_j \delta_{x_i + y_j}$$

The entropy-analog of the small doubling condition $|A + A| \leq C|A|$ is the inequality $H(\mu * \mu) \leq H(\mu) + C'$, where $H(\mu)$ is the entropy of a measure with respect to the partition into points (remember that entropy is like cardinality, but in logarithmic scale). Alternatively we could discretize at scale ε, giving $H_\rho(\mu * \mu) \leq H_\rho(\mu) + O(1/\log(1/\varepsilon))$. Tao [Tao10] has shown that such inequalities have implications similar to Freiman's theorem. Related results were also obtained by Madiman [Mad08], see also [MMT12].

The regime that interests us is, as before, the analog of $|A + B| \leq |A|^{1+\delta}$, which by formal analogy takes the form $H_\rho(\mu * \nu) \leq (1+\delta)H_\rho(\mu)$. When μ is supported on $[0, 1]$ we have $H_\rho(\mu) \leq 1 + o(1)$ as $\rho \to 0$, and this inequality is implied (and in the cases that interest us essentially equivalent to)

$$H_\rho(\mu * \nu) \leq H_\rho(\mu) + \delta \tag{18}$$

Before stating the inverse theorem for entropy we need a few more definitions. Consider the lift of μ to a tree-measure $\widetilde{\mu}$ on the full binary tree of height h (see Sect. 2.5). Given a node $\sigma = \sigma_1 \dots \sigma_k$ and $m \in \mathbb{N}$, write $\sigma\{0,1\}^m$ for the set of descendants of σ m-generations down. Let $\widetilde{\mu}_{\sigma,m}$ denote the probability measure on $\sigma\{0,1\}^m$ that assigns to each node its normalized weight according to $\widetilde{\mu}$. Since $\sum_{\eta\in\sigma\{0,1\}^m} \widetilde{\mu}(\eta) = \widetilde{\mu}(\sigma)$, this measure is given by $\widetilde{\mu}_{\sigma,m}(\eta) = \widetilde{\mu}(\eta)/\widetilde{\mu}(\sigma)$.

We say that $\widetilde{\mu}$ is δ-concentrated at σ for m generations if $H(\widetilde{\mu}_{\sigma,m}) < \delta$, that is, if

$$-\frac{1}{m} \sum_{\eta\in\sigma\{0,1\}^m} \frac{\widetilde{\mu}(\eta)}{\widetilde{\mu}(\sigma)} \log \frac{\widetilde{\mu}(\eta)}{\widetilde{\mu}(\sigma)} < \delta.$$

6 In general there is a similar formula: $\mu*\nu = \int\int \delta_{x+y} d\mu(x) d\nu(y)$, where the integral is interpreted as a measure by integrating against Borel functions.

For a tree measure μ_T associated to a tree T and for fixed m, this is equivalent to T being δ'-concentrated for m generations at σ for an appropriate δ' which tends to 0 together with δ. We say that $\widetilde{\mu}$ is δ-uniform at σ for m generations if $H(\widetilde{\mu}_{\sigma,m}) > \log m - \delta$. Note that for m fixed, when δ is small enough this implies that $\widetilde{\mu}(\eta) > 0$ for all $\eta \in \sigma\{0,1\}^m$, so this indeed generalizes full branching. We can now state the inverse theorem:

Theorem 4.1 (Theorem 2.7 of [Hoc12]) *For every $\varepsilon > 0$ and $m \geq 1$, there is a $\delta > 0$ such that for sufficiently small $\rho > 0$ the following holds. Let μ, ν be probability measures on $[0, 1]$ and suppose that*

$$H_\rho(\mu * \nu) \leq H_\rho(\mu) + \delta$$

Let $\widetilde{\mu}, \widetilde{\nu}$ denote the lifts of μ, ν to the full binary trees of height $h = \lceil \log_2(1/\rho) \rceil$. Then there is a partition of the levels $\{0, \ldots, h\}$ into three sets $U \cup V \cup W$ such that

1. *For $i \in U$,*

$$\mathbb{P}_{\sigma\sim\widetilde{\mu}}(\widetilde{\mu} \textit{ is } \varepsilon\textit{-uniform at } \sigma \textit{ for } m \textit{ generations} \mid \sigma \textit{ is in level } i) > 1 - \varepsilon$$

2. *For $j \in V$,*

$$\mathbb{P}_{\sigma\sim\widetilde{\mu}}(\widetilde{\nu} \textit{ is } \varepsilon\textit{-concentrated at } \sigma \textit{ for } m \textit{ generations} \mid \sigma \textit{ is in level } i) > 1 - \varepsilon$$

3. *$|W| < \delta n$.*

4.3 Reduction of Theorem 1.2 to a Convolution Inequality

We return to our IFS Φ with attractor $0 \in X \subseteq [0, 1]$, as in Sect. 3. Define measures $\mu^{(n)}$ analogous to X_n by

$$\mu^{(n)} = \frac{1}{|\Lambda|^n} \sum_{\underline{i}\in\Lambda^n} \delta_{f_{\underline{i}}(0)}$$

Write $S_t\mu(A) = \mu(t^{-1}A)$ (this is the usual push-forward of μ by S_t). Then the analog of the sumset relation $X_{m+n} = X_m + r^m X_n$ is

$$\mu^{(m+n)} = \mu^{(m)} * S_{r^m}\mu^{(n)}$$

The derivation is elementary, using the definition of convolution, equation (6) and the identity $S_t\delta_y = \delta_{ty}$. Next, as in Sect. 3.1, if we define $s_m = H(\mu^{(m)}, \mathcal{I}_{r^m})$ then the sequence s_n is almost super-additive in the sense that $s_{m+n} \geq s_m + s_n - O(1)$. This is proved by a similar argument to the covering number case but in the language of entropy. It follows that the limit

$$\alpha = \lim_{m\to\infty} H_{r^m}(\mu^{(m)})$$

exists. Since $\mu^{(m)}$ is supported on X, by (17) we have $\alpha \le \dim X$.

Turning to Theorem 1.2, write

$$\beta = \min\{1, \operatorname{sdim} X\}$$

and assume for the sake of contradiction that $\dim X < \beta$ and $\Delta_n \ge 2^{-kn}$ for some k. Since $\alpha \le \dim X$, we can choose $\varepsilon > 0$ so that $\alpha < \beta - \varepsilon$. Arguing analogously to Proposition 3.3 one obtains the analogous result:

Proposition 4.2 *There is a constant c (depending on β, ε) such that for large enough m,*

$$\mu^{(m)}\left(\bigcup\left\{I \in \mathcal{I}_{r^m} \;:\; H_{r^{m+n}}(\mu_I^{(m)}) > cm\right\}\right) > c$$

This lemma does not appear explicitly in [Hoc12], since that is a direct proof. Ours is a proof by contradiction, and the contradiction can be interpreted as showing that the lemma above is false. This falsehood is demonstrated directly in the last displayed equation of Sect. 5.3 of [Hoc12].

Next, for a probability measure ν and set E with $\nu(E) > 0$, write ν_E for the conditional measure on E, that is, $\nu_E(A) = \frac{1}{\nu(E)}\nu(E \cap A)$. The analog of Proposition 3.4 then holds, again with an analogous proof:

Proposition 4.3 (See Equation (40) of [Hoc12]) *For every $\delta > 0$, as $m \to \infty$*

$$\mu^{(m)}\left(\bigcup\left\{I \in \mathcal{I}_{r^m} \;:\; \frac{1}{n}H_{r^{m+n}}(\mu_I^{(m)} * S_{r^m}\mu^{(n)}) \le \alpha + \delta\right\}\right) \ge 1 - o(1)$$

From the last two propositions one sees that for given $\delta > 0$ and large enough m, there are intervals $I = I_m \in \mathcal{I}_{r^m}$ that appear in the unions in the conclusions of both propositions. Taking ν_m to be the re-scaling of $\mu_I^{(m)}$ by r^{-m} (translated back to $[0, 1)$), we have the rigorous analog of "Proposition" 3.5:

Proposition 4.4 *There is a $\tau > 0$ such that for every $\delta > 0$, for all sufficiently large m, there is a measure ν_m supported on $[0, 1)$ with*

$$\frac{1}{m}H_{r^m}(\nu_m) > \tau \tag{19}$$

$$\frac{1}{m}H_{r^m}(\mu^{(m)} * \nu_m) < \frac{1}{m}H_{r^m}(\mu^{(m)}) + \delta \tag{20}$$

4.4 Getting a Contradiction

The missing ingredient in Sect. 3.3 was the ability to estimate the number of leaves of a tree from the average amount of branching of its nodes. This is where entropy really comes in handy, because of the following (easy!) lemma. Recall that given a tree-measure θ, we write $\theta_{\sigma,m}$ for the normalized weights on the nodes m generations down from σ.

Lemma 4.5 (Lemma 3.4 of [Hoc12]) *Let $\widetilde{\theta}$ be a tree-measure on the full binary tree T of height h. Write $\partial\widetilde{\theta}$ for the measure induced by $\widetilde{\theta}$ on the leaves of the tree. Then for any m,*

$$\frac{1}{h}H(\partial\widetilde{\theta}) = \mathbb{E}_{\sigma\sim\widetilde{\theta}}(\frac{1}{m}H(\widetilde{\theta}_{\sigma,m})) + O(\frac{m}{h})$$

From here the argument proceeds exactly as in Sect. 3.3. Let $\tau > 0$ be the constant provided by Proposition 4.4. Choose a small parameter $\varepsilon > 0$. Choose m large enough that

$$H_{r^m}(\mu^{(m)}) \geq (1-\varepsilon)\alpha$$

Apply the inverse theorem 4.1 with parameters ε, m and let $\delta > 0$ be the resulting number. Applying Proposition 4.4 with this δ, there exist probability measures ν_n on $[0, 1]$ satisfying (19) and (20). Write $\widetilde{\mu}^{(n)}$, $\widetilde{\nu}_n$ for the lift of $\mu^{(n)}$, ν_n, respectively, to the binary tree of height $h_n = \lceil 1/\log(r^n)\rceil$. By the inverse theorem there is a partition $U_n \cup V_n \cup W_n$ of the levels $\{1, \ldots, h_n\}$ such that

(I) At scales $i \in U_n$, the $\widetilde{\mu}^{(n)}$-mass of nodes at level i that are ε-uniform for m generations is at least $1-\varepsilon$.
(II) At scales $j \in V_n$, the $\widetilde{\nu}_n$-mass of nodes at level i that are ε-concentrated for m generations is at least $1-\varepsilon$
(III) $|W_n| \leq \varepsilon h_n$.

If $|V_n| > (1-\tau/2)h_n$ and ε is small enough compared to τ, then sufficiently many nodes (with respect to $\widetilde{\nu}_n$) would have $H(\widetilde{\nu}^n_{\sigma,m}) < \varepsilon$ that we could invoke Lemma 4.5 and conclude that the entropy $H_{r^n}(\nu_n) \approx \frac{1}{n\log(1/r)}H(\widetilde{\nu}^n) < \tau$, contradicting (19). Therefore $|V_n| \leq (1-\tau/2)h_n$. In particular, assuming ε is small enough compared to τ,

$$|U_n| \geq h_n - |V_n| - |W_n| \geq \frac{\tau}{3}h_n$$

Next, suppose that m is large enough so that $H_{r^n}(\mu^{(n)}) > (1-\varepsilon)\alpha$. Using self-similarity of X and an argument analogous to the one outlined in Proposition 3.7, we get the analogous result:

Lemma 4.6 (Lemma 5.4 of [Hoc12]) *For all large enough n,*

$$\frac{1}{h_n+1}\sum_{\sigma}\left\{\widetilde{\mu}^{(n)}(\sigma)\,:\,\frac{1}{m}H(\widetilde{\mu}^{(n)}_{\sigma,m})>(1-2\varepsilon)\alpha\right\}>1-\varepsilon$$

Now, from the definition of U_n and our bound $|U_n|\geq\frac{\tau}{3}h_n$, we know that at least a $(1-\varepsilon)\tau/3$-fraction of the nodes of $\widetilde{\mu}^{(n)}$ satisfy $\frac{1}{m}H(\widetilde{\mu}^{(n)}_{\sigma,m})>(1-\varepsilon)m$. Of the remaining nodes, by the last lemma all but a ε-fraction satisfy $\frac{1}{m}H(\widetilde{\mu}^{(n)}_{\sigma,m})\geq(1-2\varepsilon)\alpha$. Therefore by Lemma 4.5 again, for all large enough n,

$$H_{r^n}(\mu^{(n)})\approx\frac{1}{h_n}H(\widetilde{\mu}^{(n)})>(1-\varepsilon)^2\frac{\tau}{3}+(1-(1-\varepsilon)\frac{\tau}{3}-\varepsilon)\alpha>\alpha+\varepsilon$$

assuming ε is small compared to τ. This contradicts the definition of α.

Acknowledgments Many thanks to Boris Solomyak for his coments on the paper.

References

[Bou03] Bourgain, J.: On the Erdős-Volkmann and Katz-Tao ring conjectures. Geom. Funct. Anal. **13**(2), 334–365 (2003)

[Bou10] Bourgain, J.: The discretized sum-product and projection theorems. J. Anal. Math. **112**, 193–236 (2010)

[CT06] Thomas, M.C., Joy A.T.: Elements of Information Theory. 2nd edn. Wiley, Hoboken (2006)

[EV66] Erdős, P., Volkmann, B.: Additive Gruppen mit vorgegebener Hausdorffscher Dimension. J. Reine Angew. Math. **221**, 203–208 (1966)

[Fal89] Falconer, K.J.: Dimensions and measures of quasi self-similar sets. Proc. Amer. Math. Soc. **106**(2), 543–554 (1989)

[Gar02] Gardner, R.J.: The Brunn-Minkowski inequality. Bull. Amer. Math. Soc. (N.S.), **39**(3), 355–405 (2002)

[Gar62] Garsia, A.M.: Arithmetic properties of Bernoulli convolutions. Trans. Amer. Math. Soc. **102**, 409–432 (1962)

[Hoc12] Michael H.: Self-similar sets with overlaps and sumset phenomena for entropy. preprint (2012)

[Hoc13] Michael H.: Hausdorff dimension of sums and convolutions. in preparation (2013)

[Hut81] Hutchinson, J.E.: Fractals and self-similarity. Indiana Univ. Math. J. **30**(5), 713–747 (1981)

[Ken97] Kenyon, R.: Projecting the one-dimensional Sierpinski gasket. Israel J. Math. **97**, 221–238 (1997)

[Kör08] Körner, T.W.: Hausdorff dimension of sums of sets with themselves. Studia Math. **188**(3), 287–295 (2008)

[Mad08] Madiman, M.: On the entropy of sums. In: Information Theory Workshop, 2008. ITW '08. IEEE, pp. 303–307 (2008)

[MMT12] Madiman, M., Marcus, A.W., Tetali, P.: Entropy and set cardinality inequalities for partition-determined functions. Random Struct. Algorithms **40**(4), 399–424 (2012)

[Orp13] Tuomas O.: On the packing measure of self-similar sets. preprint (2013). http://arxiv.org/abs/1212.4813

[PSS00] Yuval, P., Wilhelm, S., Boris, S.: Sixty years of Bernoulli convolutions. In: Fractal Geometry and Stochastics, II (Greifswald/Koserow, 1998), Progr. Probab., vol. 46, pp. 39–65. Birkhäuser, Basel (2000)

[PS00] Yuval, P., Boris, S.: Problems on self-similar sets and self-affine sets: an update. In: Fractal Geometry and stochastics, II (Greifswald/Koserow, 1998), Progr. Probab., vol. 46, pp. 95–106. Birkhäuser, Basel (2000)

[PS95] Pollicott, M., Simon, K.: The Hausdorff dimension of λ-expansions with deleted digits. Trans. Amer. Math. Soc. **347**(3), 967–983 (1995)

[SS10] Jörg, S., Pablo, S.: On the dimension of iterated sumsets. In: Recent Developments in Fractals and Related Fields, Appl. Numer. Harmon. Anal., pp. 55–72. Birkhäuser Boston Inc., Boston (2010)

[Shm13] Pablo, S.: On the exceptional set for absolute continuity of bernoulli convolutions. preprint (2013). http://arxiv.org/abs/1303.3992

[Tao10] Tao, T.: Sumset and inverse sumset theory for Shannon entropy. Combin. Probab. Comput. **19**(4), 603–639 (2010)

[TV06] Terence, T., Van V.: Additive Combinatorics, Cambridge Studies in Advanced Mathematics. vol. 105, Cambridge University Press, Cambridge (2006)

Quasisymmetric Modification of Metrics on Self-Similar Sets

Jun Kigami

Abstract Using the notions of scales and their gauge functions associated with self-similar sets, we give a necessary and sufficient condition for two metrics on a self-similar set being quasisymmetric to each other. As an application, we construct metrics on the Sierpinski carpet which is quasisymmetric with respect to the Euclidean metrics and obtain an upper estimate of the conformal dimension of the Sierpinski carpet.

1 Introduction

The main purpose of this paper is to give a characterization of quasisymmetry for self-similar sets in terms of scales and related notions introduced in [Kig09]. As an application, we will construct a series of metrics on the Sierpinski carpet which are quasisymmetric to the restriction of the Euclidean metric and give an upper estimate of the quasiconformal dimension of the Sierpinski carpet (Fig. 1).

Quasisymmetric maps have been introduced by Tukia and Väisälä in [TV80] as a generalization of quasiconformal mappings in the complex plane.

Definition 1.1 (*Quasisymmetry*).

(1) Let (X, d) and (X, ρ) be metric spaces. ρ is said to be quasisymmetric, or QS for short, with respect to d if and only if there exists a homeomorphism h from $[0, +\infty)$ to itself such that $h(0) = 0$ and, for any $t > 0$, $\rho(x, z) < h(t)\rho(x, y)$ whenever $d(x, z) < td(x, y)$. We write $\rho \underset{\mathrm{QS}}{\sim} d$ if ρ is quasisymmetric with respect to d.

(2) Let (X, d) be a metric space. A homeomorphism $f : X \to X$ is called quasisymmetric if and only if $d \underset{\mathrm{QS}}{\sim} d_f$, where $d_f(x, y)$ is defined by $d_f(x, y) = d(f(x), f(y))$.

The above definition immediately implies the following facts.

J. Kigami (✉)
Graduate School of Informatics, Kyoto University, Kyoto 606-8501, Japan
e-mail: kigami@i.kyoto-u.ac.jp

D.-J. Feng and K.-S. Lau (eds.), *Geometry and Analysis of Fractals*, Springer Proceedings in Mathematics & Statistics 88,
DOI 10.1007/978-3-662-43920-3_9

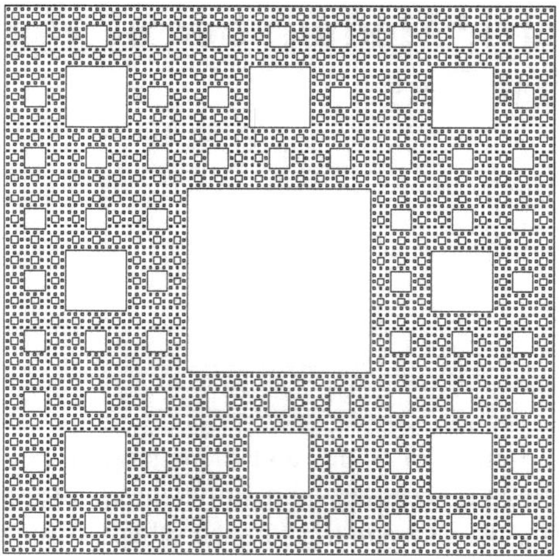

Fig. 1 The Sierpinski carpet

Proposition 1.2 *Let (X, d) and (X, ρ) be metric spaces.*

(1) *If $\rho \underset{\text{QS}}{\sim} d$, then the identity may of X is a homeomorphism from (X, d) to (X, ρ).*
(2) *The relation $\underset{\text{QS}}{\sim}$ is an equivalence relation among metrics on X. In particular, $\rho \underset{\text{QS}}{\sim} d$ if and only if $d \underset{\text{QS}}{\sim} \rho$.*

Associated with the notion of quasisymmetry, the quasiconformal dimension of a metric space has been introduced by Pansu in [Pan89] as an invariant under quasi-symmetric modification of a metric.

Definition 1.3 (*Quasiconformal dimension*) Let (X, d) be a metric space. We define the conformal dimension of (X, d), $\dim_C(X, d)$, by

$$\dim_C(X, d) = \inf\{\dim_H(X, \rho) | \rho \text{ is a metric on } X \text{and } d \underset{\text{QS}}{\sim} \rho\},$$

where $\dim_H(X, \rho)$ is the Hausdorff dimension of (X, ρ).

Quasisymmetric maps on self-similar sets have been paid much attentions in recent years as well as their conformal dimensions. For example, Bonk and Merenkov have shown that any quasisymmetric homeomorphism from the Sierpinski carpet to itself is a composition of rotations and reflections in [BM00]. About the conformal dimensions, Tyson and Wu have proven that the conformal dimension of the Sierpinski gasket is one in [TW06]. For the Sierpinski carpet, it is known that

$$1 + \frac{\log 3}{\log 2} \leq \dim_C(\text{SC}, d_E) < \dim_H(\text{SC}, d_E) = \frac{\log 8}{\log 3}, \tag{1.1}$$

where SC is the Sierpinski carpet and d_E is the restriction of the Euclidean metric. The strict inequality between the Hausdorff and the quasiconformal dimensions in (1.1) has shown by Keith and Laakso [KL04]. See [MT10] for details.

The first problem we are going to study is to obtain a verifiable characterization of quasisymmetric metrics. It will turn out that scales and related notions introduced in [Kig09] are useful in dealing with such a problem. Let K be a connected self-similar set associated with the family of contractions $\{F_1, \ldots, F_N\}$, i. e. $K = F_1(K) \cup \ldots \cup F_N(K)$. Define $F_{w_1 \ldots w_m} = F_{w_1} \circ \ldots \circ F_{w_m}$ and $K_{w_1 \ldots w_m} = F_{w_1 \ldots w_m}(K)$ for any $w_1, \ldots, w_m \in \{1, \ldots, N\}$. The notion of scales has been introduced in order to study how to find a metric under which the contraction mappings $\{F_1, \ldots, F_N\}$ have prescribed values of contraction ratios. A scale essentially gives "diameters" of $K_{w_1 \ldots w_m}$'s and induces a family of assumed "balls" $U_s(x)$ around $x \in K$ with radius $s > 0$. See Sect. 2 for precise definitions. In the language of scales, we are going to present an equivalent condition in Theorem 3.4 for metrics being quasisymmetric to each other which is easy to verify for concrete examples, in particular, in the case of "self-similar" metrics.

As an application, we will present a systematic way of constructing a self-similar metric on the Sierpinski carpet which is quasisymmetric to d_E and Ahlfors regular. The main idea is to find an "invisible" set introduced in Sect. 4. Roughly speaking, an invisible set is a collection of places where the shortest paths between two separated boundary points will not visit. (We define the "boundary" of the Sierpinski carpet by the union of four line segments, namely, the most upper, lower, right and left line segments of the square which is the convex hull of the Sierpinski carpet.) Putting an arbitrary weight on an invisible set, we will obtain a self-similar metric having the desired properties mentioned above with an explicit formula for its Hausdorff dimension in Theorem 5.3. Constructing series of invisible sets and taking advantage of the associated metrics, we will show that

$$\dim_C(\mathrm{SC}, d_E) \leq \frac{\log\left(\frac{9+\sqrt{41}}{2}\right)}{\log 3} = 1.858183\ldots < \frac{\log 8}{\log 3} = 1.892789\ldots.$$

in Sect. 6.[1]

Note that the conformal dimension in the above inequality can be replaced by the Ahlfors regular conformal dimension since our metrics are Ahlfors regular. See [MT10] for the definition of the Ahlfors regular conformal dimension.

The following is a convention in notations in this paper.

Let f and g be functions with variables $x_1, \ldots, x_n$. We use "$f \asymp g$ for any $(x_1, \ldots, x_n) \in A$" if and only if there exist positive constants c_1 and c_2 such that

[1] After completion of the preliminary version of this paper, B. Kleiner informed me that he had obtained better upper bound of $\dim_C(\mathrm{SC}, d_E)$ around 1999 by a different method in [Kle00]. His upper bound is about 1.856685 .... The author would like to express his gratitude to Professor Bruce Kleiner for his detailed comments.

$$c_1 f(x_1, \ldots, x_n) \leq g(x_1, \ldots, x_n) \leq c_2 f(x_1, \ldots, x_n)$$

for any $(x_1, \ldots, x_n) \in A$.

2 Basic Notions

This section is devoted to introducing fundamental notions and results regarding scales and self-similar sets and scales.

The following is the standard definitions on (finite and infinite) sequences of finite symbols.

Definition 2.1 Let S be a finite set. For $m \geq 0$, define $W_m(S) = S^m = \{w|w = w_1 \ldots w_m, w_i \in S\}$, where $W_0(S) = \{\emptyset\}$. Define $W_* = \cup_{m \geq 0} W_m$. Also $\Sigma(S) = S^{\mathbb{N}} = \{\omega|\omega = \omega_1\omega_2 \ldots, \omega_i \in S\}$. For $w = w_1 \ldots w_m \in W_*(S)$, the length $|w|$ of w is defined by $|w| = m$. For $w = w_1 \ldots w_m$ and $v = v_1 \ldots v_n \in W_*(S)$, we define $w \cdot v$ (or wv for short) by $w \cdot v = w_1 \ldots w_m v_1 \ldots v_n$. For a subseteq $A, B \in W_*(S)$, $A \cdot B$ (or AB for short) is defined by $A \cdot B = \{wv|w \in A, v \in B\}$.

Remark The notion of "gauge function" given in the above definition is not related to the notion of "conformal gauge" which is commonly used in literatures concerning the conformal dimension, for example, [MT10].

With the product topology, $\Sigma(S)$ is compact, perfect and totally disconnected. In other words, $\Sigma(S)$ is a Cantor set. A scale is defined by a gauge function which assign a "diameter" to every $w \in W_*(S)$.

Definition 2.2 (*Scale*). Let S be a finite set.

(1) A function $g : W_*(S) \to (0, 1]$ is called a gauge function if and only if $g(\emptyset) = 1$, $g(w_1 \ldots w_m) \leq g(w_1 \ldots w_{m-1})$ and $\max_{w \in W_m(S)} g(w) \to 0$ as $m \to 0$. A gauge function g is said to be elliptic if and only if there exists $c \in (0, 1)$ and n such that $g_{wi} \geq cg(w)$ for any $i \in S$ and any $w \in W_*(S)$ and $g_{wv} \leq cg(w)$ for any $w \in W_*(S)$ and $v \in W_n$.
(2) Let g be a gauge function. Define

$$\Lambda_s^g = \{w = w_1 \ldots w_m | g(w_1 \ldots w_{m-1}) \geq s > g(w_1 \ldots w_m)\}$$

We call $\mathcal{S}^g = \{\Lambda_s^g\}_{s \in (0,1]}$ a scale on Σ associated with the gauge function g.

If no confusion may occur, we omit S in $W_m(S)$, $W_*(S)$ and $\Sigma(S)$ and simply write W_m, W_* and Σ respectively.

The notion of self-similar structure describes topological feature of self-similar sets.

Definition 2.3 $(K, S, \{F_i\}_{i \in S})$ is called a self-similar structure if the following four conditions (S1), (S2), (S3) and (S4) are satisfied:

(S1) K is a compact metrizable set.
(S2) S is a finite set.
(S3) $F_s : K \to K$ is continuous for any $s \in S$.
(S4) There exists a continuous surjection $\pi : \Sigma(S) \to K$ such that $F_s \circ \pi = \pi \circ \sigma_s$ for any $s \in K$, where $\sigma_s : \Sigma(S) \to \Sigma(S)$ is defined by $\sigma_s(\omega_1\omega_2 \ldots) = s\omega_1\omega_2 \ldots$.

Hereafter in this paper, $(K, S, \{F_s\}_{s \in S})$ is always a self-similar structure.

Notation Define $F_{w_1 \ldots w_m} = F_{w_1} \circ \cdots \circ F_{w_m}$ and $K_w = F_w(K)$. Moreover, define $K(A) = \cup_{w \in A} K_w$ for a subset $A \subseteq W_*$.

A scale $\mathcal{S}$ on $\Sigma(S)$ induces a family of "balls" $U^{(n)}(x, s)$ around $x \in X$ with "radius" s. One of the main concerns is the existence of a metric under which those "balls" are really balls, in other words, the existence of adapted metric according to the following definition.

Definition 2.4 Let $\mathcal{S} = \{\Lambda_s\}_{s \in (0,1]}$ be a scale on Σ associated with a gauge function g.

(1) For $x \in K$, define $(\Lambda_s)_x^{(n)}$ and $U_{\mathcal{S}}^{(n)}(x, s)$ inductively by

$$(\Lambda_s)_x^{(0)} = \{w | w \in \Lambda_s, x \in K_w\}$$
$$U_{\mathcal{S}}^{(n)}(x, s) = K\left((\Lambda_s)_x^{(n)}\right)$$
$$(\Lambda_s)_x^{(n)} = \{w | w \in \Lambda_s, K_w \cap U_{\mathcal{S}}^{(n-1)}(x, s) \neq \emptyset\}$$

(2) A metric d on K is said to be adapted to the scale $\mathcal{S}$ if and only if there exist $\alpha, \beta > 0$ and $n \geq 1$ such that

$$B_d(x, \alpha s) \subseteq U_{\mathcal{S}}^{(n)}(x, s) \subseteq B_d(x, \beta s)$$

for any $x \in K$ and any s.

The notion of gentleness between scales is introduced in [Kig09] as a part of the equivalence condition for a measure being volume doubling with respect to a scale. Roughly, if two scales are gentle with respect to each other, then the transition to one scale to the other is "smooth".

Definition 2.5 Let $\mathcal{S}^g$ and $\mathcal{S}^l$ be scales on Σ associated with gauge functions g and l respectively. We say $\mathcal{S}^l$ is gentle with respect to $\mathcal{S}^g$ if and only if there exists $c > 0$ such that $l(w) \leq cl(v)$ whenever $w, v \in \Lambda_s$ for some $s > 0$ and $K_w \cap K_v \neq \emptyset$. We write $\mathcal{S}^g \underset{\text{GE}}{\sim} \mathcal{S}^l$ if $\mathcal{S}^l$ is gentle with respect to $\mathcal{S}^g$.

Proposition 2.6 *Among elliptic scales, i.e. scales whose gauge functions are elliptic,* $\underset{\text{GE}}{\sim}$ *is an equivalent relation. In particular, if g and l are elliptic, then* $\mathcal{S}^g \underset{\text{GE}}{\sim} \mathcal{S}^l$ *implies* $\mathcal{S}^l \underset{\text{GE}}{\sim} \mathcal{S}^g$.

There exists a natural "pseudo"metric associated with a scale which is defined by the infimum of the "length" of paths between two points.

Definition 2.7 (1) A sequence $(w(1), \ldots, w(n))$ is called a path in K if and only if $w(1), \ldots, w(n) \in W_*$, $K_{w(i)} \cap K_{w(i+1)} \neq \emptyset$ for any $i = 1, \ldots, n-1$. The collection of all the paths is denoted by $\mathcal{CH}$. For $U, V \subseteq K$, a path $(w(1), \ldots, w(n))$ is called a path between U and V if and only if $K_{w(1)} \cap U \neq \emptyset$ and $K_{w(n)} \cap V \neq \emptyset$. We use $\mathcal{CH}(U, V)$ to denote the collection of paths between U and V. For two paths $\mathbf{p}_1 = (w(1), \ldots, w(n))$ and $\mathbf{p}_2 = (v(1), \ldots, v(m))$, if $K_{w(n)} \cap K_{v(1)} \neq \emptyset$, we define $\mathbf{p}_1 \vee \mathbf{p}_2 \in \mathcal{CH}$ by $\mathbf{p}_1 \vee \mathbf{p}_2 = (w(1), \ldots, w(n), v(1), \ldots, v(m))$.

(2) Let $\mathcal{S}$ be a scale on Σ associated with a gauge function g. For any $x, y \in K$, we define

$$D_{\mathcal{S}}(x, y) = \inf\{\sum_{i=1}^{n} g(w(i)) | (w(1), \ldots, w(n)) \in \mathcal{CH}(x, y)\}.$$

Remark We identify a point $x \in X$ and a set $\{x\}$ if no confusion may occur.

Remark We often use D_g instead of $D_{\mathcal{S}}$ if $\mathcal{S}$ is the scale associated with a gauge function g.

Proposition 2.8 *$D_{\mathcal{S}}$ is a pseudometric, i.e. $D_{\mathcal{S}}(x, y) = D_{\mathcal{S}}(y, x)$, $D_{\mathcal{S}}(x, y) \geq 0$, $D_{\mathcal{S}}(x, x) = 0$ and $D_{\mathcal{S}}(x, y) \leq D_{\mathcal{S}}(x, z) + D_{\mathcal{S}}(z, y)$.*

By [Kig09, Lemma 2.3.10], we have the following theorem, which says that a metric adapted to a scale $\mathcal{S}$, if such a metric exists at all, is essentially $D_{\mathcal{S}}$.

Theorem 2.9 *Let $\mathcal{S}$ be a scale. There exists a metric d on K such that d is adapted to $\mathcal{S}$ if and only if $D_{\mathcal{S}}$ is a metric on K which is adapted to $\mathcal{S}$.*

3 Quasisymmetric Metrics and Scales

In this section, we give an equivalent condition for two metrics on a self-similar set being quasisymmetric in terms of scales and related notions introduced in Sect. 2.

Let $(K, S, \{F_i\}_{i \in S})$ be a self-similar structure. Assume that $K \neq \overline{V_0}$. Hereafter in this section, every metric on K is assumed to satisfy the following two properties:

1. It produces the same topology as the original topology of K.
2. The diameter of K under it equals one.

The next lemma can be verified immediately by the definitions in the previous section.

Lemma 3.1 *Let $\mathcal{S}_1 = \{\Lambda_s^1\}$ and $\mathcal{S}_2 = \{\Lambda_s^2\}$ be scales. If $\mathcal{S}_1 \underset{\text{GE}}{\sim} \mathcal{S}_2$, then for any $n \geq 1$, there exists $c_n \in (0, 1)$ such that*

$$U_1^{(n)}(x, c_n t) \subseteq U_2^{(n)}(x, s) \subseteq U_1^{(n)}(x, t/c_n)$$

for any $x \in K$, *any* (s, t) *with* $\Lambda_{t,x}^1 \cap \Lambda_{s,y}^2 \neq \emptyset$, *where* $U_i^{(n)}(x, s) = U_{S_i}^{(n)}(x, s)$ *and* $\Lambda_{t,x}^i = (\Lambda_s^i)_x^{(0)}$ *for* $i = 1, 2$.

First we define a scale associated with a metric.

Definition 3.2 Let d be a metric on K with $\mathrm{diam}(K, d) = 1$. Define $\mathcal{S}^d = \{\Lambda_s^d\}$ be the scale with the gauge function $d_w = \mathrm{diam}(K_w, d)$.

Lemma 3.3 *Let* $\mathcal{S} = \{\Lambda_s\}$ *be an elliptic scale and let* d *be a metric on* K *which is adapted to* $\mathcal{S}$. *Let* $l(w)$ *be the gauge function of* $\mathcal{S}$. *Then*

(1) $d_w \asymp l(w)$ *for any* $w \in W_*$.
(2) *The pseudometric* $D_{\mathcal{S}}$ *associated with* $\mathcal{S}$ *is a metric and* $\mathcal{D}_{\mathcal{S}}(x, y) \asymp d(x, y)$ *for any* $x, y \in K$.
(3) $\mathcal{S}^d$ *is elliptic and* d *is adapted to* $\mathcal{S}^d$.

Proof Write $U^{(n)}(x, r) = U_{\mathcal{S}}^{(n)}(x, r)$. Since d is adapted to $\mathcal{S}$, we have

$$U^{(n)}(x, \beta s) \subseteq B_d(x, s) \subseteq U^{(n)}(x, \alpha r) \tag{3.2}$$

(1) For $w \in W_*$, $U^{(n)}(x, l(w)) \subseteq B_d(x, \alpha l(w))$. Hence $d_w \leq \alpha l(w)$. Now by [Kig09, Lemma 1.3.12], there exists $y \in K_w$ and $\gamma \in (0, 1)$ such that $U^{(n)}(y, \gamma l(w)) \subseteq K_w$. Hence $B_d(x, \beta\gamma l(w)) \subseteq K_w$. Since K is connected, we have $\beta\gamma l(w) \leq d_w$.
(2) This is immediate from Theorem 2.9.
(3) These claims are immediate from (1) and Lemma 3.1. □

Now we present one of the main results of this paper. The following theorem gives an equivalent condition for certain metrics on a self-similar set being quasisymmetic. It plays a crucial role in the proof of Theorem 5.3.

Theorem 3.4 *Let* d *be a metric on* K *and let* $\mathcal{S} = \{\Lambda_s\}$ *be an elliptic scale. Assume that* d *is adapted to* $\mathcal{S}$. *Let* ρ *be a metric on* K. *Then* $d \underset{\text{QS}}{\sim} \rho$ *if and only if* $\mathcal{S}^\rho$ *is elliptic,* $\mathcal{S} \underset{\text{GE}}{\sim} \mathcal{S}^\rho$ *and* ρ *is adapted to* $\mathcal{S}^\rho$.

The rest of this section is devoted to the proof of Theorem 3.4.

Proof First we show ⇒. Assume $d \underset{\text{QS}}{\sim} \rho$. By Lemma 3.3, we may regard the gauge function of $\mathcal{S}$ is d_w and hence $\mathcal{S} = \mathcal{S}^d$.

By the results in [Kig00, Part 2], $d \underset{\text{QS}}{\sim} \rho$ is equivalent to the facts that there exists $\delta \in (0, 1)$ such that

$$\begin{aligned} B_d(x,r) &\supseteq B_\rho\left(x, \delta\overline{\rho}_d(x,s)\right) \\ B_\rho(x,r) &\supseteq B_d\left(x, \delta\overline{d}_\rho(x,r)\right) \end{aligned} \tag{3.3}$$

and

$$\begin{aligned} \overline{\rho}_d(x,r/2) &\geq \delta\overline{\rho}_d(x,r) \\ \overline{d}_\rho(x,r/2) &\geq \delta\overline{d}_\rho(x,r), \end{aligned} \tag{3.4}$$

where $\overline{\rho}_d(x,r) = \sup_{y \in B_d(x,r)} \rho(x,y)$ and $\overline{d}_d(x,r) = \sup_{y \in B_\rho(x,r)} d(x,y)$.

First we show the following claim.

Claim 1 *Let $w \in \Lambda_s^d$. Then there exists $z \in K_w$ such that $\rho_w \geq c\overline{\rho}_d(z,s)$, where c is a constant which is independent of w and s.*

Proof of Claim 1 By [Kig09, Lemma 1.3.12] and (3.3), we may find $z \in K_w$ such that

$$K_w \supseteq U_d^{(n)}(z,\gamma s) \supseteq B_d(z,\gamma s/\alpha) \supseteq B_\rho\left(z, \delta\overline{\rho}_d(z,\gamma s/\alpha)\right)$$

Hence by (3.4)

$$\rho_w \geq \delta\overline{\rho}_d(z,\gamma s/\alpha) \geq c\overline{\rho}_d(z,s).$$

□

Step 1: $\mathcal{S}^\rho$ is elliptic.

Proof of "$\rho_{wi} \geq c\rho_w$ for any $w \in W_*$ and any $i \in S$:

By Claim 1, it follows that

$$\rho_{wi} \geq c'\overline{\rho}_d(z,d_{wi}) \geq c'\overline{\rho}_d(z,2d_{wi}). \tag{3.5}$$

for some $z \in K_{wi}$". On the other hand,

$$K_w \subseteq B_d(z,2d_w) \subseteq B_\rho(z,\rho_d(x,2d_w)).$$

Hence

$$\rho_w \leq \overline{\rho}_d(z,2d_w).$$

This with (3.5) suffices.

Proof of "there exists $c \in (0,1)$ and m such that $\rho_{wv} \leq c\rho_w$ for any $w \in W_*$ and any $v \in W_m$".

Since $K_{wv} \subseteq B_\rho(x,\overline{\rho}_d(x,2d_{wv}))$, we have

$$\rho_{wv} \leq \overline{\rho}_d(x,2d_{wv}) \leq \delta\overline{\rho}_d(x,d_{wv}), \tag{3.6}$$

where $x \in K_{wv}$. On the other hand, by [Kig09, Lemma 1.3.12], there exists $z \in K_w$ such that

$$K_w \supseteq U_d^{(n)}(x, \gamma d_w) \supseteq B_\rho\left(x, \delta\overline{\rho}_d(x, \gamma d_w)\right).$$

Hence

$$\rho_w \geq \delta\overline{\rho}_d(x, \gamma d_w) \geq \delta'\overline{\rho}_d(x, d_w) \tag{3.7}$$

Now, since $\mathcal{S}^d$ is elliptic, there exists $a \in (0, 1)$ such that

$$d_{wv} \leq ca^{|v|}d_w$$

for any w and v. Hence by (3.6) and (3.7), the uniform decay of ρ with respect to d, (See [Kig00, Proposition 10.7]),

$$\rho_{wv} \leq \delta\overline{\rho}_d(x, d_{wv}) \leq \delta\overline{\rho}_d(x, ca^{|v|}d_w) \leq cb^{|v|}\overline{\rho}_d(x, d_w) \leq c'b^{|v|}\rho_w,$$

where $b \in (0, 1)$. Hence choosing sufficiently large $m = |v|$, we obtain the desired inequality.

Thus we have shown that $\mathcal{S}^\rho$ is elliptic.

Step 2: $\mathcal{S} \underset{\text{GE}}{\sim} \mathcal{S}^\rho$

Let $w, v \in \Lambda_s^d$ with $K_w \cap K_v \neq \emptyset$. Choose $x \in K_w$ and $y \in K_v$. Then $d(x, y) \leq 2s$ and hence $B_d(x, 3s) \supseteq B_d(y, s)$. This implies $\overline{\rho}_d(x, 3s) \geq \overline{\rho}_d(y, s)$. By (3.4),

$$\overline{\rho}_d(x, s) \asymp \overline{\rho}_d(y, s).$$

By Claim 1, choosing $y \in K_v$ properly, we see that $\rho_v \geq c\overline{\rho}_d(y, s)$. Since $\overline{\rho}_d(x, 2s) \geq \rho_w$, (3.4) shows that $\mathcal{S}^d \underset{\text{GE}}{\sim} \mathcal{S}$.

Step 3: ρ is adapted to $\mathcal{S}^\rho$.

Let $x \in K$ and let $w \in \Lambda_{r,x}^d \cap \Lambda_{s,x}^\rho$. Then by Lemma 3.1, (3.3) and (3.4),

$$\begin{aligned} U_\rho^{(n)}(x, cs) \supseteq U_d^{(n)}(x, r) \supseteq B_d(x, r/\alpha) \supseteq B_\rho\left(x, \delta\overline{\rho}_d(x, r/\alpha)\right) \\ \supseteq B_\rho\left(x, \delta'\overline{\rho}_d(x, 2r)\right) \supseteq B_\rho(x, \delta'\rho_w) \supseteq B_\rho(x, \delta''s). \end{aligned}$$

On the other hand, let $x \in K$ and let $w \in \Lambda_s^\rho \cap \Lambda_t^d$. Then

$$B_\rho(x, s) \supseteq B_d\left(x, \delta\overline{d}_\rho(x, s)\right) \supseteq U_d^{(n)}\left(x, \beta\delta\overline{d}_\rho(x, s)\right) \supseteq U_\rho^{(n)}(x, c'r), \tag{3.8}$$

where $wv \in \Lambda^d_{\beta\delta\overline{d}_\rho(x,s),x} \cap \Lambda^\rho_{r,x}$. Since $B_\rho(x, 2s) \supseteq K_w$, we see that $\overline{d}_\rho(x, 2s) \geq d_w$. Hence $\overline{d}_\rho(x, s) \geq c_1 d_w$. Consequently, $d_{wv} \geq c_2 d_w$, where c_2 is independent of w and v. This implies that $|v|$ is uniformly bounded. Since $\mathcal{S}^\rho$ is elliptic, $\rho_{wv} \geq c_3\rho_w$. This implies $U_\rho^{(n)}(x, c'r) \supseteq U_\rho^{(n)}(x, c_4 s)$. By (3.8), it follows that $B_\rho(x, s) \supseteq U_\rho^{(n)}(x, c_5 s)$. Thus we have shown that ρ is adapted to $\mathcal{S}^\rho$.

This concludes the proof of $\Rightarrow$. □

To show the converse direction of Theorem 3.4, we need the following lemma.

Lemma 3.5 *Assume that d is adapted to $\mathcal{S}^d$. Then, for any n and k, there exists $\lambda \in (0, 1)$ such that*

$$U_d^{(n)}(x, r) \supseteq U_d^{(n+k)}(x, \lambda r)$$

for any $x \in K$ and any r.

Proof Since d is adapted to $\mathcal{S}^d$, there exists $c > 0$ such that $U_d^{(n)}(x, r) \supseteq B_d(x, cr)$. Then $B_d(x, cr) \supseteq U_d^{(n+k)}(x, cr/(n + k + 2))$. □

Proof (of $\Leftarrow$ of Theorem 3.4) Since d and ρ are adapted to $\mathcal{S}^d$ and $\mathcal{S}^\rho$ respectively,

$$U_d^{(n)}(x, \beta_1 r) \subseteq B_d(x, r) \subseteq U_d^{(n)}(x, \alpha_1 r)$$
$$U_\rho^{(m)}(x, \beta_2 r) \subseteq B_\rho(x, r) \subseteq U_\rho^{(m)}(x, \alpha_2 r).$$

First we show (3.3). By Lemma 3.1,

$$B_d(x, r) \subseteq U_d^{(n)}(x, \alpha_1 r) \subseteq U_\rho^{(n)}(x, c\rho_w), \tag{3.9}$$

where $w \in \Lambda^d_{\alpha_1 r, x}$. Using Lemma 3.5 if necessary, we obtain

$$B_d(x, r) \subseteq U_\rho^{(m)}(x, c_1\rho_w) \subseteq B_\rho(x, c_2\rho_w).$$

Hence $\overline{\rho}_d(x, r) \leq c_2\rho_w$. Now by Lemma 3.1,

$$B_d(x, r) \supseteq U_d^{(n)}(x, \beta_1 r) \supseteq U_\rho^{(n)}(x, c'\rho_{wv}), \tag{3.10}$$

where $wv \in \Lambda^d_{\beta_1 r, x}$. By making use of Lemma 3.5 if necessary, we have

$$B_d(x, r) \supseteq U_\rho^{(m)}(x, c''\rho_{wv}) \supseteq B_\rho(x, c''\beta_2\rho_{wv}).$$

Since $\mathcal{S}^d$ is elliptic, the fact that $w \in \Lambda^d_{\alpha_1 r, x}$ and $wv \in \Lambda^d_{\beta_1 r, x}$ implies that $|v|$ is uniformly bounded with respect to x and r. Since $\mathcal{S}^\rho$ is also elliptic, we see that $\rho_w v \geq c_3\rho_w \geq c_4\overline{\rho}_d(x, r)$. Hence (3.3) holds. (By exchanging ρ and d, we also obtain the other one.)

Next we show (3.4). By (3.9),

$$\overline{\rho}_d(x, r) \leq c(n + 1)\rho_w,$$

where $w \in \Lambda^d_{\alpha_1 r}$. Replacing r by λr for $\lambda \in (0, 1)$ in (3.10), we have

$$B_d(x, \lambda r) \supseteq U^{(n)}(x, c'\rho_{wv}),$$

where $wv \in \Lambda^d_{\lambda\beta_1 r,x}$. This implies $\overline{\rho}_d(x, \lambda r) \geq c'\rho_{wv}$. The same arguments as above show that $|v|$ is uniformly bounded and $\rho_{wv} \geq c\rho_w$. Combining all these, we obtain

$$\overline{\rho}_d(x, \lambda r) \geq c'\rho_{wv} \geq c''\rho_w \geq c'''\overline{\rho}_d(x, r).$$

Again the other one is obtained by exchanging d and ρ. Thus we have obtained (3.4). □

4 Sierpinski Carpet and Its Invisible Sets

In this and the following sections, we are going to apply Theorem 3.4 to the Sierpinski carpet. First we give the definition of the Sierpinski carpet.

Definition 4.1 Let $S = \{\swarrow, \downarrow, \searrow, \rightarrow, \nearrow, \uparrow, \nwarrow, \leftarrow\}$. Define $p_{\swarrow} = -1 - \sqrt{-1}$, $p_{\downarrow} = -\sqrt{-1}$, $p_{\searrow} = 1 - \sqrt{-1}$, $p_{\rightarrow} = 1$, $p_{\nearrow} = 1 + \sqrt{-1}$, $p_{\uparrow} = \sqrt{-1}$, $p_{\nwarrow} = -1 + \sqrt{-1}$ and $p_{\leftarrow} = -1$. Moreover, define $F_s : \mathbb{C} \rightarrow \mathbb{C}$ for $s \in S$ by

$$F_s(z) = \frac{(z - p_s)}{3} + p_s.$$

The Sierpinski carpet K is the unique non-empty compact set which satisfies

$$K = \bigcup_{s \in S} F_s(K).$$

Let d_E be the restriction of the Euclidean metric on the Sierpinski carpet K.

We consider d_E as the standard metric on K and are going to construct metrics which is quasisymmetric with respect to d_E. Obviously, the scale $\mathcal{S}_{d_E}$ associated with d_E is elliptic and d_E is adapted to the scale $\mathcal{S}_{d_E}$. In fact, the gauge function associated with d_E is given by $3^{-|w|}$ for any $w \in W_*$.

Next we introduce notions and notations regarding the boundary of the Sierpinski carpet (Fig. 2).

Definition 4.2

(1) Define $L = K \cap \{z | \mathrm{Re}\, z = -1\}$, $R = K \cap \{z | \mathrm{Re}\, z = 1\}$, $T = K \cap \{z | \mathrm{Im}\, z = 1\}$ and $B = K \cap \{z | \mathrm{Im}\, z = -1\}$. Let $H_w = F_w(H)$ for any $w \in W_*$ and any $H \in \{L, R, T, B\}$. Moreover define $\partial_m = \{L_w, R_w, T_w, B_w | w \in W_m\}$.
(2) Define $L^m = \{\swarrow, \leftarrow, \nwarrow\}^m$, $R^m = \{\searrow, \rightarrow, \nearrow\}^m$, $T^m = \{\nwarrow, \uparrow, \nearrow\}^m$, $B^m = \{\swarrow, \downarrow, \searrow\}^m$ and $\delta_m = L^m \cup R^m \cup T^m \cup B^m$.

Remark Recall that $K(A) = \cup_{w \in A} K_w$ for a subset $A \subseteq W_*$. The map $A \rightarrow K(A)$ can be regarded as a map from the subsets of W_* to the subsets of K. In the case

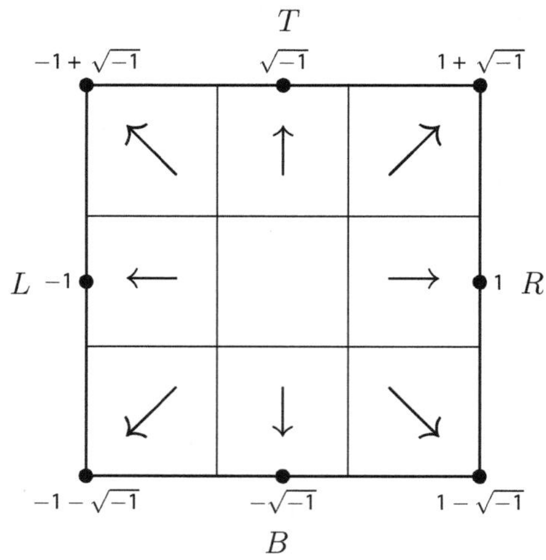

Fig. 2 Generation of the Sierpinski carpet

of the Sierpinski carpet, this map is injective, i.e. if $A \neq B$, then $K(A) \neq K(B)$. Therefore, if no confusion may occur, we identify $A \subseteq W_*$ with $K(A) \subseteq K$.

Note that $D_{d_E}(x, y) \geq 1$ for any $(x, y) \in (L \times R) \cup (T \times B)$. This fact may remain true even if you put 0 as weights (length) of some pieces of w's. Such a collection of w's is called an invisible set, whose precise definition is given below.

Definition 4.3

(1) Let

$$\mathcal{CH}_m = \{(w(1), \ldots, w(n)) | (w(1), \ldots, w(n)) \in \mathcal{CH}, w(i) \in W_m\}$$

and let $\mathcal{CH}_m(U, V) = \mathcal{CH}(U, V) \cap \mathcal{CH}_m$ for $U, V \subseteq K$.

(2) Let $A \subseteq W_m$. For $\mathbf{p} = (w(1), \ldots, w(n)) \in \mathcal{CH}_m$, define

$$\ell_A(\mathbf{p}) = \frac{\#\{i | i = 1, \ldots, n, w(i) \notin A\}}{3^m}$$

(3) Let $A \subseteq W_m$. A is said to be an invisible set if and only if

$$\inf_{\mathbf{p} \in \mathcal{CH}_m(L,R) \cup \mathcal{CH}_m(T,B)} \ell_A(\mathbf{p}) \geq 1$$

(4) Let $A \subseteq W_m$. A is said to be $+$-invariant if and only if $K(A)$ is symmetric with respect to both the real and imaginary axes.

Since L^m, R^m, T^m and B_m are the shortest paths, we have the following proposition.

Proposition 4.4 *Let $A \subseteq W_m$. If A is invisible, then $A \cap \delta_m = \emptyset$.*

The next theorem is one of the fundamental property of an invisible set. It will play a key role in constructing a metric associated with an invisible set in the next section.

Theorem 4.5 *Let $A \subseteq W_m$ be an invisible set and let $X \subseteq W_n$ be an invisible and $+$-invariant set. Then $AW_n \cup W_m X$ is an invisible set.*

The rest of this section is devoted to the proof of Theorem 4.5.

Definition 4.6

(1) Let $A \subseteq W_m$. Define $\partial_m A = \{F | F \in \partial_m, F \subseteq K(A) \cap \overline{K \backslash K(A)}\}$.
(2) Define $f_{m,\rightarrow}(z) = z + 3^{-m}$, $f_{m,\leftarrow}(z) = z - 3^{-m}$, $f_{m,\uparrow}(z) = z + 3^{-m}\sqrt{-1}$ and $f_{m,\downarrow}(z) = z - 3^{-m}\sqrt{-1}$. Moreover, let $f_{m,\swarrow} = f_{m,\downarrow} \circ f_{m,\leftarrow}$, $f_{m,\searrow} = f_{m,\downarrow} \circ f_{m,\rightarrow}$, $f_{m,\nwarrow} = f_{m,\uparrow} \circ f_{m,\leftarrow}$ and $f_{m,\nearrow} = f_{m,\uparrow} \circ f_{m,\rightarrow}$.
(3) Let $w \in W_m$. For $s \in S$, if there exists $w' \in W_m$ such that $f_{m,s}(K_w) = K_{w'}$, then define $(w)_s = w'$. Otherwise define $(w)_s = \%$, where % is used as the symbol which represents non-existence (Fig. 3).

Lemma 4.7 *Let $F \in \partial_m$ and let $G \in \partial_m(W_m(F))$. If $X \subseteq W_n$ is invisible and $+$-invariant, then $\ell_{W_m X}(\mathbf{p}) \geq 3^{-m}$ for any $\mathbf{p} \in \mathcal{CH}_{m+n}(F, G)$.*

Proof Note that $\#(W_m(F)) \leq 6$. Up to parallel translations, the reflections in the real and the imaginary axes and the $\pi/2$-rotation, we may assume that $F = B_w$ for some $w \in W_m$. Then $W_m(F) \subseteq \{w, (w)_\leftarrow, (w)_\swarrow, (w)_\downarrow, (w)_\searrow, (w)_\rightarrow\}$, where some of them may be %. In fact there are 7 cases. (See Fig. 4.)

Case 1 $\#(W_m(F)) = 6$.
Case 2 $\#(W_m(F)) = 5$ and $(w)_\searrow = \%$.
Case 3 $\#(W_m(F)) = 5$ and $(w)_\downarrow = \%$.
Case 4 $\#(W_m(F)) = 4$ and $(w)_\downarrow = (w)_\searrow = \%$.
Case 5 $\#(W_m(F)) = 3$ and $(w)_\downarrow = (w)_\searrow = (w)_\swarrow = \%$.
Case 6 $\#(W_m(F)) = 3$ and $(w)_\leftarrow = (w)_\swarrow = (w)_\searrow = \%$.
Case 7 $\#(W_m(F)) = 2$ and $(w)_\downarrow = (w)_\swarrow = (w)_\searrow = (w)_\leftarrow = \%$.

We consider the first case. The other cases can be treated by the similar discussion. If $D = \cup_{U \in \partial_m(W_m(F))} U$, then $D = \partial K(W_m(F))$. Let $\mathbf{p} = (w(1), \ldots, w(k)) \in \mathcal{CH}_{m+n}(F, G)$. The reflection in the line containing F induces a natural bijection from $W_m(F) \cdot W_n$ to itself, which is denoted by η. Define $\theta : W_m(F) \cdot W_n \to \{(w)_\leftarrow, w, (w)_\rightarrow\} \cdot W_n$ by

$$\theta(uv) = \begin{cases} uv & \text{if } u \in \{(w)_\leftarrow, w, (w)_\rightarrow\} \text{ and } v \in W_n, \\ \eta(uv) & \text{if } u \in \{(w)_\swarrow, (w)_\downarrow, (w)_\searrow\} \text{ and } v \in W_n. \end{cases}$$

Define $v(i) = \theta(w(i))$ and $\widetilde{\mathbf{p}} = (v(1), \ldots, v(k))$. Then the $+$-invariant property of X implies that $\widetilde{\mathbf{p}} \in \mathcal{CH}_{m+n}(F, D_1)$, where $D_1 = L_{(w)_\leftarrow} \cup T_{(w)_\leftarrow} \cup T_w \cup T_{(w)_\rightarrow} \cup R_{(w)_\rightarrow}$, and

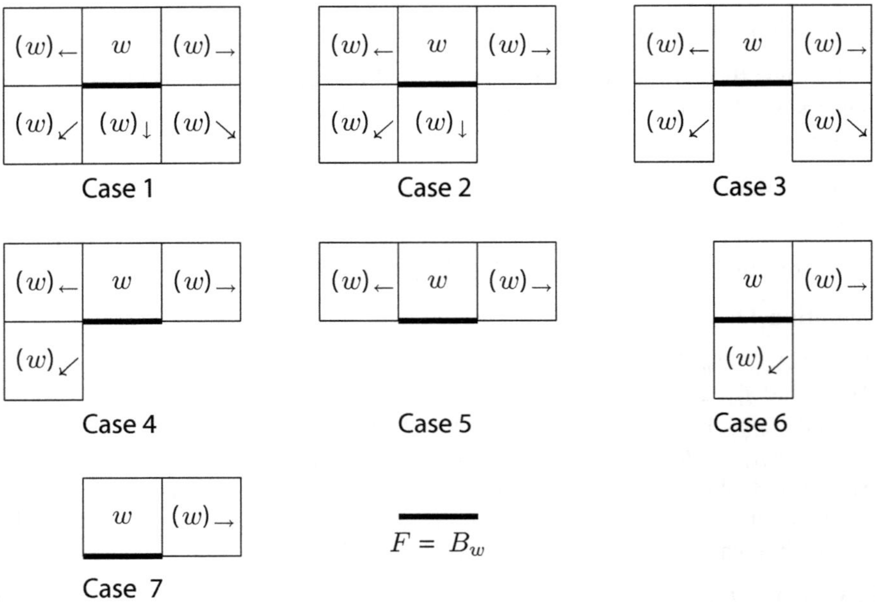

Fig. 3 Structures of $W_m(F)$

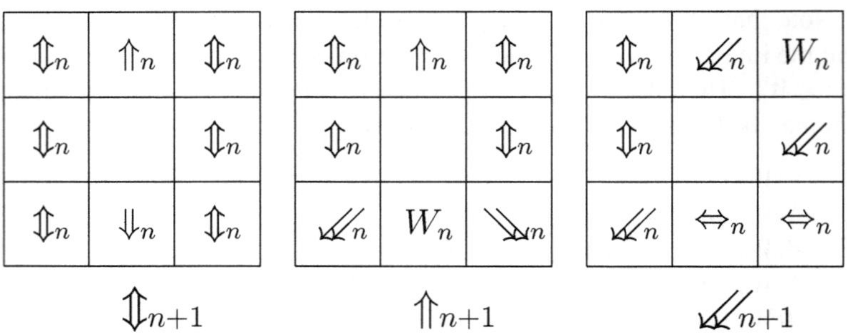

Fig. 4 Construction of $\Updownarrow_n$, $\Uparrow_n$ and $\swarrow\!\!\swarrow_n$

$$\ell_{W_m X}(\mathbf{p}) = \ell_{W_m X}(\widetilde{\mathbf{p}}),$$

If $v(k) \cap L_{(w)_\leftarrow} \neq \emptyset$, then there exists j such that $(v(j), v(j+1), \ldots, v(k)) \in \mathcal{CH}_m(R_{(w)_\leftarrow}, L_{(w)_\leftarrow})$ and $K_{v(i)} \subseteq (w)_\leftarrow \cdot W_n$ for any $i \in \{j, j+1, \ldots, k\}$. Since X is invisible, it follows that

$$\ell_{W_m X}(\mathbf{p}) \geq \ell_{W_m X}((v(j), \ldots, v(k))) \geq 3^{-m}.$$

The same discussion shows that $\ell_{W_m X}(\mathbf{p}) \geq 3^{-m}$ if $K_{v(k)} \cap R_{(w)_\rightarrow} \neq \emptyset$.

Next suppose $v(K) \cap (T_{(w)_{\leftarrow}} \cup T_w \cup T_{(w)_{\rightarrow}}) \neq \emptyset$. Then using the reflections in the lines containing L_w and R_w, we may construct $(u(1), \ldots, u(k)) \in \mathcal{CH}_{m+n}(B_w, T_w)$ which satisfies $u(i) \in w \cdots W_n$ for any i and $\ell_{W_m X}(\mathbf{p}) = \ell_{W_m X}((u(1), \ldots, u(k)))$. Since X is invisible, it follows that $\ell_{W_m X}(\mathbf{p}) \geq 3^{-m}$. □

Lemma 4.8 *Let $F, G \in \partial_m$ with $F \cap G = \emptyset$ and let $\mathbf{p} = (w(1), \ldots, w(k)) \in \mathcal{CH}_{m+n}(F, G)$. If $\{w(i)\}_{i=1}^{k} \cap AW_n = \emptyset$, then there exists $\mathbf{p}_* \in \mathcal{CH}_m(F, G)$ such that $\ell_A(\mathbf{p}_*) \leq \ell_{AW_n \cup W_m X}(\mathbf{p})$.*

Proof Let $k_1 = \max\{j | \{w(i)\}_{i=1}^{j} \subseteq W_{m+n}(F)\}$. Define $v(1) = [w(k_1)]_m$. Note that $v(1) \notin A$. There exists a unique $F_1 \in \{L_{v(1)}, R_{v(1)}, T_{v(1)}, B_{v(1)}\} \cap \partial_m(W_m(F))$ such that $K_{w(k_1)} \subseteq F_1$. By Lemma 4.7,

$$\ell_{AW_n \cup W_m X}((w(1), \ldots, w(k_1))) \geq 3^{-m} = \ell_A((v(1))).$$

Now, if $F_1 \cap G \neq \emptyset$, $(v(1)) \in \mathcal{CH}_m(F, G)$ and $\ell_{AW_n \cup W_m X}(\mathbf{p}) \geq \ell_A((v(1)))$. Hence we have constructed $\mathbf{p}_* = (v(1))$. Otherwise, replacing $(w(1), \ldots, w(k))$ and F by $(w(k_1), \ldots, w(k))$ and F_1 respectively, we repeat the same procedure as above and obtain k_2, $v(2)$ and F_2. Inductively, we have $\mathbf{p}_* = (v(1), \ldots, v(l))$ with the desired properties. □

Lemma 4.9 *Let $F, G \in \partial_m$ with $F \cap G = \emptyset$. Then for any $\mathbf{p} \in \mathcal{CH}_{m+n}(F, G)$, there exists $\mathbf{p}_* \in \mathcal{CH}_m(F, G)$ such that $\ell_A(\mathbf{p}_*) \leq \ell_{AW_n \cup W_m X}(\mathbf{p})$.*

Proof Let $\mathbf{p} = (w(1), \ldots, w(k))$. If $w(i) \notin AW_n$ for any i, then Lemma 4.8 suffices. Hence we assume that there exists i such that $w(i) \in AW_n$.

Claim 1 *Without loss of generality, we may assume that there exists $p_1 \geq 1$ and $G_1 \in \partial_m$ such that $w(1), \ldots, w(p_1) \in W_{m+n} \backslash AW_n$, $w(p_1+1) \in AW_n$, $G_1 \cap F = \emptyset$, $G_1 \subseteq K_{[w(p_1+1)]_m}$ and $(w(1), \ldots, w(p_1)) \in \mathcal{CH}_{m+n}(F, G_1)$.*

Proof of Claim 1 Case 1: $F \cap K(A) = \emptyset$

In this case, define

$$p_1 = \min\{i | w(i) \in AW_n\} - 1$$

and choose $G_1 \in \partial_m$ so that $G_1 \cap K_{w(p_1)} \cap K_{w(p_1+1)} \neq \emptyset$ and $G_1 \subseteq K_{[w(p_1)]_m}$.
Case 2: $F \cap K(A) \neq \emptyset$

In this case, F intersects at most two connected components of $K(A)$. Let C_1 and C_2 be those connected components of $K(A)$ (It is possible that $C_1 = C_2$).
Case 2.1: $\{i | K_{w(i)} \subseteq C_1 \cup C_2\} = \emptyset$.

Define p_1 and choose G_1 as in Case 1. Then p_1 and G_1 satisfies the desired property.
Case 2.2: $\{i | K_{w(i)} \subseteq C_1 \cup C_2\} \neq \emptyset$.
Define

$$q = \max\{i | K_{w(i)} \in C_1 \cup C_2\}.$$

We may choose $F_0 \in \partial_m$ so that $F_0 \cap K_{w(q)} \cap K_{w(q+1)} \neq \emptyset$ and $F_0 \subseteq K_{[w(q)]_m}$. Moreover, we may choose $\mathbf{p}^0 = (v(1), \ldots, v(k_0)) \in \mathcal{CH}_m(F, F_0)$ so that $v(i) \in AW_n$ for any $i = 1, \ldots, k_0$ and $v(k_0) = [w(q)]_m$. Note that $\ell_A(\mathbf{p}^0) = 0$. If $F_0 \cap G \neq \emptyset$, then $K_{v(k_0)} \cap G \neq \emptyset$ and $\mathbf{p}^0 \in \mathcal{CH}_m(F, G)$. Hence letting $\mathbf{p}_* = \mathbf{p}^0$, we have constructed $\mathbf{p}_*$ which satisfies all the conditions. Assume that $F_0 \cap G = \emptyset$. Since $(w(1), \ldots, w(q)) \in \mathcal{CH}_{m+n}(F, F_0)$ corresponds $\mathbf{p}^0 \in \mathcal{CH}_m(F, F_0)$, it is enough to show the statement of the lemma in the case where F and $\mathbf{p}$ are replaced by F_0 and $(w(q+1), \ldots, w(k))$ respectively. In this situation, the counterpart of Case 2.1 holds and so does Claim 1 (End of Proof of Claim 1). □

Claim 2 *Without loss of generality, we may assume that there exists k_* and $F_* \in \partial_m$ such that $w(k_*), \ldots, w(k) \in W_{m+n} \backslash AW_n$, $w(k_* - 1) \in AW_n$, $F_* \cap G = \emptyset$, $F_* \subseteq K_{w(k_*)}$ and $(w(k_*), \ldots, w(k)) \in \mathcal{CH}_{m+n}(F_*, G)$.*

Proof of Claim 2 By considering the chain $(w(k), w(k-1), \ldots, w(1)) \in \mathcal{CH}_{m+n}(G, F)$, the same argument as in the proof of Claim 1 yields this claim (End of Proof of Claim 2).

Now under Claim 1 and Claim 2, we may choose $p_1, \ldots, p_{j+1}$ and $q_0, q_1, \ldots, q_j$ which satisfy the following conditions (A), (B), (C) and (D):

(A) $q_0 = 0$, $p_{j+1} = k$, $q_i < p_{i+1} < q_{i+1}$ for any i.
(B) $\{(w(q_{i-1}+1), \ldots, w(p_i)\} \cap AW_n = \emptyset$ for any $i = 1, \ldots, j+1$
(C) $K_{w(p_i+1)}$ and $K_{w(q_i)}$ belong to the same connected component of $K(A)$ for any $i = 1, \ldots, j$.
(D) $K_{w(q_i)}$ and $K_{w(p_{i+1}+1)}$ belong to the different connected components of $K(A)$ for any $i = 1, 2, \ldots, j-1$

Let $\mathbf{p}_i = (w(q_{i-1}+1), \ldots, w(p_i))$ for $i = 1, \ldots, j+1$. Define $F_1 = F$. For $i \geq 2$, we may choose $F_i \in \partial_m$ so that $F_i \cap K_{w(q_{i-1})} \cap K_{w(q_{i-1}+1)} \neq \emptyset$ and $F \subseteq K_{[w(q_{i-1})]_m}$. Moreover, for $i = 1, \ldots, j$, we may choose $G_i \in \partial_m$ so that $G_i \cap K_{w(p_i)} \cap K_{w(p_i+1)} \neq \emptyset$ and $G_i \subseteq K_{[w(p_i+1)]_m}$. Also let $F_{j+1} = G$. By the condition (D), $F_i \cap G_i = \emptyset$ for any $i = 1, \ldots, j+1$. Hence letting $F = F_i$, $G = G_i$ and $\mathbf{p} = \mathbf{p}_i$ and applying Lemma 4.8, we obtain $\mathbf{p}_{*,i} = (v(i,1), \ldots, v(i,k_i)) \in \mathcal{CH}_m(F_i, G_i)$ which satisfies $\ell_A(\mathbf{p}_{*,i}) \leq \ell_{AW_n \cup W_m X}(\mathbf{p}_i)$.

Note that G_i and F_i belong to the same connected component of $K(A)$ by the condition (C). Hence there exists $\mathbf{p}_i^1 = (u(i,1), \ldots, u(i,l_i)) \in \mathcal{CH}_m(G_i, F_i)$ such that $u(i,1), \ldots, u(i,l_i) \in A$.

Finally let $\mathbf{p}_* = (\mathbf{p}_{*,1}, \mathbf{p}_1^1, \mathbf{p}_{*,2}, \ldots, \mathbf{p}_j^1, \mathbf{p}_{*,j+1})$. Then $\mathbf{p}_* \in \mathcal{CH}_m(F, G)$ and $\ell_A(\mathbf{p}_*) \leq \ell_{AW_n \cup W_m X}(\mathbf{p})$. □

Proof of Theorem 4.5 Let $\mathbf{p} = (w(1), \ldots, w(k)) \in \mathcal{CH}_{m+n}(L, R)$. Set $F = L_{[w(1)]_m}$ and $G = R_{[w(k)]_m}$. By Lemma 4.9, there exists $\mathbf{p}_* \in \mathcal{CH}_m(F, G)$ such that $\ell_A(\mathbf{p}_*) \leq \ell_{AW_n \cup W_m X}(\mathbf{p})$. Since A is invisible, we have $\ell_A(\mathbf{p}_*) \geq 1$. Hence $\ell_{AW_n \cup W_m X}(\mathbf{p}) \geq 1$. In the same way, if $\mathbf{p}' \in \mathcal{CH}_{m+n}(T, B)$, it follows that $\ell_{AW_n \cup W_m X}(\mathbf{p}') \geq 1$. Thus $AW_n \cup W_m X$ is invisible. □

5 Metric Associated with Invisible Set

In this section, we construct a metric associated with a +-invariant invisible set and characterize the Hausdorff dimension and the Hausdorff measure with respect to the metric.

Throughout this section, we fix a +-invariant invisible set $A \subseteq W_m$.

Notation We write $W_{m,n} = (W_m)^n = W_{mn}$, $W_{m,*} = \cup_{n\geq 0} W_{m,n}$ and $\Sigma^{(m)} = (W_m)^{\mathbb{N}}$.

Naturally $W_{m,*}$ is regarded as a subset of W_* and $\Sigma^{(m)}$ is identified with Σ.

Definition 5.1

(1) Let $\epsilon > 0$. Define $D_\epsilon^A(w)$ for $w \in W_m$ by

$$D_\epsilon^A(w) = \begin{cases} 3^{-m} & \text{if } w \notin A, \\ \epsilon & \text{if } w \in A. \end{cases}$$

and $D_\epsilon^A(\emptyset) = 1$ for $\emptyset \in W_0$. For any $w = w^{(1)} \cdots w^{(n)} \in W_{m,n}$, where $w^{(i)} \in W_m$, define

$$D_\epsilon^A(w) = D_\epsilon^A(w^{(1)}) D_\epsilon^A(w^{(2)}) \cdots D_\epsilon^A(w^{(n)}).$$

(2) Define

$$\mathcal{CH}^{(m)} = \{(w(1), \ldots, w(k)) | \\ (w(1), \ldots, w(k)) \in \mathcal{CH},\ w(i) \in W_{m,*} \text{ for any } i = 1, \ldots, k\}.$$

and $\mathcal{CH}^{(m)}(U, V) = \mathcal{CH}(U, V) \cap \mathcal{CH}^{(m)}$ for $U, V \subseteq K$. Moreover, define $\ell^{A,\epsilon}(\mathbf{p}) = \sum_{i=1}^k D_\epsilon^A(w(i))$ for any $\mathbf{p} = (w(1), \ldots, w(k)) \in \mathcal{CH}^{(m)}$ and, for $x, y \in K$,

$$d_\epsilon^A(x, y) = \inf\{\ell^{A,\epsilon}(\mathbf{p}) | \mathbf{p} \in \mathcal{CH}^{(m)}(x, y)\}.$$

$D_\epsilon^A(\cdot)$ is a gauge function on $\Sigma^{(m)}$ and d_ϵ^A is the associated pseudometric.

The next fact is obvious from the definition.

Proposition 5.2 *$d_0^A(x, y) \leq d_\epsilon^A(x, y)$ for any $x, y \in K$ and any $\epsilon > 0$.*

The next theorem shows that d_ϵ^a is really a metric and $d_\epsilon^A \underset{\text{QS}}{\sim} d_E$.

Theorem 5.3 *For any $\epsilon > 0$, d_ϵ^A is a metric on K which is quasisymmetric with respect to d_E. The Hausdorff dimension of K with respect to the metric d_ϵ^A, $\dim_H(K, d_\epsilon^A)$ is given by the unique α which satisfies*

$$(8^m - \#(A))3^{-m\alpha} + \#(A)\epsilon^\alpha = 1. \tag{5.1}$$

Furthermore, let $\mathcal{H}^\alpha$ be the α-dimensional Hausdorff measure on (X, d_ϵ^A). Then the metric measure space $(X, d_\epsilon^A, \mathcal{H}^\alpha)$ is Ahlfors α-regular, i.e.

$$\mathcal{H}^\alpha(B_d(x, r)) \asymp r^\alpha$$

for any $x \in K$ and $r \in [0, \mathrm{diam}(X, d_\epsilon^A))$.

Letting $\epsilon \downarrow 0$ in (5.1), we obtain the following corollary.

Corollary 5.4

$$\dim_C(K, d_E) \leq \frac{\log 8}{\log 3} + \frac{1}{m \log 3} \log\left(1 - \frac{\#(A)}{8^m}\right).$$

In the rest of this section, we are going to prove the above theorem. Hereafter, we omit A in the notations $D_\epsilon^A(w)$, $\ell^{A,\epsilon}(\mathbf{p})$ and $d_\epsilon^A(x, y)$ and write $D_\epsilon(w)$, $\ell^\epsilon(\mathbf{p})$ and $d_\epsilon(x, y)$ respectively.

Lemma 5.5 *Define $A_n \subseteq W_{mn}$ inductively by $A_1 = A$ and*

$$A_{n+1} = AW_{mn} \cup W_m A_n.$$

Then A_n is $+$-invariant and invisible.

Proof Letting $X = A_n$ and applying Theorem 4.5, we see inductively that A_{n+1} is $+$-invariant and invisible. □

Lemma 5.6 $d_0^A(x, y) \geq 1$ *for any* $(x, y) \in (L \times R) \cup (T \times B)$.

Proof Define $I(\mathbf{p}) = \max_{i=1,\ldots,k} |w(i)|/m$ for any $\mathbf{p} = (w(1), \ldots, w(k)) \in \mathcal{CH}^{(m)}(L, R)$. We are going to show that $\ell^0(\mathbf{p}) \geq 1$ by an induction in $I(\mathbf{p})$. If $I(\mathbf{p}) = 0$, then $\mathbf{p} = (\emptyset)$ and $\ell^0(\mathbf{p}) = D_0(\emptyset) = 1$. Let $J = \{i | i = 1, \ldots, k, |w(i)| = I(\mathbf{p})m\}$. Then there exists $k_1, \ldots, k_l$ and $j_1, \ldots, j_l$ such that $k_i \leq j_i < k_{i+1}$ and $J = \cup_{i=1,\ldots,l}\{j | k_i \leq j \leq j_i\}$. Let $\mathbf{p}^i = (w(k_i), \ldots, w(j_i))$. Since $|w(k_i - 1)| \leq (I(\mathbf{p}) - 1)m$ and $|w(j_i + 1) \leq (I(\mathbf{p}) - 1)m$, there exist $F, G \in \partial_M$, where $M = (I(\mathbf{p}) - 1)m$, such that $F \subseteq K_{w(k_i - 1)}$, $F \cap K_{w(k_i)} \neq \emptyset$, $G \subseteq K_{w(j_i+1)}$ and $G \cap K_{w(j_i)} \neq \emptyset$. If $F \cap G = \emptyset$, then $K_{w(k_i-1)} \cap K_{w(j_i+1)} \neq \emptyset$. Hence if $\mathbf{p}' = (w(1), \ldots, w(k_i - 1), w(j_i + 1), \ldots, w(k)) \in \mathcal{CH}^{(m)}(L, R)$, then we define $\mathbf{p}_*^i$ as the empty sequence. Note that $\ell^0(\mathbf{p}) \geq \ell^0(\mathbf{p}')$. Now assume that $F \cap G = \emptyset$. Set $X = A_M$. Lemma 5.5 shows that X is $+$-invariant and invisible. Then by Lemma 4.9, there exists $\mathbf{p}_*^i = (v(1), \ldots, v(l)) \in \mathcal{CH}_M(F, G)$ such that $\ell_{A_M}(\mathbf{p}_*^i) \leq \ell_{A_M W_m \cup W_M A}(\mathbf{p}^i)$. Note that $A_M W_m \cup W_M A = A_{I(\mathbf{p})m}$, that $\ell_{A_M}(\mathbf{p}_*^i) = \ell^0(\mathbf{p}_*^i)$ and that $\ell_{A_M W_m \cup W_M A}(\mathbf{p}^i) = \ell^0(\mathbf{p}^i)$. Let $\mathbf{p}_*$ be the chain where $\mathbf{p}^i$ is replaced by $\mathbf{p}_*^i$ for all i. Then $\mathbf{p}_* \in \mathcal{CH}^{(m)}(L, R)$, $I(\mathbf{p}_*) < I(\mathbf{p})$ and $\ell^0(\mathbf{p}) \geq \ell^0(\mathbf{p}_*)$. Now we have $\ell^0(\mathbf{p}) \geq \ell^0(\mathbf{p}_*) \geq 1$ by induction hypothesis.

Now, $d_0^A(x, y) \geq \inf\{\ell^0(\mathbf{p}) | \mathbf{p} \in \mathcal{CH}^{(m)}(x, y)\} \geq 1$ for any $x \in L$ and any $y \in R$. In the same manner, it follows that $d_0^A(x, y) \geq 1$ for any $x \in T$ and any $y \in B$ as well. □

Lemma 5.7 *$d_\epsilon^A(\cdot,\cdot)$ is a metric on K for any $\epsilon > 0$.*

Proof Let $x, y \in K$ with $x \neq y$. Then $\mathrm{Re}\, x \neq \mathrm{Re}\, y$ or $\mathrm{Im}\, x \neq \mathrm{Im}\, y$. Suppose $\mathrm{Re}\, x < \mathrm{Re}\, y$. Then there exist n and $i \in \{0, 1, \ldots, 3^{mn} - 1\}$ such that $\mathrm{Re}\, x \leq (2i - 3^{mn})3^{-mn} < (2i + 2 - 3^{mn})3^{-mn} \leq \mathrm{Re}\, y$.

Claim $d_\epsilon^A(x, y) \geq \min\{D_\epsilon^A(w)|w \in W_{m,k}, k = 0, 1, \ldots, n\}$.

Proof of Claim Define $W_{m,n}^i = \{w|w \in W_{m,n}, K_w \subseteq \{z|(2i - 3^{mn})3^{-mn} \leq \mathrm{Re}\, z \leq (2i + 2 - 3^{mn})3^{-mn}\}$. Let $D_{mn,i} = \min\{D_\epsilon(w)|w \in W_{m,n}^i\}$. We also define $L_{mn,i} = \cup_{w \in W_{m,n}^i} L_w$ and $R_{mn,i} = \cup_{w \in W_{m,n}^i} R_w$. Let $\mathbf{p} = (w(1), \ldots, w(k)) \in \mathcal{CH}^{(m)}(x, y)$. If $|w(i)| \leq mn$ for some i, then the claim is trivial. Hence assume that $|w(i)| < mn$ for any $i = 1, \ldots, k$. Then $\mathbf{p}$ contains $(w(p), w(p+1), \ldots, w(q)) \in \mathcal{CH}(L_{mn,i}, R_{mn,i})$ which satisfies $w(i) \in \cup_{w \in W_{m,n}^i} w W_{m,*}$. Let $w(i) = u(i)v(i)$ for $i = p, \ldots, q$, where $u(i) \in W_{m,n}^i$ and $v(i) \in W_{m,*}$. It follows that

$$\ell^{A,\epsilon}(\mathbf{p}) \geq \ell^{A,\epsilon}((w(p), w(p+1), \ldots, w(q))) \geq D_{mn,i} \sum_{i=p}^{q} D_\epsilon(v(i)). \tag{5.2}$$

Now the reflection ψ in the real axis induces a natural bijection $\varphi_\leftrightarrow : W_* \to W_*$ defined by $\psi(K_w) = K_{\varphi_\leftrightarrow(w)}$ which satisfies $\varphi_\leftrightarrow(\varphi_\leftrightarrow(w)) = w$. Hereafter in this section, we write $\varphi = \varphi_\leftrightarrow$. There exist $p_1, p_2, \ldots p_l$ such that $p_1 = p$, $p_l = q + 1$, $p_i < p_{i+1}$, $u(p_i) = u(p_i + 1) = \ldots = u(p_{i+1} - 1)$ and $u(p_i) \neq u(p_{i+1})$ for any i. Let $\bar{v}(j) = \varphi^i(v(j))$ for $j = p_i, p_i + 1, \ldots, p_{i+1} - 1$, where φ^j is the j-th iteration of φ. Then $(\bar{v}(p), \bar{v}(p+1), \ldots, \bar{v}(q)) \in \mathcal{CH}^{(m)}(L, R)$. Since A is $+$-invariant, $\sum_{i=p}^q D_\epsilon(v(i)) = \sum_{i=p}^q D_\epsilon(\bar{v}(i))$. Hence Lemma 5.6 implies that

$$\sum_{i=p}^{q} D_\epsilon(v(i)) = \sum_{i=p}^{q} D_\epsilon(\bar{v}(i)) \geq \sum_{i=p}^{q} D_0(\bar{v}(i)) \geq 1.$$

Combining this with (5.2), we have $\ell^{A,\epsilon}(\mathbf{p}) \geq D_{mn,i}$. Hence the claim holds (End of Proof of Claim).

The claim shows that $d_\epsilon^A(x, y) > 0$ if $\mathrm{Re}\, x \neq \mathrm{Re}\, y$. Similar discussion implies $d_\epsilon^A(x, y) > 0$ if $\mathrm{Im}\, x \neq \mathrm{Im}\, y$. □

Proof of Theorem 5.3 Let $\mathcal{S}^{(m)}(A, \epsilon) = \{\Lambda_s^{(m)}(A, \epsilon)\}_{s \in (0,1]}$ be the scale on $\Sigma^{(m)}$ whose gauge function is D_ϵ^A and let $\mathcal{S}^{(m)}$ by the scale on $\Sigma^{(m)}$ whose gauge function g is given by $g(w(1)\ldots w(k)) = 3^{-mk}$ for $w(1)\ldots w(k) \in W_{m,*}$ with $w(1), \ldots, w(k) \in W_m$. Obviously $\Sigma^{(m)}$ is adapted to the Euclidean metric on K. Also since $\mathcal{S}^{(m)}(A, \epsilon)$ and $\mathcal{S}^{(m)}$ are self-similar, they are elliptic.

Note that $(K, W_m, \{F_w\}_{w \in W_m})$ is a rationally ramified self-similar structure. (See [Kig09, Sect. 1.5] for the definition of rationally ramified self-similar structures.) In fact, define $h : L^1 \to R^1$ by $h(\nwarrow) = \nearrow, h(\leftarrow) = \rightarrow, h(\swarrow) = \searrow$ and $g : T^1 \to B^1$ by $g(\nwarrow) = \swarrow, g(\uparrow) = \downarrow, g(\nearrow) = \searrow$. Then define $h_m : L^m \to R^m$

by $h_m(w_1 \dots w_m) = h(w_1) \dots h(w_m)$ for $w_1 \dots w_m \in L^m$ and $g_m : T^m \to B^m$ by $g_m(w_1 \dots w_m) = g(w_1) \dots g(w_m)$ for $w_1 \dots w_m \in T^m$. Then a relation set $\mathcal{R}_m$ of $(K, W_m, \{F_w\}_{w \in W_m})$ is given by

$$\mathcal{R}_m = \{(L^m, R^m, h_m, w, v) | w, v \in W_m, L_w = R_v\} \cup \{(T^m, B^m, g_m, w, v) | w, v \in W_m, T_w = B_v\}$$

By Proposition 4.4, $D_\epsilon^A(w) = 3^{-m}$ for any $w \in L^m \cup R^m \cup T^m \cup B^m$. Using [Kig09, Theorem 1.6.6], we see that $\mathcal{S}^{(m)}(A, \epsilon) \underset{\text{GE}}{\sim} \mathcal{S}^{(m)}$.

Theorems 1.6.1 and 2.2.7 in [Kig09] imply that $\mathcal{S}^{(m)}(A, \epsilon)$ is intersection type finite. Since d_ϵ^A is a metric on K by Lemma 5.7, we may apply [Kig09, Theorem 2.3.16] and show that d_ϵ^A is adapted to the scale $\mathcal{S}^{(m)}(A, \epsilon)$. Thus we have obtained all the conditions in Theorem 3.4 and hence shown that d_ϵ^A is quasisymmetric with respect to the Euclidean metric.

The Hausdorff dimension and Ahlfors regularity of the Hausdorff measure of (K, d_ϵ^A) are immediately obtained by [Kig01, Theorem 1.5.7]. □

6 Construction of Invisible Sets

Under the existence of an invisible set, we have constructed a corresponding metric which is quasisymmetric with respect to d_E and characterized the associated Hausdorff dimension in the previous two sections. In this section, it is shown that invisible sets do exist. In fact, we construct a series of invisible sets inductively.

Definition 6.1 Let $\psi_\updownarrow$ and $\psi_\leftrightarrow$ be the reflections in the real and complex axes respectively. Then $\psi_\updownarrow$ induces a natural bijection $\varphi_\updownarrow$ from W_* to itself defined by $\psi_\updownarrow(K_w) = K_{\varphi_\updownarrow(w)}$. In the same way, we define a bijection $\varphi_\leftrightarrow$ from W_* to itself by $\psi_\leftrightarrow(K_w) = K_{\varphi_\leftrightarrow(w)}$. Moreover, let R be the $\pi/2$-rotation around the origin 0 and let $\rho : W_* \to W_*$ be the bijection defined by $R(K_w) = K_{\rho(w)}$.

The idea to have invisible sets is to divide the notion of a invisible set into a vertically invisible set and a horizontally invisible set. The final existence of invisible sets are established by taking intersections of vertically invisible set and horizontally invisible set in Theorem 6.4.

Definition 6.2 Define $\Updownarrow_n$, $\Uparrow_n$ and $\Swarrow_n$ as subsets of W_n inductively by

$$\Updownarrow_{n+1} = \{\nwarrow, \leftarrow, \swarrow, \nearrow, \rightarrow, \searrow\} \cdot \Updownarrow_n \cup \uparrow \cdot \Uparrow_n \cup \downarrow \cdot \Downarrow_n \tag{6.1}$$

$$\Uparrow_{n+1} = \{\nearrow, \leftarrow, \nwarrow, \rightarrow\} \cdot \Updownarrow_n \cup \uparrow \cdot \Uparrow_n \cup \swarrow \cdot \Swarrow_n \cup \downarrow \cdot W_n \cup \searrow \cdot \Searrow_n \tag{6.2}$$

$$\Swarrow_{n+1} = \{\nwarrow, \leftarrow\} \cdot \Updownarrow_n \cup \{\downarrow, \searrow\} \cdot \Leftrightarrow_n \cup \{\uparrow, \rightarrow, \swarrow\} \cdot \Swarrow_n \cup \nearrow \cdot W_n \tag{6.3}$$

and $\Updownarrow_0 = \Uparrow_0 = \Swarrow_0 = \emptyset$, where $\Downarrow_n = \varphi_\updownarrow(\Uparrow_n)$, $\Searrow_n = \varphi_\leftrightarrow(\Swarrow_n)$ and $\Leftrightarrow_n = \rho(\Updownarrow_n)$.

Lemma 6.13 will show that $\Updownarrow_n$ and $\Leftrightarrow_n$ are vertically and horizontally invisible sets respectively.

Lemma 6.3

$$\#(\Updownarrow_n) = 8^n - \frac{7+\sqrt{41}}{2\sqrt{41}}\left(\frac{9+\sqrt{41}}{2}\right)^n + \frac{7-\sqrt{41}}{2\sqrt{41}}\left(\frac{9-\sqrt{41}}{2}\right)^n$$

Proof Write $a_n = \#(\Updownarrow_n)$, $b_n = \#(\Uparrow_n)$ and $c_n = \#(\Swarrow_n)$. By (6.1), (6.2) and (6.3), it follows that

$$\begin{aligned} a_{n+1} &= 6a_n + 2b_n \\ b_{n+1} &= 4a_n + b_n + 2c_n + 8^n \\ c_{n+1} &= 4a_n + 3c_n + 8^n. \end{aligned}$$

Solving these with $a_0 = b_0 = c_0 = 0$, we obtain a_n as in the statement of the lemma. □

Now we have the main theorem of this section.

Theorem 6.4 *Let $A_n = \Updownarrow_n \cap \Leftrightarrow_n$. Then A_n is a $+$-invariant invisible set and*

$$\alpha_n \leq 8^n - \#(A_n) \leq 2\alpha_n,$$

where

$$\alpha_n = \frac{7+\sqrt{41}}{2\sqrt{41}}\left(\frac{9+\sqrt{41}}{2}\right)^n - \frac{7-\sqrt{41}}{2\sqrt{41}}\left(\frac{9-\sqrt{41}}{2}\right)^n.$$

Example 6.5 $A_0 = A_1 = A_2 = A_3 = \emptyset$.

$$A_4 = \{\uparrow\downarrow\leftarrow\rightarrow, \uparrow\downarrow\rightarrow\leftarrow, \downarrow\uparrow\leftarrow\rightarrow, \downarrow\uparrow\leftarrow\rightarrow, \leftarrow\rightarrow\uparrow\downarrow, \leftarrow\rightarrow\downarrow\uparrow, \rightarrow\leftarrow\uparrow\downarrow, \rightarrow\leftarrow\downarrow\uparrow\}.$$

Applying Corollary 5.4 and letting $n \to \infty$, we obtain the following upper estimate of the conformal dimension of the Sierpinski carpet.

Corollary 6.6

$$\dim_C(K, d_E) \leq \frac{\log\left(\frac{9+\sqrt{41}}{2}\right)}{\log 3} = 1.858183... < \frac{\log 8}{\log 3} = 1.892789....$$

Remark The known lower bound of $\dim_C(K, d_E)$ given in (1.1) is $\frac{\log 6}{\log 3} = 1.630929....$

The rest of this section is devoted to proving Theorem 6.4.

Lemma 6.7 (1) $\varphi_{\leftrightarrow}(\Updownarrow_n) = \Updownarrow_n$ *and* $\varphi_{\updownarrow}(\Updownarrow_n) = \Updownarrow_n$.
(2) $\varphi_{\leftrightarrow}(\Uparrow_n) = \Uparrow_n$.
(3) $\varphi_{\leftrightarrow} \circ \rho(\swarrow_n) = \swarrow_n$.

Definition 6.8 Define the vertical index $I^n_{\updownarrow} : W_n \to \{1, \ldots, 3^n\}$ by

$$I^n_{\updownarrow}(w) = \frac{3^n(\mathrm{Im}\,(F_w(\sqrt{-1})) + 1)}{2}$$

For $H \in \{L, R\}$, define $w^n_H(i)$ for $i = 1, \ldots, 3^n$ as the unique $w \in H^n$ which satisfies $I^n_{\updownarrow}(w) = i$. Moreover, for $w, v \in W_n$, define $\mathbf{p}^n_H(w, v) \in \mathcal{CH}_n$ by

$$\mathbf{p}^n_H(w, v) = \begin{cases} (w^n_H\left(I^n_{\updownarrow}(w)\right), w^n_H\left(I^n_{\updownarrow}(w)+1\right), \ldots, w^n_H\left(I^n_{\updownarrow}(v)\right) & \text{if } I^n_{\updownarrow}(w) \le I^n_{\updownarrow}(v), \\ (w^n_H\left(I^n_{\updownarrow}(w)\right), w^n_H\left(I^n_{\updownarrow}(w)-1\right), \ldots, w^n_H\left(I^n_{\updownarrow}(v)\right) & \text{if } I^n_{\updownarrow}(v) \le I^n_{\updownarrow}(w). \end{cases}$$

In the same way, we define the horizontal index $I^n_{\leftrightarrow} : W_n \to \{1, \ldots, 3^n\}$, $w^n_T(i)$, $w^n_B(i)$, $\mathbf{p}^n_T(w, v)$ and $\mathbf{p}^n_B(w, v)$.

Lemma 6.9 *Let $A \subseteq W_n$. Assume that*

$$\inf\{\ell_A(\mathbf{p}_*)|\mathbf{p}_* \in \mathcal{CH}_n(T, p_{\swarrow})\} \ge 1 \tag{6.4}$$

Let $\mathbf{p} = (w(1), \ldots, w(k)) \in \mathcal{CH}_n$. If $(w(1), w(k)) \in (T^n \cup L^n) \times L^n$, then

$$\ell_A(\mathbf{p}) \ge \frac{|I^n_{\updownarrow}(w(1)) - I^n_{\updownarrow}(w(k))| + 1}{3^n} = \ell_A(\mathbf{p}^n_L(w(1), w(k))). \tag{6.5}$$

Remark Using the symmetries, we may exchange $(T, L, p_{\swarrow})$ in the statement of Lemma 6.9 by $(T, R, p_{\searrow})$, $(B, L, p_{\nwarrow})$ and $(B, R, p_{\nearrow})$.

Proof Since $\ell_A((w^n_L(i))_{i=1,\ldots,3^n}) \ge 1$, it follows that $\{w^n_L(i)|i = 1, \ldots, 3^n\} \cap A = \emptyset$. Let $\mathbf{p} = (w(1), \ldots, w(k)) \in \mathcal{CH}_n$.

Suppose that $(w(1), w(k)) \in T^n \times L^n$. Note that $w(k) = w^n_L(i)$ for some i. Then $\mathbf{p}^n_L(w(1), w(k)) = (w^n_L(3^n), w^n_L(3^n-1), \ldots, w^n_L(i))$ and $\ell_A(\mathbf{p}^n_L(w(1), w(k)))$ $= 1 - (i-1)/3^n$. Since $\mathbf{p} \vee \mathbf{p}^n_L(w^n_L(i-1), w^n_L(1)) \in \mathcal{CH}_n(T, p_{\swarrow})$, (6.4) implies

$$\ell_A(\mathbf{p}) + \ell_A(\mathbf{p}^n_L(w^n_L(i-1), w^n_L(1))) = \ell_A(\mathbf{p} \vee \mathbf{p}^n_L(w^n_L(i-1), w^n_L(1))) \ge 1.$$

This shows (6.5) in this case.

Suppose that $(w(1), w(k)) \in L^n \times L^n$. Set $w(1) = w^n_L(j)$ and $w(k) = w^n_L(i)$. If $j < i$, then we consider $(w(k), \ldots, w(1))$ in place of $(w(1), \ldots, w(k))$. In this way, we may assume that $j \ge i$ without loss of generality. Let $\widetilde{\mathbf{p}} = \mathbf{p}^n_L\left(w^n_L(3^n), w^n_L(j+1)\right) \vee \mathbf{p} \vee \mathbf{p}^n_L\left(w^n_L(j-1), w^n_L(1)\right)$. Since $\widetilde{\mathbf{p}} \in \mathcal{CH}_n(T, p_{\swarrow})$, we have

$$\frac{3^n - j}{3^n} + \ell_A(\mathbf{p}) + \frac{i-1}{3^n} = \ell_A(\widetilde{\mathbf{p}}) \geq 1$$

This immediately implies (6.5) in this case. □

Lemma 6.10 *Let $X, Y \subseteq W_n$. Assume that*

$$\inf\{\ell_X(\mathbf{p})|\mathbf{p} \in \mathcal{CH}_n(T, B)\} \geq 1$$

and that

$$\inf\{\ell_Y(\mathbf{p})|\mathbf{p} \in \mathcal{CH}_n(T, p_{\swarrow})\} \geq 1.$$

Define $A = \nwarrow \cdot X \cup \uparrow \cdot Y$. If $\mathbf{p} = (w(1), \dots, w(k)) \in \mathcal{CH}_{n+1}(T, B_{\nwarrow})$ and $\{w(1), \dots, w(k)\} \subseteq \{\nwarrow, \uparrow\} \cdot W_n$, then

$$\ell_A(\mathbf{p}) \geq \frac{1}{3}.$$

Proof Let $w(i) = s(i)v(i)$, where $s(i) \in \{\nwarrow, \uparrow\}$ and $v(i) \in W_n$.

First assume that $s(1) = \nwarrow$. Then there exist $j_1, j_2, \dots, j_{2p+2}$ which satisfies the following three conditions (C1), (C2) and (C3):

(C1) $j_1 = 1$, $j_{2p+2} = k + 1$ and $j_1 < j_2 < \dots < j_{2p+2}$
(C2) $s(i) = \nwarrow$ for $i = j_{2q-1}, \dots, j_{2q} - 1$ and $q = 1, \dots, p + 1$
(C3) $s(i) = \uparrow$ for $i = j_{2q}, \dots, j_{2q+1} - 1$ and $q = 1, \dots, p$.

Set $\mathbf{p}_{1,q} = \big(w(j_{2q-1}), \dots, w(j_{2q} - 1)\big)$ and $\widetilde{\mathbf{p}}_{1,q} = \big(v(j_{2q-1}), \dots, v(j_{2q} - 1)\big)$. Since $\big(v(j_{2q-1}), v(j_{2q} - 1)\big) \in (T^n \times R^n) \cup (R^n \times R^n) \cup (R^n \times B^n)$, Lemma 6.9 and its variants explained in the remark imply

$$\ell_A\left(\mathbf{p}_{1,q}\right) = \frac{1}{3}\ell_X\left(\widetilde{\mathbf{p}}_{1,q}\right) \geq \frac{1}{3}\ell_X\left(\mathbf{p}^n_R\left(v\left(j_{2q-1}\right), v\left(j_{2q} - 1\right)\right)\right). \tag{6.6}$$

Set $\mathbf{p}_{2,q} = \left(w\left(j_{2q}\right), \dots, w\left(j_{2q+1} - 1\right)\right)$ and $\widetilde{\mathbf{p}}_{2,q} = \left(v\left(j_{2q}\right), \dots, v\left(j_{2q+1} - 1\right)\right)$. Since $\left(v\left(j_{2q}\right), v\left(j_{2q+1} - 1\right)\right) \in L^n \times L^n$, Lemma 6.9 shows that

$$\ell_A\left(\mathbf{p}_{2,q}\right) = \frac{1}{3}\ell_Y\left(\widetilde{\mathbf{p}}_{2,q}\right) \geq \frac{1}{3}\ell_Y\left(\mathbf{p}^n_L\left(v\left(j_{2q}\right), v\left(j_{2q+1} - 1\right)\right)\right). \tag{6.7}$$

Note that for any $i = 1, \dots, 3^n$, there exists $l = 1, 2, \dots, 2q + 1$ such that $I^n_{\updownarrow}(v(j_l)) \leq i \leq I^n_{\updownarrow}(v(j_{l+1} - 1))$ or $I^n_{\updownarrow}(v(j_l)) \geq i \geq I^n_{\updownarrow}(v(j_{l+1} - 1))$. Hence

$$\sum_{q=1}^{p+1} \ell_X\left(\mathbf{p}^n_R\left(v\left(j_{2q-1}\right), v\left(j_{2q} - 1\right)\right)\right) + \sum_{q=1}^{p} \ell_Y\left(\mathbf{p}^n_L\left(v\left(j_{2q}\right), v\left(j_{2q+1} - 1\right)\right)\right) \geq 1.$$

Combining this with (6.6) and (6.7), we obtain

$$\ell_A(\mathbf{p}) = \sum_{q=1}^{p+1} \ell_A\left(\mathbf{p}_{1,q}\right) + \sum_{q=1}^{p} \ell_A\left(\mathbf{p}_{2,q}\right) \geq \frac{1}{3}.$$

Thus we have shown the desired statement in the case when $s(1) = \nwarrow$.

If $s(1) = \uparrow$, slight modification of the above arguments yields the lemma as well. □

Definition 6.11 Define $\pi : W_* \to W_*$ by

$$\pi(w) = \begin{cases} w & \text{if } \mathrm{Re}\, F_w(0) \leq 0, \\ \varphi_{\leftrightarrow}(w) & \text{if } \mathrm{Re}\, F_w(0) > 0. \end{cases}$$

For $\mathbf{p} = (w(1), \ldots, w(k)) \in \mathcal{CH}_n$, we define $\pi_n(\mathbf{p}) = (\pi(w(1)), \ldots, \pi(w(k))$. Also define $\xi : W_* \to W_*$ by

$$\xi(w) = \begin{cases} w & \text{if } \mathrm{Re}\, F_w(0) \leq \mathrm{Im}\, F_w(0), \\ \varphi_{\leftrightarrow}(\rho(w)) & \text{if } \mathrm{Re}\, F_w(0) > \mathrm{Im}\, F_w(0). \end{cases}$$

For $\mathbf{p} = (w(1), \ldots, w(k)) \in \mathcal{CH}_n$, we define $\xi_n(\mathbf{p}) = (\xi(w(1)), \ldots, \xi(w(k))$.

By the symmetry of $\Updownarrow_n$, $\Uparrow_n$ and $\swarrow\!\!\swarrow_n$ given in Lemma 6.7, we have the following lemma.

Lemma 6.12 (1) $\pi_n : \mathcal{CH}_n \to \mathcal{CH}_n$, $\ell_{\Updownarrow_n}(\pi_n(\mathbf{p})) = \ell_{\Updownarrow_n}(\mathbf{p})$ *and* $\ell_{\Uparrow_n}(\pi_n(\mathbf{p})) = \ell_{\Uparrow_n}(\mathbf{p})$.

(2) $\xi_n : \mathcal{CH}_n \to \mathcal{CH}_n$ *and* $\ell_{\swarrow\!\!\swarrow_n}(\xi_n(\mathbf{p})) = \ell_{\swarrow\!\!\swarrow_n}(\mathbf{p})$.

Lemma 6.13 *Suppose that*

$$\inf\{\ell_{\Updownarrow_n}(\mathbf{p}) | \mathbf{p} \in \mathcal{CH}_n(T, B)\} \geq 1 \tag{6.8}$$

and

$$\inf\{\ell_{\Uparrow_n}(\mathbf{p}) | \mathbf{p} \in \mathcal{CH}_n(T, \{p_{\swarrow}, p_{\searrow}\}) \geq 1. \tag{6.9}$$

If $\mathbf{p} = (w(1), \ldots, w(k)) \in \mathcal{CH}_{n+1}(T, B_{\nwarrow} \cup B_{\nearrow})$ *and* $\{w(i)\}_{i=1}^k \subseteq \{\nwarrow, \uparrow, \nearrow\}\cdot W_n$, *then* $\ell_{\Updownarrow_{n+1}}(\mathbf{p}) \geq 1/3$.

Proof Replacing $\mathbf{p}$ by $\pi_{n+1}(\mathbf{p})$, we may assume that $w(1), \ldots, w(k) \in \{\nwarrow, \uparrow\}\cdot W_n$ and $w(k) \in \nwarrow \cdot B^n$ without loss of generality. Set $X = \Updownarrow_n$ and $Y = \Uparrow_n$. Then the assumptions (6.8) and (6.9) of Lemma 6.10 follows. Hence $\ell_{\Updownarrow_{n+1}}(\mathbf{p}) \geq 1/3$. □

Lemma 6.14 *Suppose that* (6.8) *holds and that*

$$\inf\{\ell_{\Swarrow_n}(\mathbf{p})|\mathbf{p} \in \mathcal{CH}_n(T \cup R, p_{\swarrow}) \geq 1. \tag{6.10}$$

Let $\mathbf{p} = (w(1), \ldots, w(k)) \in \mathcal{CH}_{n+1}(T, B_{\swarrow})$. *If* $\{w(1), \ldots, w(k)\} \subseteq \nwarrow, \uparrow, \nearrow \cdot W_n$, *then* $\ell_{\Swarrow_{n+1}}(\mathbf{p}) \geq 1/3$.

Proof First assume that $\{w(1), \ldots, w(k)\} \subseteq \{\nwarrow, \uparrow\} \cdot W_n$. By (6.3), applying Lemma 6.10 with $X = \Updownarrow_n$ and $Y = \Swarrow_n$, we have $\ell_{\Swarrow_{n+1}}(\mathbf{p}) \geq 1/3$.

Next, suppose that $w(i) \in \nearrow \cdot W_n$ for some i. Let

$$i_* = \max\{i|w(i) \in \nearrow \cdot W_n\} + 1.$$

and let

$$j_* = \min\{j|w(j) \in \nwarrow \cdot W_n, j \geq i_*\} - 1.$$

Then, for $i = \{i_*, \ldots, j_*\}$, there exists $v(i) \in W_n$ such that $w(i) = \uparrow \cdot v(i)$. Define

$$\mathbf{p}_* = (\uparrow \cdot \xi (v (i_*)), \uparrow \cdot \xi (v (i_* + 1)), \ldots, \uparrow \cdot \xi (v (j_*)), w (j_* + 1), \ldots, w (k)).$$

By (6.3) and Lemma 6.12,

$$\begin{aligned}
\ell_{\Swarrow_{n+1}} (w (i_*), \ldots, w (j_*)) &= \frac{1}{3} \ell_{\Swarrow_n} (v (i_*), \ldots, v (j_*)) \\
&= \frac{1}{3} \ell_{\Swarrow_n} (\xi (v (i_*)), \ldots, \xi (v (j_*))) \\
&= \ell_{\Swarrow_{n+1}} (\uparrow \cdot \xi (v (i_*)), \ldots, \uparrow \cdot \xi (v (j_*))).
\end{aligned}$$

Hence $\ell_{\Swarrow_{n+1}}(\mathbf{p}) \geq \ell_{\Swarrow_{n+1}}(\mathbf{p}_*)$. Let $\mathbf{p}_* = (w_*(1), w_*(2), \ldots, w_*(l))$. Then $w^*(i) \in \{\nwarrow, \uparrow\} \cdot W_n$ for any $i = 1, \ldots, l$. Now replacing $\mathbf{p}$ by $\mathbf{p}_*$, we are exactly in the first case and hence the desired inequality is satisfied. □

Lemma 6.15 (6.8), (6.9) *and* (6.10) *hold for any* $n \geq 0$.

Proof We use induction on n. Obviously (6.8), (6.9) and (6.10) holds for $n = 0$ since $\Updownarrow_n$, $\Uparrow_n$ and $\Swarrow_n$ are the empty sets. Assume that (6.8), (6.9) and (6.10) are true for $n = m$.

First we show (6.8) holds for $n = m + 1$. Let $\mathbf{p} = (w(1), \ldots, w(k)) \in \mathcal{CH}_{m+1}(T, B)$. Note that by Lemma 6.7-(1), $\pi_{n+1}(\mathbf{p}) \in \mathcal{CH}_{m+1}(T, B)$ and $\ell_{\Updownarrow_{m+1}}(\mathbf{p}) = \ell_{\Updownarrow_{m+1}}(\pi_{n+1}(\mathbf{p}))$. Hence replacing $\mathbf{p}$ by $\pi_{n+1}(\mathbf{p})$, we may assume that $w(i) \in \{\nwarrow, \uparrow, \leftarrow, \swarrow, \downarrow\} \cdot W_m$ for any $i = 1, \ldots, k$ without loss of generality. Set $w(i) = s(i)v(i)$, where $s(i) \in \{\nwarrow, \uparrow, \leftarrow, \swarrow, \downarrow\}$ and $v(i) \in W_m$. We may choose i_1, i_2, i_3 and i_4 which satisfies $i_1 < i_2 < i_3 < i_4$ and the following tree conditions (a1), (b1) and (c1):

(a1) $s(1), \ldots, s(i_1) \in \{\nwarrow, \uparrow\}$, $(v(1), \ldots, v(i_1)) \in \mathcal{CH}_m(T, B_{\swarrow})$,

(b1) $s(i) = \leftarrow$ for $i = i_2, \ldots, i_3$, $(v(i_2), \ldots, v(i_3)) \in \mathcal{CH}_m(T, B)$,
(c1) $s(i_4), \ldots, s(k) \in \{\swarrow, \downarrow\}$, $w_*(i_4) \in \swarrow \cdot T^m$.

Let $\mathbf{p}_1 = (w(1), \ldots, w(i_1))$. Then by the induction hypothesis, we may apply Lemma 6.13 and see that $\ell_{\Updownarrow_{m+1}}(\psi_1) \geq 1/3$.

Let $\mathbf{p}_2 = (w(i_2), \ldots, w(i_3))$. Since $(v(i_2), \ldots, v(i_3)) \in \mathcal{CH}_m(T, B)$, the induction hypothesis implies

$$\ell_{\Updownarrow_{m+1}}(\mathbf{p}_2) = \frac{1}{3}\ell_{\Updownarrow_m}(v(i_2), \ldots, v(i_3)) \geq \frac{1}{3}.$$

Set $\mathbf{p}_3 = (w(i_3), \ldots, w(k))$ and $\widetilde{\mathbf{p}}_3 = \left(\varphi_\updownarrow (w(k)), \varphi_\updownarrow (w(k-1)), \ldots, \varphi_\updownarrow (w(i_3))\right)$. Then $\ell_{\Updownarrow_{m+1}}(\mathbf{p}_3) = \ell_{\Updownarrow_{m+1}}(\widetilde{\mathbf{p}}_3)$. As $\mathbf{p}_1$, we may apply Lemma 6.13 to $\widetilde{\mathbf{p}}_3$ and obtain $\ell_{\Updownarrow_{m+1}}(\widetilde{\mathbf{p}}_3) \geq 1/3$. Combining all the estimates on $\ell_{\Updownarrow_{m+1}}(\mathbf{p}_i)$ for $i = 1, 2, 3$, we have $\ell_{\Updownarrow_{m+1}}(\mathbf{p}) \geq 1$.

Secondly, we show that (6.9) holds for $n = m + 1$. Let $\mathbf{p} = (w(1), \ldots, w(k)) \in \mathcal{CH}_{m+1}(T, \{p_\swarrow, p_\searrow\})$. As in the first case, we may assume that $w(i) \in \{\nwarrow, \uparrow, \leftarrow, \swarrow, \downarrow\}$ for any $i = 1, \ldots, k$ without loss of generality. Set $w(i) = s(i)v(i)$, where $s(i) \in \{\nwarrow, \uparrow, \leftarrow, \swarrow, \downarrow\}$ and $v(i) \in W_m$. We may choose i_1, i_2, i_3 and i_4 which satisfies $i_1 < i_2 < i_3 < i_4$ and the following tree conditions (a2), (b2) and (c2):

(a2) $s(1), \ldots, s(i_1) \in \{\nwarrow, \uparrow\}$, $(v(1), \ldots, v(i_1)) \in \mathcal{CH}_m(T, B_\swarrow)$,
(b2) $s(i) = \leftarrow$ for $i = i_2, \ldots, i_3$, $(v(i_2), \ldots, v(i_3)) \in \mathcal{CH}_m(T, B)$,
(c2) $s(i) = \swarrow$ for $i = i_4, \ldots, k$. $(v(i_4), \ldots, v(k)) \in \mathcal{CH}_m(T \cup R, p_\swarrow)$.

Define $\mathbf{p}_1$, $\mathbf{p}_2$ and $\mathbf{p}_3$ as in the first case. Then using the same discussion as in the first case, we obtain $\ell_{\Uparrow_{m+1}}(\mathbf{p}_j) \geq 1/3$ for $j = 1, 2$. Since $(v(i_4), \ldots, v(k)) \in \mathcal{CH}_m(T \cup R, p_\swarrow)$, The induction hypothesis and Lemma 6.14 yield that $\ell_{\Swarrow_m}((v(i_4), \ldots, v(k))) \geq 1$. By (6.2), it follows that

$$\ell_{\Uparrow_{m+1}}(\mathbf{p}_3) = \frac{1}{3}\ell_{\Swarrow_m}((v(i_4), \ldots, v(k))) \geq 1/3.$$

Thus, we have shown that $\ell_{\Uparrow_{m+1}}(\mathbf{p}) \geq 1$.

Finally we show that (6.10) holds for $n = m + 1$. Let $\mathbf{p} = (w(1), \ldots, w(k)) \in \mathcal{CH}_{m+1}(T \cup R, p_\swarrow)$. Note that $\xi_{m+1}(\mathbf{p}) \in \mathcal{CH}_{m+1}(T, p_\swarrow)$ and $\ell_{\Swarrow_{m+1}}(\mathbf{p}) = \ell_{\Swarrow_{m+1}}(\xi_{m+1}(\mathbf{p}))$. Hence replacing $\mathbf{p}$ by $\xi_{m+1}(\mathbf{p})$, we may assume that $w(i) \in \{\nwarrow, \uparrow, \nearrow, \leftarrow, \swarrow\} \cdot W_m$ for any $i = 1, \ldots, k$ without loss of generality (Fig. 5). Set $w(i) = s(i)v(i)$, where $s(i) \in \{\nwarrow, \uparrow, \nearrow, \leftarrow, \swarrow\}$ and $v(i) \in W_m$. We may choose i_1, i_2, i_3 and i_4 which satisfies $i_1 < i_2 < i_3 < i_4$ and the following tree conditions (a3), (b3) and (c3):

(a3) $s(1), \ldots, s(i_1) \in \{\nwarrow, \uparrow\}$, $(v(1), \ldots, v(i_1)) \in \mathcal{CH}_m(T, B_\swarrow)$,
(b3) $s(i) = \leftarrow$ for $i = i_2, \ldots, i_3$, $(v(i_2), \ldots, v(i_3)) \in \mathcal{CH}_m(T, B)$,
(c3) $s(i) = \swarrow$ for $i = i_4, \ldots, k$. $(v(i_4), \ldots, v(k)) \in \mathcal{CH}_m(T \cup R, p_\swarrow)$.

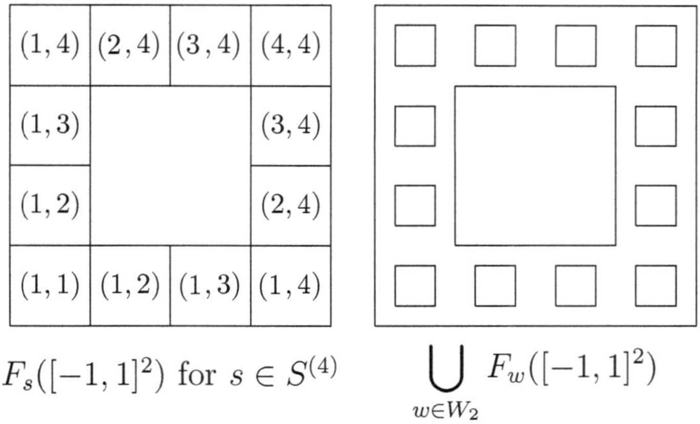

Fig. 5 Construction of $K^{(4)}$

Define $\mathbf{p}_1, \mathbf{p}_2$ and $\mathbf{p}_3$ as in the above two cases. Then by the induction hypothesis and (6.3), it follows that $\ell_{\Swarrow_{m+1}}(\mathbf{p}_j) \geq 1/3$ for $j = 2, 3$. Furthermore, Lemma 6.14 implies $\ell_{\Swarrow_{m+1}}(\mathbf{p}_1) \geq 1/3$. Hence we have $\ell_{\Swarrow_{m+1}}(\mathbf{p}) \geq 1$.

Thus we have obtained (6.8), (6.9) and (6.10) for $n = m + 1$. □

Proof of Theorem 6.4 Since $A_n \subseteq \Updownarrow_n$, $\ell_{A_n}(\mathbf{p}) \geq 1$ for any $\mathbf{p} \in \mathcal{CH}_n(T, B)$. By the fact that $\Leftrightarrow_n = \rho(\Updownarrow_n)$, it follows that $\ell_{\Leftrightarrow_n}(\mathbf{p}) \geq 1$ for any $\mathbf{p} \in \mathcal{CH}_n(L, R)$. Hence $\ell_{A_n}(\mathbf{p}) \geq 1$ for any $\mathbf{p} \in \mathcal{CH}_n(L, R)$. Thus A_n is invisible. By Lemma 6.7-(1), it follows that A_n is $+$-invariant.

Lemma 6.3 shows that $8^n - \#(\Updownarrow_n) = \#(W_n \backslash \Updownarrow_n) = \alpha_n$. Since $W_n \backslash \Updownarrow_n \subseteq W_n \backslash A_n \subseteq (W_n \backslash \Updownarrow_n) \cup (W_n \backslash \Leftrightarrow_n)$, we have $\alpha_n \leq 8^n - \#(A_n) \leq 2\alpha_n$. □

7 Generalized Sierpinski Carpet

The idea of invisible sets can be exploited for the generalized Sierpinski carpets. We will present results for a special class of the generalized Sierpinski carpet in this section. We fix $N \geq 3$. The complex plane $\mathbb{C}$ is identified with $\mathbb{R}^2$ in the usual manner.

Definition 7.1

(1) For any $(i, j) \in \{1, \ldots, N\}^2$, we define $J_{(i,j)} = [-1+2(i-1)/N, -1+2i/N] \times [-1 + 2(j-1)/N, -1 + 2j/N]$ and $F_{(i,j)} : \mathbb{R}^2 \to \mathbb{R}^2$ by $F_{(i,j)}(x, y) = (x/N + a_{(i,j)}, y/N + b_{(i,j)})$, where $a_{(i,j)} = -1 + (2i-1)/N$ and $b_{(i,j)} = -1 + (2j-1)/N$.

(2) Define $S^{(N)} = \{(i, j) | (i, j) \in \{1, \ldots, N\}^2, i \in \{1, N\}$ or $j \in \{1, N\}\}$. Let $K^{(N)}$ be the unique compact set which satisfies

$$K^{(N)} = \bigcup_{(i,j)\in S^{(N)}} F_{(i,j)}\left(K^{(N)}\right).$$

When $N = 3$, $K^{(3)}$ is the Sierpinski carpet.

Proposition 7.2 $\#(S^{(N)}) = 4N-4$ *and* $\dim_H\left(K^{(N)}, d_E\right) = \dfrac{\log(4N-4)}{\log N}$, *where* d_E *is the restriction of the Euclidean metric.*

In the following, we occasionally omit N in $S^{(N)}$ and $K^{(N)}$ and write them S and K respectively. Also we use W_m, W_* and Σ in place of $W_m\left(S^{(N)}\right)$, $W_*\left(S^{(N)}\right)$ and $\Sigma\left(S^{(N)}\right)$.

Definition 7.3 Let $A \subseteq W_m$.

(1) Let $A \subseteq W_m$. For $\mathbf{p} = (w(1), \ldots, w(k)) \in \mathcal{CH}_m$, define

$$\ell_A(\mathbf{p}) = \frac{\#(\{i|i = 1, \ldots, k, w(i) \notin A\})}{N^m}.$$

(2) A is called an invisible set if and only if

$$\inf_{\mathbf{p}\in\mathcal{CH}_m(T,B)\cup\mathcal{CH}_m(L,R)} \ell_A(\mathbf{p}) \geq 1,$$

where T, B, L and R are the same as in the last tree sections.

We also define the notion of $+$-invariance exactly same as in the previous sections. Then the analogous results as Theorems 4.5 and 5.3 hold. As a consequence we have the following statement.

Theorem 7.4 *Let $A \subset W_m$ be a $+$-invariant invisible set. Then*

$$\dim_C(K^{(N)}, d_E) \leq \frac{\log((4N-4)^m - \#(A))}{m \log N}.$$

A procedure which is similar to that in Sect. 6 produces a sequence of invisible sets. We assume $N \geq 4$ hereafter. The maps $\varphi_{\leftrightarrow}$, $\varphi_{\updownarrow}$ and ρ from W_* to itself associated with symmetries can be defined in the same way as in the last section.

Definition 7.5 Define $\Updownarrow_n \subseteq W_n$ and $\searrow_n \subseteq W_n$ inductively by

$$\begin{aligned}\Updownarrow_{n+1} = {}& \{(i,j)|(i,j) \in S.i \in \{1,N\}\}\cdot \Updownarrow_n \\ & \cup (2,1)\cdot \searrow_n \cup (2,N)\cdot \nearrow_n \cup (N-1,1)\cdot \swarrow_n \cup (N-1,N)\cdot \nwarrow_n \\ & \cup \{(i,j)|(i,j) \in S, j \in \{1,N\}, i \notin \{1,2,N-1,N\}\}\cdot W_n,\end{aligned}$$

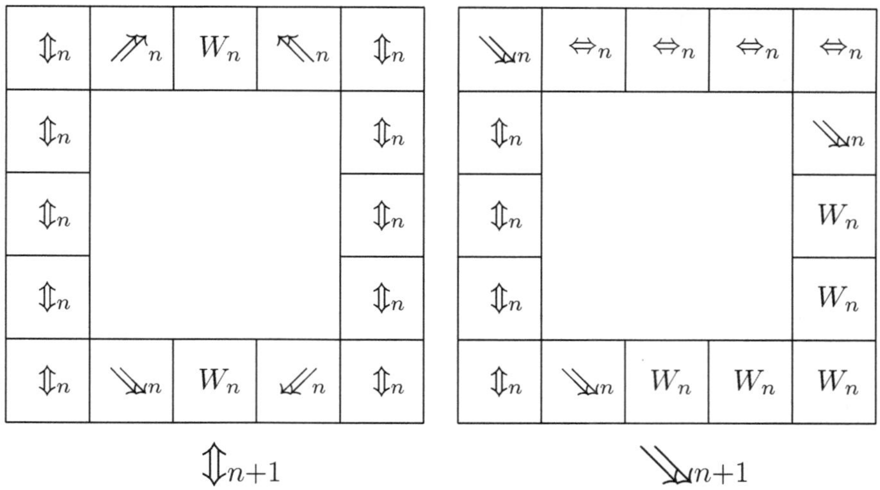

Fig. 6 Construction of $\Updownarrow_n$ and $\Searrow_n$ for $N = 5$

$$\begin{aligned}\Searrow_{n+1} &= \{(1, N), (2, 1), (N-1, N)\}\cdot \Searrow_n \\ &\quad \cup \{(1, j) | j = 1, \dots, N-1\}\cdot \Updownarrow_n \cup \{(i, N) | i = 2, \dots, N\}\cdot \Leftrightarrow_n \\ &\qquad \cup \{(1, j) | j = 3, \dots, N\}\cdot W_n \cup \{(i, N) | i = 1, \dots, N-2\}\cdot W_n,\end{aligned}$$

$\Updownarrow_0 = \emptyset$ and $\Searrow_0 = \emptyset$, where $\Leftrightarrow_n = \rho(\Updownarrow_n)$, $\Swarrow_n = \varphi_{\leftrightarrow}(\Searrow_n)$, $\Nearrow_n = \varphi_{\updownarrow}(\Searrow_n)$ and $\Nwarrow_n = \varphi_{\leftrightarrow}(\Nearrow_n)$ (Fig. 6).

By the above definition, it follows that

$$\begin{aligned} x_{n+1} &= 2N x_n + 4 y_n + 2(N-4)(4N-4)^n \\ y_{n+1} &= 2(N-1) x_n + 3 y_n + (2N-5)(4N-4)^n, \end{aligned}$$

where $x_n = \#(\Updownarrow_n)$ and $y_n = \#(\Searrow_n)$. Define

$$\tau_N = \sqrt{4N^2 + 20N - 23}.$$

Then we have

$$\begin{aligned} x_n = (4N-4)^n &- \Big(\frac{2N+5}{2\tau_N} + \frac{1}{2}\Big)\Big(\frac{2N+3+\tau_N}{2}\Big)^n \\ &+ \Big(\frac{2N+5}{2\tau_N} - \frac{1}{2}\Big)\Big(\frac{2N+3-\tau_N}{2}\Big)^n \end{aligned}$$

The same discussion as in the last section shows

$$\inf_{\mathbf{p} \in \mathcal{CH}_n(T,B)} \ell_{\Updownarrow_n}(\mathbf{p}) \geq 1.$$

Hence we obtain the counterpart of Theorem 6.4.

Theorem 7.6 *Let* $A_n = \Updownarrow_n \cap \Leftrightarrow_n$. *Then* A_n *is* $+$*-invariant invisible set and there exist* $c_1, c_2 > 0$ *such that*

$$c_1 \left(\frac{2N+3+\tau_N}{2}\right)^n \leq (4N-4)^n - \#(A_n) \leq c_2 \left(\frac{2N+3+\tau_N}{2}\right)^n$$

for sufficiently large n.

As an corollary, we obtain the following estimate of the conformal dimension of $K^{(N)}$. The lower estimate is shown by applying [MT10, Example 4.1.9].

Corollary 7.7

$$\begin{aligned}\frac{\log(2N)}{\log N} \leq \dim_C\left(K^{(N)}, d_E\right) &\leq \frac{\log\frac{2N+3+\tau_N}{2}}{\log N} \\ &< \frac{\log(4N-4)}{\log N} = \dim_H\left(K^{(N)}, d_E\right).\end{aligned}$$

Remark

$$2N+3 \leq \frac{2N+3+\tau_n}{2} < 2N+4.$$

References

[BM00] Bonk, M., Merenkov, S.: Rigidity of square Sierpinski carpets, preprint.

[KL04] Keith, S., Laakso, T.: Conformal Assouad dimension and modulus. Geom. Funct. Anal. **14**, 1278–1321 (2004)

[Kig00] Kigami, J.: Resistance Forms, Quasisymmetric Maps and Heat Kernel Estimates. Memoirs of the American Mathematical Society. American Mathematical Society, Providence, RI (2012)

[Kle00] Kleiner, B.: Personal communication

[Kig01] Kigami, J.: Analysis on Fractals, Cambridge Tracts in Mathematics. Cambridge University Press, Cambridge (2001)

[Kig09] Kigami, J.: Volume doubling measures and heat kernel estimates on self-similar sets. Mem. Am. Math. Soc. **199**(932), XXX (2009)

[MT10] Mackay, J.M., Tyson, J.T.: Conformal Dimension, Theory and Application. University Lecture Series. American Mathematical Society, Providence, RI (2010)

[Pan89] Pansu, P.: Dimension conforme et sphère á l'infini des variétés à courbure nègative. Ann. Acad. Sci. Fenn. Ser. A I Math. **14**, 177–212 (1989)

[TV80] Tukia, P., Väisälä, J.: Quasisymmetric embeddings of metric spaces. Ann. Acad. Sci. Fenn. Ser. A I Math. **5**, 97–114 (1980)

[TW06] Tyson, J.T., Wu, J.-M.: On quasiconformal dimensions of self-similar fractals. Rev. Mat. Iberoamericana **22**, 205–258 (2006)

Recent Progress on Dimensions of Projections

Pertti Mattila

Abstract This is a survey on recent progress on the question: how do projections effect dimensions generically? I shall also discuss briefly dimensions of plane sections.

Keywords Hausdorff dimension · Projections · Heisenberg group

2000 Mathematics subject Classification Primary 28A75

1 Introduction

I give a survey on the question how projection-type transformations change dimensions of sets. I shall mainly discuss Hausdorff dimension but packing and Minkowski dimensions will also be briefly looked at. First I review classical Marstrand's projection theorem and give Kaufman's proof for it. Then I present recent partial analogues of Marstrand's projection theorem in Heisenberg groups due to Balogh, Durand-Cartagena, Fässler, Tyson and myself. After that I discuss generalized projections of Peres and Schlag. In Heisenberg groups and other situations one encounters small, restricted, families of transformations. I review recent results of E. Järvenpää, M. Järvenpää, Ledrappier, Leikas and Keleti and of Fässler and Orponen on them. Then I mention briefly older results of Falconer and Howroyd and recent results of Fässler and Orponen on packing and Minkowski dimensions. For them one has generally only inequalities, but Falconer and Howroyd proved also a constancy theorem. I present this and some recent constancy theorem of Fässler and Orponen on Hausdorff dimension for a particular restricted family of projections. Finally we shall have

The author was supported by the Academy of Finland.

P. Mattila (✉)
Department of Mathematics and Statistics, University of Helsinki, P.O. Box 68,
00014 Helsinki, Finland
e-mail: pertti.mattila@helsinki.fi

D.-J. Feng and K.-S. Lau (eds.), *Geometry and Analysis of Fractals*,
Springer Proceedings in Mathematics & Statistics 88,
DOI 10.1007/978-3-662-43920-3_10

a look at Marstrand's classical result on Hausdorff dimension of plane sections and recent analogues of it in Heisenberg groups.

Background on this topic can be found in the books [Fa85, Map2]. Recent related surveys are [Map3, Map4].

I would like to thank Katrin Fässler and Tuomas Orponen for several useful comments.

2 Marstrand's Projection Theorem

Marstrand proved in 1954 the following theorem in [Ma54]:

Theorem 2.1 *Suppose* $A \subset \mathbb{R}^2$ *is a Borel set and denote by* $P_\theta, \theta \in [0, \pi)$, *the orthogonal projection onto the line* $L_\theta = \{t(\cos\theta, \sin\theta) : t \in \mathbb{R}\}$: $P_\theta(x, y) = (\cos\theta)x + (\sin\theta)y$.

(1) *If* $\dim A \leq 1$, *then* $\dim P_\theta(A) = \dim A$ *for almost all* $\theta \in [0, \pi)$.

(2) *If* $\dim A > 1$, *then* $\mathcal{L}^1(P_\theta(A)) > 0$ *for almost all* $\theta \in [0, \pi)$.

Here dim means Hausdorff dimension and $\mathcal{L}^1$ is the one-dimensional Lebesgue measure.

Marstrand's original proof was based on the definition and basic properties of Hausdorff measures. Kaufman used in [Ka68] potential theoretic and Fourier analytic methods to give a different proof. To present Kaufman's proof let us first look at the required preliminaries.

The Hausdorff dimension of a Borel set $A \subset \mathbb{R}^n$ can be determined by looking at the behaviour of Borel measures μ with compact support $\operatorname{spt}\mu \subset A$. Denote the family of such measures μ with $0 < \mu(A) < \infty$ by $\mathcal{M}(A)$. By the well-known Frostman's lemma $\dim A$ is the supremum of the numbers s such that there exists $\mu \in \mathcal{M}(A)$ for which

$$\mu(B(x, r)) \leq r^s \quad \text{for } x \in \mathbb{R}^n. \tag{2.1}$$

This is easily transformed into an integral condition. Let

$$I_s(\mu) = \iint |x - y|^{-s} d\mu x d\mu y$$

be the s-energy of μ. Then $\dim A$ is the supremum of the numbers s such that there exists $\mu \in \mathcal{M}(A)$ for which

$$I_s(\mu) < \infty. \tag{2.2}$$

For a fixed μ (2.1) and (2.2) may not be equivalent, but they are closely related: (2.2) implies that the restriction of μ to a suitable set with positive μ measure satisfies (2.1), and (2.1) implies that μ satisfies (2.2) for any $s' < s$. Defining the Riesz kernel $k_s, k_s(x) = |x|^{-s}$, the s-energy of μ can written as

$$I_s(\mu) = \int k_s * \mu d\mu.$$

For $0 < s < n$ the Fourier transform of k_s is (in the sense of distributions) $\widehat{k_s} = c(s,n)k_{n-s}$. Thus we have by Plancherel's theorem

$$I_s(\mu) = \int \widehat{k_s}|\widehat{\mu}|^2 = c(s,n)\int |x|^{s-n}|\widehat{\mu}(x)|^2 dx.$$

Consequently, dim A is the supremum of the numbers $s \leq n$ such that there exists $\mu \in \mathcal{M}(A)$ for which

$$\int |x|^{s-n}|\widehat{\mu}(x)|^2 dx < \infty. \tag{2.3}$$

To prove (1) of Theorem 2.1 let $0 < s < \dim A$ and choose by (2.2) a measure $\mu \in \mathcal{M}(A)$ such that $I_s(\mu) < \infty$. Let $\mu_\theta \in \mathcal{M}(P_\theta(A))$ be the push-forward of μ under P_θ: $\mu_\theta(B) = \mu(P_\theta^{-1}(B))$. Then

$$\begin{aligned}\int_0^\pi I_s(\mu_\theta)d\theta &= \int_0^\pi \iint |P_\theta(x-y)|^{-s} d\mu x d\mu y d\theta \\ &= \iint \int_0^\pi |P_\theta(\tfrac{x-y}{|x-y|})|^{-s} d\theta |x-y|^{-s} d\mu x d\mu y \\ &= c(s)I_s(\mu) < \infty,\end{aligned}$$

where for $v \in S^1$, $c(s) = \int_0^\pi |P_\theta(v)|^{-s} d\theta < \infty$ as $s < 1$. Referring again to (2.2) we see that dim $P_\theta(A) \geq s$ for almost all $\theta \in [0,\pi)$. By the arbitrariness of $s, 0 < s < \dim A$, we obtain dim $P_\theta(A) \geq \dim A$ for almost all $\theta \in [0,\pi)$. The opposite inequality follows from the fact that the projections P_θ are Lipschitz.

To prove (2) choose by (2.3) a measure $\mu \in \mathcal{M}(A)$ such that $\int |x|^{-1}|\widehat{\mu}(x)|^2 dx < \infty$. Directly from the definition of the Fourier transform we see that $\widehat{\mu_\theta}(t) = \widehat{\mu}(t(\cos\theta, \sin\theta))$ for $t \in \mathbb{R}, \theta \in [0,\pi)$. Integrating in polar coordinates we obtain

$$\int_0^\pi \int_{-\infty}^\infty |\widehat{\mu_\theta}(t)|^2 dt d\theta = 2\int_0^\pi \int_0^\infty |\widehat{\mu}(t(\cos\theta, \sin\theta))|^2 dt d\theta$$
$$= \int |x|^{-1}|\widehat{\mu}(x)|^2 dx < \infty.$$

Thus for almost all $\theta \in [0, \pi)$, $\widehat{\mu_\theta} \in L^2(\mathbb{R})$ which means that μ_θ is absolutely continuous with L^2-density and hence $\mathcal{L}^1(P_\theta(A)) > 0$.

It is not difficult to prove (2) without Fourier transform: application of Fubini's theorem and some simple estimates yield

$$\int_0^\pi \int \liminf_{\delta\to 0} \delta^{-1}\mu_\theta(x-\delta, x+\delta) d\mu_\theta x d\theta \le C I_1(\mu), \tag{2.4}$$

from which (2) follows by standard results on differentiation of measures.

Theorem 2.1 has the following generalization:

Theorem 2.2 *Suppose $A \subset \mathbb{R}^2$ is a Borel set.*

(1) *If $0 \le t \le \dim A \le 1$, then*

$$\dim\{\theta \in [0, \pi) : \dim P_\theta(A) < t\} \le t.$$

(2) *If $\dim A > 1$, then*

$$\dim\{\theta \in [0, \pi) : \mathcal{L}^1(P_\theta(A)) = 0\} \le 2 - \dim A.$$

Part (1) was proved by Kaufman with a similar method as above; one uses Frostman's lemma also for the exceptional set of directions. Part (2) was proved by Falconer with a Fourier-analytic method.

To formulate the higher dimensional version of Theorem 2.2, denote by $G(n, m)$ the Grassmannian manifold of linear m-dimensional subspaces of $\mathbb{R}^n$. For $V \in G(n, m)$, let

$$P_V : \mathbb{R}^n \to V$$

be the orthogonal projection. As above, we shall often write also $P_V : \mathbb{R}^n \to \mathbb{R}^m$ in a natural way. Identifying V with P_V, $G(n, m)$ becomes a smooth submanifold of dimension $m(n-m)$ of $\mathbb{R}^{n^2}$.

Theorem 2.3 *Suppose $A \subset \mathbb{R}^n$ is a Borel set.*

(1) *If $\dim A \le m$, then*

$$\dim\{V \in G(n, m) : \dim P_V(A) < t\} \le m(n-m) - m + t.$$

(2) *If* $\dim A > m$, *then*

$$\dim\{V \in G(n,m) : \mathcal{L}^1(P_V(A)) = 0\} \leq m(n-m) + m - \dim A.$$

Part (1) was proved in [Map5] and (2) in [Fa82]. The bound in (1) is sharp when $t = \dim A$. This was shown by Kaufman and myself in [KM75] with examples based on Jarnik's results on dimension and diophantine approximation, see also [Fa85], Sect. 8.5. Similar examples work also for (2). As far as I know the sharp bound in (1) for $t < \dim A$ is unkown. Anyway, the one given in Theorem 2.2 is not always sharp due to the following result of Bourgain in [Bo10] and Oberlin in [Ob12]:

Theorem 2.4 *Suppose* $A \subset \mathbb{R}^2$ *is a Borel set. Then*

$$\dim\{\theta \in [0,\pi) : \dim P_\theta(A) < \dim A/2\} = 0.$$

The construction in [KM75] can be used to get for any $0 < t \leq s < 2$ a compact set $A \subset \mathbb{R}^2$ with $\dim A = s$ such that

$$\dim\{\theta \in [0,\pi) : \dim P_\theta(A) \geq t\} \geq 2t - s.$$

Could $2t - s$ be the sharp upper bound in the range $s/2 \leq t \leq \min\{1, s\}$? In any case this shows that to get dimension 0 for the exceptional set, the bound $\dim A/2$ is the best possible.

Bourgain's estimate is somewhat stronger than the above. He obtained his result as part of deep investigations in additive combinatorics, whereas Oberlin's proof is much simpler and more direct. Oberlin also had another exceptional set estimate in [Ob13].

Some improvements on part (2) of Theorem 2.3 will be given soon in Sect. 4 in a more general setting.

3 Projection Theorems in Heisenberg Groups

Heisenberg group $\mathbb{H}^n$ is $\mathbb{R}^{2n+1}$ equipped with a non-abelian group structure, with a left invariant metric and with natural dilations. The first Heisenberg group $\mathbb{H}^1$ is the simplest of these. We can write $\mathbb{H}^1 = \mathbb{C} \times \mathbb{R}$, where the points are written as $p = (w,s), q = (z,t) \in \mathbb{H}^1$. Then the product of p and q is

$$p \cdot q = (w + z, s + t + 2Im(w\bar{z})).$$

To define the distance between p and q, set first

$$\|p\| = (|z|^4 + t^2)^{1/4},$$

and then

$$d(p,q) = \|p^{-1} \cdot q\| = (|z-w|^4 + |t-s-2Im(w\bar{z})|^2)^{1/4}.$$

It is easy to check that d really is a metric and that it is left invariant. We have the dilations

$$\delta_r(p) = (rz, r^2 t)$$

for which

$$d(\delta_r(p), \delta_r(q)) = r d(p,q).$$

When the distance is restricted to the t-axis $\{0\} \times \mathbb{R}$ it is just the square root distance. Essentially because of this the Heisenberg Hausdorff dimension of $\mathbb{H}^1$ is

$$\dim_H \mathbb{H}^1 = 4.$$

Here $\dim_H$ refers to the Hausdorff dimension with respect to the Heisenberg metric. Always dim will refer to the Hausdorff dimension with respect to the Euclidean metric. It is easy to check that

$$\dim(A) \le \dim_H(A) \le 2\dim(A),\ A \subset \mathbb{H}^1.$$

These inequalities are sharp. For example, if A is a subset of the x-axis, $\dim_H(A) = \dim(A)$, and if A is a subset of the t-axis, $\dim_H(A) = 2\dim(A)$. However, one can improve them for sets A with $\dim(A) > 1$. Very precise inequalities were obtained by Balogh et al. [BT09].

We define the projections in $\mathbb{H}^1$ in the group sense. Good subgroups of $\mathbb{H}^1$ for this purpose are those which are invariant under the dilations and have a complementary subgroup in the sense described below. They are precisely the horizontal lines

$$V_\theta = \{te_\theta : t \in \mathbb{R}\},\ e_\theta = (\cos\theta, \sin\theta, 0),\ 0 \le \theta < \pi,$$

and the vertical planes

$$W_\theta = V_\theta^\perp.$$

The horizontal lines V_θ are Euclidean, the distance restricted to them is the Euclidean distance, whereas the vertical planes W_θ are non-Euclidean; for them $\dim_H W_\theta = \dim W_\theta + 1$. We have the splitting

$$\mathbb{H}^1 = W_\theta \cdot V_\theta,$$

that is, for $p \in \mathbb{H}^1$ we have the unique factorization

$$p = Q_\theta(p) \cdot P_\theta(p),\ \ P_\theta(p) \in V_\theta, Q_\theta(p) \in W_\theta.$$

Thus we get the group projections

$$P_\theta : \mathbb{H}^1 \to V_\theta,\ Q_\theta : \mathbb{H}^1 \to W_\theta, 0 \le \theta < \pi.$$

Writing $p = (z,t) = (x+iy,t) \in \mathbb{H}^1$ we have the explicit formulas

$$P_\theta(p) = ((x\cos\theta + y\sin\theta)e_\theta, 0),$$

$$Q_\theta(p) = ((y\cos\theta - x\sin\theta)e_\theta^\perp, t - 2\cos(2\theta)xy + \sin(2\theta)(x^2 - y^2)),$$

where $e_\theta^\perp = (-\sin\theta, \cos\theta)$. So P_θ is the standard linear projection, essentially the one we considered above in $\mathbb{R}^2$, but Q_θ is a non-linear projection. P_θ is nice, it is Lipschitz and group homomorphism, but Q_θ is neither of those, it is only Hölder continuous with exponent $1/2$.

Now we have the following analogue for horizontal projections of Marstrand's projection theorem from [BD13]:

Theorem 3.1 *Let $A \subset \mathbb{H}^1$ be a Borel set. Then for almost all $\theta \in [0,\pi)$,*

$$\dim_H P_\theta(A) \ge \dim_H A - 2 \text{ if } \dim_H A \le 3,$$

$$\mathcal{H}^1(P_\theta(A)) > 0 \text{ if } \dim_H A > 3.$$

This is sharp: consider $A = \{(x,0,t) : x \in C, t \in [0,1]\}, C \subset \mathbb{R}$. Then $\dim_H A = \dim C + 2$ and

$$\dim_H P_\theta(A) = \dim P_\theta(A) = \dim P_\theta(C) = \dim C$$

for all but one θ.

Theorem 3.1 follows easily applying Marstrand's projection theorem to the projection of A on $\mathbb{C} \times \{0\}$.

For the vertical projections we have:

Theorem 3.2 *Let $A \subset \mathbb{H}^1$ be a Borel set. If $\dim_H A \le 1$, then for almost all $\theta \in [0,\pi)$,*

$$\dim_H A \le \dim_H Q_\theta(A) \le 2\dim_H A.$$

For A with $\dim_H A \le 1$ this is sharp:
if $A \subset t$-axis, $\dim_H Q_\theta(A) = \dim_H A$ for all θ,
if $A \subset x$-axis, $\dim_H Q_\theta(A) = 2\dim_H A$ for all but one θ.

The upper bound $2\dim_H A$ follows from the Hölder continuity of Q_θ. For the lower bound we use again the energy integrals. Let

$$p = (z,t), q = (\zeta,\tau) \in \mathbb{H}^1$$

and denote

$$\varphi_1 = \arg(z - \zeta), \varphi_2 = \arg(z + \zeta).$$

Then one can check that

$$d(p,q)^4 = |z - \zeta|^4 + (t - \tau + |z^2 - \zeta^2| \sin(\varphi_1 - \varphi_2))^2$$

and

$$d(Q_\theta(p), Q_\theta(q))^4 = |z - \zeta|^4 \sin^4(\varphi_1 - \theta) \\ + (t - \tau - |z^2 - \zeta^2| \sin(\varphi_2 + \varphi_1 - 2\theta))^2$$

To get $\int_0^\pi d(Q_\theta(p), Q_\theta(q))^{-s} d\theta \lesssim d(p,q)^{-s}$, one needs for $a \in \mathbb{R}$,

$$\int_0^\pi \frac{d\theta}{|a + \sin\theta|^{s/2}} \lesssim 1, \tag{3.1}$$

which is easy to check when $s < 1$.

If $\dim_H A > 1$, we have some estimates which quite likely are not sharp. For example, we do not know if $\dim_H A > 3$ implies $\mathcal{H}^2(Q_\theta(A)) > 0$ for almost all $\theta \in [0, \pi)$. Here $\mathcal{H}^2$ is the Euclidean two-dimensional Hausdorff measure. When restricted to a vertical plane it agrees with the three-dimensional Heisenberg Hausdorff measure, both give the Haar measure for this subgroup.

A related Euclidean question is: does $\dim A > 2$ imply $\mathcal{H}^2(Q_\theta(A)) > 0$ for almost all $\theta \in [0, \pi)$?

Let us now consider higher dimensions, these were treated in [BF12]. Then the basic notions and facts are

- $\mathbb{H}^n = \mathbb{C}^n \times \mathbb{R}$, $p = (w, s)$, $q = (z, t) \in \mathbb{H}^n$,
- $\omega(w, z) = 2Im(w \cdot z) = 2\sum_{j=1}^n (v_j x_j - u_j y_j)$, $w = (u_j + iv_j)$, $z = (x_j + iy_j)$,
- $p \cdot q = (w + z, s + t + \omega(w, z))$,
- $||p|| = (|z|^4 + t^2)^{1/4}$,
- $d(p,q) = ||p^{-1} \cdot q|| = (|z - w|^4 + |t - s - \omega(w, z)|^2)^{1/4}$,
- $\delta_r(p) = (rz, r^2 t)$,
- $d(\delta_r(p), \delta_r(q)) = rd(p,q)$,
- $d(p \cdot q_1, p \cdot q_2) = d(q_1, q_2)$,
- $\dim_H \mathbb{H}^n = 2n + 2$.

The subgroups invariant under dilations split again to horizontal and vertical subgroups. The horizontal ones are those m-dimensional linear subspaces of $\mathbb{R}^{2n}$, $0 < m \leq n$, on which the bilinear form ω vanishes. That is, the elements of

$$G_h(n, m) = \{V \in G(2n, m) : \omega(w, z) = 0\ \forall w, z \in V\}.$$

They are called isotropic subspaces. The unitary group $U(n) \subset O(2n)$ acts transitively on $G_h(n,m)$; by definition $g \in U(n)$ if $\omega(g(w), g(z)) = \omega(w,z)$ for all $w, z \in \mathbb{C}^n$. The vertical subgroups are all linear subspaces of $\mathbb{R}^{2n+1}$ which contain the t-axis. The horizontal subgroups are again Euclidean and for the vertical subgroups W we have $\dim_H W = \dim W + 1$. Then

$$\mathbb{H}^n = V^\perp \cdot V, V^\perp \subset \mathbb{R}^{2n+1}, V \in G_h(n,m),$$

$$p = Q_V(p) \cdot P_V(p),\ P_V(p) \in V, Q_V(p) \in V^\perp,\ \text{for } p \in \mathbb{H}^n.$$

Again $P_V : \mathbb{H}^n \to V$ is the standard linear projection, but $Q_V : \mathbb{H}^n \to V^\perp$,

$$Q_V(z,t) = (P_{V^\perp}(z), t - \omega((P_{V^\perp}(z), P_V(z))),$$

is a non-linear projection.

Notice that in the above splitting the linear dimension of V is always at most n. The vertical subgroups W of linear dimension $1 \le \dim W \le n$ have no complementary subgroups in the above sense.

We have the following horizontal projection theorem in $\mathbb{H}^n$:

Theorem 3.3 *Let $A \subset \mathbb{H}^n$ be a Borel set. If $\dim_H A \le m + 2$, then*

$$\dim P_V(A) \ge \dim_H A - 2$$

for $\mu_{n,m}$ almost all $V \in G_h(n,m)$. Furthermore, if $\dim_H A > m + 2$, then

$$\mathcal{H}^m(P_V(A)) > 0 \textit{ for } \mu_{n,m} \textit{ almost } V \in G_h(n,m).$$

This is again sharp. Above $\mu_{n,m}$ is the unique $U(n)$-invariant Borel probability measure on $G_h(n,m)$.

For the vertical projections we have

Theorem 3.4 *Let $A \subset \mathbb{H}^n$ be a Borel subset with $\dim_H A \le 1$. Then for $\mu_{n,m}$ almost $V \in G_h(n,m)$,*

$$\dim_H A \le \dim_H Q_V A \le 2 \dim_H A.$$

This is sharp when $\dim_H A \le 1$. Some, probably rather imprecise, partial results are known when $\dim_H A > 1$. One might expect that the methods would yield this theorem for $\dim_H A \le m$, but there are some serious obstacles. Let us see what they are. We can now write

$$d_H(p,q) = \sqrt[4]{|z - w|^4 + (t - s - 2\omega(\zeta, z))^2},$$

and

$$d_H(Q_V(p), Q_V(q))^4 = |P_{V^\perp}(z-w)|^4 + (t - s - \omega(P_{V^\perp}(z), P_V(z)) + \omega(P_{V^\perp}(w), P_V(w)) - \omega(P_{V^\perp}(w), P_{V^\perp}(z)))^2.$$

The key estimate in the proof is

$$\int_{G_h(n,m)} |a - \omega(v, P_V(w))|^{-s/2} d\mu_{n,m} V \lesssim 1$$

for all $0 < s < 1$, $a \in \mathbb{R}$ and $v, w \in S^{2n-1}$. In local coordinates for V the expression $a - 2\omega(v, P_V(w))$ is a second degree polynomial which can vanish to second order. Because of this the above estimate is false for $s \geq 1$ and it seems to be difficult to find anything to replace it.

There are various other results in the papers [BD13, BF12]. In particular, quite precise information is obtained on inequalities that hold for all projections.

4 Generalized Projections

Studying Kaufman's proof of Marstrand's projection theorem one notices quickly that it applies to much more general families of mappings than orthogonal projections onto lines and planes. Peres and Schlag developed this idea in [PS00] much farther. The following is still a special case of their general setting:

Let (Ω, d) be a compact metric space, $Q \subset \mathbb{R}^k$ an open connected set. We have mappings

$$\pi_\lambda : \Omega \to \mathbb{R}^m, \quad \lambda \in Q,$$

such that the mapping $\lambda \mapsto \pi_\lambda(x)$ is in $C^\infty(Q)$ for every fixed $x \in \Omega$, and to every compact $K \subset Q$ and any multi-index $\eta = (\eta_1, \ldots, \eta_n) \in \mathbb{N}^n$ there corresponds a finite constant $C_{\eta,K} > 0$ such that

$$|\partial_\lambda^\eta \pi_\lambda(x)| \leq C_{\eta,K}, \qquad \lambda \in K. \tag{4.1}$$

Definition 4.1 Define

$$\Phi_\lambda(x, y) = \frac{\pi_\lambda(x) - \pi_\lambda(y)}{d(x, y)}.$$

The family $\{\pi_\lambda, \lambda \in Q\}$ is said to be *transversal*, if there exists a finite constant $C_0 > 0$ such that

$$|\Phi_\lambda(x, y)| \leq C_0 \implies \det(D_\lambda \Phi_\lambda(x, y)(D_\lambda \Phi_\lambda(x, y)^t)) \geq C_0 \tag{4.2}$$

for $\lambda \in Q$ and $x, y \in \Omega, x \neq y$. The family $\{\pi_\lambda, \lambda \in Q\}$, is said to be *regular*, if to every multi-index $\eta = (\eta_1, \dots, \eta_n) \in \mathbb{N}^n$ there correspond a finite constant $C_\eta > 0$ such that

$$|\Phi_\lambda(x, y)| \leq C_0 \implies \left|\partial_\lambda^\eta \Phi_\lambda(x, y)\right| \leq C_\eta \tag{4.3}$$

for $\lambda \in Q$ and $x, y \in \Omega, x \neq y$.

Orthogonal projections when restricted to some compact set are easily seen to form transversal and regular families of mappings. When considering projections from $\mathbb{R}^n$ onto m-planes, we can take $k = \dim G(n, m) = m(n - m)$.

Since we are looking for lower bounds for dimensions of projections, the bad pairs of points are such x and y which are mapped close to each other. The point in transversality is that if (x, y) is a pair of bad points for some λ, then it becomes quickly better when λ moves a bit. For real-valued maps ($m = 1$), such as projections onto lines, the transversality means

$$|\Phi_\lambda(x, y)| \leq C_0 \implies |\nabla_\lambda \Phi_\lambda(x, y)| \geq C_0.$$

Here is a special case of a theorem of Peres and Schlag:

Theorem 4.2 *Suppose the above transversality and regularity conditions hold. Let $A \subset \Omega$ be a Borel set and $s = \dim A$.*

(a) *If $s \leq m$ and $t \in (0, s)$, then*

$$\dim\{\lambda \in Q : \dim \pi_\lambda(A) < t\} \leq k - m + t.$$

(b) *If $s > m$, then*

$$\dim\{\lambda \in Q : \dim \pi_\lambda(A) < t\} \leq k - s + t$$

and

$$\dim\{\lambda \in Q : \mathcal{L}^m(\pi_\lambda(A)) = 0\} \leq k - s + m.$$

(c) *If $s > 2m$, then*

$$\dim\{\lambda \in Q : \textit{the interior of } \pi_\lambda(A) \textit{ is empty}\} \leq n - s + 2.$$

In addition to being applicable to many families of mappings, this theorem also improves Theorem 2.3 in the case of orthogonal projections. As Peres and Schlag showed it can be applied in many interesting situations, for example to Bernoulli convolutions, sum sets and pinned distance sets.

In $\mathbb{R}^{2n}$ the horizontal Grassmannian, the Grassmannian of isotropic subspaces, $G_h(n, m)$, discussed before in the case of Heisenberg groups, is a proper lower dimensional submanifold of the full Grassmannian $G(2n, m)$ when $1 < m \leq n$. Nevertheless Marstrand's projection theorem holds for this submanifold. We proved this in [BF12]. Hovila established it in [Ho18] by verifying that the family $P_V : \mathbb{R}^{2n} \to V$, $V \in G_h(n,m)$, is transversal. This has two further consequnces: exceptional set estimates and Besicovitch-Federer projection theorem. The first follows from the above results of Peres and Schlag, the second from Hovila's joint work with Järvenpää et al. in [HLJ2]. There they proved Besicovitch-Federer projection theorem for transversal families of generalized projections. The classical Besicovitch-Federer projection theorem says that an $\mathcal{H}^m$ measurable set $A \subset \mathbb{R}^n$ with $\mathcal{H}^m(A) < \infty$ p! rojects into zero $\mathcal{H}^m$ measure in almost all m-planes $V \in G(n,m)$ if an and only if it meets every m-dimensional C^1-surface in a set of zero $\mathcal{H}^m$ measure, see [Fe69] or [Map2].

Neither the vertical nor the horizontal projections in Heisenberg groups satisfy transversality; these families are too small for that.

5 Restricted Families of Projections

The reason that it is not possible to get precise almost everywhere equalities for dimensions of projections in Heisenberg groups is that we have too few projections. It is of interest to search projection theorems for such restricted families of projections also in Euclidean spaces. That is, one considers a proper lower dimensional submanifold G of the Grassmannian $G(n,m)$ and the projections P_V, $V \in G$. This splits into two cases: G is general allowing flat submanifolds or G is required to possess some curvature properties. What these mean becomes clearer below. In the first case less can be said and it is completely solved by E. Järvenpää, M. Järvenpää and Keleti as we shall see soon. The second case is extremely difficult and some partial results have been obtained by Fässler and Orponen.

One motivation for studying restricted families of projection-type transformations comes from the work of E. Järvenpää, M. Järvenpää, Ledrappier and their co-workers on measures invariant under geodesic flows on manifolds, see [HJL2] and the references given there.

A simple restricted family of projections in $\mathbb{R}^3$ is given by the horizontal projections, or the projections onto the lines $L_\theta = \{t(\cos\theta, \sin\theta, 0) : t \in \mathbb{R}\}$,

$$P_\theta : \mathbb{R}^3 \to \mathbb{R},\ P_\theta(x,y,z) = x\cos\theta + y\sin\theta,\ 0 \le \theta < \pi. \tag{5.1}$$

Since $P_\theta(A) = P_\theta((\pi(A))$ where $\pi(x,y,z) = (x,y)$, and $\dim A \le \dim \pi(A) + 1$, it is easy to conclude using Marstand's projection theorem that for any Borel set $A \subset \mathbb{R}^3$, for almost all $\theta \in [0,\pi)$,

$$\dim P_\theta(A) \ge \dim A - 1 \text{ if } \dim A \le 2,$$
$$\dim P_\theta(A) = 1 \text{ if } \dim A \ge 2.$$

This is sharp by trivial examples; consider product sets $A = B \times C, B \subset \mathbb{R}^2$, $C \subset \mathbb{R}$.

A simple example of projections onto planes is given by

$$\Pi_\theta : \mathbb{R}^3 \to \mathbb{R}^2, \Pi_\theta(x, y, z) = (x \sin\theta - y \cos\theta, z), 0 < \theta < \pi. \tag{5.2}$$

These are essentially orthogonal projections onto the orthogonal complements of the lines L_θ.

Also now it is easy to prove that for any Borel set $A \subset \mathbb{R}^3$, for almost all $\theta \in [0, \pi)$,

$$\begin{aligned}
&\dim \Pi_\theta(A) \geq \dim A \text{ if } \dim A \leq 1,\\
&\dim \Pi_\theta(A) \geq 1 \text{ if } 1 \leq \dim A \leq 2,\\
&\dim \Pi_\theta(A) \geq \dim A - 1 \text{ if } \dim A \geq 2.
\end{aligned}$$

Again by easy examples these inequalities are sharp.

Järvenpää et al. proved in [JJ05] that the above sets of inequalities remain in force for any smooth, in a suitable sense non-degenerate, one-dimensional families of orthogonal projections onto lines and planes in $\mathbb{R}^3$. In fact, they proved such inequalities in more general dimensions and in [JJ13] Järvenpää et al. found the complete solution in all dimensions; sharp inequalities for smooth non-degenerate families of orthogonal projections onto m-planes in $\mathbb{R}^n$.

Consider now a slightly modified family of one-dimensional projections; let $p_\theta, \theta \in [0, 2\pi)$, be the orthogonal projection onto the line l_θ spanned by $(\cos\theta, \sin\theta, 1)$. The previous lines L_θ spanned a plane, but the lines l_θ span a cone. The trivial counter-examples do not work anymore and in fact one can now improve the above estimates for p_θ's relatively easily by showing that if $A \subset \mathbb{R}^3$ is a Borel set with $\dim A \leq 1/2$, then

$$\dim p_\theta(A) \geq \dim A \text{ for almost all } \theta \in [0, 2\pi).$$

The restriction $1/2$ comes because using Kaufman's method one is now lead to estimate integrals of the type

$$\int_0^{2\pi} \frac{d\theta}{|a + \sin\theta|^s}$$

for $s < \dim A$, and they are bounded only if $s < 1/2$. So this is the best one get without new ideas. Introducing some geometric arguments Fässler and Orponen were able to prove in [FO50] the following theorem for the packing dimensions, $\dim_p$, of the projections:

Theorem 5.1 *Let $U \subset \mathbb{R}$ be an open interval and $\gamma : U \to S^2$ be a C^3 curve such that for all $\theta \in U$ the vectors $\gamma(\theta)$, $\gamma'(\theta)$ and $\gamma''(\theta)$ span $\mathbb{R}^3$. Let*

$$p_\theta(x) = \gamma(\theta) \cdot x$$

be the orthogonal projection onto the line l_θ spanned by $\gamma(\theta)$ and

$$\pi_\theta(x) = x - (\gamma(\theta) \cdot x)\gamma(\theta)$$

the orthogonal projection onto the orthogonal complement of l_θ. Suppose $A \subset \mathbb{R}^3$ is a Borel set with $\dim A = s$.

(1) *If $s > 1/2$, there exists a number $\sigma_1(s) > 1/2$ such that*

$$\dim_p p_\theta(A) \geq \sigma_1(s) \textit{ for almost all } \theta \in U.$$

(2) *If $s > 1$, there exists a number $\sigma_1(s) > 1$ such that*

$$\dim_p \pi_\theta(A) \geq \sigma_2(s) \textit{ for almost all } \theta \in U.$$

It is not known if here the packing dimension could be replaced by the Hausdorff dimension. In [Orp3] Orponen was able to do this for the special family of orthogonal projections onto the lines l_θ spanned by $(\cos\theta, \sin\theta, 1)$ and their orthogonal complements.

It would be very interesting to find similar results for some non-linear families of mappings, for example for the vertical projections Q_θ of the Heisenberg group $\mathbb{H}^1$ considered just as mappings in $\mathbb{R}^3$:

$$Q_\theta(x, y, t) = (y\cos\theta - x\sin\theta, t - 2\cos(2\theta)xy + \sin(2\theta)(x^2 - y^2)), \theta \in [0, \pi).$$

Although, as said before, for the corresponding linear projections Π_θ as in (5.2) nothing more can be said than what we get from Marstrand's theorem, the non-linear mappings might be better.

6 Minkowski and Packing Dimensions

The analogue of Marstrand's theorem fails for Minkowski and packing dimensions; the dimension of the set does not prescribe the dimensions of the typical projections. However, Falconer and Howroyd proved in [FH96] the following sharp inequalities:

Theorem 6.1 *Let $A \subset \mathbb{R}^n$ be a Borel set. Then*

$$\dim_p P_V(A) \geq \frac{\dim_p A}{1 + (1/m - 1/n)\dim_p A} \textit{ for almost all } V \in G(n, m).$$

The same inequality holds also for upper and lower Minkowski dimensions. Examples of Järvenpää in [Jm94] show that the lower bound is sharp. In these examples the Hausdorff dimension of A is 0. Later on we proved with Falconer in [Fm96] a version of this result which gives a sharp lower bound for the packing dimension of the typical projections given both Hausdorff and packing dimension of A.

Finding good dimension estimates for exceptional sets in packing dimension projection theorems has turned out to be a very delicate question, Rams obtained some results in [Ram2] for self-conformal sets. Orponen proved in [Ort2] several such estimates and constructed many illustrative examples. He also established Baire category results.

7 Constancy Results for Projections

Although there is no dimension preservation for the packing and Minkowski dimensions under projections, Falconer and Howroyd proved in [FH97] that given the set A, the dimensions equal almost surely a constant called $\mathrm{Dim}_m A$. The number $\mathrm{Dim}_m A$ comes from certain potentials. More precisely, we first define it for measures $\mu \in \mathcal{M}(A)$:

$$\mathrm{Dim}_m \mu = \sup\{t \geq 0 : \liminf_{r \to 0} r^{-t} F_m^{\mu}(x, r) = 0 \text{ for } \mu \text{ almost all } x \in \mathbb{R}^n\},$$

where

$$F_m^{\mu}(x, r) = \int \min\{1, r^m |x - y|^{-m}\} d\mu y.$$

For sets we define

$$\mathrm{Dim}_m A = \sup\{\mathrm{Dim}_m \mu : \mu \in \mathcal{M}(A)\}.$$

The theorem of Falconer and Howroyd now reads

Theorem 7.1 *Let $A \subset \mathbb{R}^n$ be a Borel set. Then*

$$\dim_p P_V(A) = \mathit{Dim}_m A \textit{ for almost all } V \in G(n, m).$$

The relation of the potentials $F_m^{\mu}(x, r)$ to projections comes from the following observation:

$$\begin{aligned} F_m^{\mu}(x, r) &\approx \int \gamma_{n,m}(\{V \in G(n, m) : |P_V(x - y)| \leq r\}) d\mu y \\ &= \int P_V \mu(B(P_V(x), r)) d\gamma_{n,m} V, \end{aligned}$$

where $P_V \mu$ is the push forward of μ under P_V.

Similar tools were also used in [FH96, Fm96].

Perhaps such constancy results hold also in Heisenberg groups. This is not known but Fässler and Orponen proved in [FO13] constancy results for some restricted families of projections in $\mathbb{R}^3$. They did it in general dimensions but for simplicity I state their result only in $\mathbb{R}^3$:

Theorem 7.2 *Let $A \subset \mathbb{R}^3$ be a Borel set. Then for the projections Π_θ as in (5.2)*

$$\dim \Pi_\theta(A) = \sup\{\Pi_\theta(A) : \theta \in (0, \pi)\} \text{ for almost all } \theta \in (0, \pi).$$

Notice that for the projections P_θ onto lines given in (5.1) the constancy is trivial by Marstrand's projection theorem: for almost all $\theta \in (0, \pi)$,

$$\dim P_\theta(A) = \dim P_\theta(\pi(A)) = \min\{\dim \pi(A), 1\},$$

where $\pi(x, y, z) = (x, y)$.

Fässler and Orponen proved analogous results also for packing and Minkowski dimensions.

8 Slicing Theorems

Marstrand's line intersection theorem says that if A is a Borel subset of the plane with $\dim A > 1$, then the typical lines which intersect A intersect it in dimension $\dim A - 1$. Here is a way to state it more precisely and in higher dimensions:

Theorem 8.1 *Let $A \subset \mathbb{R}^n$ be a Borel set, $s > m$ and $0 < \mathcal{H}^s(A) < \infty$. Then for $\gamma_{n,m}$ almost $V \in G(n, m)$,*

$$\mathcal{H}^m(\{v \in V : \dim(A \cap (V^\perp + v)) = s - m\}) > 0.$$

Here $\gamma_{n,m}$ is the orthogonally invariant Borel probability measure on $G(n, m)$.

With Balogh et al. we proved in [BF12] the analogous result in Heisenberg groups:

Theorem 8.2 *Let $A \subset \mathbb{H}^n$ be a Borel set, $s > m + 2$ and $0 < \mathcal{H}^s_H(A) < \infty$. Then for $\mu_{n,m}$ almost $V \in G_h(n, m)$,*

$$\mathcal{H}^m(\{v \in V : \dim_H(A \cap (V^\perp \cdot v)) = s - m\}) > 0.$$

Here for $v \in V$, $V^\perp \cdot v$ is the coset $\{p \cdot v : p \in V^\perp\}$. The assumption $\dim_H A > m + 2$ is necessary.

Another way to formulate such a result, actually the one Marstrand used, is

Theorem 8.3 *Let $A \subset \mathbb{H}^n$ be a Borel set, $s > m + 2$ and $\mathcal{H}^s_H(A) < \infty$. Then for $\mathcal{H}^s_H$ almost all $p \in A$ we have*

$$\dim_H(A \cap (V^{\perp} \cdot p)) = s - m \text{ for } \mu_{n,m} \text{ almost all } V \in G_h(n,m).$$

Here $\mathcal{H}_H^s$ is the s-dimensional Hausdorff measure with respect to the Heisenberg metric.

Orponen studied in [Or12] the problem of the dimension of exceptional sets for line sections. A higher dimensional version of his results is

Theorem 8.4 *Let $A \subset \mathbb{R}^n$ be a Borel set, $s > m$ and $0 < \mathcal{H}^s(A) < \infty$. Then there is a Borel set $E \subset G(n,m)$ such that*

$$\dim E \leq m(n-m) + m - s$$

and

$$\mathcal{H}^m(\{a \in V : \dim A \cap (V^{\perp} + a) = s - m\}) > 0 \text{ for all } V \in G(n,m) \setminus E.$$

The upper bound is again sharp. Observe that this strengthens part (2) of Theorem 2.3.

To get his result, Orponen proved the following inequality: if $s > m$, $\mu \in \mathcal{M}(\mathbb{R}^n)$ and $V \in G(n,m)$, then

$$\int_V I_{s-m}(\mu_{V,a})\, \mathrm{d}\mathcal{H}^m a \lesssim \int_{\mathbb{R}^n} |P_{V^{\perp}}(x)|^{s-n} |\widehat{\mu}(x)|^2 dx.$$

Here the measures $\mu_{V,a}$ are natural slices of μ with planes $V+a$. They have supports in $\mathrm{spt}\mu \cap (V+a)$ and they disintegrate μ if the projection of μ on V is absolutely continuous with respect to the m-dimensional Hausdorff measure on V.

Fraser et al. found in [FOS3] another interesting application for this inequality: they showed that any one-dimensional graph has Fourier dimension 1. More precisely, in general dimensions

Theorem 8.5 *For any function $f : A \to \mathbb{R}^{n-m}$, $A \subset \mathbb{R}^m$, and for its graph $G_f = \{(x, f(x) : x \in A\}$, if $\mu \in \mathcal{M}(G_f)$, $s > 0$ and*

$$|\widehat{\mu}(x)| \leq |x|^{-s/2} \text{ for } x \in \mathbb{R}^n,$$

then $s \leq m$.

Notice that we make no assumptions on f, not even measurability. Still, before the work of Fraser, Orponen and Sahlsten this question was open even for Brownian graphs.

9 Final Comments

I have restricted here to general sets. Many of the above results are formulated, and are more natural and general, for measures and their dimensions. I have completely ignored the very interesting question on what can be said in various special cases, for example for self-similar and related sets and measures. For these one can often obtain results which hold for all directions or the exceptional directions can be specified. Outstanding work on self-similar and other dynamically generated sets has been recently done by Furstenberg in [Fu08], by Peres and Shmerkin in [PSh9], by Hochman and Shmerkin in [HS12], and by Hochman in [Ho16]. There have also been many results on dimensions of sections in various special cases, for example in [BFS12, LXZ7, MS03, WWXl, WX10].

References

[BD13] Balogh, Z.M., Cartagena, D.E., Fässler, K., Mattila, P., Tyson, J.T.: The effect of projections on dimension in the Heisenberg group, Revista Mat. Iberoamericana **29**, 381–432 (2013)
[BF12] Balogh, Z.M., Fässler, K., Mattila, P., Tyson, J.T.: Projection and slicing theorems in Heisenberg groups. Adv. Math. **231**, 569–604 (2012)
[BT09] Balogh, Z.M., Tyson, J.T., Warhurst, B.: Sub-Riemannian versus Euclidean dimension comparision and fractal geometry on Carnot groups. Adv. Math. **220**, 560–619 (2009)
[BFS12] Barany, B., Ferguson, A., Simon, K.: Slicing the Sierpinski gasket. Nonlinearity **25**, 1753–1770 (2012)
[Bo10] Bourgain, J.: Discretized sum-product and projection theorems. J. Anal. Math. **112**, 193–236 (2010)
[Fa82] Falconer, K.J.: Hausdorff dimension and the exceptional set of projections. Mathematika **29**, 90–96 (1982)
[Fa85] Falconer, K.J.: Geometry of Fractal Sets. Cambridge University Press, Cambridge (1985)
[FH96] Falconer, K.J., Howroyd, J.D.: Projection theorems for box and packing dimensions. Math. Proc. Camb. Phil. Soc. **119**, 287–295 (1996)
[FH97] Falconer, K.J., Howroyd, J.D.: Packing dimension of projections and dimension profiles. Math. Proc. Camb. Phil. Soc. **121**, 269–286 (1997)
[Fm96] Falconer, K.J., Mattila, P.: Packing dimension of projections and sections of measures. Math. Proc. Camb. Phil. Soc. **119**, 695–713 (1996)
[FO13] Fässler, K., Orponen, T.: Constancy results for special families of projections. Math. Proc. Camb. Philos. Soc. **154**, 549–568 (2013). arXiv:1208.1876
[FO50] Fässler, K., Orponen, T.: On restricted families of projections in $\mathbb{R}^3$. arXiv:1302.6550
[Fe69] Federer, H.: Geometric Measure Theory. Springer, New York (1969)
[FOS3] Fraser, J., Orponen, T., Sahlsten, T.: On Fourier analytic properties of graphs. IMRN. arXiv:1212.4813 (to appear)
[Fu08] Furstenberg, H.: Ergodic fractal measures and dimension conservation. Ergodic Theo. Dynam. Syst. **28**, 405–422 (2008)
[Ho16] Hochman, M.: On self-similar sets with overlaps and inverse theorems for entropy. arXiv:1212.1873
[HS12] Hochman, M., Shmerkin, P.: Local entropy averages and projections of fractal measures. Ann. Math. **175**, 1001–1059 (2012)
[Ho18] Hovila, R.: Transversality of isotropic projections, unrectifiability and Heisenberg groups. Revista Mat. Iberoamericana (to appear)

[HLJ2] Hovila, R., Järvenpää, E., Järvenpää, M., Ledrappier, F.: Besicovitch-Federer projection theorem and geodesic flows on Riemann surfaces. Geom. Dedicata. **161**, 51–61 (2012)

[HJL2] Hovila, R., Järvenpää, E., Järvenpää, M., Ledrappier, F.: Singularity of projections of 2-dimensional measures invariant under the geodesic flow. Comm. Math. Phys. **312**, 127–136 (2012)

[JJ13] Järvenpää, E., Järvenpää, M., Keleti, T.: Hausdorff dimension and non-degenerate families of projections. J. Geom. Anal. (2013)

[JJ05] Järvenpää, E., Järvenpää, M., Ledrappier, F., Leikas, M.: (Non)regularity of projections of measures invariant under geodesic flow. Comm. Math. Phys. **254**, 695–717 (2005)

[JJ28] Järvenpää, E., Järvenpää, M., Leikas, M.: One-dimensional families of projections. Nonlinearity **21**, 453–463 (2008)

[Jm94] Järvenpää, M.: On the upper Minkowski dimension, packing dimension, and orthogonal projections. Ann. Acad. Sci. Fenn. Ser. A I Math. Dissertationes **99**, 34 (1994)

[Ka68] Kaufman, R.: On Hausdorff dimension of projections. Mathematika **15**, 153–155 (1968)

[KM75] Kaufman, R., Mattila, P.: Hausdorff dimension and exceptional sets of linear transformations. Ann. Acad. Sci. Fenn. Ser. A I Math. **1**, 387–392 (1975)

[LXZ7] Liu, Q.H., Xi, L.F., Zhao, Y.F.: Dimensions of intersections of the Sierpinski carpet with lines of rational slopes. Proc. Edinb. Math. Soc. **50**, 411–428 (2007)

[MS03] Manning, A., Simon, K.: Dimension of slices through the Sierpinski carpet. Trans. Amer. Math. Soc. **365**, 213–250 (2013)

[Ma54] Marstrand, J.M.: Some fundamental geometrical properties of plane sets of fractional dimensions. Proc. London Math. Soc. **3**(4), 257–302 (1954)

[Map5] Mattila, P.: Hausdorff dimension, orthogonal projections and intersections with planes. Ann. Acad. Sci. Fenn. Ser. A I Math. **1**, 227–244 (1975)

[Map2] Mattila, P.: Geometry of Sets and Measures in Euclidean Spaces. Cambridge University Press, Cambridge (1995)

[Map3] Mattila, P.: Hausdorff dimension, projections, and the Fourier transform. Publ. Mat. **46**, 3–48 (2004)

[Map4] Mattila, P.: Marstrand's theorems. In: Córdoba, A., Férnandez, J.L., Férnandez, P. (eds.) All That Math, Portraits of Mathematicians as Young Readers, pp. 235–245. Revista Mathematica Iberoamericana (2011)

[Ob12] Oberlin, D.M.: Restricted Radon transforms and projections of planar set. Canad. Math. Bull. **55**, 815–820 (2012)

[Ob13] Oberlin, D.M.: Exceptional sets of projections, unions of k-planes, and associated transforms. arXiv:1107.4913

[Or12] Orponen, T.: Slicing sets and measures, and the dimension of exceptional parameters. J. Geom. Anal. **24**, 47–80 (2012)

[Ort2] Orponen, T.: On the packing dimension and category of exceptional sets of orthogonal projections. arXiv:1204.2121

[Orp3] Orponen, T.: Hausdorff dimension estimates for some restricted families of projections in $\mathbb{R}^3$. arXiv:1304.4955

[PS00] Peres, Y., Schlag, W.: Smoothness of projections, Bernoulli convolutions, and the dimension of exceptions. Duke Math. J. **102**, 193–251 (2000)

[PSh9] Peres, Y., Shmerkin, P.: Resonance between Cantor sets. Ergodic Theor. Dynam. Syst. **29**, 201–221 (2009)

[Ram2] Rams, M.: Packing dimension estimation for exceptional parameters. Israel J. Math. **130**, 125–144 (2002)

[WWXl] Wen, Z., Wu, W., Xi, L.: Dimension of slices through a self-similar set with initial cubical pattern. Ann. Acad. Sci. Fenn. Ser. A I (to appear)

[WX10] Wen, Z.-Y., Xi, L.: On the dimension of sections for the graph-directed sets. Ann. Acad. Sci. Fenn. Ser. A **I**(35), 515–535 (2010)

The Geometry of Fractal Percolation

Michał Rams and Károly Simon

Abstract A well studied family of random fractals called fractal percolation is discussed. We focus on the projections of fractal percolation on the plane. Our goal is to present stronger versions of the classical Marstrand theorem, valid for almost every realization of fractal percolation. The extensions go in three directions:

- the statements work for all directions, not almost all,
- the statements are true for more general projections, for example radial projections onto a circle,
- in the case $\dim_H > 1$, each projection has not only positive Lebesgue measure but also has nonempty interior.

Keywords Random fractals · Hausdorff dimension · Processes in random environment

2000 Mathematics Subject Classification: Primary 28A80 Secondary 60J80 · 60 J85

Rams was partially supported by the MNiSW grant N201 607640 (Poland). The research of Simon was supported by OTKA Foundation # K 104745.

M. Rams
Institute of Mathematics, Polish Academy of Sciences, ul. Śniadeckich 8, Warsaw 00-956, Poland
e-mail: rams@impan.gov.pl

K. Simon (✉)
Institute of Mathematics, Technical University of Budapest, Budapest H-1529 B.O.box 91, Hungary
e-mail: simonk@math.bme.hu

D.-J. Feng and K.-S. Lau (eds.), *Geometry and Analysis of Fractals*, Springer Proceedings in Mathematics & Statistics 88,
DOI 10.1007/978-3-662-43920-3_11

1 Introduction

To model turbulence, Mandelbrot [Ma74, Ma83] introduced a statistically self-similar family of random Cantor sets. Since that time this family has got at least three names in the literature: fractal percolation, Mandelbrot percolation and canonical curdling, among which we will use the first one.

In 1996 Lincoln Chayes [Ch96] published an excellent survey giving an account about the most important results known in that time. His survey focused on the percolation related properties while we place emphasis on the geometric measure theoretical properties (projections and slices) of fractal percolation sets.

About the projections of a general Borel set the celebrated Marstrand Theorem gives the following information:

Theorem 1 ([Ma54]) *Let $E \subset \mathbb{R}^2$ be a Borel set.*

- *If* $\dim_H(E) < 1$ *then for Lebesgue almost all θ* $\dim_H(\mathrm{proj}_\theta(E)) = \dim_H(E)$.
- *If* $\dim_H(E) > 1$ *then for Lebesgue almost all θ we have* $\mathcal{L}\mathrm{eb}(\mathrm{proj}_\theta(E)) > 0$.

where proj_θ *is the orthogonal projection in direction θ.*

In this paper we review some recent results which give more precise information in the special case of the projections of fractal percolation Cantor sets.

2 The Construction and Its Immediate Consequences

The construction consists of the infinite iteration of two steps. We start from the unit cube in $\mathbb{R}^d$.

- All cubes we have after the n-th iteration of the process (they will be called level n cubes) we subdivide into smaller cubes of equal size,
- Among them some are retained and some are discarded. Retaining or discarding of different cubes are independent random events. The cubes that were retained are the level $n+1$ cubes.

Those points that have never been discarded form the fractal percolation set.

Please note that in literature the term fractal percolation is often used to denote object which we call homogeneous fractal percolation. That is, the fractal percolation for which all squares have equal probabilities of being retained.

2.1 An Informal Description of Fractal Percolation

We fix integer $M \geq 2$. We partition the unit cube $Q \subset \mathbb{R}^d$ into M^d congruent cubes of side length M^{-1} and we assign a probability to each of the cubes in this partition (Fig. 1a). We retain each of the cubes of this partition with the corresponding

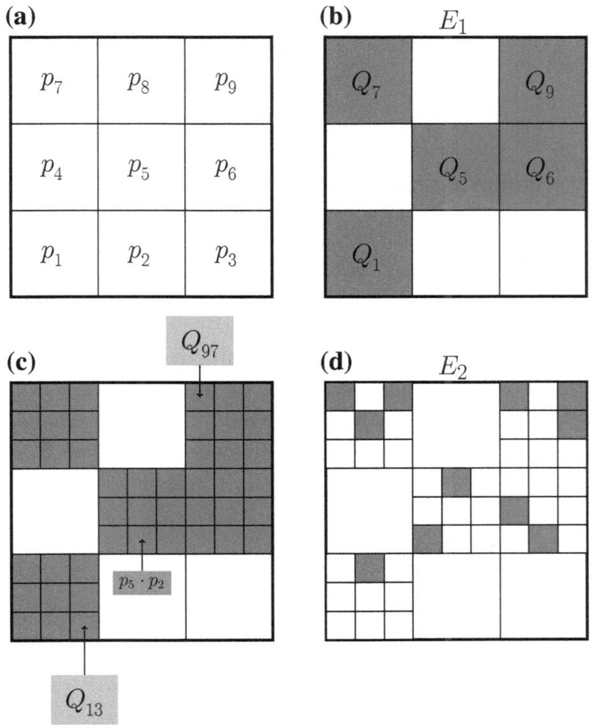

Fig. 1 The first two steps of the construction. $\mathbb{P}(Q_{52} \text{ retained}) = p_5 \cdot p_2$. For this realization $\mathcal{E}_1 = \{1, 5, 6, 7, 9\}$, $\mathcal{E}_2 = \{17, 51, 58, 62, 64, 75, 77, 79, 96, 97, 99\}$

probability independently and discard it with one minus the corresponding probability. The union of the retained squares is the first approximation of the random set to be constructed (Fig. 1b). We obtain the second approximation by repeating this process independently of everything in each of the retained squares (Fig. 1c, d). We continue this process at infinitum.

The object of our investigation is the collection of those points which have not been discarded. It will be called **fractal percolation set** and denoted by $E = (d, M, \mathbf{p})$, where $\mathbf{p}$ is the chosen vector of the probabilities $\{p_i\}$. In the special case when all p_i are equal we obtain the **homogeneous fractal percolation set** which is denoted by $E^h = E^h(d, M, p)$.

2.2 Fractal Percolation Set in More Details

For simplicity we give the construction on the plane but the definition works with obvious modifications in $\mathbb{R}^d$ for all $d \geq 1$. Besides the dimension of the ambient space the two other parameters of the construction are: the natural number $M \geq 2$ and a vector of probabilities $\mathbf{p} \in [0,1]^{M^2}$ (note: not a probabilistic vector). To shorten the notation we write $\mathcal{I}$ for the set of indices of $\mathbf{p}$:

$$\mathcal{I} := \left\{1, \ldots, M^2\right\}$$

The statistically self-similar random set which is the object of our study is defined as

$$E := \bigcap_{n=1}^{\infty} E_n, \tag{2.1}$$

where E_n is the n-th approximation of E. The inductive definition of E_n will occupy the rest of this subsection. Actually E_n is the union of a random collection of level n squares. First we define the level n squares and then we introduce the random rule with which those level n squares are selected whose union form E_n.

2.2.1 The Process of Subdivision

We divide the unit square $Q = [0,1]^2$ into M^2 congruent squares $Q_1, \ldots, Q_{M^2}$ of size M^{-1} numbered according to lexicographical order (or any other order). These squares are the level one M-adic squares. Let

$$\mathcal{N}_1 := \{x_i\}_{i \in \mathcal{I}}$$

be the set of midpoints of the level one squares (Fig. 2). For each midpoint x_i we define the homothetic map $\varphi_i : Q \to Q_i$:

$$\varphi_i(y) := x_i + M^{-1} \cdot \left(y - \left(\frac{1}{2}, \frac{1}{2}\right)\right).$$

For every $\mathbf{i} \in \mathcal{I}^n$, $\mathbf{i} = (i_1, \ldots, i_n)$ we write

$$x_{\mathbf{i}} := \varphi_{\mathbf{i}}\left(\frac{1}{2}, \frac{1}{2}\right).$$

and we define the map

$$\varphi_{\mathbf{i}}(y) := x_{\mathbf{i}} + M^{-n} \cdot \left(y - \left(\frac{1}{2}, \frac{1}{2}\right)\right).$$

To simplify the notation, we will not distinguish the set of the centers of level n squares

$$\mathcal{N}_n := \left\{\varphi_{\mathbf{i}}\left(\frac{1}{2}, \frac{1}{2}\right) : \mathbf{i} \in \mathcal{I}^n\right\}$$

and the family of level n-squares:

$$\left\{Q_{\mathbf{i}} := \varphi_{\mathbf{i}}(Q) : \mathbf{i} \in \mathcal{I}^n\right\}. \tag{2.2}$$

2.2.2 The Process of Retention

The square $Q = Q_\emptyset$ is retained. For any $\mathbf{i} \in \mathcal{I}^n$ for which the square $Q_{\mathbf{i}}$ is retained and for each $j \in \mathcal{I}$, the square $Q_{\mathbf{i}j}$ is retained with probability p_j. The events '$Q_{\mathbf{i}j}$ is retained' and '$Q_{\mathbf{i}'j'}$ is retained' are independent whenever $\mathbf{i} \neq \mathbf{i}'$ or $j \neq j'$.

We define E_1 as the union of retained squares $Q_i, i \in \mathcal{I}$. Similarly, E_n is the union of retained squares $Q_{\mathbf{i}}, \mathbf{i} \in \mathcal{I}^n$. We write

$$\mathcal{E}_n := \left\{\mathbf{i} \in \mathcal{I}^n : Q_{\mathbf{i}} \text{ retained }\right\}.$$

2.3 *The Corresponding Probability Space and Statistical Self-similarity*

The probability space corresponding to this random construction is best described by Dekking [De09]. For the convenience of the reader we repeat it here. Let $\mathcal{T}$ be the M^d array tree that is

$$\mathcal{T} := \bigcup_{n=0}^{\infty} \mathcal{I}^n,$$

where $\mathcal{I}^0 := \emptyset$ is the root of three . Let $\Omega := \{0, 1\}^{\mathcal{T}}$ that is Ω is the set of labeled trees where we label every node of $\mathcal{T}$ by 0 or 1. The probability measure $\mathbb{P}_{\mathbf{p}}$ on Ω is define in such a way that the family of labels $X_{\mathbf{i}} \in \{0, 1\}$ of nodes $\mathbf{i} \in \mathcal{T}$ satisfy:

- $\mathbb{P}_{\mathbf{p}}(X_\emptyset = 1) = 1$
- $\mathbb{P}_{\mathbf{p}}(X_{i_1,\dots,i_n}) = p_{i_n}$
- $\{X_{\mathbf{i}}\}_{\mathbf{i}\in\mathcal{T}}$ are independent.

Following [De09] we define the survival set of level n by

$$S_n := \left\{\mathbf{i} \in \mathcal{I}^n : X_{i_1\dots,i_k} = 1,\ \forall 1 \leq k \leq n\right\}.$$

Then

$$E_n = \bigcup_{\mathbf{i}\in S_n} Q_{\mathbf{i}},\quad E = \bigcap_{n=1}^{\infty} E_n.$$

It follows from the construction that generalized fractal percolation set is statistically self-similar and the number of retained cubes form a branching process:

Lemma 2 *(a)* $\{\#\mathcal{E}_n\}$ *is a branching process with average number of offsprings* $\sum_{i \in \mathcal{I}} p_i$. *In particular if* $p_i \equiv p$ *then the offspring distribution is* `Binomial` (M^d, p).

(b) For every $n \geq 1$ *and* $\mathbf{i} \in \mathcal{E}_n$ *the rescaled copy* $\varphi_{\mathbf{i}}^{-1}(E \cap Q_{\mathbf{i}})$ *has the same distribution as E itself.*

(c) The sets $\{E \cap Q_{\mathbf{i}}\}_{\mathbf{i} \in \mathcal{E}_n}$ *are independent.*

Using this it is not hard to prove that

$$E \neq \emptyset \text{ implies that } \dim_{\mathrm{H}}(E) = \dim_{\mathrm{B}}(E) = \frac{\log \sum_{i \in \mathcal{I}} p_i}{\log M} \text{ a.s.} \tag{2.3}$$

This was proved by Kahane and Peyriere [KP76], Hawkes [Ha81], Falconer [Fa86], Mauldin and Williams [MW86] independently. A canonical example of the inhomogeneous fractal percolation set is:

Example 3 [Random Sierpiński Carpet] Let $SC_p := E(2, 3, \mathbf{p})$, where using the notation of Fig. 1c:

$$p_5 = 0 \, and \; for \, i \in \{1, \ldots, 9\} \setminus \{5\} : \; p_i = p.$$

□

3 Percolation and Projection to Coordinate Axes

In this section we work on the plane so $Q = [0, 1]^2$. The connectivity properties of $E^h(2, M, p)$ for an arbitrary $M \geq 2$ was first investigated by Chayes, Chayes and Durrett [CCD88]. Dekking and Meester [DM90] gave a simpler proof and extended the scope of the theorem for some inhomogeneous fractal percolation sets like the random Sierpiński carpet SC_p. Here we summarize briefly some of the most interesting results of this area. For a much more detailed account see by Chayce [Ch96].

We say that E **percolates** if E contains a connected set which intersects both the left and the right sides of Q. If E percolates then E has a large connected component.

3.1 The Homogeneous Case

The following very important result was proved by Chayce,Chayce, Durrett.

Theorem 4 ([CCD88]) *Fix an arbitrary* $M \geq 2$. *Then there is a critical probability* $\frac{1}{M} < p_c < 1$ *such that*

(1) If $p < p_c$ *then* $E^h(2, M, p)$ *is a random dust that is totally disconnected almost surely.*

(2) *If $p \geq p_c$ then $E^h(2, M, p)$ percolates with positive probability. This implies that $E^h(2, M, p)$ is not totally disconnected almost surely.*

This shows a remarkable difference in between the fractal percolation and the usual percolation: in the latter case, the probability of percolation at critical parameter $p = p_c$ is 0.

3.2 The Inhomogeneous Case

Using some earlier works of Dekking and Grimmett [DG88], the results above were extended by Dekking and Meester [DM90]. They proved that by changing the components of $\mathbf{p}$ the inhomogeneous fractal percolation set $E(2, M, \mathbf{p})$ can go through the six stages below. Here the projection to the x-axis is denoted by proj_x. That is $\text{proj}_x(a, b) = a$.

The DM stages of $\mathbf{E(2, M, p)}$:

I: $E = \emptyset$ almost surely.
II: $\mathbb{P}(E \neq \emptyset) > 0$ but $\dim_{\mathrm{H}}\left(\text{proj}_x E\right) = \dim_{\mathrm{H}}(E)$ almost surely.
III: $\dim_{\mathrm{H}}\left(\text{proj}_x E\right) < \dim_{\mathrm{H}}(E)$ if $E \neq \emptyset$ but $\mathcal{L}\text{eb}\left(\text{proj}_x E\right) = 0$ almost surely.
IV: $0 < \mathcal{L}\text{eb}\left(\text{proj}_x E\right) < 1$ almost surely.
V: $\mathcal{L}\text{eb}\left(\text{proj}_x E\right) = 1$ holds with positive probability but E does not percolate almost surely.
VI: E percolates with positive probability.

It was proved in [DM90] that the random Sierpiński Carpet SC_p goes through all of these stages as we increase the value of p. The following theorem gives the precise answer when exactly a system appears in stages I, II, III.

Theorem 5 ([DG88, Fa86]) *Let m_r be the sum of the probabilities in the r-th column, that is the expected number of squares in column r. Then*

(1) *$E = \emptyset$ almost surely iff $\sum_{i=1}^{M^2} p_i \leq 1$. Except when $\exists i$ such that $p_i = 1$ and $p_j = 0$ for all $i \neq j$. In this case E is a singleton.*

(2) *$\dim_{\mathrm{H}}(\text{proj}_x(E)) = \dim_{\mathrm{H}}(E)$ holds almost surely, iff $\sum_{r=1}^{M} m_r \log m_r \leq 0$.*

(3) *$\mathcal{L}\text{eb}(\text{proj}_x E) = 0$ holds almost surely iff $\sum_{r=1}^{M} \log m_r \leq 0$.*

This result was strengthened by Falconer and Grimmett:

Theorem 6 ([FG92, FG94]) *Assume that $m := \min\{m_r\} > 1$. Then $\text{proj}_x(E)$ contains an interval almost surely, conditioned on non-extinction.*

We will present the proof in the fifth section.

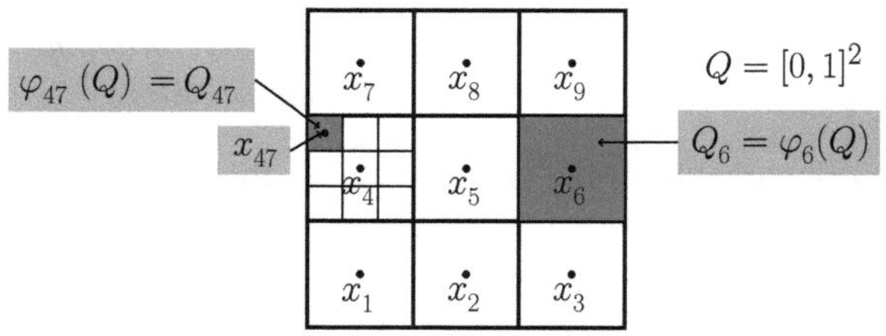

Fig. 2 Definition of level n squares

3.3 *The DM Stages for The Homogeneous Case*

For the homogeneous case $m_r = M \cdot p$ Hence we obtain that almost surely:

- If $0 < p \leq \frac{1}{M^2}$ then $E = \emptyset$.
- If $\frac{1}{M^2} < p \leq \frac{1}{M}$ then the system is in stage II.
- If $\frac{1}{M} < p < p_c$ then the system is in stage V.

Stages III and IV do not appear in the homogeneous case.

4 The Arithmetic Sum/Difference of Two Fractal Percolations

There is a very nice and more detailed survey of this field due to M. Dekking [De09]. In the previous section we studied the connectivity properties and the 90° projections of random Cantor sets. In this section we consider sets which are products of inhomogeneous fractal percolation sets and we take their 45°, (−45°) projections in order to study the arithmetic difference (arithmetic sum) respectively of independent copies of $E(1, M, \mathbf{p})$.

4.1 *The Arithmetic Sum and Its Visualization*

Let $A, B \subset \mathbb{R}$ be arbitrary. Then the arithmetic sum $A + B := \{a + b : a \in A, b \in B\}$ is the −45°-projection of $A \times B$ to the x-axis (this is the direction of the line ℓ_a on Fig. 3). Similarly, we can visualize the arithmetic difference by taking the projection of the product set with the line of +45° angle.

The motivation for studying the arithmetic difference (or sum) of random Cantor sets comes from a conjecture of Palis which states that typically (in a natural sense which depends on the actual setup), the arithmetic difference of two dynamically defined Cantors is either small in the sense that it has Lebesgue measure zero or big in the sense that it contains some intervals, but at least typically, it does not occur that the arithmetic difference set is a set of positive Lebesgue measure with empty interior. This conjecture does not hold for the algebraic difference of inhomogeneous

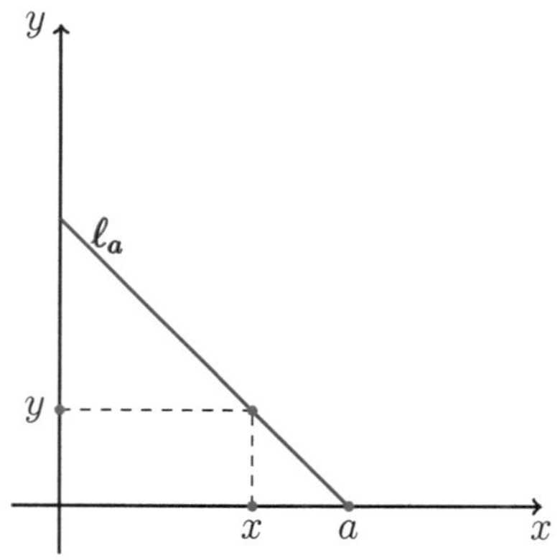

Fig. 3 Algebraic sum as $-45°$ projection: $a = x + y = \text{proj}_{-45°}(x, y)$

fractal percolation sets, but it holds in the homogeneous case. The way to prove this is via the 45°-projections of $E(1, M, \mathbf{p}) \times E(1, M, \mathbf{p})$.

4.2 The Product of Two One Dimensional Fractal Percolation Versus a Two Dimensional Fractal Percolation

We explain this relation in the case when $M = 3$. Assume that we are given the inhomogeneous fractal percolations $E(1, 3, \mathbf{a})$, and $E(1, 3, \mathbf{b})$, where $\mathbf{a} = (a_1, a_2, a_3)$, $\mathbf{b} = (b_1, b_2, b_3)$ are the vectors of probabilities. We define the vector $\mathbf{p} \in [0, 1]^9$ as their product $\mathbf{p} = \mathbf{a} \bigotimes \mathbf{b}$ in the natural way which is suggested by looking at Fig. 1a. That is:

$$p_i := a_u \cdot b_v \text{ if } i - 1 = 3 * (v - 1) + (u - 1), \ 1 \le u, v \le 3.$$

The reason that $E(2, 3, \mathbf{p})$ and $E(1, 3, \mathbf{a}) \times E(1, 3, \mathbf{b})$ are similar is explained in **(a)** and the essential difference between them is pointed out in **(b)** below:

(a) Let $\mathbf{i} \in \{1, \dots, 9\}^n$. Then the probability that $Q_{\mathbf{i}}$ is retained is the same during the construction of $E(2, 3, \mathbf{p})$ and the construction of $E(1, 3, \mathbf{a}) \times E(1, 3, \mathbf{b})$.
(b) Let K and L be level n squares for some n. Assume that both K and L are retained during the construction of $E(2, 3, \mathbf{p})$ and $E(1, 3, \mathbf{a}) \times E(1, 3, \mathbf{b})$. Then

- In the construction of $E = E(2, 3, \mathbf{p})$ the sets $E \cap K$ and $E \cap L$ are independent.
- In the construction of $E(1, 3, \mathbf{a}) \times E(1, 3, \mathbf{b})$ the sets $E \cap K$ and $E \cap L$ are independent iff $\text{proj}_x K \neq \text{proj}_x L$ and $\text{proj}_y K \neq \text{proj}_y L$ hold.

In dimension $d \geq 2$ the analogy is the same: the probability of the retention of a level n cube is the same for the d-dimensional percolation and for the d-fold product of the corresponding one dimensional percolations. On the other hand, the future of what ever happens in two distinct retained level n cubes is:

- always independent in the d-dimensional percolation case,
- independent for the d-fold product of the corresponding one dimensional fractal percolations iff the two cubes do not share any common projections to coordinate axes.b

4.3 The Existence of An Interval in the Arithmetic Difference Set

Let $E_1 := E(1, M, \mathbf{p})$ and $E_2 := E(1, M, \mathbf{q})$. We define the cyclic cross correlation coefficients:

$$\gamma_k := \sum_{i=1}^{M} p_i q_{i-k(\text{mod } M)} \text{ for } k = 1, \dots, M. \tag{4.1}$$

Theorem 7 ([DS08]) *Assuming that $E_1, E_2 \neq \emptyset$, we have*

(a) If $\forall i = 1, \dots, M : \gamma_i > 1$ *then almost surely*

$$E_2 - E_1 \; contains\; an\; interval.$$

(b) If $\exists i \in \{1, \dots, M\} : \gamma_i, \gamma_{i+1 \text{ mod } M} < 1$ *then almost surely*

$$E_2 - E_1 \; does\; not\; contain\; any\; interval.$$

In the homogeneous case and in the case when $M = 3$ this gives complete characterization. Otherwise we can change to higher order Cantor sets (collapsing $n \geq 2$ steps of the construction into one) and we can apply the same theorem in that case. The fact that this can be done is not trivial because higher order fractal percolations are correlated. That is the way as the random set develops in one level n square is dependent how it develops in some other squares. Nevertheless, M. Dekking and H. Don proved that this can be done by pointing out that the proof of the theorem above can be carried out for more general, correlated random sets than the inhomogeneous fractal percolations. This more general family includes the higher order fractal percolation sets.

4.4 The Lebesgue Measure of the Arithmetic Difference Set

Let E_2, E_2 be two independent realizations of $E(1, M, \mathbf{p})$. Then

$$\gamma_k := \sum_{i=1}^{M} p_i p_{i-k(\text{mod } M)} \text{ for } k = 1, \dots, M.$$

Let $\Gamma := \gamma_1 \cdots \gamma_M$.

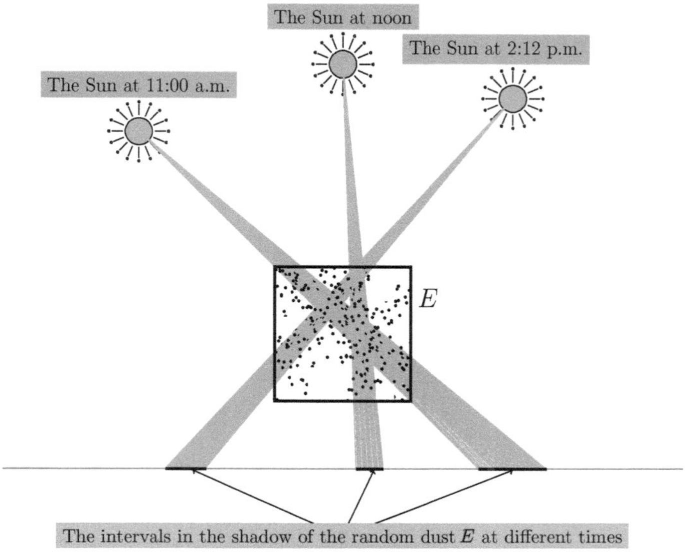

Fig. 4 Projections of fractal percolation

Theorem 8 ([MSS09]) *If* $\Gamma > 1$ *then*
$\mathcal{L}\text{eb}(E_2 - E_1) > 0.$

Combined application of Theorems 7 and 8 yields that the Palis conjecture does not hold in the case when for $M = 3$ and $\mathbf{p} = (0.52, 0.5, 0.72)$. Namely, in this case $\gamma_1 = 1.0388$ and $\gamma_2 = \gamma_3 = 0.941$. Let E_1, E_2 be two independent copies of $E(1, 3, \mathbf{p})$. Then by Theorem 7 there is no interval in $E_1 - E_2$ (since there are two consecutive γ's that are smaller than one) and by Theorem 8 we have $\mathcal{L}\text{eb}\,(E_1 - E_2) > 0$ since $\gamma_1 \cdot \gamma_2 \cdot \gamma_3 = 1.0272 > 1$.

5 General Projections: The Opaque Case

In this and in the following sections we study the projections of fractal percolation sets in general directions. In this section we consider the case $\dim_{\mathrm{H}}(E) > 1$. Under some mild assumption, almost surely projections of E have not only positive Lebesgue measure, as per Marstrand theorem, but also non-empty interior. Furthermore it holds for all and not only almost all directions. Moreover, this remains valid if we replace the orthogonal projection with a much more general family of projections.

One practical application of our result is shown above (Fig. 4). One does not need to rotate such a set to use it as an umbrella.

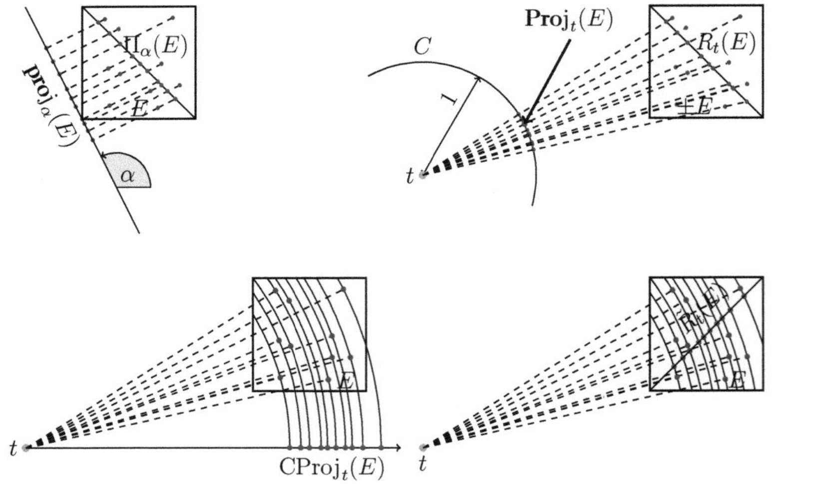

Fig. 5 The orthogonal proj_α, radial Proj_t, co-radial CProj_t projections and the auxiliary projections Π_α, R_t, and $\tilde{R}_t$

We have already studied the horizontal and vertical projections. So we can restrict our attention to the directions $\alpha \in \mathcal{D} := (0, 90°)$ A condition $A(\alpha)$, $\alpha \in \mathcal{D}$ on the vector of probabilities **p** will be defined below.

Theorem 9 ([RS00b]) *Let* $\alpha \in \mathcal{D}$. *If* $A(\alpha)$ *holds and* E *is nonempty then almost surely* $\text{proj}_\alpha(E)$ *contains an interval.*

Theorem 10 ([RS00b]) *If* $A(\alpha)$ *holds for all* $\alpha \in \mathcal{D}$ *and* $E \neq \emptyset$ *then almost surely all projections* $\text{proj}_\alpha(E)$ *contain an interval.*

Remark 11 The assertions of Theorems 9 and 10 remain valid if we replace proj_α with more general families of projections, see [RS00b, Sect. 6]. In particular, radial or co-radial projections (see Fig. 5) are included. □

Example 12 If either

(1) Homogeneous case: $p_i = p > M^{-1}$ for all i, or
(2) Generalized random Sierpiński Carpet: $M = 3$, $p_5 = q$, $p_i = p$ for $i \neq 5$, and

$$p > \max\left(\frac{1}{3}, \frac{1-q}{2}\right)$$

then Condition A(α) is satisfied for all $\alpha \in \mathcal{D}$. Note that (1) is equivalent to $\dim_{\text{H}}(E) > 1$ almost surely. □

5.1 Horizontal and Vertical Projections

Let us start by presenting the large deviation argument (by Falconer and Grimmett) working for horizontal and vertical projections. If $\dim_H \Lambda > 1$ then from the dimension formula for some n one can find a level n column with (exponentially) many squares. We prove inductively that in its every N-th level sub-column, $N > n$, we typically have exponentially many squares on each level (probability of existence of $N > n$ and an N-th level subcolumn which does not have exponentially many squares is super-exponentially small). When we move from level n column to its level $n+1$ subcolumns, each square in the column gives birth to an expected number of $pM > 1$ number of level $n+1$ squares in each of the subcolumns. By large deviation theorem there is only a superexponentially small probability that the number of level $n+1$ squares in a subcolumn is smaller than a fixed $\alpha \in (1, pM)$ multiple of the level n squares in the column. By induction, if this exceptional situation does not happen (or happens only finitely many times), for each $N > n$ the number of squares of level N in each subcolumn will be at least of order α^{N-n}.

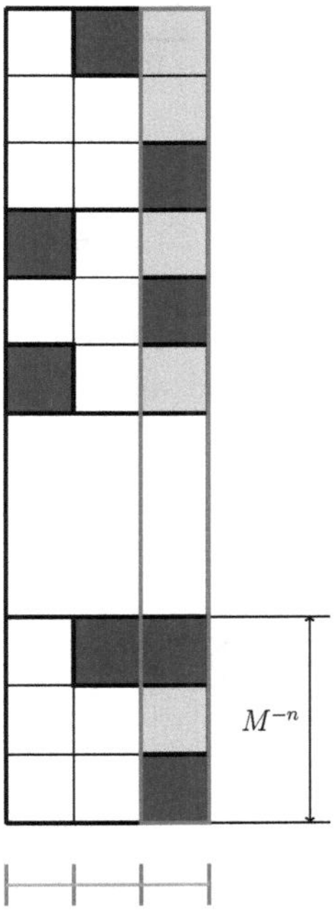

5.2 Condition A

Our goal in this subsection is to modify this argument to work in a more complicated situation of projections in general directions. Indeed, contrary to the horizontal/vertical projections case, here it is in general not true that if a line intersects a square of level n then the expected number of squares of level $n+1$ it intersects is greater than 1. It is still true if the line intersects 'central' part of the square, but not if it hits it close to the corners.

Nevertheless, we are able to find a modified version of the argument. We fix $\alpha \in \mathcal{D}$. We are going to consider Π_α instead of proj_α, i.e. we are projecting onto a diagonal Δ_α of Q, see Fig. 5. For any $\mathbf{i} \in \mathcal{I}^n$ the map $\Pi_\alpha \circ \varphi_{\mathbf{i}} : \Delta_\alpha \to \Delta_\alpha$ is a linear contraction of ratio M^{-n}. We will use its inverse: a map $\psi_{\alpha,\mathbf{i}} : \Pi_\alpha(Q_{\mathbf{i}}) \to \Delta_\alpha$. It is a linear expanding map (of ratio M^n) and it is onto.

Consider the class of nonnegative real functions on Δ_α, vanishing on the endpoints. There is a natural random inverse Markov operator G_α defined as

$$G_\alpha f(x) = \sum_{i \in \mathcal{E}_1; x \in \Pi_\alpha(Q_i)} f \circ \psi_{\alpha,i}(x).$$

The corresponding operator on the n-th level is

$$G_\alpha^{(n)} f(x) = \sum_{\mathbf{i} \in \mathcal{E}_n; x \in \Pi_\alpha(Q_{\mathbf{i}})} f \circ \psi_{\alpha,\mathbf{i}}(x).$$

In particular for any $H \subset \Delta^\alpha$ we have

$$G_\alpha^{(n)} \mathbf{1}_H(x) = \# \{ \mathbf{i} \in \mathcal{E}_n : x \in \Pi_\alpha(\varphi_{\mathbf{i}}(H)) \}.$$

Although $G_\alpha^{(n)}$ should not be thought of as the n-th iterate of G_α, the expected value of $G_\alpha^{(n)}$ is the n-th iterate of the expected value of G_α. Namely, let

$$F_\alpha = \mathbb{E}[G_\alpha] \text{ and } F_\alpha^n = \mathbb{E}\left[G_\alpha^n\right]$$

We then have the formulas

$$F_\alpha f(x) = \sum_{i \in \mathcal{I}; x \in \Pi_\alpha(Q_i)} p_i \cdot f \circ \psi_{\alpha,i}(x)$$

and

$$F_\alpha^n f(x) = \sum_{\mathbf{i}; x \in \Pi_\alpha(Q_{\mathbf{i}})} p_{\mathbf{i}} \cdot f \circ \psi_{\alpha,\mathbf{i}}(x),$$

where

$$p_{\mathbf{i}} = \prod_{k=1}^{n} p_{i_k}.$$

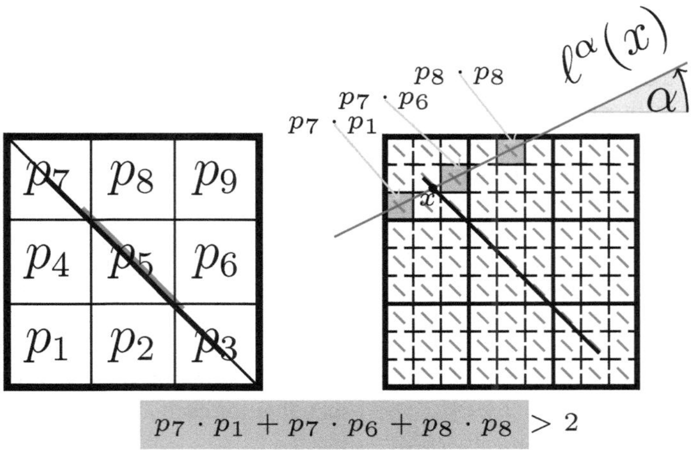

Fig. 6 Condition $A(\alpha)$: I_1^α is the small red, I_2^α is the big blue interval on the left. $r_\alpha = 2$ and the small red intervals are the scaled copies of I_1^α

Definition 13 We say the percolation model satisfies **Condition A**(α) if there exist closed intervals $I_1^\alpha, I_2^\alpha \subset \Delta_\alpha$ and a positive integer r_α such that

(i) $I_1^\alpha \subset \text{int} I_2^\alpha$, $I_2^\alpha \subset \text{int}\Delta_\alpha$,
(ii) $F_\alpha^{r_\alpha} \mathbf{1}_{I_1^\alpha} \geq 2 \cdot \mathbf{1}_{I_2^\alpha}$.

It will be convenient to use additional notation. For $x \in \Delta_\alpha$, $\alpha \in \mathcal{D}$, and $I \subset \Delta_\alpha$ we denote

$$D_n(x, I, \alpha) = \{\mathbf{i} \in \mathcal{I}^n; x \in \Pi_\alpha \circ \varphi_{\mathbf{i}}(I)\}.$$

That is, if we write $\ell^\alpha(x)$ for the line segment through $x \in \Delta_\alpha$ in direction α, $D_n(x, I, \alpha)$ is the set of $\mathbf{i}$ for which $\ell^\alpha(x)$ intersects $\varphi_{\mathbf{i}}(I)$.

The point ii) of Definition 13 can then be written as

$$\forall_{x \in I_2^\alpha} \sum_{\mathbf{i} \in D_{r_\alpha}(x, I_1^\alpha, \alpha)} p_{\mathbf{i}} \geq 2.$$

In other words, Condition $A(\alpha)$ is satisfied if for given α one can define 'small central' and 'large central' part of each square in such a way that for some $r \in \mathbb{N}$ if a line in direction α intersects the 'large central' part of some n-th level square then the expected number of 'small central' parts of its $n + r$-th level subsquares it intersects is uniformly greater than 1 (Fig. 6).

5.3 Consequences of Condition $A(\alpha)$

It is clear that if $A(\alpha)$ holds then one can apply the large deviation argument for projection in direction α-modulo a minor technical problem that the random variables in the large deviations theorem are not identically distributed.

A bit more complicated is the proof that almost surely all the projections contain intervals. It is based on the following robustness properties:

Proposition 14 *If condition $A(\alpha)$ holds for some $\alpha \in \mathcal{D}$ for some I_1^α, I_2^α and r_α then it will also hold in some neighbourhood $J \ni \alpha$. Moreover, for all $\theta \in J$ we can choose $I_1^\theta = I_1'$, $I_2^\theta = I_2$, $r_\theta = r_\alpha$ not depending on θ.*

A natural corollary is that the whole range $\mathcal{D}$ can be presented as a countable union of closed intervals $J_i = [\alpha_i^-, \alpha_i^+]$ such that for each i Condition A(α) holds for all $\alpha \in J_i$ with the same I_1^i, I_2^i, r_i.

Proposition 15 *Let $I \subset B(I, \ell) \subset J \subset \Delta_\alpha$. If $\mathbf{i} \in D_n(x, I, \alpha)$ then $\mathbf{i} \in D_n(x, J, \beta)$ for all $\beta \in (\alpha - \ell M^{-n}, \alpha + \ell M^{-n})$.*

Hence, inside each J_i one does not need to repeat the large deviation argument separately for each α. At level n it is enough to check it for approximately M^n directions. As the number of directions one needs to check grows only exponentially fast with n, the proof goes through.

5.4 Checking Condition $A(\alpha)$

One last thing needed is an efficient way to check whether $A(\alpha)$ holds.

Definition 16 We say that the fractal percolation model satisfies **Condition B**(α) if there exists a nonnegative continuous function $f : \Delta_\alpha \to \mathbb{R}$ such that f is strictly positive except at the endpoints of Δ_α and that

$$F_\alpha f \geq (1 + \varepsilon) f \tag{5.1}$$

for some $\varepsilon > 0$.

Proposition 17 *$B(\alpha)$ implies $A(\alpha)$.*

In particular, for homogeneous case $p_i = p > M^{-1}$ for any α one can choose $f(x)$ as the length of the intersection of Q with the line in direction α passing through x. It is easy to check that this function satisfies (5.1) for $\varepsilon = pM - 1$.

5.5 Application: Visibility

For a given set E, we define the visible subset (from direction α) as the set of points $x \in E$ such that the half-line starting at x and going in direction α does not meet any other point $y \in E$. Similarly, given $z \in \mathbb{R}^2$, the visible subset (from z) is the set of points $x \in E$ such that the interval $\overline{xz}$ does not meet any other point $y \in E$.

Let E be a homogeneous fractal percolation with $p > M^{-1}$. By Theorem 9, E is quite opaque: the orthogonal projection in any direction almost surely contain intervals. In particular, with large probability it contains large intervals. By stochastical self-similarity of E, the same is true for each $E \cap Q_{\mathbf{i}}$. Hence, not many points can be visible:

Theorem 18 ([Ar12]) *If E is nonempty, almost surely the visible set from direction α has finite one-dimensional Hausdorff measure for each α and the visible set from point z has Hausdorff dimension 1 for each $z \in \mathbb{R}^2$.*

6 General Projections: The Transparent Case

In this section we present results analogous to the second part of the Marstrand theorem. For homogeneous fractal percolation with Hausdorff dimension smaller than 1 almost surely $\dim_{\mathrm{H}}(\mathrm{proj}_{\alpha}(E)) = \dim_{\mathrm{H}} E$ for all α. Together with the results of the previous section, it implies

Theorem 19 ([RS00]) *In the homogeneous case, that is $E = E^h(2, M, p)$ for almost all realizations of E*

$$\forall \alpha, \ \dim_{\mathrm{H}}(\mathrm{proj}_{\alpha} E) = \min\{1, \dim_{\mathrm{H}}(E)\}. \tag{6.1}$$

Principal Assumption for this Section: In this section we always work in the homogeneous case:

$$E = E^h(2, M, p),$$

where

$$M^{-2} < p \leq M^{-1}. \tag{6.2}$$

That is p is chosen to ensure that $E \neq \emptyset$ with positive probability and $\dim_{\mathrm{H}}(E) \leq 1$ almost surely conditioned on non-extinction. To prove Theorem 19 one needs to analyze the structure of the slices of E_n:

Informal description of the structure of slices of $\mathbf{E_n}$ (which was defined as the n-th approximation of E): Namely, for almost all realizations of E and for **all** straight lines ℓ: the number of level n squares having nonempty intersection with E is at most const $\cdot\, n$. On the other hand, almost surely for n big enough, we can find some line of 45° angle which intersects const $\cdot\, n$ level n squares.

Let $\mathcal{L}^{\varepsilon}$ be the set of lines on the plane whose angle is separated both from 0° and 90° at least by ε. Further for a line ℓ let $\mathcal{E}_n(\ell)$ be the set of retained level n squares that intersect ℓ. That is,

$$\mathcal{E}_n(\ell) := \{\mathbf{i} \in \mathcal{E}_n : Q_{\mathbf{i}} \cap \ell \neq \emptyset\}.$$

Theorem 20 ([RS00]) *For almost all realizations of E we have*

$$\forall \varepsilon \in \left(0, \frac{\pi}{2}\right), \ \exists N, \ \forall n \geq N, \ \forall \ell \in \mathcal{L}^{\varepsilon}; \quad \#\mathcal{E}_n(\ell) \leq \text{const} \cdot n. \tag{6.3}$$

For simplicity, the proof in horizontal/vertical direction only (for general directions one needs to apply techniques presented in previous subsection). The proof is once again based on the large deviation argument, but working in the opposite direction. This time the expected number of squares in a subcolumn is smaller (by a constant bounded away from 1) than the number of squares in the column (and not greater, like in the opaque case). Hence, we can guarantee that if the column has sufficiently many squares for the large deviation theory to work, the number of squares in all subcolumns will shrink. This leads to an estimation on the possible rate of growth.

This estimation is sharp:

Proposition 21 ([RS00]) *There exists a constant $0 < \lambda < 1$ such that for almost all realizations, conditioned on $E \neq \emptyset$, there exists an N such that for all $n > N$ there exists a line ℓ with*

$$\#\mathcal{E}_n(\ell) > \lambda n. \tag{6.4}$$

Theorem 19 is an immediate consequence of Theorem 20.

7 The Arithmetic Sum of at Least Three Fractal Percolations

To study arithmetic sums of more than two fractal percolations we need to combine results of the previous three sections. Like in Sect. 4, we look at the projection $(x_1, \ldots, x_d) \to \sum x_i$ from the cartesian product of fractal percolations to the real line. The proof is based on the large deviation argument presented in Sect. 5. However, the main technical difficulty is the presence of dependencies. We will use the results from Sect. 6 to bound their impact.

Let

$$E^i := E^h(1, M, p_i), \ i = 1, 2, 3, \quad p := p_1 \cdot p_2 \cdot p_3 \text{ and } E := E^h(3, M, p).$$

Then

$$\dim_{\mathrm{H}}\left(E^1 \times E^2 \times E^3\right) = \dim_{\mathrm{H}}(E) = \frac{\log M^3 \cdot p}{\log M}.$$

Moreover, the probability that a level n cube C is contained in any of the two random Cantor sets above is equal to p^n.

Let S_a be the plane $\{\sum x_i = a\}$. We can write

$$E^{\text{sum}} := E^1 + E^2 + E^3 = \left\{a : S_a \cap \left(E^1 \times E^2 \times E^3\right) \neq \emptyset\right\}.$$

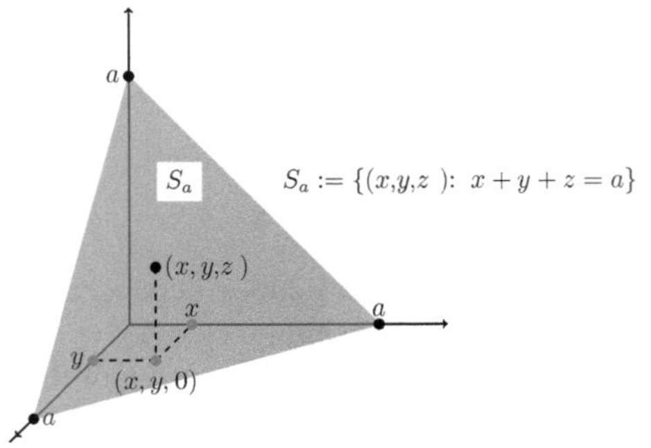

That is we can consider E^{sum} as the projection of $E_1 \times E_2 \times E_3$ to the x-axis with planes orthogonal the vector $(1, 1, 1)$. So, E^{sum} can contain an interval only if its dimension is greater than one, that is $p > M^{-2}$. It is a sufficient condition as well:

Theorem 22 ([RS00]) *Let $d \geq 2$ and for $i = 1, \ldots, d$ let $E^i := E^h(1, M, p_i)$ satisfying*

$$p := \prod_{i=1}^{d} p_i > M^{-d+1}. \tag{7.1}$$

Then for every $\mathbf{b} = (b_1, \ldots, b_d) \in \mathbb{R}^d$, $b_i \neq 0$ for all $i = 1, \ldots, d$ the sum $E^{\text{sum}}_{\mathbf{b}} = \sum_{i=1}^{d} b_i E^i$ contains an interval almost surely, conditioned on all E^i being nonempty.

We explain the proof of this theorem in the special case when $d = 3$ and $\mathbf{b} = (1, 1, 1)$. To verify that a certain $a \in E^{\text{sum}}$ we need to prove that the n approximation of the product intersects S_a, that is $(E^1 \times E^2 \times E^3)_n \cap S_a \neq \emptyset$ for every n. It follows from the dimension formula and (7.1) that we have $M^{n(1+\tau)}$ retained level n cubes for some $\tau > 0$. By the pigeon hole principle for at least one $k = 0, \ldots, 3M^n$ the plane $S_{kM^{-n}}$ intersects at least $M^{n\tau}$ retained level n cubes. For such a k we write $a = kM^{-n}$. So, $\#\{\mathcal{E}_n \cap S_a\} \geq M^{n\tau}$.

Fix an $0 \leq m \leq M$. How many level $n+1$ retained cubes intersect $S_{a+mM^{-(n+1)}}$? If the way $E^1 \times E^2 \times E^3$ develops in every level n cube was independent then we could get that the answer by the large deviation argument: exponentially many except for an event with a super exponentially small probability.

We remind that the cubes are dependent if they have the same x_1, x_2 or x_3 coordinate. Figure 7 shows the geometric position of (some of: we consider only

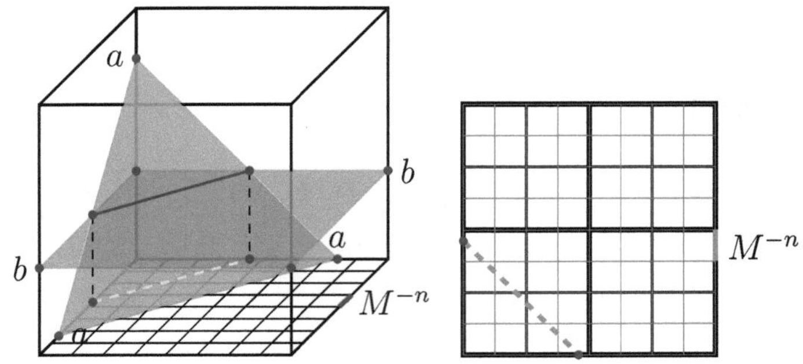

Fig. 7 The cubes intersecting the *red line* are not independent

the cubes with the same x_3 coordinate) cubes dependent on one chosen cube: $x_1 + x_2 + x_3 =$ const and $x_3 =$ const imply $x_1 + x_2 =$ const. Potentially there could be exponentially many such cubes. The key step of the proof is that using a theorem analogous to Theorem 20 for $E^1 \times E^2$ instead of $E^h(2, M, p_1 \cdot p_2)$ one can check that on the red dashed line on Fig. 7 there are only constant times n retained squares, consequently the $M^{n\tau}$ level n cubes having non-empty intersection with S_a (the blue plane on Fig. 7) can be divided into const $\cdot\, n$ classes such that the coordinate axes projection of any two cubes in a class are different. The events inside each class are independent, hence we can use the large deviation theory separately for each class. A technical comment: in order to be able to go with this procedure we may have to decrease p_1, p_2, p_3 in such a way that for the modified values we have

$$p_1 \cdot p_2 \cdot p_3 > M^{-2} \text{ but } p_i \cdot p_j < M^{-1} \text{ for distinct } i, j \in \{1, 2, 3\}.$$

That is, $E^1 \times E^2 \times E^3$ is a big set in the sense that it has dimension greater than one but its all coordinate plane projections should be small sets having dimension smaller than one—only then the n-th approximates of the coordinate plane projections intersect every line in at most const $\cdot\, n$ retained squares. However, the property of almost surely having intervals in the algebraic sum is monotonous with respect to $\{p_i\}$.

Hence among those level n retained cubes that intersect the blue plane S_a there cannot be more than const $\cdot\, n$ on the red line (any coordinate plane parallel line) which imply that the number of cubes dependent on any one cube is polynomial (const $\cdot\, n$). This bound on the dependency matrix lets us control the dependencies.

References

[Ar12] Arhosalo, I., Järvenpää, E., Järvenpää, M., Rams, M., Shmerkin, P.: Visible parts of fractal percolation. Proc. Edinburgh Math. Soc. (Series 2), **55**(02):311–331 (2012)

[CCD88] Chayes, J.T., Chayes, L., Durrett, R.: Connectivity properties of mandelbrot's percolation process. Probability theory and related fields **77**(3), 307–324 (1988)

[Ch96] Chayes, L.: On the length of the shortest crossing in the super-critical phase of mandelbrot's percolation process. Stochastic processes and their applications **61**(1), 25–43 (1996)

[DG88] Dekking, F.M., Grimmett, G.R.: Superbranching processes and projections of random cantor sets. Probability theory and related fields **78**(3), 335–355 (1988)

[De09] Dekking, M.: Random cantor sets and their projections. Fractal Geom. Stochast. IV, 269–284 (2009)

[DM90] Dekking, M., Meester, R.W.J.: On the structure of mandelbrot's percolation process and other random cantor sets. J. Stat. Phys. **58**(5), 1109–1126 (1990)

[DS08] Dekking, M., Simon, K.: On the size of the algebraic difference of two random cantor sets. Random Struct. Algorithms **32**(2), 205–222 (2008)

[Fa86] Falconer, K.J.: Random fractals. Math. Proc. Cambridge Philos. Soc **100**(3), 559–582 (1986)

[FG92] Falconer, K.J., Grimmett, G.R.: On the geometry of random cantor sets and fractal percolation. J. Theor. Probab. **5**(3), 465–485 (1992)

[FG94] Falconer, K.J., Grimmett, G.R.: Correction: On the geometry of random cantor sets and fractal percolation. J. Theor. Probab. **7**(1), 209–210 (1994)

[Ha81] Hawkes, J.: Trees generated by a simple branching process. J. Lond. Math. Soc. **2**(2), 373–384 (1981)

[KP76] Kahane, J.-P., Peyriere, J.: Sur certaines martingales de benoit mandelbrot. Adv. math. **22**(2), 131–145 (1976)

[Ma74] Mandelbrot, B.B.: Intermittent turbulence in self-similar cascades- divergence of high moments and dimension of the carrier. J. Fluid Mech. **62**(2), 331–358 (1974)

[Ma83] Mandelbrot, B.B: The fractal geometry of nature/revised and enlarged edition. p. 495. WH Freeman and Co., New York (1983)

[Ma54] Marstrand, J.M.: Some fundamental geometrical properties of plane sets of fractional dimensions. Proc. Lond. Math. Soc. **3**(1), 257–302 (1954)

[MW86] Mauldin, R.D., Williams, S.C.: Random recursive constructions: asymptotic geometric and topological properties. Trans. Amer. Math. Soc **295**(1), 325–346 (1986)

[MSS09] Mora, P., Simon, K., Solomyak, B.: The lebesgue measure of the algebraic difference of two random cantor sets. Indagationes Mathematicae **20**(1), 131–149 (2009)

[RS00] Rams, M., Simon, K.: The dimension of projections of fractal percolations. J. Stat. Phys **154**(3), 633–655 (2014)

[RS00b] Rams, M., Simon, K.: Projections of fractal percolations. To appear in Ergodic Theor. Dyn.Syst. (2013). doi:10.1017/etds.2013.45

Self-affine Sets and the Continuity of Subadditive Pressure

Pablo Shmerkin

Abstract The affinity dimension is a number associated to an iterated function system of affine maps, which is fundamental in the study of the fractal dimensions of self-affine sets. De-Jun Feng and the author recently solved a folklore open problem, by proving that the affinity dimension is a continuous function of the defining maps. The proof also yields the continuity of a topological pressure arising in the study of random matrix products. I survey the definition, motivation and main properties of the affinity dimension and the associated SVF topological pressure, and give a proof of their continuity in the special case of ambient dimension two.

Keywords Topological pressure · Self-affine sets · Affinity dimension · Subadditive thermodynamic formalism

2010 Mathematics Subject Classification Primary 37C45 · 37D35 · 37H15

1 Introduction

Let $\mathcal{F} = (f_1, \ldots, f_m)$ be a collection of contractive affine maps on some Euclidean space $\mathbb{R}^d$. That is, $f_i(x) = A_i x + t_i$, where $A_i \in \mathbb{R}^{d\times d}$ are linear maps, $t_i \in \mathbb{R}^d$ are translations, and $\|A_i\| < 1$, where $\|\cdot\|$ denotes Euclidean operator norm (although any other operator norm would work equally well). It is well known that there exists a unique nonempty compact set $E = E(\mathcal{F})$ such that

$$E = \bigcup_{i=1}^{m} f_i(E) = \bigcup_{i=1}^{m} A_i E + t_i. \tag{1.1}$$

The author was partially supported by a Leverhulme Early Career Fellowship.

P. Shmerkin (✉)
University of Surrey, Guildford, UK
e-mail: pshmerkin@utdt.edu

P. Shmerkin
Torcuato Di Tella University, Buenos Aires, Argentina

D.-J. Feng and K.-S. Lau (eds.), *Geometry and Analysis of Fractals*,
Springer Proceedings in Mathematics & Statistics 88,
DOI 10.1007/978-3-662-43920-3_12

Such sets are called *self-affine*. The tuple $\mathcal{F}$ is termed as an *iterated function system*, and E is the *attractor* or *invariant set* of $\mathcal{F}$.

An important special case is that in which the maps f_i are all similarities; in this case E is known as a *self-similar set*. It is known that for self-similar sets, Hausdorff, lower and upper box counting dimensions all agree. Moreover, if s is the only real solution to $\sum_{i=1}^{m} r_i^s = 1$, where r_i is the similarity ratio of f_i, then s is an upper bound for the Hausdorff dimension of E, and equals the Hausdorff dimension of E under a number of "controlled overlapping" conditions, the strongest and simplest being the strong separation condition, which requires that the basic pieces $f_i(E)$ are mutually disjoint. The number s is called the *similarity dimension* of the system $\mathcal{F}$, and is clearly continuous, and indeed real-analytic, as a function of the maps f_i (identified with the Euclidean space of the appropriate dimension).

The situation is dramatically more complex for general self-affine sets. It is well-known that the Hausdorff and box counting dimensions of self-affine sets may differ, and that each of them is a discontinuous function of the defining maps, even under the strong separation condition. Strikingly, it is not even known whether lower and upper box dimensions always coincide for self-affine sets. No general formula for either the Hausdorff or box counting dimension is known or expected to exist, again even in the strongly separated case. However, although the dimension theory of self-affine sets may appear at first sight like a bleak subject, many interesting and deep results have been obtained. Among these, Falconer's seminal paper [Fal92] has been highly influential. There, Falconer introduces a number $s = s(\mathcal{F})$ associated to an affine IFS $\mathcal{F}$, which we will term *affinity dimension* (no standard terminology exists; the term singularity dimension is also often used). As a matter of fact, s depends only on $A = (A_1, \ldots, A_m)$, i.e. the linear parts of the affine maps f_i, and is independent of the translations.

Falconer proved that the affinity dimension is always an upper bound for the upper box-counting, and therefore the Hausdorff, dimension of E, and in some sense, "typically" equals the Hausdorff dimension of E; his result is described in more detail in Sect. 2.3 below. The definition of the affinity dimension is rather more involved than the definition of similarity dimension, which it extends, and is postponed to Sect. 2.3.

The question of whether the affinity dimension is continuous as a function of the generating maps has been a folklore open question in the fractal geometry community for well over a decade (I learned it from B. Solomyak around 2000), and was raised explicitly in [FS09]. Recently, together with Feng and Shmerkin [FS13] we proved that the answer is affirmative:

Theorem 1.1 *The affinity dimension s is a continuous function of the linear maps $(A_1, \ldots, A_m)$.*

A related but in some sense simpler result concerns the norms of matrix products. Again let $A = (A_1, \ldots, A_m)$ be a finite collection of invertible linear maps on $\mathbb{R}^d$. Given $s \geq 0$, define

$$M(A, s) = \lim_{n\to\infty} \frac{1}{n} \log \left(\sum_{(i_1 \cdots i_k)} \|A_{i_k} \cdots A_{i_1}\|^s \right), \qquad (1.2)$$

where $\|\cdot\|$ is the standard Euclidean norm (the limit is easily seen to exist from subadditivity). The reader familiar with the thermodynamic formalism may note that this definition resembles the definition of topological pressure of a continuous potential on the full shift on m symbols, except that here we consider norms of matrix products instead of Birkhoff sums; this is an instance of the topological pressure in the setting of the *subadditive* thermodynamic formalism. This will be discussed in Sect. 3.1.

The quantity $M(A, s)$ is rather natural. On one hand, in the thermodynamic setting it is closely linked to the Lyapunov exponent of an IID random matrix product (with respect to ergodic measures under the shift). On the other hand, the "zero temperature limit" as $s \to \infty$ is the joint spectral radius of the matrices $(A_1, \ldots, A_m)$, which is an important quantity in a wide variety of fields. Although the joint spectral radius is well-known to be continuous, it is far from clear from the definition whether $M(A, s)$ is always continuous. Together with Feng [FS13], we have proved that it is:

Theorem 1.2 *$M(A, s)$ is jointly continuous in (A, s).*

Although Theorems 1.1 and 1.2 are in effect linear algebraic statements, the proofs make heavy use of dynamical systems theory, and in particular the variational principle for sub-additive potentials.

The goal of this survey is twofold. On one hand, it is an overview of the definition and main properties of the affinity dimension and the closely related singular value pressure, and the geometric reasons why it comes up naturally in the study of self-affine sets. On the other hand, it contains a full proof of Theorems 1.1 and 1.2 in the case of ambient dimension $d = 2$ (for $d = 1$, both results are trivial). The two-dimensional case captures many of the main ideas of the general case, while being technically much simpler.

I note that De-Jun Feng [private communication] has observed that a result of Bocker-Neto and Viana [BV10] on continuity of Lyapunov exponents for IID $\mathbb{R}^2$ matrix cocycles can be used to give a short alternative proof of Theorems 1.1 and 1.2 in the case $d = 2$. However, that proof does not generalize to any other dimensions.

2 SVF, Topological Pressure, and Affinity Dimension

2.1 Definition and Basic Properties of the SVF

Recall that given a linear map $A \in \mathrm{GL}_d(\mathbb{R})$, its *singular values* $\alpha_1(A) \geq \cdots \geq \alpha_d(A) > 0$ are the lengths of the semi-axes of the ellipsoid $A(B^d)$, where B^d is the unit ball of $\mathbb{R}^d$. Alternatively, the singular values are the square roots of the eigenvalues of A^*A (where A^* is the adjoint of A). In particular, $\alpha_1(A)$ is nothing else than the Euclidean norm of A:

$$\alpha_1(A) = \sup\{\|Av\| : \|v\| = 1\},$$

where $\|v\|$ denotes the Euclidean norm of $v \in \mathbb{R}^d$. Likewise,

$$\alpha_d(A) = \inf\{\|Av\| : \|v\| = 1\} = \|A^{-1}\|^{-1}.$$

Also,

$$\det(A) = \det(A^*A)^{1/2} = \prod_{i=1}^{d} \alpha_i(A).$$

Given $s \in [0, d)$, we define the *singular value function* (SVF) $\varphi^s : \mathrm{GL}_d(\mathbb{R}) \to (0, \infty)$ as follows. Let $m = \lfloor s \rfloor$. Then

$$\varphi^s(A) = \alpha_1(A) \cdots \alpha_m(A)\alpha_{m+1}(A)^{s-m}.$$

An alternative way of expressing this is:

$$\varphi^s(A) = \|A\|_m^{m+1-s} \cdot \|A\|_{m+1}^{s-m}, \tag{2.1}$$

where

$$\|A\|_k = \alpha_1(A) \cdots \alpha_k(A).$$

The reason why (2.1) is useful is that $\|A\|_k$ is a sub-multiplicative seminorm ($\|AB\|_k \le \|A\|_k \|B\|_k$). Indeed, $\|A\|_k$ is the operator norm of A when acting on the space of exterior k-forms. Alternatively, $\|A\|_k = \sup\{\det(A|_\pi) : \pi \in G(d, k)\}$ where $G(d, k)$ is the Grassmanian of k-dimensional linear subspaces of $\mathbb{R}^d$. As an immediate consequence of (2.1), we get the following key property of the SVF:

Lemma 2.1 (Sub-multiplicativity of the SVF) *$\varphi^s(AB) \le \varphi^s(A)\varphi^s(B)$ for all $A, B \in GL_d(\mathbb{R})$ and $s \in [0, d)$.*

It is also clear that $\varphi^s(A)$ is jointly continuous in s and A; it is also jointly real-analytic for non-integer s, but in general it is not even differentiable at $s = 1, \ldots, d-1$ for a fixed A. We note also that $\varphi^1(A) = \alpha_1(A) = \|A\|$ and $\lim_{s \to d} \varphi^s(A) = \det(A)$. For completeness we define $\varphi^s(A) = \det(A)^{s/d}$ for $s \ge d$, and note that this definition preserves all of the previous properties when $s \ge d$.

2.2 SVF Topological Pressure

Let $I = \{1, \ldots, m\}$. We denote by I^* the family of finite words with symbols in I, and write $|\mathtt{i}|$ for the length of $\mathtt{i} \in I^*$. The space $X := I^{\mathbb{N}}$ of right-infinite sequences is endowed with the left-shift operator σ, i.e. $\sigma(i_1 i_2 \cdots) = (i_2 i_3 \cdots)$. Given $\mathtt{i} \in X$, the restriction of $\mathtt{i}$ to its first k coordinates is denoted by $\mathtt{i}|k$. Finally, if $\mathtt{j} \in I^*$, then $[\mathtt{j}] \subset X$ is the family of all infinite sequences which start with $\mathtt{j}$.

Given $A = (A_1, \ldots, A_m) \in (\mathrm{GL}_d(\mathbb{R}))^m$ and $\mathtt{i} = (i_1 \cdots i_n) \in I^*$, we denote $A(\mathtt{i}) = A_{i_n} \cdots A_{i_1}$. The next lemma introduces the main concept of this article.

Lemma 2.2 *Given $A \in (GL_d(\mathbb{R}))^m$ and $s \geq 0$, let*

$$S_n(A, s) = \log \sum_{\mathtt{i} \in I^n} \varphi^s(A(\mathtt{i})).$$

Then the limit

$$P(A, s) := \lim_{n \to \infty} \frac{S_n(A, s)}{n} \tag{2.2}$$

exists and equals $\inf_{n \geq 1} S_n(A, s)/n > -\infty$.

Proof Lemma 2.1 implies that the sequence $S_n = S_n(A, s)$ is subadditive, i.e. $S_{n+k} \leq S_n + S_k$. But it is well known that for any subadditive sequence S_n, the limit of S_n/n exists and equals $\inf_{n \geq 1} S_n/n$. Finally, since $\varphi^s(A) \geq \det(A)^s$ and we are assuming that the maps A_i are invertible, one can easily check that

$$\lim_{n \to \infty} \frac{S_n}{n} \geq \log \left(\sum_{i \in I} \det(A(\mathtt{i}))^s \right) > -\infty.$$

□

Definition 2.1 The function $P(A, s)$ defined in (2.2) is called the *SVF topological pressure*.

It is instructive to compare the definitions of $P(A, s)$ and $M(A, s)$ given in (1.2). Both quantities coincide for $0 \leq s \leq 1$; for $s > 1$, the definition of $P(A, s)$ takes into account different singular values of the matrix products involved. Since $\varphi^s(A) \leq \|A\|^s$, one always has $P(A, s) \leq M(A, s)$.

The following lemma summarizes some elementary but important continuity properties of the topological pressure.

Lemma 2.3 *The following hold:*

(1) *Given $A = (A_1, \ldots, A_m) \in (GL_d(\mathbb{R}))^m$, let*

$$\alpha_* = \min_{i \in I}\{\alpha_d(A_i)\}, \quad \alpha^* = \max_{i \in I}\{\alpha_1(A_i)\}.$$

Then

$$(\log \alpha_*)\varepsilon \le P(A, s+\varepsilon) - P(A, s) \le (\log \alpha^*)\varepsilon.$$

(2) *For fixed* $A = (A_1, \ldots, A_m) \in (GL_d(\mathbb{R}))^m$, *the function* $s \to P(A, s)$ *is Lipschitz continuous; the Lipschitz constant is uniform in a neighborhood of A.*
(3) $P(A, s)$ *is upper semicontinuous (as a function of both A and s).*

Proof Note that $\varphi^s(B)\alpha_d(B)^\varepsilon \le \varphi^{s+\varepsilon}(B) \le \varphi^s(B)\alpha_1(B)^\varepsilon$ for any $s, \varepsilon > 0$ and $B \in \mathrm{GL}_d(\mathbb{R})$. Also, $\alpha_1(B) = \|B\|$ is sub-multiplicative and $\alpha_d(B) = \|B^{-1}\|^{-1}$ is super-multiplicative. Combining these facts yields

$$n\varepsilon \log \alpha_* + S_n(A, s) \le S_n(A, s+\varepsilon) \le n\varepsilon \log \alpha^* + S_n(A, s),$$

which yields the first claim. The second claim is immediate from the first, and the fact that α^*, α_* are continuous functions of A.

Finally, upper semicontinuity follows since $P(A, s) = \inf_{n\ge 1} S_n(A, s)/n$ is an infimum of continuous functions. □

In light of the previous lemma, it seems natural to ask whether $P(A, s)$ is not just upper semicontinuous but in fact continuous. As we will see in the next section, this question is closely linked to Theorem 1.1. Falconer and Sloan [FS09] proved continuity of P at tuples of linear maps satisfying certain assumptions, and raised the general continuity problem. Feng and the author [FS13] recently proved that continuity always holds:

Theorem 2.1 *The map* $(A, s) \to P(A, s)$ *is continuous on* $(GL_d(\mathbb{R}))^m \times [0, +\infty)$.

A proof of this theorem in dimension $d = 2$ will be presented in Sect. 4.

2.3 Affinity Dimension and Self-affine Sets

So far, no assumptions have been made on the maps A_i, other than invertibility. However, the motivation for the study of the SVF topological pressure came from the theory of self-affine sets, and in this context the maps A_i are strict contractions.

Lemma 2.4 *If* $A = (A_1, \ldots, A_m) \in (GL_d(\mathbb{R}))^m$ *and* $\|A_i\| < 1$ *for all* $i \in I$, *then* $s \to P(A, s)$ *is a continuous, strictly decreasing function of s on* $[0, \infty)$, *and has a unique zero on* $(0, \infty)$.

Proof That $P(A, s)$ is continuous and strictly decreasing in s follows immediately from Lemma 2.3, since $\alpha^* < 1$ when the maps are strict contractions. By definition $P(A, 0) = \log m > 0$. On the other hand,

$$P(A, s) \le \log\left(\sum_{i\in I} \varphi^s(A_i)\right) \to -\infty \quad \text{as } s \to \infty.$$

Hence $P(A, \cdot)$ has a unique zero. □

Definition 2.2 Given $A = (A_1, \dots, A_m) \in (\mathrm{GL}_d(\mathbb{R}))^m$ with $\|A_i\| < 1$ for all $i \in I$, its *affinity dimension* is the unique positive root s of the pressure equation $P(A, s) = 0$.

Note that Theorem 1.1 is in fact an immediate corollary of Theorem 1.2.

In the rest of this section we indicate why this number is relevant in the study of self-affine sets; this material is by now standard. To begin, let us recall the definition of Hausdorff dimension in terms of Hausdorff content: $\dim_H(E) = \inf\{s : \mathcal{H}^s_\infty(E) = 0\}$, where

$$\mathcal{H}^s_\infty(E) = \inf\left\{\sum_i r_i^s : E \subset B(x_i, r_i)\right\}.$$

Now suppose $E = \bigcup_{i\in I} A_i E + t_i$ is the invariant set of the IFS $\{A_i x + t_i\}_{i\in I}$. Since the maps $f_i(x) = A_i x + t_i$ are strict contractions, for all large enough R the closed ball B_R of radius R and center at the origin is mapped into its interior by all the maps f_i. Let $E_0 = B_R$ and define inductively $E_{k+1} = \bigcup_{i\in I} f_i(E_k)$. By our choice of R, it is easy to see inductively that E_k is a decreasing sequence of compact sets; moreover, if we call $F = \bigcap_{k=0}^\infty E_k$, then one can check that $F = \bigcup_{i\in I} f_i(F)$. Thus $E = F$ by uniqueness.

The above discussion shows that E is covered by E_k for any k; moreover, by construction

$$E_k = \bigcup_{i\in I^k} f_{i_1} \cdots f_{i_k}(B_R) =: \bigcup_{i\in I^k} U_{i_1\cdots i_k},$$

where $U_{i_1\cdots i_k}$ is an ellipsoid with semi-axes $R\alpha_1(A_{i_1} \cdots A_{i_k}) \geq \cdots \geq R\alpha_d(A_{i_1} \cdots A_{i_k})$. This shows that there is a natural cover of the self-affine sets by *ellipsoids*. In order to estimate Hausdorff content (and hence Hausdorff dimension) effectively, one needs to find efficient coverings by *balls*. What Falconer observed is that we can cover each ellipsoid efficiently by balls, in a way that depends on the dimension of the Hausdorff content we are trying to estimate. Namely, for each integer $1 \leq m < d$, we can cover an ellipsoid in $\mathbb{R}^d$ with semi-axes $\alpha_1 \geq \cdots \geq \alpha_d$ by a parallelepiped with sides $2\alpha_1 \geq \cdots \geq 2\alpha_d$. In turn, this can be covered by at most

$$(4R\alpha_1/\alpha_m)\cdots(4R\alpha_{m-1}/\alpha_m)(4R)^{d-m+1}$$

cubes of side length α_m, each of which is contained in a ball of radius $\sqrt{d}\alpha_m$. It turns out that if we want to estimate $\mathcal{H}^s_\infty(E)$ by covering each of the ellipsoids that make up E_k in this way, independently of each other, the optimal choice of m is $\lfloor s \rfloor$. This particular choice yields (after some straightforward calculations) a bound

$$\mathcal{H}^s_\infty(E) \leq C_{R,d} \sum_{(i_1,\dots,i_k)\in I^k} \varphi^s(A_{i_1} \cdots A_{i_k}).$$

From here one can deduce that if $P(A,t) < 0$, then $\mathcal{H}^t_\infty(E) = 0$, whence $\dim_H(E) \le t$. Letting $t \to s$, the affinity dimension, finally shows that $\dim_H(E) \le s$.

The argument above can be modified to reveal that the affinity dimension is also an upper bound for the upper box counting dimension of E. Thus, we can say that the affinity dimension is a *candidate* to the (Hausdorff, or box-counting) dimension of a self-affine set, obtained by using the most natural coverings, and is always an upper bound for both the box-counting and Hausdorff dimension. In general, these natural coverings may be far from optimal. For example, many of the ellipsoids making up E_k may overlap substantially or be aligned in such a way that it is far more efficient to cover them together rather than separately. Also, most of the cubes that we employed to cover each ellipsoid might not intersect E at all. And indeed, it may happen that the Hausdorff dimension, and/or the box-counting dimension are strictly smaller than the affinity dimension; this is the case for many kinds of self-affine carpets, see e.g. [Ba07] and references therein. However, it is perhaps surprising that, as discovered by Falconer [Fal92], *typically* the Hausdorff and box-counting dimensions of self-affine sets do coincide with the affinity dimension, in the following precise way:

Theorem 2.2 *Let $A = (A_1, \ldots, A_m)$, with the A_i invertible linear maps on $\mathbb{R}^d$. Assume further that $\|A_i\| < 1/2$ for all $i \in I$. Given $t_1, \ldots, t_m \in \mathbb{R}^d$, denote by $E(t_1, \ldots, t_m)$ the self-affine set corresponding to the IFS $\{A_i x + t_i\}_{i\in I}$.*

Then for Lebesgue-almost all $(t_1, \ldots, t_m) \in \mathbb{R}^{md}$, the Hasudorff dimension of $E(t_1, \ldots, t_m)$ equals the affinity dimension of A.

We remark that Falconer proved the second part under the assumption $\|A_i\| < 1/3$. Solomyak [Sol98] later pointed out a modification in the proof that allows to replace $1/3$ by $1/2$. By an observation of Edgar [Edg92], the bound $1/2$ is optimal. Since Falconer's pioneering work, many advances have been obtained in this direction. A natural question is whether one can give *explicit* conditions under which the Hausdorff and/or box-counting dimensions equal the affinity dimension; this was achieved in [Fal92, HL95, KS09]. In a different direction, Falconer and Miao [FM08] provided a bound on the dimension of exceptional parameters $(t_1, \ldots, t_m)$. For other recent directions in the study of the dimension of self-affine sets, see the survey [Fal13].

3 Further Background

3.1 Subadditive Thermodynamic Formalism

The topological pressure $P(\varphi)$ of a Hölder continuous potential φ is a key component of the thermodynamic formalism, which in turn, as discovered by Bowen, is a formidable tool in the dimension theory of conformal dynamical systems. In the classical setting, the functional P is continuous as a function of φ in the appropriate topology.

It is well-known that the dimension theory of *non-conformal* dynamical systems is far more difficult, and the classical thermodynamic formalism is no longer the appropriate tool. Instead, starting with the insights of Barreira [Ba96] and Falconer [Fal88], a *sub-additive* thermodynamic formalism has been developed. Both the thermodynamic formalism itself and its application to the dimension of invariant sets and measures is far more difficult in the non-conformal case. The proofs of Theorems 2.1 and 1.2 depend crucially on this subadditive thermodynamic formalism, and hence we review the main elements in this section.

We limit ourselves to potentials defined on the full shift on m symbols $X = I^{\mathbb{N}}$. Let $\Phi = \{\varphi_n\}_{n=1}^{\infty}$ be a collection of continuous real-valued maps on X. We say that Φ is *subadditive* if

$$\varphi_{k+n}(\mathtt{i}) \le \varphi_k(\mathtt{i}) + \varphi_n(\sigma^k \mathtt{i}) \quad \text{for all } \mathtt{i} \in X. \tag{3.1}$$

Important examples of subadditive potentials, which will be relevant in the proofs of Theorems 2.1 and 1.2, are

$$\varphi_n(\mathtt{i}) = s \log \|A_{i_n} \cdots A_{i_1}\|, \tag{3.2}$$

$$\varphi_n(\mathtt{i}) = \log \varphi^s(A_{i_n} \cdots A_{i_1}). \tag{3.3}$$

We note that the order of the products is the reverse of the order usually considered in the IFS literature; the reason for this will become apparent later when we apply Oseledets' Theorem. Let $\mathcal{E}$ denote the set of probability measures ergodic and invariant under the shift σ. The thermodynamic formalism consists of three main pieces: the *entropy* h_μ of a measure $\mu \in \mathcal{E}$, the *topological pressure* $P(\Phi)$ of a subadditive potential Φ, and the *energy* or *Lyapunov exponent* $E_\mu(\Phi)$ of Φ with respect to a measure $\mu \in \mathcal{E}$. These are defined as follows:

$$h_\mu = \lim_{n\to\infty} \frac{1}{n} \sum_{\mathtt{i}\in I^n} -\mu[\mathtt{i}] \log \mu[\mathtt{i}],$$

$$P(\Phi) = \lim_{n\to\infty} \frac{1}{n} \log \sum_{\mathtt{i}\in I^n} \sup_{\mathtt{j}\in\mathtt{i}} \varphi_n(\mathtt{j}),$$

$$E_\mu(\Phi) = \lim_{n\to\infty} \frac{1}{n} \log \int \varphi_n \, d\mu.$$

By standard subadditivity arguments, all the limits exist. Moreover, if $\mu \in \mathcal{E}$, then for μ-almost all $\mathtt{i}$,

$$h_\mu = \lim_{n\to\infty} \frac{-\log \mu[\mathtt{i}|n]}{n}, \tag{3.4}$$

$$E_\mu(\Phi) = \lim_{n\to\infty} \frac{\log \varphi_n(\mathtt{i})}{n}. \tag{3.5}$$

The first equality is a particular case of the Shannon-McMillan-Breiman, while the second is a consequence of Kingman's subadditive ergodic theorem.

These quantities are related via the following *variational principle* due to Cao, Feng, and Huang [CF08]:

Theorem 3.1 ([CF08], Theorem 1.1) *If Φ is a subadditive potential on X, then*

$$P(\Phi) = \sup\left\{h_\mu + E_\mu(\Phi) : \mu \in \mathcal{E}\right\}.$$

Particular cases of the above, under stronger assumptions on the potentials, were previously obtained by many authors, see for example [Kae04, Mum06, Ba10] and references therein.

It follows from the semicontinuity of the entropy with respect to the shift map that the supremum in Theorem 3.1 is in fact a maximum; measures which attain the supremum are known as *equilibrium measures* or *equilibrium states* (for the potential Φ). The variational principle and the existence of equilibrium measures for the potentials given in (3.2) and (3.3) go back to [Kae04]. We remark that, unlike the classical setting, equilibrium measures do not need to be unique in the subadditive setting, not even in the locally constant case. Feng and Käenmäki [FK11] characterize all equilibrium measures for potentials of the form $\varphi_n(\mathtt{i}) = s\log\|A_{i_1}\cdots A_{i_n}\|$.

3.2 Oseledets' Multiplicative Ergodic Theorem

We recall a version of the Multiplicative Ergodic Theorem of Oseledets. For simplicity we state it only in dimension $d = 2$; see e.g. [Kre185, Theorem 5.7] for the full version.

Theorem 3.2 *Let $B_1, \ldots, B_m \in GL_2(\mathbb{R})$, and for $\mathtt{i} \in X$ write*

$$B(\mathtt{i}, n) = B_{\mathtt{i}_n} B_{\mathtt{i}_{n-1}} \cdots B_{\mathtt{i}_1}.$$

Further, let μ be a σ-invariant and ergodic measure on X. Then, one of the two following situations occur:

(A) *(Equal Lyapunov exponents). There exists $\lambda \in \mathbb{R}$ such that for μ-almost all i,*

$$\lim_{n\to\infty} \frac{\log|B(\mathtt{i}, n)v|}{n} = \lambda \quad \textit{uniformly for } |v| = 1.$$

(B) *(Distinct Lyapunov exponents). There exist $\lambda_1 > \lambda_2$ and measurable families $\{E_1(\mathtt{i})\}, \{E_2(\mathtt{i})\}$ of one-dimensional subspaces such that for μ-almost all $\mathtt{i}$:*

(a) $\mathbb{R}^2 = E_1(\mathtt{i}) \oplus E_2(\mathtt{i})$.

(b) $B_{i_1} E_j(\mathtt{i}) = E_j(\sigma\mathtt{i})$, $j = 1, 2$.

(c) *For all* $v \in E_j(\mathtt{i}) \setminus \{0\}$, $j = 1, 2$,

$$\lim_{n\to\infty} \frac{\log |B(\mathtt{i}, n)v|}{n} = \lambda_j.$$

3.3 The Cone Condition

The proof of Theorems 1.2 and 2.1 rely on finding a subsystem (after iteration) which is better behaved than the original one and captures almost all of its topological pressure. In the case of distinct Lyapunov exponents (with respect to a measure chosen from an application of the variatonal principle), the good behavior of this subsystem will consist in satisfying the (strict) cone condition: all the maps will send some fixed cone into its interior (except for the origin). Recall that a *cone* $K \subset \mathbb{R}^d$ is a closed set such that $K \cap -K = \{0\}$ and $tx \in K$ whenever $t > 0, x \in K$ (here $-K = \{-x : x \in K\}$).

This kind of cone condition is ubiquitous in the study of dynamical systems and associated matrix cocycles. In our situation, its usefulness will be derived from the following lemma.

Lemma 3.1 *Let* $K', K \subset \mathbb{R}^d$ *be cones such that* $K' \setminus \{0\} \subset$ *interior*(K). *There exists a constant* $c > 0$ *(depending on the cones) such that*

$$\|A\| \geq c\frac{\|Aw\|}{\|w\|} \tag{3.6}$$

for all $w \in K$ *and all* $A \in \mathbb{R}^{d\times d}$ *such that* $AK \subset K'$.

In particular, there is $c' > 0$ *such that if* $A_1, A_2 \in \mathbb{R}^{d\times d}$ *are such that* $A_jK \subset (K' \cup -K')$, $j = 1, 2$, *then*

$$\|A_1A_2\| \geq c'\|A_1\|\|A_2\|.$$

Proof Suppose (3.6) does not hold. Then, for all n we can find a linear map A_n of norm 1 with $A_n(K) \subset K'$, and $w_n \in K'$ also of norm 1, such that $\|A_nw_n\| < 1/n$. By compactness, this implies that there are a linear map A on V of norm 1 (in particular nonzero) such that $A(K) \subset K'$, and a vector $w \in K'$ such that $Aw = 0$. Now pick $u \in K$ such that $Au \neq 0$ and $w - u \in K$; this is possible since $K' \setminus \{0\} \subset$ interior(K). It follows that $A(w - u) = -Au \in -K'$, whence $Au \in K' \cap -K'$, contradicting that K' is a cone.

For the second claim, we may assume (replacing A_j by $-A_j$ if needed) that $A_jK \subset K'$ for $j = 1, 2$. The claim now follows from the first one, since for fixed $w \in K'$ of norm 1,

$$\|A_1 A_2\| \geq c\|A_1(A_2 w)\| \geq c^2\|A_1\|\|A_2 w\| \geq c^3\|A_1\|\|A_2\|.$$

□

A tuple $A = (A_1, \dots, A_m)$ is said to satisfy the *cone condition* if there exist cones K', K such that $K' \setminus \{0\}$ is contained in the interior of K, and $A_i K \subset (K' \cup -K')$ for all $i \in I$. The relevance of this condition can be seen from the following lemma.

Lemma 3.2 *If A satisfies the cone condition, then M is continuous on $\mathcal{U} \times [0, +\infty)$ for some neighborhood $\mathcal{U}$ of A, and the same holds for P if $d = 2$.*

Proof We know from Lemma 2.3 that P is upper semicontinuous and Lipschitz continuous in s, with the Lipschitz constant locally uniformly bounded. The same arguments show that the same is true for M. Hence the task is to prove the claim with "lower continuous" in place of "continuous", with the value of s fixed.

A trivial but key observation is that the cone condition is robust, in the following sense: if $A = (A_1, \dots, A_m)$ satisfies the cone condition with cones K, K', then there are a neighborhood $\mathcal{U}$ of A and cones $\widetilde{K}, \widetilde{K}'$ such that any $B \in \mathcal{U}$ satisfies the cone condition with cones $\widetilde{K}, \widetilde{K}'$. In particular, applying Lemma 3.1 we find that there exists a constant $c = c(\mathcal{U}) \in (0, 1)$, such that if $B = (B_1, \dots, B_m) \in \mathcal{U}$, then

$$\|B_{\mathtt{i}} B_{\mathtt{j}}\| \geq c\,\|B_{\mathtt{i}}\|\|B_{\mathtt{j}}\| \quad \text{for all } \mathtt{i}, \mathtt{j} \in I^*. \tag{3.7}$$

Now, for this constant c, let

$$\widetilde{S}_n(B, s) = c \sum_{\mathtt{i} \in I^n} \|B_{\mathtt{i}}\|^s,$$

and observe that if $B \in \mathcal{U}$ then, thanks to (3.7), $\widetilde{S}_{n+k}(B, s) \geq \widetilde{S}_n(B, s)\widetilde{S}_k(B, s)$. Therefore, for $B \in \mathcal{U}$,

$$M(B, s) = \lim_{n \to \infty} \frac{1}{n} \log \widetilde{S}_n(B, s) = \sup_n \frac{1}{n} \widetilde{S}_n(B, s).$$

Since a supremum of continuous functions is lower semicontinuous, this yields the claim for M.

Suppose now $d = 2$. Since $\alpha_2(B) = \|B^{-1}\|^{-1}$ for $B \in \mathrm{GL}_2(\mathbb{R})$, we have that $\alpha_2(B_1 B_2) \geq \alpha_2(B_1)\alpha_2(B_2)$ for any $B_1, B_2 \in \mathrm{GL}_2(\mathbb{R})$. Let $\widetilde{K}, \widetilde{K}', \mathcal{U}, c$ be as before. Then

$$\varphi^s(B_{\mathtt{i}} B_{\mathtt{j}}) \geq c\,\varphi^s(B_{\mathtt{i}})\varphi^s(B_{\mathtt{j}}) \quad \text{for all } \mathtt{i}, \mathtt{j} \in I^*.$$

Thus, arguing as before,

$$P(B, s) = \sup_n \frac{1}{n} \log\left(\sum_{\mathtt{i}\in I^n} c\,\varphi^s(B_{\mathtt{i}})\right),$$

which is lower semicontinuous as a supremum of continuous functions. □

Although we will not use this result directly, the ideas in its proof will arise in our proof of continuity of M and P in dimension $d = 2$.

4 Proof of the Continuity of Subadditive Pressure in $\mathbb{R}^2$

4.1 General Strategy and the Case of Equal Lyapunov Exponents

In this section we prove Theorems 1.2 and 2.1 in dimension $d = 2$ (recall that Theorem 1.1 is an immediate corollary of Theorem 2.1). We are going to give the details of the continuity of $P(A, s)$; the proof for $M(A, s)$ is essentially identical. We have already observed that P is upper semicontinuous, hence it is enough to prove it is lower continuous. Moreover, by the second part of Lemma 2.3, it is enough to prove continuity in A for a fixed value of s.

Fix $\varepsilon > 0$ for the course of the proof. Consider the potential $\Phi = \{\varphi_n\}$ where $\varphi_n(\mathtt{i}) = \varphi^s(A(\mathtt{i}, n))$ (this is the potential given in (3.3)). Thanks to the variational principle for subadditive potentials (Theorem 3.1), we know that there exists an ergodic measure μ on X, such that

$$h_\mu + E_\mu(\Phi) \geq P(\Phi) - \varepsilon = P(A, s) - \varepsilon. \tag{4.1}$$

(In fact, by the remark after Theorem 3.1, we can take $\varepsilon = 0$ in the above, but we do not need this). The potential Φ and the measure μ will remain fixed for the rest of the proof; we underline that they depend on s and A.

We apply Oseledets' Theorem (Theorem 3.2) to the the matrices $(A_1, \ldots, A_m)$ and the measure μ. The proof splits depending on whether the resulting Lyapunov exponents are equal or distinct. However, in both cases we will rely on the general scheme given in the next lemma.

Lemma 4.1 *Suppose there are $n = n(\varepsilon)$, $Y_n \subset I^n$, and a neighborhood $\mathcal{U}$ of A such that the following hold:*

(1) $\log |Y_n| \geq n(h_\mu - \varrho_1(\varepsilon))$,

(2) *If $\mathtt{i}$ is a juxtaposition of k words from Y_n, and $B \in \mathcal{U}$, then*

$$\log \varphi^s(B_{\mathtt{i}}) \geq nk(E_\mu(\Phi) - \varrho_2(\varepsilon)).$$

Then $P(B, s) \geq P(A, s) - \varepsilon - \varrho_1(\varepsilon) - \varrho_2(\varepsilon)$ for all $B \in \mathcal{U}$.

Proof Let Y_n^k denote the family of juxtapositions of k words from Y_n. If $B \in \mathcal{U}$, then

$$\begin{aligned}
P(B,s) &\ge \limsup_{k\to\infty} \frac{1}{nk} \log \sum_{\mathtt{i}\in Y_n^k} \varphi^s(B_{\mathtt{i}}) \\
&\ge \lim_{k\to\infty} \frac{1}{nk}\left(k\log|Y_n| + \min_{\mathtt{i}\in Y_n^k} \log \varphi^s(B_{\mathtt{i}}) \right) \\
&\ge h_\mu + E_\mu(\Phi) - \varrho_1(\varepsilon) - \varrho_2(\varepsilon) \\
&\ge P(A,s) - \varepsilon - \varrho_1(\varepsilon) - \varrho_2(\varepsilon),
\end{aligned}$$

where in the last line we have used (4.1). □

In practice we will take $\varrho_i(\varepsilon)$ to be a multiple of ε, so that in the limit as $\varepsilon \to 0$ we obtain the required lower semicontinuity. Finding a set Y_n such that (4.1) holds is not difficult, and likewise if we also require (4.1) only for $\mathtt{i} \in Y_n$ (rather than Y_n^k). The challenge is to make (4.1) stable under compositions of the maps $B_{\mathtt{j}}$, $\mathtt{j} \in Y_n$ as well, and for this we will need geometric and ergodic-theoretic arguments depending on Oseledets' Theorem.

First we deal with the simpler case in which the Lyapunov exponents are equal; the case of different exponents is addressed in the next subsection.

Suppose then that there is a single Lyapunov exponent λ. It follows easily from (3.5) and Theorem 3.2 that

$$\lim_{n\to\infty} \frac{1}{n} \int \log \varphi^s(A(\mathtt{i},n))d\mu(\mathtt{i}) = s\lambda. \tag{4.2}$$

By Theorem 3.2, (3.4) and Egorov's Theorem, we can find a set $Y \subset X$ such that $\mu(Y) \ge 1/2$, and $n_0 \in \mathbb{N}$ such that if $\mathtt{i} \in Y$ and $n \ge n_0$ then

$$\begin{aligned}
|A(\mathtt{i},n)v| &\ge e^{\lambda n} e^{-\varepsilon n}|v| \quad \text{for all } v \in \mathbb{R}^2 \setminus \{0\}, \\
\mu[\mathtt{i}|n] &\le e^{-n(h_\mu-\varepsilon)}.
\end{aligned} \tag{4.3}$$

Fix some $n \ge n_0$ and write $Y_n = \{i|n : \mathtt{i} \in Y\}$. Note that

$$\frac{1}{2} \le \sum_{\mathtt{j}\in Y_n} \mu[\mathtt{j}] \le |Y_n| e^{-n(h_\mu-\varepsilon)},$$

whence (if n is large enough)

$$|Y_n| \ge e^{n(h_\mu-2\varepsilon)}.$$

On the other hand, since Y_n is finite, we can find a neighborhood $\mathcal{U}$ of A such that if $B = (B_1, \ldots, B_m) \in \mathcal{U}$, then

$$|B_{\mathtt{j}} v| \ge e^{n\lambda} e^{-2\varepsilon n}|v| \quad \text{for all } v \in \mathbb{R}^2 \setminus \{0\}, \mathtt{j} \in Y_n.$$

Therefore

$$|B_{\mathtt{i}} v| \geq e^{kn\lambda} e^{-2\varepsilon kn} |v| \quad \text{for all } v \in \mathbb{R}^2 \setminus \{0\}, \mathtt{i} \in Y_n^k,$$

where $Y_n^k \subset I^{kn}$ is the set of all juxtapositions of k words from Y_n. In particular,

$$\varphi^s(B_{\mathtt{i}}) \geq e^{kns\lambda} e^{-2\varepsilon kns} \quad \text{for all } \mathtt{i} \in Y_n^k.$$

We have therefore established the hypotheses of Lemma 4.1, with $\varrho_1(\varepsilon) = 2\varepsilon$ and $\varrho_2(\varepsilon) = 2s\varepsilon$. Applying that lemma and letting $\varepsilon \to 0$ establishes lower semicontinuity when the Lyapunov exponents are equal.

4.2 The Case of Distinct Lyapunov Exponents

Suppose now that the Lyapunov exponents are $\lambda_1 > \lambda_2$. We will again construct sets Y_n so that we can apply Lemma 4.1; this is trickier in this case, and the main idea is to use Oseledet's Theorem, Egorov's Theorem and recurrence, to find sets Y_n (for n arbitrarily large) so that the hypotheses of Lemma 4.1 hold when $k = 1$, *and in addition* $\{A_{\mathtt{i}} : \mathtt{i} \in Y_n\}$ *satisfies the cone condition.* The cone condition will allow us to pass to a neighborhood of A first, and to iterates of the $B_{\mathtt{i}}$, $\mathtt{i} \in Y_n$, later.

Recall that for μ-almost all $\mathtt{i}$ there is an Oseledets splitting $\mathbb{R}^2 = E_1(\mathtt{i}) \oplus E_2(\mathtt{i})$. The family of splittings $\mathbb{R}^2 = E_1 \oplus E_2$ has a natural separable metrizable topology; for example, we can take $d(E_1 \oplus E_2, E_1' \oplus E_2') = \max(\angle(E_1, E_1'), \angle(E_2, E_2'))$, where $\angle$ is the angle between two lines. We can then find a fixed splitting $\mathbb{R}^2 = F_1 \oplus F_2$ which is in the support of the push-forward of μ under the Oseledets splitting or, in other words,

$$\mu(\mathtt{i} : d(E_1(\mathtt{i}) \oplus E_2(\mathtt{i}), F_1 \oplus F_2) < \eta) > 0 \quad \text{for all } \eta > 0.$$

We write $F_i^\gamma = \{E : \angle(E, F_i) < \gamma\}$.

Lemma 4.2 *There are $R, \eta > 0$ and two cones $K', K \subset \mathbb{R}^2$ with $K' \setminus \{0\} \subset K$, such that the following holds. Suppose that $A \in GL_2(\mathbb{R})$ is such that $Av_j \in F_j^\eta$ for some $v_j \in F_j^\eta$ of unit norm, $j = 1, 2$, and moreover $|Av_1| > R|Av_2|$. Then $AK \subset (K' \cup -K')$.*

Proof The lemma is essentially a consequence of compactness. Let v be a unit vector in F_1, and let K, K' be any cones such that

$$v \in \text{interior}(K') \setminus \{0\} \subset K' \setminus \{0\} \subset K \subset \mathbb{R}^2 \setminus F_2.$$

Suppose the claim fails with this choice of cones. Then for each n there are $A_n \in \mathrm{GL}_d(2)$ and $v_{n,j} \in F_j^{1/n}$ such that

$$1 = |A_n v_{n,1}| \geq n|A_n v_{n,2}|, \tag{4.4}$$

and $A_nK \not\subset K' \cup -K'$. By passing to a subsequence (and replacing $v_{n,1}$ by $-v_{n,1}$ whenever needed), we may assume that $v_{n,1} \to v$, $v_{n,2} \to w$ and $A_n \to A$ for some $w \in F_2$ of unit norm, and $A \in \mathbb{R}^{2\times 2}$. Moreover, there is $z \in K$ of unit norm such that $Az \notin \text{interior}(K' \cup -K')$. However, (4.4) implies that Az is a non-zero multiple of v for any $z \notin F_2$, which contradicts our choice of cones. This contradiction finishes the proof of the lemma. □

From now on let R, η, K, K' be as in the statement of the Lemma. By our choice of F_1, F_2, we have $\mu(\Delta) > 0$, where

$$\Delta = \{\mathtt{i} \in X : E_j(\mathtt{i}) \in F_j^{\eta}, \quad j = 1, 2\}.$$

At this point we recall the following quantitative version of Poincaré recurrence due to Khintchine, see [Pet89, Theorem 3.3] for a proof. Although it applies to any set of positive measure in a measure-preserving system, we state only the special case we will require.

Lemma 4.3 *For every $\delta > 0$, the set $\{n : \mu(\sigma^{-n}\Delta \cap \Delta) > \mu(\Delta)^2 - \delta\}$ is infinite (and it even has bounded gaps).*

In particular, if we set $\kappa := \mu(\Delta)^2/2 > 0$, then the set $S := \{n : \mu(\sigma^{-n}\Delta \cap \Delta) > \kappa\}$ is infinite. On the other hand, by (3.4), (3.5), the last part of Oseledets' Theorem, and Egorov's Theorem, we may find $n_0 \in \mathbb{N}$ and a set $\Sigma \subset X$ with $\mu(\Sigma) > 1 - \kappa/2$, such that if $n \geq 0$ and $\mathtt{i} \in \Sigma$, then:

$$\mu[\mathtt{i}|n] \leq e^{-n(h_\mu - \varepsilon)}, \tag{4.5}$$

$$\log \varphi^s(A(\mathtt{i}, n)) \geq n(E_\mu(\Phi) - \varepsilon), \tag{4.6}$$

$$|A(\mathtt{i}, n)\widehat{E}_1(\mathtt{i})| \geq R\,|A(\mathtt{i}, n)\widehat{E}_2(\mathtt{i})|,$$

where $\widehat{E}_j(\mathtt{i})$ is a unit vector in $E_j(\mathtt{i})$. This is the point where we use that the Lyapunov exponents are different.

Taking stock, we have seen that if $n \geq n_0$, and $\mathtt{i} \in \Delta \cap \sigma^{-n}\Delta \cap \Sigma$ then, by Lemma 4.2, the map $A(\mathtt{i}, n)$ satisfies

$$A(\mathtt{i}, n)K \subset (K' \cup -K').$$

Hence for $n \in S \cap [n_0, \infty)$, we define $Y_n = \{\mathtt{i}|n : \mathtt{i} \in \Delta \cap \sigma^{-n}\Delta \cap \Sigma\}$. We will show that these sets meet the conditions of Lemma 4.1, with suitable functions $\varrho_j(\varepsilon)$. Firstly, note that

$$\begin{aligned}\kappa/2 &\leq \mu(\Delta \cap \sigma^{-n}\Delta \cap \Sigma) \\ &\leq \sum_{\mathtt{j} \in Y_n} \mu[\mathtt{j}] \\ &\leq |Y_n| e^{-n(h_\mu - \varepsilon)}.\end{aligned}$$

Hence $\log|Y_n| \geq n(h_\mu - 2\varepsilon)$, provided n is taken large enough that $e^{-\varepsilon n} < \kappa/2$.

On the other hand, $\{A_{\mathtt{j}} : \mathtt{j} \in Y_n\}$ satisfies the cone condition with cones K, K' (these cones are independent of n). Then there are a neighborhood $\mathcal{U}$ of A in $(\mathrm{GL}_d(\mathbb{R}))^m$ and cones $\widetilde{K}, \widetilde{K'}$ such that if $B \in \mathcal{U}$, then $\{B_{j_n} \cdots B_{j_1} : (j_1 \dots j_n) \in Y_n\}$ satisfies the cone condition with cones $\widetilde{K}, \widetilde{K'}$. In particular, by Lemma 3.1, there exists $c > 0$ (depending only on $\mathcal{U}$, and not on n) such that

$$\|B_{\mathtt{i}}\| \geq c^{k-1}\left(\min_{\mathtt{j}\in Y_n}\|B_{\mathtt{j}}\|\right)^k \quad \text{for all } \mathtt{i} \in Y_n^k.$$

Arguing as in the proof of Lemma 3.2,

$$\varphi^s(B_{\mathtt{i}}) \geq c^{k-1}\left(\min_{\mathtt{j}\in Y_n}\varphi^s(B_{\mathtt{j}})\right)^k \quad \text{for all } \mathtt{i} \in Y_n^k.$$

By taking n large enough, we may assume that $\log c/n > -\varepsilon$. Furthermore, in light of (4.6) we may find a neighborhood $\mathcal{V} \subset \mathcal{U}$ containing A, such that if $B \in \mathcal{V}$ and $\mathtt{j} \in Y_n$, then $\log\varphi^s(B_{\mathtt{j}}) > n(E_\mu(\Phi) - 2\varepsilon)$. We conclude that if $B \in \mathcal{V}$ and $\mathtt{i} \in Y_n^k$, then

$$\log\varphi^s(B_{\mathtt{i}}) > kn(E_\mu(\Phi) - 3\varepsilon).$$

We are now able to apply Lemma 4.1 to conclude that if $B \in \mathcal{V}$, then

$$P(B, s) \geq P(A, s) - 6\varepsilon.$$

As $\varepsilon > 0$ was arbitrary, this yields the desired lower semicontinuity.

4.3 Some Remarks on the Higher-Dimensional Case

We finish the paper with some brief remarks on the proof of the continuity of M and P in any dimension.

The proof of the continuity of M in dimension 2 extends fairly easily to arbitrary dimension d (using the general version of Oseledets' Theorem): if all d Lyapunov exponents are equal, then the proof is identical to the two-dimensional case. If not all exponents are equal, let $1 \leq k < d$ be the multiplicity of the largest Lyapunov exponent. Then the argument is very similar, except that one uses cones around k-planes.

For $d \geq 3$, one cannot reduce φ^s to a quantity involving only matrix norms (of the given maps and their inverses), so it is clear that some new tools are required to prove continuity of P in general dimension. The proof follows the same outline, but it involves working with higher exterior powers of the maps A_i, and proving cone conditions for two different exterior powers simultaneously. Altough passing to exterior products is a common trick in the area, this makes the general proof far more technical. The reader is referred to [FS13] for further details, as well as consequences and generalizations of these results.

Acknowledgments I am grateful to De-Jun Feng for useful comments on an earlier version of the article.

References

[Ba07] Barański, K.: Hausdorff dimension of the limit sets of some planar geometric constructions. Adv. Math. **210**(1), 215–245 (2007)

[Ba10] Barreira, L.: Almost additive thermodynamic formalism: some recent developments. Rev. Math. Phys. **22**(10), 1147–1179 (2010)

[Ba96] Barreira, L.M.: A non-additive thermodynamic formalism and applications to dimension theory of hyperbolic dynamical systems. Ergodic Theor. Dyn. Syst. **16**(5), 871–927 (1996)

[BV10] Bocker-Neto, C., Viana, M.: Continuity of lyapunov exponents for random 2d matrices (preprint) (2010). http://arxiv.org/abs/1012.0872

[CF08] Cao, Y.-L., Feng, D.-J., Huang, W.: The thermodynamic formalism for sub-additive potentials. Discrete Contin. Dyn. Syst. **20**(3), 639–657 (2008)

[Edg92] Edgar, G.A.: Fractal dimension of self-affine sets: some examples. Rend. Circ. Mat. Palermo **2**(28), 341–358 (1992) (Measure theory (Oberwolfach, 1990))

[Fal88] Falconer, K.J.: A subadditive thermodynamic formalism for mixing repellers. J. Phys. A **21**(14), L737–L742 (1988)

[Fal13] Falconer, K.: Dimensions of self-affine sets: a survey. In Barral, J., Seuret, S. (eds.) Further Developments in Fractals and Related Fields, pp. 115–134. Springer, Berlin (2013)

[FM08] Falconer, K., Miao, J.: Exceptional sets for self-affine fractals. Math. Proc. Camb. Philos. Soc. **145**(3), 669–684 (2008)

[FS09] Falconer, K., Sloan, A.: Continuity of subadditive pressure for self-affine sets. Real Anal. Exchange **34**(2), 413–427 (2009)

[Fal92] Falconer, K.J.: The Hausdorff dimension of self-affine fractals. Math. Proc. Cambridge Philos. Soc. **103**(2), 339–350 (1988)

[FK11] Feng, D.-J., Käenmäki, A.: Equilibrium states of the pressure function for products of matrices. Discrete Continuous Dyn. Syst. (DCDS-A) **30**(3), 699–708 (2011)

[FS13] Feng, D.-J., Shmerkin, P.: Non-conformal repellers and the continuity of pressure for matrix cocycles. Geom. Funct. Anal. (2014)

[HL95] Hueter, I., Lalley, S.P.: Falconer's formula for the Hausdorff dimension of a self-affine set in R^2. Ergodic Theor. Dyn. Syst. **15**(1), 77–97 (1995)

[Kae04] Käenmäki, A.: On natural invariant measures on generalised iterated function systems. Ann. Acad. Sci. Fenn. Math. **29**(2), 419–458 (2004)

[KS09] Käenmäki, A., Shmerkin, P.: Overlapping self-affine sets of Kakeya type. Ergodic Theor. Dyn. Syst. **29**(3), 941–965 (2009)

[Krel85] Krengel, U.: Ergodic theorems, vol. 6 of de Gruyter Studies in Mathematics. Walter de Gruyter & Co., Berlin (1985) (with a supplement by Antoine Brunel)

[Mum06] Mummert, A.: The thermodynamic formalism for almost-additive sequences. Discrete Continuous Dyn. Syst. **16**(2), 435–454 (2006)

[Pet89] Petersen, K.: Ergodic theory, volume 2 of Cambridge Studies in Advanced Mathematics. Cambridge University Press, Cambridge (1989) (corrected reprint of the 1983 original)

[Sol98] Solomyak, B.: Measure and dimension for some fractal families. Math. Proc. Camb. Philos. Soc. **124**(3), 531–546 (1998)

Stability Properties of Fractal Curvatures

Martina Zähle

Abstract Lipschitz-Killing curvatures of singular sets are known from geometric measure theory. These are extensions of classical notions from convex and differential geometry. In recent years their fractal versions have been introduced via approximation by parallel sets of small distances. In the present paper stability properties of the corresponding limits under small perturbations of these neighborhoods are studied. The well-known Minkowski content may be considered as marginal case.

1 Introduction

Fractal versions of the Lipschitz-Killing curvatures of singular sets known from geometric measure theory [Fed59] have been introduced via approximation by parallel sets of small distances. This construction generalizes that of the Minkowski content considered by many authors, which is a marginal case. A survey on related developments over the last years as well as on some background from classical convex and differential geometry is given in [Zah13]. Moreover, in this paper we outline an approach for self-similar sets by means of an associated dynamical system, which leads to short proofs for the total curvatures, their densities and the curvature measures.

An extension of the average limits of the curvature-direction measures to self-conformal sets is given in [Boh13]. There the tools of thermodynamic formalism are more involved, but the proofs are shorter and the results are more general than in the literature.

An advantage of the approaches via renewal theory, e.g. in [Gat00, Win08, Win11, WZ13, Zah11, Zah13], or gap structures [FK12, KK12, Kom11, LPW11, LPW13,

M. Zähle (✉)
Institute of Mathematics, Friedrich Schiller University Jena, Jena, Germany
e-mail: martina.zaehle@uni-jena.de

D.-J. Feng and K.-S. Lau (eds.), *Geometry and Analysis of Fractals*,
Springer Proceedings in Mathematics & Statistics 88,
DOI 10.1007/978-3-662-43920-3_13

PW12], are the more practicable integral representations of the fractal limits for special classes of self-similar or self-conformal sets.

The aim of the present paper is to show that the difference between the approximation by parallel sets and by neighborhoods arising from certain small perturbations of these sets is asymptotically vanishing. To this aim we use suitable Lipschitz mappings between the unit normal bundles of the parallel sets and those of the close neighborhoods describing the perturbations. They have to be chosen in such a manner that the geometry is asymptotically the same. In this approach the representation of the curvature measures by means of associated normal cycles, i.e. by integration of the Lipschitz-Killing curvature forms over the unit normal bundles, appears as a useful tool. We present here only the main ideas. More precise error estimates can be derived.

In Chap. 2, Lemma 2, a general estimate of the difference between the curvature-direction measures of two subsets of $\mathbb{R}^d$ admitting normal cycles which are close to each other is derived. It is assumed that the second normal cycle is the pushforward of the first one under a Lipschitz mapping f, and there distance is measured in terms of the distance between the orthogonally transformed approximate differential of f and the identity in $\mathbb{R}^d \times S^{d-1}$.

If we are only interested in Lebesgue measure or surface area, which corresponds to the Minkowski content in applications to fractals, Lemma 1 provides evident tools. The main result is formulated in Theorem 1, where we prove stability of the above mentioned fractal curvatures under asymptotically vanishing perturbations. Here we specify Lemma 2 to this situation. Finally we demonstrate on the example of the Sierpinski gasket how the theorem can be applied.

2 Some Background from Geometric Measure Theory

2.1 Normal Cycles and Curvatures

Throughout the paper we use notions and notations from geometric integration theory in the sense of Federer [Fed69] (see also Morgan [Mor88], Krantz and Parks [KP08]). In particular, we consider multivector fields and differential forms in $\mathbb{R}^d \times \mathbb{R}^d$ in the language of the exterior algebra and the alternating algebra, respectively. The symbol $\langle \eta, \varphi \rangle$ denotes dual pairing of a multivector η and a differential form φ. Currents are continuous linear functionals on spaces of differential forms. In our case they are given by integrating differential $(d-1)$-forms over certain Hausdorff-$(d-1)$-rectifiable subsets of $\mathbb{R}^d \times S^{d-1}$, namely the unit normal bundles of geometric sets in $\mathbb{R}^d$. $\mathcal{H}^{d-1}$ denotes the corresponding Hausdorff measure.

We next recall some results from [RZ01, Zah86], and [RZ05].

For general X with reach $X > 0$ there is an associated rectifiable current called the *unit normal cycle* of X which is given by

$$N_X(\varphi) := \int_{\mathrm{nor}X} \langle a_X(x,n), \varphi(x,n)\rangle \mathscr{H}^{d-1}(d(x,n))$$

for an appropriate unit simple $(d-1)$-vector field $a_X = a_1 \wedge \ldots \wedge a_{d-1}$ associated a.e. with the approximate tangent spaces of the unit normal bundle nor$X$ and for integrable differential $(d-1)$-forms φ. In these terms for $k \leq d-1$ the curvature measure may be represented by

$$C_k(X,B) = N_X \llcorner \mathbf{1}_{B\times\mathbb{R}^d}(\varphi_k) = \int_{\mathrm{nor}X\cap(B\times\mathbb{R}^d)} \langle a_X(x,n), \varphi_k(n)\rangle \mathscr{H}^{d-1}(d(x,n))$$

for any bounded Borel set $B \subset \mathbb{R}^d$, where the *k-th Lipschitz-Killing curvature form* φ_k does not depend on the points x and is defined by its action on a simple $(d-1)$-vector $\eta = \eta_1 \wedge \ldots \wedge \eta_{d-1}$ as follows: Let

$$\pi_0(y,z) := y \text{ and } \pi_1(y,z) := z$$

be the *coordinate projections* in $\mathbb{R}^d \times \mathbb{R}^d$, $e_1', \ldots, e_d'$ be the *dual basis* of the standard basis $e_1, \ldots, e_d$ in $\mathbb{R}^d$ and $\mathscr{O}_k$ be the surface area of the k-dimensional unit sphere. Then we have

$$\langle \eta, \varphi_k(n)\rangle := \mathscr{O}_{d-k}^{-1} \sum_{\varepsilon_i\in\{0,1\},\ \sum\varepsilon_i=d-1-k} \langle \pi_{\varepsilon_1}\eta_1 \wedge \ldots \wedge \pi_{\varepsilon_{d-1}}\eta_{d-1} \wedge n, e_1' \wedge \ldots \wedge e_d'\rangle. \tag{1}$$

Moreover, in these notations the sign of the above unit simple $(d-1)$-vector field a_X is a.e. determined by

$$\left\langle \Lambda^{d-1}(\pi_0 + \varepsilon\pi_1)a_T(x,n) \wedge n, e_1' \wedge \ldots \wedge e_d'\right\rangle > 0$$

for sufficiently small ε. It can be represented in the form $a_X = a_1 \wedge \ldots \wedge a_{d-1}$ with

$$a_i(x,n) = \left(\frac{1}{\sqrt{1+\varkappa_i(x,n)^2}} b_i(x,n), \frac{\varkappa_i(x,n)}{\sqrt{1+\varkappa_i(x,n)^2}} b_i(x,n)\right), \tag{2}$$

where the $\varkappa_i(x,n) \in (-\infty, \infty]$ are the *generalized principal curvatures* of X at (x,n) (with convention $\frac{\infty}{\infty} = 1$), and their *direction vectors* $b_i(x,n)$ form together with the unit normal vector n a positively oriented orthonormal basis in $\mathbb{R}^d$, i.e., $\langle b_1 \wedge \ldots \wedge b_{d-1} \wedge n, e_1' \wedge \ldots \wedge e_d'\rangle = 1$. (Note that for convex sets all $\varkappa_i(x,n)$ are either 0 or ∞.)

From this one obtains the integral representation of the *curvature-direction measures* $\widetilde{C_k}(X,\widetilde{B}) := N_X \llcorner \mathbf{1}_{\widetilde{B}}(\varphi_k)$ on $\mathbb{R}^d \times S^{d-1}$:

$$\widetilde{C}_k(X, A) = \mathcal{O}_{d-k}^{-1} \int_{\mathrm{nor}X} \mathbf{1}_A(x, n) \sum_{i_1 < \cdots < i_{d-1-k}} \frac{\varkappa_{i_1}(x, n) \cdots \varkappa_{i_{d-1-k}}(x, n)}{\prod_{i=1}^{d-1} \sqrt{1 + \varkappa_i(x, n)^2}} \mathcal{H}^{d-1}(d(x, n))$$

for any bounded Borel set A in $\mathbb{R}^d \times S^{d-1}$. We define its *strong variation measure* by

$$\widetilde{C}_k^{svar}(X, A) := \int_{\mathrm{nor}X} \mathbf{1}_A ||\varphi_k|| \, d\mathcal{H}^{d-1} ,$$

where $||\varphi(x, n)||$ denotes here the comass norm of an alternating $(d-1)$-form $\varphi(x, n)$ on the tangent space $\mathrm{Tan}(\mathrm{nor} F_\varepsilon, (x, n))$, defined for a.a. $(x, n) \in \mathrm{nor}X$. Then one infers

$$\widetilde{C}_k^{svar}(X, A) := \mathcal{O}_{d-1-k}^{-1} \int_{\mathrm{nor}X} \mathbf{1}_A \sum_{i_1 < \cdots < i_{d-1-k}} \frac{\left|\varkappa_{i_1} \cdots \varkappa_{i_{d-1-k}}\right|}{\prod_{i=1}^{d-1} \sqrt{1 + \varkappa_i^2}} \, d\mathcal{H}^{d-1} .$$

$C_k^{svar}(X, \cdot) := \widetilde{C}_k^{svar}(X, (\cdot) \times S^{d-1})$ is called *strong variation measure* of the curvature measure $C_k(X, \cdot)$.

For d-dimensional domains X denote the *closure of the complement* of X by $\widetilde{X}$. If the latter has positive reach then the normal cycle of X may be introduced by the pushforward of that of $\widetilde{X}$ under the normal reflection $\rho(x, n) := (x, -n)$:

$$\mathrm{nor}X := \rho(\mathrm{nor}\widetilde{X}) \text{ and } N_X := \rho_\# N_{\widetilde{X}} ,$$

which yields

$$C_k(X, \cdot) = (-1)^{d-1-k} C_k(\widetilde{X}, \cdot) ,$$

and

$$C_k^{svar}(X, \cdot) = C_k^{svar}(\widetilde{X}, \cdot)$$

for the strong variation measures. Note that for such X, $C_{d-1}(X, \cdot)$ agrees with half the surface are measure $\mathcal{H}^{d-1}$ on the boundary of X.

Normal cycles and curvatures have been introduced for various other classes of singular sets X. (These are $(d-1)$ rectifiable currents without boundaries vanishing on the contact 1-form.) Then the above representations remains valid, but in general, the orienting $(d-1)$-vector field a_X has to be multiplied be an associated integer-valued topological index function i_X, i.e.,

$$N_X(\varphi) := \int_{\mathrm{nor}X} \langle i_X(x, n) a_X(x, n), \varphi(x, n) \rangle \mathcal{H}^{d-1}(d(x, n))$$

and the strong variation measure of the k-th curvature (-direction) measure is given by

$$\int_{\mathrm{nor}X} \mathbf{1}_{(\cdot)}(x,n)|i_X(x,n)|\; ||\varphi_k||(x,n)\, \mathscr{H}(d(x,n))\,.$$

The *index function* is determined by

$$i_X(x,n) := \mathbf{1}_X(x)\left(1 - \lim_{\varepsilon\to 0}\lim_{\delta\to 0}\chi\big(X\cap B(x+(\varepsilon+\delta)n,\varepsilon)\big)\right)$$

for the Euler-Poincaré characteristic χ in the sense of singular homology, where $B(z,r)$ denotes the closed ball with centre z and radius r.

In the general case one introduces additionally $C_d(X,\cdot) := \mathscr{L}^d(X\cap(\cdot))$ for the *Lebesgue measure.*

Then the *total values* $C_i(X) := C_i(X,\mathbb{R}^d)$, $i = 0,\dots d$, for certain classes of sets X form a complete system of motion invariant Euclidean valuations, which are continuous with respect to the flat seminorms of the associated normal cycles N_X, see [Zah90]. (In the convex setting this is a well-known theorem of Hadwiger.)

Below we will choose for X the parallel sets F_ε of small distances ε or different neighborhoods F^ε of self-similar and other fractals F with normal cycles N_{F_ε} and N_{F^ε}, respectively. We are seeking for conditions that guarantee the same fractal curvatures arising from the average limits if in the approximation procedure the parallel sets F_ε are replaced by the F^ε. Here we need a stronger continuity version than that of flat convergence.

2.2 Distance Estimates for Curvature Measures

Let us first mention some simple distance estimate for the Hausdorff measures of rectifiable sets which can be applied to the above cases $k = d-1$ and $k = d$. They easily follow from the Area theorem [Fed69, 3.2.22]. Here and in the following I denotes the *identity map* and ap Df the *approximate differential* of a Lipschitz mapping f.

Lemma 1 *If X and Y are $(\mathscr{H}^m, m)$-rectifiable and $\mathscr{H}^m$-measurable subsets of $\mathbb{R}^d$ and there exists a Lipschitz mapping f from X onto Y such that for $\mathscr{H}^m$-a.a. $y\in Y$ the set $f^{-1}(y)$ is a singleton, then we get for any Borel set B,*

$$\left|\mathscr{H}^m(X\cap f^{-1}(B)) - \mathscr{H}^m(Y\cap B)\right| \le \int_{X\cap f^{-1}(B)} \sum_{i=1}^{m}\binom{m}{i}\Delta_f^i\, d\mathscr{H}^m\,,$$

where $\Delta_f(x) = ||I - ap\, Df(x)||$ at a.a. $x\in X$, for the operator norm in the corresponding approximate tangent spaces.

A similar version for the curvature measures can be formulated as follows. For a rectifiable set $Z\subset\mathbb{R}^d\times\mathbb{R}^d$ in the above sense, a Lipschitz mapping $f: Z\to$

$\mathbb{R}^d \times \mathbb{R}^d$, and any family of orthogonal mappings $O(n) : \mathbb{R}^d \to \mathbb{R}^d$, measurable in $n \in S^{d-1}$, such that

$$\pi_1(f(x,n)) = O(n)n \text{ at a.a.}(x,n) \in \text{nor}X \tag{3}$$

we use the notation

$$\widetilde{D}f(x,n) := \widetilde{O}(n)^{-1} \circ \text{ap } Df(x,n) \tag{4}$$

for a.a. $(x,n) \in Z$, where the approximate differential of f is extended to the whole $\mathbb{R}^d \times \mathbb{R}^d$ by the identity on the orthogonal complement of the approximate tangent space of Z at (x,n). The mappings $\widetilde{O}(n) : \mathbb{R}^d \times \mathbb{R}^d \to \mathbb{R}^d \times \mathbb{R}^d$ are given by

$$\widetilde{O}(n)(u,v) := (O(n)u, O(n)v), \quad u, v \in \mathbb{R}^d .$$

Lemma 2 *If X and Y admit normal cycles N_X and N_Y and there exists a Lipschitz mapping f : nor$X \to$ nor$Y$ such that*

$$f_\# N_X = N_Y ,$$

then we get for $k = 0, \ldots, d-1$ and any bounded Borel set A in $\mathbb{R}^d \times S^{d-1}$,

$$\left|\widetilde{C}_k(X, f^{-1}(A)) - \widetilde{C}_k(Y, A)\right| \le \sum_{l=0}^{d-1} \int_{f^{-1}(A)} \sum_{m=\max(k-l,1)}^{d-1} c(d,k,l,m)\, \Delta_f^m \, d\widetilde{C}_l^{svar}(X, \cdot) ,$$

for certain constants $c(d,k,l,m)$ and

$$\Delta_f(x,n) := ||I - \widetilde{D}f(x,n)||$$

at a.a. $(x,n) \in$ nor$X$, in the above notations.

Proof Since $N_Y = f_\# N_X$ we obtain (in short notation ap $Df = Df$)

$$\widetilde{C}_k(Y,A) = \int_{\text{nor}X} \mathbf{1}_A(f(x,n)) 1_X(x,n) \langle \Lambda^{d-1} Df(x,n) a_X(x,n), \varphi_k(f(x,n))\rangle \, \mathscr{H}^{d-1}(d(x,n)) .$$

Moreover,

$$\widetilde{C}_k(X, f^{-1}(A)) = \int_{\text{nor}X} \mathbf{1}_A(f(x,n)) 1_X(x,n) \langle a_X(x,n), \varphi_k(x,n)\rangle \, \mathscr{H}^{d-1}(d(x,n))$$

Below we shall show that for $\mathscr{H}^{d-1}$-a.a. $(x,n) \in$ nor$X$,

$$\left|\langle a_X(x,n),\varphi_k(x,n)\rangle - \langle \Lambda^{d-1}Df(x,n)a_X(x,n),\varphi_k(f(x,n))\rangle\right| \le \prod_{j=1}^{d-1}\left(1+\varkappa_j(x,n)^2\right)^{-1/2}$$

$$\sum_{i_1<\ldots<i_{d-1-k}}\sum_{m=1}^{d-1}\Delta_f^m\sum_{j_1<\ldots<j_m}\prod_{l=1}^{m}(1+|\varkappa_{j_l}(x,n)|)\prod_{j\in\{i_1,\ldots,i_{d-1-k}\}\setminus\{j_1,\ldots,j_m\}}|\varkappa_j(x,n)|\,, \qquad (5)$$

where all indices are positive and do not exceed $d-1$, and the last product over the empty set equals 1. Recalling that

$$|1_X|\prod_{j=1}^{d-1}\left(1+\varkappa_j^2\right)^{-1/2}\sum_{i_1<\ldots<i_{d-1-l}}|\varkappa_{i_1}\cdots\varkappa_{i_{d-1-l}}|$$

equals the density of the strong variation measure $\widetilde{C}_l^{svar}(X,\cdot)$ with respect to Hausdorff measure $\mathscr{H}^{d-1}$ and rearranging the summands in (5) according to the number of $\varkappa_j$ in the corresponding products the assertion follows by integration. (Note that the constants $c(d,k,l,m)$ can be determined by combinatorial arguments.)

In order to prove inequality (5) we rewrite the operators $\widetilde{D}f$ in $\mathbb{R}^d\times\mathbb{R}^d$ in matrix representation as

$$\widetilde{D}f = \begin{pmatrix} I+D_{00} & D_{01} \\ D_{10} & I+D_{11}\end{pmatrix},$$

for linear operators $D_{ij}:\mathbb{R}^d\to\mathbb{R}^d$ and note that $||D_{ij}||\le\Delta_f$, $i,j=0,1$. In this notation we obtain for the basis vectors $a_1,\ldots,a_{d-1}$ in the approximate tangent spaces of nor$X$ chosen according to (2)

$$\pi_0(\widetilde{D}fa_i) = \frac{1}{\sqrt{1+\varkappa_i^2}}(I+D_{00}+\varkappa_j D_{01})b_i$$

$$\pi_1(\widetilde{D}fa_i) = \frac{1}{\sqrt{1+\varkappa_i^2}}(\varkappa_j I+D_{10}+\varkappa_j D_{11})b_i\,.$$

Setting $I_i^0:=I$ and $I_i^1:=\varkappa_i I$, $D_i^0:=D_{00}+\varkappa_i D_{01}$, $D_i^1:=D_{10}+\varkappa_i D_{11}$ we get

$$\pi_{\varepsilon_i}(\widetilde{D}fa_i) = \frac{1}{\sqrt{1+\varkappa_i^2}}(I_i^{\varepsilon_i}+D_i^{\varepsilon_i})b_i\,,\quad \varepsilon_i=0,1\,,$$

and

$$||D_i^{\varepsilon_i}||\le(1+|\varkappa_i|)\,\Delta_f\,,\quad ||I_i^1||=|\varkappa_i|\,. \qquad (6)$$

Furthermore, recalling $a_X=a_1\wedge\ldots a_{d-1}$ and the definition of the Lipschitz-Killing curvature form φ_k we infer (for the sum running over $\varepsilon_i=0,1$ with $\sum_{i=1}^{d-1}\varepsilon_i=$

$d-1-k$ and the for volume form $\Omega_d := e_1' \wedge \ldots \wedge e_d'$:

$$\begin{aligned}
&\langle \Lambda^{d-1} Df(x,n) a_X(x,n), \varphi_k(f(x,n))\rangle \\
&= \sum \langle \pi_{\varepsilon_1}(Df(x,n)a_1(x,n)) \wedge \ldots \wedge \pi_{\varepsilon_{d-1}}(Df(x,n)a_{d-1}(x,n)) \wedge \pi_1(f(x,n)), \Omega_d\rangle \\
&= \sum \langle \pi_{\varepsilon_1}(\widetilde{D}f(x,n)a_1(x,n)) \wedge \ldots \wedge \pi_{\varepsilon_{d-1}}(\widetilde{D}f(x,n)a_{d-1}(x,n)) \wedge n, \Omega_d\rangle,
\end{aligned}$$

because of the invariance of Ω_d under the orientation preserving orthogonal mapping $O(n)$ and the definition of $\widetilde{D}f(x,n)$. Using the above representation of $\pi_{\varepsilon_i}(\widetilde{D}f(x,n)a_i)$ (and omitting the arguments (x,n) for brevity) the last expression may be rewritten as

$$\begin{aligned}
&\prod_{i=1}^{d-1}(1+\varkappa_i^2)^{-1/2} \sum \langle (I_1^{\varepsilon_1} + D_1^{\varepsilon_1}))b_1 \wedge \ldots \wedge (I_{d-1}^{\varepsilon_{d-1}} + D_{d-1}^{\varepsilon_{d-1}})b_{d-1} \wedge n, \Omega_d\rangle \\
&\quad = \prod_{i=1}^{d-1}(1+\varkappa_i^2)^{-1/2} \sum \langle I_1^{\varepsilon_1} b_1 \wedge \ldots \wedge I_{d-1}^{\varepsilon_{d-1}} b_{d-1} \wedge n, \Omega_d\rangle \\
&\quad + \prod_{i=1}^{d-1}(1+\varkappa_i^2)^{-1/2} \sum\sum \langle E_1^{\delta_1\varepsilon_1} b_1 \wedge \ldots \wedge E_{d-1}^{\delta_{d-1}\varepsilon_{d-1}} b_{d-1} \wedge n, \Omega_d\rangle\,,
\end{aligned}$$

where the last inner sum runs over $\delta_j = 0, 1$ with $\sum_{j=1}^{d-1}\delta_j \geq 1$, and $E_j^{\delta_j\varepsilon_j} := I_j^{\varepsilon_j}$ if $\delta_j = 0$, $E_j^{\delta_j\varepsilon_j} := D_j^{\varepsilon_j}$ if $\delta_j = 1$.

The first summand of the last expression agrees with $\langle a_X, \varphi_k\rangle$. The absolute value of the second summand coincides with the left hand side of the asserted inequality. In view of the above operator norm estimates (6) it does not exceed

$$\prod_{i=1}^{d-1}(1+\varkappa_i^2)^{-1/2} \sum_{\varepsilon_i}\sum_{\delta_j}\prod_{i=1}^{d-1} |E_i^{\delta_i\varepsilon_i} b_i| \leq \prod_{i=1}^{d-1}(1+\varkappa_i^2)^{-1/2}$$

$$\sum_{i_1<\ldots<i_{d-1-k}} \sum_{\sum\delta_j\geq 1} \prod_{j\in\{i_1,\ldots,i_{d-1-k}\}} \big((1-\delta_j)|\varkappa_j| + \delta_j(1+|\varkappa_j|)\Delta_f\big)$$

$$\prod_{j\notin\{i_1,\ldots,i_{d-1-k}\}} \big(1-\delta_j + \delta_j(1+|\varkappa_j|)\Delta_f\big) = \prod_{i=1}^{d-1}(1+\varkappa_i^2)^{-1/2}$$

$$\sum_{i_1<\ldots<i_{d-1-k}} \sum_{m=1}^{d-1} \Delta_f^m \sum_{j_1<\ldots<j_m} \prod_{l=1}^{m}(1+|\varkappa_{j_l}(x,n)|) \prod_{j\in\{i_1,\ldots,i_{d-1-k}\}\setminus\{j_1,\ldots,j_m\}} |\varkappa_j(x,n)|\,.$$

The last expression equals the right hand side of the asserted inequality (5), thus it is proved.

Note that the formal multiplication by $\prod_{i=1}^{d-1}(1+\varkappa_i^2)^{-1/2}$ everywhere has to be understood in the commutative sense, i.e., the factors with $\varkappa_i = \infty$ should be canceled with the corresponding $|\varkappa_i|$ in the following products. □

3 Stability of Fractal Curvatures Under Approximate Perturbations

We now turn back to the problem of approximating fractal curvatures by different neighborhoods. For simplicity we here consider only the total values. Recall that for certain classes of compact fractal sets F the *average Minkowski content* ($k = d, d-1$) and *fractal curvatures* of order $0 \leq k \leq d-2$ have been introduced by means of the parallel sets F_ε as

$$C_k^{frac}(F) = \operatorname{ess}\lim_{\delta\to 0} \frac{1}{|\ln\delta|}\int_\delta^{\delta_0} \varepsilon^{D-k} C_k(F_\varepsilon)\,\varepsilon^{-1}d\varepsilon\,, \tag{7}$$

where D is the Hausdorff or Minkowski dimension of F and the essential limit is taken with respect to the Lebesgue measure. Under additional assumptions (non-arithmetic cases) the ordinary limits are shown to exist. Then the question arises whether one can choose other approximating sets, say F^ε, in order to obtain the same limits. For example, for practical purposes, it would be useful to replace the parallel sets by close polyhedra. The above estimates can be applied to this and more general situations, where the F_ε are not necessarily parallel sets:

Theorem 1 *For Lebesgue a.a. $\varepsilon < \varepsilon_0$ let F_ε and F^ε be any compact subsets of $\mathbb{R}^d$ admitting normal cycles N_{F_ε} and N_{F^ε}, resp., such that there exists a Lipschitz mapping $f_\varepsilon : \operatorname{nor} F_\varepsilon \to \operatorname{nor} F^\varepsilon$ satisfying*

$$(f_\varepsilon)_\# N_{F_\varepsilon} = N_{F^\varepsilon} \text{ and } \|I - \widetilde{D} f_\varepsilon\| \leq \alpha(\varepsilon)\varepsilon \tag{8}$$

for some bounded function $\alpha(\varepsilon)$, where $\widetilde{D} f_\varepsilon$ is defined as in (4)*. Then we get for $k = 0, \ldots, d-1$,*

$$\left|\frac{1}{|\ln\delta|}\int_\delta^{\delta_0} \varepsilon^{D-k} C_k(F_\varepsilon)\,\varepsilon^{-1}d\varepsilon - \frac{1}{|\ln\delta|}\int_\delta^{\delta_0} \varepsilon^{D-k} C_k(F^\varepsilon)\,\varepsilon^{-1}d\varepsilon\right|$$

$$\leq \text{const} \sum_{l=0}^{d-1} \frac{1}{|\ln\delta|}\int_\delta^{\delta_0} \alpha(\varepsilon)\varepsilon^{D-l} C_l^{svar}(F_\varepsilon)\,\varepsilon^{-1}d\varepsilon \tag{9}$$

provided the first two integrals exist for some $D > 0$. If

$$\max_{0\le l\le d-1}\sup_{0<\delta<\delta_0}\frac{1}{|\ln\delta|}\int_{\delta}^{\delta_0}\varepsilon^{D-l}C_l^{svar}(F_\varepsilon)\,\varepsilon^{-1}d\varepsilon<\infty \text{ and } \operatorname{ess\,lim}_{\varepsilon\to o}\alpha(\varepsilon)=0 \quad (10)$$

then we get the same limits

$$\operatorname{ess\,lim}_{\delta\to 0}\frac{1}{|\ln\delta|}\int_{\delta}^{\delta_0}\varepsilon^{D-k}C_k(F^\varepsilon)\,\varepsilon^{-1}d\varepsilon=\operatorname{ess\,lim}_{\delta\to 0}\frac{1}{|\ln\delta|}\int_{\delta}^{\delta_0}\varepsilon^{D-k}C_k(F_\varepsilon)\,\varepsilon^{-1}d\varepsilon \quad (11)$$

provided one of them exists.

Analogous statements hold for the non-averaged versions.

Proof We substitute in Lemma 2 the set X by F_ε, the set Y by F^ε and the mapping f by f_ε. Using the estimate

$$||\Delta_{f_\varepsilon}||\le\varepsilon\alpha(\varepsilon)$$

we then infer for $\varepsilon\le\delta_0$ and varying constant factors independent of F_ε, F^ε,

$$|C_k(F_\varepsilon)-C_k(F^\varepsilon)|\le \text{const}\left(\sum_{l=0}^{k-1}C_l^{svar}(F_\varepsilon)\sum_{m=k-l}^{d-1}\varepsilon^m\alpha(\varepsilon)^m+\sum_{l=k}^{d-1}C_l^{svar}(F_\varepsilon)\sum_{m=1}^{d-1}\varepsilon^m\alpha(\varepsilon)^m\right)$$

$$\le \text{const}\left(\sum_{l=0}^{k-1}C_l^{svar}(F_\varepsilon)\varepsilon^{k-l}\alpha(\varepsilon)^{l-l}\sum_{m=0}^{d-1-(k-l)}\varepsilon^m\alpha(\varepsilon)^m+\sum_{l=k}^{d-1}C_l^{svar}(F_\varepsilon)\varepsilon\alpha(\varepsilon)\sum_{m=0}^{d-2}\varepsilon^m\alpha(\varepsilon)^m\right)$$

$$\le \text{const}\left(\varepsilon^k\alpha(\varepsilon)\left(\sum_{l=0}^{k-1}\varepsilon^{-l}C_l^{svar}(F_\varepsilon)+\varepsilon\sum_{l=k}^{d-1}\varepsilon^{-l}C_l^{svar}(F_\varepsilon)\le \text{const}\,\varepsilon^k\alpha(\varepsilon)\sum_{l=0}^{k-1}\varepsilon^{-l}C_l^{svar}(F_\varepsilon)\,.\right.\right.$$

This implies

$$\left|\frac{1}{|\ln\delta|}\int_{\delta}^{\delta_0}\varepsilon^{D-k}C_k(F_\varepsilon)\,\varepsilon^{-1}d\varepsilon-\frac{1}{|\ln\delta|}\int_{\delta}^{\delta_0}\varepsilon^{D-k}C_k(F^\varepsilon)\,\varepsilon^{-1}d\varepsilon\right|$$

$$\le \text{const}\sum_{l=0}^{k-1}\frac{1}{|\ln\delta|}\int_{\delta}^{\delta_0}\alpha(\varepsilon)\varepsilon^{D-l}C_l^{svar}(F_\varepsilon)\,\varepsilon^{-1}d\varepsilon\,,$$

i.e. (9). The limit equality (11) is a consequence. Finally note that the above estimates also work without averaging over ε. □

Remark 1 For F_ε we can take the parallel sets of a *self-similar or self-conformal fractal* with the open set condition and Hausdorff dimension D. If the parallel sets are regular in the above mentioned sense and the integrability condition in (10) is satisfied then the average limits

$$\operatorname*{ess\,lim}_{\delta\to 0} \frac{1}{|\ln\delta|}\int_{\delta}^{\delta_0} \varepsilon^{D-k} C_k(F_\varepsilon)\,\varepsilon^{-1}d\varepsilon$$

do exist (see [Boh13, BZ13, RZ12, Win08, WZ13]). Note that the integrability condition (10) can be checked for many classical examples, where even existence of the average limits of the strong variation measures can be proved. Then the theorem states that one obtains the same limits if one considers small perturbations F^ε of F_ε in the sense of (8).

For example, in the case of a *general Sierpinski gasket* in the plane only the interior triangular holes of the parallel sets determine the limit behaviour. If one investigates small affine perturbations of these triangles such that distances between the corresponding vertices are of order $o(\varepsilon)$, then the mappings $f_\varepsilon : \operatorname{nor} F_\varepsilon \to \operatorname{nor} F^\varepsilon$ can locally be chosen as follows

$$f_\varepsilon(x,n) := \left(\Pi_{\widetilde{F^\varepsilon}}\big(x-\beta(\varepsilon)n\big), -\frac{x-\Pi_{\widetilde{F^\varepsilon}}\big(x-\beta(\varepsilon)n\big)}{|x-\Pi_{\widetilde{F^\varepsilon}}\big(x-\beta(\varepsilon)n\big)|}\right).$$

Here $\Pi_{\widetilde{F^\varepsilon}}$ denotes the metric projection onto $\widetilde{F^\varepsilon} = \overline{(F^\varepsilon)^c}$ and the constant $\beta(\varepsilon)$ depends on the corresponding affine transformation. (The parallel set of distance $\beta(\varepsilon)$ of the triangle of $\widetilde{F}_\varepsilon$ under consideration must contain the corresponding perturbed triangle of $\widetilde{F^\varepsilon}$.) Note that in view of convexity in such a case the estimates are simpler than in general.

Remark 2 The error estimates in Lemma 2 and Theorem 1 are rough. Our method of proof shows that they can be improved and the corresponding constants can be computed. For the cases $k = d-1, d$, which yield up to some constant the same limit, namely the (average) *Minkowski content* (see [RW13]), Lemma 1 provides better tools than Lemma 2 and the above conditions are not needed. The remaining arguments in this case are similar and we do not expose the details.

References

[Boh13] T. Bohl: Fractal curvatures and Minkowski content of self-conformal sets. Preprint, arXiv:1211.3421v1, PhD thesis, University of Jena (2013)

[BZ13] Bohl, T., Zähle, M.: Curvature-direction measures of self-similar sets. Geom. Dedicata **167**, 215–231 (2013)

[Fed59] Federer, H.: Curvature measures. Trans. Amer. Math. Soc. **93**, 418–491 (1959)

[Fed69] Federer, H.: Geometric Measure Theory. Springer, Berlin (1969)

[FK12] Freiberg, U., Kombrink, S.: Minkowski content and local Minkowski content for a class of self-conformal sets. Geom. Dedicata **159**, 307–325 (2012)

[Gat00] Gatzouras, D.: Lacunarity of self-similar and stochastically self-similar sets. Trans. Amer. Math. Soc. **352**, 1953–1983 (2000)

[KK12] Kesseböhmer, M., Kombrink, S.: Fractal curvature measures and Minkowski content for self-conformal subsets of the real line. Adv. Math. **230**, 2474–2512 (2012)

[Kom11] S. Kombrink: Fractal Curvature Measures and Minkowski Content for Limit Sets of Conformal Function Systems. PhD thesis, University of Bremen (2011) http://nbn-resolving.de/urn:nbn:de:gbv:46–00102477-19
[KP08] Krantz, S.G., Parks, H.R.: Geometric Integration Theory. Birkhäuser, Boston (2008)
[LPW11] Lapidus, M.L., Pearse, E.P.J., Winter, S.: Pointwise tube formulas for fractal sprays and self-similar tilings with arbitrary generators. Adv. Math. **227**, 1349–1398 (2011)
[LPW13] Lapidus, M.L., Pearse, E.P.J., Winter, S.: Minkowski measurability results for self-similar tilings and fractals with monophase generators. Contemp. Math. **600**, 185–203 (2013)
[Mor88] Morgan, F.: Geometric Measure Theory–A Beginner's Guide. Academic Press, Boston (1988)
[PW12] Pearse, E.P.J., Winter, S.: Geometry of canonical self-similar tilings. Rocky Mt. J. Math. **42**, 1327–1357 (2012)
[RW13] Rataj, J., Winter, S.: Characterization of Minkowski measurability in terms of surface area. J. Math. Anal. Appl. **400**, 120–132 (2013)
[RZ01] Rataj, J., Zähle, M.: Curvatures and currents for unions of sets with positive reach. Ann. Glob. Anal. Geom. **20**, 1–21 (2001)
[RZ05] Rataj, J., Zähle, M.: General normal cycles and Lipschitz manifolds of bounded curvature. Ann. Glob. Anal. Geom. **27**, 135–156 (2005)
[RZ12] Rataj, J., Zähle, M.: Curvature densities of self-similar sets. Indiana Univ. Math. J. **61**, 1425–1449 (2012)
[Win08] Winter, S.: Curvature Measures and Fractals. Diss. Math. **453**, 1–66 (2008)
[Win11] Winter, S.: Curvature bounds for neighborhoods of self-similar sets. Comment. Math. Univ. Carolin. **52**, 205–226 (2011)
[WZ13] Winter, S., Zähle, M.: Fractal curvature measures of self-similar sets. Adv. Geom. **13**, 229–244 (2013)
[Zah86] Zähle, M.: Integral and current representation of Federer's curvature measures. Arch. Math. **46**, 557–567 (1986)
[Zah90] Zähle, M.: Approximation and characterization of generalized Lipschitz-Killing curvatures. Ann. Global Anal. Geom. **8**, 249–260 (1990)
[Zah11] Zähle, M.: Lipschitz-Killing curvatures of self-similar random fractals. Trans. Amer. Math. Soc. **363**, 2663–2684 (2011)
[Zah13] Zähle, M.: Curvature measures of fractal sets. Contemp. Math. **600**, 381–399 (2013)

Participants

- Shigeki Akiyama, Tsukuba University
- Patricia Alonso Ruiz, University of Siegen
- Peter Arzt, University of Siegen
- Jung-Chao Ban, National Dong Hwa University
- Christoph Bandt, Arndt University
- Michael Barnsley, Australian National University
- Julien Barral, Université Paris 13
- Yann Bugeaud, Université de Strasbourg
- Yongluo Cao, Suzhou University
- Joe Po-Chou Chen, Cornell University
- Zhen-Qing Chen, University of Washington
- Thierry Coulhon, The Australian National University
- Eva Curry, Acadia University
- Xinrong Dai, Sun Yat-Sen University
- Guotai Deng, Central China Normal University
- Qirong Deng, Fujian Normal University
- Xinhan Dong, Hunan Normal University
- Yangge Du, The Chinese University of Hong Kong
- Kenneth J. Falconer, University of St. Andrews
- Ai-Hua Fan, Université de Picardie Jules Verne
- Chun Fang, University of Helsinki
- Ábel Farkas, University of St Andrews
- De-Jun Feng, The Chinese University of Hong Kong
- Axel Françoise, Université Paris-Sud
- Jonathan Fraser, University of St Andrews
- Uta Freiberg, University of Siegen
- Xiaoye Fu, The Chinese University of Hong Kong
- Jean-Pierre Gabardo, McMaster University
- Peter Grabner, Technische Universität Graz
- Alexander Grigor'yan, Universität Bielefeld
- Deguang Han, University of Central Florida

D.-J. Feng and K.-S. Lau (eds.), *Geometry and Analysis of Fractals*,
Springer Proceedings in Mathematics & Statistics 88,
DOI 10.1007/978-3-662-43920-3

- Douglas Hardin, Vanderbilt University
- Xinggang He, Central China Normal University
- Joachim Heinze, Springer-Verlag
- Masanori Hino, Kyoto University
- Ching Wei Ho, The Chinese University of Hong Kong
- Mike Hochman, The Hebrew University
- Jiaxin Hu, Tsinghua University
- Wen Huang, University of Science and Technology of China
- Satoshi Ishiwata, Yamagata University
- Yu Ito, Kyoto University
- Stéphane Jaffard, Université Paris-Est-Créteil Val-de-Marne
- Esa Järvenpää, University of Oulu
- Maarit Järvenpää, University of Oulu
- Xiong Jin, University of St Andrews
- Timo Jolivet, Université Paris Diderot
- Thomas Jordan, University of Bristol
- Palle E. T. Jorgensen, University of Iowa
- Antti Käenmäki, University of Jyväskylä
- Naotaka Kajino, Universität Bielefeld
- Charlene Kalle, Leiden University
- Tamás Keleti, Eötvös Loránd University
- Tom Kempton, Utrecht University
- Jun Kigami, Kyoto University
- Henna Koivusalo, University of Oulu
- Shilei Kong, The Chinese University of Hong Kong
- Takashi Kumagai, Kyoto University
- Chun Kit Lai, McMaster University
- Wai Kit Lam, The Chinese University of Hong Kong
- Ka-Sing Lau, The Chinese University of Hong Kong
- Gregory F. Lawler, University of Chicago
- François Ledrappier, University of Notre Dame
- Woon Yin Paul Lee, The Chinese University of Hong Kong
- Chi Wai Leung, The Chinese University of Hong Kong
- King Shun Leung, The Hong Kong Institute of Education
- Bing Li, South China University of Technology & University of Oulu
- Wenxia Li, East China Normal University
- Xiangyu Liang, Université Paris-Sud
- Lingmin Liao, Université Paris-Est-Créteil Val-de-Marne
- Yong Lin, People's University of China
- Jingcheng Liu, Hunan Normal University
- Yuan Liu, Chinese Academy of Sciences
- Qinghui Liu, Beijing Institute of Technology
- Quansheng Liu, Université de Bretagne-Sud
- Chiu Hong Lo, The Chinese University of Hong Kong
- Fan Lü, Huazhong University of Science and Technology

- Jun Luo, Zhongshan University
- Jun Jason Luo, Shatou University
- Jihua Ma, Wuhan University
- Peter Massopust, Technische Universität München & Helmholtz Zentrum München
- Pertti Mattila, University of Helsinki
- John Ling-Lei Meng, Springer-Verlag
- Tuen Wai Ng, University of Hong Kong
- Sze-Man Ngai, Georgia Southern University
- Jacques Peyrière, Université Paris-Sud
- Mark Pollicott, University of Warwick
- Hua Qiu, Nanjing University
- Yanhui Qu, Tsinghua University
- Hui Rao, Central China Normal University
- John A. Rock, Cal Poly Pomona
- Luke Rogers, University of Connecticut
- Huojun Ruan, Zhejiang University
- Tuomas Sahlsten, University of Helsinki
- Jörg Schmeling, Lund University
- Stéphane Seuret, Université Paris-Est-Créteil Val-de-Marne
- Narn-Rueih Shieh, National Taiwan University
- Pablo Shmerkin, University of Surrey
- Lin Shu, Peking University
- Károly Simon, Technical University of Budapest
- Örjan Stenflo, Uppsala University
- Weiyi Su, Nanjing University
- Qiyu Sun, University of Central Florida
- Yuhua Sun, Universität Bielefeld
- Ville Suomala, University of Oulu
- Hanako Takahashi, Kyoto University
- Bo Tan, Huazhong University of Science and Technology
- Donglei Tang, Nanjing Audit University
- Tai Man Tang, Xiangtan University
- András Telcs, University of Pannonia
- Alexander Teplyaev, University of Connecticut
- Jonathan Tsai, The University of Hong Kong
- Kam Piu Tsang, The Chinese University of Hong Kong
- Baris Ugurcan, Cornell University
- Baowei Wang, Huazhong University of Science and Technology
- Xiangyang Wang, Sun Yat-Sen University
- Yang Wang, Michigan State University
- Zhi-Ying Wen, Tsinghua University
- Shengyou Wen, Hubei University
- Ting Kam Wong, University of Washington
- Dongsheng Wu, University of Alabama in Huntsville

- Haihua Wu, Hunan Normal University
- Jun Wu, Huazhong University of Science and Technology
- Min Wu, South China University of Technology
- Chang Xiao, The Chinese University of Hong Kong
- Yimin Xiao, Michigan State University
- Jiansheng Xie, Fudan University
- Ying Xiong, South China University of Technology
- Jian Xu, Huazhong University of Science and Technology
- S. C. P. Yam, The Chinese University of Hong Kong
- Dawei Yang, Jilin University
- Xiangdong Ye, University of Science and Technology of China
- Yuanling Ye, South China Normal University
- Zuguo Yu, Xiangtan University
- Martina Zähle, University of Jena
- Rüdiger Zeller, Universität Greifswald
- Guohua Zhang, Fudan University
- Pengfei Zhang, The Chinese University of Hong Kong
- Xinghua Zheng, Hong Kong University of Science and Technology
- Ding-Xuan Zhou, City University of Hong Kong
- Zuoling Zhou, Zhongshan University

MIX
Papier aus verantwortungsvollen Quellen
Paper from responsible sources
FSC® C105338

If you have any concerns about our products,
you can contact us on
ProductSafety@springernature.com

In case Publisher is established outside the EU,
the EU authorized representative is:
Springer Nature Customer Service Center GmbH
Europaplatz 3, 69115 Heidelberg, Germany

Printed by Libri Plureos GmbH
in Hamburg, Germany